FINITE
MATHEMATICS

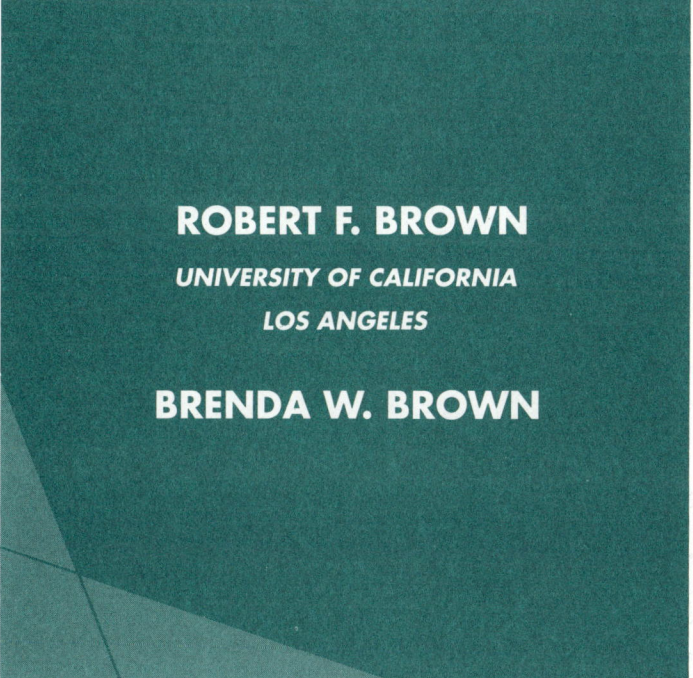

ROBERT F. BROWN
UNIVERSITY OF CALIFORNIA
LOS ANGELES

BRENDA W. BROWN

ARDSLEY HOUSE PUBLISHERS, INC., NEW YORK

Address orders and editorial
correspondence to:
Ardsley House, Publishers, Inc.
320 Central Park West
New York, NY 10025

Copyright © 1992, by Ardsley House, Publishers, Inc.

All rights reserved.

Reproduction or translation of any part of this work
beyond that permitted by Sections 107 and 108 of
the 1976 United States Copyright Act without the
permission of the copyright owner is unlawful.
Requests for permission or further information
should be addressed to the Permissions Department,
Ardsley House.

ISBN: 0-912675-96-9

Printed in the United States of America

10 9 8 7 6 5 4 3 2 1

To The Memory of I.B. and H.G.W.

Contents

Preface ix

Chapter 1   **MATRICES AND SYSTEMS OF LINEAR EQUATIONS**   **1**

Economic Planning: The Leontief Model   1
- **1.1** Matrices   *3*
- **1.2** Matrix Multiplication   *17*
- **1.3** Nonsingular Two-by-Two Linear Systems   *34*
- **1.4** Nonsingular n-by-n Linear Systems   *47*
- **1.5** The Inverse of a Matrix   *64*
- **1.6** Matrix Methods for Linear Systems   *80*
- **1.7** Matrix Applications   *96*
- **1.8** The Leontief Model   *106*

Review Exercises for Chapter 1   *115*

Chapter 2   **LINEAR PROGRAMMING**   **118**

Farm Management   *118*
- **2.1** Linear-Programming Problems   *119*
- **2.2** The Geometry of Linear Programming   *129*
- **2.3** Matrix Forms and Duality   *146*
- **2.4** Pivoting and Bases   *163*
- **2.5** The Simplex Algorithm for Maximization Problems   *178*
- **2.6** Consequences of Duality   *197*
- **2.7** A Farm-Management Problem   *212*

Review Exercises for Chapter 2   *216*

Chapter 3 PROBABILITY 219

Probability and the Weather 219
3.1 Probability and Odds 222
3.2 Counting 229
3.3 Permutations and Factorials 237
3.4 Combinations 244
3.5 Computing Probability by Counting 253
3.6 Union of Events 258
3.7 Disjoint Events 267
3.8 Conditional Probability 275
3.9 Intersection of Events 286
Review Exercises for Chapter 3 293
Appendix: The Standard Deck of Cards 296

Chapter 4 MORE PROBABILITY 297

Drug Testing 297
4.1 Partitions 299
4.2 Bayes' Theorem 306
4.3 Random Variables and Probability Distributions 314
4.4 Expected Value and Variance 321
4.5 Binomial Experiments 331
4.6 The Normal Distribution 340
4.7 Normal Approximation for Binomial Distributions 351
Review Exercises for Chapter 4 361

Chapter 5 MARKOV CHAINS 364

Brand-Share Prediction 364
5.1 Matrices and Probability 365
5.2 Markov Chain Processes 376
5.3 Equilibrium 389
5.4 Absorbing Markov Chains 399
5.5 The Fundamental Matrix 412
Review Exercises for Chapter 5 424

Chapter 6 GAME THEORY 427

Political Campaigns 427
6.1 Matrix Games 428
6.2 Strictly Determined Games 437
6.3 Mixed Strategies 446

- **6.4** Two-by-Two Games *461*
- **6.5** Games and Linear Programming *474*
- **6.6** Sensitivity Analysis of Games *490*
  Review Exercises for Chapter 6 503

Chapter 7 GRAPHS AND NETWORKS 506

A Trading Network 506
- **7.1** Graphs *507*
- **7.2** Digraphs *523*
- **7.3** Digraph Matrices *533*
- **7.4** Networks *546*
  Review Exercises for Chapter 7 559

Chapter 8 MATHEMATICS OF FINANCE 563

Home Mortgages 563
- **8.1** Simple Interest *564*
- **8.2** Discount *572*
- **8.3** Compound Interest *581*
- **8.4** Annuities *590*
- **8.5** Amortization *600*
  Review Exercises for Chapter 8 612

Appendices

Set Theory and Logic *A1*
Tables *A13*

Answers to Odd-Numbered Exercises A23

Index A53

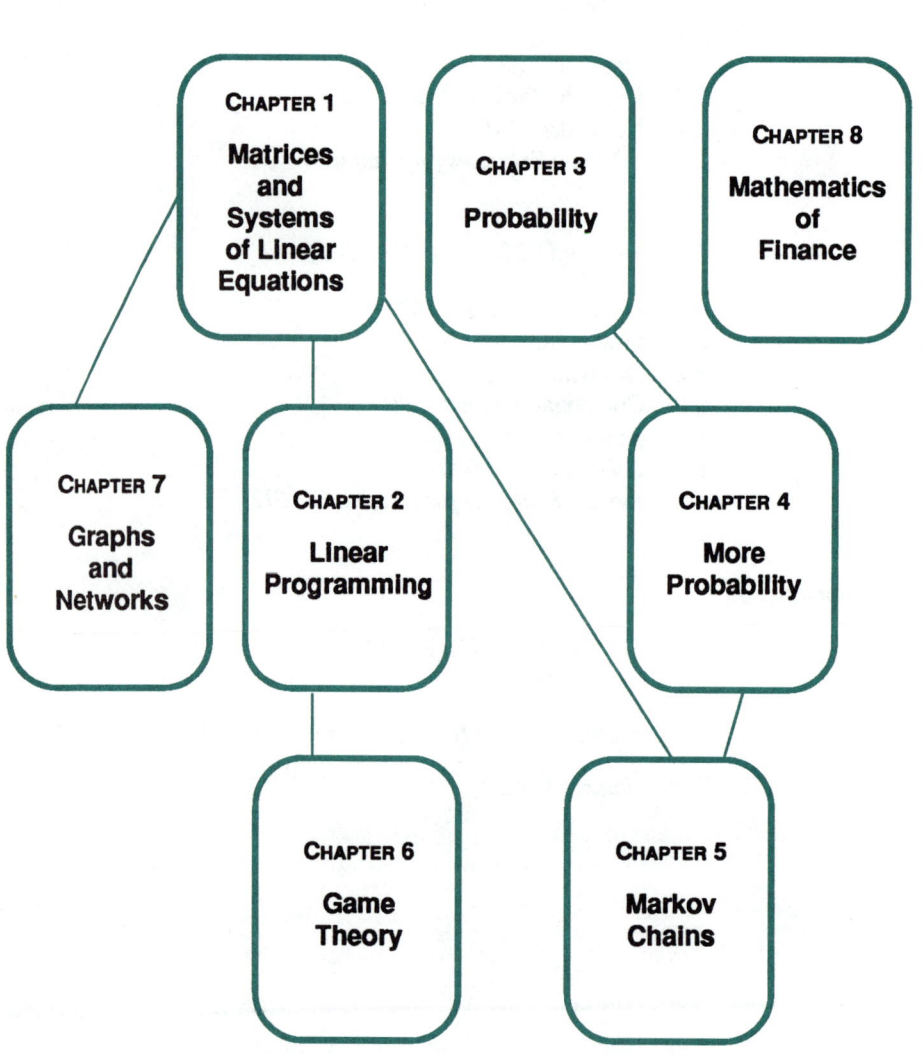

Preface

Finite Mathematics contains the topics that are customarily taught in a course with this title, intended for majors in the social and biological sciences as well as business. The structure of the book is shown in the chart on the facing page.

Each chapter begins with a fairly detailed, but nontechnical, description of a significant application of mathematics. As the chapter progresses, we refer to this application whenever the mathematics presented relates to it. By the end of the chapter, the reader will have seen how the mathematical topics discussed make a significant contribution to the application. In this way and through numerous briefer examples and exercises based on applications, the reader is assured that finite mathematics is not only accessible and attractive, but useful as well.

Each section of the text is broken into several subsections. At the end of the section there is a summary that reviews the mathematical content in a condensed form. Instructors can use the summaries as an aid in planning lectures, and students will find them helpful for reviewing. Definitions are boxed, as are important statements — but in a distinguishably different style — so as to be easily located for reference. An end-of-solution symbol, ■, clarifies when an example has been completed and when the text proceeds to the next idea.

One of the most attractive characteristics of finite mathematics is its capacity for being supported by elementary explanations. Whenever possible, a convincing, if informal, justification is presented for the mathematics taught. We also include a review of high school material, when appropriate. Above all, we wish to encourage the students to read this book, and not just do exercises from it. Therefore, our development features thorough mathematical discussions and numerous examples with detailed solutions.

The book contains many innovations—for instance:

- Unknowns head the columns of an augmented matrix so that the student can easily move back and forth between the augmented matrix and the corresponding linear system (*Section 1.4*). This anticipates the same feature of the linear-programming tableau.

- We present an efficient "pivoting without fractions" method for implementing the simplex algorithm of linear programming, with a complete explanation of why the method works and an immediate application to the linear systems studied in Chapter 1 (*Section 2.4*).

- Ordered k-tuples are used as a counting tool to complement counting trees (*Section 3.2*).

- We reinforce the Gauss-Jordan method of Chapter 1 as we use it to find the equilibrium of a regular Markov chain (*Section 5.3*) and the fundamental matrix of an absorbing Markov chain (*Section 5.5*).

- Sensitivity analysis, from linear programming, is applied to parametrized matrix games (*Section 6.6*).

- Matrices are used extensively in the study of directed graphs (*Section 7.3*).

=:=:=:=:=:=:=:=:=:=

We thank the reviewers: Barbara Bohannon, Hofstra University; Subramanian Ganesan, University of Portland; V. Rao Potluri, Reed College; Walter Johnson, Fayetteville State University; and Martin Zuckerman, City College of the City University of New York. We received many helpful suggestions from James Caballero, David Cohen, Shelly Cook, Charles Lange and Jon Rogawski, all of UCLA, and also from Peter Wong of Bates College. Marc Darr served as our computer consultant and Geoff Brown was our consultant on the mathematics of finance. Glenn Appleby and Jacqui Ramagge checked exercise answers for us. We are very grateful to Laura Jones for her patient and skillful production of the text.

ROBERT F. BROWN
BRENDA W. BROWN

Matrices and Systems of Linear Equations

Economic Planning: The Leontief Model

In 1973, Wassily Leontief won the Nobel prize in Economic Science for his development of input-output economics. Leontief first expressed his ideas on the workings of the economy in a paper he wrote while he was still a graduate student in Berlin during the 1920s. His basic work, *The Structure of the American Economy 1919–29*, was published in 1941. The theory was put to use during World War II when Leontief worked with the U. S. Bureau of Labor Statistics to develop useful input-output tables for the American economy.

In an article that appeared in *Scientific American* in 1951,[†] Leontief defined input-output economics as

> a method of analysis that takes advantage of the relatively stable pattern of the flow of goods and services among the elements of our economy to bring a much more detailed statistical picture of the system into the range of manipulation by economic theory. As such, the method has had to await the modern high-speed computing machine as well as the present propensity of government and private agencies to accumulate mountains of data.

The central concept of Leontief's theory states that there is a fundamental relationship between the volume of the output from an

---

[†]Wassily Leontief, "Input-Output Economics," *Scientific American* (Oct. 1, 1951).

industry and the size of the input to it. These fundamental relationships are expressed in input-output analysis as the ratio of each input to the total output of which it becomes part. For example, for each million dollars of production, the automobile industry uses so many dollars' worth of iron and steel, industrial and heating equipment, electrical equipment, and so on. The dollar amounts rise or fall with changing prices, but the actual amount of steel going into a car varies only when the technology changes. The ratio of input of material to output of cars remains relatively stable over a period of several years. Barring any drastic technological change, the needs of the industry can be predicted quite accurately, assuming we can estimate how many cars will be produced in a given year.

The relationships among all industries can be presented in a systematic way by displaying them in tabular form in what is called an "input-output table." The input-output ratio that describes the relationship between the two industries (known as "sectors" in this theory) occupies a specific location in the table.

The Leontief model also uses a "bill of demands," a table that represents exports, surpluses, government and individual consumption, and the like. The problem of predicting the number of units needed in order to run the economy and fill the bill of demands is complicated because each sector must produce not only the units required by the bill of demands, but also units of input to the other sectors so that they, in turn, can produce what the first sector needs to operate. The more each sector produces, the more input it needs; but then the more input from it the other sectors need in order to satisfy its demands. This interdependence among the various sectors of the economy is the heart of input-output economic theory.

The mathematics of input-output analysis is not difficult, but it requires a lot of arithmetic. Also the input-output models have grown larger as the economy has become more complex. Fortunately, the increasing complexities of the economy have been matched by the increasing efficiency of computing equipment.

The theory *does* work. In 1945, a study of the postwar economy was conducted based on a list of economic demands that assumed full employment in 1950. The demands were inserted into an input-output model based on a table of ratios for the year 1939, the last year of normal peacetime production. The model indicated, for example, that there would be a need for 98 million ingot tons of steel. Steel production, at full capacity in 1950, in fact, turned out to be 96.8 million tons, and there was a shortage as predicted.

Another use of input-output economics is in producing models of the economies of developing nations. These models give planners a clearer picture of the strengths and weaknesses of the developing economy. Also, input-output theory is useful in managing the environment. This application involves associating the amount of pollutants generated (foul air, polluted water, solid waste, noise, etc.) to every unit of output of each economic sector.

Individual corporations have also constructed input-output tables to aid in planning. From this kind of analysis, corporations discover indirect costs of which they had been previously unaware. For example, the Celanese Plastics Company found that in order to turn out products that sell for $100, they use $43.20 in chemicals bought directly from a manufacturer. But in producing their $100 worth of products, they actually consume $61.30 in chemicals. The extra $18.10 represents chemicals bought by the packaging concerns, printers, and other companies whose products are used in the manufacture of plastics. The importance of this indirect cost for the plastics company is that if the price of chemicals changed, the change must be figured on the indirect cost of chemicals, as well as on the direct cost, when estimating the necessary price changes for the plastics.

The value of input-output economics in planning—whether it is used by a single company, a nation, or an ecologist studying global problems—depends on the accuracy of the data and the skill and intelligence of those who interpret the information the models yield. Thus, the proper use of input-output theory requires skills from business and economics as well as from mathematics. The mathematics involved in Leontief's model is based on the theory of matrices, which we will study in this chapter.

1.1 Matrices

Definition of a Matrix

A rectangular array of numbers is called a **matrix** (*plural:* **matrices**). There is no restriction on the size of the matrix or on the numbers in it. Thus, all of the following are matrices

$$\begin{bmatrix} 2 & 3 & 1 \\ 2 & 1 & 0 \\ 0 & 0 & 1 \end{bmatrix} \begin{bmatrix} -1 & -1 & 0 \\ 0 & 0 & 0 \end{bmatrix} \begin{bmatrix} 2 \\ \frac{3}{2} \\ -5 \\ -\frac{1}{2} \end{bmatrix} \begin{bmatrix} 1 & 1 \end{bmatrix} \begin{bmatrix} 0.47 & 0.12 \\ -3.1 & -6 \end{bmatrix}$$

The types of matrices that we will discuss in connection with input-output theory are:

1. The "input-output matrix." (The following is based on the transportation activity of the 1958 U. S. economy.)

$$\begin{bmatrix} 0.41 & 0 & 0.06 \\ 0 & 0.22 & 0 \\ 0.19 & 0.04 & 0.25 \end{bmatrix}$$

4 MATRICES AND SYSTEMS OF LINEAR EQUATIONS

2. The demand matrix. (The following matrix relates to basic industries in the Western European economy of 1953.)

$$\begin{bmatrix} 90 \\ 320 \\ 154 \\ 43 \end{bmatrix}$$

3. A matrix of unknowns, called the "intensity matrix,"

$$\begin{bmatrix} x \\ y \\ z \end{bmatrix}$$

The matrices that economists actually use are much larger (those given are summarized from the full tables.) For instance, the complete input-output matrix for the 1958 U. S. economy has 81 rows.

Rows and Columns

A matrix is described by means of **rows** and **columns**. Rows are always numbered from top to bottom. For example, consider

$$A = \begin{bmatrix} 2 & 3 & -1 \\ 2 & 1 & 0 \\ 0 & 0 & 1 \end{bmatrix} \begin{matrix} \text{First Row} \\ \text{Second Row} \\ \text{Third Row} \end{matrix}$$

Columns are always numbered from left to right; thus, for the matrix A,

$$\begin{bmatrix} 2 & 3 & -1 \\ 2 & 1 & 0 \\ 0 & 0 & 1 \end{bmatrix}$$

First Column, Second Column, Third Column

For the matrix

$$B = \begin{bmatrix} 2 \\ \frac{3}{2} \\ -5 \\ -\frac{1}{2} \end{bmatrix}$$

there are four rows, each consisting of a single number, but there is only one column, the matrix B itself.

Size and Location

The number of rows and columns of a matrix is called its **size**. The number of rows is always given first, and the numbers are separated by the word "by" or by the symbol "×." Thus, the matrix A is a 3-by-3 (or 3×3-) matrix, and the matrix B is 4×1. An $n \times n$-matrix is called a **square matrix**, a $1 \times n$-matrix is a **row matrix**, and an $n \times 1$-matrix, such as B, is a **column matrix**.

The numbers in a matrix are called its **entries**. Each entry has a specific **location**, described by first stating the row and then the column in which it lies. There is a convenient shorthand for naming locations. We will illustrate it with the matrix

$$A = \begin{bmatrix} 2 & 3 & -1 \\ 2 & 1 & 0 \\ 0 & 0 & 1 \end{bmatrix}$$

The entry in row number i and column number j is represented by the symbol a_{ij}, where the letter "a" is used to tell us that the number is in the matrix A. For instance, instead of saying that the number -1 is in the upper right-hand corner of the matrix A, that is, in the first row and third column, we just write $a_{13} = -1$. For the number in the third row and the first column we write $a_{31} = 0$. Since in matrix notation the row always comes before the column, we know that in a_{12} the "1" refers to the first row; thus, $a_{12} = 3$.

Example 1. Given the "input-output" matrix,

$$A = \begin{bmatrix} 0.41 & 0 & 0.06 \\ 0 & 0.22 & 0 \\ 0.19 & 0.04 & 0.25 \end{bmatrix}$$

identify:

(a) the second row
(b) the third column
(c) the size of the matrix
(d) the entry in the third row and second column
(e) the location of the number 0.06

Solution

(a) $\begin{bmatrix} 0 & 0.22 & 0 \end{bmatrix}$ (b) $\begin{bmatrix} 0.06 \\ 0 \\ 0.25 \end{bmatrix}$ (c) 3×3

(d) 0.04 (e) first row, third column; $a_{13} = 0.06$ ∎

6 MATRICES AND SYSTEMS OF LINEAR EQUATIONS

Equality

A matrix is characterized not only by its size and the numbers in it, but also by the location of the numbers. Thus, two matrices are **equal** if they are of the same size and have equal entries in every corresponding location

> Matrices A, with entries a_{ij}, and B, with entries b_{ij}, are **equal**, written $A = B$, if A and B are of the same size and $a_{ij} = b_{ij}$ for every row i and every column j.

Example 2. In each of the following, determine whether or not $A = B$.

(a) $A = \begin{bmatrix} 2 & 1 & 3 \\ 4 & -1 & 0 \end{bmatrix}$ and $B = \begin{bmatrix} 1+1 & 1 & 4-1 \\ 4 & -1 & 2-2 \end{bmatrix}$

(b) $A = \begin{bmatrix} -1 & 0 \\ 1 & 1 \end{bmatrix}$ and $B = \begin{bmatrix} 0 & -1 \\ 1 & 1 \end{bmatrix}$

(c) $A = \begin{bmatrix} 0 & 0 & 0 \\ 0 & 0 & 0 \end{bmatrix}$ and $B = \begin{bmatrix} 0 & 0 \\ 0 & 0 \end{bmatrix}$

Solution

(a) Yes. Both matrices are 2×3 and certainly $a_{12} = b_{12}$, $a_{21} = b_{21}$, and $a_{22} = b_{22}$. Moreover,
$$b_{11} = 1 + 1 = 2 = a_{11}, \quad b_{13} = 4 - 1 = 3 = a_{13}, \quad b_{23} = 2 - 2 = 0 = a_{23}$$

(b) No, even though both A and B are 2×2-matrices. In fact, $a_{11} = -1$, whereas $b_{11} = 0$; so $a_{11} \neq b_{11}$. If matrices differ in any location, then they cannot be equal.

(c) No, because even though all the numbers in both matrices are zeros, the matrices are of different sizes; A is the 2×3-**zero matrix** and B is the 2×2-**zero matrix**. ∎

Zero Matrices

Matrices, such as the ones of Example 2(c), in which the entries in every location are zeros are called **zero matrices**.

Example 3. Solve for x, y, and z.

$$\begin{bmatrix} x & -1 \\ 0 & y+1 \\ 2z & 2 \end{bmatrix} = \begin{bmatrix} 1 & 1-2 \\ 0 & 2 \\ 4 & 1+1 \end{bmatrix}$$

Solution

If we call the matrices A and B, where

$$A = \begin{bmatrix} x & -1 \\ 0 & y+1 \\ 2z & 2 \end{bmatrix} \quad \text{and} \quad B = \begin{bmatrix} 1 & 1-2 \\ 0 & 2 \\ 4 & 1+1 \end{bmatrix}$$

then we are told that $A = B$. Since $a_{11} = x$ and $b_{11} = 1$, we know, by the definition of matrix equality, that $a_{11} = b_{11}$, that is, that $x = 1$. Similarly, $a_{22} = y + 1$, whereas $b_{22} = 2$; thus, $y + 1 = 2$, so that $y = 1$. Finally, $a_{31} = 2z$ and $b_{31} = 4$; therefore, $2z = 4$ and thus, $z = 2$. ■

Addition and Subtraction of Matrices

Addition or subtraction of two matrices to produce a third matrix is possible only if the two matrices are of the same size, and then their sum or difference will also be of that size. Otherwise, the sum or difference is just not defined—that is, there is no such thing. To add or subtract matrices of the same size, simply add (or subtract) the corresponding entries of the two matrices.

> To add matrices A and B of the same size, form the sum matrix $C = A + B$ by defining $c_{ij} = a_{ij} + b_{ij}$ for every row i and every column j. The difference matrix $D = A - B$ is defined by $d_{ij} = a_{ij} - b_{ij}$ in every location.

Example 4. Perform the indicated addition or subtraction, if possible.

(a) $\begin{bmatrix} 1 & 2 & -1 \\ 3 & 2 & 0 \\ 1 & 1 & 2 \end{bmatrix} + \begin{bmatrix} 3 & 1 & 3 \\ -1 & -1 & -2 \\ 1 & 1 & -1 \end{bmatrix}$

(b) $\begin{bmatrix} 1 & -1 & 2 \end{bmatrix} + \begin{bmatrix} 0 & 1 & 1 \end{bmatrix}$

(c) $\begin{bmatrix} -1 & 0 \\ 1 & 2 \\ 2 & 0 \end{bmatrix} + \begin{bmatrix} 1 \\ 1 \\ 3 \end{bmatrix}$

(d) $\begin{bmatrix} 2 & 3 & 2 \\ -1 & 4 & 2 \end{bmatrix} - \begin{bmatrix} 3 & 0 & 1 \\ -2 & -2 & 2 \end{bmatrix}$

(e) $\begin{bmatrix} 2 & -1 \\ 0 & 0 \\ 2 & 1 \end{bmatrix} + \begin{bmatrix} 0 & 0 \\ 0 & 0 \\ 0 & 0 \end{bmatrix}$

Solution

(a) $$\begin{bmatrix} 1 & 2 & -1 \\ 3 & 2 & 0 \\ 1 & 1 & 2 \end{bmatrix} + \begin{bmatrix} 3 & 1 & 3 \\ -1 & -1 & -2 \\ 1 & 1 & -1 \end{bmatrix} = \begin{bmatrix} 1+3 & 2+1 & -1+3 \\ 3+(-1) & 2+(-1) & 0+(-2) \\ 1+1 & 1+1 & 2+(-1) \end{bmatrix}$$

$$= \begin{bmatrix} 4 & 3 & 2 \\ 2 & 1 & -2 \\ 2 & 2 & 1 \end{bmatrix}$$

(b) $$\begin{array}{r} \begin{bmatrix} 1 & -1 & 2 \end{bmatrix} \\ + \begin{bmatrix} 0 & 1 & 1 \end{bmatrix} \\ \hline \begin{bmatrix} 1 & 0 & 3 \end{bmatrix} \end{array}$$

(c) The sum is not defined because the left-hand matrix is 3×2, but the right-hand matrix is a different size, namely, 3×1.

(d) $$\begin{bmatrix} 2 & 3 & 2 \\ -1 & 4 & 2 \end{bmatrix} - \begin{bmatrix} 3 & 0 & 1 \\ -2 & -2 & 2 \end{bmatrix} = \begin{bmatrix} 2-3 & 3-0 & 2-1 \\ (-1)-(-2) & 4-(-2) & 2-2 \end{bmatrix}$$

$$= \begin{bmatrix} -1 & 3 & 1 \\ 1 & 6 & 0 \end{bmatrix}$$

(e) $$\begin{bmatrix} 2 & -1 \\ 0 & 0 \\ 2 & 1 \end{bmatrix} + \begin{bmatrix} 0 & 0 \\ 0 & 0 \\ 0 & 0 \end{bmatrix} = \begin{bmatrix} 2+0 & (-1)+0 \\ 0+0 & 0+0 \\ 2+0 & 1+0 \end{bmatrix} = \begin{bmatrix} 2 & -1 \\ 0 & 0 \\ 2 & 1 \end{bmatrix}$$ ∎

Properties of Addition

As Part (e) of Example 4 illustrates, zero matrices behave like 0 does in ordinary arithmetic: just as $1 + 0 = 1$ and $0 + 14 = 14$, so too, a matrix does not change when a zero matrix is added to it. However, in matrix theory there are many zero matrices—one for each size. If A is a matrix and O is a zero matrix of the same size, then

$$A + O = A$$

Addition of matrices behaves just like addition of ordinary numbers in the sense that it does not matter in what order matrices are added. This is the *commutative property* of matrix addition.

> If A and B are matrices of the same size, then addition is **commutative**, that is,
>
> $$A + B = B + A$$

Example 5. Given the matrices

$$A = \begin{bmatrix} -1 & 0 \\ 2 & 1 \end{bmatrix} \quad \text{and} \quad B = \begin{bmatrix} 0 & 2 \\ 2 & 2 \end{bmatrix}$$

compute $A + B$ and $B + A$.

Solution

$$A + B = \begin{bmatrix} -1 & 0 \\ 2 & 1 \end{bmatrix} + \begin{bmatrix} 0 & 2 \\ 2 & 2 \end{bmatrix} = \begin{bmatrix} (-1)+0 & 0+2 \\ 2+2 & 1+2 \end{bmatrix}$$

$$= \begin{bmatrix} -1 & 2 \\ 4 & 3 \end{bmatrix}$$

$$B + A = \begin{bmatrix} 0 & 2 \\ 2 & 2 \end{bmatrix} + \begin{bmatrix} -1 & 0 \\ 2 & 1 \end{bmatrix} = \begin{bmatrix} 0+(-1) & 2+0 \\ 2+2 & 2+1 \end{bmatrix}$$

$$= \begin{bmatrix} -1 & 2 \\ 4 & 3 \end{bmatrix}$$ ∎

It is clear from the example why we always have the commutative property, $A + B = B + A$. For instance, just as $(-1) + 0 = 0 + (-1)$ in the first row and first column of both $A + B$ and $B + A$, we will always have $a_{ij} + b_{ij} = b_{ij} + a_{ij}$ because we are simply adding ordinary numbers in every location. It may seem strange that we are making such a point of the fact that matrix addition is commutative. The reason is that, in contrast:

> Matrix multiplication is not commutative.

That is, $2 \cdot 3 = 3 \cdot 2$ and, in general, $ab = ba$ for all numbers a and b. But for matrices A and B, the product AB is not necessarily the same as BA, as we shall see in the next section.

Solution of Equations

With these rules of arithmetic for matrices, it is possible to solve simple matrix equations.

Example 6. Solve the following equation for the matrix X:

$$\begin{bmatrix} 1 & 2 \\ 3 & -1 \end{bmatrix} + X = \begin{bmatrix} 0 & 2 \\ 2 & 0 \end{bmatrix}$$

Solution

The problem is to find a matrix X, which must be of size 2 by 2, so that when it is added to the matrix on its left, we obtain the matrix to the right of the equals sign. Just as the ordinary equation $3 + x = 2$ is solved by subtracting 3 from both sides to obtain $x = 2 - 3$, so too, the matrix equation is solved by subtracting the left-hand matrix from both sides:

$$X = \begin{bmatrix} 0 & 2 \\ 2 & 0 \end{bmatrix} - \begin{bmatrix} 1 & 2 \\ 3 & -1 \end{bmatrix} = \begin{bmatrix} -1 & 0 \\ -1 & 1 \end{bmatrix}$$

We can check the solution by adding:

$$\begin{bmatrix} 1 & 2 \\ 3 & -1 \end{bmatrix} + \begin{bmatrix} -1 & 0 \\ -1 & 1 \end{bmatrix} = \begin{bmatrix} 0 & 2 \\ 2 & 0 \end{bmatrix}$$

■

Scalar Multiplication

There are two kinds of multiplication that involve matrices: scalar multiplication and matrix multiplication. Here, we introduce scalar multiplication; the next section is devoted to matrix multiplication. In **scalar multiplication** we multiply a matrix A by a single number k (called a "scalar" to emphasize that we are thinking of k as a constant number and not as a 1-by-1 matrix) to form a matrix kA. To multiply the matrix A by k, we just multiply each number in A by k.

If A is an $m \times n$-matrix with entries a_{ij} and k is a number, then the **scalar product** kA is the $m \times n$-matrix with entry ka_{ij} in the ith row and jth column.

For instance, when we multiply the matrix

$$A = \begin{bmatrix} 1 & 0 & -2 & 1 \\ 2 & 1 & -3 & 2 \\ -1 & 1 & 4 & 0 \end{bmatrix}$$

by 3, the scalar product is

$$3A = \begin{bmatrix} 3(1) & 3(0) & 3(-2) & 3(1) \\ 3(2) & 3(1) & 3(-3) & 3(2) \\ 3(-1) & 3(1) & 3(4) & 3(0) \end{bmatrix} = \begin{bmatrix} 3 & 0 & -6 & 3 \\ 6 & 3 & -9 & 6 \\ -3 & 3 & 12 & 0 \end{bmatrix}$$

Example 7. Given the matrices

$$A = \begin{bmatrix} 1 & -2 \\ -1 & 1 \\ 0 & 2 \end{bmatrix} \quad \text{and} \quad B = \begin{bmatrix} -1 & 3 \\ 0 & 2 \\ -2 & -1 \end{bmatrix}$$

find the following matrices:

(a) $-2A$ (b) $4A - 3B$ (c) $\frac{1}{2}(A + B)$

Solution

(a)
$$-2A = \begin{bmatrix} -2(1) & -2(-2) \\ -2(-1) & -2(1) \\ -2(0) & -2(2) \end{bmatrix} = \begin{bmatrix} -2 & 4 \\ 2 & -2 \\ 0 & -4 \end{bmatrix}$$

(b) We first compute

$$4A = \begin{bmatrix} 4 & -8 \\ -4 & 4 \\ 0 & 8 \end{bmatrix} \quad \text{and} \quad 3B = \begin{bmatrix} -3 & 9 \\ 0 & 6 \\ -6 & -3 \end{bmatrix}$$

and then subtract

$$4A - 3B = \begin{bmatrix} 4 & -8 \\ -4 & 4 \\ 0 & 8 \end{bmatrix} - \begin{bmatrix} -3 & 9 \\ 0 & 6 \\ -6 & -3 \end{bmatrix} = \begin{bmatrix} 7 & -17 \\ -4 & -2 \\ 6 & 11 \end{bmatrix}$$

(c)
$$A + B = \begin{bmatrix} 1 & -2 \\ -1 & 1 \\ 0 & 2 \end{bmatrix} + \begin{bmatrix} -1 & 3 \\ 0 & 2 \\ -2 & -1 \end{bmatrix} = \begin{bmatrix} 0 & 1 \\ -1 & 3 \\ -2 & 1 \end{bmatrix}$$

and therefore,

$$\frac{1}{2}(A + B) = \frac{1}{2}\begin{bmatrix} 0 & 1 \\ -1 & 3 \\ -2 & 1 \end{bmatrix} = \begin{bmatrix} 0 & \frac{1}{2} \\ -\frac{1}{2} & \frac{3}{2} \\ -1 & \frac{1}{2} \end{bmatrix}$$ ∎

Using Matrix Operations

When we present data from a rectangular table of numbers, regarded as a matrix, we can use matrix addition, matrix subtraction, and scalar multiplication to obtain further information about the data. Here is a simplified version of a typical application. Suppose a manufacturer of sailboats makes just two models, small and large, and sells them through retailers in three locations—Seattle, San Diego, and Honolulu. The manufacturer records the sales for the months of May and June as shown in Table 1-1. Thus, for instance, the San Diego retailer sold 4 small sailboats in May and the Seattle retailer sold 3 large sailboats in June.

12   MATRICES AND SYSTEMS OF LINEAR EQUATIONS

	May				June		
	Seattle	San Diego	Honolulu		Seattle	San Diego	Honolulu
Small	6	4	9	Small	8	6	6
Large	2	1	2	Large	3	0	4

Table 1-1

We represent the tables by matrices named for the months:

$$M = \begin{bmatrix} 6 & 4 & 9 \\ 2 & 1 & 2 \end{bmatrix} \text{ and } J = \begin{bmatrix} 8 & 6 & 6 \\ 3 & 0 & 4 \end{bmatrix}$$

Now if we wish to calculate the total sales for the two months, classified by location and size of boat, we add

$$M + J = \begin{bmatrix} 6 & 4 & 9 \\ 2 & 1 & 2 \end{bmatrix} + \begin{bmatrix} 8 & 6 & 6 \\ 3 & 0 & 4 \end{bmatrix} = \begin{bmatrix} 14 & 10 & 15 \\ 5 & 1 & 6 \end{bmatrix}$$

If we wish to calculate the change in sales from May to June, we should subtract. If we use $J - M$ (rather than $M - J$), then improvements in sales will be recorded as positive numbers in the matrix:

$$J - M = \begin{bmatrix} 8 & 6 & 6 \\ 3 & 0 & 4 \end{bmatrix} - \begin{bmatrix} 6 & 4 & 9 \\ 2 & 1 & 2 \end{bmatrix} = \begin{bmatrix} 2 & 2 & -3 \\ 1 & -1 & 2 \end{bmatrix}$$

Suppose September sales are expected to be only about $\frac{1}{3}$ those of June. To predict September sales, the manufacturer can compute

$$\frac{1}{3}J = \frac{1}{3}\begin{bmatrix} 8 & 6 & 6 \\ 3 & 0 & 4 \end{bmatrix} = \begin{bmatrix} \frac{8}{3} & 2 & 2 \\ 1 & 0 & \frac{4}{3} \end{bmatrix}$$

Remember that to find the *average* of two numbers, we add them and divide the sum by 2 or, what is the same thing, multiply the sum by $\frac{1}{2}$. Thus, the average of 3 and 11 is

$$\tfrac{1}{2}(3 + 11) = \tfrac{1}{2}(14) = 7$$

We can compute the average sales during the months of May and June, according to the location and size of sailboat, by computing the average of the matrices M and J, that is,

$$\frac{1}{2}(M+J) = \frac{1}{2}\begin{bmatrix} 14 & 10 & 15 \\ 5 & 1 & 6 \end{bmatrix} = \begin{bmatrix} 7 & 5 & \frac{15}{2} \\ \frac{5}{2} & \frac{1}{2} & 3 \end{bmatrix}$$

Summary

A **matrix** is any rectangular array of numbers. **Rows** are numbered from top to bottom, **columns** from left to right. A **location** in a matrix is specified by giving the row number and the column number, in that order. If the matrix is called A, the **entry** in row i and column j is represented by the symbol a_{ij}. The **size** of a matrix is $m \times n$ if it has m rows and n columns. An $n \times n$-matrix is called a **square matrix**, a $1 \times n$-matrix is a **row matrix**, and an $n \times 1$-matrix is a **column matrix**. Two matrices are **equal** provided that they are the same size and have equal entries in each location, that is,

$A = B$ means that $a_{ij} = b_{ij}$ for every i and j

Two matrices can be added (or subtracted) only if they are the same size. Then the numbers in the same locations in the two matrices are added (or subtracted) to produce their sum (or difference) in that same location. The result is a matrix of that same size. Matrix addition has the **commutativity** property:

$$A + B = B + A$$

A **zero matrix** is a matrix with 0 in every location. A zero matrix, when added to another matrix, does not change that matrix. **Scalar multiplication** means to multiply a matrix A by a single number k to form a matrix kA by the rule: the number a_{ij} in the ith row and jth column of A is multiplied by k to obtain the number ka_{ij} in the same location in kA.

Exercises

1. For the matrix

   $$\begin{bmatrix} 2 & -1 & 1 & 1 \\ 3 & 1 & 0 & 4 \end{bmatrix}$$

   identify:
   (a) The second column
   (b) The size of the matrix
   (c) The location of the number 4

2. For the matrix

   $$\begin{bmatrix} 0.05 & 2 & 1 \\ 1 & -1 & -1.3 \\ 0.1 & 0 & 0.02 \end{bmatrix}$$

   identify:
   (a) The third row
   (b) The size of the matrix
   (c) a_{23}

14   MATRICES AND SYSTEMS OF LINEAR EQUATIONS

3. For the matrix
$$\begin{bmatrix} -1 \\ 2 \\ -3 \\ 4 \\ 0 \end{bmatrix}$$
identify:
(a) The fifth row
(b) The size of the matrix
(c) The location of the number -3

4. For the matrix
$$\begin{bmatrix} 2 & 1 \\ -1 & -1 \\ 0 & 1 \end{bmatrix}$$
identify:
(a) The first column
(b) The size of the matrix
(c) The number in the second row and the third column

5. Find examples of matrices of the following sizes:
   (a) 1×2 (b) 3×4 (c) 2×1 (d) 5×2

In Exercises 6 through 10, determine whether $A = B$. If the matrices are not equal, in what way do they fail to be equal?

6. $A = \begin{bmatrix} 2 & 2^2 \\ -1 & 3+2 \end{bmatrix}$ and $B = \begin{bmatrix} 3-1 & 4 \\ -1 & 5 \end{bmatrix}$

7. $A = \begin{bmatrix} \frac{3}{2} & -2 & 0.1 \end{bmatrix}$ and $B = \begin{bmatrix} 1+\frac{1}{2} & 1-3 & \frac{1}{10} \end{bmatrix}$

8. $A = \begin{bmatrix} 2 \\ 0 \end{bmatrix}$ and $B = \begin{bmatrix} 2 & 0 \end{bmatrix}$

9. $A = \begin{bmatrix} 2 & 1 \\ 0 & 0 \\ 0 & 1 \end{bmatrix} + \begin{bmatrix} -1 & -1 \\ 0 & -1 \\ 2 & 2 \end{bmatrix}$ and $B = \begin{bmatrix} 1 & 0 \\ 0 & -1 \\ 2 & 3 \end{bmatrix}$

10. $A = \begin{bmatrix} 2 & 0 \\ 1 & 0 \end{bmatrix}$ and $B = \begin{bmatrix} 2 & 0 \\ 2-1 & -3 \end{bmatrix}$

In Exercises 11 through 15, solve for x and y, if possible.

11. $\begin{bmatrix} -1 & x \\ 2 & y \end{bmatrix} = \begin{bmatrix} -1 & 4 \\ 1+1 & -1 \end{bmatrix}$

12. $\begin{bmatrix} 1 & x \\ x & 2y \end{bmatrix} = \begin{bmatrix} 1 & 4 \\ 3+2 & 6 \end{bmatrix}$

13. $\begin{bmatrix} x \\ y \end{bmatrix} = \begin{bmatrix} 2x+1 \\ -y+2 \end{bmatrix}$

14. $\begin{bmatrix} x+1 & y-2 \end{bmatrix} = \begin{bmatrix} 2 & -3 \end{bmatrix} + \begin{bmatrix} x & 1-y \end{bmatrix}$

15. $\begin{bmatrix} x \\ y \\ 0 \end{bmatrix} = \begin{bmatrix} 2x \\ 4 \\ 2 \end{bmatrix} + \begin{bmatrix} 1 \\ 1 \\ -2 \end{bmatrix}$

In Exercises 16 through 20, perform the arithmetic operations, if possible.

16. $\begin{bmatrix} -1 & 0 \\ 0 & 0 \end{bmatrix} - \begin{bmatrix} 2 & 2 \\ -1 & 2 \end{bmatrix}$

17. $\begin{bmatrix} 2 & \frac{1}{2} & -1 \end{bmatrix} + \begin{bmatrix} -1 & -1 & -1 \end{bmatrix}$

18. $\begin{bmatrix} 0 \\ 1 \\ 1 \end{bmatrix} + \begin{bmatrix} 5 \\ 2 \\ 0 \end{bmatrix}$

19. $\begin{bmatrix} 2 & -1 \\ 1 & 1 \end{bmatrix} + \begin{bmatrix} 0 \\ 0 \end{bmatrix}$

20. $\begin{bmatrix} -1 & 2 \\ 1 & 1 \\ 1 & 0 \end{bmatrix} - \begin{bmatrix} 2 & 1 \\ 3 & 1 \\ 0 & 0 \end{bmatrix}$

In Exercises 21 through 26, find the unknown matrix X.

21. $X + \begin{bmatrix} 2 \\ 1 \end{bmatrix} = \begin{bmatrix} -1 \\ -1 \end{bmatrix}$

22. $X - \begin{bmatrix} 2 & 1 \\ -1 & 2 \end{bmatrix} = \begin{bmatrix} 3 & -1 \\ 0 & 1 \end{bmatrix}$

23. $\begin{bmatrix} -1 & 0 \\ 0 & 1 \\ 1 & 1 \end{bmatrix} - X = \begin{bmatrix} 2 & 1 \\ 0 & 0 \\ 0 & 0 \end{bmatrix}$

24. $\left(\begin{bmatrix} 2 & 1 & 4 \end{bmatrix} + X \right) + \begin{bmatrix} -3 & 1 & 1 \end{bmatrix} = \begin{bmatrix} 0 & 0 & 2 \end{bmatrix}$

25. $\begin{bmatrix} 2 \\ 1 \\ 2 \end{bmatrix} + X = \begin{bmatrix} -2 \\ -2 \\ -2 \end{bmatrix}$

26. $\begin{bmatrix} 2 & 1 \\ 3 & 2 \end{bmatrix} - X = \begin{bmatrix} 0 & 0 \\ 0 & 0 \end{bmatrix}$

27. Given the matrices
$$A = \begin{bmatrix} 0 & 2 & -2 \\ -1 & 2 & 1 \\ 1 & 0 & 0 \end{bmatrix} \text{ and } B = \begin{bmatrix} 1 & 1 & -1 \\ 2 & 0 & 0 \\ -1 & -2 & -3 \end{bmatrix}$$
calculate:
(a) $4A$ (b) $A + 2B$ (c) $3B - 5A$

28. Given the matrices
$$A = \begin{bmatrix} 2 & 1 & 0 & 2 \\ -1 & 1 & 2 & 1 \end{bmatrix} \text{ and } B = \begin{bmatrix} 3 & 0 & -1 & 4 \\ -2 & 1 & 0 & 2 \end{bmatrix}$$
calculate:
(a) $-3B$ (b) $2A - B$ (c) $\frac{1}{2}(A + B)$

16 MATRICES AND SYSTEMS OF LINEAR EQUATIONS

29. Solve for x and y:

$$\begin{bmatrix} 0 & x \\ y & 1 \end{bmatrix} = \begin{bmatrix} 2 & 2 \\ 2y & -3 \end{bmatrix} + 2\begin{bmatrix} -1 & -4 \\ 1 & 2 \end{bmatrix}$$

30. Find the unknown matrix X:

$$X + 2\begin{bmatrix} 1 & -1 \\ 0 & 2 \\ 1 & 0 \end{bmatrix} = 3\begin{bmatrix} 2 & 0 \\ 1 & -1 \\ -1 & -2 \end{bmatrix}$$

31. A regional tire distributor has retail stores in Boston, Providence, Hartford, and New Haven. The value of sales of auto and truck tires in November and December are given (in units of $1000) in Table 1-2.

November	Auto	Truck
Boston	40	15
Providence	55	30
Hartford	82	45
New Haven	80	32

December	Auto	Truck
Boston	42	21
Providence	53	26
Hartford	80	50
New Haven	78	33

Table 1-2

Representing these tables by 4×2-matrices N and D:

(a) Calculate total sales for the two months, that is, find $N + D$.

(b) Calculate $D - N$ to determine the change in sales from November to December.

(c) Calculate the average sales, $\frac{1}{2}(N + D)$.

(d) Suppose that sales of tires in January are usually 10% higher than in December. Predict January sales by calculating $D + (0.1)D$.

32. A manufacturer has three plants located in Austin, Dallas, and Houston. Each plant has a production line, a packaging/shipping division, and an office staff. The total salaries at each plant for a quarter of the year (three-month period) are summarized in Table 1-3, in millions of dollars:

First Quarter	Austin	Dallas	Houston
Production	5.1	4.2	7.0
Packaging	1.0	0.7	1.2
Office	2.0	1.8	2.0

Second Quarter	Austin	Dallas	Houston
Production	5.0	4.3	7.2
Packaging	1.1	0.7	1.3
Office	2.0	1.9	2.1

Table 1-3

Representing the table of first-quarter salaries by a matrix A and that for the second-quarter salaries by B:

(a) Calculate $A + B$, the total salaries in each division at each plant for the entire two-quarter (six-month) period.
(b) Calculate $B - A$, which describes the changes in total salaries between the two quarters.
(c) Calculate $\frac{1}{2}(A + B)$, the average salary costs by category.
(d) Suppose that because of declining sales, the manufacturer must reduce its salary expenses by 2% from those of the second quarter. Calculate $(0.02)B$ to determine how much salary savings each plant must achieve in the next quarter at each division, expressed in *thousands* of dollars. (*Note*: 1% of one million dollars is $10,000.)

1.2 Matrix Multiplication

Row-Column Multiplication

To explain the rules for multiplying matrices, we begin with the simplest case. Suppose that A is a row matrix, B a column matrix, and the number of entries in A equals the number of entries in B. The rules for multiplication are:

1. Start at the left of A and at the top of B.
2. Moving to the right in A and down in B, multiply corresponding (first, second, etc.) numbers.
3. Add up the result.

Example 1. Find the product AB, where

$$A = \begin{bmatrix} 4 & -2 & 1 \end{bmatrix} \quad \text{and} \quad B = \begin{bmatrix} 0 \\ -1 \\ -3 \end{bmatrix}$$

Solution

Notice that A and B each have 3 entries. Applying the rules yields

$$AB = 4(0) + (-2)(-1) + 1(-3) = 0 + 2 + (-3) = -1$$

Thus, the product of the 1×3-matrix A and the 3×1-matrix B is the *number* -1. We may also write -1 as $[-1]$ and think of it as a *matrix* with only one row and one column. In this way the product of these matrices may be thought of either as a number or as another matrix, depending on how it will be used. ■

A general formula for this simplest case of the multiplication of matrices is shown in the following box. Notice that we are using a simpler notation than in the first section, where for the row matrix

18   MATRICES AND SYSTEMS OF LINEAR EQUATIONS

$$A = \begin{bmatrix} -3 & -2 & 0 \end{bmatrix}$$

we would have written

$$a_{11} = -3, a_{12} = -2, \text{ and } a_{13} = 0$$

Since there is just one row, row number 1, we will omit the row number. Similarly, we will omit the column number (always "1") from the locations in a column matrix.

Writing a $1 \times n$-(row) matrix A as

$$A = \begin{bmatrix} a_1 & a_2 & a_3 & \ldots & a_n \end{bmatrix}$$

and an $n \times 1$-(column) matrix B with the same number, n, of entries as

$$B = \begin{bmatrix} b_1 \\ b_2 \\ b_3 \\ \cdot \\ \cdot \\ b_n \end{bmatrix}$$

the multiplication rule for such matrices is

$$AB = a_1 b_1 + a_2 b_2 + a_3 b_3 + \ldots + a_n b_n$$

which may be thought of as a 1×1-matrix.

An Illustration of Matrix Multiplication

The definition of matrix multiplication may seem somewhat unnatural. There are, however, mathematical problems that are quite readily and usefully expressed in the language of matrix multiplication.

Example 2. A shopper buys four half-gallons of milk, two loaves of bread, three pounds of coffee, two 5-pound sacks of flour, and one 5-pound bag of sugar at the prices given in Table 1-4. What is the total cost of the order?

Item	Cost
Milk	$0.95 per half-gallon
Bread	$1.05 a loaf
Coffee	$3.50 a pound
Flour	$1.10 for a 5-pound sack
Sugar	$1.60 for a 5-pound bag

Table 1-4

Solution

The cost of four half-gallons of milk is ($0.95)(4) = $3.80. Similarly, two loaves of bread cost ($1.05)(2) = $2.10, and so on. Let c stand for the total cost of the purchases. Then

$$c = (\$0.95)(4) + (\$1.05)(2) + (\$3.50)(3) + (\$1.10)(2) + (\$1.60)(1)$$

which has the form of a row matrix times a column matrix. Let us see which matrices are involved. Define a 1×5-(row) matrix P of prices

$$P = [\$0.95 \quad \$1.05 \quad \$3.50 \quad \$1.10 \quad \$1.60]$$

and a 5×1-(column) matrix Q of quantities

$$Q = \begin{bmatrix} 4 \\ 2 \\ 3 \\ 2 \\ 1 \end{bmatrix}$$

Notice that the prices of the food items are in the order: milk, bread, coffee, flour, sugar, from left to right in P, and the quantities are in the same order from top to bottom in Q. By the definition of matrix multiplication,

$$PQ = \begin{bmatrix} \$0.95 & \$1.05 & \$3.50 & \$1.10 & \$1.60 \end{bmatrix} \begin{bmatrix} 4 \\ 2 \\ 3 \\ 2 \\ 1 \end{bmatrix}$$

$$= \left[(\$0.95)(4) + (\$1.05)(2) + (\$3.50)(3) + (\$1.10)(2) + (\$1.60)(1) \right]$$

$$= \$20.20$$

Thus, the formula for the cost of the groceries is given by the matrix equation

$$[c] = PQ \qquad \blacksquare$$

The same procedure can be applied to any list of purchases. Letting P be the matrix of prices and Q the matrix of quantities purchased, then $[c] = PQ$ will determine the total cost. Thus, if P is known, several possible shopping lists (matrices for Q) can be multiplied by P to find which combinations of purchases are possible within a fixed budget (that is, keeping c less than or equal to a given amount.)

The Size Restriction

Although the rule for multiplying a $1 \times n$-row matrix and an $n \times 1$-column matrix may be unfamiliar, it is really the only new idea you will have to deal with. The rules for multiplying in general can be reduced to performing this special case over and over again. However, before we can state the rules for computing the matrix product AB in general, we have to explain the way in which the sizes of the matrices enter into the picture. You have seen size restrictions when we add and subtract matrices, so you will not be surprised to learn that the product of matrices A and B, written AB, is also defined only for matrices A and B of the appropriate sizes. However, the size restriction is different here.

> In order to form the product AB, the number of columns of A, the matrix on the left, must equal the number of rows of B, the matrix on the right. Thus, A must be of size $m \times k$ and B of size $k \times n$, for some whole numbers m, k, and n. The product matrix, AB, will have as many rows as A and as many columns as B, that is, AB will be an $m \times n$-matrix.

Example 3. Is the product AB defined in each of the following cases? If so, what size will it be?

(a)
$$A = \begin{bmatrix} 1 & 2 & -1 \end{bmatrix} \quad \text{and} \quad B = \begin{bmatrix} 0 \\ -2 \\ 3 \end{bmatrix}$$

(b)
$$A = \begin{bmatrix} 2 & 1 & -1 & 2 \\ 0 & 0 & 1 & 2 \end{bmatrix} \quad \text{and} \quad B = \begin{bmatrix} 2 & 1 & 0 \\ 1 & 2 & -1 \\ 0 & 2 & -1 \\ 0 & 1 & 0 \end{bmatrix}$$

(c) $A = \begin{bmatrix} 2 & 1 & 2 \end{bmatrix}$ and $B = \begin{bmatrix} 3 & 2 & 1 \end{bmatrix}$

Solution

(a) Yes, because A has 3 columns and B has 3 rows; AB be will be 1×1 because A has only one row and B only one column.
(b) Yes, because A is of size 2×4 and B is 4×3, so A has 4 columns and B has 4 rows. The product AB will be 2×3.
(c) No, because A and B are both of size 1×3, so A has 3 columns but B has only 1 row. ∎

Order of Multiplication

In the last section we learned that addition of matrices is commutative, just like addition of ordinary numbers: If $A + B$ makes sense, so does $B + A$ and the sum is the same: $A + B = B + A$. In Part (b) of Example 3 we saw that

$$AB = \begin{bmatrix} 2 & 1 & -1 & 2 \\ 0 & 0 & 1 & 2 \end{bmatrix} \begin{bmatrix} 2 & 1 & 0 \\ 1 & 2 & -1 \\ 0 & 2 & -1 \\ 0 & 1 & 0 \end{bmatrix}$$

is a 2 × 3-matrix. We claim that

$$BA = \begin{bmatrix} 2 & 1 & 0 \\ 1 & 2 & -1 \\ 0 & 2 & -1 \\ 0 & 1 & 0 \end{bmatrix} \begin{bmatrix} 2 & 1 & -1 & 2 \\ 0 & 0 & 1 & 2 \end{bmatrix}$$

is not defined because the left-hand matrix (now B) has 3 columns, but there are only 2 rows in the right-hand matrix (now A). Thus, the existence of AB does not automatically imply the existence of BA. So, it is important to keep track of the order in which matrices are multiplied.

Multiplication in General

The general rule for the multiplication of matrices is:

If A is an $m \times k$-matrix and B is a $k \times n$-matrix, then to compute the entry in the ith row and the jth column of the $m \times n$-product matrix AB, multiply the entire ith row of A by the entire jth column of B.

Example 4. Compute AB, where

$$A = \begin{bmatrix} -3 & -2 & 0 \end{bmatrix} \quad \text{and} \quad B = \begin{bmatrix} 4 & 2 \\ -1 & 1 \\ 3 & 5 \end{bmatrix}$$

Solution

Since A has 1 row and B has 2 columns, we know that AB must be a 1×2-matrix, that is, it is of the form

$$AB = \begin{bmatrix} x_1 & x_2 \end{bmatrix}$$

The problem is then to define the numbers x_1 and x_2. The procedure is to think of each column of B as a 3×1-matrix in its own right and multiply that matrix by the single row of A according to the rule we used before. That is, let

$$B_1 = \begin{bmatrix} 4 \\ -1 \\ 3 \end{bmatrix} \quad \text{and} \quad B_2 = \begin{bmatrix} 2 \\ 1 \\ 5 \end{bmatrix}$$

22 MATRICES AND SYSTEMS OF LINEAR EQUATIONS

and compute

$$AB_1 = \begin{bmatrix} -3 & -2 & 0 \end{bmatrix} \begin{bmatrix} 4 \\ -1 \\ 3 \end{bmatrix} = (-3)(4) + (-2)(-1) + (0)(3) = -10$$

and

$$AB_2 = \begin{bmatrix} -3 & -2 & 0 \end{bmatrix} \begin{bmatrix} 2 \\ 1 \\ 5 \end{bmatrix} = (-3)(2) + (-2)(1) + (0)(5) = -8$$

In this case the definition gives

$$x_1 = AB_1 = -10, \qquad x_2 = AB_2 = -8$$

Therefore,

$$AB = \begin{bmatrix} -3 & -2 & 0 \end{bmatrix} \begin{bmatrix} 4 & 2 \\ -1 & 1 \\ 3 & 5 \end{bmatrix} = \begin{bmatrix} -10 & -8 \end{bmatrix}$$

∎

Example 5. Compute AB, where

$$A = \begin{bmatrix} 0 & 1 & 4 \\ -1 & 3 & -2 \end{bmatrix} \quad \text{and} \quad B = \begin{bmatrix} 1 & -2 \\ -3 & 5 \\ -1 & 0 \end{bmatrix}$$

Solution

Since A is a 2×3-matrix and B is 3×2, we know that AB must be a 2×2-matrix, that is, AB is of the form

$$AB = \begin{bmatrix} w & x \\ y & z \end{bmatrix}$$

We think of B as a collection of columns,

$$B = \begin{bmatrix} B_1 & B_2 \\ \begin{bmatrix} 1 \\ -3 \\ -1 \end{bmatrix} & \begin{bmatrix} -2 \\ 5 \\ 0 \end{bmatrix} \end{bmatrix}$$

and of A as a collection of rows,

$$A = \begin{bmatrix} 0 & 1 & 4 \\ -1 & 3 & -2 \end{bmatrix} \begin{matrix} A_1 \\ A_2 \end{matrix}$$

Now, each row of A is itself a 1×3-matrix, and can be multiplied by each column of B, a 3×1-matrix, to produce a number. There are four such calculations in this example:

$$A_1 B_1 = \begin{bmatrix} 0 & 1 & 4 \end{bmatrix} \begin{bmatrix} 1 \\ -3 \\ -1 \end{bmatrix} = 0 + (-3) + (-4) = -7$$

$$A_1B_2 = \begin{bmatrix} 0 & 1 & 4 \end{bmatrix} \begin{bmatrix} -2 \\ 5 \\ 0 \end{bmatrix} = 0 + 5 + 0 = 5$$

$$A_2B_1 = \begin{bmatrix} -1 & 3 & -2 \end{bmatrix} \begin{bmatrix} 1 \\ -3 \\ -1 \end{bmatrix} = (-1) + (-9) + 2 = -8$$

$$A_2B_2 = \begin{bmatrix} -1 & 3 & -2 \end{bmatrix} \begin{bmatrix} -2 \\ 5 \\ 0 \end{bmatrix} = 2 + 15 + 0 = 17$$

Since w is in the first row and first column of AB, w is the product of the first row, A_1, of A and the first column, B_1, of B. Thus,

$$w = A_1B_1 = \begin{bmatrix} 0 & 1 & 4 \end{bmatrix} \begin{bmatrix} 1 \\ -3 \\ -1 \end{bmatrix} = -7$$

Similarly, since x is in the first row and the second column of AB, $x = A_1B_2 = 5$. Also, $y = A_2B_1 = -8$ and $z = A_2B_2 = 17$. We conclude that

$$AB = \begin{bmatrix} 0 & 1 & 4 \\ -1 & 3 & -2 \end{bmatrix} \begin{bmatrix} 1 & -2 \\ -3 & 5 \\ -1 & 0 \end{bmatrix} = \begin{bmatrix} -7 & 5 \\ -8 & 17 \end{bmatrix}$$ ∎

Example 6. Compute AB, where

(a) $A = \begin{bmatrix} -1 & 1 \\ 2 & 0 \\ 3 & 2 \end{bmatrix}$, $B = \begin{bmatrix} 1 & 0 \\ 1 & -1 \end{bmatrix}$

(b) $A = \begin{bmatrix} 0 \\ 2 \\ 1 \end{bmatrix}$, $B = \begin{bmatrix} 1 & 2 & -1 \end{bmatrix}$

Solution

(a) Since A is 3×2 and B is 2×2, then AB will be 3×2. Writing the row-column multiplications in a more compact way, we find that

$$AB = \begin{bmatrix} -1 & 1 \\ 2 & 0 \\ 3 & 2 \end{bmatrix} \begin{bmatrix} 1 & 0 \\ 1 & -1 \end{bmatrix} = \begin{bmatrix} (-1)(1) + (1)(1) & (-1)(0) + (1)(-1) \\ (2)(1) + (0)(1) & (2)(0) + (0)(-1) \\ (3)(1) + (2)(1) & (3)(0) + (2)(-1) \end{bmatrix} = \begin{bmatrix} 0 & -1 \\ 2 & 0 \\ 5 & -2 \end{bmatrix}$$

(b) Here, A is 3×1 and B is 1×3, so that AB must be 3×3. Each entry in A is a distinct row and each entry in B is a distinct column. The rule for multiplying implies that

$$AB = \begin{bmatrix} 0 \\ 2 \\ 1 \end{bmatrix} \begin{bmatrix} 1 & 2 & -1 \end{bmatrix} = \begin{bmatrix} (0)(1) & (0)(2) & (0)(-1) \\ (2)(1) & (2)(2) & (2)(-1) \\ (1)(1) & (1)(2) & (1)(-1) \end{bmatrix} = \begin{bmatrix} 0 & 0 & 0 \\ 2 & 4 & -2 \\ 1 & 2 & -1 \end{bmatrix}$$ ∎

24 MATRICES AND SYSTEMS OF LINEAR EQUATIONS

Multiplication by Matrices of Unknowns

A particularly important kind of matrix multiplication occurs when one of the matrices being multiplied consists of letters that stand for unknown quantities. We will encounter this situation when we apply matrices to the solution of systems of equations later in the chapter.

Example 7. Compute AX, where

$$A = \begin{bmatrix} 2 & 1 & -1 \\ 1 & 3 & 0 \end{bmatrix} \quad \text{and} \quad X = \begin{bmatrix} x \\ y \\ z \end{bmatrix}$$

Solution

$$\begin{bmatrix} 2 & 1 & -1 \end{bmatrix} \begin{bmatrix} x \\ y \\ z \end{bmatrix} = (2)(x) + (1)(y) + (-1)(z) = 2x + y - z$$

$$\begin{bmatrix} 1 & 3 & 0 \end{bmatrix} \begin{bmatrix} x \\ y \\ z \end{bmatrix} = 1x + 3y + 0z = x + 3y$$

and so we have computed

$$AX = \begin{bmatrix} 2x + y - z \\ x + 3y \end{bmatrix}$$

Multiplying the 2×3-matrix A times the 3×1-matrix X does give us a 2×1-matrix AX; but in each location we have an algebraic expression, such as $2x + y - z$, rather than a single number. ■

Another Use of Matrix Multiplication

We saw in Example 2 how row-column multiplication can be used to compute the cost of a shopping list. The next example illustrates how the more general matrix-multiplication procedure can be applied.

Example 8. A baker specializes in one type of cake and one kind of cookie. Each cake requires 1.5 (one and one-half) cups of sugar and 2 eggs, and each batch of cookies uses 0.5 cup of sugar and 1 egg. If the baker plans to bake 15 cakes and 10 batches of cookies, use matrix multiplication to compute how much sugar and how many eggs will be required.

Solution

Because the baker uses 1.5 cups of sugar for each cake, $(1.5)(15) = 22.5$ cups are needed for all the cakes. The sugar requirement for the 10 batches of cookies is $(0.5)(10) = 5$ cups, so the total amount of sugar needed is

$$\text{Sugar} = (1.5)(15) + (0.5)(10) = 27.5 \text{ cups}$$

1.2 MATRIX MULTIPLICATION 25

Similarly, the egg requirement is

$$\text{Eggs} = 2 \cdot 15 + 1 \cdot 10 = 40$$

That really answers the question, but we are also asked to express the answer in terms of matrices. To do that, we describe the baker's requirements for sugar and eggs per unit product as shown in Table 1-5.

Ingredient	Cake	Cookies
Sugar	1.5	0.5
Eggs	2	1

Table 1-5

Let A be the matrix corresponding to the table

$$A = \begin{bmatrix} 1.5 & 0.5 \\ 2 & 1 \end{bmatrix}$$

We also form a column matrix Q of the quantities of cakes and cookies being produced:

$$Q = \begin{bmatrix} 15 \\ 10 \end{bmatrix} \begin{matrix} \text{Cakes} \\ \text{Batches of Cookies} \end{matrix}$$

Finally, we let N be the matrix of the quantities of each ingredient needed:

$$N = \begin{bmatrix} \text{Sugar} \\ \text{Eggs} \end{bmatrix}$$

We claim that the two equations

$$\text{Sugar} = (1.5)(15) + (0.5)(10)$$
$$\text{Eggs} = 2 \cdot 15 + 1 \cdot 10$$

form the single matrix equation

$$N = \begin{bmatrix} \text{Sugar} \\ \text{Eggs} \end{bmatrix} = \begin{bmatrix} 1.5 & 0.5 \\ 2 & 1 \end{bmatrix} \begin{bmatrix} 15 \\ 10 \end{bmatrix} = AQ$$

The equation $N = AQ$ means (by the definition of equality of matrices) that the top number in N (Sugar) is equal to the top number in AQ, and the bottom number in N (Eggs) equals the bottom number in AQ. The top number of the 2×1-matrix AQ is

$$\begin{bmatrix} 1.5 & 0.5 \end{bmatrix} \begin{bmatrix} 15 \\ 10 \end{bmatrix} = (1.5)(15) + (0.5)(10) = 27.5$$

and the bottom number is

$$\begin{bmatrix} 2 & 1 \end{bmatrix} \begin{bmatrix} 15 \\ 10 \end{bmatrix} = 2 \cdot 15 + 1 \cdot 10 = 40$$

which verifies our claim. ■

Noncommutativity of Matrix Multiplication

Our next example illustrates the fact that AB may be different from BA, even when both are defined and of the same size.

Example 9. Compute AB and BA, where

$$A = \begin{bmatrix} 1 & 2 \\ 0 & -1 \end{bmatrix} \quad \text{and} \quad B = \begin{bmatrix} -2 & 0 \\ 1 & -1 \end{bmatrix}$$

Solution

Multiplying the rows of A times the columns of B to compute AB, we have

$$AB = \begin{bmatrix} 1 & 2 \\ 0 & -1 \end{bmatrix}\begin{bmatrix} -2 & 0 \\ 1 & -1 \end{bmatrix} = \begin{bmatrix} (1)(-2)+(2)(1) & (1)(0)+(2)(-1) \\ (0)(-2)+(-1)(1) & (0)(0)+(-1)(-1) \end{bmatrix} = \begin{bmatrix} 0 & -2 \\ -1 & 1 \end{bmatrix}$$

To compute BA, we use the rows of B on the left and the columns of A on the right:

$$BA = \begin{bmatrix} -2 & 0 \\ 1 & -1 \end{bmatrix}\begin{bmatrix} 1 & 2 \\ 0 & -1 \end{bmatrix} = \begin{bmatrix} (-2)(1)+(0)(0) & (-2)(2)+(0)(-1) \\ (1)(1)+(-1)(0) & (1)(2)+(-1)(-1) \end{bmatrix} = \begin{bmatrix} -2 & -4 \\ 1 & 3 \end{bmatrix}$$

Clearly,

$$AB = \begin{bmatrix} 0 & -2 \\ -1 & 1 \end{bmatrix} \neq \begin{bmatrix} -2 & -4 \\ 1 & 3 \end{bmatrix} = BA$$

even though both products are 2×2-matrices. ∎

Identity Matrices

Remember that a zero matrix (a matrix with zeros for all entries) behaves like the number 0 in ordinary addition—when it is added to a matrix of the same size, the matrix remains unchanged. There are also matrices, called *identity matrices*, that behave like the number 1 in ordinary multiplication. Just as

$$1 \cdot x = x \cdot 1 = x$$

for any number x, so too, when an identity matrix is multiplied by any matrix (of the appropriate size), the matrix remains unchanged. Identity matrices are square, that is, they have the same number of rows and columns. There is one identity matrix of each possible size. The first few, in order of size, are

$$[1], \quad \begin{bmatrix} 1 & 0 \\ 0 & 1 \end{bmatrix}, \quad \text{and} \quad \begin{bmatrix} 1 & 0 & 0 \\ 0 & 1 & 0 \\ 0 & 0 & 1 \end{bmatrix}$$

In general, an **identity matrix** is defined to be a square matrix that has a 1 in every location on the *main diagonal* of the matrix and 0s in all other locations. The **main diagonal** of a square matrix consists of the "diagonal" locations from upper left to lower right, where the row number and the column number are the same. Identity matrices are usually identified by the letter I.

> If I is the $n \times n$-identity matrix, we can multiply AI provided that the matrix A has n columns. The identity property of I is that in this case,
> $$AI = A$$

For instance, if I is the 4×4-identity matrix and A is a 2×4-matrix, we can multiply and the product will be given by $AI = A$.

> We can multiply the $n \times n$-identity matrix I on the right by a matrix B if B has n rows, and the result will be that
> $$IB = B$$

The next example verifies this property when I is the 2×2-identity matrix and B is of size 2×3.

Example 10. Perform the multiplication.

$$IB = \begin{bmatrix} 1 & 0 \\ 0 & 1 \end{bmatrix} \begin{bmatrix} 2 & -1 & 1 \\ 3 & 0 & 1 \end{bmatrix}$$

Solution

$$IB = \begin{bmatrix} 1 & 0 \\ 0 & 1 \end{bmatrix} \begin{bmatrix} 2 & -1 & 1 \\ 3 & 0 & 1 \end{bmatrix} = \begin{bmatrix} (1)(2) + (0)(3) & (1)(-1) + (0)(0) & (1)(1) + (0)(1) \\ (0)(2) + (1)(3) & (0)(-1) + (1)(0) & (0)(1) + (1)(1) \end{bmatrix}$$

$$= \begin{bmatrix} 2 & -1 & 1 \\ 3 & 0 & 1 \end{bmatrix} \quad \blacksquare$$

The Distributive Law

Another important property of matrix arithmetic, which it shares with ordinary arithmetic, is called the **distributive law**. We know that given the arithmetic problem of finding

$$3 \cdot 2 + 3 \cdot 4$$

either we can solve directly,

$$3 \cdot 2 + 3 \cdot 4 = 6 + 12 = 18$$

or we can first factor out the 3:

$$3 \cdot 2 + 3 \cdot 4 = 3(2 + 4) = 3 \cdot 6 = 18$$

28   MATRICES AND SYSTEMS OF LINEAR EQUATIONS

The algebraic formula for factoring is expressed by the distributive law; it is

$$ab + ac = a(b + c)$$

Using capital letters to denote matrices, the same distributive law works for matrices, that is,

$$\boxed{AB + AC = A(B + C)}$$

Thus, there is a factoring process for matrices. However, we must be more careful in factoring matrices than in factoring numbers. The arithmetic calculation of

$$3 \cdot 2 + 4 \cdot 3$$

allows for factoring because

$$3 \cdot 2 + 4 \cdot 3 = 3 \cdot 2 + 3 \cdot 4 = 3(2 + 4)$$

but, in general, $AB + CA$ *cannot be factored* because it may be that $CA \neq AC$, so that $AB + CA$ is not the same as $AB + AC$.

We can factor sums and differences just as in ordinary arithmetic, so long as the term factored out is on a single side. The formulas are:

$$\boxed{\begin{array}{ll} AB + AC = A(B + C) & AB - AC = A(B - C) \\ BA + CA = (B + C)A & BA - CA = (B - C)A \end{array}}$$

As the next example illustrates, when factoring is possible, it provides a welcome computational shortcut in the multiplication of matrices.

Example 11.   Compute $BA - CA$ given the matrices

$$A = \begin{bmatrix} -1 & -2 \\ 0 & 2 \end{bmatrix}, \quad B = \begin{bmatrix} 1 & 2 \\ -1 & -2 \\ 0 & 1 \end{bmatrix}, \quad \text{and} \quad C = \begin{bmatrix} 2 & 3 \\ -1 & 2 \\ 1 & -1 \end{bmatrix}$$

Solution

The most efficient procedure is to factor out the A-matrix, this time on the right. That is, use the formula $BA - CA = (B - C)A$ and first compute

$$B - C = \begin{bmatrix} 1 & 2 \\ -1 & -2 \\ 0 & 1 \end{bmatrix} - \begin{bmatrix} 2 & 3 \\ -1 & 2 \\ 1 & -1 \end{bmatrix} = \begin{bmatrix} -1 & -1 \\ 0 & -4 \\ -1 & 2 \end{bmatrix}$$

Then

$$BA - CA = (B - C)A = \begin{bmatrix} -1 & -1 \\ 0 & -4 \\ -1 & 2 \end{bmatrix} \begin{bmatrix} -1 & -2 \\ 0 & 2 \end{bmatrix}$$

$$= \begin{bmatrix} (-1)(-1) + (-1)(0) & (-1)(-2) + (-1)(2) \\ (0)(-1) + (-4)(0) & (0)(-2) + (-4)(2) \\ (-1)(-1) + (2)(0) & (-1)(-2) + (2)(2) \end{bmatrix} = \begin{bmatrix} 1 & 0 \\ 0 & -8 \\ 1 & 6 \end{bmatrix} \blacksquare$$

Summary

Matrices A and B can be multiplied to form the product AB only if the number of columns of A is equal to the number of rows of B. If A is of size $m \times k$ and B is of size $k \times n$, then AB is $m \times n$. To obtain the entry in the ith row and the jth column of the matrix AB, multiply the entire ith row of A by the entire jth column of B. If the ith row of A is

$$\begin{bmatrix} a_1 & a_2 & a_3 & \ldots & a_k \end{bmatrix}$$

and the jth column of B is

$$\begin{bmatrix} b_1 \\ b_2 \\ b_3 \\ \vdots \\ b_k \end{bmatrix}$$

then their product is defined to be the number

$$a_1 b_1 + a_2 b_2 + a_3 b_3 + \ldots + a_k b_k$$

An **identity matrix** I is a square ($n \times n$) matrix with 1s on the **main diagonal**, where the row number and the column number are the same, and 0s elsewhere. It has the property $AI = A$ and $IB = B$ for any matrices A and B for which the products are defined. Matrices obey the distributive laws

$$AB + AC = A(B + C) \qquad AB - AC = A(B - C)$$
$$BA + CA = (B + C)A \qquad BA - CA = (B - C)A$$

Exercises

In each of Exercises 1 through 9, decide whether AB or BA exist and, if they do, what size they must be. Do not actually carry out the multiplication.

1. $A = \begin{bmatrix} -1 & 2 & \frac{1}{2} \end{bmatrix}, B = \begin{bmatrix} 2 & 0 & 0 & 1 \\ 3 & 0 & 1 & -1 \\ 2 & \frac{1}{2} & 0 & 0 \end{bmatrix}$

2. $A = \begin{bmatrix} 2 \\ 1 \end{bmatrix}, B = \begin{bmatrix} -1 & 0 & 0 \\ 1 & 1 & 2 \\ 0 & 0 & 0 \end{bmatrix}$

3. $A = \begin{bmatrix} 2 & 1 \\ 1 & -1 \end{bmatrix}, B = \begin{bmatrix} 0 & 3 \\ -2 & 1 \end{bmatrix}$

4. $A = \begin{bmatrix} 1 & -1 & 2 \end{bmatrix}, B = \begin{bmatrix} 3 \end{bmatrix}$

5. $A = \begin{bmatrix} 1 & 0 & 1 & -1 \\ 2 & -1 & 1 & 1 \\ 2 & 0 & 1 & 2 \end{bmatrix}, B = \begin{bmatrix} 1 & 2 & 1 \end{bmatrix}$

6. $A = \begin{bmatrix} 1 & 2 & 2 & -1 \\ 0 & 0 & 1 & 0 \end{bmatrix}, B = \begin{bmatrix} -1 & -1 \\ 0 & 0 \\ 1 & -2 \\ 2 & -1 \end{bmatrix}$

7. $A = \begin{bmatrix} 0 & 1 & 2 \end{bmatrix}, B = \begin{bmatrix} 2 & -1 & 1 & 2 \end{bmatrix}$

8. $A = \begin{bmatrix} \frac{1}{2} \end{bmatrix}, B = \begin{bmatrix} 2 & -1 & 1 & 2 \end{bmatrix}$

9. $A = \begin{bmatrix} 1 \\ 2 \\ -1 \\ -2 \end{bmatrix}, B = \begin{bmatrix} 1 & 0 & 2 \end{bmatrix}$

In Exercises 10 through 24, compute AB.

10. $A = \begin{bmatrix} 2 & 1 \end{bmatrix}, B = \begin{bmatrix} 0 \\ 1 \end{bmatrix}$

11. $A = \begin{bmatrix} 0 & 1 & -1 & -1 \end{bmatrix}, B = \begin{bmatrix} -2 \\ 1 \\ 0 \\ 0 \end{bmatrix}$

12. $A = \begin{bmatrix} -1 & 2 & 1 \end{bmatrix}, B = \begin{bmatrix} x \\ y \\ z \end{bmatrix}$

13. $A = \begin{bmatrix} -1 & -1 \end{bmatrix}, B = \begin{bmatrix} 0 & 1 & -1 & 2 \\ 0 & 0 & -1 & 0 \end{bmatrix}$

14.
$$A = \begin{bmatrix} 0 & \frac{1}{2} & 2 & -1 & -3 & -2 \end{bmatrix}, B = \begin{bmatrix} 1 & \frac{1}{2} \\ -1 & 2 \\ 0 & -1 \\ 0 & 0 \\ 2 & 1 \\ -1 & 2 \end{bmatrix}$$

15.
$$A = \begin{bmatrix} a & b & c \end{bmatrix}, B = \begin{bmatrix} 2 & 1 & -1 \\ 1 & 0 & 0 \\ -1 & 1 & 0 \end{bmatrix}$$

16.
$$A = \begin{bmatrix} 1 & 0 \\ 0 & 1 \end{bmatrix}, B = \begin{bmatrix} -1 & 3 & 1 \\ \frac{1}{2} & -1 & 2 \end{bmatrix}$$

17.
$$A = \begin{bmatrix} 1 \\ 2 \\ -1 \\ -2 \end{bmatrix}, B = \begin{bmatrix} 1 & 0 & 2 \end{bmatrix}$$

18.
$$A = \begin{bmatrix} \frac{1}{2} & \frac{1}{2} & 2 & 1 \\ 1 & 2 & \frac{1}{2} & 2 \end{bmatrix}, B = \begin{bmatrix} -1 \\ 2 \\ -1 \\ 0 \end{bmatrix}$$

19.
$$A = \begin{bmatrix} -1 & 2 & 0 \\ 3 & 0 & 0 \\ -1 & 0 & -1 \end{bmatrix}, B = \begin{bmatrix} x \\ y \\ z \end{bmatrix}$$

20.
$$A = \begin{bmatrix} 1 & 2 & -1 \\ 0 & 1 & 1 \end{bmatrix}, B = \begin{bmatrix} 1 & 0 & 0 \\ 0 & 1 & 0 \\ 0 & 0 & 1 \end{bmatrix}$$

21.
$$A = \begin{bmatrix} 0 & -1 \\ 2 & 1 \\ -3 & 0 \end{bmatrix}, B = \begin{bmatrix} 3 \\ -2 \end{bmatrix}$$

22.
$$A = \begin{bmatrix} 2 & \frac{1}{2} & 0 & 1 \\ -1 & 1 & -\frac{1}{3} & 1 \end{bmatrix}, B = \begin{bmatrix} 0 & 2 & 0 & -1 \\ -4 & 0 & 1 & 0 \\ 0 & -3 & 1 & 0 \\ 2 & 1 & 0 & \frac{1}{2} \end{bmatrix}$$

23.
$$A = \begin{bmatrix} -1 & \frac{1}{2} & 1 & 0 \\ \frac{1}{3} & 2 & 0 & -1 \\ 0 & -\frac{2}{3} & 1 & \frac{1}{4} \end{bmatrix}, B = \begin{bmatrix} w \\ x \\ y \\ z \end{bmatrix}$$

24.
$$A = \begin{bmatrix} -1 & \frac{1}{2} \\ -1 & -\frac{1}{2} \\ 0 & 1 \end{bmatrix}, B = \begin{bmatrix} \frac{1}{2} & 1 \\ 1 & -1 \end{bmatrix}$$

MATRICES AND SYSTEMS OF LINEAR EQUATIONS

In Exercises 25 through 27, compute both AB and BA.

25. $A = \begin{bmatrix} -1 & 2 \\ 0 & 1 \end{bmatrix}, B = \begin{bmatrix} 0 & 2 \\ 0 & 0 \end{bmatrix}$ 26. $A = \begin{bmatrix} -4 & 0 \\ 0 & 3 \end{bmatrix}, B = \begin{bmatrix} -2 & -1 \\ 1 & 3 \end{bmatrix}$

27. $A = \begin{bmatrix} 4 & 2 \\ -2 & 0 \end{bmatrix}, B = \begin{bmatrix} -3 & 0 \\ 0 & -3 \end{bmatrix}$

28. Write the $n \times n$-identity matrix where   (a) n is 4,   (b) n is 5,   (c) n is 6.

In Exercises 29 and 30, carry out the indicated multiplication, showing your work in detail.

29. $\begin{bmatrix} 1 & 0 & 0 \\ 0 & 1 & 0 \\ 0 & 0 & 1 \end{bmatrix} \begin{bmatrix} 1 & 2 \\ -1 & 0 \\ 2 & 3 \end{bmatrix}$ 30. $\begin{bmatrix} 2 & -1 \end{bmatrix} \begin{bmatrix} 1 & 0 \\ 0 & 1 \end{bmatrix}$

In Exercises 31 through 34, compute using a distributive law.

31. $\begin{bmatrix} -1 & 0 \\ 2 & 3 \end{bmatrix} \begin{bmatrix} -1 & 2 & 1 \\ 1 & 0 & 1 \end{bmatrix} + \begin{bmatrix} 0 & 1 \\ -1 & -2 \end{bmatrix} \begin{bmatrix} -1 & 2 & 1 \\ 1 & 0 & 1 \end{bmatrix}$

32. $\begin{bmatrix} 2 & -1 & 1 & 2 \\ 3 & 1 & 4 & 0 \end{bmatrix} \begin{bmatrix} 1 \\ 2 \\ 0 \\ 1 \end{bmatrix} - \begin{bmatrix} 2 & -1 & 1 & 2 \\ 3 & 1 & 4 & 0 \end{bmatrix} \begin{bmatrix} 2 \\ 1 \\ 1 \\ 2 \end{bmatrix}$

33. $\begin{bmatrix} 1 \\ -1 \end{bmatrix} \begin{bmatrix} 2 & -1 & 0 \end{bmatrix} + \begin{bmatrix} 1 \\ -1 \end{bmatrix} \begin{bmatrix} -2 & 1 & 0 \end{bmatrix}$

34. $\begin{bmatrix} 2 & -1 & \frac{1}{2} \end{bmatrix} \begin{bmatrix} 2 \\ 1 \\ 0 \end{bmatrix} - \begin{bmatrix} 2 & -1 & \frac{1}{2} \end{bmatrix} \begin{bmatrix} 3 \\ 0 \\ 2 \end{bmatrix}$

35. An investor plans to buy 100 shares of telephone stock, 200 shares of oil stock, 400 shares of automobile stock, and 100 shares of airline stock. The telephone stock is selling for $46 a share, the oil stock for $34 a share, the automobile stock for $15 a share, and the airline stock for $10 a share.
    (a) Express the numbers of shares as a row matrix.
    (b) Express the prices of the stocks as a column matrix.
    (c) Use matrix multiplication to compute the total cost of the investor's purchase.

36. A manufacturer of custom-designed jewelry has orders for two rings, three pairs of earrings, five pins, and one necklace. The manufacturer estimates that it takes 1 work-hour of labor to produce a ring, 1.5 work-hours for each pair of earrings, 0.5 work-hour for each pin, and 2 work-hours for each necklace.
   (a) Express the manufacturer's orders as a row matrix.
   (b) Express the work-hour requirements for the various types of jewelry as a column matrix.
   (c) Use matrix multiplication to calculate the total number of work-hours required to complete all the orders.

37. A company pays its executives a salary and gives them shares of its stock as an annual bonus. Last year, the president of the company received $160,000 and 50 shares of stock, each of the three vice-presidents were paid $90,000 and 20 shares of stock, and the treasurer was paid $80,000 and 10 shares of stock.
   (a) Express the payments to the executives in money and stocks by means of a 2×3-matrix.
   (b) Express the number of executives of each rank by means of a column matrix.
   (c) Use matrix multiplication to calculate the total amount of money and the total number of shares of stock the company paid these executives last year.

38. One day, an ice-cream shop sold 80 ice-cream sodas and 150 milk shakes. The ingredients in a soda are 1 ounce of syrup, 1 ounce of milk, 3 ounces of soda water, and 4 ounces of ice cream. The recipe for a milk shake calls for blending 1 ounce of syrup, 4 ounces of milk, and 3 ounces of ice cream.
   (a) Express the syrup, milk, and ice cream in each product by means of a matrix.
   (b) Express the number of sodas and milk shakes sold that day as a column matrix.
   (c) Use matrix multiplication to compute the amount of syrup, milk, and ice cream used by the shop that day.

39. A family consists of two adults, one teenager, and three young children. Each adult consumes $\frac{1}{5}$ loaf of bread, no milk, $\frac{1}{10}$ pound of coffee, and $\frac{1}{8}$ pound of cheese on a typical day. The teenager eats $\frac{2}{5}$ loaf of bread, drinks 1 quart of milk but no coffee, and eats $\frac{1}{8}$ pound of cheese. Each child eats $\frac{1}{5}$ loaf of bread, drinks $\frac{1}{2}$ quart of milk and no coffee, and eats $\frac{1}{16}$ pound of cheese.
   (a) Use a matrix to express the daily consumption of bread, milk, coffee, and cheese by the various types of family members.
   (b) Express the number of family members of the various types by means of a column matrix.
   (c) Use matrix multiplication to calculate the total amount of bread, milk, coffee, and cheese consumed by this family on a typical day.

1.3 Nonsingular Two-by-Two Linear Systems

Matrices can be used to solve systems of linear equations. In this section we will discuss the case of two linear equations in two unknowns.

Linear Equations

We begin by recalling what a linear equation is. We will usually consider linear equations in several unknowns, which we represent by letters like x, y, and so on. An equation in these unknowns is called **linear** if it is possible to write it as a sum of expressions of the form "constant times an unknown" set equal to a constant. Examples of linear equations are:

$$2x + 5y = -1 \quad \text{and} \quad -2x + 3y + 2z = 3$$

Equations such as

$$3x^2 + 2y = 1 \quad \text{and} \quad 4x + 2yz = 2$$

are not linear because $3x^2 = 3 \cdot x \cdot x$ involves a repeated unknown as a factor and $2yz$ involves multiplying two unknowns. On the other hand,

$$w + 3x - 4y - z = 2$$

is a linear equation because we can rewrite it as

$$1w + 3x + (-4)y + (-1)z = 2$$

in which every term in the sum on the left is in "constant times an unknown" form. The equations

$$x = 2y \quad \text{and} \quad y = 2x + 1$$

are also linear. For instance, we can rewrite the first as

$$1x + (-2)y = 0$$

and the second as

$$(-2)x + 1y = 1$$

Systems and Solutions

A **solution of a linear equation** means a collection of values of the unknowns that make the equation true. For instance, a solution of

$$2x + 5y = -1$$

is $x = 2$, $y = -1$ because

$$2(2) + 5(-1) = -1$$

Another solution is $x = -3$, $y = 1$, and there are infinitely many others. Usually, however, we will not be concerned with the solutions of a single linear equation, but rather, with the solution of a collection of such equations, all in the same unknowns. Such a collection is called a **system of linear equations** and a **solution** consists of values of the unknowns for which all the equations in the system are true. For example, the system of linear equations in x and y:

$$x + 3y = 1$$
$$2x - y = -5$$

has the values $x = -2$ and $y = 1$ as a solution, as you can check by substituting these values:

$$(-2) + 3(1) = 1$$
$$2(-2) - 1 = -5$$

There are a number of techniques available for solving systems of linear equations. The systems of linear equations used in fields such as economics are often very large, sometimes involving hundreds, or even thousands, of equations. Although techniques using matrices are not necessarily the most efficient for solving small systems, such as two equations in two unknowns, they work well for large systems because they can easily be carried out on a computer.

Matrix Form of a System

We begin with the problem: Find numbers x and y for which both of the following equations are true.

$$3x + 4y = 1$$
$$2x + 5y = 2$$

Since you are expecting to use matrices at this point, you may have thought of this one right away:

$$\begin{bmatrix} 3 & 4 \\ 2 & 5 \end{bmatrix}$$

It consists of the numbers, to the left of the equal signs, that are multiplied by the x's and y's. Such numbers are called the "coefficients" of the x's and y's, so the matrix is called the **coefficient matrix**. A system of two linear equations in two unknowns is called a "two-by-two system," in which the first two refers to the number of equations and the second to the number of unknowns. The fact that the coefficient matrix is also "2-by-2" is no accident. The languages of matrices and of linear systems are meant to correspond to one another.

There is another matrix encoded in this two-by-two system, made up of the numbers on the right-hand side of the equal signs:

$$\begin{bmatrix} 1 \\ 2 \end{bmatrix}$$

But there is a third matrix in the system, though it is a bit harder to find— namely,

$$\begin{bmatrix} x \\ y \end{bmatrix}$$

The role of this matrix becomes clear if we multiply it on the left by the coefficient matrix:

$$\begin{bmatrix} 3 & 4 \\ 2 & 5 \end{bmatrix} \begin{bmatrix} x \\ y \end{bmatrix} = \begin{bmatrix} 3x + 4y \\ 2x + 5y \end{bmatrix}$$

So, we see that the entire left-hand side of the linear system can be thought of as the result of a matrix multiplication: the coefficient matrix times the matrix of unknowns.

If we give our matrices the names

$$A = \begin{bmatrix} 3 & 4 \\ 2 & 5 \end{bmatrix}, \quad B = \begin{bmatrix} 1 \\ 2 \end{bmatrix}, \quad \text{and} \quad X = \begin{bmatrix} x \\ y \end{bmatrix}$$

then since

$$AX = \begin{bmatrix} 3 & 4 \\ 2 & 5 \end{bmatrix} \begin{bmatrix} x \\ y \end{bmatrix} = \begin{bmatrix} 3x + 4y \\ 2x + 5y \end{bmatrix}$$

we find that the left-hand side of the system is the product AX and the right-hand side is the matrix B. Furthermore, because of the way equality of matrices was defined in Section 1.1, the two equations

$$3x + 4y = 1$$
$$2x + 5y = 2$$

can be thought of as the single matrix equation

$$AX = B$$

The Inverse

Once we write the problem in this way, we can see that there is a natural way to try to solve it. If someone gave you the ordinary equation

$$2x = 6$$

to solve, you would just divide both sides by 2 to find the solution

$$x = 3$$

In general, to solve the ordinary equation

$$ax = b$$

where a is not zero, you always divide both sides by a. Our method for solving the matrix equation

$$AX = B$$

will do much the same thing. To make sense of an instruction such as "divide both sides by 2" when we are talking about matrices, we must first take a closer look at what "divide by" means in ordinary arithmetic. We can always think of division in terms of multiplication. Instead of *dividing* 6 by 2, for instance, we can *multiply* 6 by $\frac{1}{2}$ with the same result since

$$\frac{1}{2} \cdot 6 = 3$$

1.3 NONSINGULAR TWO-BY-TWO LINEAR SYSTEMS

The number $\frac{1}{2}$ is called the **reciprocal** or **inverse of** 2. Arithmetically, the relationship between the number 2 and its inverse is given by

$$2 \cdot \frac{1}{2} = \frac{1}{2} \cdot 2 = 1$$

In general, we can represent division as multiplication by writing

$$b \div a = b \cdot a^{-1}$$

where a^{-1} (read "a **inverse**") is that number for which

$$a \cdot a^{-1} = a^{-1} \cdot a = 1$$

For instance, the inverse of $a = 2$ is given by $a^{-1} = \frac{1}{2}$. We can, to some extent, carry this point of view over to the subject of matrices. Note that the solution of the equation $ax = b$ is

$$x = a^{-1} \cdot b$$

because

$$a \cdot a^{-1} b = 1 \cdot b = b$$

> A matrix A is said to have an **inverse**, written A^{-1}, (again, read "A **inverse**") if the matrix A^{-1} has the property that
>
> $$AA^{-1} = A^{-1}A = I$$
>
> where I is an identity matrix.

Singular and Nonsingular Matrices

Not quite all ordinary numbers have an inverse: there is no number a^{-1} when $a = 0$. Many matrices fail to have inverses, and we will have more to say about such matrices later. For now, we just notice that A cannot possibly have an inverse unless it is a square matrix, that is, unless it has the same number of rows and columns. The reason is as follows: An identity matrix I is square. The requirement $AA^{-1} = I$ implies that A has the same number of rows as I, whereas $A^{-1}A = I$ implies that A has the same number of columns as I. Thus, A is the same size as I and is therefore square. But not all square matrices have inverses. For instance, a square zero matrix has no inverse. A square matrix is called **singular** if it has no inverse and **nonsingular** if it has an inverse.

Solving the Matrix Equation

Let us return to the matrix equation $AX = B$, now assuming that A is a nonsingular matrix. We will be thinking of A as a 2-by-2 matrix; but in fact, the solution of the equation is the same for a nonsingular matrix A of *any* size. To solve the equation, we multiply both sides by A^{-1}. (Remember the instruction, "divide both sides of $2x = 6$ by 2," which is the same as "multiply both sides of the equation by $2^{-1} = \frac{1}{2}$.") That gives us

$$A^{-1}AX = A^{-1}B$$

But $A^{-1}A = I$, so on the left-hand side we have[†]

$$A^{-1}AX = IX = X$$

by the basic property of I.

> The solution of the matrix equation $AX = B$ is
>
> $$X = A^{-1}B$$
>
> In words, to compute the solution X of the matrix equation $AX = B$, where A is an n-by-n nonsingular matrix, multiply B on the left by the inverse of A.

The 2-by-2 Inverse

The solution of $AX = B$ requires us to find the inverse A^{-1} of the nonsingular matrix A. We will discuss a general procedure for doing this in Section 1.5. For now, it will be convenient to remember that we are supposed to be concentrating on 2-by-2 systems, so we will suppose that A is a 2-by-2 nonsingular matrix. It happens that there is an easy way to find A^{-1} in this case, but we must emphasize from the start that nothing like it will work for any larger matrix, even for a 3-by-3 matrix.

We will illustrate this special technique using the matrix

$$A = \begin{bmatrix} 3 & 4 \\ 2 & 5 \end{bmatrix}$$

We need to find the inverse matrix A^{-1}, which is also 2 by 2. Since we do not know A^{-1}, we will write it as a matrix of unknowns:

$$A^{-1} = \begin{bmatrix} s & t \\ u & v \end{bmatrix}$$

Now when we write out in detail what it means to say that $AA^{-1} = I$, here is what we get:

[†]When we multiply AX on the left by A^{-1}, the product is $A^{-1}(AX)$. As in ordinary multiplication, matrix multiplication has the associative property:

$$A(BC) = (AB)C$$

Thus,

$$A^{-1}(AX) = (A^{-1}A)X$$

From now on, we will omit parentheses when we multiply matrices.

1.3 NONSINGULAR TWO-BY-TWO LINEAR SYSTEMS

$$AA^{-1} = \begin{bmatrix} 3 & 4 \\ 2 & 5 \end{bmatrix} \begin{bmatrix} s & t \\ u & v \end{bmatrix} = \begin{bmatrix} 3s+4u & 3t+4v \\ 2s+5u & 2t+5v \end{bmatrix} = \begin{bmatrix} 1 & 0 \\ 0 & 1 \end{bmatrix} = I$$

The definition of equality of matrices then tells us that the matrix equation $AA^{-1} = I$ is the same as the following four equations

$$(*) \begin{cases} 3s + 4u = 1 \\ 2s + 5u = 0 \end{cases} \qquad (**) \begin{cases} 3t + 4v = 0 \\ 2t + 5v = 1 \end{cases}$$

If you have not forgotten that the purpose of finding A^{-1} was supposed to be to solve the single system of two linear equations in two unknowns

$$3x + 4y = 1$$
$$2x + 5y = 2$$

you may be wondering if we are not moving away from our objective, rather than towards it. In fact, we now have two systems of just this type to solve—system (∗) in the unknowns s and u, and system (∗∗) in the unknowns t and v. But keep in mind that the purpose of what we are doing is not just to solve this single system, but to develop a method for solving *all* 2-by-2 systems $AX = B$, where the coefficient matrix A is nonsingular. Furthermore, these new systems are rather special, as we will see.

Let us focus on this equation from system (∗)

$$2s + 5u = 0$$

We put both the constants on a single side by these steps:

$$2s = -5u$$

$$s = -\frac{5}{2}u$$

$$\frac{s}{u} = -\frac{5}{2}$$

There is no *single* solution of the equation

$$\frac{s}{u} = -\frac{5}{2}$$

since it contains two unknowns. But there is an obvious solution that is very simple: let $s = 5$ and $u = -2$. Although there is no guarantee in mathematics that the simplest approach to a problem will be the right one, it is usually a good idea to try it. For this problem it works, as we will see. You may be wondering why we did not choose $s = -5$ and $u = 2$, which is just as simple. We could have done so, and it would give us the same solution for A^{-1}.

If we look at the corresponding equation in system (∗∗), that is $3t + 4v = 0$, the same steps lead us to

$$\frac{t}{v} = -\frac{4}{3}$$

which also has two "simplest" solutions: $t = 4$, $v = -3$, and $t = -4$, $v = 3$. This will give us two candidates for A^{-1}:

and
$$\begin{bmatrix} s & t \\ u & v \end{bmatrix} = \begin{bmatrix} 5 & 4 \\ -2 & -3 \end{bmatrix}$$

$$\begin{bmatrix} s & t \\ u & v \end{bmatrix} = \begin{bmatrix} 5 & -4 \\ -2 & 3 \end{bmatrix}$$

Let us see what happens when we multiply each of these by A:

$$\begin{bmatrix} 3 & 4 \\ 2 & 5 \end{bmatrix} \begin{bmatrix} 5 & 4 \\ -2 & -3 \end{bmatrix} = \begin{bmatrix} 7 & 0 \\ 0 & -7 \end{bmatrix}$$

and

$$\begin{bmatrix} 3 & 4 \\ 2 & 5 \end{bmatrix} \begin{bmatrix} 5 & -4 \\ -2 & 3 \end{bmatrix} = \begin{bmatrix} 7 & 0 \\ 0 & 7 \end{bmatrix}$$

The results are very similar, and neither of them gives the identity matrix. But it is the second choice, $t = -4$, $v = 3$, which gives us the answer we have been looking for. We can see that the product with A is not the identity matrix I only because the numbers on the main diagonal, though equal, are "seven times too large." Let us see what happens when we divide each of the numbers in our second attempt to find A^{-1} by the number 7 that turned up when we multiplied by A:

$$\begin{bmatrix} 3 & 4 \\ 2 & 5 \end{bmatrix} \begin{bmatrix} \frac{5}{7} & -\frac{4}{7} \\ -\frac{2}{7} & \frac{3}{7} \end{bmatrix} = \begin{bmatrix} 1 & 0 \\ 0 & 1 \end{bmatrix} = I$$

Now we have found the inverse of A:

$$A^{-1} = \begin{bmatrix} \frac{5}{7} & -\frac{4}{7} \\ -\frac{2}{7} & \frac{3}{7} \end{bmatrix}$$

and we can solve the original system

$$X = \begin{bmatrix} x \\ y \end{bmatrix} = A^{-1} B = \begin{bmatrix} \frac{5}{7} & -\frac{4}{7} \\ -\frac{2}{7} & \frac{3}{7} \end{bmatrix} \begin{bmatrix} 1 \\ 2 \end{bmatrix} = \begin{bmatrix} -\frac{3}{7} \\ \frac{4}{7} \end{bmatrix}$$

We can check that the answer is correct by substituting $x = -\frac{3}{7}$, $y = \frac{4}{7}$ into the system:

$$3\left(-\frac{3}{7}\right) + 4\left(\frac{4}{7}\right) = \frac{7}{7} = 1$$

$$2\left(-\frac{3}{7}\right) + 5\left(\frac{4}{7}\right) = \frac{14}{7} = 2$$

Of course, all that was too much work just to solve one problem; but now we can go through exactly the same steps to obtain an easy technique for solving *all* 2-by-2 systems in which the coefficient matrix is nonsingular (and, as we shall see later, that means all such systems that have single solutions).

The General 2-by-2 Case

Since we want a formula that will work for any nonsingular 2-by-2 matrix, we will write A in a way that represents any 2-by-2 matrix—namely,

$$A = \begin{bmatrix} a & b \\ c & d \end{bmatrix}$$

We could also have used the symbols a_{11}, a_{12}, and so on, that we introduced in Section 1.1, but they are not too helpful in this problem because they would make the equations look unnecessarily complicated. We can still use the same unknowns as before in A^{-1}. Then we have

$$AA^{-1} = \begin{bmatrix} a & b \\ c & d \end{bmatrix} \begin{bmatrix} s & t \\ u & v \end{bmatrix} = \begin{bmatrix} as+bu & at+bv \\ cs+du & ct+dv \end{bmatrix} = \begin{bmatrix} 1 & 0 \\ 0 & 1 \end{bmatrix} = I$$

and thus, we have the four equations

$$as + bu = 1 \qquad at + bv = 0$$
$$cs + du = 0 \qquad ct + dv = 1$$

The equations $cs + du = 0$ and $at + bv = 0$ can be rewritten, just as before, with the unknowns, s, t, u, v, on one side:

$$\frac{s}{u} = -\frac{d}{c} \qquad \text{and} \qquad \frac{t}{v} = -\frac{b}{a}$$

If we choose the "simplest" solutions: $s = d$, $u = -c$, $t = -b$, $v = a$ and multiply by A, we find

$$\begin{bmatrix} a & b \\ c & d \end{bmatrix} \begin{bmatrix} d & -b \\ -c & a \end{bmatrix} = \begin{bmatrix} ad-bc & 0 \\ 0 & -cb+da \end{bmatrix} = \begin{bmatrix} \Delta & 0 \\ 0 & \Delta \end{bmatrix}$$

where

$$\Delta = ad - bc = -cb + da$$

The Determinant

The number Δ in the preceding matrix equation is called the **determinant** of the matrix A. As long as Δ is not zero, we can divide each number in the simplest solution by Δ, just as in the numerical example, to obtain the general formula.

42 MATRICES AND SYSTEMS OF LINEAR EQUATIONS

> If
> $$A = \begin{bmatrix} a & b \\ c & d \end{bmatrix}$$
> then
> $$A^{-1} = \begin{bmatrix} \dfrac{d}{\Delta} & -\dfrac{b}{\Delta} \\ -\dfrac{c}{\Delta} & \dfrac{a}{\Delta} \end{bmatrix}$$
> where $\Delta = ad - bc$ is the determinant of the matrix A.

We can summarize this formula in words: Switch the two numbers on the main diagonal, change the signs of the remaining two numbers (but leave them where they are), then divide everything by the determinant.

Example 1. Find values x and y that solve the linear system

$$x + 2y = -1$$
$$-3x - 4y = 3$$

Solution

We will turn this into the matrix equation $AX = B$ in which

$$X = \begin{bmatrix} x \\ y \end{bmatrix} \text{ and } B = \begin{bmatrix} -1 \\ 3 \end{bmatrix}$$

but we have to be more careful about the coefficient matrix A. If we think of A as being of the form

$$A = \begin{bmatrix} a & b \\ c & d \end{bmatrix}$$

we see that we must write the left-hand side of the linear system so that each of the unknowns x and y has a number (coefficient) in front of it and the only arithmetic operation is addition:

$$x + 2y = 1 \cdot x + 2 \cdot y$$
$$-3x - 4y = (-3)x + (-4)y$$

Now it is easy to find A:

$$A = \begin{bmatrix} 1 & 2 \\ -3 & -4 \end{bmatrix}$$

and, computing the determinant

1.3 NONSINGULAR TWO-BY-TWO LINEAR SYSTEMS

$$\Delta = 1(-4) - 2(-3) = (-4) - (-6) = 2$$

we see from the general formula that

$$A^{-1} = \begin{bmatrix} -\frac{4}{2} & -\frac{2}{2} \\ \frac{3}{2} & \frac{1}{2} \end{bmatrix} = \begin{bmatrix} -2 & -1 \\ \frac{3}{2} & \frac{1}{2} \end{bmatrix}$$

The solution is given by

$$X = A^{-1}B = \begin{bmatrix} -2 & -1 \\ \frac{3}{2} & \frac{1}{2} \end{bmatrix} \begin{bmatrix} -1 \\ 3 \end{bmatrix} = \begin{bmatrix} -1 \\ 0 \end{bmatrix}$$

In other words, $x = -1$ and $y = 0$. ∎

Matrix Codes

Nonsingular matrices can be used to create secret codes. First, the message is converted to numerical form, using Table 1-6.

A	B	C	D	E	F	G	H	I	J	K	L	M
1	2	3	4	5	6	7	8	9	10	11	12	13
N	O	P	Q	R	S	T	U	V	W	X	Y	Z
14	15	16	17	18	19	20	21	22	23	24	25	26

Table 1-6

For instance, the message NEED HELP becomes 14, 5, 5, 4   8, 5, 12, 16. Our message is now in a simple substitution code. Even if we rearranged the numbers in Table 1-6, a simple substitution code is rather easy to break. For instance, the fact that the number 5 appears three times in the coded message would suggest that 5 represents E, the most common letter in English.

To construct the matrix code, we select a nonsingular matrix, called the *encoding matrix* to convert our message to code. For an example, we will use the encoding matrix

$$E = \begin{bmatrix} 2 & 1 \\ 1 & 1 \end{bmatrix}$$

We encode the message by forming matrices of two letters from the message, converting them to numerical form using Table 1-6, and then multiplying them by E:

44 MATRICES AND SYSTEMS OF LINEAR EQUATIONS

$$\begin{bmatrix} 2 & 1 \\ 1 & 1 \end{bmatrix} \begin{bmatrix} N \\ E \end{bmatrix} = \begin{bmatrix} 2 & 1 \\ 1 & 1 \end{bmatrix} \begin{bmatrix} 14 \\ 5 \end{bmatrix} = \begin{bmatrix} 33 \\ 19 \end{bmatrix}$$

$$\begin{bmatrix} 2 & 1 \\ 1 & 1 \end{bmatrix} \begin{bmatrix} E \\ D \end{bmatrix} = \begin{bmatrix} 2 & 1 \\ 1 & 1 \end{bmatrix} \begin{bmatrix} 5 \\ 4 \end{bmatrix} = \begin{bmatrix} 14 \\ 9 \end{bmatrix}$$

$$\begin{bmatrix} 2 & 1 \\ 1 & 1 \end{bmatrix} \begin{bmatrix} H \\ E \end{bmatrix} = \begin{bmatrix} 2 & 1 \\ 1 & 1 \end{bmatrix} \begin{bmatrix} 8 \\ 5 \end{bmatrix} = \begin{bmatrix} 21 \\ 13 \end{bmatrix}$$

$$\begin{bmatrix} 2 & 1 \\ 1 & 1 \end{bmatrix} \begin{bmatrix} L \\ P \end{bmatrix} = \begin{bmatrix} 2 & 1 \\ 1 & 1 \end{bmatrix} \begin{bmatrix} 12 \\ 16 \end{bmatrix} = \begin{bmatrix} 40 \\ 28 \end{bmatrix}$$

The encoded message is now: 33, 19, 14, 9, 21, 13, 40, 28. Notice that although the letter E appears three times in the original message, it is now represented by three different numbers—19, 14, and 13—in the newly encoded message.

If we know the encoding matrix E, we can decode any message we receive by using the inverse matrix E^{-1}. To see how this is done, we let T be one of the column matrices obtained from the message in simple substitution code. For instance, in the first multiplication in the preceding message we let

$$T = \begin{bmatrix} 14 \\ 5 \end{bmatrix}$$

Let C represent the corresponding column matrix of the encoded message we obtained by matrix multiplication, that is,

$$C = ET$$

For this T we obtained

$$C = ET = \begin{bmatrix} 2 & 1 \\ 1 & 1 \end{bmatrix} \begin{bmatrix} 14 \\ 5 \end{bmatrix} = \begin{bmatrix} 33 \\ 19 \end{bmatrix}$$

To decode the message, we multiply the encoded message C by E^{-1}, thus obtaining

$$E^{-1}C = E^{-1}(ET) = IT = T$$

that is the original uncoded message. Thus, if we multiplied

$$C = \begin{bmatrix} 33 \\ 19 \end{bmatrix}$$

by E^{-1}, we would obtain

$$T = \begin{bmatrix} 14 \\ 5 \end{bmatrix}$$

again. The following example illustrates the decoding procedure.

Example 2. The message 41, 23, 6, 5, 74, 49 is encoded using the matrix

$$E = \begin{bmatrix} 2 & 1 \\ 1 & 1 \end{bmatrix}$$

Decode this message.

Solution

To find the inverse matrix E^{-1}, we calculate that the determinant is

$$\Delta = 2 \cdot 1 - 1 \cdot 1 = 1$$

and therefore, by the formula for computing the inverse of a 2-by-2 nonsingular matrix, we have

$$E^{-1} = \begin{bmatrix} 1 & -1 \\ -1 & 2 \end{bmatrix}$$

Performing the multiplication $E^{-1}C$ and then referring back to Table 1-6 to find the letter equivalents, we have

$$\begin{bmatrix} 1 & -1 \\ -1 & 2 \end{bmatrix} \begin{bmatrix} 41 \\ 23 \end{bmatrix} = \begin{bmatrix} 18 \\ 5 \end{bmatrix} = \begin{bmatrix} R \\ E \end{bmatrix}$$

$$\begin{bmatrix} 1 & -1 \\ -1 & 2 \end{bmatrix} \begin{bmatrix} 6 \\ 5 \end{bmatrix} = \begin{bmatrix} 1 \\ 4 \end{bmatrix} = \begin{bmatrix} A \\ D \end{bmatrix}$$

$$\begin{bmatrix} 1 & -1 \\ -1 & 2 \end{bmatrix} \begin{bmatrix} 74 \\ 49 \end{bmatrix} = \begin{bmatrix} 25 \\ 24 \end{bmatrix} = \begin{bmatrix} Y \\ X \end{bmatrix}$$

Ignoring the last X (that made the number of letters even), we see that the message is

READY ■

Summary

A system of two linear equations in two unknowns:

$$ax + by = h$$
$$cx + dy = k$$

can be written as the matrix equation $AX = B$, where A is the **coefficient matrix**

$$A = \begin{bmatrix} a & b \\ c & d \end{bmatrix}$$

$$X = \begin{bmatrix} x \\ y \end{bmatrix} \quad \text{and} \quad B = \begin{bmatrix} h \\ k \end{bmatrix}$$

MATRICES AND SYSTEMS OF LINEAR EQUATIONS

A matrix A is **nonsingular** if it has an **inverse** A^{-1}, that is, if there is a matrix A^{-1} such that
$$AA^{-1} = A^{-1}A = I$$
where I is an identity matrix. Otherwise, it is **singular**. If A is nonsingular, then the solution of $AX = B$ is $X = A^{-1}B$. The inverse of A is

$$A^{-1} = \begin{bmatrix} \dfrac{d}{\Delta} & -\dfrac{b}{\Delta} \\ -\dfrac{c}{\Delta} & \dfrac{a}{\Delta} \end{bmatrix}$$

where $\Delta = ad - bc$ is called the **determinant** of A.

Exercises

In Exercises 1 through 6, the system can be written as a matrix equation $AX = B$. Find the coefficient matrix A.

1. $2x + 2y = 12$
   $5x + 2y = -1$

2. $3x + y = 2$
   $4x + 5y = 2$

3. $2x - 3y = 0$
   $-2x + 4y = 1$

4. $x + 2y = -1$
   $2x - y = 3$

5. $-x + 3y = 4$
   $x - 2y = 1$

6. $\dfrac{1}{3}x - 2y = \dfrac{1}{2}$
   $-3x - \dfrac{1}{2}y = 1$

In Exercises 7 through 13, find the inverse of the given matrix

7. $\begin{bmatrix} 1 & 1 \\ 3 & 4 \end{bmatrix}$

8. $\begin{bmatrix} 5 & 2 \\ 1 & 1 \end{bmatrix}$

9. $\begin{bmatrix} 3 & 1 \\ 1 & 1 \end{bmatrix}$

10. $\begin{bmatrix} 1 & -1 \\ 1 & 1 \end{bmatrix}$

11. $\begin{bmatrix} -2 & 1 \\ 1 & 2 \end{bmatrix}$

12. $\begin{bmatrix} 3 & 2 \\ 2 & 1 \end{bmatrix}$

13. $\begin{bmatrix} 2 & -1 \\ 3 & -2 \end{bmatrix}$

In Exercises 14 through 16, solve the matrix equation $AX = B$ for X, where the matrices A and B are as given.

14. $A = \begin{bmatrix} 2 & 1 \\ 1 & 1 \end{bmatrix}$, $B = \begin{bmatrix} 1 \\ 1 \end{bmatrix}$

15. $A = \begin{bmatrix} 1 & -1 \\ 2 & 1 \end{bmatrix}$, $B = \begin{bmatrix} 0 \\ 2 \end{bmatrix}$

16. $A = \begin{bmatrix} -1 & 1 \\ 3 & 2 \end{bmatrix}$, $B = \begin{bmatrix} -2 \\ 1 \end{bmatrix}$

In Exercises 17 through 22, solve the given system for x and y, using the matrix method explained in this section.

17. $3x + 2y = 1$
$\quad\;\, 2x + y = 2$

18. $x - y = 3$
$\quad\;\, -x + 2y = 1$

19. $-2x + y = 0$
$\quad\;\, -3x + 2y = 1$

20. $-x + 2y = -1$
$\quad\;\, 2x - y = 1$

21. $3x - y = -2$
$\quad\;\, 2x - 2y = 1$

22. $-2x + 3y = 3$
$\quad\;\, x + y = 2$

In Exercises 23 and 24, the message was encoded using the given matrix. Decode the message.

23. Message: 13, 34, 0, 18, –1, 23

$E = \begin{bmatrix} -1 & 1 \\ -1 & 2 \end{bmatrix}$

24. Message: 57, 29, 85, 45, 35, 25

$E = \begin{bmatrix} 3 & 1 \\ 1 & 1 \end{bmatrix}$

1.4 Nonsingular *n*-by-*n* Linear Systems

We solved the system of two linear equations in two unknowns:

$$3x + 4y = 1$$
$$2x + 5y = 2$$

by finding the inverse A^{-1} of the coefficient matrix

$$A = \begin{bmatrix} 3 & 4 \\ 2 & 5 \end{bmatrix}$$

and computing the solution $X = A^{-1}B$, where

$$B = \begin{bmatrix} 1 \\ 2 \end{bmatrix}$$

A 3-by-3 Example

Now we will look at a larger problem, with three equations in three unknowns:

$$x - y + 2z = 3$$
$$2x - 2y + z = -2$$
$$-x \quad\;\; + z = 1$$

48 MATRICES AND SYSTEMS OF LINEAR EQUATIONS

If we write out the left-hand side with all the coefficients made explicit, as we did in the last section, we obtain

$$\begin{aligned} x - y + 2z &= 1x + (-1)y + 2z \\ 2x - 2y + z &= 2x + (-2)y + 1z \\ -x + z &= (-1)x + 0y + 1z \end{aligned}$$

We can easily see that the coefficient matrix is

$$A = \begin{bmatrix} 1 & -1 & 2 \\ 2 & -2 & 1 \\ -1 & 0 & 1 \end{bmatrix}$$

The 3-by-3 system can be written as the matrix equation $AX = B$ if we write:

$$X = \begin{bmatrix} x \\ y \\ z \end{bmatrix} \quad \text{and} \quad B = \begin{bmatrix} 3 \\ -2 \\ 1 \end{bmatrix}$$

We chose this problem so that A would be a nonsingular matrix, that is, so that A would have an inverse A^{-1}. In fact, all the coefficient matrices in this section will be of this type, as we indicated by the title of the section. If we knew A^{-1}, we could solve the system by multiplying: $X = A^{-1}B$, because the rule for the solution had nothing to do with the size of the matrices. However, there is no formula for the inverse, as simple as the one we found in the last section for 2-by-2 matrices, that works for larger matrices. In this section, we will use some simple facts about solving equations to present a procedure, called the *Gauss-Jordan method*, for solving a larger system of linear equations directly, without computing A^{-1}. Later in this chapter, in Section 1.6, we will extend the same method to solve matrix equations $AX = B$, where A is any matrix. The method is easiest to describe when A is a nonsingular matrix, so that is why we do it first.

Operations that Preserve the Solution

Everything we will do in this section is based on the idea that you can make certain simple changes in a system of linear equations so that it looks different, but has the same solution. The simplest change of all is just to write the equations in a different order. For instance, if we start with the preceding problem, of finding values of x, y, and z for which

$$\begin{aligned} x - y + 2z &= 3 \\ 2x - 2y + z &= -2 \\ -x + z &= 1 \end{aligned}$$

and write the second equation before the first:

$$\begin{aligned} 2x - 2y + z &= -2 \\ x - y + 2z &= 3 \\ -x + z &= 1 \end{aligned}$$

1.4 NONSINGULAR n-BY-n LINEAR SYSTEMS 49

this is the same problem, that is, the same values of x, y, and z, whatever they are, will still work.

The second idea is almost as simple, and is certainly very familiar. If you multiply or divide both sides of an equation by the same nonzero constant, the equation is the same as before in the sense that it has the same solution. When we discussed the equation $2x = 6$ in the last section, it was natural to solve it, that is, to find out what value of x makes the two sides equal, by dividing both sides by 2 to find that $x = 3$. The same principle works for an equation with more than one unknown. For instance, if we multiply both sides of the equation

$$x - y + 2z = 3$$

by 3, to obtain the equation

$$3x - 3y + 6z = 9$$

any numbers x, y, and z that made the left-hand side of the first equation add up to 3 will make the left-hand side of the second add up to 9. Therefore, if we change the original linear system

$$\begin{aligned} x - y + 2z &= 3 \\ 2x - 2y + z &= -2 \\ -x + z &= 1 \end{aligned}$$

by replacing the first equation by 3 times itself:

$$\begin{aligned} 3x - 3y + 6z &= 9 \\ 2x - 2y + z &= -2 \\ -x + z &= 1 \end{aligned}$$

the system looks different, but it has the same solution.

The third, and final, thing we have to observe about equations concerns what happens when we add the left-hand sides of two of them—for instance the first two:

$$(x - y + 2z) + (2x - 2y + z)$$

The system tells us that $x - y + 2z = 3$ and $2x - 2y + z = -2$, so it is correct to write that

$$(x - y + 2z) + (2x - 2y + z) = 3 + (-2) = 1$$

We can simplify the left-hand side of the sum by combining the x-terms, the y-terms and the z-terms:

$$(x + 2x) + (-y + (-2y)) + (2z + z) = 3x - 3y + 3z = 1$$

but a much neater way to do this is to line them up as now indicated and add:

$$\begin{array}{rl} x - y + 2z = 3 & \text{(FIRST EQUATION)} \\ + \ 2x - 2y + z = -2 & \text{(SECOND EQUATION)} \\ \hline 3x - 3y + 3z = 1 & \text{(SUM)} \end{array}$$

That is what we will do from now on.

If we replaced both the first and second equations by their sum, that would change the problem because then there would be only two equations, rather than three. This would indicate that we must have lost some information. But if, for instance, we keep the first equation and replace the second with the sum:

$$x - y + 2z = 3$$
$$3x - 3y + 3z = 1$$
$$-x + z = 1$$

then again we have only changed the appearance of the system, not its solution.

To summarize, we have discussed three operations on the equations in a system: changing the order of equations, multiplying or dividing both sides of an equation in the system, adding two equations. The operations change the appearance of the system, but they do not change the values of the unknowns that solve it. However, you may wonder what good it would do to perform these operations since it would appear to be as difficult to find the solutions of the new systems as it was to solve the original problem. The point is that you have to perform these operations in the right way. If in the problem $2x = 6$ we had multiplied both sides by 3 to produce $6x = 18$, we would have been no closer to the solution. Instead, by dividing both sides by 2, we were handed the solution: $x = 3$. We propose to do exactly the same thing with the system of 3 equations in 3 unknowns: rewrite it in such a way that the solution is just as obvious as $x = 3$.

The Augmented Matrix

We will simplify the description of how to rewrite the system by taking advantage of matrix notation. If we write the problem so that the coefficient matrix is easy to see:

$$1x + (-1)y + 2z = 3$$
$$2x + (-2)y + 1z = -2$$
$$(-1)x + 0y + 1z = 1$$

we can preserve all that information in a matrix form like this:

$$\begin{array}{ccc} x & y & z \end{array}$$
$$\left[\begin{array}{ccc|c} 1 & -1 & 2 & 3 \\ 2 & -2 & 1 & -2 \\ -1 & 0 & 1 & 1 \end{array} \right]$$

The letters at the top of the columns record the fact that the numbers in that column are all multiplied by that unknown, and the vertical line substitutes for the equal signs. If we think of the system of three linear equations as the matrix equation $AX = B$, as we did at the beginning of this section, then we see that the A- and B-matrices appear in this matrix, which is now of the form $[A \mid B]$. The matrix $[A \mid B]$ is called the **augmented matrix** corresponding to the matrix equation $AX = B$, or what amounts to the same thing, to the system of linear equations. The augmented matrix preserves all the information given in the linear system because from this matrix we can easily reconstruct the linear system from which it comes.

1.4 NONSINGULAR n-BY-n LINEAR SYSTEMS

Example 1. Write the system of linear equations whose augmented matrix is

$$\begin{array}{cccc} w & x & y & z \end{array}$$
$$\left[\begin{array}{cccc|c} 0 & 0 & -2 & 0 & 3 \\ -1 & 0 & 3 & -1 & -2 \\ 2 & 4 & 0 & 0 & 1 \\ 1 & 3 & -3 & 1 & 0 \end{array}\right]$$

Solution

In every row each of the first four numbers is multiplied by the unknown above it. These products are then added and equated to the number lying to the right of the vertical line. Thus, we can write

$$\begin{aligned} 0w + 0x + (-2)y + 0z &= 3 \\ (-1)w + 0x + 3y + (-1)z &= -2 \\ 2w + 4x + 0y + 0z &= 1 \\ 1w + 3x + (-3)y + 1z &= 0 \end{aligned}$$

or, more briefly,

$$\begin{aligned} -2y &= 3 \\ -w + 3y - z &= -2 \\ 2w + 4x &= 1 \\ w + 3x - 3y + z &= 0 \end{aligned}$$
■

Operations on Augmented Matrices

Now we can represent the operations we discussed earlier on the equations of the system as operations on the rows of the augmented matrix

$$\begin{array}{ccc} x & y & z \end{array}$$
$$\left[\begin{array}{ccc|c} 1 & -1 & 2 & 3 \\ 2 & -2 & 1 & -2 \\ -1 & 0 & 1 & 1 \end{array}\right]$$

Interchanging the first and second equations does the same to the first and second rows.

$$\begin{array}{ccc} x & y & z \end{array}$$
$$\left[\begin{array}{ccc|c} 2 & -2 & 1 & -2 \\ 1 & -1 & 2 & 3 \\ -1 & 0 & 1 & 1 \end{array}\right]$$

Multiplying the first row of the original system by 3 has this effect:

$$\begin{array}{ccc} x & y & z \end{array}$$
$$\left[\begin{array}{ccc|c} 3 & -3 & 6 & 9 \\ 2 & -2 & 1 & -2 \\ -1 & 0 & 1 & 1 \end{array}\right]$$

52   MATRICES AND SYSTEMS OF LINEAR EQUATIONS

Adding the first two equations of the system the way we did before yields

$$\begin{array}{rl} x - y + 2z = & 3 \quad \text{(FIRST EQUATION)} \\ +\ 2x - 2y + z = & -2 \quad \text{(SECOND EQUATION)} \\ \hline 3x - 3y + 3z = & 1 \quad \text{(SUM)} \end{array}$$

This looks much the same when we use the rows of the augmented matrix:

$$\begin{array}{c} \begin{array}{ccc} x & y & z \end{array} \\ \begin{bmatrix} 1 & -1 & 2 \mid 3 \end{bmatrix} \quad \text{(first row)} \\ +\begin{bmatrix} 2 & -2 & 1 \mid -2 \end{bmatrix} \quad \text{(second row)} \\ \hline \begin{bmatrix} 3 & -3 & 3 \mid 1 \end{bmatrix} \quad \text{(sum)} \end{array}$$

We then substitute that sum for the second row:

$$\begin{array}{ccc} x & y & z \end{array}$$
$$\begin{bmatrix} 1 & -1 & 2 & \mid & 3 \\ 3 & -3 & 3 & \mid & 1 \\ -1 & 0 & 1 & \mid & 1 \end{bmatrix}$$

In the procedure for rewriting the augmented matrix, instead of just adding two rows, usually we will want to add to a row another row after we have multiplied both sides of that other row by a constant. This operation combines two steps that we have already discussed and that do not change the solutions of the system. For instance, we will see that we will want to add to the second row of the original augmented matrix

$$\begin{array}{ccc} x & y & z \end{array}$$
$$\begin{bmatrix} 1 & -1 & 2 & \mid & 3 \\ 2 & -2 & 1 & \mid & -2 \\ -1 & 0 & 1 & \mid & 1 \end{bmatrix}$$

the first row multiplied by -2, that is,

$$[-2 \ \ 2 \ \ -4 \ \mid \ -6] \quad \text{(-2 times first row)}$$

with this result:

$$\begin{array}{c} \begin{array}{ccc} x & y & z \end{array} \\ \begin{bmatrix} 2 & -2 & 1 \mid -2 \end{bmatrix} \quad \text{(second row)} \\ +\begin{bmatrix} -2 & 2 & -4 \mid -6 \end{bmatrix} \quad \text{(-2 times first row)} \\ \hline \begin{bmatrix} 0 & 0 & -3 \mid -8 \end{bmatrix} \quad \text{(sum)} \end{array}$$

When we substitute this sum in place of the second row, we obtain

$$\begin{array}{ccc} x & y & z \end{array}$$
$$\begin{bmatrix} 1 & -1 & 2 & \mid & 3 \\ 0 & 0 & -3 & \mid & -8 \\ -1 & 0 & 1 & \mid & 1 \end{bmatrix}$$

This time the matrix really does look somewhat simpler.

Example 2. In the following augmented matrix, add 3 times the second row to the third row and substitute the sum for the third row:

$$\begin{array}{cccc} w & x & y & z \end{array}$$
$$\left[\begin{array}{cccc|c} 1 & 0 & -1 & 0 & 2 \\ 0 & 1 & 2 & -1 & 0 \\ 0 & -3 & 1 & 2 & 1 \\ 0 & 2 & 1 & 0 & -1 \end{array}\right]$$

Solution

We first write out the sum

$$\begin{array}{cccc} w & x & y & z \end{array}$$
$$\begin{array}{r} \left[\begin{array}{cccc|c} 0 & -3 & 1 & 2 & 1 \end{array}\right] \\ +\left[\begin{array}{cccc|c} 0 & 3 & 6 & -3 & 0 \end{array}\right] \\ \hline \left[\begin{array}{cccc|c} 0 & 0 & 7 & -1 & 1 \end{array}\right] \end{array} \begin{array}{l} \text{(third row)} \\ \text{(3 times second row)} \\ \text{(sum)} \end{array}$$

and then we substitute it for the third row of the augmented matrix:

$$\begin{array}{cccc} w & x & y & z \end{array}$$
$$\left[\begin{array}{cccc|c} 1 & 0 & -1 & 0 & 2 \\ 0 & 1 & 2 & -1 & 0 \\ 0 & 0 & 7 & -1 & 1 \\ 0 & 2 & 1 & 0 & -1 \end{array}\right]$$ ∎

The Gauss-Jordan Method

We have discussed three operations on the rows of the augmented matrix that change the appearance of the matrix, but do not change the solutions of the corresponding system of linear equations. These were:

1. Interchange two rows of the matrix.
2. Multiply (or divide) all the numbers in a row by the same nonzero constant.
3. Add a constant times a row to another row and substitute the sum for one of these rows.

Now we will describe a procedure that uses these operations to solve a system of n linear equations in n unknowns with a nonsingular coefficient matrix A. Recall that a column of an identity matrix, say,

$$\begin{bmatrix} 1 & 0 & 0 \\ 0 & 1 & 0 \\ 0 & 0 & 1 \end{bmatrix}$$

consists of a 1 in the **diagonal location**, in which the number of the row is the same as the number of that column, and a 0 in every other location.

54 MATRICES AND SYSTEMS OF LINEAR EQUATIONS

> The **Gauss-Jordan method** transforms the columns to the left of the vertical line in the augmented matrix $[\,A \mid B\,]$ into columns of an identity matrix. The rules for transforming each column are:
>
> 1. Change the number in the diagonal location of the column to a 1.
> 2. Then change the number in every other location of the column to 0.
>
> These rules are applied column by column from left to right.

It is crucial that we work from left to right because otherwise, at some stage we may undo the progress we made earlier in turning columns of the augmented matrix into columns of an identity matrix.

To execute Step 1 of the rules, first notice whether the number in the diagonal location of the column is 0. If it is 0, interchange the row containing that element with some row *below* it that contains a nonzero number in the column we are working on. Once there is a nonzero number d in the diagonal location, divide the row containing d by d itself. This step will change the number originally in the diagonal location to a 1 and so complete Step 1.

To carry out Step 2 of the rules, consider any nonzero number of the column other than the number (now 1) in the diagonal location. Let us call that number a. Add $-a$ times the row containing the diagonal location to the row containing a and replace the row containing a with the sum. The effect of this operation is to change a to 0. Repeat this procedure, in any order, to complete Step 2.

If A were a singular matrix, the Gauss-Jordan method would not be able to convert it to the identity matrix as it does when A is nonsingular, though it would still give us valuable information about the linear system, as we shall see in Section 1.6. We do not have to be concerned about this possibility for now because, as we mentioned before, all the coefficient matrices in this section are nonsingular.

Solving the Example

We will now apply the Gauss-Jordan method to the problem with which we started this section:

$$\begin{aligned} x - y + 2z &= 3 \\ 2x - 2y + z &= -2 \\ -x \quad\;\;\; + z &= 1 \end{aligned}$$

Looking at the corresponding augmented matrix

1.4 NONSINGULAR n-BY-n LINEAR SYSTEMS   55

$$\begin{array}{c} x   y   z \\ \left[\begin{array}{rrr|r} 1 & -1 & 2 & 3 \\ 2 & -2 & 1 & -2 \\ -1 & 0 & 1 & 1 \end{array}\right] \end{array}$$

we see that the number 1 is already in the upper left-hand corner, the diagonal location of the first column; so for now we can go on to Step 2. The number just below the 1 is 2, so the procedure calls for adding -2 times the first row to the second row, which will make the left-hand number of the sum a 0:

$$\begin{array}{r} x   y   z \\ \left[\begin{array}{rrr|r} 2 & -2 & 1 & -2 \end{array}\right] \quad \text{(second row)} \\ +\left[\begin{array}{rrr|r} -2 & 2 & -4 & -6 \end{array}\right] \quad \text{(-2 times first row)} \\ \hline \left[\begin{array}{rrr|r} 0 & 0 & -3 & -8 \end{array}\right] \quad \text{(sum)} \end{array}$$

We then substitute this sum for the original second row. That is what we did just before Example 2, with the following result:

$$\begin{array}{c} x   y   z \\ \left[\begin{array}{rrr|r} 1 & -1 & 2 & 3 \\ 0 & 0 & -3 & -8 \\ -1 & 0 & 1 & 1 \end{array}\right] \end{array}$$

Continuing down the first column, we come to -1. So we repeat Step 2, this time multiplying the first row by $-(-1) = 1$, which just gives us the first row again. We then add that to the third row, the one we wish to change:

$$\begin{array}{r} x   y   z \\ \left[\begin{array}{rrr|r} -1 & 0 & 1 & 1 \end{array}\right] \quad \text{(third row)} \\ +\left[\begin{array}{rrr|r} 1 & -1 & 2 & 3 \end{array}\right] \quad \text{(first row)} \\ \hline \left[\begin{array}{rrr|r} 0 & -1 & 3 & 4 \end{array}\right] \quad \text{(sum)} \end{array}$$

Substituting the sum for the third row causes all the nondiagonal numbers in the first column to equal zero.

$$\begin{array}{c} x   y   z \\ \left[\begin{array}{rrr|r} 1 & -1 & 2 & 3 \\ 0 & 0 & -3 & -8 \\ 0 & -1 & 3 & 4 \end{array}\right] \end{array}$$

Since the first column on the left is now the first column of the 3-by-3 identity matrix, we move on to the next column. There is a 0 in the diagonal location; so, according to Step 1, we must interchange the second row with the one below it, in order to move a nonzero number into that location.

$$\begin{bmatrix} x & y & z & \\ 1 & -1 & 2 & | & 3 \\ 0 & -1 & 3 & | & 4 \\ 0 & 0 & -3 & | & -8 \end{bmatrix}$$

We have still not completed Step 1 because the number in the diagonal location of the second column is −1. We change it to a 1 by dividing that row by −1 itself.

$$\begin{bmatrix} x & y & z & \\ 1 & -1 & 2 & | & 3 \\ 0 & 1 & -3 & | & -4 \\ 0 & 0 & -3 & | & -8 \end{bmatrix}$$

We return to Step 2 to change the −1 in the second column to a 0. Recalling the instructions, we will change the first row by adding the second row (which contains the diagonal location for this column) to it:

$$\begin{array}{r} \begin{bmatrix} x & y & z & \\ 1 & -1 & 2 & | & 3 \end{bmatrix} \text{ (first row)} \\ +\begin{bmatrix} 0 & 1 & -3 & | & -4 \end{bmatrix} \text{ (second row)} \\ \hline \begin{bmatrix} 1 & 0 & -1 & | & -1 \end{bmatrix} \text{ (sum)} \end{array}$$

We substitute this sum for the first row:

$$\begin{bmatrix} x & y & z & \\ 1 & 0 & -1 & | & -1 \\ 0 & 1 & -3 & | & -4 \\ 0 & 0 & -3 & | & -8 \end{bmatrix}$$

Now the second column looks like that column of the identity matrix, so we can work on the third column. According to Step 1, we should divide the third row by −3 to obtain a 1 in the diagonal location of the third column.

$$\begin{bmatrix} x & y & z & \\ 1 & 0 & -1 & | & -1 \\ 0 & 1 & -3 & | & -4 \\ 0 & 0 & 1 & | & \frac{8}{3} \end{bmatrix}$$

By Step 2, we will change the −1 in the first row to a 0 by adding the third row to it.

$$\begin{array}{r} \begin{bmatrix} x & y & z & \\ 1 & 0 & -1 & | & -1 \end{bmatrix} \text{ (first row)} \\ +\begin{bmatrix} 0 & 0 & 1 & | & \frac{8}{3} \end{bmatrix} \text{ (third row)} \\ \hline \begin{bmatrix} 1 & 0 & 0 & | & \frac{5}{3} \end{bmatrix} \text{ (sum)} \end{array}$$

1.4 NONSINGULAR n-BY-n LINEAR SYSTEMS

Substituting this sum results in a new first row:

$$\begin{array}{c} \begin{array}{ccc} x & y & z \end{array} \\ \left[\begin{array}{ccc|c} 1 & 0 & 0 & \frac{5}{3} \\ 0 & 1 & -3 & -4 \\ 0 & 0 & 1 & \frac{8}{3} \end{array}\right] \end{array}$$

The last number we have to change is the -3, and Step 2 tells us how to do it: Add to the second row 3 times the third row.

$$\begin{array}{c} \begin{array}{ccc} x & y & z \end{array} \\ \begin{array}{r} \left[\begin{array}{ccc|c} 0 & 1 & -3 & -4 \end{array}\right] \\ +\left[\begin{array}{ccc|c} 0 & 0 & 3 & 8 \end{array}\right] \\ \hline \left[\begin{array}{ccc|c} 0 & 1 & 0 & 4 \end{array}\right] \end{array} \end{array} \begin{array}{l} \text{(second row)} \\ \text{(3 times third row)} \\ \text{(sum)} \end{array}$$

We substitute the sum for the previous second row so that the remaining non-diagonal number in the third column becomes a zero:

$$\begin{array}{c} \begin{array}{ccc} x & y & z \end{array} \\ \left[\begin{array}{ccc|c} 1 & 0 & 0 & \frac{5}{3} \\ 0 & 1 & 0 & 4 \\ 0 & 0 & 1 & \frac{8}{3} \end{array}\right] \end{array}$$

We have completed the procedure and have turned the matrix to the left of the vertical line into an identity matrix.

The Solution

Now making use of the letters at the top of the columns, we can see that this augmented matrix corresponds to the following very simple system of three equations in three unknowns:

$$\begin{aligned} x &= \frac{5}{3} \\ y &= 4 \\ z &= \frac{8}{3} \end{aligned}$$

and there certainly is no question what the solution of that system is. We arrived at this system by the use of three operations: interchanging rows, multiplying or dividing a row by a constant, adding (a constant times) a row to another row and substituting the sum for one of these rows. We have seen that the solution to a problem remains the same after each application of any of these operations; so the solution to the original problem

58   MATRICES AND SYSTEMS OF LINEAR EQUATIONS

$$x - y + 2z = 3$$
$$2x - 2y + z = -2$$
$$-x + z = 1$$

must be the same one, that is $x = \frac{5}{3}$, $y = 4$, and $z = \frac{8}{3}$. To see that this is so, and as a check that we did not make any arithmetic mistakes, we compute

$$\frac{5}{3} - 4 + 2\left(\frac{8}{3}\right) = 3$$

$$2\left(\frac{5}{3}\right) - 2(4) + \frac{8}{3} = -2$$

$$-\frac{5}{3} + \frac{8}{3} = 1$$

Before we complete this section, we should notice that the unknowns above the columns of the augmented matrix were not used until the very end, to write down the solution. The operations never change the order of the terms within a row, so the labelling of rows will be the same at the end of the procedure as at the beginning. Consequently, there was no need to keep writing x, y, and z at the top of the matrices. We can label the first augmented matrix and, when we have an identity matrix to the left of the vertical line, look back and label this final augmented matrix in the same way.

Example 3. Use the Gauss-Jordan method to find the values of x, y, and z that solve the following system of linear equations:

$$2x + y - z = 0$$
$$-2x - z = 2$$
$$-2y + z = -1$$

Solution

We begin with the augmented matrix

$$\begin{array}{ccc} x & y & z \end{array}$$
$$\begin{bmatrix} 2 & 1 & -1 & | & 0 \\ -2 & 0 & -1 & | & 2 \\ 0 & -2 & 1 & | & -1 \end{bmatrix}$$

Step 1 tells us to divide the first row by 2, the number in the diagonal location of the first column, so that the number in that location becomes a 1.

$$\begin{bmatrix} 1 & \frac{1}{2} & -\frac{1}{2} & | & 0 \\ -2 & 0 & -1 & | & 2 \\ 0 & -2 & 1 & | & -1 \end{bmatrix}$$

Going on to Step 2, we see that $a_{21} = -2$, so we must add 2 times the first row to the second row.

1.4 NONSINGULAR n-BY-n LINEAR SYSTEMS

$$\begin{array}{r}[-2 \quad 0 \quad -1 \mid 2] \\ +[2 \quad 1 \quad -1 \mid 0] \\ \hline [0 \quad 1 \quad -2 \mid 2]\end{array}$$ (second row)
(2 times first row)
(sum)

We substitute this sum for the second row so that the number in the first column below the 1 is now a zero.

$$\begin{bmatrix} 1 & \frac{1}{2} & -\frac{1}{2} & 0 \\ 0 & 1 & -2 & 2 \\ 0 & -2 & 1 & -1 \end{bmatrix}$$

The diagonal location in the second column is already occupied by a 1, so we next change the number above it to a zero by adding $-\frac{1}{2}$ times the second row to the first row, to produce a new first row.

$$\begin{array}{r}\left[1 \quad \frac{1}{2} \quad -\frac{1}{2} \mid 0\right] \\ +\left[0 \quad -\frac{1}{2} \quad 1 \mid -1\right] \\ \hline \left[1 \quad 0 \quad \frac{1}{2} \mid -1\right]\end{array}$$ (first row)
$\left(-\frac{1}{2} \text{ times second row}\right)$
(sum)

The matrix now looks like this:

$$\begin{bmatrix} 1 & 0 & \frac{1}{2} & -1 \\ 0 & 1 & -2 & 2 \\ 0 & -2 & 1 & -1 \end{bmatrix}$$

We complete the process of making the nondiagonal numbers in the second column all zeros by adding 2 times the second row to the third row.

$$\begin{array}{r}[0 \quad -2 \quad 1 \mid -1] \\ +[0 \quad 2 \quad -4 \mid 4] \\ \hline [0 \quad 0 \quad -3 \mid 3]\end{array}$$ (third row)
(2 times second row)
(sum)

This yields:

$$\begin{bmatrix} 1 & 0 & \frac{1}{2} & -1 \\ 0 & 1 & -2 & 2 \\ 0 & 0 & -3 & 3 \end{bmatrix}$$

For the third column, we first divide the third row by -3 to change the diagonal entry to a 1.

$$\begin{bmatrix} 1 & 0 & \frac{1}{2} & -1 \\ 0 & 1 & -2 & 2 \\ 0 & 0 & 1 & -1 \end{bmatrix}$$

We change a_{13} to a 0 by adding $-\frac{1}{2}$ times the third row to the first row—

$$\begin{array}{r}\begin{bmatrix} 1 & 0 & \frac{1}{2} & | & -1 \end{bmatrix} \text{ (first row)} \\ +\begin{bmatrix} 0 & 0 & -\frac{1}{2} & | & \frac{1}{2} \end{bmatrix} \left(-\frac{1}{2} \text{ times third row}\right) \\ \hline \begin{bmatrix} 1 & 0 & 0 & | & -\frac{1}{2} \end{bmatrix} \text{ (sum)} \end{array}$$

obtaining

$$\begin{bmatrix} 1 & 0 & 0 & | & -\frac{1}{2} \\ 0 & 1 & -2 & | & 2 \\ 0 & 0 & 1 & | & -1 \end{bmatrix}$$

For the final operation, which will make the matrix to the left of the vertical line an identity matrix, we add 2 times the third row to the second row.

$$\begin{array}{r}\begin{bmatrix} 0 & 1 & -2 & | & 2 \end{bmatrix} \text{ (second row)} \\ +\begin{bmatrix} 0 & 0 & 2 & | & -2 \end{bmatrix} \text{ (2 times third row)} \\ \hline \begin{bmatrix} 0 & 1 & 0 & | & 0 \end{bmatrix} \text{ (sum)} \end{array}$$

This time, when we substitute into the augmented matrix, we will recall the unknowns labelling the columns, just as at the beginning of the problem.

$$\begin{array}{ccc} x & y & z \end{array}$$
$$\begin{bmatrix} 1 & 0 & 0 & | & -\frac{1}{2} \\ 0 & 1 & 0 & | & 0 \\ 0 & 0 & 1 & | & -1 \end{bmatrix}$$

This is the augmented matrix corresponding to the system

$$\begin{aligned} x &= -\frac{1}{2} \\ y &= 0 \\ z &= -1 \end{aligned}$$

We conclude that the solution to the system

$$\begin{aligned} 2x + y - z &= 0 \\ -2x \quad\quad - z &= 2 \\ -2y + z &= -1 \end{aligned}$$

is $x = -\frac{1}{2}$, $y = 0$, $z = -1$. We should verify that we did not make any arithmetic mistakes by substituting these numbers into the given equations:

$$2\left(-\frac{1}{2}\right) + 0 - (-1) = 0$$
$$-2\left(-\frac{1}{2}\right) \quad - (-1) = 2$$
$$-2(0) + (-1) = -1$$

∎

A Look Ahead

We would like you to notice a feature of the solution to Example 3. Suppose we change the right-hand side of the system. For instance, suppose we replace the number 0 on the right-hand side of the first equation with -2:

$$2x + y - z = -2$$
$$-2x \quad - z = 2$$
$$-2y + z = -1$$

To solve this new problem, we will have to start the Gauss-Jordan method all over again with a slightly different augmented matrix. It will have a -2 at the top of the column to the *right* of the vertical line instead of a 0. But in the solution to Example 3, all the decisions we made about which operations to perform, and in which order, depended on the numbers to the *left* of the vertical line. The reason is that the left is where both the numbers in the diagonal location and the numbers that have to be converted to 0s are found.

If we alter Example 3 by changing only the numbers on the right-hand sides of the equations (that is, the numbers to the right of the vertical line of the augmented matrix), then the steps of the new solution will be exactly the same as those of the solution just given. However, the numbers to the right of the vertical line will usually be different at every stage of the solution. Thus, the numbers to the right and those to the left of the vertical line of the augmented matrix play rather different roles in the Gauss-Jordan method. This observation will be the key to the way we use the Gauss-Jordan method to calculate the inverses of nonsingular matrices with more than two rows, as we will see in the next section.

Summary

If a system of n linear equations in n unknowns is written as the matrix equation $AX = B$ and if the coefficient matrix A is nonsingular (that is, has an inverse), then the system can be solved by the **Gauss-Jordan method**. This method consists of applying to the **augmented matrix** $[\,A\,|\,B\,]$ the following operations: interchange of rows, multiplication or division of all the numbers of a row by a nonzero constant, addition of (a constant times) a row to another row, and substitution of this sum for one of these rows. We work on columns from left to right. If the number in the **diagonal location** (where row number equals column number) is 0, interchange the row containing that element with some row below it that contains a nonzero number in the relevant column. Once there is a nonzero number d in the diagonal location, divide the row containing d by d itself to change d to a 1. Take any nonzero number a in the column and add $-a$ times the row containing the diagonal location to the row containing a. Replace the row containing a with the sum to change a to 0. When this process is completed, the augmented matrix has the form $[\,I\mid B^*\,]$, and the solution of the original system is given by the numbers in the column B^* to the right of the vertical line.

Exercises

In Exercises 1 through 4, write the augmented matrix corresponding to the given system of linear equations, but do not attempt to solve it.

1. $3x + 2y + z = 1$
   $x + y - 3z = -2$
   $2x - 2y + z = 1$

2. $2x \quad\quad + z = 2$
   $x + y - z = 0$
   $-y + z = 1$

3. $w + x + y + z = 1$
   $-w + \quad\quad y - z = 0$
   $\quad\quad 2x + 3y + z = -1$
   $w + x \quad\quad - z = 1$

4. $\quad\quad 2x + 3y \quad\quad = -2$
   $w + x \quad\quad - 2z = 1$
   $\quad\quad -2y + z = 3$
   $2w \quad\quad\quad - z = 1$

5. Multiply the first row of
$$\begin{bmatrix} -1 & 2 & 0 & | & -2 \\ \frac{1}{3} & 1 & 2 & | & 0 \\ -1 & 1 & 2 & | & -1 \end{bmatrix}$$
by -1.

6. Multiply the second row of
$$\begin{bmatrix} -1 & 2 & 0 & | & -2 \\ \frac{1}{3} & 1 & 2 & | & 0 \\ -1 & 1 & 2 & | & -1 \end{bmatrix}$$
by $\frac{1}{2}$.

7. Add 3 times the first row of
$$\begin{bmatrix} 1 & -2 & 1 & | & 1 \\ -3 & 4 & 0 & | & 2 \\ 2 & -1 & 1 & | & -1 \end{bmatrix}$$
to the second row, and substitute the sum for the second row.

8. Add -2 times the second row of
$$\begin{bmatrix} 1 & 2 & 1 & | & 1 \\ 0 & 1 & -\frac{1}{2} & | & 1 \\ 0 & -1 & 0 & | & 1 \end{bmatrix}$$
to the first row, and substitute the sum for the first row.

1.4 NONSINGULAR n-BY-n LINEAR SYSTEMS

9. Add $-\frac{1}{3}$ times the third row of

$$\begin{bmatrix} 1 & 0 & 0 & \Big| & \frac{1}{3} \\ 0 & 1 & \frac{1}{3} & \Big| & -2 \\ 0 & 0 & 1 & \Big| & 2 \end{bmatrix}$$

to the second row, and substitute the sum for the second row.

10. Add $\frac{1}{2}$ times the third row of

$$\begin{bmatrix} 1 & 0 & 0 & \frac{1}{2} & \Big| & \frac{1}{2} \\ 0 & 1 & 0 & 2 & \Big| & 1 \\ 0 & 0 & 1 & 3 & \Big| & -1 \\ 0 & 0 & -\frac{1}{2} & 0 & \Big| & 1 \end{bmatrix}$$

to the fourth row, and substitute the sum for the fourth row.

In Exercises 11 through 18, a stage of the solution of a system of linear equations by the Gauss-Jordan method is shown. State in words what the next step should be, but do not actually carry out the arithmetic.

11. $\begin{bmatrix} 1 & 0 & -1 & \Big| & 2 \\ 0 & 1 & 2 & \Big| & 0 \\ 0 & 0 & 3 & \Big| & 1 \end{bmatrix}$

12. $\begin{bmatrix} 1 & 0 & 2 & \Big| & -1 \\ 0 & 1 & 1 & \Big| & -1 \\ 0 & -2 & 1 & \Big| & 0 \end{bmatrix}$

13. $\begin{bmatrix} 1 & 2 & 1 & \Big| & 1 \\ 0 & 0 & 1 & \Big| & 2 \\ 0 & -3 & 1 & \Big| & -1 \end{bmatrix}$

14. $\begin{bmatrix} 1 & 2 & 1 & \Big| & 1 \\ 3 & 2 & 0 & \Big| & -1 \\ 0 & 2 & 0 & \Big| & -1 \end{bmatrix}$

15. $\begin{bmatrix} 0 & 1 & -2 & \Big| & 0 \\ 0 & 2 & 1 & \Big| & 1 \\ 2 & 3 & -1 & \Big| & 1 \end{bmatrix}$

16. $\begin{bmatrix} 1 & 0 & \frac{1}{2} & \Big| & 1 \\ 0 & 1 & 0 & \Big| & 1 \\ 0 & 0 & \frac{1}{2} & \Big| & 2 \end{bmatrix}$

17. $\begin{bmatrix} 1 & 0 & 1 & -2 & \Big| & 1 \\ 0 & 0 & 2 & 0 & \Big| & 1 \\ 0 & -2 & -1 & 0 & \Big| & -1 \\ 0 & 3 & -2 & 1 & \Big| & 0 \end{bmatrix}$

18. $\begin{bmatrix} 1 & 0 & 0 & 2 & \Big| & 1 \\ 0 & 1 & 0 & 1 & \Big| & \frac{1}{2} \\ 0 & 0 & 1 & \frac{1}{2} & \Big| & -1 \\ 0 & 0 & -\frac{1}{3} & 1 & \Big| & 0 \end{bmatrix}$

In Exercises 19 through 28, solve the given system of linear equations by the Gauss-Jordan method.

19. $\begin{aligned} y + z &= 2 \\ x + y + z &= -1 \\ -x + z &= -1 \end{aligned}$

20. $\begin{aligned} x - y &= 2 \\ -2x + 2y + z &= 0 \\ y - z &= 2 \end{aligned}$

64   MATRICES AND SYSTEMS OF LINEAR EQUATIONS

21. $x + y - z = 1$
    $2y = -1$
    $-x - y + 2z = 0$

22. $2y + z = 0$
    $y + z = 1$
    $x - y = -1$

23. $-x + y + 2z = 1$
    $2x - 2z = 0$
    $-x - y - 2z = 2$

24. $2x - 4z = 0$
    $y + 2z = 1$
    $-x + 2y + z = -1$

25. $2x + y - z = 0$
    $-x + y = 1$
    $2x + y + 2z = 0$

26. $w - z = 2$
    $-w + x + z = 1$
    $w + y - z = 0$
    $x + z = -1$

27. $-w + x + 2z = 1$
    $w + y + z = -1$
    $-x + y + z = 2$
    $-2w + y + 3z = 0$

28. $-x + y = 2$
    $w + 2x + z = 2$
    $2w + x - y - z = 0$
    $-w - x + y + 2z = 0$

1.5 The Inverse of a Matrix

In Section 1.3 we introduced the inverse, A^{-1}, of a matrix A, which was defined by the property

$$AA^{-1} = A^{-1}A = I$$

where I is an identity matrix. Only square matrices, that is, $n \times n$-matrices, can have inverses, but not all square matrices have them. A matrix that has an inverse is called **nonsingular** and a square matrix with no inverse is called **singular**. In that same section we learned that the inverse can be used to solve matrix equations of the form $AX = B$, where A is a nonsingular matrix. A slightly more complicated matrix equation must be solved as a part of input-output economic theory, and its solution also involves the inverse, as we shall now see.

The Input-Output Equation

The matrix equation of input-output theory concerns three matrices: an $n \times n$-matrix A called the "input-output matrix," an $n \times 1$-matrix D called the "demand matrix," and another $n \times 1$-matrix X, this time consisting of unknowns, called the "intensity matrix." We will show you in Section 1.8 that, in input-output economics, the equation that relates these matrices is

$$X = AX + D$$

Notice that AX is an $n \times 1$-matrix, so the equation at least makes sense. Let us use the matrix algebra we have learned so far to solve this equation for the unknown

1.5 THE INVERSE OF A MATRIX

matrix X, and thus produce the formula the economist (with considerable help from a computer) uses to compute the "intensities" of X.

It is generally a good idea to get all the terms of the equation that involve the unknown on the same side, so we will subtract AX from both sides of the equation:

$$X - AX = D$$

The next step is to "factor" the unknown matrix X out of the left hand side of the equation. We have to be careful that we use only procedures that we know work properly for matrices. The way to do this is first to write X as IX, where I is the $n \times n$-identity matrix. That is, by the property of I, we know that $X = IX$, so we can substitute back into our equation with this result:

$$IX - AX = D$$

"Factoring" means using the distributive law to obtain

$$(I - A)X = D$$

Now we notice that the equation is of the form $BX = D$, and we already know how to solve a matrix equation of this type, provided B is a nonsingular matrix. The solution is given by $X = B^{-1}D$. It will be more convenient to write this solution in terms of the original matrices, A, D, and X. In that case, we say that if the matrix $I - A$ is nonsingular, then the intensity matrix can be computed by the formula

$$X = (I - A)^{-1}D$$

In a typical computation of this type (but with matrices much smaller than in a real economics application), we are given

$$A = \begin{bmatrix} 0.474 & 0.026 & 0.079 \\ 0.018 & 0.358 & 0.025 \\ 0.105 & 0.173 & 0.234 \end{bmatrix}, \quad D = \begin{bmatrix} 223{,}133 \\ 23{,}527 \\ 263{,}985 \end{bmatrix}, \quad \text{and} \quad X = \begin{bmatrix} x \\ y \\ z \end{bmatrix}$$

If we subtract A from the 3×3-identity matrix, the result is

$$I - A = \begin{bmatrix} 1 & 0 & 0 \\ 0 & 1 & 0 \\ 0 & 0 & 1 \end{bmatrix} - \begin{bmatrix} 0.474 & 0.026 & 0.079 \\ 0.018 & 0.358 & 0.025 \\ 0.105 & 0.173 & 0.234 \end{bmatrix} = \begin{bmatrix} 0.526 & -0.026 & -0.079 \\ -0.018 & 0.642 & -0.025 \\ -0.105 & -0.173 & 0.766 \end{bmatrix}$$

With the aid of a computer, we find the inverse (actually, a close approximation of it) of $I - A$ to be

$$(I - A)^{-1} = \begin{bmatrix} 1.941 & 0.133 & 0.205 \\ 0.017 & 1.577 & 0.059 \\ 0.277 & 0.375 & 1.348 \end{bmatrix}$$

multiply it by the demand matrix D to compute X, with this result:

$$X = (I-A)^{-1}D = \begin{bmatrix} 1.941 & 0.133 & 0.205 \\ 0.017 & 1.577 & 0.059 \\ 0.277 & 0.375 & 1.348 \end{bmatrix} \begin{bmatrix} 233{,}133 \\ 23{,}527 \\ 263{,}985 \end{bmatrix} = \begin{bmatrix} 490{,}347 \\ 56{,}470 \\ 425{,}143 \end{bmatrix}$$

We will discuss the economic interpretation of this calculation in Section 1.8.

In Section 1.3, we presented a formula for the inverse of a two-by-two nonsingular matrix; but we remarked that it is not practical to write down a formula for finding the inverse A^{-1} of a nonsingular n-by-n matrix A when n is larger than 2. In input-output theory a size such as $n = 81$ is more what we expect. Instead, A^{-1} is computed from A using a systematic procedure based on the Gauss-Jordan method of the preceding section, which we will next describe in detail. This is the procedure that the computer used to produce the matrix $(I - A)^{-1}$ just discussed.

Systems for the Inverse

We have seen how the problem of finding the inverse of the two-by-two matrix

$$A = \begin{bmatrix} 3 & 4 \\ 2 & 5 \end{bmatrix}$$

became a problem in solving systems of linear equations. We wrote the inverse as a matrix of unknowns

$$A^{-1} = \begin{bmatrix} s & t \\ u & v \end{bmatrix}$$

The definition of the inverse means that $AA^{-1} = I$, and when we wrote out what that means, here was the result:

$$AA^{-1} = \begin{bmatrix} 3 & 4 \\ 2 & 5 \end{bmatrix} \begin{bmatrix} s & t \\ u & v \end{bmatrix} = \begin{bmatrix} 3s + 4u & 3t + 4v \\ 2s + 5u & 2t + 5v \end{bmatrix} = \begin{bmatrix} 1 & 0 \\ 0 & 1 \end{bmatrix} = I$$

By the definition of matrix equality, we see that

$$(*) \begin{cases} 3s + 4u = 1 \\ 2s + 5u = 0 \end{cases} \qquad (**) \begin{cases} 3t + 4v = 0 \\ 2t + 5v = 1 \end{cases}$$

We should again notice that the two equations on the left concern only the unknowns s and u, whereas the equations on the right have t and v as the unknowns. We think of these equations as two systems, (*) and (**), each consisting of two linear equations in two unknowns. The solution of (*) would tell us the numbers s and u in the first column of A^{-1}, and the solution of (**) would give the numbers in the second column of A^{-1}. Furthermore, the two systems have something very important in common. The coefficient matrix is the same for both systems: it is the matrix A whose inverse we wish to compute.

Let us see what happens when we go through exactly the same steps with a larger matrix. Our example will be

$$A = \begin{bmatrix} 1 & -1 & 1 \\ 2 & 1 & 0 \\ -1 & -1 & 1 \end{bmatrix}$$

The inverse is a 3-by-3 matrix of unknowns, and since there are so many unknowns, it will be convenient to take advantage of the notation we introduced at the beginning of the chapter to indicate where these unknowns are located in the matrix A^{-1}:

$$A^{-1} = \begin{bmatrix} x_{11} & x_{12} & x_{13} \\ x_{21} & x_{22} & x_{23} \\ x_{31} & x_{32} & x_{33} \end{bmatrix}$$

Now when we write out what $AA^{-1} = I$ means, we have

$$A(A^{-1}) = \begin{bmatrix} 1 & -1 & 1 \\ 2 & 1 & 0 \\ -1 & -1 & 1 \end{bmatrix} \begin{bmatrix} x_{11} & x_{12} & x_{13} \\ x_{21} & x_{22} & x_{23} \\ x_{31} & x_{32} & x_{33} \end{bmatrix} = \begin{bmatrix} 1 & 0 & 0 \\ 0 & 1 & 0 \\ 0 & 0 & 1 \end{bmatrix} = I$$

that is, when we multiply out these matrices, we obtain nine linear equations. But these equations may again be grouped into systems that share the same unknowns:

$$\begin{aligned} x_{11} - x_{21} + x_{31} &= 1 \\ 2x_{11} + x_{21} &= 0 \\ -x_{11} - x_{21} + x_{31} &= 0 \end{aligned} \qquad \begin{aligned} x_{12} - x_{22} + x_{32} &= 0 \\ 2x_{12} + x_{22} &= 1 \\ -x_{12} - x_{22} + x_{32} &= 0 \end{aligned} \qquad \begin{aligned} x_{13} - x_{23} + x_{33} &= 0 \\ 2x_{13} + x_{23} &= 0 \\ -x_{13} - x_{23} + x_{33} &= 1 \end{aligned}$$

The Augmented Matrices

If we look at the three augmented matrices that correspond to these three systems of linear equations,

$$\begin{array}{c} \begin{matrix} x_{11} & x_{21} & x_{31} \end{matrix} \\ \left[\begin{array}{ccc|c} 1 & -1 & 1 & 1 \\ 2 & 1 & 0 & 0 \\ -1 & -1 & 1 & 0 \end{array}\right] \end{array}, \quad \begin{array}{c} \begin{matrix} x_{12} & x_{22} & x_{32} \end{matrix} \\ \left[\begin{array}{ccc|c} 1 & -1 & 1 & 0 \\ 2 & 1 & 0 & 1 \\ -1 & -1 & 1 & 0 \end{array}\right] \end{array}, \text{ and } \begin{array}{c} \begin{matrix} x_{13} & x_{23} & x_{33} \end{matrix} \\ \left[\begin{array}{ccc|c} 1 & -1 & 1 & 0 \\ 2 & 1 & 0 & 0 \\ -1 & -1 & 1 & 1 \end{array}\right] \end{array}$$

we note that the "matrix" to the left of the vertical line is always the same; it is the matrix A that we started with. As we remarked at the end of the last section, if we solved these systems by the Gauss-Jordan method, we would go through exactly the same sequence of row operations in solving each of these systems; only the columns to the right of the vertical lines would be different. Therefore, what we should do is to combine these three matrices into one, with the matrix A to the left of the vertical line, and all three right-hand columns to the right. Since these columns are just those of the 3-by-3 identity matrix, this larger augmented matrix is of the form $[\,A\,|\,I\,]$. We should then use the Gauss-Jordan method, but should

68 MATRICES AND SYSTEMS OF LINEAR EQUATIONS

carry each step through the entire row (now consisting of six numbers, rather than four) to change the matrix to the left of the vertical line into an identity matrix. The numbers to the right of the vertical line will be the solutions of the three systems of linear equations. If we look back at the unknowns at the top of the three matrices, we see that for each column of the identity matrix on the right, the solution will be the corresponding column of the matrix A^{-1}, just as in the 2-by-2 case. When we have changed the 3-by-3 matrix to the left of the vertical line in the augmented matrix $[\,A\,|\,I\,]$ to the identity matrix, we will have changed the 3-by-3 matrix to the right to the inverse matrix A^{-1}. Consequently, the matrix will have the form $[\,I\,|\,A^{-1}\,]$.

An Illustration

We will next apply the procedure based on the Gauss-Jordan method to the example

$$A = \begin{bmatrix} 1 & -1 & 1 \\ 2 & 1 & 0 \\ -1 & -1 & 1 \end{bmatrix}$$

We form the augmented matrix $[\,A\,|\,I\,]$:

$$\begin{bmatrix} 1 & -1 & 1 & | & 1 & 0 & 0 \\ 2 & 1 & 0 & | & 0 & 1 & 0 \\ -1 & -1 & 1 & | & 0 & 0 & 1 \end{bmatrix}$$

Since the number in the upper left-hand corner, the diagonal element of the first column, is already a 1, we move right on to changing the number 2 below it to a zero. For this purpose, we multiply the first row by -2 and add it to the second row.

$$\begin{array}{r} \begin{bmatrix} 2 & 1 & 0 & | & 0 & 1 & 0 \end{bmatrix} \text{ (second row)} \\ +\begin{bmatrix} -2 & 2 & -2 & | & -2 & 0 & 0 \end{bmatrix} \text{ (-2 times first row)} \\ \hline \begin{bmatrix} 0 & 3 & -2 & | & -2 & 1 & 0 \end{bmatrix} \text{ (sum)} \end{array}$$

This sum then becomes the new second row.

$$\begin{bmatrix} 1 & -1 & 1 & | & 1 & 0 & 0 \\ 0 & 3 & -2 & | & -2 & 1 & 0 \\ 1 & -1 & 1 & | & 0 & 0 & 1 \end{bmatrix}$$

In order to change the -1 at the bottom of the first column to a zero, we just have to add the first row to the third row, and substitute the sum for the third row:

$$\begin{array}{r} \begin{bmatrix} -1 & -1 & 1 & | & 0 & 0 & 1 \end{bmatrix} \text{ (third row)} \\ +\begin{bmatrix} 1 & -1 & 1 & | & 1 & 0 & 0 \end{bmatrix} \text{ (first row)} \\ \hline \begin{bmatrix} 0 & -2 & 2 & | & 1 & 0 & 1 \end{bmatrix} \text{ (sum)} \end{array}$$

1.5 THE INVERSE OF A MATRIX

Thus, we have

$$\begin{bmatrix} 1 & -1 & 1 & | & 1 & 0 & 0 \\ 0 & 3 & -2 & | & -2 & 1 & 0 \\ 0 & -2 & 2 & | & 1 & 0 & 1 \end{bmatrix}$$

This completes the process of changing the first column to that of the identity matrix, and we move on to the diagonal element of the second column, which is a 3. We divide the second row by 3 so that the diagonal number in the second column becomes a 1.

$$\begin{bmatrix} 1 & -1 & 1 & | & 1 & 0 & 0 \\ 0 & 1 & -\frac{2}{3} & | & -\frac{2}{3} & \frac{1}{3} & 0 \\ 0 & -2 & 2 & | & 1 & 0 & 1 \end{bmatrix}$$

We change the top number in the second row from −1 to zero by adding the second row (which contains the diagonal element) to the first row, and substituting the sum for the first row:

$$\begin{array}{r} \begin{bmatrix} 1 & -1 & 1 & | & 1 & 0 & 0 \end{bmatrix} \text{ (first row)} \\ +\begin{bmatrix} 0 & 1 & -\frac{2}{3} & | & -\frac{2}{3} & \frac{1}{3} & 0 \end{bmatrix} \text{ (second row)} \\ \hline \begin{bmatrix} 1 & 0 & \frac{1}{3} & | & \frac{1}{3} & \frac{1}{3} & 0 \end{bmatrix} \text{ (sum)} \end{array}$$

This yields

$$\begin{bmatrix} 1 & 0 & \frac{1}{3} & | & \frac{1}{3} & \frac{1}{3} & 0 \\ 0 & 1 & -\frac{2}{3} & | & -\frac{2}{3} & \frac{1}{3} & 0 \\ 0 & -2 & 2 & | & 1 & 0 & 1 \end{bmatrix}$$

We change −2 to zero by multiplying the second row by 2 and adding it to the third row.

$$\begin{array}{r} \begin{bmatrix} 0 & -2 & 2 & | & 1 & 0 & 1 \end{bmatrix} \text{ (third row)} \\ +\begin{bmatrix} 0 & 2 & -\frac{4}{3} & | & -\frac{4}{3} & \frac{2}{3} & 0 \end{bmatrix} \text{ (2 times second row)} \\ \hline \begin{bmatrix} 0 & 0 & \frac{2}{3} & | & -\frac{1}{3} & \frac{2}{3} & 1 \end{bmatrix} \text{ (sum)} \end{array}$$

The sum then becomes the new third row. This completes the process of changing the second column to that of the identity matrix.

$$\begin{bmatrix} 1 & 0 & \frac{1}{3} & | & \frac{1}{3} & \frac{1}{3} & 0 \\ 0 & 1 & -\frac{2}{3} & | & -\frac{2}{3} & \frac{1}{3} & 0 \\ 0 & 0 & \frac{2}{3} & | & -\frac{1}{3} & \frac{2}{3} & 1 \end{bmatrix}$$

To put a 1 in place of the present $\frac{2}{3}$ in the diagonal location of the third column, we divide the third row by that number, $\frac{2}{3}$, or what is the same thing, we multiply the row by $\frac{3}{2}$. (Remember that dividing by a fraction is the same thing as multiplying by its reciprocal—the fraction with the numerator and denominator interchanged.)

$$\begin{bmatrix} 1 & 0 & \frac{1}{3} & \bigm| & \frac{1}{3} & \frac{1}{3} & 0 \\ 0 & 1 & -\frac{2}{3} & \bigm| & -\frac{2}{3} & \frac{1}{3} & 0 \\ 0 & 0 & 1 & \bigm| & -\frac{1}{2} & 1 & \frac{3}{2} \end{bmatrix}$$

To change the number $\frac{1}{3}$ at the top of the third row to a zero, we will multiply the third row by $-\frac{1}{3}$ and add it to the first row.

$$\begin{array}{r} \begin{bmatrix} 1 & 0 & \frac{1}{3} & \bigm| & \frac{1}{3} & \frac{1}{3} & 0 \end{bmatrix} \quad \text{(first row)} \\ +\begin{bmatrix} 0 & 0 & -\frac{1}{3} & \bigm| & \frac{1}{6} & -\frac{1}{3} & -\frac{1}{2} \end{bmatrix} \quad \left(-\frac{1}{3} \text{ times second row}\right) \\ \hline \begin{bmatrix} 1 & 0 & 0 & \bigm| & \frac{1}{2} & 0 & -\frac{1}{2} \end{bmatrix} \quad \text{(sum)} \end{array}$$

Then we substitute the sum for the first row:

$$\begin{bmatrix} 1 & 0 & 0 & \bigm| & \frac{1}{2} & 0 & -\frac{1}{2} \\ 0 & 1 & -\frac{2}{3} & \bigm| & -\frac{2}{3} & \frac{1}{3} & 0 \\ 0 & 0 & 1 & \bigm| & -\frac{1}{2} & 1 & \frac{3}{2} \end{bmatrix}$$

The final step changes $-\frac{2}{3}$ to zero by multiplying the third row by $\frac{2}{3}$ and adding it to the second row.

$$\begin{array}{r} \begin{bmatrix} 0 & 1 & -\frac{2}{3} & \bigm| & -\frac{2}{3} & \frac{1}{3} & 0 \end{bmatrix} \quad \text{(second row)} \\ +\begin{bmatrix} 0 & 0 & \frac{2}{3} & \bigm| & -\frac{1}{3} & \frac{2}{3} & 1 \end{bmatrix} \quad \left(\frac{2}{3} \text{ times second row}\right) \\ \hline \begin{bmatrix} 0 & 1 & 0 & \bigm| & -1 & 1 & 1 \end{bmatrix} \quad \text{(sum)} \end{array}$$

We obtain

$$\begin{bmatrix} 1 & 0 & 0 & \bigm| & \frac{1}{2} & 0 & -\frac{1}{2} \\ 0 & 1 & 0 & \bigm| & -1 & 1 & 1 \\ 0 & 0 & 1 & \bigm| & -\frac{1}{2} & 1 & \frac{3}{2} \end{bmatrix}$$

1.5 THE INVERSE OF A MATRIX

The inverse is therefore given by

$$A^{-1} = \begin{bmatrix} \frac{1}{2} & 0 & -\frac{1}{2} \\ -1 & 1 & 1 \\ -\frac{1}{2} & 1 & \frac{3}{2} \end{bmatrix}$$

It is a good idea to multiply this matrix by A to be sure that the product really is I, to check that we did not make any arithmetic errors in finding the inverse.

$$A(A^{-1}) = \begin{bmatrix} 1 & -1 & 1 \\ 2 & 1 & 0 \\ -1 & -1 & 1 \end{bmatrix} \begin{bmatrix} \frac{1}{2} & 0 & -\frac{1}{2} \\ -1 & 1 & 1 \\ -\frac{1}{2} & 1 & \frac{3}{2} \end{bmatrix} = \begin{bmatrix} 1 & 0 & 0 \\ 0 & 1 & 0 \\ 0 & 0 & 1 \end{bmatrix} = I$$

You will certainly have noticed that it requires quite a few individual steps even to invert a 3 × 3-matrix. As the size of the matrix increases, the number of individual steps required to find the inverse increases rapidly. As we have seen, input-output economics requires the inversion of matrices, and since the matrices are quite large, it is a process that involves a tremendous number of steps. However, because the procedure for inverting a matrix is so straightforward, and each of the three types of row operations can be carried out very rapidly on a computer, there are very efficient computer programs for doing it.

Now that we know how to find the inverses of larger matrices, we can use this technique to solve matrix equations of the form $AX = B$, where A is a nonsingular square matrix larger than two-by-two. The solution is $X = A^{-1}B$, regardless of the size of A, because the solution of $AX = B$ that we found in Section 1.3 did not depend on size. We can also solve slightly more complicated matrix equations in this way, as the first example illustrates:

Example 1. Solve the matrix equation

$$AX + B = C$$

for the matrix X of unknowns, where

$$A = \begin{bmatrix} -1 & 0 & 2 \\ 0 & 0 & -1 \\ 2 & 1 & 0 \end{bmatrix}, \quad B = \begin{bmatrix} 2 \\ 1 \\ -1 \end{bmatrix}, \quad \text{and} \quad C = \begin{bmatrix} 0 \\ 1 \\ 1 \end{bmatrix}$$

Solution
We will first find a general solution of the matrix equation

$$AX + B = C$$

72 MATRICES AND SYSTEMS OF LINEAR EQUATIONS

and then apply the formula we obtain to this particular problem. We wish to rewrite the matrix equation with the X on one side of the equals sign and the known matrices on the other side. Just as with equations of elementary algebra, we subtract the matrix B from both sides and the equation becomes

$$AX = C - B$$

In an algebraic equation we would then divide both sides of this equation by A. We recall from Section 1.3 that the corresponding step with matrices is to multiply both sides of the equation by A^{-1} (we are still assuming that the given square matrices are all nonsingular). The result is given by the formula

$$X = A^{-1}(C - B)$$

In order to use this formula, we need to find A^{-1}. Therefore, we write the augmented matrix

$$\begin{bmatrix} -1 & 0 & 2 & | & 1 & 0 & 0 \\ 0 & 0 & -1 & | & 0 & 1 & 0 \\ 2 & 1 & 0 & | & 0 & 0 & 1 \end{bmatrix}$$

We divide the first row by -1 to make the diagonal entry of the first column a 1.

$$\begin{bmatrix} 1 & 0 & -2 & | & -1 & 0 & 0 \\ 0 & 0 & -1 & | & 0 & 1 & 0 \\ 2 & 1 & 0 & | & 0 & 0 & 1 \end{bmatrix}$$

We will complete the process of changing the first column to that of an identity matrix by adding -2 times the first row to the third row. From now on, we will not show the addition step because we will assume that you are familiar with this process. When we substitute the sum in place of the third row, the augmented matrix becomes

$$\begin{bmatrix} 1 & 0 & -2 & | & -1 & 0 & 0 \\ 0 & 0 & -1 & | & 0 & 1 & 0 \\ 0 & 1 & 4 & | & 2 & 0 & 1 \end{bmatrix}$$

We must interchange the second and third rows to obtain a nonzero number as the diagonal entry of column two. This puts the second column immediately into the desired form.

$$\begin{bmatrix} 1 & 0 & -2 & | & -1 & 0 & 0 \\ 0 & 1 & 4 & | & 2 & 0 & 1 \\ 0 & 0 & -1 & | & 0 & 1 & 0 \end{bmatrix}$$

Going on to the third column, we divide the third row by -1 so that we have a 1 in the diagonal location of column three.

$$\begin{bmatrix} 1 & 0 & -2 & | & -1 & 0 & 0 \\ 0 & 1 & 4 & | & 2 & 0 & 1 \\ 0 & 0 & 1 & | & 0 & -1 & 0 \end{bmatrix}$$

To change the -2 on the top of the column to a zero, we multiply the third row by 2 and add it to the first row.

1.5 THE INVERSE OF A MATRIX

$$\begin{bmatrix} 1 & 0 & 0 & | & -1 & -2 & 0 \\ 0 & 1 & 4 & | & 2 & 0 & 1 \\ 0 & 0 & 1 & | & 0 & -1 & 0 \end{bmatrix}$$

The final step, to change the 4 to a zero, is accomplished by multiplying the third row by -4 and adding it to the second row. Substituting the sum for the second row completes the computation of the inverse because the identity matrix is now to the left of the vertical line.

$$\begin{bmatrix} 1 & 0 & 0 & | & -1 & -2 & 0 \\ 0 & 1 & 0 & | & 2 & 4 & 1 \\ 0 & 0 & 1 & | & 0 & -1 & 0 \end{bmatrix}$$

It is a good idea at this point to check that you did not make any mistakes—by multiplying this answer by the original matrix:

$$\begin{bmatrix} -1 & -2 & 0 \\ 2 & 4 & 1 \\ 0 & -1 & 0 \end{bmatrix} \begin{bmatrix} -1 & 0 & 2 \\ 0 & 0 & -1 \\ 2 & 1 & 0 \end{bmatrix} = \begin{bmatrix} 1 & 0 & 0 \\ 0 & 1 & 0 \\ 0 & 0 & 1 \end{bmatrix}$$

Thus, we have found that

$$A^{-1} = \begin{bmatrix} -1 & -2 & 0 \\ 2 & 4 & 1 \\ 0 & -1 & 0 \end{bmatrix}$$

Recalling the formula $X = A^{-1}(C - B)$ that we developed at the beginning of the solution, we compute

$$C - B = \begin{bmatrix} 0 \\ 1 \\ 1 \end{bmatrix} - \begin{bmatrix} 2 \\ 1 \\ -1 \end{bmatrix} = \begin{bmatrix} -2 \\ 0 \\ 2 \end{bmatrix}$$

and so, the solution of the matrix equation $AX + B = C$ is

$$X = A^{-1}(C - B) = \begin{bmatrix} -1 & -2 & 0 \\ 2 & 4 & 1 \\ 0 & -1 & 0 \end{bmatrix} \begin{bmatrix} -2 \\ 0 \\ 2 \end{bmatrix} = \begin{bmatrix} 2 \\ -2 \\ 0 \end{bmatrix} \quad \blacksquare$$

In solving a matrix equation $AX = B$, it is quicker to form the augmented matrix $[\,A \mid B\,]$ and to use the Gauss-Jordan method as it was presented in the last section than to compute A^{-1} and then multiply to obtain $X = A^{-1}B$. However, in many applications the matrix B changes from time to time, whereas the matrix A stays the same. (Matrices used in input-output economics are of this type.) For such applications, it is more efficient to calculate A^{-1} once and for all because we can use it over and over again to calculate $X = A^{-1}B$ for the different matrices B. We next present a simple example of this approach.

74   MATRICES AND SYSTEMS OF LINEAR EQUATIONS

Example 2. Solve the following systems of linear equations

(a) $x + 2y + z = 1$
$-2x - 2y - z = 3$
$x - y = 2$

(b) $x + 2y + z = -2$
$-2x - 2y - z = 3$
$x - y = -1$

Solution

The coefficient matrix is the same in both parts, namely

$$A = \begin{bmatrix} 1 & 2 & 1 \\ -2 & -2 & -1 \\ 1 & -1 & 0 \end{bmatrix}$$

so we will use the inverse A^{-1} rather than treating these as two separate problems. We write the linear systems as matrix equations:

(a) $AX = B$ and   (b) $AX = C$,

where,

$$X = \begin{bmatrix} x \\ y \\ z \end{bmatrix}, \quad B = \begin{bmatrix} 1 \\ 3 \\ 2 \end{bmatrix}, \quad \text{and} \quad C = \begin{bmatrix} -2 \\ 3 \\ -1 \end{bmatrix}$$

The solutions are:

(a) $X = A^{-1}B$ and   (b) $X = A^{-1}C$.

Therefore, the next step is to find A^{-1}. Write the augmented matrix

$$\begin{bmatrix} 1 & 2 & 1 & | & 1 & 0 & 0 \\ -2 & -2 & -1 & | & 0 & 1 & 0 \\ 1 & -1 & 0 & | & 0 & 0 & 1 \end{bmatrix}$$

Multiply the first row by 2, add it to the second row, and then substitute the sum to form the new second row.

$$\begin{bmatrix} 1 & 2 & 1 & | & 1 & 0 & 0 \\ 0 & 2 & 1 & | & 2 & 1 & 0 \\ 1 & -1 & 0 & | & 0 & 0 & 1 \end{bmatrix}$$

Multiply the first row by -1 and add it to the third row.

$$\begin{bmatrix} 1 & 2 & 1 & | & 1 & 0 & 0 \\ 0 & 2 & 1 & | & 2 & 1 & 0 \\ 0 & -3 & -1 & | & -1 & 0 & 1 \end{bmatrix}$$

Going on to the second column, divide the second row by 2.

$$\begin{bmatrix} 1 & 2 & 1 & | & 1 & 0 & 0 \\ 0 & 1 & \frac{1}{2} & | & 1 & \frac{1}{2} & 0 \\ 0 & -3 & -1 & | & -1 & 0 & 1 \end{bmatrix}$$

1.5 THE INVERSE OF A MATRIX 75

Multiply the second row by −2 and add it to the first row.

$$\begin{bmatrix} 1 & 0 & 0 & | & -1 & -1 & 0 \\ 0 & 1 & \frac{1}{2} & | & 1 & \frac{1}{2} & 0 \\ 0 & -3 & -1 & | & -1 & 0 & 1 \end{bmatrix}$$

Multiply the second row by 3 and add it to the third row.

$$\begin{bmatrix} 1 & 0 & 0 & | & -1 & -1 & 0 \\ 0 & 1 & \frac{1}{2} & | & 1 & \frac{1}{2} & 0 \\ 0 & 0 & \frac{1}{2} & | & 2 & \frac{3}{2} & 1 \end{bmatrix}$$

The procedure to change the diagonal element in the third column to a 1 is to divide the third row 3 by $\frac{1}{2}$, that is, to multiply it by the reciprocal, 2.

$$\begin{bmatrix} 1 & 0 & 0 & | & -1 & -1 & 0 \\ 0 & 1 & \frac{1}{2} & | & 1 & \frac{1}{2} & 0 \\ 0 & 0 & 1 & | & 4 & 3 & 2 \end{bmatrix}$$

We multiply the third row 3 by $-\frac{1}{2}$ and add it to the second row to complete the computation of A^{-1}.

$$\begin{bmatrix} 1 & 0 & 0 & | & -1 & -1 & 0 \\ 0 & 1 & 0 & | & -1 & -1 & -1 \\ 0 & 0 & 1 & | & 4 & 3 & 2 \end{bmatrix}$$

The solution of the matrix equation (a) is

$$X = \begin{bmatrix} x \\ y \\ z \end{bmatrix} = A^{-1}B = \begin{bmatrix} -1 & -1 & 0 \\ -1 & -1 & -1 \\ 4 & 3 & 2 \end{bmatrix} \begin{bmatrix} 1 \\ 3 \\ 2 \end{bmatrix} = \begin{bmatrix} -4 \\ -6 \\ 17 \end{bmatrix}$$

and therefore, the solution of the system is

$$x = -4 \qquad y = -6 \qquad z = 17$$

We may check that it is correct by substituting this solution into the system (a):

(a) $\quad x + 2y + z = -4 \quad + 2(-6) + 17 = 1$
$\quad\quad -2x - 2y - z = -2(-4) - 2(-6) - 17 = 3$
$\quad\quad x - y \quad\quad = -4 - \quad (-6) \quad\quad = 2$

It takes very little additional effort to solve matrix equation (b) because we can use the same matrix A^{-1}.

$$X = A^{-1}C = \begin{bmatrix} -1 & -1 & 0 \\ -1 & -1 & -1 \\ 4 & 3 & 2 \end{bmatrix} \begin{bmatrix} -2 \\ 3 \\ -1 \end{bmatrix} = \begin{bmatrix} -1 \\ 0 \\ -1 \end{bmatrix}$$

76 MATRICES AND SYSTEMS OF LINEAR EQUATIONS

Therefore, the solution of the system (b) is

$$x = -1 \qquad y = 0 \qquad z = -1$$

We check that this is correct:

$$\begin{array}{rll}
(b) & x + 2y + z = & -1 + 2(0) + (-1) = -1 \\
& -2x - 2y - z = & -2(-1) - 2(0) - (-1) = 3 \\
& x - y = & -1 - 0 = -1
\end{array}$$

■

A Singular Example

Since the determinant of the matrix

$$\begin{bmatrix} 1 & 2 \\ 2 & 4 \end{bmatrix}$$

is

$$\Delta = 1 \cdot 4 - 2 \cdot 2 = 0$$

we cannot find its inverse by the formula of Section 1.3, because we cannot divide by 0. Let us see what happens if we apply the Gauss-Jordan method to this matrix. We form the augmented matrix

$$\left[\begin{array}{cc|cc} 1 & 2 & 1 & 0 \\ 2 & 4 & 0 & 1 \end{array}\right]$$

and attempt to change the matrix to the left of the vertical line to a 2-by-2 identity matrix. Since the diagonal entry of the first column is already 1, we proceed immediately to the next stage, which is to add -2 times the first row to the second and substitute the sum for the second row.

$$\left[\begin{array}{cc|cc} 1 & 2 & 1 & 0 \\ 0 & 0 & -2 & 1 \end{array}\right]$$

According to the Gauss-Jordan method, the next step should be to change the diagonal entry of the second column into a 1. But a 0 occupies that location and there is no number below it in the column to interchange with it. Therefore, we cannot continue. *This outcome will always occur if the matrix we started with was singular.* The rules will force us to come to a dead end before we succeed in changing the matrix to the left of the vertical line into an identity matrix. In fact, this is a way to show that the original matrix has no inverse.

The Determinant

We did not have to apply the Gauss-Jordan method to

$$\begin{bmatrix} 1 & 2 \\ 2 & 4 \end{bmatrix}$$

1.5 THE INVERSE OF A MATRIX

to discover that it has no inverse because we calculated that its determinant is zero, and that told us the same thing. There is a number called the "determinant" of an $n \times n$-matrix that gives the same information. That is, a square matrix is singular if its determinant is zero, whereas if the determinant is nonzero, the matrix has an inverse. However, in contrast to our use of the 2×2-determinant in calculating inverses in Section 1.3, for larger matrices, the determinant concept does not provide a useful method for calculating the inverse.

Summary

The **inverse**, A^{-1}, of a nonsingular matrix A is a matrix such that

$$AA^{-1} = A^{-1}A = I$$

Given a nonsingular matrix A, its inverse A^{-1} is obtained by transforming the **augmented matrix** $[\,A\,|\,I\,]$ to the form $[\,I\,|\,A^{-1}\,]$ by means of the Gauss-Jordan method as follows: Work column by column from left to right. If the **diagonal entry** of the column (where the row number equals the column number) is 0, change it to a nonzero number by interchanging the row containing the diagonal location with a lower row. Once a nonzero number d occupies this diagonal location, make it a 1 by dividing the row by d. Then change all the other numbers in the column to zeros. Specifically, for any other nonzero number a in the column, add $-a$ times the row containing the diagonal location to the row containing a, and substitute the sum for the row containing a. When A is a nonsingular matrix, then the equation

$$AX = B$$

can be solved for the unknown matrix X with solution

$$X = A^{-1}B$$

If a square matrix A is singular, the procedure will fail: at some stage the diagonal entry and all the numbers below it will be zeros.

Exercises

1. For the matrix

$$\left[\begin{array}{cc|cc} 1 & 2 & 1 & 0 \\ 2 & 1 & 0 & 1 \end{array}\right]$$

add -2 times the first row to the second row and substitute the sum for the second row.

MATRICES AND SYSTEMS OF LINEAR EQUATIONS

2. For the matrix

$$\begin{bmatrix} 1 & -\frac{1}{2} & 3 & \bigm| & -\frac{1}{2} & 0 & 0 \\ 0 & 1 & 1 & \bigm| & -\frac{1}{4} & 1 & 0 \\ 0 & 2 & -1 & \bigm| & 1 & 0 & 1 \end{bmatrix}$$

add $\frac{1}{2}$ times the second row to the first row and substitute the sum for the first row.

3. For the matrix

$$\begin{bmatrix} 1 & 0 & 0 & 1 & \bigm| & \frac{1}{4} & -\frac{1}{2} & 2 & 0 \\ 0 & 1 & 3 & 0 & \bigm| & \frac{1}{2} & 2 & 0 & 0 \\ 0 & 0 & 1 & 1 & \bigm| & \frac{1}{8} & -\frac{1}{2} & 1 & 0 \\ 0 & 0 & 1 & 2 & \bigm| & 1 & -1 & 0 & 1 \end{bmatrix}$$

add -3 times the third row to the second row and substitute the sum for the second row.

Find the inverses of the matrices in Exercises 4–11 by the Gauss-Jordan method.

4. $\begin{bmatrix} 0 & -2 \\ 1 & 4 \end{bmatrix}$
5. $\begin{bmatrix} 2 & 3 \\ -1 & -2 \end{bmatrix}$

6. $\begin{bmatrix} 2 & 0 & 0 \\ 2 & -1 & 0 \\ 0 & -1 & 1 \end{bmatrix}$
7. $\begin{bmatrix} 2 & -1 & 0 \\ 1 & -1 & -2 \\ 1 & 0 & 1 \end{bmatrix}$

8. $\begin{bmatrix} 0 & 1 & -1 \\ 2 & -1 & 1 \\ 3 & -2 & -1 \end{bmatrix}$
9. $\begin{bmatrix} 2 & -1 & 0 \\ 0 & 3 & 1 \\ 2 & 1 & 2 \end{bmatrix}$

10. $\begin{bmatrix} 1 & -1 & \frac{1}{2} & -3 \\ 0 & 1 & 0 & 3 \\ -1 & 0 & \frac{1}{2} & 0 \\ 0 & 0 & 0 & 3 \end{bmatrix}$
11. $\begin{bmatrix} 0 & 0 & -2 & 0 \\ -1 & 0 & 3 & -1 \\ 2 & 4 & 0 & 0 \\ 1 & 3 & -3 & 1 \end{bmatrix}$

1.5 THE INVERSE OF A MATRIX

In Exercises 12 through 15, solve the given equation for the matrix X using the matrix A^{-1}, calculated by the Gauss-Jordan method.

12. Solve $AX = B$, where

$$A = \begin{bmatrix} -1 & 0 & 2 \\ 0 & 0 & -1 \\ 2 & 1 & 0 \end{bmatrix} \quad \text{and} \quad B = \begin{bmatrix} -2 \\ 1 \\ 2 \end{bmatrix}$$

13. Solve $AX = B + C$, where

$$A = \begin{bmatrix} 2 & 4 \\ 3 & 1 \end{bmatrix}, \quad B = \begin{bmatrix} 1 \\ -1 \end{bmatrix}, \quad \text{and} \quad C = \begin{bmatrix} -2 \\ 2 \end{bmatrix}$$

14. Solve $AX + B = C$, where

$$A = \begin{bmatrix} 2 & 0 & 1 \\ -2 & 2 & 0 \\ 2 & 1 & 0 \end{bmatrix}, \quad B = \begin{bmatrix} -1 \\ -2 \\ 0 \end{bmatrix}, \quad \text{and} \quad C = \begin{bmatrix} 4 \\ 0 \\ 1 \end{bmatrix}$$

15. Solve $B - AX = C$, where

$$A = \begin{bmatrix} -1 & -1 & -1 \\ 0 & 2 & 1 \\ 1 & 0 & 1 \end{bmatrix}, \quad B = \begin{bmatrix} 1 \\ 3 \\ -1 \end{bmatrix}, \quad \text{and} \quad C = \begin{bmatrix} 0 \\ 2 \\ 1 \end{bmatrix}$$

In Exercises 16 through 21, solve the given system of linear equations by finding the inverse of the coefficient matrix by the Gauss-Jordan method and using it in both parts of the exercise:

16. (a) $\quad 2x \quad\quad + z = 2$
    $\quad\quad -x + y - z = -2$
    $\quad\quad\quad\quad -y + z = 1$

    (b) $\quad 2x \quad\quad + z = -1$
    $\quad\quad -x + y - z = 0$
    $\quad\quad\quad\quad -y + z = 2$

17. (a) $\quad x + y - z = 1$
    $\quad\quad\quad 2y + z = 0$
    $\quad\quad 2x \quad\quad - z = -1$

    (b) $\quad x + y - z = 0$
    $\quad\quad\quad 2y + z = 2$
    $\quad\quad 2x \quad\quad - z = -1$

18. (a) $\quad -x + 2y - z = 1$
    $\quad\quad 2x - y + z = -1$
    $\quad\quad x + y - 2z = 3$

    (b) $\quad -x + 2y - z = 0$
    $\quad\quad 2x - y + z = -2$
    $\quad\quad x + y - 2z = 2$

19. (a) $\quad w + x + y + z = 1$
    $\quad -w + \quad\quad y - z = 0$
    $\quad\quad\quad 2x + 3y + z = -1$
    $\quad w + x \quad\quad - z = 1$

    (b) $\quad w + x + y + z = 1$
    $\quad -w \quad\quad + y - z = -1$
    $\quad\quad\quad 2x + 3y + z = 0$
    $\quad w + x \quad\quad - z = -2$

20. (a) $y - 2z = -3$
    $x - y + z = 2$
    $-2x + 2y - z = -1$

    (b) $y - 2z = -1$
    $x - y + z = \frac{1}{2}$
    $-2x + 2y - z = \frac{1}{2}$

21. (a) $-w + y + z = 2$
    $w - x - z = -4$
    $-w + 2x + z = 1$
    $2w + x - y = 0$

    (b) $-w + y + z = 0$
    $w - x - z = 0$
    $-w + 2x + z = 1$
    $2w + x - y = 1$

1.6 Matrix Methods for Linear Systems

If we try to solve the system
$$x + 2y = 1$$
$$2x + 4y = 3$$
by the methods we have discussed so far, we will be unsuccessful because, as we saw at the end of the last section, the coefficient matrix
$$A = \begin{bmatrix} 1 & 2 \\ 2 & 4 \end{bmatrix}$$
does not have an inverse.

The purpose of this section is to use the Gauss-Jordan method to study any linear system or, in matrix terms, any matrix equation $AX = B$, not just those in which A is nonsingular. Before we do this, we will take a brief look at some systems of two linear equations in two unknowns, such as the one just given. These small systems exhibit all the types of behavior of solutions that occur in systems of any size. If we keep these pictures in mind, they will help us visualize what matrix methods will tell us about larger systems.

Graphing Linear Equations

To understand what is happening in a two-by-two system, we will draw a picture of the equations. To review how this is done, we begin with the equation
$$3x + 4y = 1$$
The graph of this equation consists of the points (x, y) on the plane for which the equation is true. If we solve the equation for y, it becomes
$$4y = -3x + 1$$
and then

1.6 MATRIX METHODS FOR LINEAR SYSTEMS

$$y = -\frac{3}{4}x + \frac{1}{4}$$

We choose different values of x and compute the corresponding values of y, as shown in Table 1-7.

x	y
-1	1
0	$\frac{1}{4}$
1	$-\frac{1}{2}$
2	$-\frac{5}{4}$

Table 1-7

Notice the pattern: For each unit increase in x, the value of y decreases by the constant amount $\frac{3}{4}$. If we plot the corresponding points on the x,y-plane, as in Figure 1-8, we see that they lie on a straight line, which is why such equations are called "linear equations." Once we realize that the pictures of the equations we discuss will be straight lines, we do not have to compute so many points on them. We will just find any two points and they will determine the line.

Intersecting and Parallel Lines

Let us look at the following system that we solved in Section 1.3:

$$3x + 4y = 1$$
$$2x + 5y = 2$$

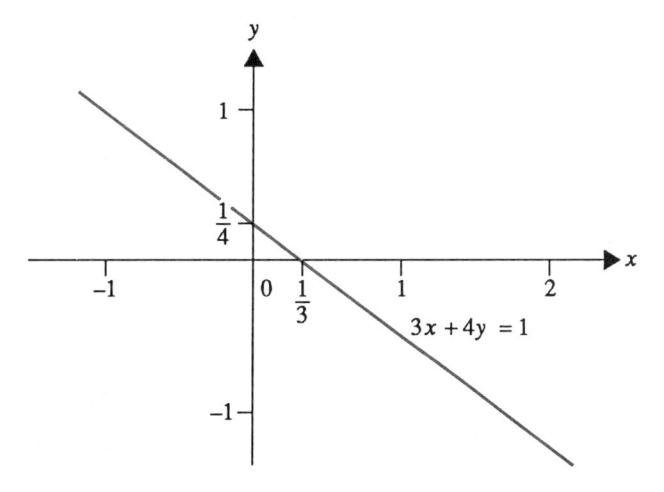

Figure 1-8

Figure 1-8 shows us the graph of the first equation. For the second, we will use the simple procedure of locating two points on the line that is the graph of $2x + 5y = 2$. We do not even have to solve this equation for y as we did before. If we choose $x = 0$, for instance, the equation becomes

$$2(0) + 5y = 2$$

which is the same as

$$5y = 2$$

Thus, $y = \frac{2}{5}$. Taking another convenient value, $x = 1$, and substituting, the equation becomes

$$2(1) + 5y = 2$$

so that

$$5y = 0 \quad \text{and} \quad y = 0$$

In Section 1.3, we found that the system has the solution $x = -\frac{3}{7}$ and $y = \frac{4}{7}$. We graph the two equations in Figure 1-9 and observe that the two lines meet at a single point. Clearly, this point corresponds to the solution of the system, $(-\frac{3}{7}, \frac{4}{7})$.

Returning to the system

$$x + 2y = 1$$
$$2x + 4y = 3$$

that introduced this section, we use the same method to graph these two equations

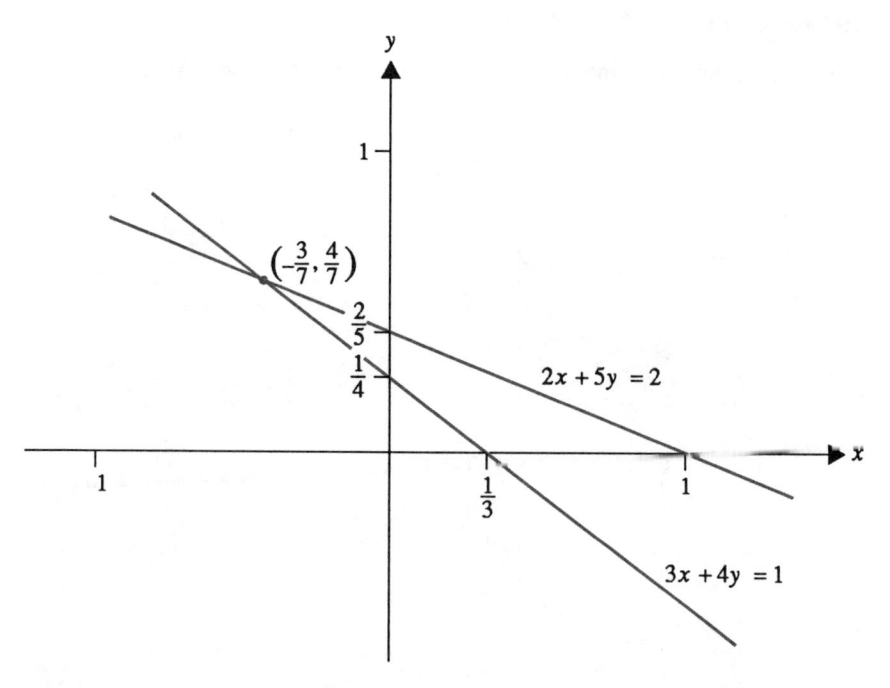

Figure 1-9

1.6 MATRIX METHODS FOR LINEAR SYSTEMS

Figure 1-10

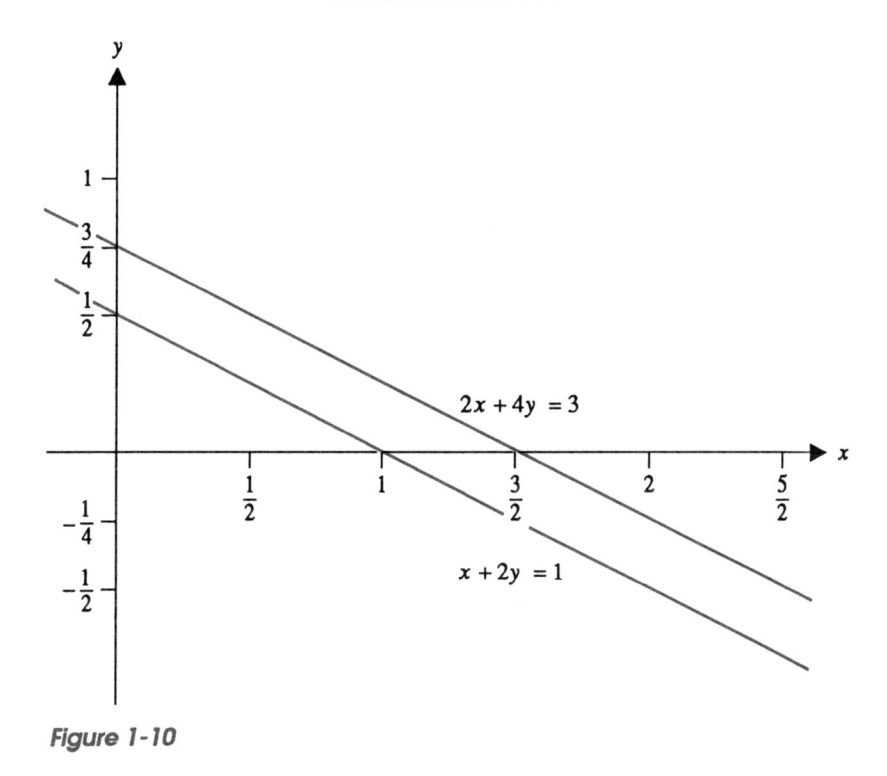

in Figure 1-10. We see that the equations are represented by two parallel lines. We could not find the inverse of the coefficient matrix of this system

$$A = \begin{bmatrix} 1 & 2 \\ 2 & 4 \end{bmatrix}$$

by either of the methods we have discussed. Nor would we be able to do it by any other method. If we could, the matrix $X = A^{-1}B$ would produce a solution of the system and that would amount to giving the coordinates of a point where the two parallel lines meet. Thus, this matrix A has no inverse; it is what we called previously a *singular* matrix. We can decide whether or not a 2-by-2 matrix is singular by computing its determinant. If it is not zero, then we can solve the corresponding system by the method of Section 1.3.

A Single Line

However, if in the matrix equation $AX = B$ corresponding to a two-by-two system, the matrix A is singular, it does not necessarily follow that there are no solutions of the system. An example is the slightly different system

$$x + 2y = 1$$
$$2x + 4y = 2$$

84 MATRICES AND SYSTEMS OF LINEAR EQUATIONS

In terms of the matrix equation $AX = B$, the coefficient matrix A has not changed, but the matrix B has. So A is still singular, but now the graph is as shown in Figure 1-11.

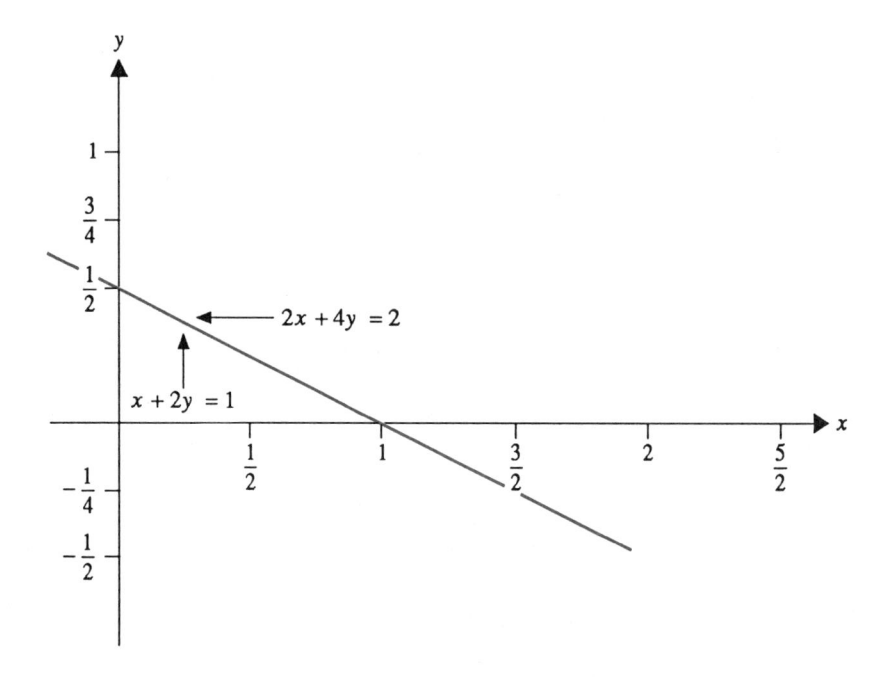

Figure 1-11

We cannot see the distinct lines corresponding to the two equations because one line lies right over the other. In this case, thinking of the solutions of the system as the points of intersection of the two lines, there are as many solutions as there are points on the lines, that is, an infinite number of them. We can produce solutions by solving the equation

$$x + 2y = 1$$

for y to obtain the general formula

$$y = -\frac{1}{2}x + \frac{1}{2}$$

Thus, three solutions are given by $x = 0, y = \frac{1}{2}$; $x = 1, y = 0$; and $x = 2, y = -\frac{1}{2}$.

The Three Possibilities

We have now seen three different types of geometric behavior exhibited by two lines in the plane. The lines may meet in a single point, they may be parallel, or they may be identical. Because there are no other geometric possibilities, there are no other possibilities for the behavior of the solutions of two-by-two linear systems. Let us summarize these possibilities in terms of matrices. Write the system as the matrix equation $AX = B$.

1.6 MATRIX METHODS FOR LINEAR SYSTEMS

> If the matrix A is nonsingular, then the system has exactly one solution, given by
>
> $$X = A^{-1}B$$
>
> If the matrix A is singular, then there are two possibilities:
>
> either there are no solutions
> or
> there are an infinite number of solutions

These three possibilities—one solution, no solutions, an infinite number of solutions—describe the possible sets of solutions, not just of two-by-two systems, but of any system of linear equations.

Simplifying Augmented Matrices

The system of linear equations

$$x - 2y = 3$$
$$2x - y = 0$$
$$-y = 2$$

cannot be analyzed by the methods we have presented so far. There are two unknowns, but there are three linear equations. We can still write this as the matrix equation $AX = B$ if we let

$$A = \begin{bmatrix} 1 & -2 \\ 2 & -1 \\ 0 & -1 \end{bmatrix}, \quad X = \begin{bmatrix} x \\ y \end{bmatrix}, \quad \text{and} \quad B = \begin{bmatrix} 3 \\ 0 \\ 2 \end{bmatrix}$$

However, notice that in this case the coefficient matrix A is a 3×2-matrix. Nevertheless, we can use the Gauss-Jordan method to understand this linear system or any other matrix equation of the form $AX = B$. Just as in Section 1.4, we form the augmented matrix $[\,A \mid B\,]$. We apply the row operations to this augmented matrix to transform the columns of A into the columns of an identity matrix, working from left to right. The result is a new augmented matrix $[\,A' \mid B'\,]$. However, in general, A' will not be an identity matrix now because it is the same size as A, and therefore, it may not be square. If A has more rows than columns, then A' usually looks like an identity matrix with some columns missing—for instance,

$$A' = \begin{bmatrix} 1 & 0 \\ 0 & 1 \\ 0 & 0 \end{bmatrix}$$

(Compare this with Example 1, which follows.) If A has more columns than rows, then the "extra" columns are not necessarily made up of 0s and 1s. An example of this is

$$A' = \begin{bmatrix} 1 & 0 & \frac{3}{5} \\ 0 & 1 & -\frac{1}{5} \end{bmatrix}$$

as we will see in Example 4. Even when A is square, A' still may not be the identity matrix, as Example 3 will illustrate, where

$$A' = \begin{bmatrix} 1 & 0 & -\frac{5}{3} \\ 0 & 1 & 2 \\ 0 & 0 & 0 \end{bmatrix}$$

Furthermore, the Gauss-Jordan method will not always produce a matrix A' in which the columns of the identity matrix are collected at the left, as we will see in Example 5, where

$$A' = \begin{bmatrix} 1 & -\frac{1}{2} & 0 & 0 \\ 0 & 0 & 1 & 0 \\ 0 & 0 & 0 & 1 \end{bmatrix}$$

We observed in Section 1.4 what happened to a system of linear equations when we interchanged the order of the equations, or multiplied an equation by a constant, or added (a multiple of) an equation to another and substituted the sum for one of them. The appearance of the system changed, but the solution was the same. Consequently, when we represented the system by an augmented matrix $[\ A\ |\ B\ ]$ and applied the corresponding operations to the rows, the same thing happened. Even though applying these operations by means of the Gauss-Jordan method does not necessarily convert A to the identity matrix, as it did in Section 1.4, it makes the augmented matrix $[\ A'\ |\ B'\ ]$ simple enough so that we can see what its solutions are. (They may be infinite in number or there may be none at all.) Therefore, we can understand the solutions of the original linear system.

The examples that follow illustrate the various things that can happen.

Example 1. Use the Gauss-Jordan method to find values x and y that solve the system

$$\begin{aligned} x - 2y &= 3 \\ 2x - y &= 0 \\ -y &= 2 \end{aligned}$$

Solution

The corresponding augmented matrix is

1.6 MATRIX METHODS FOR LINEAR SYSTEMS

$$[A \mid B] = \begin{array}{c} x y \\ \left[\begin{array}{cc|c} 1 & -2 & 3 \\ 2 & -1 & 0 \\ 0 & -1 & 2 \end{array}\right] \end{array}$$

Since the diagonal entry of the first column is already 1, we go right to the next step and add −2 times the first row to the second row. We then substitute this sum for the second row:

$$\left[\begin{array}{cc|c} 1 & -2 & 3 \\ 0 & 3 & -6 \\ 0 & -1 & 2 \end{array}\right]$$

Then we divide the second row by 3:

$$\left[\begin{array}{cc|c} 1 & -2 & 3 \\ 0 & 1 & -2 \\ 0 & -1 & 2 \end{array}\right]$$

We add 2 times the second row to the first row and substitute the sum for the first row:

$$\left[\begin{array}{cc|c} 1 & 0 & -1 \\ 0 & 1 & -2 \\ 0 & -1 & 2 \end{array}\right]$$

Finally, we add the second row to the third row and substitute for the third row. This is the last possible step because it will convert the second column to that of an identity matrix and because there are no more columns to the left of the vertical line. Recalling the column labels of the original matrix $[A \mid B]$, we have

$$\begin{array}{c} x y \\ \left[\begin{array}{cc|c} 1 & 0 & -1 \\ 0 & 1 & -2 \\ 0 & 0 & 0 \end{array}\right] = [A' \mid B'] \end{array}$$

The corresponding linear system is

$$\begin{aligned} x & = -1 \\ y & = -2 \\ 0x + 0y & = 0 \end{aligned}$$

which tell us that $x = -1$ and $y = -2$ is the solution to the problem. We should check our arithmetic by substituting these values back into the original system.

$$\begin{aligned} -1 - 2(-2) &= 3 \\ 2(-1) - (-2) &= 0 \\ -(-2) &= 2 \end{aligned}$$

∎

You may have noticed in this example that the last rows of A and B seem not to have made a contribution to the solution of the problem because they just reduce to the obvious statement $0 = 0$. You might wonder whether we could delete the last equation and solve the matrix equation

$$AX = \begin{bmatrix} 1 & -2 \\ 2 & -1 \end{bmatrix} \begin{bmatrix} x \\ y \end{bmatrix} = \begin{bmatrix} 3 \\ 0 \end{bmatrix} = B$$

by calculating $X = A^{-1}B$ as in Section 1.3. This procedure would indeed produce the correct answer in this case. But it is really a very poor idea to throw away equations in order to turn the problem into one with a convenient form, as the next example demonstrates.

Example 2. Use the Gauss-Jordan method to find values x, y, and z that solve the system of linear equations

$$\begin{aligned} y - z &= 2 \\ x + 2y + 2z &= 0 \\ -x \quad\quad + z &= -1 \\ 2y + z &= -1 \end{aligned}$$

Solution

Beginning with

$$[A \mid B] = \begin{matrix} & x & y & z & \\ & \begin{bmatrix} 0 & 1 & -1 & 2 \\ 1 & 2 & 2 & 0 \\ -1 & 0 & 1 & -1 \\ 0 & 2 & 1 & -1 \end{bmatrix} \end{matrix}$$

we interchange the first two rows:

$$\begin{bmatrix} 1 & 2 & 2 & 0 \\ 0 & 1 & -1 & 2 \\ -1 & 0 & 1 & -1 \\ 0 & 2 & 1 & -1 \end{bmatrix}$$

Then we add the first row to the third row.

$$\begin{bmatrix} 1 & 2 & 2 & 0 \\ 0 & 1 & -1 & 2 \\ 0 & 2 & 3 & -1 \\ 0 & 2 & 1 & -1 \end{bmatrix}$$

Adding multiples of the second row to the other rows, we obtain

$$\begin{bmatrix} 1 & 0 & 4 & -4 \\ 0 & 1 & -1 & 2 \\ 0 & 0 & 5 & -5 \\ 0 & 0 & 3 & -5 \end{bmatrix}$$

Next, we divide the third row by 5.

$$\begin{bmatrix} 1 & 0 & 4 & | & -4 \\ 0 & 1 & -1 & | & 2 \\ 0 & 0 & 1 & | & -1 \\ 0 & 0 & 3 & | & -5 \end{bmatrix}$$

Adding multiples of the third row to the other rows gives the following result, with the column labels recalled.

$$\begin{matrix} x & y & z & \end{matrix}$$
$$\begin{bmatrix} 1 & 0 & 0 & | & 0 \\ 0 & 1 & 0 & | & 1 \\ 0 & 0 & 1 & | & -1 \\ 0 & 0 & 0 & | & -2 \end{bmatrix} = [A' \mid B']$$

This is the augmented matrix corresponding to the system of linear equations

$$\begin{aligned} x & = 0 \\ y & = 1 \\ z & = -1 \\ 0x + 0y + 0z & = -2 \end{aligned}$$

But this last equation means that $0 = -2$, which is obviously false. Therefore, the original system of four equations was "false" in the sense that the four statements about the numbers x, y, and z contradict each other. There are no numbers x, y, and z that can solve all four equations at once, so the system has no solutions at all. ■

If we threw away the last equation in Example 2, there would be no contradiction. Therefore, we would have obtained the wrong answer.

An n-by-n linear system cannot be solved by computing an inverse, as in Section 1.4, if the coefficient matrix A is singular. However, we can analyze such a system by the Gauss-Jordan method, as the next example illustrates.

Example 3. Use the Gauss-Jordan method to find values x, y, and z that solve the system of linear equations

$$\begin{aligned} 3x + 2y - z & = 2 \\ y + 2z & = -1 \\ 3x + y - 3z & = 3 \end{aligned}$$

Solution

Begin with the augmented matrix

$$\begin{matrix} & & x & y & z & \end{matrix}$$
$$[A \mid B] = \begin{bmatrix} 3 & 2 & -1 & | & 2 \\ 0 & 1 & 2 & | & -1 \\ 3 & 1 & -3 & | & 3 \end{bmatrix}$$

Dividing the first row by 3, then adding −3 times the first row to the third gives us

$$\begin{bmatrix} 1 & \frac{2}{3} & -\frac{1}{3} & \bigg| & \frac{2}{3} \\ 0 & 1 & 2 & \bigg| & -1 \\ 0 & -1 & -2 & \bigg| & 1 \end{bmatrix}$$

Adding multiples of the second row to the other two produces

$$\begin{array}{ccc} x & y & z \end{array}$$
$$\begin{bmatrix} 1 & 0 & -\frac{5}{3} & \bigg| & \frac{4}{3} \\ 0 & 1 & 2 & \bigg| & -1 \\ 0 & 0 & 0 & \bigg| & 0 \end{bmatrix}$$

We now have 0 in the a_{33} location, which is at the bottom of its column, directly to the left of the vertical line. Therefore, we cannot perform any more steps and this matrix is $[\,A'\,|\,B'\,]$. It is the augmented matrix of the linear system

$$x - \frac{5}{3}z = \frac{4}{3}$$
$$y + 2z = -1$$
$$0 = 0$$

When we solve these equations for x and for y, we get

$$x = \frac{4}{3} + \frac{5}{3}z \qquad y = -1 - 2z$$

which shows us that the values of x and y will depend on what z is. For example, if $z = 0$, then

$$x = \frac{4}{3} + \frac{5}{3}(0) = \frac{4}{3} \qquad y = -1 - 2(0) = -1$$

so $x = \frac{4}{3}$, $y = -1$, $z = 0$ is a solution of the system of linear equations. If $z = 1$, then

$$x = \frac{4}{3} + \frac{5}{3}(1) = \frac{9}{3} = 3 \qquad y = -1 - 2(1) = -3$$

and we have another solution: $x = 3$, $y = -3$, $z = 1$. There is a solution for each value of z, so there are an infinite number of solutions. ∎

We have seen examples of linear systems with

1. a single solution;
2. no solution;
3. infinitely many solutions.

These are the only possible types of solutions.

In the systems of linear equations that arise in applications, it often happens that the number of unknowns is greater than the number of equations. In such systems if there are any solutions at all, then the set of solutions must be infinite. Our last two examples are of this type.

1.6 MATRIX METHODS FOR LINEAR SYSTEMS

Example 4. Use the Gauss-Jordan method to find values of x, y, and z that solve the system of linear equations

$$2x + y + z = 3$$
$$x - 2y + z = 4$$

Solution

$$[A \mid B] = \begin{bmatrix} \overset{x}{2} & \overset{y}{1} & \overset{z}{1} & 3 \\ 1 & -2 & 1 & 4 \end{bmatrix}$$

We divide the first row by 2 and then subtract that row from the second row:

$$\begin{bmatrix} 1 & \frac{1}{2} & \frac{1}{2} & \frac{3}{2} \\ 0 & -\frac{5}{2} & \frac{1}{2} & \frac{5}{2} \end{bmatrix}$$

Next we divide the second row by $-\frac{5}{2}$, that is, multiply it by $-\frac{2}{5}$.

$$\begin{bmatrix} 1 & \frac{1}{2} & \frac{1}{2} & \frac{3}{2} \\ 0 & 1 & -\frac{1}{5} & -1 \end{bmatrix}$$

We then add $-\frac{1}{2}$ times the second row to the first, which gives us the matrix

$$\begin{bmatrix} \overset{x}{1} & \overset{y}{0} & \overset{z}{\frac{3}{5}} & 2 \\ 0 & 1 & -\frac{1}{5} & -1 \end{bmatrix} = [A' \mid B']$$

This time the corresponding linear system is

$$x \quad\;\; + \frac{3}{5}z = 2$$
$$y - \frac{1}{5}z = -1$$

The solutions can be found from the equivalent equations

$$x = 2 - \frac{3}{5}z \qquad y = -1 + \frac{1}{5}z$$

Among the infinitely many solutions we have $x = 2$, $y = -1$, $z = 0$ and $x = \frac{7}{5}$, $y = -\frac{4}{5}$, $z = 1$. ∎

Example 5. Use the Gauss-Jordan method to find values of w, x, y, and z that solve the system of linear equations

$$2w - x + y \quad\;\; = 2$$
$$-2w + x \quad\;\; + z = -1$$
$$4w - 2x - y + z = 0$$

Solution

$$[A \mid B] = \begin{matrix} w & x & y & z & \\ \begin{bmatrix} 2 & -1 & 1 & 0 & 2 \\ -2 & 1 & 0 & 1 & -1 \\ 4 & -2 & -1 & 1 & 0 \end{bmatrix} \end{matrix}$$

Dividing the first row by 2 and then adding multiples of it to the other rows gives

$$\begin{bmatrix} 1 & -\frac{1}{2} & \frac{1}{2} & 0 & 1 \\ 0 & 0 & 1 & 1 & 1 \\ 0 & 0 & -3 & 1 & -4 \end{bmatrix}$$

Examining the second column, we find a zero in the a_{22}-location and a zero below it, so we cannot apply the Gauss-Jordan method to this column. However, unlike Example 3, this does not mean that it is impossible to continue with the method in order to produce an augmented matrix $[A' \mid B']$ that will clearly show us the solutions of the original system of linear equations. In the present example we still have two columns to the left of the vertical line that we have not used. So we go to the third column and attempt to convert it to the form of the *second* column of the 3×3-identity matrix, since we have not yet succeeded in transforming a column into this form. We therefore add $-\frac{1}{2}$ times the second row to the first.

$$\begin{bmatrix} 1 & -\frac{1}{2} & 0 & -\frac{1}{2} & \frac{1}{2} \\ 0 & 0 & 1 & 1 & 1 \\ 0 & 0 & -3 & 1 & -4 \end{bmatrix}$$

We then add 3 times the second row to the third row.

$$\begin{bmatrix} 1 & -\frac{1}{2} & 0 & -\frac{1}{2} & \frac{1}{2} \\ 0 & 0 & 1 & 1 & 1 \\ 0 & 0 & 0 & 4 & -1 \end{bmatrix}$$

The third column is now of the desired form, and we can go on to work with the fourth column. We want to convert it to the form of the last column of the identity matrix. Dividing the third row by 4 and then adding multiples of that row to the other rows completes the application of the Gauss-Jordan method:

$$[A' \mid B'] = \begin{matrix} w & x & y & z & \\ \begin{bmatrix} 1 & -\frac{1}{2} & 0 & 0 & \frac{3}{8} \\ 0 & 0 & 1 & 0 & \frac{5}{4} \\ 0 & 0 & 0 & 1 & -\frac{1}{4} \end{bmatrix} \end{matrix}$$

The corresponding linear system is

$$w - \frac{1}{2}x = \frac{3}{8}$$
$$y = \frac{5}{4}$$
$$z = -\frac{1}{4}$$

which we can rewrite in the form

$$w = \frac{3}{8} + \frac{1}{2}x, \qquad y = \frac{5}{4}, \qquad z = -\frac{1}{4}$$

to see that there are an infinite number of solutions. In fact, we obtain a different one for every value of x. For instance, when $x = 0$, then $w = \frac{3}{8}$, so that the corresponding solution is $w = \frac{3}{8}$, $x = 0$, $y = \frac{5}{4}$, $z = -\frac{1}{4}$. Another solution is $w = \frac{7}{8}$, $x = 1$, $y = \frac{5}{4}$, $z = -\frac{1}{4}$. Notice that y and z remain the same in all solutions since they, unlike w, do not depend on the value of x. ■

Row-reduced Form

If you carry out the Gauss-Jordan method as we have described it in this section, when you arrive at the final augmented matrix $[\,A'\mid B'\,]$, the matrix A' will be so simple that you can easily determine whether or not the linear system has solutions and, if it does, what they are. However, the last example illustrates the danger that you might terminate the Gauss-Jordan method too soon because, in that example, the method would not apply to the second column, though it did to the third, as we showed. There is a simple test that you can carry out just by looking at the augmented matrix that will tell you whether you have carried out the Gauss-Jordan method as far as it can go. When the Gauss-Jordan method is completed, the augmented matrix will be in *row-reduced form*, which involves only the numbers to the left of the vertical line.

> The **row-reduced form** is described by the following rules:
>
> 1. Looking from left to right along each row, either the first nonzero number is a 1 (called the **leading 1**) or the row (to the left of the vertical line) consists only of zeros.
> 2. The rows of zeros all lie below the rows with leading 1s.
> 3. Every leading 1 is the only nonzero number in its column.
> 4. Every leading 1 is to the right of the leading 1s in the rows above it.

The final matrices of Examples 2 and 3 each have rows of zeros (to the left of the vertical line) as their last rows, even though there is a nonzero number to the right of the vertical line in Example 2. If you look back at the second matrix in the solution to Example 5, you will see that this matrix is not in row-reduced form because although there is a leading 1 in the second row, it is not the only nonzero number in the third column. Therefore, we know we should continue the Gauss-Jordan method.

Efficiency of the Method

The Gauss-Jordan method is not the most efficient method of solving systems of linear equations. High-speed methods that are well suited to computers have been developed to solve large systems relatively quickly. On the other hand, the Gauss-Jordan method is a simple, matrix-based procedure that makes a good foundation for understanding these more rapid, specialized techniques. It is also flexible: we can use it to find matrix inverses as well.

Summary

To solve a matrix equation of the form $AX = B$ by the **Gauss-Jordan method**, where A is a given $m \times n$-matrix, B is a given $m \times 1$-matrix, and X is an $n \times 1$-matrix of unknowns, first form the augmented matrix $[\,A \mid B\,]$. To the extent possible, transform columns of A, working from left to right, to columns of an identity matrix. The allowable operations are interchanges of rows, multiplication and division of a row by a nonzero constant, and addition of a nonzero multiple of one row to another row. After these operations have been used to produce a matrix $[\,A' \mid B'\,]$, the solutions of the corresponding linear system are the same as the solutions of the original system. If one of the equations in the system corresponding to $[\,A' \mid B'\,]$ is of the form $0 = k$ for some $k \neq 0$, then the system has no solution. Otherwise, $AX = B$ has either a single solution or an infinite number of solutions. When the Gauss-Jordan method is completed, the augmented matrix will be in **row-reduced form**, which means that looking from left to right along each row, up to the vertical line, either the first nonzero number is a 1 (called the **leading 1**) or the row consists only of zeros. The rows of zeros all lie below the rows with leading 1s; every leading 1 is the only nonzero number in its column; and every leading 1 is to the right of the leading 1s in the rows above it.

Exercises

In Exercises 1 through 20, apply the Gauss-Jordan method to the corresponding matrix equation of the form $AX = B$ to determine whether or not the system of linear equations has a solution. If there is one solution, find out what it is. If there are an infinite number of solutions, write down any two of them.

1.6 MATRIX METHODS FOR LINEAR SYSTEMS

1. $x + y = -1$
   $4x + y = -3$
   $x + 4y = -2$

2. $2x - y = 0$
   $3x - y = 4$

3. $x + 2y = -1$
   $-x - 2y = -2$
   $2x + 2y = -3$

4. $2x - y = -2$
   $-x + 3y + 2z = 1$
   $x + 2y + 2z = -1$

5. $y - z = 1$
   $y + z = 1$
   $x + 2y + z = 0$
   $-2x + z = 4$
   $2x + y = -1$

6. $w + 2x - y - 2z = -1$
   $x - y = -1$
   $-w + 2x + z = -1$

7. $y + 2z = -1$
   $w - x + 2y + z = 2$
   $-w + 2x = 0$
   $x + 2y = 1$
   $y + z = 3$

8. $w + 2x + 3y + z = 0$
   $v + 3w + 2x + 2z = 1$
   $v + 2w - 3y + z = -1$

9. $2x - y = 2$
   $-2x + 2y + z = -1$
   $2x + y = 1$

10. $2x - y + z = 1$
    $2y - z = 2$
    $2x + z = -2$

11. $2y + z = 2$
    $-x + y = 0$
    $2x - 2y + z = -1$
    $x + y + 2z = 1$

12. $x - y = 2$
    $x + 2y + 2z = 5$

13. $-x - y = 1$
    $2x + y + 3z = -1$
    $y + 3z = 2$
    $x + y = -2$

14. $x + z = 2$
    $y - 2z = 1$
    $-x + y + z = -1$
    $-2x - 2z = -4$
    $2x + y = 1$
    $y = 1$

15. $w + x + 2z = 2$
    $y = 1$
    $w + x + 2y - z = 0$

16. $-2x + y + z = 1$
    $2w - y + z = -1$
    $-w + x + 2z = 2$
    $-w - x - 2y = -2$

17. $-w + x + 2y = 0$
    $2w - x - 3y + z = 1$
    $2w - 2x - 4y + z = -1$

18. $x + 2y - z = 0$
    $-2x - 4y = 1$

19. $2x + y = 1$
    $2x + y + z = -2$
    $4x + 2y + 3z = -1$

20. $2w - 2x + y = -1$
    $-3w + 3x - 2y + z = 2$

1.7 Matrix Applications

We now discuss some applications of matrix techniques. The applications we present illustrate, in a simplified form, how matrices are used. Although, as a general rule, it can be difficult to fit equations to "story problems," in this section we have the advantage that we know we are looking for *linear* equations—in order to apply matrix techniques. Thus, we will be limited to problems that involve only equations like $3x + 2y = 4$ or $x - 2y + z = -2$.

Example 1. An ice-cream store sells ice-cream sodas and milk shakes. An ice-cream soda contains 4 ounces of ice cream and 1 ounce of syrup and a milk shake contains 3 ounces of ice cream and 1 ounce of syrup. The store used the amounts of ice cream and syrup (in ounces) shown in Table 1-12.

	Ice Cream	Syrup
Friday	512	160
Saturday	485	145
Sunday	600	185

Table 1-12

How many ice-cream sodas and how many milk shakes did the store sell on Friday, on Saturday, and on Sunday?

Solution

We will begin with Friday's sales. What is the problem asking us to find out? The last sentence tells us that we are looking for two numbers: the number of ice-cream sodas and the number of milk shakes sold on Friday. There are two unknowns in the problem. Let x be the number of sodas and y the number of milk shakes. Each soda contains 4 ounces of ice cream, so x of them would use $4x$ ounces. Similarly, the y milk shakes require $3y$ ounces of ice cream since each shake takes 3 ounces. Altogether, x sodas and y shakes will account for a total of $4x + 3y$ ounces. We are told that the store used 512 ounces of ice cream on Friday. Therefore,

$$4x + 3y = 512$$

which is a linear equation, just as we expect. In the same way, x sodas and y shakes require $x + y$ ounces of syrup, since each soda and each shake require 1 ounce. Referring to Table 1-12, we see that

$$x + y = 160$$

If we write this system of two linear equations in matrix form as $AX = B$, then

$$A = \begin{bmatrix} 4 & 3 \\ 1 & 1 \end{bmatrix} \text{ and } B = \begin{bmatrix} 512 \\ 160 \end{bmatrix}$$

Before trying to solve the matrix equation $AX = B$, let us look at the problem for Saturday. We are still looking for x, the number of sodas sold that day, and y, the number of shakes. Furthermore, the number of ounces of ice cream per soda or shake does not depend on which day it is, so the consumption can still be written as $4x + 3y$. The only difference is in the total number of ounces consumed, 485. The same is true for the syrup, so for Saturday the system of equations is

$$4x + 3y = 485$$
$$x + y = 145$$

This time we have the matrix equation $AX = B$ with

$$A = \begin{bmatrix} 4 & 3 \\ 1 & 1 \end{bmatrix} \quad \text{and} \quad B = \begin{bmatrix} 485 \\ 145 \end{bmatrix}$$

Notice that only B changes, but A remains as before. In the same way the information for Sunday gives us the equations

$$4x + 3y = 600$$
$$x + y = 185$$

and again A does not change. Note that A is nonsingular (its determinant is $\Delta = 4 - 3 = 1$). As we suggested earlier, rather than treating this as three separate problems, we will use the solution $X = A^{-1}B$, with the *same* A^{-1} each time and only the B's changing. By the formula for finding the inverse of a 2×2-matrix,

$$A^{-1} = \begin{bmatrix} 1 & -3 \\ -1 & 4 \end{bmatrix}$$

so for Friday the solution is given by

$$X = \begin{bmatrix} x \\ y \end{bmatrix} = A^{-1}B = \begin{bmatrix} 1 & -3 \\ -1 & 4 \end{bmatrix} \begin{bmatrix} 512 \\ 160 \end{bmatrix} = \begin{bmatrix} 32 \\ 128 \end{bmatrix}$$

This means that on Friday the store sold 32 ice-cream sodas and 128 milk shakes. For Saturday, the computation is

$$\begin{bmatrix} 1 & -3 \\ -1 & 4 \end{bmatrix} \begin{bmatrix} 485 \\ 145 \end{bmatrix} = \begin{bmatrix} 50 \\ 95 \end{bmatrix}$$

so sales were 50 sodas and 95 shakes. Finally, for Sunday,

$$\begin{bmatrix} 1 & -3 \\ -1 & 4 \end{bmatrix} \begin{bmatrix} 600 \\ 185 \end{bmatrix} = \begin{bmatrix} 45 \\ 140 \end{bmatrix}$$

which means that the store sold 45 ice-cream sodas and 140 milk shakes. ∎

Example 2. An historian who is studying a voyage of exploration learns that a religious charity donated one medal for each participant in the voyage and that 40 medals were donated. Further, the quartermaster ordered 348 ounces of wine per day of the voyage. The historian estimates that seamen were allowed a ration of 8 ounces of wine a day and officers 12 ounces. A final piece of information uncovered by the historian is that the owner of the ship put on deposit 2 shillings for each seaman and 5 shillings for each officer as a fund to help the dependents of any men who did not sur-

98 MATRICES AND SYSTEMS OF LINEAR EQUATIONS

vive. The total deposit was 101 shillings. How many seamen and how many officers made the voyage?

Solution

The historian wishes to know x, the number of seamen, and y, the number of officers. Since 40 medals were donated for the voyage—one for each participant—we conclude that 40 sailors were on board; that is,

$$x + y = 40$$

Wine consumption was $8x$ ounces for all seamen and $12y$ ounces a day for all officers. The quartermaster ordered 348 ounces of wine per day, so

$$8x + 12y = 348$$

Applying the same reasoning to the information about the amount of money on deposit produces the equation

$$2x + 5y = 101$$

This system of three linear equations is equivalent to the matrix equation $AX = B$, where

$$A = \begin{bmatrix} 1 & 1 \\ 8 & 12 \\ 2 & 5 \end{bmatrix} \quad \text{and} \quad B = \begin{bmatrix} 40 \\ 348 \\ 101 \end{bmatrix}$$

Since A is not a square matrix, there is no possibility of finding an inverse; and so we work with the augmented matrix

$$[A \mid B] = \begin{bmatrix} x & y & \\ 1 & 1 & 40 \\ 8 & 12 & 348 \\ 2 & 5 & 101 \end{bmatrix}$$

Using the Gauss-Jordan method, first we subtract multiples of the first row from the other two rows,

$$\begin{bmatrix} 1 & 1 & 40 \\ 0 & 4 & 28 \\ 0 & 3 & 21 \end{bmatrix}$$

and then turn the second column into that of the identity matrix, to produce

$$\begin{bmatrix} x & y & \\ 1 & 0 & 33 \\ 0 & 1 & 7 \\ 0 & 0 & 0 \end{bmatrix} = [A' \mid B']$$

The system of linear equations corresponding to this augmented matrix is

$$\begin{aligned} x \quad &= 33 \\ y &= 7 \\ 0x + 0y &= 0 \end{aligned}$$

The answer to the historian's problem is that there were 33 seamen and 7 officers on the ship. ∎

Example 3. A company invests a total of $1 million in three different forms: short-term loans, treasury notes, and municipal bonds. The short-term loans pay annual interest at a rate of 10%, the treasury notes pay 8% interest, and the bonds pay 4% interest. The total interest for the year is $60,000, or $0.06 million. If the company has three times as much money invested in municipal bonds as in treasury notes, how much money is invested in each type of investment: loans, notes, and bonds?

Solution

Let x be the amount, in millions of dollars, invested in short-term loans, y the amount invested in treasury notes, and z the amount invested in municipal bonds. Since the total is $1 million, we write

$$x + y + z = 1$$

The interest from x million dollars of loans will be 10% of x or, in decimal form, $0.10x$. Similarly, the interest on the treasury notes is $0.08y$ and on the bonds, $0.04z$. Since the total interest is $0.06 million we write

$$0.10x + 0.08y + 0.04z = 0.06$$

We can easily get rid of those awkward decimals by taking advantage of the fact that the solutions of an equation are unchanged when we multiply both sides of the equation by the same number. So, we multiply both sides by 100, in effect, moving the decimal points two places to the right. The result is a more familiar sort of equation:

$$10x + 8y + 4z = 6$$

We were told that the company has three times as much money, z, invested in municipal bonds as in treasury notes, y. Therefore, we would have to triple the amount invested in treasury notes (to $3y$) to equal the total, z, for bonds. In symbols,

$$3y = z$$

To get the usual sort of linear equation, we must have all the unknowns on one side—so we subtract z from both sides:

$$3y - z = 0$$

The system of linear equations is now

$$\begin{aligned} x + y + z &= 1 \\ 10x + 8y + 4z &= 6 \\ 3y - z &= 0 \end{aligned}$$

In matrix form $AX = B$, we see that

$$A = \begin{bmatrix} 1 & 1 & 1 \\ 10 & 8 & 4 \\ 0 & 3 & -1 \end{bmatrix} \quad \text{and} \quad B = \begin{bmatrix} 1 \\ 6 \\ 0 \end{bmatrix}$$

Although it would be possible to find A^{-1}, it will be less work to use the Gauss-Jordan method—first forming the augmented matrix

100   MATRICES AND SYSTEMS OF LINEAR EQUATIONS

$$[A \mid B] = \begin{array}{c} x y z \\ \begin{bmatrix} 1 & 1 & 1 & 1 \\ 10 & 8 & 4 & 6 \\ 0 & 3 & -1 & 0 \end{bmatrix} \end{array}$$

and then applying the usual steps

$$\begin{bmatrix} 1 & 1 & 1 & 1 \\ 0 & -2 & -6 & -4 \\ 0 & 3 & -1 & 0 \end{bmatrix}$$

$$\begin{bmatrix} 1 & 0 & -2 & -1 \\ 0 & 1 & 3 & 2 \\ 0 & 0 & -10 & -6 \end{bmatrix}$$

$$\begin{array}{c} x y z \\ \begin{bmatrix} 1 & 0 & 0 & \frac{1}{5} \\ 0 & 1 & 0 & \frac{1}{5} \\ 0 & 0 & 1 & \frac{3}{5} \end{bmatrix} \end{array} = [A' \mid B']$$

We see that there is a single answer:

$$x = \frac{1}{5}, \quad y = \frac{1}{5}, \quad z = \frac{3}{5}$$

Since we are working with units of money, it would be more natural to express the answer in decimal terms: the company has $0.2 million ($200,000) invested in short-term loans, $0.2 million invested in treasury notes, and $0.6 million invested in municipal bonds. ∎

Example 4. A stolen car, chased by a police car, strikes and damages a truck while passing it. In statements regarding the incident, the truck driver says that the stolen car was going 3 times as fast as the truck, the policeman says he was going 20 miles per hour faster than the stolen car, and the car thief estimates that the sum of the speeds of the police car and the stolen car equaled 6 times the speed of the truck. Determine the speeds of the stolen car, the police car, and the truck; or show that the statements of the three drivers contradict one another.

Solution

Let x be the speed of the stolen car, y be the speed of the police car, and z be the speed of the truck. The truck driver's statement is that x is 3 times as great as z, that is, $x = 3z$. The policeman's statement indicates that you would have to add 20 miles per hour to the stolen car's speed to reach his car's speed, y; thus, $y = x + 20$. The car thief describes $x + y$, the sum of the police car's speed and the stolen car's speed, and estimates that $x + y = 6z$. To summarize, we have the three equations

$$x = 3z$$
$$y = x + 20$$
$$x + y = 6z$$

When we put all of the unknowns on the left sides of the equations, we obtain

$$x - 3z = 0$$
$$-x + y = 20$$
$$x + y - 6z = 0$$

It is easy to see that the corresponding augmented matrix is

$$\begin{array}{ccc} x & y & z \end{array}$$
$$\begin{bmatrix} 1 & 0 & -3 & | & 0 \\ -1 & 1 & 0 & | & 20 \\ 1 & 1 & -6 & | & 0 \end{bmatrix}$$

We add multiples of the first row to the two others to change the first column to that of an identity matrix.

$$\begin{bmatrix} 1 & 0 & -3 & | & 0 \\ 0 & 1 & -3 & | & 20 \\ 0 & 1 & -3 & | & 0 \end{bmatrix}$$

But when we subtract the second row from the third, in order to change the second column to that of the identity matrix, we have this result:

$$\begin{array}{ccc} x & y & z \end{array}$$
$$\begin{bmatrix} 1 & 0 & -3 & | & 0 \\ 0 & 1 & -3 & | & 20 \\ 0 & 0 & 0 & | & -20 \end{bmatrix}$$

The last row means

$$0x + 0y + 0z = -20, \quad \text{that is,} \quad 0 = -20$$

which is false. Since there is no solution possible from the given information, we conclude that not all the statements can be correct; indeed, they contradict one another. ∎

Example 5. The annual report of a company contains a summary of personnel figures. It states that the company has a total of 900 employees in three classifications—clerical, factory, and trainee—and that it budgets a total of 2450 weeks for paid vacations for these employees during the year and 10,450 days for paid sick leave. The report also states that a clerical employee receives two weeks of paid vacation and 8 days of sick leave each year, that a factory employee receives 3 weeks of vacation and 13 days of sick leave, and that a trainee gets one week of vacation and 3 days of sick leave. The report does not state how many employees the company has in each of the three categories. If possible, determine this from the information given.

Solution

Let x be the number of clerical employees, y the number of factory employees, and z the number of trainees. We know the total number of employees is given by

$$x + y + z = 900$$

Each clerical employee receives 2 weeks of vacation, so this category requires $2x$ weeks. The vacation requirements for the other employees amount to $3y$ for the factory employees and z for the trainees. Since the company requires 2450 weeks in all, we see that

$$2x + 3y + z = 2450$$

Similarly, the sick-leave requirements for each type of employee tell us that

$$8x + 13y + 3z = 10{,}450$$

The augmented matrix is

$$\begin{array}{ccc} x & y & z \end{array}$$
$$\begin{bmatrix} 1 & 1 & 1 & | & 900 \\ 2 & 3 & 1 & | & 2450 \\ 8 & 13 & 3 & | & 10{,}450 \end{bmatrix}$$

When we apply the Gauss-Jordan method to the first column, the result is

$$\begin{bmatrix} 1 & 1 & 1 & | & 900 \\ 0 & 1 & -1 & | & 650 \\ 0 & 5 & -5 & | & -3250 \end{bmatrix}$$

After we do the same to the second column, this is the result:

$$\begin{array}{ccc} x & y & z \end{array}$$
$$\begin{bmatrix} 1 & 0 & 2 & | & 250 \\ 0 & 1 & -1 & | & 650 \\ 0 & 0 & 0 & | & 0 \end{bmatrix}$$

This tells us that the system of three linear equations has the same infinite set of solutions as the linear system

$$\begin{aligned} x + 2z &= 250 \\ y - z &= 650 \end{aligned}$$

We conclude that there was not enough information obtained from the report to determine the number of employees in each classification. ∎

As a final application, we return to the matrix codes we discussed in Section 1.3. We will again need Table 1-6, which we repeat here for convenience.

A	B	C	D	E	F	G	H	I	J	K	L	M
1	2	3	4	5	6	7	8	9	10	11	12	13
N	O	P	Q	R	S	T	U	V	W	X	Y	Z
14	15	16	17	18	19	20	21	22	23	24	25	26

Table 1-6

1.7 MATRIX APPLICATIONS

Example 6. The message: 38, 43, 50, 89, 114, 108, 58, 59, 76, 116, 140, 140 is in matrix code with encoding matrix

$$E = \begin{bmatrix} 1 & 2 & 2 \\ 1 & 3 & 2 \\ 1 & 2 & 3 \end{bmatrix}$$

Decode the message.

Solution

The encoded message was produced by using Table 1-6 to convert the original message to column matrices T, now of size 3×1, and again,

$$C = ET$$

gave the encoded message. Therefore, to decode the message, we must multiply by E^{-1} to find the original message, that is,

$$T = E^{-1}C$$

Using the Gauss-Jordan method, as in Section 1.5, we find that the inverse is

$$E^{-1} = \begin{bmatrix} 5 & -2 & -2 \\ -1 & 1 & 0 \\ -1 & 0 & 1 \end{bmatrix}$$

We multiply column matrices from the encoded message by E^{-1}.

$$\begin{bmatrix} 5 & -2 & -2 \\ -1 & 1 & 0 \\ -1 & 0 & 1 \end{bmatrix} \begin{bmatrix} 38 \\ 43 \\ 50 \end{bmatrix} = \begin{bmatrix} 4 \\ 5 \\ 12 \end{bmatrix}$$

$$\begin{bmatrix} 5 & -2 & -2 \\ -1 & 1 & 0 \\ -1 & 0 & 1 \end{bmatrix} \begin{bmatrix} 89 \\ 114 \\ 108 \end{bmatrix} = \begin{bmatrix} 1 \\ 25 \\ 19 \end{bmatrix}$$

$$\begin{bmatrix} 5 & -2 & -2 \\ -1 & 1 & 0 \\ -1 & 0 & 1 \end{bmatrix} \begin{bmatrix} 58 \\ 59 \\ 76 \end{bmatrix} = \begin{bmatrix} 20 \\ 1 \\ 18 \end{bmatrix}$$

$$\begin{bmatrix} 5 & -2 & -2 \\ -1 & 1 & 0 \\ -1 & 0 & 1 \end{bmatrix} \begin{bmatrix} 116 \\ 140 \\ 140 \end{bmatrix} = \begin{bmatrix} 20 \\ 24 \\ 24 \end{bmatrix}$$

From Table 1-6 we have

4,	5,	12,	1,	25,	19,	20,	1,	18,	20,	24,	24
D	E	L	A	Y	S	T	A	R	T	X	X

Ignoring the X's at the end that were added to permit the last multiplication, we see that the secret message was

DELAY START

104   MATRICES AND SYSTEMS OF LINEAR EQUATIONS

Exercises

1. The historian of Example 2 applies the same technique to another voyage. He knows there were 29 men on the ship. The daily wine allowance was 252 ounces, and the monthly payroll was 30,000 maravedis. The historian still assumes the daily wine allowance per man was 8 ounces for seamen and 12 ounces for officers. The salary was 1000 maravedis a month for a seaman and 2000 for an officer. Calculate how many seamen and officers made the voyage, or show that the historian's assumptions are incorrect because the equations are inconsistent.

2. On a European vacation a student spent $30 a day for housing in England, $20 a day in France, and $20 a day in Spain. For food she spent $20 a day in England, $30 a day in France, and $20 a day in Spain. She also spent $10 a day in each country for incidental expenses. Her records show a total of $340 spent for housing, $320 for food, and $140 for incidental expenses. Calculate the number of days the traveler spent in each country, or show that the records must be incorrect because the amounts spent are incompatible with each other.

3. A man refuses to tell anyone his age, but he likes to drop hints about it. For example, he remarks that twice his age plus his mother's age adds up to 140 and also that his age plus his father's age add up to 105. Furthermore, he says that the sum of his age and his mother's age is 30 more than his father's age. Calculate the man's age or show that his hints contradict one another.

4. An intelligence agent knows that 60 aircraft, consisting of fighter planes and bombers, are stationed at a certain secret airfield. The agent wants to find out how many aircraft are fighter planes and how many are bombers. There is a type of rocket carried by both kinds of planes; the fighter carries 6 of these rockets, the bomber only 2. The agent learns that 250 rockets are required to arm every plane at the airfield. Furthermore, the agent overhears a remark that there are twice as many fighter planes as bombers at the base. Calculate the number of fighter planes and bombers at the airfield, or show that the agent's information must be incorrect because it is inconsistent.

5. A card player holds a thirteen-card bridge hand. She says that she has three more red cards than black cards, that twice the number of red cards in the hand plus three times the number of black cards adds up to 31, and that three times the number of red cards plus the number of black cards equals 29. Calculate the number of red cards and black cards in the hand, if possible.

6. A farmer feeds her cattle a mixture of two types of feed. One standard unit of type A feed supplies a steer with 10% of its minimum daily requirement of protein and 15% of its requirement of carbohydrates. Type B feed contains 12% of the requirement of protein and 8% of the requirement of carbohydrates in a standard unit. If the farmer wishes to feed her cattle exactly 100% of their minimum daily requirement of protein and carbohydrates, how many units of each type of feed should she give a steer each day?

7. Last year, a large corporation paid its vice presidents $100,000 a year in salary, 100 shares of stock, and an entertainment allowance of $20,000. A division

manager received $70,000 in salary, 50 shares of stock, and $5000 for official entertaining. An assistant manager received $40,000 in salary, but neither stock nor an entertainment allowance. Last year, the corporation paid a total of $1,600,000 in salaries, 1000 shares of stock, and $150,000 in expense allowances to its vice presidents, division managers, and assistant managers. Since the company had an unprofitable year, it decides that this year it will keep the same salary, stock, and expense-allowance schedules for these categories of employees, but will fire some of these employees, so that the totals will be $1,140,000 in salaries, 700 shares of stock, and $110,000 in expense allowances. How many vice presidents, division managers, and assistant division managers did the corporation have last year and how many will it have this year?

8. Messages intercepted from the headquarters of an army division reveal its total strength to be 10,000 soldiers. Some are combat soldiers (all male), some are male support soldiers, and the rest are female support soldiers. Another message indicates that there are 3 times as many male support soldiers as female soldiers. A third message implies that the number of combat soldiers plus 4 times the number of female soldiers equals the total division size, that is, 10,000. Are the three messages consistent with each other? If they are, determine whether there is enough information given to calculate the number of combat soldiers in the division and, if so, calculate this number.

9. An automobile service station employs mechanics and station attendants. Each works 8 hours a day. An attendant only pumps gas, but mechanics are expected to spend 6 hours repairing automobiles and 2 hours pumping gas. The station owner calculates that next Thursday, Friday, and Saturday, the station will have work-hours to sell gas and repair cars as indicated in Table 1-13:

	Gas	Repair
Thursday	32	24
Friday	46	18
Saturday	34	6

Table 1-13

How many attendants and how many mechanics will be working on each of these three days?

10. An investor remarks to a stockbroker that all her stock holdings are in three companies—Eastern Airlines, Hilton Hotels, and McDonald's—and that two days ago the value of her stocks went down $350, but yesterday the value increased by $600. The broker recalls that two days ago the price of Eastern stock dropped by $1 a share, Hilton dropped by $1.50, but the price of McDonald's stock rose by $0.50. The broker also remembers that yesterday the price of Eastern rose $1.50, there was a further drop of $0.50 a share in Hilton stock, and McDonald's rose $1 a share. Show that the broker does not have enough information to calculate the number of shares the investor owns in each company's stock; but that when the investor says she owns 200 shares of McDonald's, the broker can calculate the number of shares she owns of Eastern Airlines and Hilton Hotels.

In Exercises 11 and 12, the message was encoded using the given matrix. Decode the message.

11. Message: −28, 61, 47, −6, 25, 10, −34, 73, 48

$$E = \begin{bmatrix} -1 & 1 & -1 \\ 2 & -1 & 2 \\ 2 & -1 & 1 \end{bmatrix}$$

12. Message: 85, 107, −49, 88, 107, −51, 70, 89, −42, 32, 37, −8, 72, 84, −39

$$E = \begin{bmatrix} -1 & 2 & 3 \\ -1 & 2 & 4 \\ 1 & -1 & -2 \end{bmatrix}$$

1.8 The Leontief Model

The Leontief economic model assumes there is a simple relationship between the amount produced by each sector of an economy and the amount of input required from other sectors. We will illustrate the relationship using Leontief's analysis of the 1958 American economy. Leontief divided the economy into 81 sectors. In order to keep our discussion simple, we will consolidate the data into a 3-sector mode. The sectors are Industry (including agriculture), which we abbreviate as I, Energy (E), and Services (S). This simplified model of the American economy in 1958 is described in the input-output table (Table 1-14).

		Output		
		I	E	S
	I	0.474	0.026	0.079
Input	E	0.018	0.358	0.025
	S	0.105	0.173	0.234

Table 1-14

We will demonstrate the meaning of Table 1-14 by considering the left-hand column. The top number in this column means that one unit of industrial production requires the consumption of 0.474 unit of other industrial production. Similarly, from the next number in the column, one unit of industrial production will consume 0.018 unit of energy, and, from the bottom number, 0.105 unit of services. Since the unit of measurement that Leontief used is millions of dollars, we conclude that the production of $1 million worth of industrial production consumes $0.474 million, or in other words $474,000 worth of industrial products, $18,000 worth of energy, and $105,000 worth of services. Similarly, the entry in the column headed by E and opposite S of 0.173 means that $173,000 worth of labor and other services goes into the production of $1 million worth of energy, and the number 0.358 in the column headed by E and opposite E means that $358,000 worth of energy is consumed to produce $1 million worth of energy.

A basic assumption of Leontief's model is that the proportion of input required for output stays the same no matter what the level of production. For instance, it assumes that the production of $1 million of industrial production will consume $(0.474)(1) = 0.474$ million dollars of other industrial production, that $30 million of industrial consumption will consume $(0.474)(30) = 14.22$ million dollars worth, and that $115 million will require $(0.474)(115) = 54.51$ million, and so on. The input-output matrix is then telling us that the production of n units of industrial production consumes $0.474n$ units of industrial output, $0.018n$ units of energy, and $0.105n$ units of services. For example, the production of $50 million worth of products from the industrial sector of the 1958 American economy required $(0.474)(50) = 23.7$ units ($23.7 million) worth of industrial output, $(0.018)(50) = 0.9$ unit of energy, and $(0.105)(50) = 5.25$ units of services.

Example 1. According to the simplified input-output table for the 1958 American economy, how many dollars worth of industrial production, energy, and services are required to produce $120 million worth of services?

Solution

From the third (S) column of the input-output table, we learn that each unit ($1 million worth) of services requires 0.079 unit of industrial production because the number in that column of the table opposite I is 0.079. Thus, $120 million, or 120 units, requires $(0.079)(120) = 9.48$ units, or $9.48 million, worth of industrial products. Furthermore, 120 units of services uses $(0.025)(120) = 3$ units of energy and $(0.234)(120) = 28.08$ units of other services, or $3 million and $28.08 million worth of energy and services, respectively. ■

As we explained in the introduction to the chapter, the Leontief model also uses a "bill of demands," a table that represents the production needed to satisfy the requirements of the economy. For an individual country, the bill of demands would include enough production to meet the needs of the people in the country and of the government, to furnish exports to other countries, perhaps some planned surpluses, and so on. The bill of demands for the simplified 1958 American economy is (in millions of dollars)

I: 223,133
E: 23,527
S: 263,985

Finding the Equation to be Solved

We now turn to the problem posed in the introduction. How many units of output from each sector are needed to run the economy (internally, as described by the input-output table) and also fill the bill of demands? The number of units of industrial production required to solve the problem will be represented by the symbol x, the number of units of energy required, by y, and the number of units of services, by z. These three numbers are the quantities we are calculating.

Let the symbol s_{11} represent the number of units of industrial production required to produce x units of industrial output. We will not know how large s_{11} is until we determine x, but for the present we need the symbol. Furthermore, s_{12} will stand for the number of units of industrial output required to produce y units of energy. Similarly, s_{13} represents the industrial units used to produce z units of services. Thus, in each case, the first subscript, 1, indicates that the input consists of industrial units. The second subscript tells us to which of the three segments the industrial production will go. As in Section 1.1, the symbol s_{ij} concerns the flow of units corresponding to the location row i (here 1) and column j in the input-output table.

The total requirement x for industrial units consists of the demand for 223,133 units from the bill of demands (the first row, naturally), plus the sum of the requirements for industrial units from the three sectors. We write

$$x = s_{11} + s_{12} + s_{13} + 223{,}133$$

In just the same way, we use the s_{ij}-notation, meaning the total number of units of production from the segment of the ith row that will be required by the jth-column segment to produce the (unknown) output required from that segment, to obtain two other equations, one for energy and one for services:

$$y = s_{21} + s_{22} + s_{23} + 23{,}527$$
$$z = s_{31} + s_{32} + s_{33} + 263{,}985$$

Now, according to the input-output table, it takes 0.474 unit of industrial input to produce 1 unit of industrial output. Recalling the basic assumption underlying this kind of economic theory, it then takes $0.474x$ units of industrial input to produce x industrial units. For example, if x were 400,000 units, then it would require $(0.474)(400{,}000) = 189{,}600$ industrial units to produce 400,000 units. Since s_{11} was defined to be the number of units of industrial input required to produce x industrial units, we have the equation

$$s_{11} = 0.474x$$

Similarly, from the table, it takes 0.026 unit of industrial production to make 1 energy unit; so, it takes $0.026y$ units to make y energy units. Thus, we conclude that

$$s_{12} = 0.026y$$

We also have

$$s_{13} = 0.079z$$

Substituting into the equation

$$x = s_{11} + s_{12} + s_{13} + 223{,}133$$

provides us with

$$x = 0.474x + 0.026y + 0.079z + 223{,}133$$

In the same way, we know that
$$s_{21} = 0.018x$$
That is, in order to produce x units of industrial output, it will take 0.018 unit of energy to do the job. Substituting for s_{21} and the other s_{ij} in the equations for y and z, we now have the following equations:

$$(*) \begin{cases} x = 0.474x + 0.026y + 0.079z + 223{,}133 \\ y = 0.018x + 0.358y + 0.025z + 23{,}527 \\ z = 0.105x + 0.173y + 0.234z + 263{,}985 \end{cases}$$

The Matrix Equation

It would be possible to solve these equations by techniques of algebra for the unknowns x, y, and z. Keep in mind, however, that we want a procedure that will work for an 81-sector model, like the one Leontief actually used for the 1958 American economy, and not just for simplified 3-sector examples. Thus, the next step, in which we turn the problem of computing the numbers x, y, and z into a matrix-arithmetic problem, is of fundamental importance in obtaining a practical solution to the problem.

Let A be the 3×3 input-output matrix corresponding to the input-output table:

$$A = \begin{bmatrix} 0.474 & 0.026 & 0.079 \\ 0.018 & 0.358 & 0.025 \\ 0.105 & 0.173 & 0.234 \end{bmatrix}$$

The bill of demands gives rise to a 3×1-demand matrix:

$$D = \begin{bmatrix} 223{,}133 \\ 23{,}527 \\ 263{,}985 \end{bmatrix}$$

In addition, we have a 3×1-matrix of unknowns

$$X = \begin{bmatrix} x \\ y \\ z \end{bmatrix}$$

called the **intensity matrix** in input-output economic theory. We claim that the system of three linear equations $(*)$ is the same as the single matrix equation:

$$X = AX + D$$

If you check that

$$AX = \begin{bmatrix} 0.474 & 0.026 & 0.079 \\ 0.018 & 0.358 & 0.025 \\ 0.105 & 0.173 & 0.234 \end{bmatrix} \begin{bmatrix} x \\ y \\ z \end{bmatrix} = \begin{bmatrix} 0.474x + 0.026y + 0.079z \\ 0.018x + 0.358y + 0.025z \\ 0.105x + 0.173y + 0.234z \end{bmatrix}$$

you should have no difficulty in verifying that our claim is correct.

The general input-output model leads to the same equation. Leontief assumed that the coefficients of the input-output table for an n-sector economy have been determined on the basis of economic data, producing an $n \times n$ input-output matrix A. The bill of demands is computed from a knowledge of the economic situation, and it becomes the $n \times 1$-matrix D. The number of units required for each of the n sectors in order to run the economy and fill the bill of demands is calculated by solving the matrix equation

$$X = AX + D$$

where X is the $n \times 1$-matrix of unknowns, and where the order of sectors is the same as in the rows of A and D. That is why we solved this equation in Section 1.5, where we showed that the solution was

$$X = (I - A)^{-1}D$$

Now we can easily describe the method by which the computer solves the problem of determining the required total output for each sector. The input-output matrix A is subtracted from the identity matrix of the same size. The matrix $I - A$ is usually nonsingular. In case it is not, a very small numerical adjustment to A, which has no economic significance, will make $I - A$ nonsingular. The computer then finds the inverse of this matrix $I - A$ (essentially) by the Gauss-Jordan method. This is a very formidable task in the case of a large matrix, but one well within the capability of modern computers. Finally, the matrix $(I - A)^{-1}$ is multiplied by the $n \times 1$-demand matrix D to give the $n \times 1$-matrix X of answers.

In the simplified version of the table for the 1958 American economy

$$A = \begin{bmatrix} 0.474 & 0.026 & 0.079 \\ 0.018 & 0.358 & 0.025 \\ 0.105 & 0.173 & 0.234 \end{bmatrix} \quad \text{and} \quad D = \begin{bmatrix} 223{,}123 \\ 23{,}527 \\ 263{,}985 \end{bmatrix}$$

We subtract to obtain

$$I - A = \begin{bmatrix} 1 & 0 & 0 \\ 0 & 1 & 0 \\ 0 & 0 & 1 \end{bmatrix} - \begin{bmatrix} 0.474 & 0.026 & 0.079 \\ 0.018 & 0.358 & 0.025 \\ 0.105 & 0.173 & 0.234 \end{bmatrix} = \begin{bmatrix} 0.526 & -0.026 & -0.079 \\ -0.018 & 0.642 & -0.025 \\ -0.105 & -0.173 & 0.766 \end{bmatrix}$$

The inverse of $I - A$ (actually, an approximation since the computer has to deal with decimals) is given by

$$(I - A)^{-1} = \begin{bmatrix} 1.941 & 0.133 & 0.205 \\ 0.017 & 1.577 & 0.059 \\ 0.271 & 0.375 & 1.348 \end{bmatrix}$$

Therefore,

$$\begin{bmatrix} x \\ y \\ z \end{bmatrix} = X = (I - A)^{-1}D = \begin{bmatrix} 1.941 & 0.133 & 0.205 \\ 0.017 & 1.577 & 0.059 \\ 0.271 & 0.375 & 1.348 \end{bmatrix} \begin{bmatrix} 223{,}133 \\ 23{,}527 \\ 263{,}985 \end{bmatrix} = \begin{bmatrix} 490{,}347 \\ 56{,}470 \\ 425{,}143 \end{bmatrix}$$

1.8 THE LEONTIEF MODEL

So, we have the solution

$$x = 490,347$$
$$y = 56,470$$
$$z = 425,143$$

In other words, it would require 490,347 units ($490,347 million worth) of industrial production, 56,470 units of energy, and 425,143 units of services to run the 1958 American economy and completely fill the stated bill of demands.

Example 2. A portion of the input-output table for the Western European economy in 1953 gives us the following input-output table for some significantly interrelated industries. Here I = iron, L = lumber, C = coal, and P = paper. (See Table 1-15.) The bill of demands (in $10 million equivalents) is given in Table 1-16.

Input \ Output

	I	L	C	P
I	.20	.15	.02	.05
L	.13	.10	.05	.03
C	.40	.10	.12	.07
P	.15	.02	.03	.20

Table 1-15

I (Iron)	90
L (Lumber)	320
C (Coal)	154
P (Paper)	43

Table 1-16

(a) How many units of coal are required to produce 1 unit of iron?
(b) How many units of coal are required to produce 10,000 units of iron?
(c) How many units of paper products are required to produce 75,000 units of coal?
(d) According to the bill of demands, what was the demand on the Western European economy for production of lumber products (in the equivalent of U. S. dollars)?
(e) For A, the input-output matrix corresponding to the Western European input-output table, calculate $I - A$.
(f) Given that

$$(I - A)^{-1} = \begin{bmatrix} 1.525 & 0.263 & 0.053 & 0.109 \\ 0.371 & 1.183 & 0.078 & 0.074 \\ 0.967 & 0.296 & 1.181 & 0.175 \\ 2.905 & 0.533 & 0.146 & 1.464 \end{bmatrix}$$

how much production in each of the iron, lumber, coal, and paper industries would Western Europe have had to achieve (in millions of U. S. dollars) in 1953 in order to run the economy of the region and fill the bill of demands?

Solution

(a) 0.40

(b) (0.40)(10,000) = 4000

(c) (0.03)(75,000) = 2250

(d) $3,200,000,000

(e) $$I - A = \begin{bmatrix} 0.80 & -0.15 & -0.02 & -0.05 \\ -0.13 & 0.90 & -0.05 & -0.03 \\ -0.40 & -0.10 & 0.88 & -0.07 \\ -0.15 & -0.02 & -0.03 & 0.80 \end{bmatrix}$$

(f) We calculate that

$$\begin{bmatrix} I \\ L \\ C \\ P \end{bmatrix} = X = \begin{bmatrix} 1.525 & 0.263 & 0.053 & 0.109 \\ 0.371 & 1.183 & 0.078 & 0.074 \\ 0.967 & 0.296 & 1.181 & 0.175 \\ 2.905 & 0.533 & 0.146 & 1.464 \end{bmatrix} \begin{bmatrix} 90 \\ 320 \\ 154 \\ 43 \end{bmatrix} = \begin{bmatrix} 234.26 \\ 427.14 \\ 371.15 \\ 517.45 \end{bmatrix}$$

Therefore, $2342.6 million of iron production, $4271.4 million of lumber production, $3711.5 million of coal production, and $5174.5 million of paper production would have been necessary. ∎

Computing Changes

Returning to the general subject of input-output economics, the solution

$$X = (I - A)^{-1} D$$

tells us that if a new bill of demands D^* is calculated for a subsequent time period for the same economy, the required intensities of the various economic sectors are calculated readily by the computer. It performs the matrix multiplication

$$(I - A)^{-1} D^*$$

using the same inverse matrix $(I - A)^{-1}$. However, if new technology or new economic circumstances change the relationship between two sectors, and so change the input-output matrix A to a new matrix A^*, then the inverse $(I - A^*)^{-1}$ must be calculated all over again.

Exercises

1. A much simplified version of an input-output table for the 1958 Israeli economy divides that economy into three sectors: agriculture, manufacturing, and energy—with the result shown in Table 1-17.[†] Suppose also that the bill of demands (in millions of shekels) is

    Agriculture: 32
    Manufacturing: 14
    Energy: 10

[†]Wassily Leontief, *Input-Output Economics* (New York: Oxford University Press, 1966), pp. 54-57.

1.8 THE LEONTIEF MODEL

		Output		
		Agriculture	Manufacturing	Energy
	Agriculture	0.293	0	0
Input	Manufacturing	0.014	0.207	0.017
	Energy	0.044	0.010	0.216

Table 1-17

(a) How many units of agricultural production are required to produce 1 unit of agricultural output?
(b) How many units of agricultural production are required to produce 200,000 units of agricultural output?
(c) How many units of agricultural products go into the production of 50,000 units of energy?
(d) How many units of energy go into the production of 50,000 units of agricultural products?
(e) For A, the Israeli input-output matrix, calculate $I - A$ and check that its (approximate) inverse is

$$\begin{bmatrix} 1.414 & 0 & 0 \\ 0.027 & 1.261 & 0.027 \\ 0.080 & 0.016 & 1.276 \end{bmatrix}$$

(f) How much output of agriculture, manufacturing, and energy would be required to run the Israeli economy in 1958 and fill the bill of demands?

2. The interdependence of certain basic industries of the 1947 American economy is presented in the following input-output table for chemicals (C), petroleum (P), rubber products (R), and metals (M) (Table 1-18). Suppose the bill of demands is (in $10 million units)

$$\begin{array}{ll} C: & 132 \\ P: & 320 \\ R: & 84 \\ M: & 183 \end{array}$$

		Output			
		C	P	R	M
	C	0.184	0.015	0.213	0.010
Input	P	0.023	0.353	0.004	0.048
	R	0	0	0.014	0
	M	0.014	0.001	0.004	0.369

Table 1-18

(a) How many units of metal products are required to produce 1 unit of chemicals?
(b) How many units of rubber products are required to produce 5 units of petroleum products?

(c) How many units of chemical output are needed to produce 5 units of rubber products?
(d) How many dollars' worth of petroleum go into producing $1 million worth of metal products?
(e) How many dollars' worth of petroleum go into producing $10 million worth of metal products?
(f) How many dollars' worth of chemicals are consumed in the production of $7 million worth of chemicals?
(g) For A, the input-output matrix of these U. S. industries in 1947, calculate $I - A$ and check that its (approximate) inverse is

$$\begin{bmatrix} 1.227 & 0.028 & 0.265 & 0.022 \\ 0.046 & 1.547 & 0.017 & 0.118 \\ 0 & 0 & 1.014 & 0 \\ 0.027 & 0.003 & 0.012 & 1.585 \end{bmatrix}$$

(h) Calculate the production, in millions of dollars, of chemicals, petroleum, rubber products, and metals the United States needed to have produced in 1947 to run this portion of the economy and fill the bill of demands.

3. The interdependence among the motor-vehicle industry and other basic industries in the 1958 American economy is described by the following input-output table for motor vehicles (V), steel (S), glass (G), and rubber and plastics (R) (Table 1-19). The final demand for these products in millions of dollars is given by

$$\begin{array}{ll} V: & 5444 \\ S: & 3276 \\ G: & 119 \\ R: & 943 \end{array}$$

		\multicolumn{4}{c}{Output}
		V	S	G	R
Input	V	0.298	0.002	0	0
	S	0.088	0.212	0	0.002
	G	0.010	0	0.050	0.006
	R	0.029	0.003	0.004	0.030

Table 1-19

(a) For A, the input-output matrix of this portion of the 1958 American economy, calculate $I - A$ and check that its (approximate) inverse is

$$\begin{bmatrix} 1.425 & 0.004 & 0 & 0 \\ 0.159 & 1.269 & 0 & 0.003 \\ 0.015 & 0 & 1.053 & 0.007 \\ 0.043 & 0.004 & 0.004 & 1.031 \end{bmatrix}$$

(b) Calculate the production, in millions of dollars, of motor vehicles, steel, glass, and rubber and plastics that the United States needed to have produced in 1958 to run this portion of the economy and fill the bill of demands.

4. A study of the worldwide extraction of ores in 1970 is here simplified to three sectors: extraction of ore, production of metals from the ore, and impact on the environment. The environment input represents the pollutants emitted by the extraction and production processes, and the environment output is that of the pollution-abatement industry. (See Table 1-20.) The bill of demands (in $ billion equivalents) is given by

$$\begin{aligned} \text{Extraction:} & \quad -13 \\ \text{Production:} & \quad 2533 \\ \text{Environment:} & \quad 60 \end{aligned}$$

		Output		
		Extraction	Production	Environment
Input	Extraction	0	0.0178	0
	Production	0.3372	0.4223	0.3298
	Environment	0.0859	0.0144	0.0118

Table 1-20

(a) For A, the input-output matrix of the 1970 worldwide extraction industry, calculate $I - A$ and check that its (approximate) inverse is

$$\begin{bmatrix} 1.011 & 0.031 & 0.010 \\ 0.646 & 1.766 & 0.589 \\ 0.097 & 0.028 & 1.021 \end{bmatrix}$$

(b) Calculate the value, in billions of U. S. dollars, of the amount of ore extracted, of metals produced, and of environmental costs in 1970 to run this portion of the world's economy and fill the bill of demands.

Review Exercises for Chapter 1

In Exercises 1 through 8, let

$$A = \begin{bmatrix} 1 & -1 & -1 \\ 0 & 2 & 1 \\ 1 & 0 & 0 \end{bmatrix}, \quad B = \begin{bmatrix} 2 \\ -2 \\ 2 \end{bmatrix}, \quad C = \begin{bmatrix} 1 & 0 & 1 \end{bmatrix},$$

$$D = \begin{bmatrix} 1 \\ -1 \\ -1 \end{bmatrix}, \quad \text{and} \quad E = \begin{bmatrix} 2 & 3 & 0 \\ 1 & 2 & 1 \\ 0 & 1 & -1 \end{bmatrix}$$

Compute, if possible.

1. $A + E$
2. $B - D$
3. CD
4. $B + C$
5. BC
6. $AB + AD$
7. $2A - 3E$
8. AC

116 MATRICES AND SYSTEMS OF LINEAR EQUATIONS

In Exercises 9 and 10, find each inverse.

9. $\begin{bmatrix} -1 & 1 \\ 2 & 0 \end{bmatrix}$

10. $\begin{bmatrix} 0 & 2 & -1 \\ 1 & 1 & 0 \\ -1 & 1 & -2 \end{bmatrix}$

In Exercises 11 and 12, let

$$A = \begin{bmatrix} 2 & 1 \\ -1 & 0 \end{bmatrix}, \quad B = \begin{bmatrix} -1 \\ 3 \end{bmatrix}, \quad \text{and} \quad C = \begin{bmatrix} 2 \\ 1 \end{bmatrix}$$

Solve for the unknown matrix X.

11. $X + B = C$

12. $AX = B$

In Exercises 13 and 14, solve $AX = B$ by the Gauss-Jordan method, if possible.

13. $X = \begin{bmatrix} x \\ y \end{bmatrix}$, $A = \begin{bmatrix} 1 & 2 \\ -1 & 1 \\ 2 & 1 \end{bmatrix}$, and $B = \begin{bmatrix} 2 \\ 4 \\ -2 \end{bmatrix}$

14. $X = \begin{bmatrix} x \\ y \\ z \end{bmatrix}$, $A = \begin{bmatrix} 2 & 2 & 1 \\ -1 & 0 & -2 \\ 1 & 2 & -1 \end{bmatrix}$, and $B = \begin{bmatrix} 0 \\ 1 \\ -1 \end{bmatrix}$

15. Solve the equation $X = AX + D$ for X, where

$$A = \begin{bmatrix} 2 & 1 \\ -1 & 1 \end{bmatrix} \quad \text{and} \quad D = \begin{bmatrix} 3 \\ 2 \end{bmatrix}$$

16. If possible, use the Gauss-Jordan method to solve the equation $AX + B = C$ for X, where

$$A = \begin{bmatrix} 0 & 1 & -1 \\ 1 & 0 & 2 \\ -1 & 1 & 0 \end{bmatrix}, \quad B = \begin{bmatrix} -2 \\ 3 \\ 1 \end{bmatrix}, \quad \text{and} \quad C = \begin{bmatrix} -1 \\ 0 \\ 2 \end{bmatrix}$$

In Exercises 17 and 18, use a matrix method to find numbers x and y for which the equations are true, if such numbers exist.

17. $-x + 3y = 2$
$2x - 6y = -1$

18. $-x + 3y = 2$
$2x - y = -1$

In Exercises 19 and 20, use the Gauss-Jordan method to find numbers x, y, and z for which all the equations are true, if such numbers exist. If there are infinitely many solutions, write down any two of them.

19. $\quad x - y - z = -1$
    $\quad 2x + y = 0$

20. $\quad y + z = 2$
    $\quad x + z = -1$
    $\quad x + y - z = 0$

21. A concert hall seats 500 people in the orchestra, 240 people in the loge, and 350 people in the balcony. If ticket prices for a concert are $20 each in the orchestra, $16 each in the loge, and $12 each in the balcony, use matrix multiplication to calculate the total income from ticket sales from a concert, assuming that every seat is sold.

22. A woman travels 4 miles, partly walking and partly jogging. If she covers twice as much distance jogging as walking, use a matrix method to determine how far she walks and how far she jogs.

23. A store sells three models of mechanical pencils. The inexpensive model sells for $1, the standard model for $2, and the deluxe model for $4. One day, the store sells 15 mechanical pencils, receiving a total of $20 for the sales. If 10 more of the inexpensive model are sold than of the deluxe model, use the Gauss-Jordan method to find out how many of each model are sold.

2

Linear Programming

Farm Management

Managing a farm in the most efficient way, that is, producing the maximum profit from the available resources while preserving the fertility of the soil, is an extremely complicated business. In the 1950s, farm-management specialists discovered that a mathematical technique called *linear programming* could be very useful in their work. A good example of this approach is described in James N. Boles' article in the *Journal of Farm Economics*.[†] This article describes how linear programming was used in developing optimum resource allocations for farms in Kern County, California, where water for irrigation was limited and where the government imposed cotton-acreage restrictions.

Boles considered a typical farm with resources consisting of 150 acres of cropland, the labor and management ability of the farm operator, farm machinery and tools, working capital, and an irrigation system that could supply only a limited amount of water.

The problem was to decide how much land the farmer should devote to each crop normally grown in Kern County—cotton, potatoes, alfalfa, barley, and sugar beets—so that the profit would be as large as possible. The farm operator had sufficient working capital, labor, and equipment to undertake any crop allocation plan the mathematics proposed. To

---

[†]James N. Boles, "Linear Programming and Farm Management Analysis," *Journal of Farm Economics* 3 (Feb. 1955), pp. 251–273.

perform the mathematical analysis needed for the problem, Boles assumed that the farmer could predict accurately the profit from each crop—for instance, $207 per acre for cotton and $86 per acre for alfalfa. These predictions might not have been entirely correct, but they were considered close enough to reality to make the solution obtained from the mathematical study the best one possible with the information available. Besides the water restrictions, Boles' analysis took into account government restrictions that did not permit the farmer to devote more than 60 acres of cropland to cotton. An additional condition of the problem was that the farmer had decided not to plant more than 50 acres of potatoes because of the extreme variability of potato prices.

Boles divided the growing season into three periods with varying amounts of water available to the farmer for irrigation during each period. He then had to consider the added complication that each crop requires different amounts of water during each period. For example, each acre of cotton requires 4 inches of water during the first period, 16.6 inches during the second, and 7.8 inches during the third. But potatoes require 13.3 inches of water per acre during the first period and none for the rest of the growing season.

If we look at everything the farmer must consider to make the best use of the available resources in this relatively simple farming situation, we can understand how complex farm-management decisions are. This chapter will show how linear programming provides a tool to help solve problems of this type.

2.1 Linear-Programming Problems

Linear Polynomials

In Chapter 1 we studied systems of linear equations. These are equations such as

$$3x + y - 2z = 1$$

where the nonconstant left-hand side, $3x + y - 2z$, is expressed as a sum of terms of the form "constant times unknown."

A sum of terms of the form "constant times unknown" is called a **linear polynomial**. For an expression with many unknowns, it is convenient to label the unknowns with subscripts, such as x_1, x_2, x_3, and so on. We will also do this with constants, writing a_1, a_2, a_3, so that we can say that, in general, a **linear polynomial** is an expression of the form

$$a_1 x_1 + a_2 x_2 + \ldots + a_n x_n$$

and a **linear equation** is an equation that can be written as

$$a_1 x_1 + a_2 x_2 + \ldots + a_n x_n = b$$

where b is a constant.

Linear Inequalities

When we replace the equals sign in the preceding linear equation by the inequality signs $\geq$ and $\leq$, we obtain the **linear inequalities**

$$a_1 x_1 + a_2 x_2 + \ldots + a_n x_n \geq b$$

and

$$a_1 x_1 + a_2 x_2 + \ldots + a_n x_n \leq b$$

A **solution** of the inequality consists of numbers that when substituted for $x_1, x_2, \ldots, x_n$, yields a true statement. For instance, $x_1 = 3$, $x_2 = 2$, $x_3 = 7$ is a solution of the inequality

$$2x_1 - 4x_2 + x_3 \geq 3$$

because

$$2(3) - 4(2) + 7 = 5$$

and 5 is larger than 3. On the other hand, $x_1 = 2$, $x_2 = 1$, $x_3 = 1$ is not a solution of this inequality because

$$2(2) - 4(1) + 1 = 1$$

is less than 3.

If we replace the inequality sign in a linear inequality by an equals sign, the resulting linear equation is said to *correspond* to the inequality. Thus, the linear equation corresponding to the preceding inequality is

$$2x_1 - 4x_2 + x_3 = 3$$

Notice that a solution of the linear equation, such as $x_1 = 2$, $x_2 = 1$, $x_3 = 3$, is also a solution of this inequality.

Feasible Solutions

The farmer in the introduction had to take into account many restrictions on the acreage assigned to the five possible crops. There were limited amounts of land and water, as well as limitations due to government policies, and so on. In a linear-programming analysis of the farmer's decision, these restrictions will be described by a system of linear inequalities. For instance, in a simplified farming problem (Example 3), the farmer has 150 acres of land on which just two crops are to be grown: potatoes and barley. We will let x_1 represent the number of acres the farmer will use for potatoes and x_2 be the number of acres for barley. The restriction that the farmer has available only 150 acres in all can be written as the linear inequality

$$x_1 + x_2 \leq 150$$

Similarly, a water restriction will be expressed as the inequality

$$12x_1 + 6x_2 \leq 1200$$

The numbers of acres x_1 and x_2 of each crop will have to satisfy both inequalities.

In the terminology of linear programming, the set of all solutions of the system of linear inequalities that describes the restrictions of the problem is called the **feasible set** and a solution in that set is called a **feasible solution**.

Example 1. Determine whether each of the following choices of values

(a) $x_1 = 0, x_2 = 2, x_3 = 1$
(b) $x_1 = 1, x_2 = 1, x_3 = 1$
(c) $x_1 = 1, x_2 = 1, x_3 = 2$

is a solution of the following system of inequalities:

$$\begin{aligned} 2x_1 + x_2 + 3x_3 &\leq 6 \\ -x_1 + 2x_2 - 3x_3 &\leq 2 \\ x_1 + x_3 &\leq 3 \end{aligned}$$

Solution

(a) Substituting the values $x_1 = 0, x_2 = 2, x_3 = 1$ into the linear polynomials, we calculate that

$$\begin{aligned} 2(0) + 2 + 3(1) &= 5 \\ -0 + 2(2) - 3(1) &= 1 \\ 0 + 1 &= 1 \end{aligned}$$

This is a solution because $5 \leq 6$, $1 \leq 2$, and $1 \leq 3$. That is, all three inequalities of the system hold true.

(b)
$$\begin{aligned} 2(1) + 1 + 3(1) &= 6 \\ -1 + 2(1) - 3(1) &= -2 \\ 1 + 1 &= 2 \end{aligned}$$

This is a solution because $x_1 = 1, x_2 = 1, x_3 = 1$ is a solution of the linear equations corresponding to the first and third inequalities and, for the second inequality, $-2 \leq 2$.

(c)
$$\begin{aligned} 2(1) + 1 + 3(2) &= 9 \\ -1 + 2(1) - 3(2) &= -5 \\ 1 + 2 &= 3 \end{aligned}$$

This is not a solution because in the first inequality, 9 is larger than 6. ■

Linear Functions

Besides their role in describing the restrictions of the problem by means of linear inequalities, linear polynomials appear in another form in linear-programming problems. In Example 3, the profit per acre for potatoes is $200 and for barley is $29. If the farmer plants 40 acres of potatoes, the profit from that crop will be ($200)(40) = $8000, whereas 55 acres will bring in ($200)(55) = $11,000. In general, planting x_1 acres of potatoes will bring a profit of $200x_1$ dollars and the profit from x_2 acres of barley will be $29x_2$. We can calculate the total profit by just adding these amounts:

$$\text{Total Profit} = 200x_1 + 29x_2$$

For instance, $x_1 = 85$ acres of potatoes and $x_2 = 60$ acres of barley will earn:

$$\text{Total Profit} = 200(85) + 29(60) = 18{,}740$$

As the unknown quantities x_1 and x_2 vary, the amount of total profit varies as well. When the unknowns appear in an open-ended expression, such as the preceding one for total profit, it is customary to refer to them as **variables**. The amount of total profit can also be considered as a variable because it depends on the values of x_1 and x_2. For this reason we replace Total Profit by the single letter z and write the expression

$$z = 200x_1 + 29x_2$$

The variable z is called the **dependent variable** and x_1 and x_2 are the **independent variables**.

In general, an expression of the form

$$z = c_1x_1 + c_2x_2 + \ldots + c_nx_n$$

where $c_1, c_2, \ldots, c_n$ are constants, in which the value of the dependent variable z is computed by substituting values for the independent variables $x_1, x_2, \ldots, x_n$ is called a **linear function**.

Example 2. Calculate the value of the dependent variable z in the linear function

$$z = 2x_1 + x_2 + 3x_3$$

(a) when $x_1 = 1, x_2 = 2, x_3 = 2$;
(b) when $x_1 = 2, x_2 = 3, x_3 = 0$;
(c) when $x_1 = 1, x_2 = 0, x_3 = 1.5$.

Solution

(a) $z = 2(1) + 2 + 3(2) = 10$ (b) $z = 2(2) + 3 + 3(0) = 7$
(c) $z = 2(1) + 0 + 3(1.5) = 6.5$ ■

Maximum and Minimum

If we represent a finite set of numbers on the number line, then the number furthest to the right is called the **maximum** of the set, and the number furthest to the left is the **minimum**. For example, for the set

$$\{2.1, 1.3, -3.5, 1, -4, 0, 0.2, -1.8\}$$

the maximum is 2.1 and the minimum is -4; see Figure 2-1. Similarly, every finite set has both a maximum and a minimum.

An infinite set may not have a maximum or minimum; for instance, the set of counting numbers,

$$1, 2, 3, \ldots$$

has a minimum, 1, but no maximum.

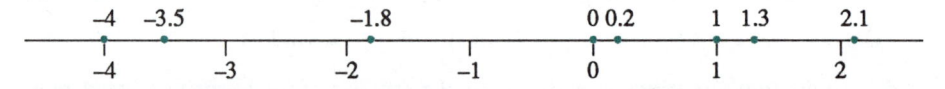

Figure 2-1

Definition of LP-Problem

To state a linear-programming problem requires

1. A system of linear inequalities.
2. A linear function, called the **objective function**, which is of the form

$$z = c_1 x_1 + c_2 x_2 + \ldots + c_n x_n$$

A **linear-programming problem** (or more briefly, an **LP-problem**) consists of determining either the maximum or the minimum value of the objective function $z = c_1 x_1 + c_2 x_2 + \ldots + c_n x_n$, where the values $x_1, x_2, \ldots, x_n$ that are substituted into the expression for z are feasible solutions, that is, solutions of the given system of linear inequalities.

Problem Formulation

Because LP-problems come from concrete situations, they are usually described in words, rather than in symbols. Expressing the problem in a form to which we can apply the mathematical techniques requires three steps:

1. Identify the variables.
2. State the objective as the maximum or minimum of a linear function.
3. List the inequalities that restrict the variables.

The examples that follow illustrate how such a verbal description is put into mathematical terms as an LP-problem. The first example is a simplified version of the agricultural problem discussed in the introduction to this chapter.

Example 3. Express the following (*Farmer's*) LP-problem in mathematical terms:

A farmer has 150 acres of land on which to grow potatoes and barley. To grow potatoes, 12 inches of water per acre are required, and 6 inches per acre are required to grow barley. The farmer has 1200 acre-inches[†] of water available for irrigation. If the farmer's profit is estimated to be $200 an acre for potatoes and $29 an acre for barley, how many acres should the farmer devote to each crop to maximize the profit?

Solution

The final sentence of the problem description asks the question that specifies the variables: the number of acres to be planted with potatoes and with barley, respectively. We use the symbol x_1 for the number of acres of potatoes and x_2 for the acres of barley. Since the farmer expects a profit of $200 from each acre planted with potatoes, he will have a profit from the x_1 acres of $200x_1$ dollars and, by the same reasoning, a

---

[†]An acre-inch of water is the amount of water needed to cover one acre to a depth of one inch.

profit of $29x_2$ dollars from x_2 acres planted with barley. Thus, the total profit to the farmer is described by the objective function

$$\text{Total Profit} = z = 200x_1 + 29x_2$$

The total acreage available is 150 acres, which, as we already observed, is expressed by the linear inequality

$$x_1 + x_2 \leq 150$$

We are told that each acre planted with potatoes requires 12 inches of water for irrigation. Thus, x_1 acres planted with potatoes will consume $12x_1$ acre-inches of water. Similarly, each acre planted with barley requires 6 inches of water for irrigation; so, if x_2 acres of barley are planted, they will use $6x_2$ acre-inches. The sum of the amount of water used to irrigate the potatoes and the amount used to irrigate the barley must be less than or equal to the total water available, which we are told is 1200 acre-inches. In symbols then,

$$12x_1 + 6x_2 \leq 1200$$

Of course the number of acres devoted to a crop certainly cannot be a negative number, but for the mathematical formulation of the problem it must be stated explicitly that both x_1 and x_2 are greater than or equal to zero. In symbols, we have $x_1 \geq 0$ and $x_2 \geq 0$. Notice that these expressions are linear inequalities; for instance, $x_1 \geq 0$ could be written as $1x_1 + 0x_2 \geq 0$. In the LP-problems of this chapter, all the variables will be required to be greater than or equal to zero. We will put these routine restrictions in a condensed form, such as $x_1, x_2 \geq 0$, at the end of the system of linear inequalities. Thus, we have the following formal mathematical description of this simplified farm problem:

Maximize: Total Profit $= z = 200x_1 + 29x_2$
Subject to: $x_1 + x_2 \leq 150$
$12x_1 + 6x_2 \leq 1200$
$x_1, x_2 \geq 0$ ■

We will solve this LP-problem in Section 2.2.

Example 4. Express the following (*Secretaries and Clerks*) LP-problem in mathematical terms:

In a certain company, a secretary can prepare 10 invoices and file 20 receipts in an hour and a clerk can prepare 4 invoices and file 30 receipts in an hour. During a peak hour, the company must have at least 100 invoices prepared and at least 300 receipts filed. Because secretaries have other responsibilities, such as making appointments, the company decides it must hire at least five secretaries. If a secretary is paid $11 an hour and a clerk $4 an hour, how many secretaries and how many clerks should the company employ to be able to cope with the volume of work described at the minimum cost to the company?

Solution

The variables are x_1, the number of secretaries to be hired, and x_2, the number of clerks. It costs the company $\$11x_1$ per hour for the x_1 secretaries and $\$4x_2$ an hour for the x_2 clerks; so, the total cost per hour to the company (in salaries) is $11x_1 + 4x_2$. The objective function is therefore

$$\text{Salaries} = z = 11x_1 + 4x_2$$

Since there must be at least five secretaries, we write $x_1 \geq 5$. There need not be any clerks, but there can hardly be a negative number of them; so $x_2 \geq 0$. Each secretary can prepare 10 invoices per hour; so, x_1 secretaries can prepare $10x_1$ invoices. In the same hour x_2 clerks can prepare $4x_2$ invoices. Thus altogether, the x_1 secretaries and x_2 clerks can prepare $10x_1 + 4x_2$ invoices per hour. Since it is sometimes necessary to prepare at least 100 invoices in a single hour,

$$10x_1 + 4x_2 \geq 100$$

The x_1 secretaries can file $20x_1$ receipts in an hour and the x_2 clerks can file $30x_2$ receipts; thus, the total filing capacity per hour is $20x_1 + 30x_2$. Because the office must be able to file at least 300 receipts in a single hour,

$$20x_1 + 30x_2 \geq 300$$

Thus, the mathematical problem is to find numbers x_1 and x_2 that

$$\begin{aligned}
&\text{Minimize:} && \text{Salaries} = z = 11x_1 + 4x_2 \\
&\text{Subject to:} && x_1 \geq 5 \\
& && 10x_1 + 4x_2 \geq 100 \\
& && 20x_1 + 30x_2 \geq 300 \\
& && x_2 \geq 0
\end{aligned}$$

■

Example 5. Express the following (*Bank Investment*) LP-problem in mathematical terms:

> A bank grants mortgages and makes business loans. It earns 8% interest on mortgages and 10% on business loans. The bank has $50 million available for these two types of investments. Suppose that state banking laws require that the bank invest at least three times as much money in mortgages as in business loans. How much money should be allotted to each type of investment in order to earn as much as possible?

Solution

The variables are x_1, the amount of money the bank should invest in mortgages, and x_2, the amount in business loans. For each $1 million the bank invests in mortgages, it will earn 8% of that amount or, in decimal form, 0.08 times the amount invested. Thus, from $\$x_1$ million invested in mortgages the bank will earn $\$0.08x_1$ million. Similarly, the earnings from $\$x_2$ million of business loans will be $0.10x_2$. Therefore, the objective function is

$$\text{Total Earnings} = z = 0.08x_1 + 0.10x_2$$

Using $1 million as the unit of measurement, the fact that the bank has $50 million to invest means that
$$x_1 + x_2 \leq 50$$
The legal requirement that the bank invest at least three times as much in mortgages as in business loans will be easier to write in symbols if we express this in a different way. The law says that even if the amount invested in business loans is multiplied by 3, that amount, $3x_2$, must still be less than x_1, the amount invested in mortgages. Therefore, the requirement is
$$3x_2 \leq x_1$$
This inequality can be put in a more familiar form by subtracting x_1 from both sides:
$$-x_1 + 3x_2 \leq 0$$
The bank cannot invest negative amounts of money; so, $x_1 \geq 0$ and $x_2 \geq 0$. Thus, the mathematical problem is to find x_1 and x_2 that

Maximize: Total Earnings = $z = 0.08x_1 + 0.10x_2$
Subject to: $x_1 + x_2 \leq 50$
$-x_1 + 3x_2 \leq 0$
$x_1, x_2 \geq 0$ ■

Summary

A **linear polynomial** is an expression of the form
$$a_1x_1 + a_2x_2 + \ldots + a_nx_n$$
a **linear equation** is an equation that can be written as
$$a_1x_1 + a_2x_2 + \ldots + a_nx_n = b$$
and **linear inequalities** have the form
$$a_1x_1 + a_2x_2 + \ldots + a_nx_n \geq b$$
$$a_1x_1 + a_2x_2 + \ldots + a_nx_n \leq b$$
where $x_1, x_2, \ldots, x_n$ are **variables**, and $a_1, a_2, \ldots, a_n$ and b are constants. A **solution** of a system of inequalities is a set of values for the variables for which the inequalities are true when we substitute them into the equations. The set of all solutions of the system of linear inequalities that describes the restrictions of a linear-programming problem is called the **feasible set** and a solution in that set is called a **feasible solution**. An expression of the form
$$z = c_1x_1 + c_2x_2 + \ldots + c_nx_n$$
in which the values of the **dependent variable** z are computed by substituting values for the **independent variables** $x_1, x_2, \ldots, x_n$ is called a **linear function**. If we represent a set of numbers on the number line, then the number furthest to the right, if there is one, is called the **maximum** of the set and the number furthest to the left, if there is one, is the **minimum**. A linear-programming

problem (more briefly, an **LP-problem**) consists of determining either the maximum or the minimum value of a linear function

$$z = c_1 x_1 + c_2 x_2 + \ldots + c_n x_n$$

called the **objective function**, where the values $x_1, x_2, \ldots, x_n$ must be solutions of a given system of linear inequalities.

Exercises

In Exercises 1 and 2, determine whether each choice of values of the variables is a solution of the given system of linear inequalities.

1. $2x_1 + 3x_2 \leq 9$
   $4x_1 - x_2 \leq 7$
   (a) $x_1 = 2, x_2 = 1$
   (b) $x_1 = 1, x_2 = 2$
   (c) $x_1 = 2, x_2 = 0$

2. $-x_1 + 2x_2 + x_3 \geq 1$
   $2x_1 + x_2 \geq 2$
   $3x_2 - x_3 \geq 0$
   (a) $x_1 = 1, x_2 = 2, x_3 = 1$
   (b) $x_1 = 1, x_2 = 0, x_3 = 1$
   (c) $x_1 = 2, x_2 = 1, x_3 = 2$

In Exercises 3 through 6, determine whether each choice of values of the variables is a feasible solution of the given system of linear inequalities for an LP-problem.

3. $x_1 + x_2 \leq 4$
   $2x_1 + 3x_2 \leq 6$
   $x_1, x_2 \geq 0$
   (a) $x_1 = 1, x_2 = 1$
   (b) $x_1 = 3, x_2 = 0$
   (c) $x_1 = 1, x_2 = 2$
   (d) $x_1 = 2, x_2 = -1$

4. $2x_1 + x_2 \geq 3$
   $-x_1 + 2x_2 \geq 0$
   $x_1 \geq 1$
   $x_2 \geq 0$
   (a) $x_1 = 1, x_2 = 2$
   (b) $x_1 = 1.5, x_2 = 0$
   (c) $x_1 = 2, x_2 = 1.5$
   (d) $x_1 = 0.5, x_2 = 2$

5. $3x_1 + x_2 \geq 2$
   $x_1 + x_3 \geq 0$
   $x_1, x_2, x_3 \geq 0$
   (a) $x_1 = 1, x_2 = -1, x_3 = 0$
   (b) $x_1 = 2, x_2 = 1, x_3 = 1$
   (c) $x_1 = 0, x_2 = 1, x_3 = 1$
   (d) $x_1 = 1.2, x_2 = 0.5, x_3 = 1.5$

6. $3x_1 + x_2 + x_3 \leq 10$
   $2x_1 - x_2 + 2x_3 \leq 8$
   $-4x_1 + x_3 \leq 5$
   $x_1, x_2, x_3 \geq 0$
   (a) $x_1 = 9, x_2 = 0.5, x_3 = 0.5$
   (b) $x_1 = 1, x_2 = 9, x_3 = 0$
   (c) $x_1 = 0, x_2 = 0, x_3 = 0$
   (d) $x_1 = 3, x_2 = 2, x_3 = 2$

For each of the linear functions of Exercises 7 through 9, calculate the values of the dependent variable corresponding to the given values of the independent variables.

7. $z = 3x_1 + 2x_2$
   (a) $x_1 = 2, x_2 = 1$
   (b) $x_1 = 1, x_2 = 0.5$
   (c) $x_1 = 0, x_2 = 2$

8. $z = x_1 + x_2 + 3x_3$
   (a) $x_1 = 2, x_2 = 1, x_3 = 1$
   (b) $x_1 = 1.5, x_2 = 0, x_3 = 2$
   (c) $x_1 = 1, x_2 = 2, x_3 = 3$

128 LINEAR PROGRAMMING

9. $z = 120x_1 + 15x_2 + 45x_3$
   (a) $x_1 = 1, x_2 = 0, x_3 = 2$
   (b) $x_1 = 10, x_2 = 100, x_3 = 20$
   (c) $x_1 = 1, x_2 = 2, x_3 = 3$

In Exercises 10 through 14, express the linear-programming problems in mathematical terms by means of a linear objective function and linear inequalities. Do not attempt to answer the questions.

10. A woman has no more than 7 hours a week to devote to exercise. She plans to jog and to ride her bicycle. She will jog at least 1 hour a week. Since she prefers bicycling, she wishes to spend at least twice as much time at that activity as she does at jogging. If jogging consumes 600 calories an hour and bicycling 350 calories an hour, how many hours should she devote to each activity in order to use up as many calories as possible while satisfying her other requirements?

11. A man promises to bake chocolate cookies and vanilla cookies for a charity bake sale. He agrees to bake at least five batches of cookies altogether and to bake at least one more batch of chocolate cookies than vanilla cookies. The man discovers that the cost of sugar will make his charitable offer pretty expensive. If a batch of chocolate cookies requires 1.25 cups of sugar and a batch of vanilla cookies requires 1 cup of sugar, how many batches of each type of cookie should he bake in order to keep his promise, and yet use as little sugar as possible?

12. A farmer's cattle are fed a mixture of two types of feed—type A and type B. One unit of type-A feed supplies a steer with 25% of its minimum daily requirement of carbohydrates and 10% of its minimum daily requirement of protein. One unit of type-B feed supplies the steer with 10% of its minimum daily requirement of carbohydrates and 50% of its requirement of protein. In order to be sure that the steer gets certain vitamins that are not present in type-A feed, the farmer must give the steer at least 1 unit of type-B feed. If 1 unit of type-A feed costs $0.10 and 1 unit of type-B feed costs $0.15, how much of each type should the farmer give a steer each day so that he gets at least 100% of his minimum daily requirement of carbohydrates and protein while at the same time, the feeding costs the farmer as little as possible?

13. A farmer has 310 acres of land on which she wishes to grow potatoes and barley. Because the price of potatoes is uncertain, she decides to devote no more than 100 acres to this crop. To grow potatoes requires 15 inches of water per acre, whereas barley needs 6 inches of water per acre. The farmer has 2000 acre-inches of water for irrigation. If the farmer's profit is estimated to be $200 an acre for potatoes and $30 an acre for barley, how many acres should she devote to each crop in order to make the maximum profit?

14. A manufacturer produces one type of chair and one type of table. The factory consists of a machine shop with 15 employees and an assembly-and-finishing line with 25 employees. Each employee works 8 hours a day. It requires 2 work-hours of machine-shop labor and 1.5 hours of assembly-and-finishing labor to produce a chair, and it takes 1.75 work-hours of machine-shop labor and 5 work-hours of assembly-and-finishing labor to make a table. The manufacturer will produce in a

day no more chairs and tables than can be sold to wholesalers, namely, no more than 60 chairs and 45 tables. If the profit is $75 for each chair and $100 for each table, how many chairs and how many tables should the manufacturer produce each day in order to make the maximum profit?

2.2 The Geometry of Linear Programming

If an LP-problem involves only the variables x_1 and x_2, as in the examples of the preceding section, we can solve the problem by drawing a picture of it. For problems involving more variables, the method of solution depends on matrix techniques. The pictures in this section will help us understand the general method.

Graphing Linear Inequalities

We have seen that the restrictions on the solution of an LP-problem are expressed as linear inequalities. To use the geometric method, we must be able to picture a linear inequality. Let us begin with the simple inequality $x_2 \geq 0$. When we label the axes of the plane by x_1 and x_2, as in Figure 2-2, the points on the plane are represented by ordered pairs (x_1, x_2) of numbers, called the **coordinates** of the point. Furthermore, the pairs in which $x_2 \geq 0$ represent points either on the horizontal axis, that is, where $x_2 = 0$, or above it, where x_2 is greater than zero (but not equal to it), that is, where $x_2 > 0$. To put it another way, if we replace the linear inequality $x_2 \geq 0$ by the corresponding linear equation $x_2 = 0$, the graph of the equation

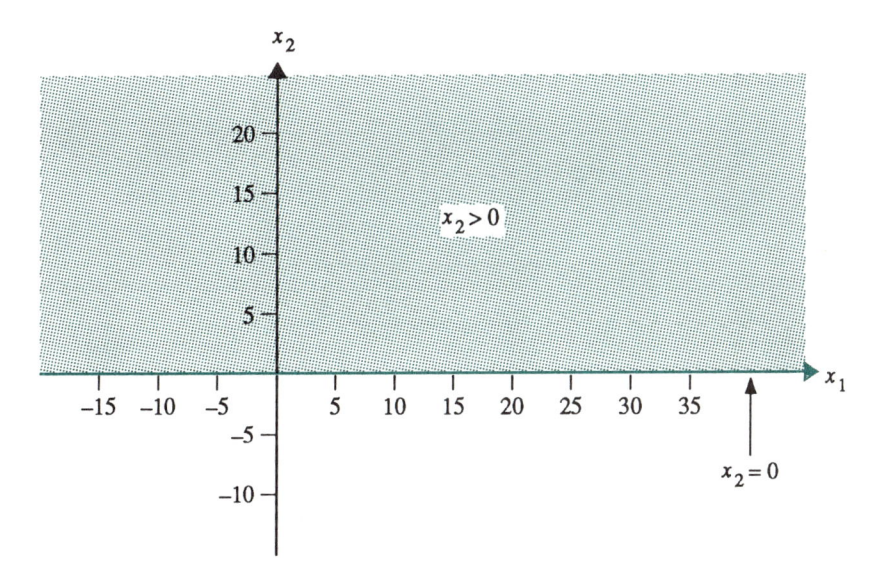

Figure 2-2

(the horizontal axis) divides the plane into two parts: above, where $x_2 > 0$, and below, where $x_2 < 0$. Notice that the set of points in the plane represented by ordered pairs (x_1, x_2) for which the inequality $x_2 \geq 0$ is true, the shaded portion of Figure 2-2, consists of the line representing $x_2 = 0$ together with the part of the plane representing $x_2 > 0$.

Recall that the graph of a linear equation $a_1x_1 + a_2x_2 = b$ consists of all points in the plane whose coordinates are solutions of the equation. Similarly, the **graph** of a linear inequality in the two variables x_1 and x_2, of the form

$$a_1x_1 + a_2x_2 \geq b \quad \text{or} \quad a_1x_1 + a_2x_2 \leq b$$

is the set of points in the plane whose coordinates make the inequality true when we substitute them into the inequality.

Example 1. Which of the points in Figure 2-3 are in the graph of $2x_1 + x_2 \geq 1$?

Solution

The inequality $2x_1 + x_2 \geq 1$ is true for the coordinates $(1, 0)$ of the point A of Figure 2-3 because if we substitute 1 for x_1 and 0 for x_2 on the left side of the equation, then $2(1) + 0 = 2 \geq 1$. Therefore, A is in the graph of $2x_1 + x_2 \geq 1$. Calculating that

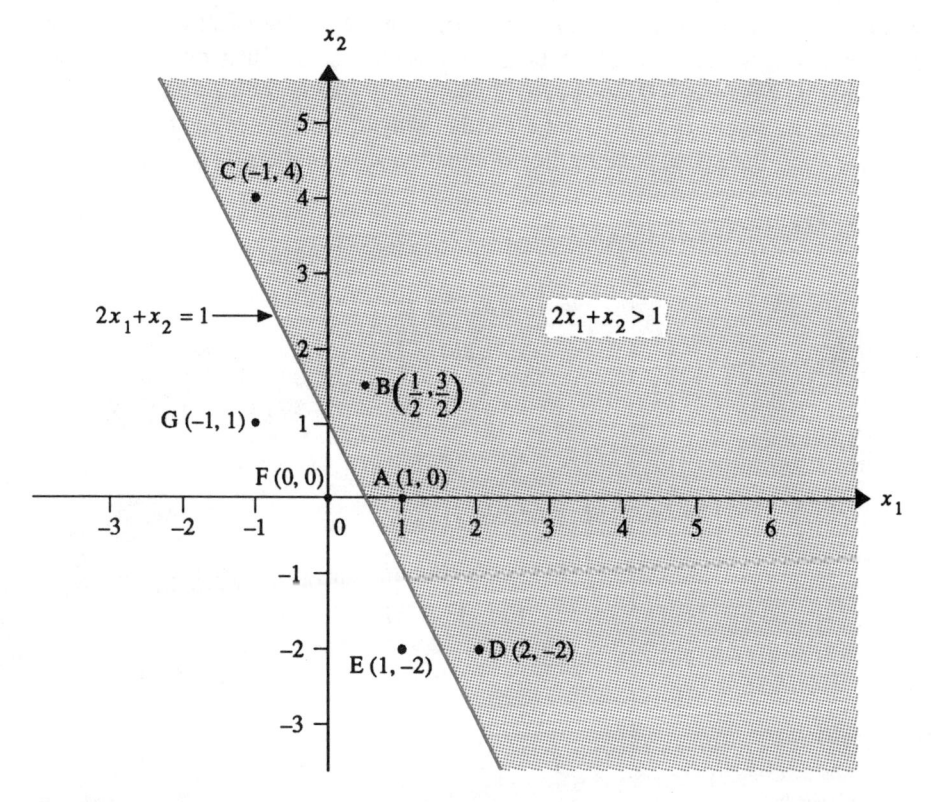

Figure 2-3

$2(\frac{1}{2}) + \frac{3}{2} = \frac{5}{2} \geq 1$ shows that B is also in the graph. You can check that points C and D are also in the graph of $2x_1 + x_2 \geq 1$. On the other hand, the point E with coordinates $(1, -2)$ does not belong to the graph of $2x_1 + x_2 \geq 1$ because $2(1) + (-2) = 0$ is not greater than or equal to 1. Neither do the points F and G lie in the graph. Notice that A, B, C, and D (the points that lie in the graph of $2x_1 + x_2 \geq 1$) are all on one side of the line that is the graph of the corresponding linear equation $2x_1 + x_2 = 1$, whereas the points E, F, and G are on the other side. ∎

We cannot picture the graph of $2x_1 + x_2 \geq 1$ just by testing random points of the plane, but Example 1 and our earlier discussion of the graph of $x_2 \geq 0$ suggest what the graph of the linear inequality should look like. First we graph the *linear equation* $2x_1 + x_2 = 1$ corresponding to the inequality, as in the first chapter, by locating two points on the graph for convenient values of x_1 and then drawing the straight line through them. Taking $x_1 = 0$, we see that $x_2 = 1$; and taking $x_1 = 1$, then x_2 must satisfy $2(1) + x_2 = 1$, so that $x_2 = -1$. We draw the line through $(0, 1)$ and $(1, -1)$. That line separates the plane into two parts. It is not hard to believe that *all* the points in the part containing the points A, B, C, and D have coordinates (x_1, x_2) for which the inequality $2x_1 + x_2 > 1$ is true. Thus, the graph of $2x_1 + x_2 \geq 1$ is the portion of the plane indicated by the shaded part of Figure 2-3.

In general, to find the graph of a linear inequality $a_1x_1 + a_2x_2 \geq b$ or $a_1x_1 + a_2x_2 \leq b$, we first draw the line that is the graph of the corresponding *linear equation* $a_1x_1 + a_2x_2 = b$. The line separates the rest of the plane into two parts, called **half-planes**. One half-plane consists of all points represented by pairs (x_1, x_2) for which

$$a_1x_1 + a_2x_2 < b$$

and the other part consists of the pairs (x_1, x_2) with

$$a_1x_1 + a_2x_2 > b$$

Therefore, the graph of the linear inequality will consist of the line and one of the half-planes. To find out which half-plane, choose any convenient ordered pair that is not on the line and substitute its values into the linear polynomial of the inequality. If the given inequality is true for that ordered pair, then the half-plane containing that point determines the graph; otherwise, use the other half-plane. For instance, in the example $2x_1 + x_2 \geq 1$, if we test the origin $(0, 0)$, we see that $2(0) + 0 = 0$ is less than 1; so, the graph consists of the line that is the graph of $2x_1 + x_2 = 1$ together with the half-plane that does not contain the origin.

Example 2. Draw the graph of $3x_1 + x_2 \leq 0$.

Solution

As we can see in Figure 2-4, the graph of the corresponding linear equation $3x_1 + x_2 = 0$ goes through the origin and the point representing $(-1, 3)$. We will test the point A with coordinates $(-1, 0)$ to determine which half-plane to use for the graph of the in-

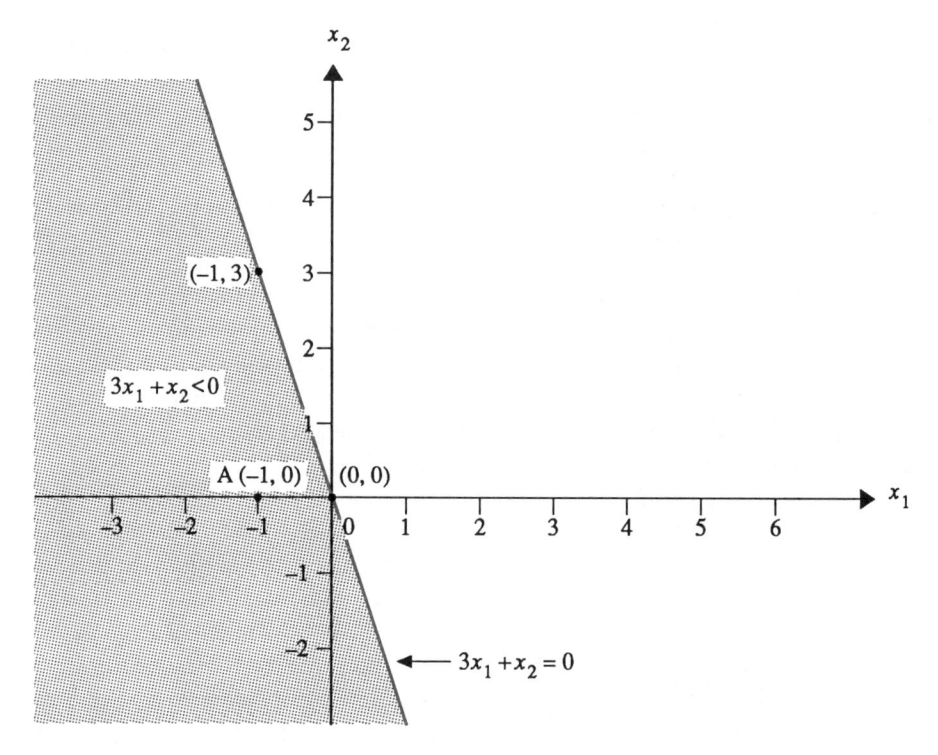

Figure 2-4

equality. When we substitute -1 for x_1 and 0 for x_2 in the linear polynomial $3x_1 + x_2$, we have $3(-1) + 0 = -3$, which is less than zero; so the half-plane containing A is in the graph. ∎

Picturing the Feasible Set

The feasible set of an LP-problem is the set of solutions of a system of linear inequalities. If the LP-problem involves just the variables x_1 and x_2, the feasible set will consist of all the pairs of numbers (x_1, x_2) for which all the inequalities are true. Recall that the solution of a system of two *linear equations* in two unknowns must lie in the graphs of both equations; so, it has to be the point where these two lines intersect. In the same way, since a solution of a system of *linear inequalities* must lie in the graph of every inequality of the system, the feasible set is the intersection of all the graphs.

The *Secretaries and Clerks* problem, Example 4 of the preceding section, has a feasible set determined by the following system of linear inequalities:

$$\begin{aligned} x_1 &\geq 5 \\ 10x_1 + 4x_2 &\geq 100 \\ 20x_1 + 30x_2 &\geq 300 \\ x_2 &\geq 0 \end{aligned}$$

2.2 THE GEOMETRY OF LINEAR PROGRAMMING 133

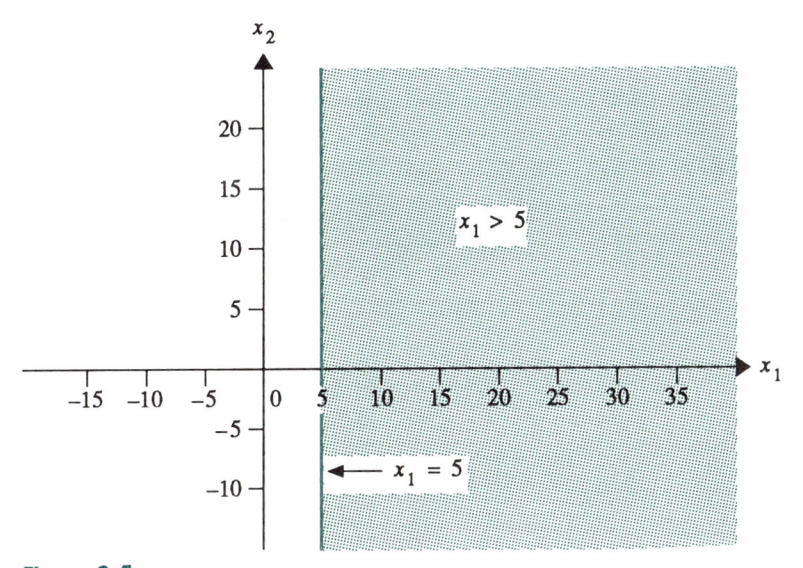

Figure 2-5

In order to draw a picture of the feasible set, we begin with the inequality $x_1 \geq 5$. The corresponding linear equation $x_1 = 5$ is satisfied by a pair of the form $(5, x_2)$, where x_2 is any number; the graph of $x_1 = 5$ is just the vertical line indicated in Figure 2-5 and the graph of $x_1 \geq 5$ consists of all the points on this line or to its right.

Looking back to Figure 2-2 on page 129, which shows the graph of the last inequality in the list on page 132, $x_2 \geq 0$, we intersect the graphs to produce Figure 2-6.

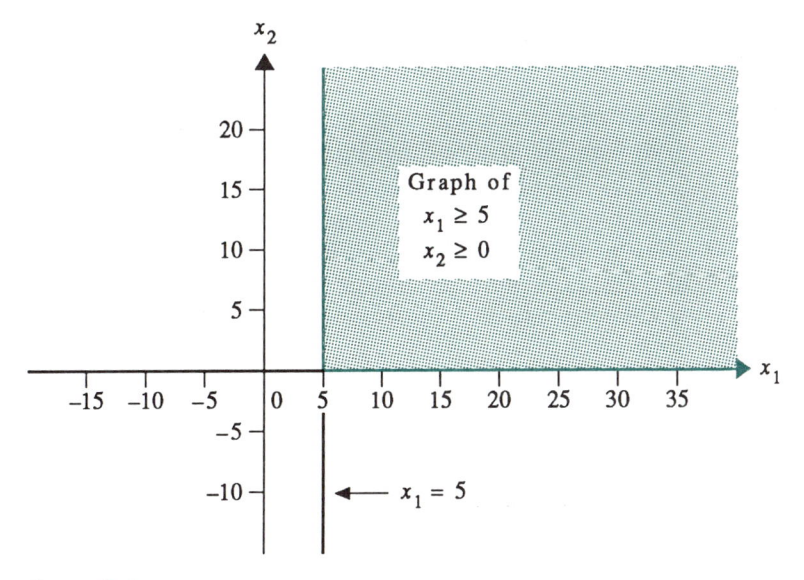

Figure 2-6

134 LINEAR PROGRAMMING

To graph $10x_1 + 4x_2 \geq 100$, we first graph the corresponding *linear equation*, which passes through (0, 25) and (10, 0), as we can see in Figure 2-7. Notice that we have superimposed this line on the region shown in Figure 2-6. Since the origin is not on the graph of $10x_1 + 4x_2 \geq 100$, we shade only the other side of the line $10x_1 + 4x_2 = 100$. The shaded portion of Figure 2-7 is the intersection of the graphs of the three linear inequalities,

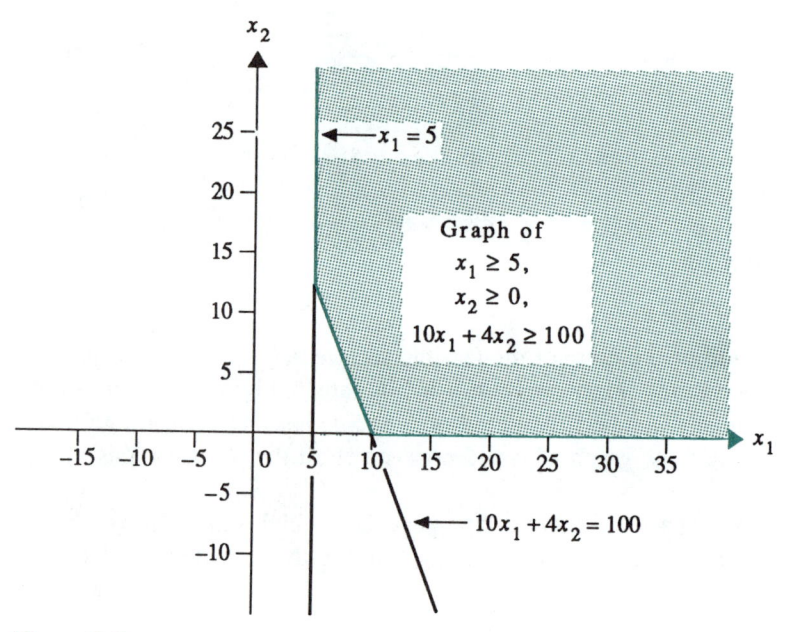

$$x_1 \geq 5, \quad x_2 \geq 0, \quad \text{and} \quad 10x_1 + 4x_2 \geq 100$$

Figure 2-7

For the final step, the intersection with the graph of $20x_1 + 30x_2 \geq 300$ in Figure 2-8, we again draw the graph of the corresponding linear equation on the set constructed up to this point. Once again, we see that the origin does not lie in the graph of the inequality, so we intersect the half-plane *not* containing the origin to complete the construction of the feasible set. In other words, every point in the shaded set of Figure 2-8 corresponds to an ordered pair of numbers whose values satisfy *all four* of the inequalities in the system of the *Secretaries and Clerks* problem.

Visualizing Linear Functions

A linear function, such as the *salary* function

$$z = 11x_1 + 4x_2$$

2.2 THE GEOMETRY OF LINEAR PROGRAMMING   135

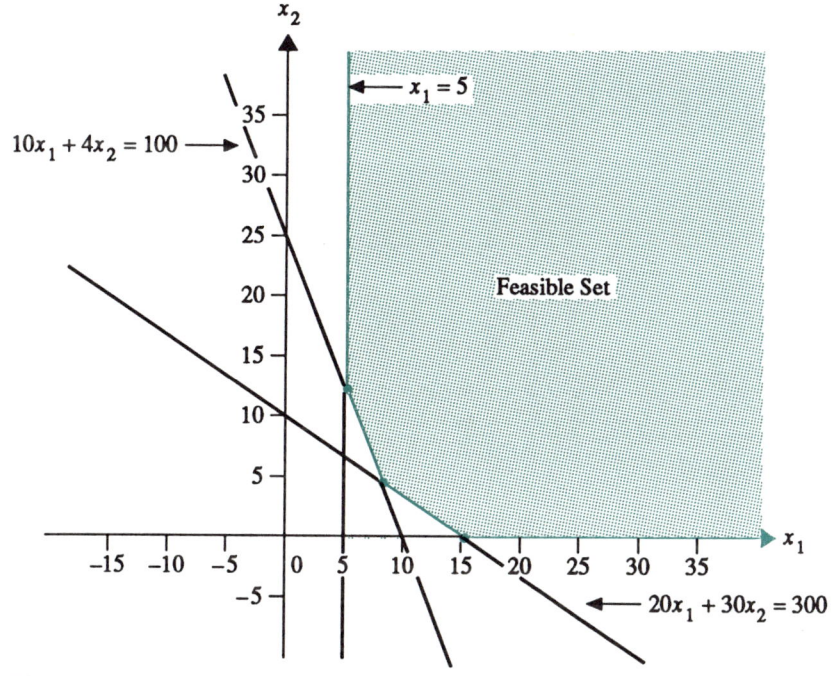

Figure 2-8

of the *Secretaries and Clerks* problem is more difficult to visualize on the plane than a linear equation or a linear inequality. But we can do it well enough to understand the LP-problem. If we choose a particular salary total such as 130, then there are many combinations of the values of x_1 and x_2 that will make $11x_1 + 4x_2$ equal 130. For example, $x_1 = 10$ and $x_2 = 5$, or $x_1 = 5$ and $x_2 = 18.75$, and so on. If we think of x_1 and x_2 as representing abstract numbers rather than numbers of secretaries and clerks, they can take on *any* decimal values, and there are an infinite number of pairs (x_1, x_2) for which $11x_1 + 4x_2 = 130$: those pairs that are represented by the points on the line $11x_1 + 4x_2$ shown in Figure 2-9.

Example 3. Draw the graphs of the following equations (for various salary totals) on one coordinate plane:

$$11x_1 + 4x_2 = 55$$
$$11x_1 + 4x_2 = 80$$
$$11x_1 + 4x_2 = 130$$
$$11x_1 + 4x_2 = 165$$

Solution

See Figure 2-9. ∎

Level Curves

Notice that choosing different total-salary levels in Example 3 produces a family of parallel lines in Figure 2-9. Notice also that as we proceed up and to the right,

136 LINEAR PROGRAMMING

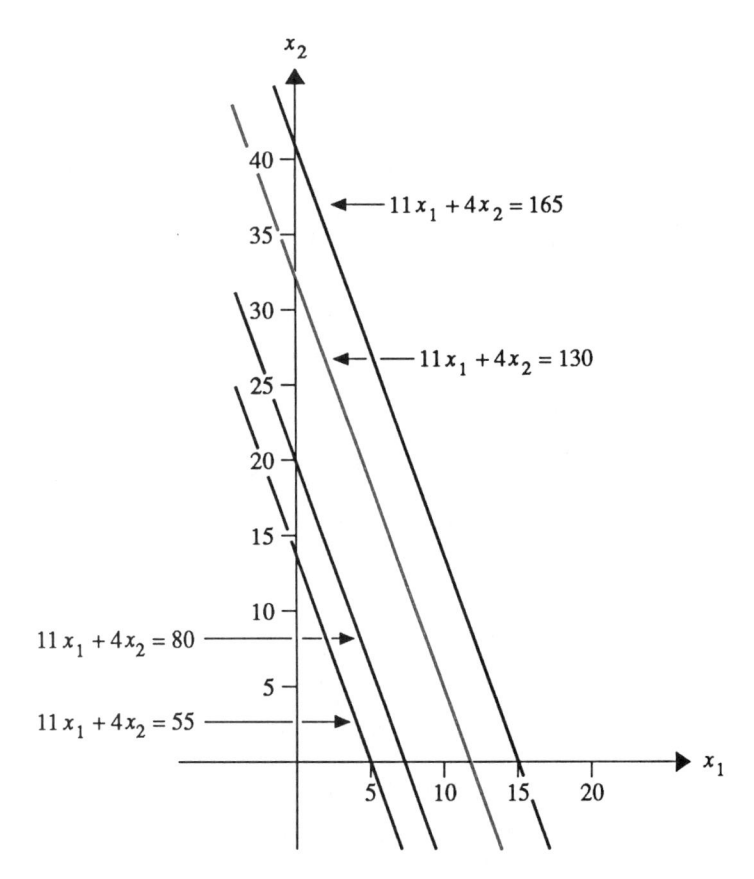

Figure 2-9

the total salary represented by the parallel lines gets larger. Moving down and to the left, the total salary becomes smaller.

The picture of the linear function

$$\text{Salary} = z = 11x_1 + 4x_2$$

in Figure 2-9 is typical of any linear function. A linear function in two independent variables is described by an expression of the form

$$z = c_1x_1 + c_2x_2$$

Choosing a total value d for the function (for instance, $d = 130$ for the total salary), we obtain a linear equation, $c_1x_1 + c_2x_2 = d$, and its graph, a straight line. As we let d vary from one value to another, we produce a family of parallel lines, the graphs of linear equations. These lines are called the **level curves** of the linear function. Let us fix one such value for the function; call it d^*. All lines parallel to the graph of $c_1x_1 + c_2x_2 = d^*$ lie to one side or the other of that line. We know from our previous discussion of the graphing of linear inequalities that all the points on one side have coordinates for which $c_1x_1 + c_2x_2 > d^*$, and for the coordinates of all the points on the other side we have $c_1x_1 + c_2x_2 < d^*$. Thus, all the parallel lines on one side are graphs of $c_1x_1 + c_2x_2 = d$, where d is greater than d^*, whereas all the parallel lines on the other side are graphs of $c_1x_1 + c_2x_2 = d$, where d is less than d^*.

Visualizing a Complete Problem

Now we are in a position to picture an entire LP-problem. Let us consider the *Secretaries and Clerks* problem, Example 4 of Section 2.1, which is to

$$\begin{aligned}
\text{Minimize:} \quad & \text{Salaries} = z = 11x_1 + 4x_2 \\
\text{Subject to:} \quad & x_1 \geq 5 \\
& 10x_1 + 4x_2 \geq 100 \\
& 20x_1 + 30x_2 \geq 300 \\
& x_2 \geq 0
\end{aligned}$$

In Figure 2-10 we superimpose the level curves of the linear function

$$z = 11x_1 + 4x_2$$

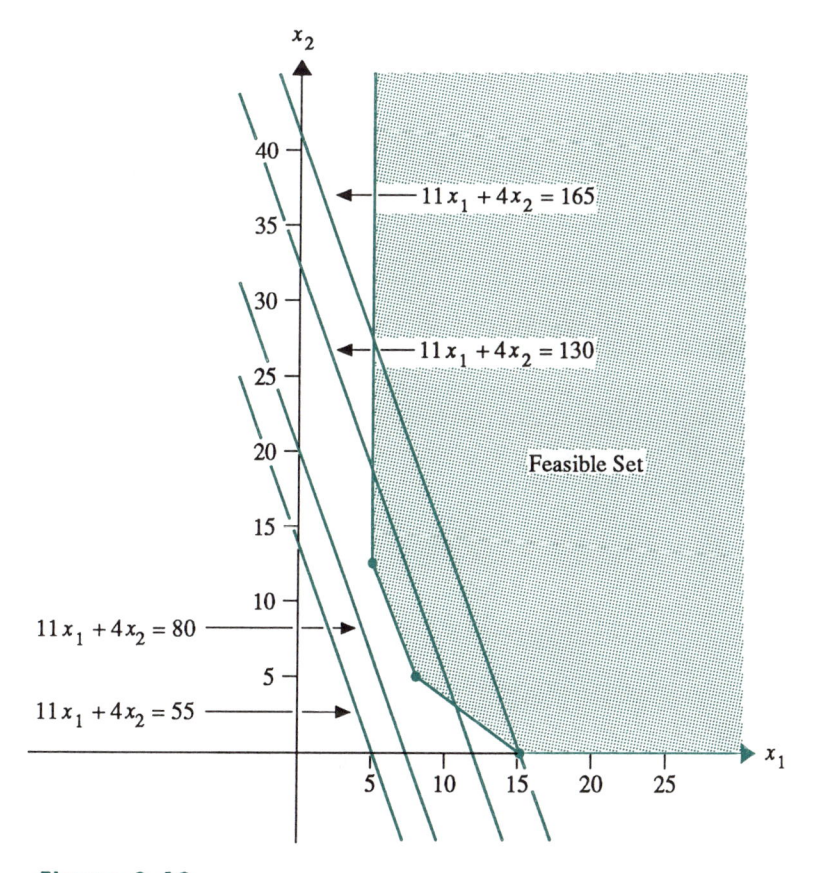

Figure 2-10

of Figure 2-9 on the feasible set of the problem that we drew in Figure 2-8. The part of the level curve $11x_1 + 4x_2 = d$ that lies in the feasible set consists of points whose coordinates satisfy all the conditions of the problem and which give a total salary of d. If a level curve misses the feasible set entirely, then no pair (x_1, x_2) will satisfy the conditions of the problem and produce a salary total equal to the constant d for that line. Thus, for example, no point in the feasible set represents a pair for which $11x_1 + 4x_2 = 80$; so, there is no solution to the LP-problem for which the total salaries will cost the company only $80 an hour.

In general, if a level curve of $c_1x_1 + c_2x_2 = d^*$ intersects the *inside* of the feasible set of the LP-problem, as in Figure 2-11, then the points in the feasible set that lie on the level curve cannot produce either a largest or smallest possible value for the function. The reason is that nearby level curves, on both sides, will also intersect the feasible set; thus, there are feasible solutions (x_1, x_2) for which $z = c_1x_1 + c_2x_2$ is larger than d^* and others for which z is smaller than d^*. However, if a level curve of $c_1x_1 + c_2x_2 = d^*$ hits the feasible set *only at a corner*, as in Figure 2-12, then all the points of the feasible set lie to one side of that line. Therefore, when we substitute the coordinates into $c_1x_1 + c_2x_2$, the sum is either always smaller than d^* or always larger than this value. Therefore, the single point at which the level curve of the equation $c_1x_1 + c_2x_2 = d^*$ intersects the feasible set has coordinates for which the function is either as large or as small as possible.

Location of Solutions

Looking at Figures 2-11 and 2-12, we see that a linear function attains its largest or smallest possible value over this feasible set at a *corner* of the set, that is, at a point at which two edges of the feasible set come together. Such points are called **extreme points** of the feasible set.

It may happen that a level curve contains one entire edge of the feasible set, as in Figure 2-13. In such a case, if the coordinates of one point on that edge solve the LP-problem, then the coordinates of every point on the edge will be a solution. Since, in particular, the extreme points A and B could be used for this purpose, it is still true even in this situation that we need only concern ourselves with extreme points.

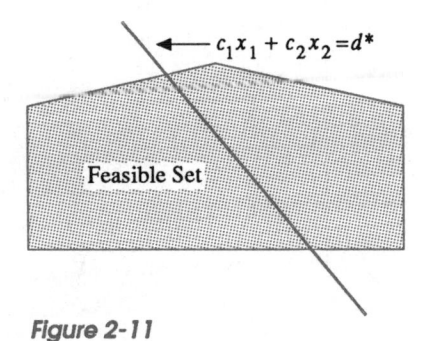

Figure 2-11

2.2 THE GEOMETRY OF LINEAR PROGRAMMING

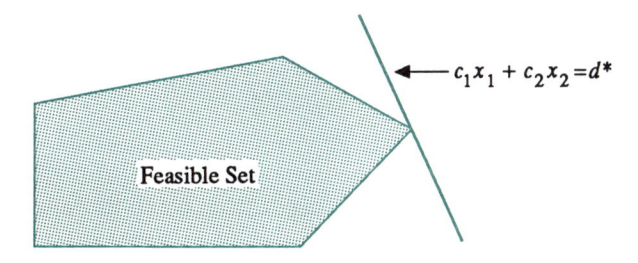

Figure 2-12

Unbounded Feasible Sets

Figures 2-11 through 2-13 have helped us focus our attention on a very important feature of the feasible set: its extreme points. However, these pictures differ from that of the feasible set of the *Secretaries and Clerks* problem (Figure 2-8) in an important way. In Figures 2-11 through 2-13, the feasible set could be pictured entirely, whereas in Figure 2-8 we understand that the feasible set extends indefinitely in the "northeast" direction. We say that the feasible set of the *Secretaries and Clerks* problem is **unbounded**; that is, we cannot draw a (complete) boundary around it. If the feasible set is bounded, as in Figures 2-11 through 2-13 or as in Figure 2-14, for Example 5, then we can find the maximum or the minimum of a linear function on the set, and this will always occur at an extreme point, just as these figures indicate. But when the feasible set is unbounded, such a maximum or minimum may not exist. For instance, suppose we replace the *Total Salary* objective function of the *Secretaries and Clerks* problem by the linear function

$$z = x_1 + x_2$$

that counts the total number of these employees. We now wish to maximize this function. Keeping the same inequalities as before, and therefore the same feasible set, the problem becomes

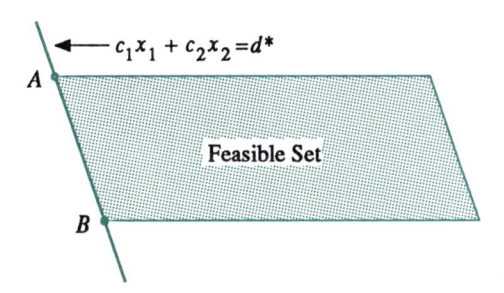

Figure 2-13

Maximize: Employees $= z = x_1 + x_2$
Subject to:
$$x_1 \geq 5$$
$$10x_1 + 4x_2 \geq 100$$
$$20x_1 + 30x_2 \geq 300$$
$$x_2 \geq 0$$

The inequalities keep the number of employees from being too small, but do not prevent x_1 and x_2 from being as large as we like. Thus, there is no restriction on how large $x_1 + x_2$ can get for (x_1, x_2) in the feasible set; this LP-problem has no solution.

What we may still say about LP-problems with unbounded feasible sets is that if there is a solution to the problem, then it will be given by the coordinates of an extreme point of the set. Furthermore, we would expect problems with just two variables that do not have solutions to be artificial: no employer would want to make the total number of employees as large as possible. Therefore, by restricting ourselves to sensible problems, we will avoid problems without solutions in this section.

For more complex LP-problems, with many variables, the matrix technique for analyzing such problems, which we will present in Section 2.5, will permit us to identify unbounded problems that have no solutions.

The Strategy of Solution

Keeping in mind that all of the LP-problems of this section will have solutions, we will take advantage of the fact that the solution is given by the coordinates of an extreme point of the feasible set. We calculate the value of the objective function at each of these coordinates. The solution to the problem is the pair that produces the largest (or, if we are seeking it, the smallest) value of the objective function. Therefore, the solution of an LP-problem involving two variables x_1 and x_2 will consist of two main steps:

1. Graph the feasible set, locate the extreme points, and find their coordinates.
2. Calculate the value of the objective function at each coordinate pair.

The solution of the LP-problem is given by the coordinates that produce the maximum (or minimum, depending on the problem) value in Step 2.

Example 4. Solve the *Secretaries and Clerks* problem (Example 4 of Section 2.1).

Minimize: Salaries $= z = 11x_1 + 4x_2$
Subject to:
$$x_1 \geq 5$$
$$10x_1 + 4x_2 \geq 100$$
$$20x_1 + 30x_2 \geq 300$$
$$x_2 \geq 0$$

Solution

We graphed the feasible set for this problem in Figure 2-8. There is an extreme point A with coordinates $(15, 0)$. There is also an extreme point B at which the graphs of the

2.2 THE GEOMETRY OF LINEAR PROGRAMMING

equations $x_1 = 5$ and $10x_1 + 4x_2 = 100$ intersect. From Chapter 1 we know that the point of intersection is represented by the solution of the system of linear equations:

$$x_1 = 5$$
$$10x_1 + 4x_2 = 100$$

The solution gives the coordinates of the point B. An easy way to solve the system would be to substitute 5 for x_1 in the second equation and solve for x_2. However, we wish to emphasize the relationship between the Gauss-Jordan method of Chapter 1 and LP-problems, because this relationship will be used in later sections to solve more difficult problems. So, we will use the matrix method even though it is less efficient. Starting with the augmented matrix

$$\begin{matrix} & x_1 & x_2 & \\ \end{matrix}$$
$$\begin{bmatrix} 1 & 0 & | & 5 \\ 10 & 4 & | & 100 \end{bmatrix}$$

we use the method of Chapter 1 to calculate that B has coordinates $\left(5, \frac{25}{2}\right)$. The remaining extreme point C is the intersection of the graphs of $10x + 4y = 100$ and $20x + 30y = 300$ and thus is the solution of the system

$$10x + 4y = 100$$
$$20x + 30y = 300$$

We solve the system to calculate that C has coordinates $\left(\frac{90}{11}, \frac{50}{11}\right)$. Substituting each pair (x_1, x_2) into the expression $11x_1 + 4x_2$, we find

A: $11(15) + 4(0) = 165$

B: $11(5) + 4\left(\frac{25}{2}\right) = 105$

C: $11\left(\frac{90}{11}\right) + 4\left(\frac{50}{11}\right) = \frac{1190}{11} = 108$ (approximately)

Therefore, the solution is $\left(5, \frac{25}{2}\right)$. That is, if the company hires 5 secretaries and $\frac{25}{2} = 12.5$ clerks (say, twelve full-time clerks and one half-time clerk), then the required amount of work will be done at the lowest possible hourly cost to the company, $105. Note that even if the pair $\left(\frac{90}{11}, \frac{50}{11}\right)$ had been the solution to the problem, we could have interpreted it in terms of full-time and part-time employees and the cost as the average total salaries per hour. ∎

Example 5. Solve the simplified *Farmer's* problem of Section 2.1, which in mathematical form was given by

Maximize: Total Profit $= z = 200x_1 + 29x_2$

Subject to: $x_1 + x_2 \leq 150$
$12x_1 + 6x_2 \leq 1200$
$x_1, x_2 \geq 0$

Solution

The feasible set is not difficult to construct; see Figure 2-14. From this figure we can see that the extreme points are the origin, A, the points B with coordinates $(100, 0)$ and C with coordinates $(0, 150)$, and the point D where the graphs of the linear equations $x_1 + x_2 = 150$ and $12x_1 + 6x_2 = 1200$ intersect. Thus, to find the coordinates of D, we must solve the system

$$x_1 + x_2 = 150$$
$$12x_1 + 6x_2 = 1200$$

Applying the technique of Chapter 1 to the augmented matrix

$$\begin{matrix} x_1 & x_2 & \\ \begin{bmatrix} 1 & 1 & | & 150 \\ 12 & 6 & | & 1200 \end{bmatrix} \end{matrix}$$

we obtain the solution $x_1 = 50$, $x_2 = 100$. Therefore, the extreme point D has coordinates $(50, 100)$.

The profits at the extreme points are:

A: $200(0) + 29(0) = 0$
B: $200(100) + 29(0) = 20,000$
C: $200(0) + 29(150) = 4350$
D: $200(50) + 29(100) = 12,900$

Thus, the most profitable plan for the farmer is to plant only potatoes, even though the lack of water limits him to just 100 acres of that crop. The total profit is then $20,000. ∎

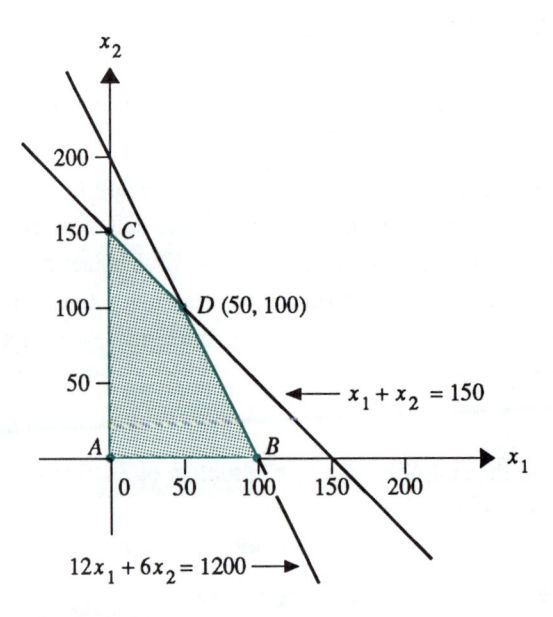

Figure 2-14

Summary

A point on the plane can be represented by an ordered pair of numbers, called the **coordinates** of that point. The **graph** of a linear inequality $a_1x_1 + a_2x_2 \leq b$ or $a_1x_1 + a_2x_2 \geq b$ is the set of points in the plane whose coordinates make the inequality true when we substitute them into the inequality. The line that is the graph of the corresponding linear equation $a_1x_1 + a_2x_2 = b$ separates the rest of the plane into two parts, called **half-planes**, and the graph of a linear inequality consists of this line and one of the half-planes. For a linear function described by an expression of the form $z = c_1x_1 + c_2x_2$, choosing a value d for the function gives a linear equation $c_1x_1 + c_2x_2 = d$. Letting d vary produces a family of parallel lines that are the graphs of linear equations, the **level curves** of the linear function. A point at which two edges of the feasible set meet is called an **extreme point**. If an LP-problem has a solution, that solution is given by the coordinates of an extreme point of the feasible set. To solve an LP-problem involving two variables x_1 and x_2, first graph the feasible set, locate the extreme points, and find their coordinates. Then calculate the value of the objective function at each coordinate pair. A solution is given by the coordinates (x_1, x_2) that produce the maximum or minimum value (depending on the problem).

Exercises

1. Determine whether each of the following pairs are the coordinates of points on the graph of the inequality $x_1 + 3x_2 \geq 4$.

   (a) $(1, 2)$
   (b) $(-1, 2)$
   (c) $(-1, -2)$
   (d) $(0, 0)$
   (e) $(1, 1)$
   (f) $(-2, 2)$

2. Determine whether each of the following pairs are the coordinates of points on the graph of the inequality $-3x_1 + x_2 \leq 2$.

   (a) $(1, 2)$
   (b) $(-1, 2)$
   (c) $(-1, -2)$
   (d) $(0, 0)$
   (e) $(1, 5)$
   (f) $(1, 0)$

In Exercises 3 through 11, draw the graphs of the given linear inequalities.

3. $x_1 \leq -1$
4. $2x_1 - x_2 \geq 0$
5. $-3x_1 + 2x_2 \leq 2$
6. $2x_1 - x_2 \geq -1$
7. $x_2 \geq -3$
8. $x_1 - x_2 \leq 1$
9. $-2x_1 - x_2 \leq 10$
10. $2x_1 + 4x_2 \leq 10$
11. $3x_1 \geq x_2$

12. Graph the points on the plane with coordinates satisfying both the inequality $x_1 \leq 4$ and the inequality $x_1 + x_2 \geq 2$.

144   LINEAR PROGRAMMING

13. Graph the points on the plane with coordinates satisfying the inequalities $x_1 \geq 0$, $2x_1 + x_2 \leq 3$, and $x_1 - 2x_2 \geq 0$.

14. Graph the points on the plane with coordinates satisfying the inequalities $x_1 \geq 0$, $x_2 \geq 0$, and $2x_1 - 3x_2 \leq 4$.

15. Graph the points on the plane with coordinates satisfying the inequalities $x_1 \geq 0$, $x_2 \geq 0$, $x_1 - x_2 \leq 1$, and $x_1 + 3x_2 \leq 3$.

In Exercises 16 through 20, draw the feasible set for the given system of linear inequalities.

16. $x_1 + x_2 \geq 2$
    $x_1 + 3x_2 \geq 3$
    $x_1, x_2 \geq 0$

17. $x_1 + x_2 \leq 4$
    $6x_1 + x_2 \leq 6$
    $x_1, x_2 \geq 0$

18. $6x_1 + 5x_2 \geq 30$
    $x_1 + 6x_2 \geq 6$
    $x_1, x_2 \geq 0$

19. $10x_1 + x_2 \geq 5$
    $3x_1 + 2x_2 \geq 6$
    $x_2 \geq 2$
    $x_1 \geq 0$

20. $x_1 + 2x_2 \leq 4$
    $-x_1 + x_2 \leq 1$
    $x_1 \leq 2$
    $x_1, x_2 \geq 0$

21. Graph the level curves of the linear function
$$z = 10x_1 + 7x_2$$
corresponding to the values $d = 30, 35, 40, 50$, and 55.

22. Graph the level curves of the linear function
$$z = 3x_1 + x_2$$
corresponding to the values $d = 0, 2, 4, 8$, and 12.

23. Graph the feasible set of the following LP-problem and draw the level curves corresponding to the values $d = 0, 3, 5$, and 8 on the same figure.

    Maximize: $z = 2x_1 + x_2$
    Subject to: $x_1 \leq 2$
    $x_1 + x_2 \leq 3$
    $x_1, x_2 \geq 0$

24. Graph the feasible set of the following LP-problem and draw the level curves corresponding to the values $d = 5, 7, 9, 14$, and 25 on the same figure.

    Minimize: $z = x_1 + 2x_2$
    Subject to: $2x_1 + x_2 \geq 10$
    $x_1 + x_2 \geq 7$
    $x_1, x_2 \geq 0$

In Exercises 25 through 28, solve the given LP-problem.

25. Maximize: $z = x_1 + 2x_2$
    Subject to: $x_1 + x_2 \leq 2$
    $10x_1 + x_2 \leq 10$
    $x_1, x_2 \geq 0$

26. Maximize: $z = 3x_1 + x_2$
    Subject to: $x_1 \leq 3$
    $x_1 + 5x_2 \leq 5$
    $x_1, x_2 \geq 0$

27. Minimize: $z = 2x_1 + 3x_2$
    Subject to: $x_1 + x_2 \geq 3$
    $4x_1 + x_2 \geq 4$
    $x_1 + 4x_2 \geq 4$
    $x_1, x_2 \geq 0$

28. Minimize: $z = 10x_1 + 5x_2$
    Subject to: $x_1 + x_2 \geq 5$
    $10x_1 + x_2 \geq 10$
    $x_1 + 15x_2 \geq 15$
    $x_1, x_2 \geq 0$

29. A copper company has two smelters. The smaller one can refine 5000 tons of copper ore per hour and the larger one refines 10,000 tons per hour. Each smelter must operate at least 8 hours a day to justify the investment the company has in the equipment. The company must process at least 150,000 tons of ore a day. If the smaller smelter consumes 6 megawatts of energy per hour and the larger one consumes 11 megawatts, how many hours should each smelter operate each day so as to consume as little energy as possible?

30. A baker plans to bake cakes and cookies. Each cake requires $\frac{5}{2}$ cups of flour and 2 cups of sugar, and each batch of cookies uses 1 cup of flour and $\frac{1}{2}$ cup of sugar. The baker wishes to use no more than 70 cups of flour and 50 cups of sugar in all. If he can sell each cake for $10 and each batch of cookies for $3, how many cakes and how many batches of cookies should he bake in order to make the greatest income?

31. A taxi with a conventional diesel engine is equipped so that it can also burn bottled natural gas. It causes twice as much pollution when it burns diesel fuel as it does when it burns natural gas. The taxi can average 30 miles per hour when it burns diesel fuel, but because of poorer engine performance with natural gas, it averages only 20 miles per hour when using this fuel. The taxi must drive at least 200 miles a day to be profitable. If the driver cannot work more than 10 hours a day, how many hours should she drive using each type of fuel in order to cause as little pollution as possible?

32. Repeat Exercise 31, but now assume that the driver can work at most 8 hours a day.

33. A bank grants mortgages and makes business loans. It has $500 million to invest. Suppose government regulations require the bank to invest at least $3 in mortgages for each $1 in business loans. If the bank charges 8% interest on mortgages and 10% interest on business loans, how much money should it devote to each type of investment in order to make the greatest profit?

34. A company owns 100 acres of land. Some of the land will be developed for low-density (single-family homes) use and some for high-density (apartment building) use. The company estimates its development costs at $0.5 million an acre for low-density use and $4 million an acre for high-density. It can raise a total of $100 million for the project. Zoning regulations require that at least 4 times as much acreage be devoted to low-density use as to high-density. The company expects its net profit to be $50,000 an acre for low-density development and $150,000 an acre for high-density. How many acres should it develop for low-density use and how many for high-density so as to obtain the maximum net profit?

2.3 Matrix Forms and Duality

Systems of Linear Inequalities

We showed in Chapter 1 that the system of linear equations

$$x + 2y = 15$$
$$8x + 20y = 132$$

could be written as the single matrix equation $AX = B$, where

$$A = \begin{bmatrix} 1 & 2 \\ 8 & 20 \end{bmatrix}, \quad X = \begin{bmatrix} x \\ y \end{bmatrix}, \quad \text{and} \quad B = \begin{bmatrix} 15 \\ 132 \end{bmatrix}$$

Observe that by the definition of matrix multiplication,

$$AX = \begin{bmatrix} x + 2y \\ 8x + 20y \end{bmatrix}$$

and the two matrices AX and B are equal because they are the same size and the numbers in each location are equal.

Matrix Inequality

The definition of inequality of matrices is similar to the definition of matrix equality.

> We say that the matrix A is **less than or equal to** the matrix B, written $A \leq B$, if A and B are the same size and if at each location, the entry in A is less than or equal to the corresponding entry in B, that is, $a_{ij} \leq b_{ij}$ for all i and j.

Example 1. Is $A \leq B$ in each of the following cases?

(a) $A = \begin{bmatrix} -1 & 2 & 1 \\ 3 & 0 & 2 \end{bmatrix}$ and $B = \begin{bmatrix} 0 & 3 & 2 \\ 3 & 1 & 2 \end{bmatrix}$

(b) $A = \begin{bmatrix} -1 & -2 \\ 0 & 1 \end{bmatrix}$ and $B = \begin{bmatrix} 0 & 0 & 1 \\ 1 & 2 & 0 \end{bmatrix}$

(c) $A = \begin{bmatrix} 1 & 2 \\ -1 & -2 \end{bmatrix}$ and $B = \begin{bmatrix} 2 & 3 \\ -2 & -1 \end{bmatrix}$

(d) $A = \begin{bmatrix} 1 & 2 \\ 2 & -1 \end{bmatrix}$ and $B = \begin{bmatrix} 1 & 2 \\ 2 & -1 \end{bmatrix}$

(e) $A = \begin{bmatrix} 1 \\ 2 \\ -1 \end{bmatrix}$ and $B = \begin{bmatrix} 2 & 3 & 0 \end{bmatrix}$

Solution

(a) Yes, because both matrices are 2×3 and we can check that $a_{11} = -1 \leq 0 = b_{11}$, $a_{12} = 2 \leq 3 = b_{12}$, $a_{13} = 1 \leq 2 = b_{13}$, $a_{21} = 3 \leq 3 = b_{21}$, $a_{22} = 0 \leq 1 = b_{22}$, and $a_{23} = 2 \leq 2 = b_{23}$.

(b) No. Here, A is 2×2 and B is 2×3 so the matrices are of different sizes.

(c) No, because $a_{21} = -1$ but $b_{21} = -2$. We do not have $a_{21} \leq b_{21}$, as required.

(d) Yes, because if $A = B$, then $A \leq B$. For each i and j, if we have $a_{ij} = b_{ij}$, then it is also true that $a_{ij} \leq b_{ij}$.

(e) No. Here, A is 3×1 and B is 1×3, so the matrices are of different sizes. ■

A matrix A is **greater than or equal to** a matrix B, written $A \geq B$, if A and B are the same size and at each location, the entry in A is greater than or equal to the corresponding entry in B, that is, $a_{ij} \geq b_{ij}$ for all i and j. If for all the numbers in a matrix A we have $a_{ij} \geq 0$, then $A \geq O$, where the zero matrix O is the same size as A.

We now turn to the matrix formulation of some important types of LP-problems.

The *Farmer's* Problem

Recall the mathematical form of the simplified *Farmer's* problem of Section 2.1.

Maximize: Total Profit = $z = 200x_1 + 29x_2$
Subject to:
$$x_1 + x_2 \leq 150$$
$$12x_1 + 6x_2 \leq 1200$$
$$x_1, x_2 \geq 0$$

The farmer's problem represents a very common type of LP-problem. The goal is to make the objective function as large as possible while the inequalities limit the sizes of the independent variables. For the objective function $z = 200x_1 + 29x_2$, the larger x_1 and x_2 are, the greater the total profit. However, the inequality restrictions, other than the standard nonnegativity condition, $x_1, x_2 \geq 0$, prevent x_1 and x_2 from being very large because the restrictions are all of the form $a_1 x_1 + a_2 x_2 \leq b$.

The Objective Function

To define this sort of maximization problem precisely, we will turn once again to matrix language. The first step is to write the objective function in this language. A linear polynomial can be written as a row matrix (of constants) times a column matrix (of variables). For instance, we can define matrices for the objective function of the *Farmer's* problem as follows:

$$C = \begin{bmatrix} 200 & 29 \end{bmatrix} \quad \text{and} \quad X = \begin{bmatrix} x_1 \\ x_2 \end{bmatrix}$$

Then we can write the objective function as

$$\text{Total Profit} = z = CX$$

In general, since every linear function is of the form

$$z = c_1 x_1 + c_2 x_2 + \ldots + c_n x_n$$

if we define the matrices

$$C = \begin{bmatrix} c_1 & c_2 & \cdots & c_n \end{bmatrix} \quad \text{and} \quad X = \begin{bmatrix} x_1 \\ x_2 \\ \cdot \\ \cdot \\ \cdot \\ x_n \end{bmatrix}$$

then we can always write the objective function in matrix form as

$$z = CX$$

Inequalities for the Feasible Set

In the *Farmer's* problem, we notice that the system of linear inequalities that defines the feasible set:

$$\begin{aligned} x_1 + x_2 &\leq 150 \\ 12x_1 + 6x_2 &\leq 1200 \\ x_1, x_2 &\geq 0 \end{aligned}$$

divides naturally into two subsystems. The first two inequalities

$$\begin{aligned} x_1 + x_2 &\leq 150 \\ 12x_1 + 6x_2 &\leq 1200 \end{aligned}$$

can be written in matrix form in the same way that we wrote systems of two linear equations. That is, if we define a coefficient matrix

$$A = \begin{bmatrix} 1 & 1 \\ 12 & 6 \end{bmatrix}$$

then the left-hand side of the system of inequalities is the matrix product AX, where X is the matrix of independent variables used for the objective function.

Defining the matrix of constants on the right-hand side of the system as

$$B = \begin{bmatrix} 150 \\ 1200 \end{bmatrix}$$

these two inequalities can be written as $AX \leq B$. We write the remaining inequalities $x_1, x_2 \geq 0$, which are just the standard restrictions on the independent variables, in matrix form as $X \geq 0$.

Maximization Problems

The type of LP-problem represented by the *Farmer's* problem is called a *maximization problem*.

> A **maximization problem** is an LP-problem that can be written in matrix form as
>
> Maximize: $z = CX$
> Subject to: $AX \leq B$
> $X \geq 0$
>
> where C is a $1 \times n$-matrix of constants, X is the $n \times 1$-matrix of independent variables, A is an $m \times n$-matrix, and therefore, B must be $m \times 1$.

Example 2. Find the matrices X, A, B, C for which the maximization problem

Maximize: $z = 8x_1 + 10x_2 + 4x_3 + 2x_4$
Subject to:
$$\begin{aligned}
x_1 &\leq 20 \\
x_2 + x_3 &\leq 100 \\
x_4 &\leq 150 \\
x_1 + 10x_2 + 3x_3 + 4x_4 &\leq 250 \\
8x_1 \quad\quad + 2x_3 + x_4 &\leq 200 \\
x_1, x_2, x_3, x_4 &\geq 0
\end{aligned}$$

can be written in the form:

Maximize: $z = CX$
Subject to: $AX \leq B$ and $X \geq 0$

Solution

From the objective function we see that we must define

$$C = \begin{bmatrix} 8 & 10 & 4 & 2 \end{bmatrix} \quad \text{and} \quad X = \begin{bmatrix} x_1 \\ x_2 \\ x_3 \\ x_4 \end{bmatrix}$$

and from the inequalities that determine the feasible set,

$$A = \begin{bmatrix} 1 & 0 & 0 & 0 \\ 0 & 1 & 1 & 0 \\ 0 & 0 & 0 & 1 \\ 1 & 10 & 3 & 4 \\ 8 & 0 & 2 & 1 \end{bmatrix} \quad \text{and} \quad B = \begin{bmatrix} 20 \\ 100 \\ 150 \\ 250 \\ 200 \end{bmatrix}$$

■

The *Secretaries and Clerks* Problem

So far in this section, the LP-problems have been of the type in which we make the objective function as large as possible. The other type of LP-problem we discussed in previous sections required the objective function to be made as small as possible. For instance, recall the mathematical form of the *Secretaries and Clerks* problem

Minimize: Salaries = $z = 11x_1 + 4x_2$
Subject to:
$$x_1 \geq 5$$
$$10x_1 + 4x_2 \geq 100$$
$$20x_1 + 30x_2 \geq 300$$
$$x_2 \geq 0$$

Matrix Form of the Problem

We can write the linear objective function as a matrix product $z = CX$, just as before, if we let

$$C = \begin{bmatrix} 11 & 4 \end{bmatrix} \quad \text{and} \quad X = \begin{bmatrix} x_1 \\ x_2 \end{bmatrix}$$

Similarly, defining matrices

$$A = \begin{bmatrix} 1 & 0 \\ 10 & 4 \\ 20 & 30 \end{bmatrix} \quad \text{and} \quad B = \begin{bmatrix} 5 \\ 100 \\ 300 \end{bmatrix}$$

the inequalities that determine the feasible set can be written as $AX \geq B$ and $X \geq 0$ (since $x_1 \geq 5$ implies that $x_1 \geq 0$, as well). Note that the inequality sign in $AX \geq B$ is in the opposite direction from that in a maximization problem like that of the farmer. The goal of this minimization problem is to make the objective function $z = 11x_1 + 4x_2$ as small as possible, and the smaller the (positive) values of x_1 and

x_2, the smaller z becomes. The inequalities limit how small the values of these independent variables can be.

Minimization Problems

> A **minimization problem** is an LP-problem that can be written in matrix form as
> $$\text{Minimize:} \quad z = CX$$
> $$\text{Subject to:} \quad AX \geq B$$
> $$X \geq 0$$

Example 3. Find the matrices X, A, B, C for which the following minimization problem

Minimize: $z = 5x_1 + 10x_2 + 3x_3$

Subject to:
$$x_3 \geq 2$$
$$x_1 + 2x_2 \geq 3$$
$$2x_1 + x_2 + 3x_3 \geq 4$$
$$3x_1 + x_3 \geq 4$$
$$x_1, x_2, x_3 \geq 0$$

can be written in the form:

Minimize: $z = CX$

Subject to: $AX \geq B$ and $X \geq 0$

Solution

From the objective function, we see that we must define

$$C = \begin{bmatrix} 5 & 10 & 3 \end{bmatrix} \quad \text{and} \quad X = \begin{bmatrix} x_1 \\ x_2 \\ x_3 \end{bmatrix}$$

and from the inequalities that determine the feasible set,

$$A = \begin{bmatrix} 0 & 0 & 1 \\ 1 & 2 & 0 \\ 2 & 1 & 3 \\ 3 & 0 & 1 \end{bmatrix} \quad \text{and} \quad B = \begin{bmatrix} 2 \\ 3 \\ 4 \\ 4 \end{bmatrix}$$

∎

The Transpose

The theory of linear programming is based on a relationship between maximization and minimization problems called *duality*. In order to explain duality, we need a new matrix concept.

152 LINEAR PROGRAMMING

> If M is a matrix, then the **transpose** matrix of M, written M^T (*read*: M transpose), is the matrix formed by reversing the row and column locations of the elements of M. That is, the entry m_{ij} in the ith row and jth column of M occupies the jth row and ith column of the matrix M^T.

It is easy to find the transpose when M is a column matrix, such as

$$M = \begin{bmatrix} 2 \\ -1 \\ 3 \\ 0 \end{bmatrix} = \begin{bmatrix} m_{11} \\ m_{21} \\ m_{31} \\ m_{41} \end{bmatrix}$$

Then, according to the definition, M^T will be a row matrix with the entries from top to bottom in M placed from left to right in M^T.

$$M^T = \begin{bmatrix} 2 & -1 & 3 & 0 \end{bmatrix}$$

Similarly, the transpose of a row matrix will be a column matrix. A convenient way to construct the transpose of a general $m \times n$-matrix M is to view M as n column matrices; transpose each column to a row, as we just did, preserving the usual matrix order. That is, the columns from left to right of M form the rows from top to bottom of M^T. Notice that this will make M^T an $n \times m$-matrix.

Example 4. Find the transpose of each of the following matrices.

$$A = \begin{bmatrix} -1 & 2 & 0 & 3 & 1 \end{bmatrix}, \quad B = \begin{bmatrix} 1 & -1 & 2 & 3 \\ 2 & 1 & -2 & 0 \end{bmatrix}, \quad C = \begin{bmatrix} 1 & 2 \\ 2 & 0 \end{bmatrix},$$

$$D = \begin{bmatrix} 0 & 2 & -1 \\ 1 & -1 & -2 \\ 2 & 4 & 0 \\ 3 & 5 & 1 \end{bmatrix}, \quad \text{and} \quad E = \begin{bmatrix} 0 & 1 & 2 & 3 \\ 2 & -1 & 4 & 5 \\ -1 & -2 & 0 & 1 \end{bmatrix}$$

Solution

$$A^T = \begin{bmatrix} -1 \\ 2 \\ 0 \\ 3 \\ 1 \end{bmatrix}, \quad B^T = \begin{bmatrix} 1 & 2 \\ -1 & 1 \\ 2 & -2 \\ 3 & 0 \end{bmatrix}, \quad C^T = \begin{bmatrix} 1 & 2 \\ 2 & 0 \end{bmatrix}, \quad D^T = \begin{bmatrix} 0 & 1 & 2 & 3 \\ 2 & -1 & 4 & 5 \\ -1 & -2 & 0 & 1 \end{bmatrix},$$

$$\text{and} \quad E^T = \begin{bmatrix} 0 & 2 & -1 \\ 1 & -1 & -2 \\ 2 & 4 & 0 \\ 3 & 5 & 1 \end{bmatrix}$$

Matrix C illustrates the fact that a matrix can be its own transpose. The matrices D and E are each the transpose of the other. Thus, if you transpose the matrix D to obtain $D^T = E$ and then transpose E, you have $E^T = D$, which is where you started. This is a general property of the transpose: if you do it twice, you return to the original matrix; in symbols,

$$(M^T)^T = M$$

for any matrix M.

The Dual of a Maximization Problem

We start with a maximization problem in matrix form: maximize: $z = CX$, subject to: $AX \leq B$ and $X \geq 0$. We will use the matrices A, B, and C to construct a minimization problem that is called its "dual" problem. For an example we will use:

Maximize: $\quad z = 2x_1 + 3x_2$
Subject to: $\quad x_1 - 3x_2 \leq 1$
$\quad\quad\quad\quad\quad 2x_1 \quad\quad\quad \leq 3$
$\quad\quad\quad\quad\quad 3x_1 + x_2 \leq 2$
$\quad\quad\quad\quad\quad x_1, x_2 \geq 0$

for which the matrices are

$$X = \begin{bmatrix} x_1 \\ x_2 \end{bmatrix}, \quad C = \begin{bmatrix} 2 & 3 \end{bmatrix}, \quad A = \begin{bmatrix} 1 & -3 \\ 2 & 0 \\ 3 & 1 \end{bmatrix}, \quad \text{and} \quad B = \begin{bmatrix} 1 \\ 3 \\ 2 \end{bmatrix}$$

The matrix C was formed from the constants of the objective function, whereas B came from the constants on the right-hand side of the inequalities that determine the feasible set. In the dual minimization problem, the roles of these two matrices will be reversed. In particular, the objective of the dual problem is: Minimize $u = B^T Y$ so that the constants of B determine the new objective function. The dependent variable is now called u just to distinguish it from the dependent variable z in the original maximization problem. The matrix of independent variables Y is entirely different from the variable matrix X in the maximization problem and may even be of a different size. Since the matrix

$$B^T = \begin{bmatrix} 1 & 3 & 2 \end{bmatrix}$$

contains three elements, the matrix product $u = B^T Y$ only makes sense if Y is a 3×1-matrix, that is,

154 LINEAR PROGRAMMING

$$Y = \begin{bmatrix} y_1 \\ y_2 \\ y_3 \end{bmatrix}$$

The objective function of the dual problem is $u = B^T Y = y_1 + 3y_2 + 2y_3$.

To complete the construction of the dual minimization problem, we use the transpose C^T of the matrix of constants C of the original objective function. Thus, C^T, a column matrix, is the right-hand side of the inequalities that determine the feasible set of the minimization problem. The left-hand side of the system of linear inequalities is $A^T Y$.

The dual of a maximization problem

Maximize: $z = CX$
Subject to: $AX \leq B$ and $X \geq 0$

is the problem

Minimize: $u = B^T Y$
Subject to: $A^T Y \geq C^T$ and $Y \geq 0$

If we apply this definition to the example, since

$$A^T = \begin{bmatrix} 1 & 2 & 3 \\ -3 & 0 & 1 \end{bmatrix} \quad \text{and} \quad C^T = \begin{bmatrix} 2 \\ 3 \end{bmatrix}$$

matrix multiplication gives us the inequalities that determine the feasible set:

$$y_1 + 2y_2 + 3y_3 \geq 2$$
$$-3y_1 + y_3 \geq 3$$
$$y_1, y_2, y_3 \geq 0$$

Example 5. Write the dual of the following LP-problem:

Maximize: $z = 2x_1 - x_2 + x_4$
Subject to:
$$4x_1 + 2x_2 + x_4 \leq 3$$
$$x_1 - 3x_3 \leq 0$$
$$x_1, x_2, x_3, x_4 \geq 0$$

Solution

The matrix form of this problem is:

Maximize: $z = CX$
Subject to: $AX \leq B$ and $X \geq 0$

where

$$C = \begin{bmatrix} 2 & -1 & 0 & 1 \end{bmatrix}, \quad A = \begin{bmatrix} 4 & 2 & 0 & 1 \\ 1 & 0 & -3 & 0 \end{bmatrix}, \quad \text{and} \quad B = \begin{bmatrix} 3 \\ 0 \end{bmatrix}$$

(We can write the objective function as $z = 2x_1 - x_2 + 0x_3 + x_4$.) To form the dual problem, we need the transposes:

$$C^T = \begin{bmatrix} 2 \\ -1 \\ 0 \\ 1 \end{bmatrix}, \quad A^T = \begin{bmatrix} 4 & 1 \\ 2 & 0 \\ 0 & -3 \\ 1 & 0 \end{bmatrix}, \quad \text{and} \quad B^T = \begin{bmatrix} 3 & 0 \end{bmatrix}$$

Since the objective function of the dual $u = B^T Y$ must have as many independent variables as there are constants in B^T, we see that there are just two independent variables, y_1 and y_2. According to the definition, the problem we are seeking has the matrix form:

Minimize: $\quad u = B^T Y$
Subject to: $\quad A^T Y \geq C^T$ and $Y \geq 0$

Multiplying the matrices, we obtain the answer.

Minimize: $\quad u = 3y_1$
Subject to:
$$\begin{aligned} 4y_1 + y_2 &\geq 2 \\ 2y_1 &\geq -1 \\ -3y_2 &\geq 0 \\ y_1 &\geq 1 \\ y_1, y_2 &\geq 0 \end{aligned}$$
■

The Dual of a Minimization Problem

Just as every maximization problem has a dual minimization problem, every problem of the type

Minimize: $\quad z = CX$
Subject to: $\quad AX \geq B$ and $X \geq 0$

has a dual maximization problem. The procedure is really the same: reverse the roles of the matrices of constants B and C, using their transposes, and form the inequalities that determine the feasible set, using the matrix A^T.

The dual of a minimization problem

Minimize: $\quad z = CX$
Subject to: $\quad AX \geq B$ and $X \geq 0$

is the problem:

Maximize: $\quad u = B^T Y$
Subject to: $\quad A^T Y \leq C^T$ and $Y \geq 0$

Again, the size of the matrix Y of independent variables is determined by the size of B^T.

156 LINEAR PROGRAMMING

Example 6. Find the dual of the LP-problem

Minimize: $\quad z = x_1 + x_2 + 4x_3$

Subject to:
$$3x_1 + x_2 + 4x_3 \geq 1$$
$$x_1 + 2x_3 \geq 2$$
$$2x_2 + x_3 \geq 2$$
$$x_1, x_2, x_3 \geq 0$$

Solution

The matrices for this problem are

$$C = \begin{bmatrix} 1 & 1 & 4 \end{bmatrix}, \quad A = \begin{bmatrix} 3 & 1 & 4 \\ 1 & 0 & 2 \\ 0 & 2 & 1 \end{bmatrix}, \quad \text{and} \quad B = \begin{bmatrix} 1 \\ 2 \\ 2 \end{bmatrix}$$

with transposes

$$C^T = \begin{bmatrix} 1 \\ 1 \\ 4 \end{bmatrix}, \quad A^T = \begin{bmatrix} 3 & 1 & 0 \\ 1 & 0 & 2 \\ 4 & 2 & 1 \end{bmatrix}, \quad \text{and} \quad B^T = \begin{bmatrix} 1 & 2 & 2 \end{bmatrix}$$

Therefore, the dual problem is

Maximize: $\quad u = y_1 + 2y_2 + 2y_3$

Subject to:
$$3y_1 + y_2 \leq 1$$
$$y_1 + 2y_3 \leq 1$$
$$4y_1 + 2y_2 + y_3 \leq 4$$
$$y_1, y_2, y_3 \geq 0$$

∎

The Duality Theorem

To form the dual of an LP-problem, the rules are essentially the same no matter whether the original is a maximization or a minimization problem. We interchange the roles of the matrix of constants of the objective function and the matrix of constants on the right-hand side of the inequalities, and we transpose the matrix of coefficients. We also adjust the size of the matrix of independent variables and we change from maximizing the objective function to minimizing it, or the reverse. Looked at in this way, and remembering that $(A^T)^T = A$, it is not hard to believe that if you start with one of these problems and form the dual problem once, then do it a second time, you return to the problem you started with. This property of the duality operation tells us that LP-problems come in pairs, with each member of the pair the dual of the other. The principal theoretical fact in linear programming, the *duality theorem*, describes how the solutions to these pairs of dual problems relate to one another.

We have seen that for every maximization problem

Maximize: $\quad z = CX$

Subject to: $\quad AX \leq B$ and $X \geq 0$

there is a dual problem, namely,

2.3 MATRIX FORMS AND DUALITY

Minimize: $u = B^T Y$

Subject to: $A^T Y \geq C^T$ and $Y \geq 0$

The **Duality Theorem** gives the key relationship between these two problems.

Suppose you compute the maximum numerical value that the linear function $z = CX$ can achieve if the values chosen for the variables in X are in the feasible set of the maximization problem. Then that number is exactly the same as the minimum value that $u = B^T Y$ can attain if the values taken in Y are in the feasible set of the minimization problem.

Moreover, we obtain the same conclusion if we start with a minimization problem and form the dual maximization problem:

The minimum value possible for $z = CX$, if X satisfies the conditions $AX \geq B$ and $X \geq 0$, is the same as the maximum value for $u = B^T Y$, if we require that $A^T Y \leq C^T$ and $Y \geq 0$.

To illustrate the duality theorem, we will look again at the simplified *Farmer's* problem, Example 5 of the preceding section.

Maximize: Total Profit $= z = 200x_1 + 29x_2$

Subject to:
$$x_1 + x_2 \leq 150$$
$$12x_1 + 6x_2 \leq 1200$$
$$x_1, x_2 \geq 0$$

We solved this problem by constructing the feasible set (see Figure 2-15) and calculating

- A: $200(0) + 29(0) = 0$
- B: $200(100) + 29(0) = 20{,}000$
- C: $200(0) + 29(150) = 4350$
- D: $200(50) + 29(100) = 12{,}900$

to see that the maximum profit occurred at the extreme point B, at which the profit was $200(100) + 29(0) = 20{,}000$.

To find the dual of this problem, we write it as a maximization problem:

Maximize: $z = CX$

Subject to: $AX \leq B$ and $X \geq 0$

where

$$C = \begin{bmatrix} 200 & 29 \end{bmatrix}, \quad X = \begin{bmatrix} x_1 \\ x_2 \end{bmatrix}, \quad A = \begin{bmatrix} 1 & 1 \\ 12 & 6 \end{bmatrix}, \quad \text{and} \quad B = \begin{bmatrix} 150 \\ 1200 \end{bmatrix}$$

Since $B^T = [150 \quad 1200]$, the dual problem contains two variables; writing

$$Y = \begin{bmatrix} y_1 \\ y_2 \end{bmatrix}, \quad A^T = \begin{bmatrix} 1 & 12 \\ 1 & 6 \end{bmatrix}, \quad \text{and} \quad C^T = \begin{bmatrix} 200 \\ 29 \end{bmatrix}$$

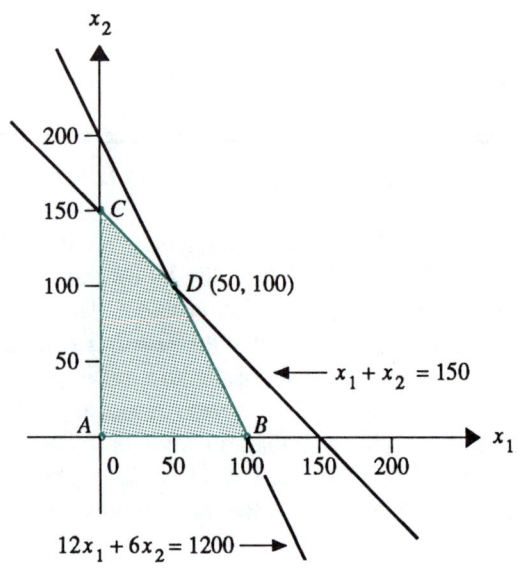

Figure 2-15

we see that the dual problem is

    Minimize: $u = 150y_1 + 1200y_2$
    Subject to: $y_1 + 12y_2 \geq 200$
                         $y_1 + 6y_2 \geq 29$
                         $y_1, y_2 \geq 0$

To solve this problem, as in Section 2.2, we first find the feasible set (see Figure 2-16). The line $y_1 + 6y_2 = 29$ lies outside the feasible set because if positive numbers y_1 and y_2 satisfy the inequality

$$y_1 + 12y_2 \geq 200$$

they must satisfy

$$y_1 + 6y_2 \geq 29$$

as well; so that inequality is not, in fact, a restriction.

To find when $u = 150y_1 + 1200y_2$ will be as small as possible, there are just two extreme points to test:

    A: $150(0) + 1200\left(\dfrac{200}{12}\right) = 20{,}000$
    B: $150(200) + 1200(0) = 30{,}000$

Thus, the minimum value of u is 20,000, just as it was the maximum value of the objective function z of the dual problem.

In the same way, in Example 6, if we find the minimum value that the linear function $z = x_1 + x_2 + 4x_3$ can attain when the variables x_1, x_2, and x_3 are restricted to the feasible set determined by the inequalities of the minimization problem, then that number is equal to the maximum value that the objective function

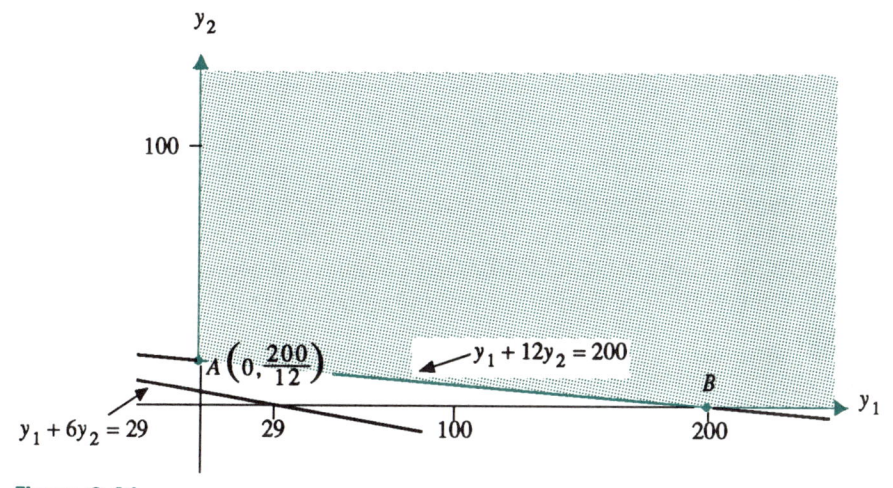

Figure 2-16

$u = y_1 + 2y_2 + 2y_3$ of the dual problem can attain when y_1, y_2, and y_3 are restricted to the feasible set of that dual problem.

Briefly, the duality theorem tells us that if z is the objective function of a minimization problem and u is the objective function of its dual maximization problem, then

$$\text{Min } z = \text{Max } u$$

It also states that if z is the objective function of a maximization problem, then

$$\text{Max } z = \text{Min } u$$

where u is the objective function of the dual minimization problem. In fact, by the Duality Theorem, when we solve a maximization or minimization problem, we are really solving two problems, the one we were given and also its dual. This theorem and the linear-programming theory that produces it have many uses, as we will demonstrate in a later section.

Summary

A matrix A is **less than or equal to** a matrix B, written $A \leq B$, if A and B are the same size and at each location, the entry in A is less than or equal to the corresponding entry in B. A matrix A **is greater than or equal to** a matrix B, written $A \geq B$, if A and B are the same size and at each location, the entry in A is greater than or equal to the corresponding entry in B. A **maximization problem** in matrix form is

Maximize: $\quad z = CX$

Subject to: $\quad AX \leq B$ and $X \geq 0$

A **minimization problem** in matrix form is

> Minimize: $z = CX$
> Subject to: $AX \geq B$ and $X \geq 0$

The **transpose** of a matrix M, written M^T, is the matrix formed by reversing the row and column locations of the entries of M. To construct the transpose of a general $m \times n$-matrix M, view M as n column matrices, transpose each column into a row; then the columns from left to right of M become the rows from top to bottom of M^T. The **dual** of the maximization problem

> Maximize: $z = CX$
> Subject to: $AX \leq B$ and $X \geq 0$

is the problem

> Minimize: $u = B^T Y$
> Subject to: $A^T Y \geq C^T$ and $Y \geq 0$

The **Duality Theorem** states that

$$\text{Max } z = \text{Min } u$$

The **dual** of the minimization problem

> Minimize: $z = CX$
> Subject to: $AX \geq B$ and $X \geq 0$

is the problem

> Maximize: $u = B^T Y$
> Subject to: $A^T Y \leq C^T$ and $Y \geq 0$

In this case, the **Duality Theorem** states that

$$\text{Min } z = \text{Max } u$$

Exercises

Determine whether $A \leq B$, $A \geq B$, or neither is true in each of Exercises 1 through 6.

1. $A = \begin{bmatrix} -1 & 0 \\ 2 & 1 \end{bmatrix}$, $B = \begin{bmatrix} 0 & 1 \\ 2 & -1 \end{bmatrix}$

2. $A = \begin{bmatrix} 5 \\ 4 \\ 0 \end{bmatrix}$, $B = \begin{bmatrix} 1 \\ 2 \\ -1 \end{bmatrix}$

3. $A = \begin{bmatrix} 4 & 5 & 8 \\ 5 & 6 & 7 \end{bmatrix}$, $B = \begin{bmatrix} 2 & 1 \\ 1 & -1 \\ 2 & 0 \end{bmatrix}$

4. $A = \begin{bmatrix} -1 & -2 \\ -3 & 0 \end{bmatrix}$, $B = \begin{bmatrix} 0 & 0 \\ 0 & 0 \end{bmatrix}$

2.3 MATRIX FORMS AND DUALITY   161

5. $A = \begin{bmatrix} 1 & 2 \\ -1 & -1 \\ 1 & 3 \end{bmatrix}$, $B = \begin{bmatrix} 1 & 2 \\ -1 & -2 \\ 1 & -3 \end{bmatrix}$ 6. $A = \begin{bmatrix} 0 & 0 & 0 \end{bmatrix}$, $B = \begin{bmatrix} 1 & 2 & 3 \end{bmatrix}$

In Exercises 7 through 12, find the transpose of the given matrix.

7. $\begin{bmatrix} -1 & -2 & 0 & \frac{1}{2} \end{bmatrix}$

8. $\begin{bmatrix} 1 & 2 & 1 \\ 0 & 1 & 1 \end{bmatrix}$

9. $\begin{bmatrix} 1 & 2 \\ -1 & 0 \\ 1 & 2 \\ 2 & -1 \end{bmatrix}$

10. $\begin{bmatrix} 0 & 1 & 2 \\ 1 & 0 & 3 \\ 2 & 3 & 4 \end{bmatrix}$

11. $\begin{bmatrix} 1 \\ 2 \\ -2 \end{bmatrix}$

12. $\begin{bmatrix} 1 & 2 & 3 & 4 \\ -3 & 2 & 1 & -1 \\ -3 & 2 & 1 & -1 \end{bmatrix}$

In Exercises 13 through 19, find the matrices X, A, B, and C for the matrix form of the LP-problem, but do not attempt to solve the problem.

13. Maximize: $z = 3x_1 + 4x_2$
    Subject to: $x_1 + x_2 \leq 4$
    $6x_1 + x_2 \leq 6$
    $x_1, x_2 \geq 0$

14. Minimize: $z = x_1 + 8x_2$
    Subject to: $6x_1 + 5x_2 \geq 30$
    $x_1 + 6x_2 \geq 6$
    $x_1, x_2 \geq 0$

15. Maximize: $z = 5x_1 + 9x_2$
    Subject to: $2x_1 + 3x_2 \leq 10$
    $x_1 + 2x_2 \leq 7$
    $3x_1 + x_2 \leq 12$
    $x_1, x_2 \geq 0$

16. Minimize: $z = 40x_1 + 35x_2$
    Subject to: $x_1 + x_2 \geq 12$
    $3x_1 + 2x_2 \geq 8$
    $5x_1 + 6x_2 \geq 15$
    $x_1, x_2 \geq 0$

17. Maximize: $z = 10x_1 + 12x_2 + 5x_3$
    Subject to: $x_1 + x_2 + x_3 \leq 100$
    $2x_1 + x_2 \leq 40$
    $2x_2 + x_3 \leq 60$
    $x_1, x_2, x_3 \geq 0$

18. Minimize: $z = 50x_1 + 150x_2 + 3000x_3$
    Subject to: $x_1 + x_2 + 30x_3 \geq 207$
    $x_2 + 15x_3 \geq 200$
    $x_1, x_2, x_3 \geq 0$

19. Maximize: $z = 2x_1 + 3x_2 + x_3 + x_4$
    Subject to: $x_1 \leq 12$
    $x_2 - x_3 \leq 0$
    $x_2 + x_3 + x_4 \leq 15$
    $x_1, x_2, x_3, x_4 \geq 0$

In Exercises 20 through 23, find the matrices $X, A, B,$ and C for the matrix form of the LP-problem, but do not attempt to solve the problem.

20. Suppose that an astronaut's ration is made up of three types of food concentrates; call them type x, type y, and type z. One unit of type x concentrate weighs 1 ounce and contains $\frac{1}{4}$ of the astronaut's minimum daily requirement (MDR) of protein, $\frac{1}{8}$ MDR of carbohydrates, and $\frac{1}{4}$ MDR of calories. One unit of type y concentrate weighs 2 ounces and contains $\frac{1}{8}$ MDR of protein, $\frac{1}{4}$ MDR of carbohydrates, and $\frac{1}{4}$ MDR of calories. One unit of type z concentrate weighs $\frac{3}{2}$ ounces and contains $\frac{3}{8}$ MDR of protein, $\frac{1}{8}$ MDR of carbohydrates, and $\frac{3}{8}$ MDR of calories. The ration must contain enough food for 3 days—that is, at least 3 MDR of protein, 3 MDR of carbohydrates, and 3 MDR of calories. How many units of each type of concentrate should be used in the ration in order to meet the nutritional requirements, but make the total weight as small as possible?

21. A university wishes to invest its $100 million endowment so as to make a good return on its money without taking too many risks. It decides to invest in mutual funds and municipal bonds, while keeping at least $15 million in savings accounts. The university estimates it can earn 9% a year from mutual funds, 7.5% a year from bonds, and 5% from savings accounts. Because of the risk associated with mutual funds (which invest in the stock market), the university decides to invest no more in mutual funds than half the amount it invests in bonds, and also to put no more in mutual funds than it puts in savings accounts. How much money should the university invest in mutual funds, bonds, and savings accounts in order to make the largest possible return on its endowment?

22. A jeweler makes rings, earrings, pins, and necklaces. He wishes to work no more than 40 hours a week. It takes him 1 hour to make a ring, $\frac{3}{2}$ hours to make a pair of earrings, $\frac{1}{2}$ hour to make a pin, and 2 hours to make a necklace. He estimates that he can sell no more than ten rings, ten pairs of earrings, fifteen pins, and three necklaces each week. The jeweler charges $50 for a ring, $80 for a pair of earrings, $30 for a pin, and $200 for a necklace. How many rings, earrings, pins, and necklaces should he make in order to maximize his earnings?

23. Suppose that the jeweler in Exercise 22 decides that if he can bring in $1500 a week, then he can afford to stay in business (that is, pay his expenses, buy materials, and make enough profit to live on). He must make at least 5 each of rings, earrings, and pins in order to have enough variety of stock in his store. How many rings, earrings, pins, and necklaces should he make so that he can stay in business, but work as few hours as possible?

In Exercises 24 through 31, write the dual of the LP-problem as a problem of maximizing or minimizing a linear function, subject to the restrictions given by the indicated linear inequalities.

24. Maximize: $z = 10x_1 + 7x_2$
    Subject to: $3x_1 + x_2 \leq 8$
    $2x_1 \leq 5$
    $x_1 + 3x_2 \leq 7$
    $x_1, x_2 \geq 0$

25. Minimize: $z = 2x_1 + 5x_2$
    Subject to: $x_1 + 4x_2 \geq 5$
    $3x_1 + x_2 \geq 4$
    $3x_1 \geq 2$
    $x_1 + x_2 \geq 1$
    $x_1, x_2 \geq 0$

26. Minimize: $z = 4x_1 + 5x_2 + x_3$
    Subject to: $x_1 + x_2 + x_3 \geq 2$
    $2x_1 + 3x_2 \geq 5$
    $3x_1 - x_3 \geq 0$
    $x_1, x_2, x_3 \geq 0$

27. Maximize: $z = 3x_2 + 4x_3$
    Subject to: $2x_1 + x_2 + 3x_3 \leq 9$
    $x_1 + 4x_2 \leq 5$
    $x_1, x_2, x_3 \geq 0$

28. Minimize: $z = 7x_1 + x_2 + x_3 + 2x_4$
    Subject to: $x_1 + x_3 - x_4 \geq 1$
    $2x_1 + x_2 + x_4 \geq 3$
    $2x_2 + x_3 \geq 4$
    $x_1, x_2, x_3, x_4 \geq 0$

29. Maximize: $z = x_1 + 2x_2 + 3x_5$
    Subject to: $x_1 + x_2 + x_3 + x_4 + x_5 \leq 50$
    $2x_1 + 3x_3 - 2x_4 + 5x_5 \leq 100$
    $x_1, x_2, x_3, x_4, x_5 \geq 0$

30. Maximize: $z = 120x_1 + 5x_2 - x_3$
    Subject to: $x_1 + x_2 + 3x_3 \leq 10$
    $x_1 - 5x_3 \leq 0$
    $2x_2 + x_3 \leq 8$
    $x_1, x_2, x_3 \geq 0$

31. Minimize: $z = x_1$
    Subject to: $-x_1 + 2x_2 + 3x_3 \geq 15$
    $3x_1 + x_2 + x_3 \geq 12$
    $x_1 - 4x_2 - 2x_3 \geq 0$
    $3x_1 + 2x_2 + x_3 \geq 9$
    $x_1, x_2, x_3 \geq 0$

2.4 Pivoting and Bases

Solving Linear Systems

In Chapter 1 we used row operations on matrices to solve systems of linear equations, such as

$$3x + y - 2z = -3$$
$$4x + 2y - 2z = -3$$
$$x + z = 1$$

164   LINEAR PROGRAMMING

by the Gauss-Jordan method. The method that we will present to solve LP-problems with more than two variables is also based on these row operations. It will be easier to explain the linear-programming method if we look at the procedure of Chapter 1 in a new way.

Remember how the Gauss-Jordan method solves the preceding system. First we form the augmented matrix

$$\begin{array}{ccc} x & y & z \end{array}$$
$$\begin{bmatrix} 3 & 1 & -2 & | & -3 \\ 4 & 2 & -2 & | & -3 \\ 1 & 0 & 1 & | & 1 \end{bmatrix}$$

Next we divide the first row by 3 so that the number 3 in the diagonal location is changed to a 1:

$$\begin{bmatrix} 1 & \frac{1}{3} & -\frac{2}{3} & | & -1 \\ 4 & 2 & -2 & | & -3 \\ 1 & 0 & 1 & | & 1 \end{bmatrix}$$

Then we wish to change the 4 below the diagonal entry to 0; so, we multiply the first row by -4 and add that to the second row.

$$\begin{array}{r} \begin{bmatrix} 4 & 2 & -2 & | & -3 \end{bmatrix} \text{ (second row)} \\ + \begin{bmatrix} -4 & -\frac{4}{3} & \frac{8}{3} & | & 4 \end{bmatrix} \text{ (}-4\text{ times first row)} \\ \hline \begin{bmatrix} 0 & \frac{2}{3} & \frac{2}{3} & | & 1 \end{bmatrix} \text{ (sum)} \end{array}$$

We substitute the sum for the second row of the augmented matrix:

$$\begin{bmatrix} 1 & \frac{1}{3} & -\frac{2}{3} & | & -1 \\ 0 & \frac{2}{3} & \frac{2}{3} & | & 1 \\ 1 & 0 & 1 & | & 1 \end{bmatrix}$$

To complete the process of changing the first column of the augmented matrix to that of a 3-by-3 identity matrix, we add -1 times the first row to the third row, thus changing the 1 in the a_{31}-location to a 0:

$$\begin{bmatrix} 1 & \frac{1}{3} & -\frac{2}{3} & | & -1 \\ 0 & \frac{2}{3} & \frac{2}{3} & | & 1 \\ 0 & -\frac{1}{3} & \frac{5}{3} & | & 2 \end{bmatrix}$$

2.4 PIVOTING AND BASES

Now, before going further, let us take a closer look at how the 2 in the a_{22}-location of the original matrix was changed to its present value of $\frac{2}{3}$. We added $-\frac{4}{3}$ to obtain

$$2 + \left(-\frac{4}{3}\right) = \frac{2}{3}$$

Suppose we write the left-hand side of the computation as a single fraction in this way:

$$2 + \left(-\frac{4}{3}\right) = \frac{3}{3} \cdot 2 - \frac{4}{3} = \frac{3 \cdot 2 - 4 \cdot 1}{3}$$

Notice that the numbers in this fraction all appear in the upper left-hand corner of the original augmented matrix:

$$\begin{bmatrix} 3 & 1 & \cdot & | & \cdot \\ 4 & 2 & \cdot & | & \cdot \\ \cdot & \cdot & \cdot & | & \cdot \end{bmatrix}$$

The Pivoting Formula

To understand why the same numbers appear in the fraction, we must examine the steps we have just carried out to convert the first column of the augmented matrix so that it became the first column of the 3-by-3 identity matrix.

We employ the terminology that is used in linear programming. The sequence of matrix operations that converts a matrix column to the form of a column of an identity matrix is called a **pivot**. The element of the matrix that is converted to the single 1 in the column is called the **pivot element**, the row in which it lies is the **pivot row**, and its column is the **pivot column**. The effect of the pivot on the pivot column is to change it to a column of an identity matrix.

Using the Gauss-Jordan method, we divide each number in the pivot row by the pivot element so that the pivot element becomes a 1. To see the effect of the pivot on the rest of the matrix, we take any element of the matrix not in either the pivot row or pivot column, call it our "target," and use the symbol t for it. We call the pivot element p. Let a be the element in the pivot column that lies in the same row as the target. We also need to keep track of the element in the pivot row that is in the same column as t, which we call b. These four numbers form a box within the matrix, for instance:

$$\begin{bmatrix} p & b & \cdot & | & \cdot \\ a & t & \cdot & | & \cdot \\ \cdot & \cdot & \cdot & | & \cdot \end{bmatrix}$$

To see the effect of the Gauss-Jordan method, we first divide the pivot row by p and note the effect on this portion of the matrix:

$$\begin{bmatrix} 1 & \frac{b}{p} & \cdot & | & \cdot \\ a & t & \cdot & | & \cdot \\ \cdot & \cdot & \cdot & | & \cdot \end{bmatrix}$$

166 LINEAR PROGRAMMING

Then we change a to 0 by adding $-a$ times the pivot row to the row containing a.

$$\begin{array}{c} \begin{bmatrix} a & t & . & | & . \end{bmatrix} \text{(target row)} \\ \begin{bmatrix} -a & (-a)\frac{b}{p} & . & | & . \end{bmatrix} \text{($-a$ times pivot row)} \\ \hline \begin{bmatrix} 0 & t+(-a)\frac{b}{p} & . & | & . \end{bmatrix} \text{(sum)} \end{array}$$

This changes the target t to an expression that we rewrite as a single fraction:

$$t + (-a)\frac{b}{p} = \frac{p}{p}(t) + (-a)\frac{b}{p} = \frac{pt-ab}{p}$$

Since in the original matrix we had

$$\begin{bmatrix} p & b & . & | & . \\ a & t & . & | & . \\ . & . & . & | & . \end{bmatrix} = \begin{bmatrix} 3 & 1 & . & | & . \\ 4 & 2 & . & | & . \\ . & . & . & | & . \end{bmatrix}$$

we see that the target, $t = 2$, changes to

$$\frac{pt-ab}{p} = \frac{3 \cdot 2 - 4 \cdot 1}{3} = \frac{2}{3}$$

just as before.

A sequence of similar steps, applied to each number of the original matrix, would complete the pivot operation with the same result:

$$\begin{bmatrix} 1 & \frac{1}{3} & -\frac{2}{3} & | & -1 \\ 0 & \frac{2}{3} & \frac{2}{3} & | & 1 \\ 0 & -\frac{1}{3} & \frac{5}{3} & | & 2 \end{bmatrix}$$

However, by analyzing what we have just done, we can simplify the process.

Pivoting Without Fractions

Although the coefficients of the original problem were all integers, that is, whole numbers, we introduced fractions into the matrix as soon as we divided the first row by the pivot element. In order to change the first column to the form of a column of the identity matrix, we introduced fractions into every row. We could eliminate all the fractions from the matrix by multiplying each row by the same number: the pivot element, $p = 3$. Recall from Chapter 1 that multiplying or dividing a row by a nonzero constant does not change the solutions of the system of linear equations. In the resulting matrix

$$\begin{bmatrix} 3 & 1 & -2 & | & -3 \\ 0 & 2 & 2 & | & 3 \\ 0 & -1 & 5 & | & 6 \end{bmatrix}$$

the first column still contains only a single nonzero number, though it is no longer a 1. We will see that this does not significantly effect our ability to read off the solution of the linear system when we have completed the matrix operations.

If we compare this matrix to the original augmented matrix of the system, we observe that the first row, that is, the pivot row, is exactly the same as at the start. Using the method of Chapter 1, we divide the pivot row by p; so, multiplying that row by p brings us back to where we started. The rest of the pivot column still consists of zeros. To see what has happened to a target element t in neither the pivot row nor the pivot column, as before, change t to the fraction

$$\frac{pt - ab}{p}$$

In order to eliminate fractions, we multiply every element by the pivot p; that cancels the denominator p, and we see that t has been changed to the simpler expression $pt - ab$. For instance, in the case of $t = a_{33} = 1$, we have

$$\begin{bmatrix} p & . & b & . \\ . & . & . & . \\ a & . & t & . \end{bmatrix} = \begin{bmatrix} 3 & . & -2 & . \\ . & . & . & . \\ 1 & . & 1 & . \end{bmatrix}$$

and t becomes

$$pt - ab = 3(1) - 1(-2) = 5$$

just as before.

Rules for Pivoting

We now summarize the rules for pivoting without introducing fractions; we assume that all the numbers in the original augmented matrix are integers.

To pivot at the pivot element $a_{ij} = p$:

1. Copy the pivot row into the new matrix without change.

2. Change all other numbers in the pivot column to zeros.

3. For a target t in neither the pivot row nor the pivot column, if a is the entry in the target row and pivot column and b is the entry in the pivot row and target column, change t to $pt - ab$.

In Step 3, it does not matter which number you call a and which b, so long as a, b, p, and t together form a "box" within the matrix. The formula just says to multiply pivot and target together and then subtract the product of the two other corners of the box.

168   LINEAR PROGRAMMING

To practice using these rules, we start with the matrix

$$\begin{bmatrix} 3 & 1 & -2 & -3 \\ 0 & 2 & 2 & 3 \\ 0 & -1 & 5 & 6 \end{bmatrix}$$

in which we have completed the pivot in the a_{11}-location. We will use this matrix to pivot at a_{22}. We begin by carrying out Steps 1 and 2, with the result:

$$\begin{bmatrix} \cdot & 0 & \cdot & \cdot \\ 0 & 2 & 2 & 3 \\ \cdot & 0 & \cdot & \cdot \end{bmatrix}$$

The missing numbers are the targets. Looking back to the matrix, we start applying Step 3, with our target $t = 3$ in the a_{11}-location. In that matrix we have a box

$$\begin{bmatrix} t & b & \cdot & \cdot \\ a & p & \cdot & \cdot \\ \cdot & \cdot & \cdot & \cdot \end{bmatrix} = \begin{bmatrix} 3 & 1 & \cdot & \cdot \\ 0 & 2 & \cdot & \cdot \\ \cdot & \cdot & \cdot & \cdot \end{bmatrix}$$

so, $t = 3$ changes to

$$tp - ab = 3 \cdot 2 - 0 \cdot 1 = 6$$

For target $t = -2$ in the a_{13}-location, the picture is

$$\begin{bmatrix} \cdot & a & t & \cdot \\ \cdot & p & b & \cdot \\ \cdot & \cdot & \cdot & \cdot \end{bmatrix} = \begin{bmatrix} \cdot & 1 & -2 & \cdot \\ \cdot & 2 & 2 & \cdot \\ \cdot & \cdot & \cdot & \cdot \end{bmatrix}$$

and t becomes

$$tp - ab = (-2)2 - 2(1) = -6$$

We continue in this way, always referring back to the matrix

$$\begin{bmatrix} 3 & 1 & -2 & -3 \\ 0 & 2 & 2 & 3 \\ 0 & -1 & 5 & 6 \end{bmatrix}$$

until we reach the a_{34}-location:

$$\begin{bmatrix} \cdot & \cdot & \cdot & \cdot \\ \cdot & p & \cdot & b \\ \cdot & a & \cdot & t \end{bmatrix} = \begin{bmatrix} \cdot & \cdot & \cdot & \cdot \\ \cdot & 2 & \cdot & 3 \\ \cdot & -1 & \cdot & 6 \end{bmatrix}$$

where $t = 6$ becomes

$$pt - ab = 2(6) - (-1)3 = 15$$

We have completed the pivot in the a_{22}-location and the matrix is:

$$\begin{bmatrix} 6 & 0 & -6 & | & -9 \\ 0 & 2 & 2 & | & 3 \\ 0 & 0 & 12 & | & 15 \end{bmatrix}$$

We will use this matrix to pivot once more, now in the a_{33}-location. The result is

$$\begin{bmatrix} 72 & 0 & 0 & | & -18 \\ 0 & 24 & 0 & | & 6 \\ 0 & 0 & 12 & | & 15 \end{bmatrix}$$

Remembering that we began with the augmented matrix of a system of linear equations in the unknowns x, y, and z, the matrix

$$\begin{array}{ccc} x & y & z \end{array}$$
$$\begin{bmatrix} 72 & 0 & 0 & | & -18 \\ 0 & 24 & 0 & | & 6 \\ 0 & 0 & 12 & | & 15 \end{bmatrix}$$

is the augmented matrix of the system

$$\begin{aligned} 72x & & & = -18 \\ & 24y & & = 6 \\ & & 12z & = 15 \end{aligned}$$

The solution is

$$x = -\frac{18}{72} = -\frac{1}{4}, \quad y = \frac{6}{24} = \frac{1}{4}, \quad \text{and} \quad z = \frac{15}{12} = \frac{5}{4}$$

Since pivoting is based on the same operations as the Gauss-Jordan method, these values of x, y, and z are also the solution of the original system

$$\begin{aligned} 3x + y - 2z &= -3 \\ 4x + 2y - 2z &= -3 \\ x \quad\quad + z &= 1 \end{aligned}$$

as we can check by substitution.

Solving Linear Systems

The method we used to solve this system of linear equations can be used to analyze any such system. The rules are just a condensed version of the Gauss-Jordan method. Moving by columns from left to right, if the diagonal element a_{ii} is nonzero, pivot in that location. If $a_{ii} = 0$, interchange the ith row with a row beneath it that contains a nonzero element in the ith column; then use the *new* a_{ii} as the pivot element. If there is no such nonzero element, pretend that column does not exist and go on to the next column to the right. When you are no longer

170 LINEAR PROGRAMMING

able to continue, you will be able to analyze the resulting system of linear equations just as you did in Chapter 1, to determine whether there is a single solution, infinitely many solutions, or no solution at all.

The use of a uniform formula to change the target elements when pivoting is very well suited to computer solution of linear systems and to the solution of linear-programming problems. However, the particular formula we are using, that avoids fractions, is suitable only for solving small problems because, otherwise, the integers involved can easily become unmanageably large.

Example 1. Find a solution of the following system of linear equations, if it has one.

$$\begin{aligned} 2w - 2x - y + z &= 1 \\ -w + 2x - 2z &= -1 \\ 3w - 2x - 2y &= 0 \end{aligned}$$

Solution

Beginning with the augmented matrix

$$\begin{array}{cccc} w & x & y & z \end{array}$$
$$\left[\begin{array}{cccc|c} 2 & -2 & -1 & 1 & 1 \\ -1 & 2 & 0 & -2 & -1 \\ 3 & -2 & -2 & 0 & 0 \end{array}\right]$$

we pivot at the a_{11}-location:

$$\left[\begin{array}{cccc|c} 2 & -2 & -1 & 1 & 1 \\ 0 & 2 & -1 & -3 & -1 \\ 0 & 2 & -1 & -3 & -3 \end{array}\right]$$

Then we use that matrix to pivot at a_{22}:

$$\begin{array}{cccc} w & x & y & z \end{array}$$
$$\left[\begin{array}{cccc|c} 4 & 0 & -4 & -4 & 0 \\ 0 & 2 & -1 & -3 & -1 \\ 0 & 0 & 0 & 0 & -4 \end{array}\right]$$

Since the last row corresponds to the equation $0w + 0x + 0y + 0z = -4$, that is, to the false statement $0 = -4$, we see that this system has no solution. ∎

Even with quite small problems, our pivoting method can quickly lead to whole numbers that are awkwardly large, especially if there is no calculator available. If this happens, a simple extra step will bring the numbers down to a convenient size, as we now demonstrate.

Example 2. Find a solution of the following system of linear equations, if it has one.

$$\begin{aligned} 2x + 3y &= 1 \\ 4x + 5y + z &= 0 \\ 3x - 2z &= -1 \\ 3x - y - z &= -3 \end{aligned}$$

Solution

Beginning with the augmented matrix

$$\begin{array}{ccc} x & y & z \end{array}$$
$$\begin{bmatrix} 2 & 3 & 0 & | & 1 \\ 4 & 5 & 1 & | & 0 \\ 3 & 0 & -2 & | & -1 \\ 3 & -1 & -1 & | & -3 \end{bmatrix}$$

we pivot at the a_{11}-location:

$$\begin{bmatrix} 2 & 3 & 0 & | & 1 \\ 0 & -2 & 2 & | & -4 \\ 0 & -9 & -4 & | & -5 \\ 0 & -11 & -2 & | & -9 \end{bmatrix}$$

Pivoting this matrix in the a_{22}-location brings us to the matrix

$$\begin{bmatrix} -4 & 0 & -6 & | & 10 \\ 0 & -2 & 2 & | & -4 \\ 0 & 0 & 26 & | & -26 \\ 0 & 0 & 26 & | & -26 \end{bmatrix}$$

At this point, you might not like the arithmetic involved in pivoting at the 26 in the a_{33}-location. In that case, recalling that each row of the matrix represents an equation, we know we can divide any row without changing the solution. In particular, we can divide the third row (and the fourth as well) by 26. While we are at it, we notice that all the numbers in the first two rows are even, so we can simplify the arithmetic further by dividing each of those rows by 2. We then have the equivalent matrix

$$\begin{bmatrix} -2 & 0 & -3 & | & 5 \\ 0 & -1 & 1 & | & -2 \\ 0 & 0 & 1 & | & -1 \\ 0 & 0 & 1 & | & -1 \end{bmatrix}$$

We use *this* matrix for the remaining pivot, in the a_{33}-location:

$$\begin{array}{ccc} x & y & z \end{array}$$
$$\begin{bmatrix} -2 & 0 & 0 & | & 2 \\ 0 & -1 & 0 & | & -1 \\ 0 & 0 & 1 & | & -1 \\ 0 & 0 & 0 & | & 0 \end{bmatrix}$$

which represents the equations

$$-2x = 2, \quad -y = -1, \quad z = -1, \quad \text{and} \quad 0 = 0$$

Therefore, the system has a single solution:

$$x = -1, \quad y = 1, \quad z = -1$$ ∎

172 LINEAR PROGRAMMING

We will be applying the pivot operations to LP-problems; we now return to the terminology of that subject with variables represented by $x_1, x_2,$ and so on.

Example 3. Find a solution of the following system of linear equations, if it has one.

$$3x_1 + 2x_2 - x_3 + x_4 = 1$$
$$-2x_1 - x_2 + 2x_3 - x_5 = -1$$

Solution

The augmented matrix is

$$\begin{array}{c} x_1 x_2 x_3 x_4 x_5 \\ \begin{bmatrix} 3 & 2 & -1 & 1 & 0 & | & 1 \\ -2 & -1 & 2 & 0 & -1 & | & -1 \end{bmatrix} \end{array}$$

We pivot first in the a_{11}-location:

$$\begin{bmatrix} 3 & 2 & -1 & 1 & 0 & | & 1 \\ 0 & 1 & 4 & 2 & -3 & | & -1 \end{bmatrix}$$

We then use the result to pivot at a_{22}:

$$\begin{array}{c} x_1 x_2 x_3 x_4 x_5 \\ \begin{bmatrix} 3 & 0 & -9 & -3 & 6 & | & 3 \\ 0 & 1 & 4 & 2 & -3 & | & -1 \end{bmatrix} \end{array}$$

We cannot continue; we see that this is the kind of augmented matrix for which the corresponding linear system

$$3x_1 - 9x_3 - 3x_4 + 6x_5 = 3$$
$$x_2 + 4x_3 + 2x_4 - 3x_5 = -1$$

has infinitely many solutions.

To find solutions, it is convenient to solve the first equation for the variable x_1,

$$3x_1 = 3 + 9x_3 + 3x_4 - 6x_5$$
$$x_1 = 1 + 3x_3 + x_4 - 2x_5$$

and then solve the second equation for x_2:

$$x_2 = -1 - 4x_3 - 2x_4 + 3x_5$$

If we wish to find just one solution, the easiest way to do this is to set each of the variables $x_3, x_4,$ and x_5 equal to zero. Then we see that $x_1 = 1$ and $x_2 = -1$. To summarize, the system has infinitely many solutions, one of which is

$$x_1 = 1, \quad x_2 = -1, \quad x_3 = 0, \quad x_4 = 0, \quad x_5 = 0 \qquad \blacksquare$$

Bases

We will use Example 3 to introduce some terminology that we will need in the following sections. Let us look again at the matrix that results from the pivots:

$$\begin{array}{c} x_1 x_2 x_3 x_4 x_5 \\ \begin{bmatrix} 3 & 0 & -9 & -3 & 6 & | & 3 \\ 0 & 1 & 4 & 2 & -3 & | & -1 \end{bmatrix} \end{array}$$

We have seen how the first row of the matrix gives us a simple way to solve for the variable x_1, and the second row does the same for x_2. The reason is that the column beneath the variable x_1 contains only one nonzero number, and the same is true for the column labeled x_2; so, each of the variables x_1 and x_2 appears in just a single equation. If for each row of the matrix we have a distinct variable that appears in that row and no other, we say that these variables form a **basis** for the matrix, and we call these variables the **basic variables**. The rest of the variables are known as **nonbasic variables**. Thus, the basis for the preceding matrix is $\{x_1, x_2\}$ and the nonbasic variables are x_3, x_4, and x_5. When a matrix has a basis, it is customary to write each basic variable to the left of the row in which it appears. Thus, in the matrix

$$\begin{array}{c} \\ x_3 \\ x_1 \\ x_4 \end{array} \begin{array}{c} x_1 \quad x_2 \quad x_3 \quad x_4 \quad x_5 \\ \left[\begin{array}{ccccc|c} 0 & 1 & 3 & 0 & -1 & 2 \\ -2 & 2 & 0 & 0 & 1 & -1 \\ 0 & 3 & 0 & 1 & 2 & 2 \end{array}\right] \end{array}$$

the basis is $\{x_1, x_3, x_4\}$. We list the basic variables in order of increasing subscripts, even though in the given matrix the x_3 is opposite the first row (since that is the row in which it appears). Of course, we could interchange rows to put the basic variables on the left in this order, but there is no reason to do so. The nonbasic variables in this case are the remaining ones, x_2 and x_5.

The Basic Solution

We noticed in Example 3 that there was a very easy way to find one solution of the given system. In the terminology we just introduced, this solution is the one in which all the nonbasic variables are set equal to zero. This solution is called the **basic solution** (corresponding to the given basis). If all we require is the basic solution, we can find it without going to the trouble of writing general expressions for all solutions, as we did in Example 3. Instead, since we know the nonbasic variables will be zero anyhow, we eliminate them from the matrix. Thus, to find the basic solution corresponding to the final matrix of Example 3,

$$\begin{array}{c} \\ x_1 \\ x_2 \end{array} \begin{array}{c} x_1 \quad x_2 \quad x_3 \quad x_4 \quad x_5 \\ \left[\begin{array}{ccccc|c} 3 & 0 & -9 & -3 & 6 & 3 \\ 0 & 1 & 4 & 2 & -3 & -1 \end{array}\right] \end{array}$$

we delete the columns of the nonbasic variables x_3, x_4, and x_5. This leaves

$$\begin{array}{c} \\ x_1 \\ x_2 \end{array} \begin{array}{c} x_1 \quad x_2 \\ \left[\begin{array}{cc|c} 3 & 0 & 3 \\ 0 & 1 & -1 \end{array}\right] \end{array}$$

corresponding to the equations $3x_1 = 3$ and $x_2 = -1$. Thus, we easily read off the basic solution:

$$x_1 = 1, \quad x_2 = -1, \quad x_3 = 0, \quad x_4 = 0, \quad x_5 = 0$$

In the matrix

174 LINEAR PROGRAMMING

$$\begin{array}{c} \\ x_3 \\ x_1 \\ x_4 \end{array} \begin{array}{c} x_1 \ \ x_2 \ \ x_3 \ \ x_4 \ \ x_5 \\ \left[\begin{array}{ccccc|c} 0 & 1 & 3 & 0 & -1 & 2 \\ -2 & 2 & 0 & 0 & 1 & -1 \\ 0 & 3 & 0 & 1 & 2 & 2 \end{array} \right] \end{array}$$

deleting the columns of the nonbasic variables x_2 and x_5 leaves

$$\begin{array}{c} \\ x_3 \\ x_1 \\ x_4 \end{array} \begin{array}{c} x_1 \ \ \ \ x_3 \ \ x_4 \\ \left[\begin{array}{ccc|c} 0 & 3 & 0 & 2 \\ -2 & 0 & 0 & -1 \\ 0 & 0 & 1 & 2 \end{array} \right] \end{array}$$

which can be written as the system

$$\begin{aligned} 3x_3 &= 2 \\ -2x_1 &= -1 \\ x_4 &= 2 \end{aligned}$$

We see that $x_3 = \frac{2}{3}, x_1 = \frac{1}{2}, x_4 = 2$. Recalling that the nonbasic variables, that is, x_2 and x_5, are zero, we conclude that the basic solution corresponding to the basis $\{x_1, x_3, x_4\}$ is

$$x_1 = \frac{1}{2}, \quad x_2 = 0, \quad x_3 = \frac{2}{3}, \quad x_4 = 2, \quad x_5 = 0$$

Changing the Basis

The geometric method for solving an LP-problem with two independent variables x_1 and x_2 depends on knowing that the solution, if it exists, is always obtained from the coordinates of a corner point of the feasible set. The matrix method for LP-problems, which works for any number of independent variables, depends on the fact that the solution is a basic solution of a system of linear equations. In the next section, we will present rules that change the basis to arrive eventually at the particular basis whose corresponding basic solution solves the LP-problem. Before we can do that, however, we must know how to change from one basis for a linear system to another.

For example, suppose we are directed to change from the matrix

$$\begin{array}{c} \\ x_3 \\ x_1 \\ x_4 \end{array} \begin{array}{c} x_1 \ \ x_2 \ \ x_3 \ \ x_4 \ \ x_5 \\ \left[\begin{array}{ccccc|c} 0 & 1 & 3 & 0 & -1 & 2 \\ -2 & 2 & 0 & 0 & 1 & -1 \\ 0 & 3 & 0 & 1 & 2 & 2 \end{array} \right] \end{array}$$

with the basis $\{x_1, x_3, x_4\}$, to the equivalent system of linear equations with the basis $\{x_1, x_4, x_5\}$. Comparing the two bases, we see that what we need to do is remove the variable x_3 from the basis and replace it by the variable x_5. To make x_5 a basic variable, there must be only one nonzero number in the column headed by x_5; so, we need to pivot in that column. We must find a row in which the basic variable x_5 can appear and, since we no longer want x_3 to be in the basis, we will

2.4 PIVOTING AND BASES 175

use the row corresponding to x_3, that is, the first row. In other words, we should pivot in the a_{15}-location.

The same reasoning gives us the general rule for changing a basis by replacing a basic variable x_r with a new basic variable x_s. We must pivot in the column of the variable x_s, that is, the sth column. There is a single row in which the variable x_r appears; call it the kth row (it is not usually row number r, but some other row). The rule is: pivot at the location a_{ks}.

To pivot at a_{15} in the preceding matrix, we begin the pivoting method by copying the first row and changing the rest of the fifth column to zeros. We also change the label on the first row from x_3 to x_5.

$$\begin{array}{c} \begin{array}{cccccc} x_1 & x_2 & x_3 & x_4 & x_5 & \end{array} \\ \begin{array}{c} x_5 \\ x_1 \\ x_4 \end{array} \left[\begin{array}{ccccc|c} 0 & 1 & 3 & 0 & -1 & 2 \\ \cdot & \cdot & \cdot & \cdot & 0 & \cdot \\ \cdot & \cdot & \cdot & \cdot & 0 & \cdot \end{array} \right] \end{array}$$

Then for each remaining target, we use the pivot element $p = -1$ and the formula

$$\text{change the target } t \text{ to } pt - ab$$

to obtain the following matrix with the basis $\{x_1, x_4, x_5\}$:

$$\begin{array}{c} \begin{array}{cccccc} x_1 & x_2 & x_3 & x_4 & x_5 & \end{array} \\ \begin{array}{c} x_5 \\ x_1 \\ x_4 \end{array} \left[\begin{array}{ccccc|c} 0 & 1 & 3 & 0 & -1 & 2 \\ 2 & -3 & -3 & 0 & 0 & -1 \\ 0 & -5 & -6 & -1 & 0 & -6 \end{array} \right] \end{array}$$

If we wish to know the corresponding basic solution, we delete the columns of the matrix corresponding to the nonbasic variables x_2 and x_3, and read off the equations: $-x_5 = 2$, $2x_1 = -1$, and $-x_4 = -6$. The basic solution is

$$x_1 = -\frac{1}{2}, \quad x_2 = 0, \quad x_3 = 0, \quad x_4 = 6, \quad x_5 = -2$$

Example 4. Find the basic solution corresponding to the basis $\{x_2, x_3, x_5\}$.

$$\begin{array}{c} \begin{array}{ccccccc} x_1 & x_2 & x_3 & x_4 & x_5 & x_6 & \end{array} \\ \begin{array}{c} x_5 \\ x_4 \\ x_2 \end{array} \left[\begin{array}{cccccc|c} 2 & 0 & 3 & 0 & -2 & 2 & 2 \\ -1 & 0 & 1 & 3 & 0 & 3 & 0 \\ 1 & -2 & -2 & 0 & 0 & -1 & 3 \end{array} \right] \end{array}$$

Solution

Before we can find the basic solution, we must change the basis from the present $\{x_2, x_4, x_5\}$ to $\{x_2, x_3, x_5\}$; so, we must add the variable x_3 to the basis in place of x_4. Since x_4 appears alone in the second row, we pivot at the location a_{23}.

$$\begin{array}{c} \begin{array}{ccccccc} x_1 & x_2 & x_3 & x_4 & x_5 & x_6 & \end{array} \\ \begin{array}{c} x_5 \\ x_3 \\ x_2 \end{array} \left[\begin{array}{cccccc|c} 5 & 0 & 0 & -9 & -2 & -7 & 2 \\ -1 & 0 & 1 & 3 & 0 & 3 & 0 \\ -1 & -2 & 0 & 6 & 0 & 5 & 3 \end{array} \right] \end{array}$$

Now, if we remove the columns headed by the nonbasic variables x_1, x_4, and x_6 from the rest of the matrix, we have the equations $-2x_5 = 2$, $x_3 = 0$, and $-2x_2 = 3$. Thus, the basic solution is

$$x_1 = 0, \quad x_2 = -\frac{3}{2}, \quad x_3 = 0, \quad x_4 = 0, \quad x_5 = -1, \quad x_6 = 0 \quad \blacksquare$$

Summary

A **pivot** modifies a column of an augmented matrix corresponding to a system of linear equations so that the solutions of the system remain the same, but the column contains only one nonzero element, the **pivot element**. The row in which the pivot element lies is the **pivot row**, and its column, the **pivot column**. To pivot at the pivot element $a_{ij} = p$, first copy the pivot row into the new matrix and change all other numbers in the pivot column to zeros. Then, for a target t neither in the pivot row nor in the pivot column, let a be the entry in the target row and pivot column and b, the entry in the pivot row and target column; change t to $pt - ab$. To analyze a system of linear equations, move from left to right in the augmented matrix. If the diagonal element a_{ii} is nonzero, pivot in that location, and if $a_{ii} = 0$, interchange the ith row with a row beneath it that contains a nonzero element in the ith column; then use the new a_{ii} as the pivot element. If there is no such nonzero element, go on to the next column to the right. When you are no longer able to continue, analyze the resulting system of linear equations, just as in the Gauss-Jordan method, to determine whether there is a single solution, infinitely many solutions, or no solution at all. If for each row of an augmented matrix we have a distinct variable that appears in that row and no other, we say that these variables form a **basis** for the matrix and we call these variables the **basic variables**. The other variables are known as **nonbasic variables**. The solution of the corresponding system of linear equations in which all the nonbasic variables are set equal to zero is called the **basic solution** (corresponding to the given basis). To change a basis by replacing a basic variable x_r appearing in the kth row with a new basic variable x_s, pivot at the location a_{ks}.

Exercises

1. Pivot at the a_{11}-location:

$$\begin{bmatrix} 2 & 1 & -2 & | & -1 \\ 1 & 0 & -1 & | & 2 \\ -1 & 2 & 2 & | & 0 \end{bmatrix}$$

2. Pivot at the a_{33}-location:

$$\begin{bmatrix} 2 & 0 & 3 & | & 0 \\ 0 & -1 & -2 & | & 2 \\ 0 & 0 & -1 & | & 2 \end{bmatrix}$$

2.4 PIVOTING AND BASES

3. Pivot at the a_{22}-location:

$$\begin{bmatrix} 3 & 1 & 2 & | & 1 \\ 0 & 1 & 1 & | & 1 \\ 0 & 2 & 1 & | & 2 \end{bmatrix}$$

4. Pivot at the a_{11}-location:

$$\begin{bmatrix} -2 & -3 & 1 & | & -1 \\ 3 & 2 & 0 & | & 2 \\ 0 & -1 & -2 & | & -3 \end{bmatrix}$$

5. Pivot at the a_{12}-location:

$$\begin{bmatrix} 2 & 1 & 0 & 1 & 1 & | & 2 \\ -1 & 3 & 2 & 0 & 1 & | & 1 \end{bmatrix}$$

6. Pivot at the a_{22}-location:

$$\begin{bmatrix} 0 & -2 & -2 & 1 & | & 0 \\ 1 & -1 & 3 & 0 & | & -1 \end{bmatrix}$$

7. Pivot at the a_{25}-location:

$$\begin{bmatrix} 1 & 0 & -1 & 1 & 2 & | & 0 \\ 1 & 2 & 2 & 0 & -2 & | & 1 \end{bmatrix}$$

8. Pivot at the a_{33}-location:

$$\begin{bmatrix} 2 & 0 & 2 & 1 & | & 0 \\ 0 & -1 & -2 & 1 & | & 1 \\ 0 & 0 & 3 & -1 & | & 1 \\ 0 & 0 & 1 & 2 & | & 2 \end{bmatrix}$$

9. Pivot at the a_{12}-location:

$$\begin{bmatrix} 0 & 2 & 2 & 0 & | & 1 \\ 1 & -1 & 0 & 0 & | & 0 \\ 0 & 1 & 0 & -2 & | & 3 \end{bmatrix}$$

10. Pivot at the a_{34}-location:

$$\begin{bmatrix} 1 & 2 & 0 & 2 & 0 & | & -2 \\ -1 & 0 & 1 & 3 & 0 & | & -1 \\ 2 & 0 & 0 & -2 & 1 & | & -2 \end{bmatrix}$$

In Exercises 11 through 20, use the pivoting method of this section to determine whether or not the system of linear equations has a solution. If there is one solution, find out what it is. If there are an infinite number of solutions, write down two of them.

11. $2x - 3y = 2$
$5x + y = -1$

12. $2x + y + z = 0$
$-x + 2y = 1$
$-2x - y + 3z = 2$

13. $x - y + z = -1$
$2x - 2y = 1$
$-x + 2y - z = 0$
$2x + 2y = 0$

14. $-3x + y - z = 0$
$x - y - 2z = -1$
$-x - y - 3z = -1$

15. $ -3y - z = 5$
$2x + y + 5z = -3$
$x + 2y + 3z = -4$

16. $2x_1 + x_2 - x_3 + 3x_4 = 1$
$-3x_1 + x_2 + 2x_3 + x_4 = -1$

17. $x_1 + x_3 - x_4 = 0$
$-x_1 - x_2 + x_4 = 1$
$2x_1 + x_4 = -1$

18. $-x_1 + 2x_2 - x_3 + x_4 = 1$
$ x_2 + 2x_3 + x_4 + x_5 = -1$
$-x_1 + 4x_2 + 3x_3 + 3x_4 + 2x_5 = -2$

19. $ x_2 + 2x_3 + 2x_4 - x_5 = 0$
$-x_1 + x_2 + x_3 - 2x_5 = -1$
$2x_1 + x_2 + 4x_3 + 6x_4 = 0$

20. $2x_1 + x_2 + x_3 - x_4 + 2x_5 = 2$
$-2x_1 + 3x_2 + x_4 = 0$
$-x_1 - x_2 + 2x_3 + 2x_4 - x_5 = -2$

In Exercises 21 through 26, find the basic solution corresponding to the given basis.

21. $\{x_2, x_3, x_4\}$:

$$\begin{array}{c} & \begin{array}{cccc} x_1 & x_2 & x_3 & x_4 \end{array} \\ \begin{array}{c} x_1 \\ x_2 \\ x_3 \end{array} & \left[\begin{array}{cccc|c} 2 & 0 & 0 & 3 & -2 \\ 0 & -2 & 0 & -1 & 1 \\ 0 & 0 & -1 & 2 & 0 \end{array}\right] \end{array}$$

22. $\{x_1, x_3, x_4\}$:

$$\begin{array}{c} & \begin{array}{ccccc} x_1 & x_2 & x_3 & x_4 & x_5 \end{array} \\ \begin{array}{c} x_1 \\ x_2 \\ x_3 \end{array} & \left[\begin{array}{ccccc|c} 2 & 0 & 0 & 2 & 1 & 2 \\ 0 & 1 & 0 & -2 & 2 & -1 \\ 0 & 0 & -3 & -2 & 3 & -1 \end{array}\right] \end{array}$$

23. $\{x_1, x_2, x_4\}$:

$$\begin{array}{c} & \begin{array}{cccc} x_1 & x_2 & x_3 & x_4 \end{array} \\ \begin{array}{c} x_2 \\ x_1 \\ x_3 \end{array} & \left[\begin{array}{cccc|c} 0 & -2 & 0 & 1 & -2 \\ 1 & 0 & 0 & 2 & -1 \\ 0 & 0 & -2 & 3 & 1 \end{array}\right] \end{array}$$

24. $\{x_2, x_5\}$:

$$\begin{array}{c} & \begin{array}{cccccc} x_1 & x_2 & x_3 & x_4 & x_5 & x_6 \end{array} \\ \begin{array}{c} x_2 \\ x_1 \end{array} & \left[\begin{array}{cccccc|c} 0 & 2 & 1 & -2 & 1 & 2 & 2 \\ -3 & 0 & 1 & 2 & 4 & -3 & 1 \end{array}\right] \end{array}$$

25. $\{x_2, x_3, x_5\}$:

$$\begin{array}{c} & \begin{array}{ccccc} x_1 & x_2 & x_3 & x_4 & x_5 \end{array} \\ \begin{array}{c} x_5 \\ x_1 \\ x_3 \end{array} & \left[\begin{array}{ccccc|c} 0 & 1 & 0 & 3 & 3 & 1 \\ -2 & 2 & 0 & 2 & 0 & 3 \\ 0 & 3 & -4 & -1 & 0 & -3 \end{array}\right] \end{array}$$

26. $\{x_1, x_2, x_3, x_4\}$:

$$\begin{array}{c} & \begin{array}{ccccc} x_1 & x_2 & x_3 & x_4 & x_5 \end{array} \\ \begin{array}{c} x_3 \\ x_1 \\ x_5 \\ x_2 \end{array} & \left[\begin{array}{ccccc|c} 0 & 0 & -2 & 2 & 0 & 2 \\ 1 & 0 & 0 & -1 & 0 & 2 \\ 0 & 0 & 0 & 3 & 3 & -1 \\ 0 & -1 & 0 & 3 & 0 & 4 \end{array}\right] \end{array}$$

2.5 The Simplex Algorithm for Maximization Problems

The Nature of the Simplex Algorithm

In Section 2.2, the solution of an LP-problem involving two variables x_1 and x_2 depended on finding the coordinates (x_1, x_2) of extreme points of the feasible set of the problem. For example, in the *Secretaries and Clerks* problem, one extreme point was the intersection of the graphs of the two linear equations

$$\begin{aligned} x_1 &= 5 \\ 10x_1 + 4x_2 &= 100 \end{aligned}$$

We found the coordinates of this extreme point by solving the matrix equation $AX = B$, where

$$A = \begin{bmatrix} 1 & 0 \\ 10 & 4 \end{bmatrix} \quad \text{and} \quad B = \begin{bmatrix} 5 \\ 100 \end{bmatrix}$$

The solution of any LP-problem is given by the coordinates of an extreme point of the feasible set. The matrix techniques of Chapter 1 provide the foundation for

2.5 THE SIMPLEX ALGORITHM FOR MAXIMIZATION PROBLEMS

a very effective method, called the **simplex algorithm**, that solves LP-problems with any number of independent variables. The method is very well suited to computer implementation, so it is a practical way to solve the very large LP-problems, often involving thousands of variables, that arise in applications.

We will describe a somewhat simplified form of the simplex algorithm that can be applied to these maximization problems.

Maximize: $\quad z = CX$

Subject to: $\quad AX \leq B, X \geq 0$, and also $B \geq 0$

Thus, the constants on the right-hand sides of the inequalities are all positive or zero. This requirement is satisfied by many important types of LP-problems found in applications. (All the examples in Section 2.3 are of this type.) In the next section we will see that this same technique also solves many minimization problems. We will point out the procedures that would be required to extend the method to produce a general simplex algorithm that could be used to analyze all LP-problems.

Slack Variables

The problem we use to explain the simplex algorithm is another simple agricultural problem based on the more realistic problem we discussed in the introduction to this chapter. This time the problem concerns a farm with three crops; so there are three variables, and therefore, we cannot solve it by the geometric method of Section 2.2.

A farmer will plant up to 60 acres of peas, up to 75 acres of beans, and, possibly, some acres of carrots on her 150-acre farm. She can use up to 3000 acre-inches of water to irrigate her crops. An acre of peas requires 30 inches of water, an acre of beans 15 inches, and an acre of carrots 6 inches. She estimates that the net profit per acre from each of the crops will be $70 for peas, $90 for beans, and $10 for carrots. How many acres of each crop should she plant to maximize her net profit?

We use the variables x_1, x_2, and x_3 for the numbers of acres devoted to peas, beans, and carrots, respectively. The acreage limitations for the first two crops are that $x_1 \leq 60$ and $x_2 \leq 75$, and the total acreage limitation for the farm is that $x_1 + x_2 + x_3 \leq 150$. The water requirement for x_1 acres of peas is $30x_1$ acre-inches, and the requirements for the other crops add $15x_2$ and $6x_3$ acre-inches, respectively; so, we know that $30x_1 + 15x_2 + 6x_3 \leq 3000$, since that is how much water is available. Referring to the profit figures above to express the total net profit from the three crops, we can write this problem as

Maximize: $\quad z = 70x_1 + 90x_2 + 10x_3$

Subject to:
$$
\begin{aligned}
x_1 &\leq 60 \\
x_2 &\leq 75 \\
x_1 + x_2 + x_3 &\leq 150 \\
30x_1 + 15x_2 + 6x_3 &\leq 3000 \\
x_1, x_2, x_3 &\geq 0
\end{aligned}
$$

180 LINEAR PROGRAMMING

This is in the form we require for the simplex algorithm, that is,

$$\text{Maximize:} \quad z = CX$$
$$\text{Subject to:} \quad AX \leq B \text{ and } X \geq 0$$

with

$$A = \begin{bmatrix} 1 & 0 & 0 \\ 0 & 1 & 0 \\ 1 & 1 & 1 \\ 30 & 15 & 6 \end{bmatrix}, \quad B = \begin{bmatrix} 60 \\ 75 \\ 150 \\ 3000 \end{bmatrix}, \quad \text{and} \quad C = \begin{bmatrix} 70 & 90 & 10 \end{bmatrix}$$

Observe that we do have the condition $B \geq 0$.

As we explained, we will make use of row operations as we did in solving systems of linear equations, but now in the efficient pivoting form of the preceding section. The system of linear equations we work with comes from the linear inequalities that describe the restrictions on the problem.

For each inequality of the form

$$a_1 x_1 + a_2 x_2 + \ldots + a_n x_n \leq b$$

we introduce a new variable, which we will temporarily call x, defined by the formula

$$x = b - (a_1 x_1 + a_2 x_2 + \ldots + a_n x_n)$$

The variable x is called a **slack variable** of the LP-problem. Notice that since b is at least as large as the expression on the left-hand side of the inequality, it follows that $x \geq 0$. Thus, the slack variables will be nonnegative, just as the other independent variables are. We introduce a different slack variable for each inequality of the given form, numbering them x_{n+1}, x_{n+2}, and so on.

Because the independent variables of the preceding example are x_1, x_2, and x_3, we begin numbering the slack variables with x_4 and define them as follows:

$$x_4 = 60 - x_1$$
$$x_5 = 75 - x_2$$
$$x_6 = 150 - (x_1 + x_2 + x_3)$$
$$x_7 = 3000 - (30x_1 + 15x_2 + 6x_3)$$

Since the inequality $x_1 + x_2 + x_3 \leq 150$ expresses the fact that the farmer has 150 acres of land available in all, the slack variable x_6 measures the amount of land that is left over if the farmer plants x_1, x_2, and x_3 acres of each of the three crops. Notice that certainly we must have $x_6 \geq 0$. Similarly, we have the water limitation of the *Farmer's* problem, $30x_1 + 15x_2 + 6x_3 \leq 3000$, and the slack variable x_7 measures the amount of available water that the farmer does not use. These new variables measure the extent to which each linear inequality fails to be an equation. An equation can be thought of as an exact or tight relationship, whereas an inequality can be thought of as the opposite of tight, that is, "slack"; this is how these variables acquired their name.

2.5 THE SIMPLEX ALGORITHM FOR MAXIMIZATION PROBLEMS

Conversion to Matrix Form

We usually write linear equations with all the variables on the left-hand side. Doing this with the equations that define the slack variables, we have

$$
\begin{aligned}
x_1 + x_4 &= 60 \\
x_2 + x_5 &= 75 \\
x_1 + x_2 + x_3 + x_6 &= 150 \\
30x_1 + 15x_2 + 6x_3 + x_7 &= 3000
\end{aligned}
$$

Now we rewrite the LP-problem to include the slack variables.

Maximize: $z = 70x_1 + 90x_2 + 10x_3$

Subject to:
$$
\begin{aligned}
x_1 + x_4 &= 60 \\
x_2 + x_5 &= 75 \\
x_1 + x_2 + x_3 + x_6 &= 150 \\
30x_1 + 15x_2 + 6x_3 + x_7 &= 3000 \\
x_1, x_2, x_3, x_4, x_5, x_6, x_7 &\geq 0
\end{aligned}
$$

The augmented matrix for this system of linear equations is

$$
\begin{array}{c}
 \\ x_4 \\ x_5 \\ x_6 \\ x_7
\end{array}
\begin{array}{c}
\begin{matrix} x_1 & x_2 & x_3 & x_4 & x_5 & x_6 & x_7 \end{matrix} \\
\left[\begin{array}{ccccccc|c}
1 & 0 & 0 & 1 & 0 & 0 & 0 & 60 \\
0 & 1 & 0 & 0 & 1 & 0 & 0 & 75 \\
1 & 1 & 1 & 0 & 0 & 1 & 0 & 150 \\
30 & 15 & 6 & 0 & 0 & 0 & 1 & 3000
\end{array} \right]
\end{array}
$$

This augmented matrix has a very special structure in terms of the matrix form of the LP-problem. The feasible set is determined by matrix inequalities $AX \geq B$ and $X \geq 0$. The augmented matrix is of the form $[A \quad I \quad | B]$, for those same matrices A and B.

Notice also that this system of 4 linear equations in 7 variables has a basis $\{x_4, x_5, x_6, x_7\}$ given by the slack variables, as we indicated on the augmented matrix. The basic solution is easy to read off the matrix since we can delete the first three columns that correspond to the nonbasic variables. The matrix B gives us the answer:

$$x_1 = 0, \quad x_2 = 0, \quad x_3 = 0, \quad x_4 = 60, \quad x_5 = 75, \quad x_6 = 150, \quad x_7 = 3000$$

The requirement that $B \geq 0$ guarantees that the basic solution consists of nonnegative numbers only. Thus, the basic solution gives the coordinates of a point in the feasible set of the original LP-problem and, for that reason, it is called a **basic feasible solution**.

The Simplex Tableau

The basic feasible solutions give the coordinates of extreme points of the feasible set of the LP-problem. The coordinates of one of those extreme points will solve the problem, that is, will give the maximum value of the objective function

$$z = 70x_1 + 90x_2 + 10x_3$$

if (x_1, x_2, x_3) must be the coordinates of a point in the feasible set. Certainly, the basic feasible solution corresponding to the basis $\{x_4, x_5, x_6, x_7\}$ given by the slack variables is not the right one because in this case x_1, x_2, and x_3 are all zero; so, $z = 0$.

The simplex algorithm is a set of rules that tells us how to change the basis of the system of linear equations to obtain a basis for which the corresponding basic feasible solution solves the LP-problem. We start with the basis furnished by the slack variables; at each step of the simplex algorithm we replace one variable in the present basis by one of the nonbasic variables, using the pivoting procedure of the preceding section. The decision as to which nonbasic variable to introduce into the basis depends on the form of the objective function. At each stage we must write the objective function only in terms of the present nonbasic variables. This condition is always satisfied at the beginning of the procedure because the initial basis is given by the slack variables and these variables come from the inequalities, and not from the objective function. For instance, the objective function

$$z = 70x_1 + 90x_2 + 10x_3$$

uses only the nonbasic variables x_1, x_2, and x_3.

In general, the objective function is

$$z = c_1 x_1 + c_2 x_2 + \ldots + c_n x_n$$

We rewrite it in the form

$$z - c_1 x_1 - c_2 x_2 - \ldots - c_n x_n = 0$$

and add the constants to the augmented matrix as a new last row. We also label the last row on the left with the dependent variable z. In the example we write the objective function as

$$z - 70x_1 - 90x_2 - 10x_3 = 0$$

and expand the augmented matrix to

$$\begin{array}{c} \\ x_4 \\ x_5 \\ x_6 \\ x_7 \\ z \end{array} \begin{array}{c} x_1 \quad x_2 \quad x_3 \quad x_4 \quad x_5 \quad x_6 \quad x_7 \\ \left[\begin{array}{ccccccc|c} 1 & 0 & 0 & 1 & 0 & 0 & 0 & 60 \\ 0 & 1 & 0 & 0 & 1 & 0 & 0 & 75 \\ 1 & 1 & 1 & 0 & 0 & 1 & 0 & 150 \\ 30 & 15 & 6 & 0 & 0 & 0 & 1 & 3000 \\ \hline -70 & -90 & -10 & 0 & 0 & 0 & 0 & 0 \end{array} \right] \end{array}$$

This expanded augmented matrix is called the *simplex tableau* of the LP-problem. In general, the objective function, written in matrix terms as $z = CX$, is rewritten in the form $z - CX = 0$ and is introduced into the augmented matrix as a new last row to form the **simplex tableau**:

$$\begin{array}{c} \\ \{x\} \\ z \end{array} \begin{array}{c} x_1 \quad \ldots \quad x_N \\ \left[\begin{array}{c|c|c} A & I & B \\ \hline -C & 0 & 0 \end{array} \right] \end{array}$$

2.5 THE SIMPLEX ALGORITHM FOR MAXIMIZATION PROBLEMS

where $\{x\}$ represents the basis and O is a row matrix of zeros. If the independent variables of the original LP-problem were $x_1, x_2, \ldots, x_n$, then we introduced slack variables called x_{n+1}, x_{n+2}, and so on, with one slack variable for each linear inequality of the system. In matrix terms, $AX \leq B$. If A is an $m \times n$-matrix, then there are m inequalities in the system, and the last slack variable is x_{n+m}. Therefore, $N = n + m$.

Example 1. Write the simplex tableau of the problem

Maximize: $\quad z = 10x_1 + 15x_2$
Subject to: $\quad x_1 + 2x_2 \leq 4$
$\quad x_1 - \frac{1}{2}x_2 \leq 0$
$\quad x_1, x_2 \geq 0$

Solution

In Section 2.4, we discussed how to pivot a matrix without introducing fractions into the calculation, provided we started out with a matrix that had no fractions. We plan to use that pivoting method later, but it will not apply if we just follow the instructions for writing the simplex tableau. The problem is that the coefficient matrix A will contain the fraction $\frac{1}{2}$ from the inequality

$$x_1 - \frac{1}{2}x_2 \leq 0$$

so that fraction will occupy the a_{22}-location in the simplex tableau. We can correct the problem by multiplying both sides of the inequality by 2, remembering that when you multiply an inequality by a positive number, the inequality sign remains unchanged. (Multiplication by a negative number reverses the direction of the inequality sign, but we will never multiply by a negative number.) Then this inequality becomes $2x_1 - x_2 \leq 0$, and the system of linear inequalities is

$$x_1 + 2x_2 \leq 4$$
$$2x_1 - x_2 \leq 0$$
$$x_1, x_2 \geq 0$$

Just as with a system of linear equations, the solutions of this system, that is, the feasible set, is exactly the same as before. Now the matrix form of the problem:

Maximize: $\quad z = CX$
Subject to: $\quad AX \leq B$ and $X \geq 0$

involves only matrices of integers. These matrices are

$$A = \begin{bmatrix} 1 & 2 \\ 2 & -1 \end{bmatrix}, \quad B = \begin{bmatrix} 4 \\ 0 \end{bmatrix}, \quad \text{and} \quad C = \begin{bmatrix} 10 & 15 \end{bmatrix}$$

Introducing the slack variables x_3 and x_4, the simplex tableau of this problem is

$$\begin{array}{c} \\ x_3 \\ x_4 \\ z \end{array} \begin{array}{c} \begin{array}{cccc} x_1 & x_2 & x_3 & x_4 \end{array} \\ \left[\begin{array}{cccc|c} 1 & 2 & 1 & 0 & 4 \\ 2 & -1 & 0 & 1 & 0 \\ \hline -10 & -15 & 0 & 0 & 0 \end{array} \right] \end{array}$$

∎

Pivot Column and Pivot Element

Although the pivot operations of the simplex method are just those we used in Section 2.4 to solve a system of linear equations, the procedure is different. Instead of always starting with the left-hand column and moving column by column to the right, we look at the numbers in the bottom row of the tableau. If there is just one negative number, then the column in which it lies will be the pivot column. If there is more than one negative number in the bottom row, we find the minimum and choose the column containing it as the pivot column. The variable that heads the pivot column will always be nonbasic because the objective function is written in terms of nonbasic variables only and the number at the bottom of the column is the negative of the coefficient of the objective function of the variable at the top.

For the tableau

$$\begin{array}{c} \\ x_4 \\ x_5 \\ x_6 \\ x_7 \\ z \end{array} \begin{array}{c} x_1 \quad x_2 \quad x_3 \quad x_4 \quad x_5 \quad x_6 \quad x_7 \\ \left[\begin{array}{ccccccc|c} 1 & 0 & 0 & 1 & 0 & 0 & 0 & 60 \\ 0 & 1 & 0 & 0 & 1 & 0 & 0 & 75 \\ 1 & 1 & 1 & 0 & 0 & 1 & 0 & 150 \\ 30 & 15 & 6 & 0 & 0 & 0 & 1 & 3000 \\ -70 & -90 & -10 & 0 & 0 & 0 & 0 & 0 \end{array}\right] \end{array}$$

the minimum number in the bottom row is -90 (from the term $90x_2$ of the objective function); so, the second column will be the pivot column, and we will add the variable x_2 to the basis in place of one of the slack variables.

We wish to replace one of the present basic variables by the nonbasic variable corresponding to the pivot column in such a way that all the numbers in the new basic solution are nonnegative. This gives us a basic feasible solution, the coordinates of an extreme point of the feasible set, though we may not have yet found the solution to the LP-problem. To do this, we divide each number in the right-hand column of the tableau by the number in the pivot column that is in the same row, provided that the number in the pivot column is positive. We place these quotients to the right of the tableau. In the example we have

$$\begin{array}{c} \\ x_4 \\ x_5 \\ x_6 \\ x_7 \\ z \end{array} \begin{array}{c} x_1 \quad x_2 \quad x_3 \quad x_4 \quad x_5 \quad x_6 \quad x_7 \\ \left[\begin{array}{ccccccc|c} 1 & 0 & 0 & 1 & 0 & 0 & 0 & 60 \\ 0 & 1 & 0 & 0 & 1 & 0 & 0 & 75 \\ 1 & 1 & 1 & 0 & 0 & 1 & 0 & 150 \\ 30 & 15 & 6 & 0 & 0 & 0 & 1 & 3000 \\ -70 & -90 & -10 & 0 & 0 & 0 & 0 & 0 \end{array}\right] \begin{array}{l} \\ 75/1 = 75 \\ 150/1 = 150 \\ 3000/15 = 200 \end{array} \end{array}$$

The quotient cannot be negative; but it will be zero if the number in the right-hand column is zero (see Example 3). The pivot row is the row with the smallest of these (positive or zero) quotients, and the number in that row in the pivot column is called the pivot element because we will pivot at that location. If there is a zero quotient, that will be the minimum. In the present example all the quotients are positive and the smallest is in the second row. Therefore, we will pivot in the

2.5 THE SIMPLEX ALGORITHM FOR MAXIMIZATION PROBLEMS

a_{22}-location and that will put x_2 in the basis in place of the slack variable x_5. We perform the pivot just as in Section 2.4, noting by an asterisk at a_{22}, the location of the pivot. To recall the steps briefly, we first copy the pivot row and change the rest of the pivot column to zeros:

$$
\begin{array}{c}
 \\ x_4 \\ x_2 \\ x_6 \\ x_7 \\ z
\end{array}
\left[\begin{array}{ccccccc|c}
x_1 & x_2 & x_3 & x_4 & x_5 & x_6 & x_7 & \\
\cdot & 0 & \cdot & \cdot & \cdot & \cdot & \cdot & \cdot \\
0 & 1^* & 0 & 0 & 1 & 0 & 0 & 75 \\
\cdot & 0 & \cdot & \cdot & \cdot & \cdot & \cdot & \cdot \\
\cdot & 0 & \cdot & \cdot & \cdot & \cdot & \cdot & \cdot \\
\hline
\cdot & 0 & \cdot & \cdot & \cdot & \cdot & \cdot & \cdot
\end{array}\right]
$$

Then, for a number t neither in the pivot row nor in the pivot column, letting a be the number in the target row and pivot column and b the number in the pivot row and target column, we change t to $pt - ab$, where, in this case, the pivot element p is 1. The result of changing all the targets, represented by dots in the preceding matrix, is

$$
\begin{array}{c}
 \\ x_4 \\ x_2 \\ x_6 \\ x_7 \\ z
\end{array}
\left[\begin{array}{ccccccc|c}
x_1 & x_2 & x_3 & x_4 & x_5 & x_6 & x_7 & \\
1 & 0 & 0 & 1 & 0 & 0 & 0 & 60 \\
0 & 1^* & 0 & 0 & 1 & 0 & 0 & 75 \\
1 & 0 & 1 & 0 & -1 & 1 & 0 & 75 \\
30 & 0 & 6 & 0 & -15 & 0 & 1 & 1875 \\
\hline
-70 & 0 & -10 & 0 & 90 & 0 & 0 & 6750
\end{array}\right]
$$

All the numbers in the last column are nonnegative; so, the basic solution consists of nonnegative numbers, as required. In addition to changing the basis to record the effect of the pivot, as we did in Section 2.4, we also multiply the symbol z below the basis by the pivot element. Since in this case the pivot element was a 1, the symbol remains a z.

A New Pivot

As we will see, if there were no negative numbers in the last row, the corresponding basic feasible solution would solve the LP-problem. The presence of negative numbers in the bottom row indicates that we have not yet found the solution to the problem. Therefore, we repeat the previous steps and locate the pivot column— the first column this time, since -70 is the minimum number in the last row. We form the quotients to the right of the tableau to determine the pivot element.

$$
\begin{array}{c}
 \\ x_4 \\ x_2 \\ x_6 \\ x_7 \\ z
\end{array}
\left[\begin{array}{ccccccc|c}
x_1 & x_2 & x_3 & x_4 & x_5 & x_6 & x_7 & \\
1 & 0 & 0 & 1 & 0 & 0 & 0 & 60 \\
0 & 1 & 0 & 0 & 1 & 0 & 0 & 75 \\
1 & 0 & 1 & 0 & -1 & 1 & 0 & 75 \\
30 & 0 & 6 & 0 & -15 & 0 & 1 & 1875 \\
\hline
-70 & 0 & -10 & 0 & 90 & 0 & 0 & 6750
\end{array}\right]
\begin{array}{l}
60/1 = 60 \\
\\
75/1 = 75 \\
1875/30 = 62.5
\end{array}
$$

We find that the pivot element is the 1 in the a_{11}-location, so we pivot there, putting x_1 in the basis and taking x_4 out:

$$\begin{array}{c} \\ x_1 \\ x_2 \\ x_6 \\ x_7 \\ z \end{array} \begin{array}{c} \begin{array}{ccccccc} x_1 & x_2 & x_3 & x_4 & x_5 & x_6 & x_7 \end{array} \\ \left[\begin{array}{ccccccc|c} 1* & 0 & 0 & 1 & 0 & 0 & 0 & 60 \\ 0 & 1 & 0 & 0 & 1 & 0 & 0 & 75 \\ 0 & 0 & 1 & -1 & -1 & 1 & 0 & 15 \\ 0 & 0 & 6 & -30 & -15 & 0 & 1 & 75 \\ \hline 0 & 0 & -10 & 70 & 90 & 0 & 0 & 10{,}950 \end{array} \right] \begin{array}{l} \\ \\ 15/1 = 15 \\ 75/6 = 12.5 \\ \\ \end{array} \end{array}$$

Again, the pivot element is a 1, so the label to the left of the bottom row is still just z. The -10 at the bottom of the third column tells us that we are not yet finished, but must bring x_3 into the basis. From the computations to the right of the matrix, we pivot at the 6 in the a_{34}-location, with this result:

$$\begin{array}{c} \\ x_1 \\ x_2 \\ x_6 \\ x_3 \\ 6z \end{array} \begin{array}{c} \begin{array}{ccccccc} x_1 & x_2 & x_3 & x_4 & x_5 & x_6 & x_7 \end{array} \\ \left[\begin{array}{ccccccc|c} 6 & 0 & 0 & 6 & 0 & 0 & 0 & 360 \\ 0 & 6 & 0 & 0 & 6 & 0 & 0 & 450 \\ 0 & 0 & 0 & 24 & 9 & 6 & -1 & 15 \\ 0 & 0 & 6 & -30 & -15 & 0 & 1 & 75 \\ \hline 0 & 0 & 0 & 120 & 390 & 0 & 10 & 66{,}450 \end{array} \right] \end{array}$$

Notice that now the label on the last row is $6z$ because it has been multiplied by the pivot element. The fact that there are no longer any negative numbers in the last row means that we have reached the extreme point whose coordinates solve the LP-problem, as we shall now demonstrate. In the first tableau of this problem, the bottom row represented the objective function

$$z = 70x_1 + 90x_2 + 10x_3$$

in the form:

$$z - 70x_1 - 90x_2 - 10x_3 = 0$$

The last row of the tableau now represents the expression

$$6z + 120x_4 + 390x_5 + 10x_7 = 66{,}450$$

If we solve this equation for z, we find that

$$z = 11{,}075 - 20x_4 - 65x_5 - \frac{5}{3}x_7$$

We still call the dependent variable z because this is the same function as the one we started with. What we have done by means of pivoting is to use the equations that define the slack variables, and so relate the several variables to each other, to express z in terms of x_4, x_5, and x_7, instead of x_1, x_2, and x_3. Remembering that each of x_4, x_5, and x_7 must be greater than or equal to zero, the minus signs on the right-hand side of the expression for z tell us that z will be as large as possible precisely when each of x_4, x_5, and x_7 is equal to zero. Any positive value for x_4, x_5, or x_7 would make z smaller than 11,075. So, if there is a set of numbers for $x_1, x_2, \ldots, x_7$ that satisfies the conditions of the problem and if each of x_4, x_5, and

2.5 THE SIMPLEX ALGORITHM FOR MAXIMIZATION PROBLEMS

x_7 is equal to zero, then this set will solve the LP-problem. The basis for the last matrix given is $\{x_1, x_2, x_3, x_6\}$; so, x_4, x_5, and x_7 are all nonbasic variables and the corresponding basic feasible solution, in which all of these are zero, is the solution we have been seeking.

To determine the basic feasible solution, we delete from the last tableau the columns corresponding to the nonbasic variables. There remains the tableau

$$\begin{array}{c|cccc|c} & x_1 & x_2 & x_3 & x_6 & \\ \hline x_1 & 6 & 0 & 0 & 0 & 360 \\ x_2 & 0 & 6 & 0 & 0 & 450 \\ x_6 & 0 & 0 & 0 & 6 & 15 \\ x_3 & 0 & 0 & 6 & 0 & 75 \\ \hline 6z & 0 & 0 & 0 & 0 & 66{,}450 \end{array}$$

We see that $x_1 = \frac{360}{6} = 60$, $x_2 = \frac{450}{6} = 75$, $x_3 = \frac{75}{6} = 12.5$, and $x_6 = \frac{15}{6} = 2.5$. We can also tell from the last row that the maximum value of z is $\frac{66{,}450}{6} = 11{,}075$, as we also know from the preceding expression for z.

We can state this solution in terms of the original problem by saying that the farmer should plant $x_1 = 60$ acres of peas, $x_2 = 75$ acres of beans, $x_3 = 12.5$ acres of carrots, and that her maximum profit is $z = \$11{,}075$. Notice that the solution does not make use of all the land, but only $x_1 + x_2 + x_3 = 147.5$ out of the 150 acres available. Looking back to the definition of the slack variables, we see that x_6 represents the amount of unused land, which explains the value $x_6 = 2.5$ in the basic feasible solution.

Checking the Answer

If there are no errors in our calculations, the solution to the original problem is

$$X = \begin{bmatrix} x_1 \\ x_2 \\ x_3 \end{bmatrix} = \begin{bmatrix} 60 \\ 75 \\ 12.5 \end{bmatrix}$$

To check this answer, recall that X must satisfy the condition $AX \leq B$, where

$$A = \begin{bmatrix} 1 & 0 & 0 \\ 0 & 1 & 0 \\ 1 & 1 & 1 \\ 30 & 15 & 6 \end{bmatrix} \quad \text{and} \quad B = \begin{bmatrix} 60 \\ 75 \\ 150 \\ 3000 \end{bmatrix}$$

By matrix multiplication we verify that X has this property:

$$AX = \begin{bmatrix} 1 & 0 & 0 \\ 0 & 1 & 0 \\ 1 & 1 & 1 \\ 30 & 15 & 6 \end{bmatrix} \begin{bmatrix} 60 \\ 75 \\ 12.5 \end{bmatrix} = \begin{bmatrix} 60 \\ 75 \\ 147.5 \\ 3000 \end{bmatrix} \leq \begin{bmatrix} 60 \\ 75 \\ 150 \\ 3000 \end{bmatrix} = B$$

188   LINEAR PROGRAMMING

As a further check, we multiply CX in the original problem:

$$z = CX = \begin{bmatrix} 70 & 90 & 10 \end{bmatrix} \begin{bmatrix} 60 \\ 75 \\ 12.5 \end{bmatrix} = 4200 + 6750 + 125 = 11{,}075$$

which agrees with the answer we obtained from the bottom row of the tableau.

Verifying by matrix multiplication that an answer X satisfies $AX \leq B$ and that the calculation $z = CX$ agrees with the last row of the tableau does not guarantee that an answer is correct. Nevertheless, we have found from experience that an arithmetic error in carrying out the simplex algorithm will usually produce an answer that fails at least one of these two tests.

The General Method

We now summarize the steps of the simplex algorithm for LP-problems of the form:

Maximize: $z = CX$
Subject to: $AX \leq B, X \geq 0$, and $B \geq 0$

where X is an $n \times 1$-matrix of independent variables, and where the matrices of constants A, B, and C are $m \times n$, $m \times 1$, and $1 \times n$, respectively.

1. Form the simplex tableau with a basis $\{x\}$ given by the slack variables $x_{n+1}, x_{n+2}, \ldots, x_{m+n} = x_N$, and label the last row with the dependent variable z.

$$\begin{array}{c} \\ \{x\} \\ z \end{array} \left[\begin{array}{ccc|c} x_1 & \cdots & x_N & \\ A & & I & B \\ \hline -C & & 0 & 0 \end{array} \right]$$

2. If there are any negative numbers in the last row, choose the column containing the minimum of that row as the pivot column. If there is a tie for the minimum, it is customary to choose the column furthest to the left.
3. Divide each positive number in the pivot column into the number in the last column in the same row, and choose as the pivot row the one in which the quotient is smallest. When there is a tie for the smallest quotient, choose as the pivot row the one that has the smallest number in the pivot column.
4. Pivot at the pivot element that lies in the pivot row and column, changing the basis accordingly and multiplying the label of the last row by the pivot element.
5. Repeat Steps 2 through 4 as long as there are any negative numbers in the last row.
6. When there are no negative numbers in the last row, the basic feasible solution for that basis solves the LP-problem. The maximum value can be computed by dividing the number in the lower right-hand corner by the constant that is the coefficient of the label z of the bottom row.

Cautions

The procedure we have just described is not foolproof. It is possible to construct examples that will cause the simplex algorithm to "cycle," that is, to repeat the same steps endlessly without reaching the solution. We can overcome this difficulty by a straightforward modification of the simplex algorithm. But, in fact, cycling seldom occurs in small problems. Therefore, rather than introducing further technicalities, we have taken care in all examples and exercises to avoid the type of problems that can cycle.

Another point we must be concerned about arose in Section 2.2: not all LP-problems have solutions. To see how the simplex algorithm responds to such a problem, we will apply it to the maximization problem

Maximize: $z = x_1 + x_2$

Subject to: $x_1 \leq 5$

$x_1, x_2 \geq 0$

If we use the method of Section 2.2, we can see that the feasible set is unbounded and contains points with coordinates (x_1, x_2), with no restriction on the size of x_2; so, there is no maximum value of z. To apply the simplex algorithm, we introduce a slack variable to rewrite the inequality as $x_1 + x_3 = 5$; the simplex tableau is

$$\begin{array}{c} \\ x_3 \\ z \end{array} \begin{array}{c} x_1 \quad x_2 \quad x_3 \\ \left[\begin{array}{ccc|c} 1 & 0 & 1 & 5 \\ -1 & -1 & 0 & 0 \end{array} \right] \end{array}$$

When we pivot in the first column, we must do it at the a_{11}-location, removing x_3 from the basis and replacing it by x_1:

$$\begin{array}{c} \\ x_3 \\ z \end{array} \begin{array}{c} x_1 \quad x_2 \quad x_3 \\ \left[\begin{array}{ccc|c} 1 & 0 & 1 & 5 \\ 0 & -1 & 1 & 5 \end{array} \right] \end{array}$$

We still have a negative number in the last row, the -1 in the second column; but we cannot pivot in the second column because there is no positive number in that column. Thus, we cannot carry out Step 3 of the method.

Suppose that in carrying out the simplex algorithm, at some stage, in one column no number is positive and the bottom number is negative. Certainly, we cannot pivot in that column; but the presence of such a column tells us much more. It means that there are points in the feasible set whose coordinates can make the objective function arbitrarily large. Thus, the second column of the original simplex tableau already showed us that this problem has no solution. Although it is easy to construct abstract problems like this one, for most applications we can be sure that this difficulty cannot arise. For instance, the farmer cannot hope to find a crop-acreage decision that will make her net profit as large as she wants. All the LP-problems in the rest of the chapter have solutions; so, it will always be possible to use the simplex algorithm, as we described it, to solve them.

Example 2. Solve the LP-problem

Maximize: $z = 10x_1 + 15x_2$
Subject to: $x_1 + 2x_2 \leq 4$
$x_1 - \frac{1}{2}x_2 \leq 0$
$x_1, x_2 \geq 0$

Solution

This is the same problem as in Example 1; from that example we know that the simplex tableau can be written entirely with integers as

$$\begin{array}{c} \\ x_3 \\ x_4 \\ z \end{array} \left[\begin{array}{cccc|c} x_1 & x_2 & x_3 & x_4 & \\ 1 & 2 & 1 & 0 & 4 \\ 2 & -1 & 0 & 1 & 0 \\ -10 & -15 & 0 & 0 & 0 \end{array} \right]$$

Since $-15 < -10$, the pivot column is the second column. The pivot row must be the first one since it contains the only positive number in the second column. Thus, we pivot at the 2 in the a_{12}-location, changing the basis to replace the variable x_3 with x_2 and remembering to multiply the symbol z to the left of the bottom row by the pivot element 2:

$$\begin{array}{c} \\ x_2 \\ x_4 \\ 2z \end{array} \left[\begin{array}{cccc|c} x_1 & x_2 & x_3 & x_4 & \\ 1 & 2 & 1 & 0 & 4 \\ 5 & 0 & 1 & 2 & 4 \\ -5 & 0 & 15 & 0 & 60 \end{array} \right]$$

The -5 at the bottom of the first column tells us we must pivot again, with this first column as the pivot column. This time we have to compute quotients to determine the pivot element, which we see is the 5 in the a_{21}-location:

$$\begin{array}{c} \\ x_2 \\ x_4 \\ 2z \end{array} \left[\begin{array}{cccc|c} x_1 & x_2 & x_3 & x_4 & \\ 1 & 2 & 1 & 0 & 4 \\ 5 & 0 & 1 & 2 & 4 \\ -5 & 0 & 15 & 0 & 60 \end{array} \right] \begin{array}{l} 4/1 = 4 \\ 4/5 = 0.8 \end{array}$$

We now pivot, changing the basis and multiplying $2z$ by the new pivot element 5:

$$\begin{array}{c} \\ x_2 \\ r_1 \\ 10z \end{array} \left[\begin{array}{cccc|c} x_1 & x_2 & x_3 & x_4 & \\ 0 & 10 & 4 & -2 & 16 \\ 5 & 0 & 1 & 2 & 4 \\ 0 & 0 & 80 & 10 & 320 \end{array} \right]$$

Since there are no negative numbers in the last row, the basic solution will give us the answer to the problem. Ignoring the columns headed by the nonbasic variables x_3 and x_4, we see that $10x_2 = 16$ and $5x_1 = 4$; the basic feasible solution that solves the problem is

$$x_1 = \frac{4}{5}, \quad x_2 = \frac{8}{5}, \quad x_3 = 0, \quad x_4 = 0$$

2.5 THE SIMPLEX ALGORITHM FOR MAXIMIZATION PROBLEMS 191

Furthermore, from the last row we have $10z = 320$; thus, the maximum value of the objective function, subject to the given inequality restrictions, is $z = 32$. As a check of the accuracy of our work, we look back at the original statement of the problem to verify that

$$AX = \begin{bmatrix} 1 & 2 \\ 2 & -1 \end{bmatrix} \begin{bmatrix} \frac{4}{5} \\ \frac{8}{5} \end{bmatrix} = \begin{bmatrix} 4 \\ 0 \end{bmatrix} \leq \begin{bmatrix} 4 \\ 0 \end{bmatrix} = B$$

and that

$$z = CX = \begin{bmatrix} 10 & 15 \end{bmatrix} \begin{bmatrix} \frac{4}{5} \\ \frac{8}{5} \end{bmatrix} = \frac{40}{5} + \frac{120}{5} = 32$$

∎

In the examples thus far, the pivoting steps have simply replaced slack variables with independent variables of the original problem until no negative numbers remained in the last row. Our next example illustrates the fact that even quite small problems may require slack variables to reenter the basis before we arrive at the solution.

Example 3. An investor has $10,000 from which she would like to make as much money as possible. She plans to invest some money in stocks, some in bonds, and to put the rest in a savings account. The investor believes she can earn 8% on the money invested in stocks and 7% on bonds. The savings bank pays 5% interest. To balance her investments, she establishes the following rules. She will limit her investment in stocks to no more than half the amount invested in bonds. Her stock investment must also be less than or equal to the amount in her savings account. Finally, the amount she invests in stocks plus the amount in bonds will be no more than $8000. How much money should she invest in stocks, how much in bonds, and how much should she put in the bank?

Solution

We let x_1 represent the amount of money to be invested in stocks, x_2 be the amount in bonds, and x_3 be the amount in the bank. There is a total of $10,000 available, so

$$x_1 + x_2 + x_3 \leq 10{,}000$$

The amount x_1 that she will invest in stocks must be less than or equal to half the amount she invests in bonds, that is,

$$x_1 \leq \frac{1}{2} x_2$$

The requirement that x_1 be less than or equal to the amount in her savings account is described by the inequality

$$x_1 \leq x_3$$

Since the amount spent on stocks plus the amount on bonds is no more than $8000, we also have the condition

$$x_1 + x_2 \leq 8000$$

To summarize, the restrictions of the problem are

$$x_1 + x_2 + x_3 \leq 10{,}000$$
$$x_1 \leq \frac{1}{2}x_2$$
$$x_1 \leq x_3$$
$$x_1 + x_2 \leq 8000$$

We can put the second inequality in a more familiar form, first by multiplying both sides of the inequality by 2 to eliminate the fraction—so that $2x_1 \leq x_2$—then by subtracting x_2 from both sides to produce $2x_1 - x_2 \leq 0$. We also subtract x_3 from both sides of the third inequality. The conditions of the problem are

$$x_1 + x_2 + x_3 \leq 10{,}000$$
$$2x_1 - x_2 \leq 0$$
$$x_1 - x_3 \leq 0$$
$$x_1 + x_2 \leq 8000$$

from which we can read off the matrices

$$A = \begin{bmatrix} 1 & 1 & 1 \\ 2 & -1 & 0 \\ 1 & 0 & -1 \\ 1 & 1 & 0 \end{bmatrix} \quad \text{and} \quad B = \begin{bmatrix} 10{,}000 \\ 0 \\ 0 \\ 8000 \end{bmatrix}$$

The investor will earn 8% on the money invested in stocks, so her earnings from one dollar would be 8 cents, or $0.08. From $$x_1$ worth of stocks her earnings will therefore be $0.08x_1$. Similarly, her earnings from bonds and from her savings account will be $0.07x_2$ and $0.05x_3$, respectively. Thus,

$$\text{Earnings} = z = 0.08x_1 + 0.07x_2 + 0.05x_3$$

We can eliminate decimals by multiplying by 100, so we have

$$100z = 8x_1 + 7x_2 + 5x_3$$

and

$$C = \begin{bmatrix} 8 & 7 & 5 \end{bmatrix}$$

However, we must remember that the bottom row of the tableau will represent $100z$, rather than z, which we can do by labeling the row accordingly.

Since A is a 4×3-matrix, there must be $4 + 3 = 7$ variables in all; so, we will introduce slack variables x_4, x_5, x_6, and x_7. We put all this information together in the simplex tableau. Since -8 is the minimum number in the last row, the first column will be the pivot column; we divide the positive numbers in that column into the corresponding numbers in the last column:

	x_1	x_2	x_3	x_4	x_5	x_6	x_7		
x_4	1	1	1	1	0	0	0	10,000	10,000/1 = 10,000
x_5	2	-1	0	0	1	0	0	0	0/2 = 0
x_6	1	0	-1	0	0	1	0	0	0/1 = 0
x_7	1	1	0	0	0	0	1	8000	8000/1 = 8000
$100z$	-8	-7	-5	0	0	0	0	0	

2.5 THE SIMPLEX ALGORITHM FOR MAXIMIZATION PROBLEMS

The minimum quotient, zero, occurs in both the second and third rows. We should choose the row that has the smaller number in the pivot column. Therefore, we choose the third row, so that the pivot element is the 1 in the a_{13}-location, rather than the 2 above it. The reason for this tie-breaking rule is that the size of the numbers in the matrix will grow more slowly if the pivot element is smaller. Pivoting in the a_{13}-location causes x_1 to replace the slack variable x_6 in the basis.

$$
\begin{array}{c}
x_1x_2x_3x_4x_5x_6x_7 \\
\begin{array}{c} x_4 \\ x_5 \\ x_1 \\ x_7 \\ 100z \end{array}
\left[\begin{array}{ccccccc|c}
0 & 1 & 2 & 1 & 0 & -1 & 0 & 10{,}000 \\
0 & -1 & 2 & 0 & 1 & -2 & 0 & 0 \\
1 & 0 & -1 & 0 & 0 & 1 & 0 & 0 \\
0 & 1 & 1 & 0 & 0 & -1 & 1 & 8000 \\
\hline
0 & -7 & -13 & 0 & 0 & 8 & 0 & 0
\end{array}\right]
\begin{array}{l} 10{,}000/2 = 5000 \\ 0/2 = 0 \\ \\ 8000/1 = 8000 \\ \end{array}
\end{array}
$$

The minimum number in the last row is -13 in the third column, and there is no question this time where to pivot: at the 2 in the a_{23}-location. Thus, x_3 replaces x_5 in the basis, and we change the label $100z$ to $200z$.

$$
\begin{array}{c}
x_1x_2x_3x_4x_5x_6x_7 \\
\begin{array}{c} x_4 \\ x_3 \\ x_1 \\ x_7 \\ 200z \end{array}
\left[\begin{array}{ccccccc|c}
0 & 4 & 0 & 2 & -2 & 2 & 0 & 20{,}000 \\
0 & -1 & 2 & 0 & 1 & -2 & 0 & 0 \\
2 & -1 & 0 & 0 & 1 & 0 & 0 & 0 \\
0 & 3 & 0 & 0 & -1 & 0 & 2 & 16{,}000 \\
\hline
0 & -27 & 0 & 0 & 13 & -10 & 0 & 0
\end{array}\right]
\begin{array}{l} 20{,}000/4 = 5000 \\ \\ \\ 16{,}000/3 = 5333 \end{array}
\end{array}
$$

The next pivot, at the 4 in the a_{12}-location brings x_2 into the basis in place of x_4, and we multiply the label $200z$ for the last row by the pivot element.

$$
\begin{array}{c}
x_1x_2x_3x_4x_5x_6x_7 \\
\begin{array}{c} x_2 \\ x_3 \\ x_1 \\ x_7 \\ 800z \end{array}
\left[\begin{array}{ccccccc|c}
0 & 4 & 0 & 2 & -2 & 2 & 0 & 20{,}000 \\
0 & 0 & 8 & 2 & 2 & -6 & 0 & 20{,}000 \\
8 & 0 & 0 & 2 & 2 & 2 & 0 & 20{,}000 \\
0 & 0 & 0 & -6 & 2 & -6 & 8 & 4000 \\
\hline
0 & 0 & 0 & 54 & -2 & 14 & 0 & 540{,}000
\end{array}\right]
\begin{array}{l} \\ 20{,}000/2 = 10{,}000 \\ 20{,}000/2 = 10{,}000 \\ 4000/2 = 2000 \end{array}
\end{array}
$$

This time, as we pivot at the 2 in the a_{45}-location, we replace the slack variable x_7 with another slack variable x_5, which left the basis earlier. It is even possible in a more complicated problem that a slack variable will return to the basis as a replacement for one of the original independent variables. The label to the left of the bottom row is multiplied once more.

$$
\begin{array}{c}
x_1x_2x_3x_4x_5x_6x_7 \\
\begin{array}{c} x_2 \\ x_3 \\ x_1 \\ x_5 \\ 1600z \end{array}
\left[\begin{array}{ccccccc|c}
0 & 8 & 0 & -8 & 0 & -8 & 16 & 48{,}000 \\
0 & 0 & 16 & 16 & 0 & 0 & -16 & 32{,}000 \\
16 & 0 & 0 & 16 & 0 & 16 & -16 & 32{,}000 \\
0 & 0 & 0 & -6 & 2 & -6 & 8 & 4000 \\
\hline
0 & 0 & 0 & 96 & 0 & 16 & 16 & 1{,}088{,}000
\end{array}\right]
\end{array}
$$

194   LINEAR PROGRAMMING

There are no negative numbers in the bottom row, so we have found the solution. Reading down the rows, ignoring the columns below the nonbasic variables x_4, x_6, and x_7, we calculate that

$$x_2 = \frac{48{,}000}{8} = 6000, \quad x_3 = \frac{32{,}000}{16} = 2000$$

$$x_1 = \frac{32{,}000}{16} = 2000, \quad x_5 = \frac{4000}{2} = 2000$$

Thus, the basic feasible solution is

$$x_1 = 2000, \ x_2 = 6000, \ x_3 = 2000, \ x_4 = 0, \ x_5 = 2000, \ x_6 = 0, \text{ and } x_7 = 0$$

Furthermore, from the last row we have $z = \frac{1{,}088{,}000}{1600} = 680$ as the maximum value of the objective function. Looking back to the original statement of the problem, if the investor invests \$2000 in stocks, \$6000 in bonds, and puts \$2000 in a savings account, then she will earn as much as possible, \$680, under the conditions of the problem. ∎

We conclude this section by applying the simplex algorithm to a maximization problem that we solved by the graphical method in Section 2.2, to see how the two methods of solution compare.

Example 4. Solve the *Farmer's* problem, Example 5 of Section 2.2 by the simplex algorithm:

Maximize: Total Profit = $z = 200x_1 + 29x_2$
Subject to:
$$x_1 + x_2 \le 150$$
$$12x_1 + 6x_2 \le 1200$$
$$x_1, x_2 \ge 0$$

Solution

The initial simplex tableau is

$$\begin{array}{c} \\ x_3 \\ x_4 \\ z \end{array} \begin{array}{c} x_1 \quad x_2 \quad x_3 \quad x_4 \\ \left[\begin{array}{cccc|c} 1 & 1 & 1 & 0 & 150 \\ 12 & 6 & 0 & 1 & 1200 \\ \hline -200 & -29 & 0 & 0 & 0 \end{array}\right] \end{array} \begin{array}{l} 150/1 = 150 \\ 1200/12 = 100 \end{array}$$

with basic feasible solution $x_1 = 0$, $x_2 = 0$, $x_3 = 150$, $x_4 = 1200$. This step corresponds to computing the value of z at the extreme point with coordinates $(0, 0)$, which was called A in Figure 2-14 of Section 2.2. That value was $z = 0$, as we can see from the lower right-hand corner of the tableau. When we pivot at the 12 in the a_{21}-location, we obtain

$$\begin{array}{c} \\ x_3 \\ x_1 \\ 12z \end{array} \begin{array}{c} x_1 \quad x_2 \quad x_3 \quad x_4 \\ \left[\begin{array}{cccc|c} 0 & 6 & 12 & -1 & 600 \\ 12 & 6 & 0 & 1 & 1200 \\ \hline 0 & 852 & 0 & 200 & 240{,}000 \end{array}\right] \end{array}$$

2.5 THE SIMPLEX ALGORITHM FOR MAXIMIZATION PROBLEMS

We see that the corresponding basic feasible solution will solve the problem. It is

$$x_1 = \frac{1200}{12} = 100, \quad x_2 = 0, \quad x_3 = \frac{600}{12} = 50, \quad x_4 = 0$$

and the maximum value of the objective function is

$$z = \frac{240{,}000}{12} = 20{,}000 \qquad \blacksquare$$

The coordinates (100, 0) correspond to the point B that solved the problem in Section 2.2. Notice, though, that the simplex algorithm went directly to the extreme point B that gave the solution, without computing the value of the objective function at the other extreme points, C and D. Thus, even in such a small problem, the simplex algorithm is a very efficient way to find the extreme point that gives the solution. Industrial-sized LP-problems often involve thousands of variables, and it is for problems of this magnitude that the efficiency of the simplex algorithm is really valuable.

Summary

To solve an LP-problem:

Maximize: $z = CX$

Subject to: $AX \leq B, X \geq 0,$ and $B \geq 0$

where X is an $n \times 1$-matrix of independent variables, and the constant matrices A, B, and C are $m \times n$, $m \times 1$, and $1 \times n$, respectively, by the **simplex algorithm**, form the **simplex tableau** with a basis $\{x\}$ given by the slack variables $x_{n+1}, x_{n+2}, \ldots, x_{m+n} = x_N$. Thus,

$$\begin{array}{c} \\ \{x\} \\ z \end{array} \begin{array}{c} \begin{array}{ccc} x_1 & \cdots & x_N \end{array} \\ \left[\begin{array}{cc|c} A & I & B \\ \hline -C & O & 0 \end{array} \right] \end{array}$$

where O is the $1 \times m$-matrix of zeros. If there are any negative numbers in the last row, choose the column containing the minimum of that row as the pivot column. Divide each positive number in the pivot column into the number in the last column of the same row, and choose as the pivot row the one in which the quotient is smallest. Pivot at the **pivot element** that lies in the pivot row and column, changing the basis accordingly and multiplying the label of the last row by the pivot element. The corresponding basic solution is a **basic feasible solution** because it gives the coordinates of a point in the feasible set. Repeat the previous steps as long as there are any negative numbers in the last row. When no negative numbers remain in the last row, the basic feasible solution for that basis solves the LP-problem. The maximum value can be computed by dividing the number in the lower right-hand corner by the coefficient of the label z of the bottom row.

Exercises

In Exercises 1 through 4, solve the maximization problem by the simplex algorithm

$$\text{Maximize:} \qquad z = CX$$
$$\text{Subject to:} \qquad AX \leq B \text{ and } X \geq 0$$

1. $X = \begin{bmatrix} x_1 \\ x_2 \end{bmatrix}$, $A = \begin{bmatrix} 2 & 3 \\ 3 & 2 \end{bmatrix}$, $B = \begin{bmatrix} 6 \\ 5 \end{bmatrix}$, and $C = \begin{bmatrix} 4 & 5 \end{bmatrix}$

2. $X = \begin{bmatrix} x_1 \\ x_2 \\ x_3 \end{bmatrix}$, $A = \begin{bmatrix} 1 & 1 & 0 \\ 0 & 1 & 0 \\ 0 & 1 & 2 \end{bmatrix}$, $B = \begin{bmatrix} 2 \\ 1 \\ 3 \end{bmatrix}$, and $C = \begin{bmatrix} \frac{1}{2} & 2 & 1 \end{bmatrix}$

3. $X = \begin{bmatrix} x_1 \\ x_2 \end{bmatrix}$, $A = \begin{bmatrix} 1 & 1 \\ 1 & 2 \\ 1 & 0 \\ 0 & 1 \end{bmatrix}$, $B = \begin{bmatrix} 3 \\ 4 \\ \frac{5}{2} \\ \frac{3}{2} \end{bmatrix}$, and $C = \begin{bmatrix} 5 & 8 \end{bmatrix}$

4. $X = \begin{bmatrix} x_1 \\ x_2 \\ x_3 \end{bmatrix}$, $A = \begin{bmatrix} 1 & 0 & 0 \\ 0 & 4 & 1 \\ 1 & -1 & 0 \\ 0 & 0 & 1 \end{bmatrix}$, $B = \begin{bmatrix} 3 \\ 2 \\ 0 \\ 1 \end{bmatrix}$, and $C = \begin{bmatrix} 5 & 1 & 3 \end{bmatrix}$

In Exercises 5 through 10, solve by the simplex algorithm.

5. Maximize: $z = 100x_1 + 150x_2$
Subject to: $3x_1 + x_2 \leq 6$
$x_1 + 2x_2 \leq 4$
$x_2 \leq 1$
$x_1, x_2 \geq 0$

6. Maximize: $z = 7x_1 + 8x_2 + 10x_3$
Subject to: $x_1 + 2x_2 + 2x_3 \leq 20$
$x_1 \leq 2x_2$
$2x_1 + 3x_2 \leq 80$
$x_1 + x_2 \leq 6$
$x_1, x_2, x_3 \geq 0$

7. Maximize: $z = 5x_1 + 4x_2$
Subject to: $x_1 \leq 2x_2$
$x_2 \leq 1$
$x_1 + x_2 \leq 5$
$2x_1 + 3x_2 \leq 9$
$x_1 \geq x_2$
$x_1, x_2 \geq 0$

8. A company can produce up to 1500 digital watches a week in two models: standard and deluxe. The company will make at least as many standard models as deluxe. Furthermore, because of limited skilled labor, the number of standard models plus 3 times the number of deluxe models must add up to no more than 2000. The company sells a standard-model watch for $50 and a deluxe model for $150. How many of each model should be made each week in order to maximize revenue?

9. A politician plans to walk the length of her state to attract attention to her candidacy, to get better acquainted with the problems of her state, and to talk with many voters. She will spend part of her time in fast walking, part in leisurely strolling, and part in stopping to talk to voters. To balance the time spent in achieving the various goals, she decides to walk as far as she can each hour, but to have no more than $\frac{3}{4}$ hour for fast-walking plus strolling time. She will also limit her fast-walking time to no more than the total of talking time plus strolling time. If her fast-walking speed is 3 miles per hour, her strolling speed is 1 mile per hour, and she stands still while she is talking to voters, what fraction of each hour should she devote to each activity in order to go as far as possible in the hour?

10. An employee of an ice-cream shop wishes to make the richest ice-cream soda (measured in calories) that he can fit in a 12-ounce glass. The ingredients of a soda are syrup, cream, soda water, and ice cream. In order to look and taste like a soda, the mixture must contain no more than 4 ounces of ice cream, at least as much soda water as the total amount of syrup and cream combined, and no more than 1 ounce more of syrup than of cream. If syrup is 75 calories an ounce, cream is 50 calories an ounce, ice cream is 40 calories an ounce, and soda water contains no calories, how many ounces of each ingredient should be used?

2.6 Consequences of Duality

Solving Minimization Problems

We have shown how the simplex algorithm is used to solve maximization problems:

Maximize: $z = CX$

Subject to: $AX \leq B, X \geq 0,$ and $B \geq 0$

Suppose now that we have a minimization problem of the form

Minimize: $z = CX$

Subject to: $AX \geq B$ and $X \geq 0$

As we saw in Section 2.3, the dual of this problem is

Maximize: $u = B^T Y$

Subject to: $A^T Y \leq C^T$ and $Y \geq 0$

If we require that $C \geq 0$ in the minimization problem, that is, that the constants of the objective function be positive (or zero), then, in the dual maximization problem, we have the condition $C^T \geq 0$, which gives that maximization problem the form we need to apply the simplex algorithm of the last section. As we will see in the examples, the condition $C \geq 0$ is a natural one for many applications.

Let us recall what the duality theorem of linear programming tells us about a minimization problem,

Minimize: $\qquad z = CX$

Subject to: $\qquad AX \geq B$ and $X \geq 0$

and its dual maximization problem,

Maximize: $\qquad u = B^T Y$

Subject to: $\qquad A^T Y \leq C^T$ and $Y \geq 0$

According to the Duality Theorem, Min z = Max u. That is, the smallest value possible for the objective function $z = CX$, for values X satisfying the given conditions, equals the largest value of $u = B^T Y$ possible provided that the values Y are similarly restricted in the dual problem. But the detailed analysis that establishes the theorem actually does much more. It not only tells us that the "best" values of the objective functions of the two dual problems are equal, but that the simplex-algorithm solution to one problem gives the complete solution to the other problem, as well. Specifically, starting with a minimization problem,

Minimize: $\qquad z = CX$

Subject to: $AX \geq B$, $X \geq 0$, where, in addition, $C \geq 0$

suppose that we form the simplex tableau of the dual problem,

Maximize: $\qquad u = B^T Y$

Subject to: $\qquad A^T Y \leq C^T$, $Y \geq 0$, and $C^T \geq 0$

This latter problem looks like this:

$$\begin{array}{c} \\ \{y\} \\ u \end{array} \begin{array}{c} \begin{array}{ccc} y_1 & \cdots & y_N \end{array} \\ \left[\begin{array}{cc|c} A^T & I & C^T \\ \hline -B^T & 0 & 0 \end{array} \right] \end{array}$$

If A is an $m \times n$-matrix, then its transpose A^T is $n \times m$, so in particular, it has m columns which, in the tableau, are headed by the independent variables $y_1, y_2, \ldots, y_m$ that make up Y, the matrix of variables. The next n columns of the tableau are headed by the slack variables $y_{m+1}, y_{m+2}, \ldots, y_{m+n} = y_N$ that form the basis $\{y\}$. The label of the last row is given by u because the objective function of the maximization problem is $u = B^T Y$. We apply the simplex algorithm to this tableau, pivoting to modify the basis, until no negative numbers remain in the bottom row. Then the basic solution corresponding to the final basis solves the maximization problem. But this final simplex tableau solves the original minimization problem, as well, in the following way. As in the preceding section, the labels above the tableau remain unchanged throughout the simplex algorithm. The

2.6 CONSEQUENCES OF DUALITY

analysis on which the proof of the duality theorem is based tells us that the n numbers that lie in the last row below the n slack variables $y_{m+1}, y_{m+2}, \ldots, y_{m+n}$ provide a solution to the given minimization problem. That is, the transpose of that $1 \times n$-matrix is the matrix X that satisfies the restrictions of the problem and that minimizes $z = CX$.

An Example

A presentation of the details of the duality-theorem analysis is beyond the scope of this book; but it is not difficult to take advantage of what it tells us. We apply it first to the minimization problem

$$\text{Minimize:} \quad z = x_1 + 2x_2 + 2x_3$$
$$\text{Subject to:} \quad x_1 + 2x_2 + 4x_3 \geq 12$$
$$x_1 + x_3 \geq 4$$
$$-x_1 + 2x_2 - x_3 \geq 0$$
$$x_1, x_2, x_3 \geq 0$$

for which we have the matrices

$$X = \begin{bmatrix} x_1 \\ x_2 \\ x_3 \end{bmatrix}, \quad A = \begin{bmatrix} 1 & 2 & 4 \\ 1 & 0 & 1 \\ -1 & 2 & -1 \end{bmatrix}, \quad B = \begin{bmatrix} 12 \\ 4 \\ 0 \end{bmatrix}, \quad \text{and} \quad C = \begin{bmatrix} 1 & 2 & 2 \end{bmatrix}$$

Notice that the condition $C \geq 0$ is satisfied. We form the transpose matrices

$$A^T = \begin{bmatrix} 1 & 1 & -1 \\ 2 & 0 & 2 \\ 4 & 1 & -1 \end{bmatrix}, \quad B^T = \begin{bmatrix} 12 & 4 & 0 \end{bmatrix}, \quad \text{and} \quad C^T = \begin{bmatrix} 1 \\ 2 \\ 2 \end{bmatrix}$$

and from them the simplex tableau

$$\{y\} \begin{bmatrix} y_1 & \cdots & y_N & & \\ A^T & & I & & C^T \\ \hline -B^T & & 0 & & 0 \end{bmatrix} = \begin{array}{c} \\ y_4 \\ y_5 \\ y_6 \\ \hline u \end{array} \begin{bmatrix} y_1 & y_2 & y_3 & y_4 & y_5 & y_6 & \\ 1 & 1 & -1 & 1 & 0 & 0 & 1 \\ 2 & 0 & 2 & 0 & 1 & 0 & 2 \\ 4 & 1 & -1 & 0 & 0 & 1 & 2 \\ \hline -12 & -4 & 0 & 0 & 0 & 0 & 0 \end{bmatrix}$$

which is the tableau of a maximization problem with slack variables y_4, y_5, and y_6. We begin the simplex algorithm by noting that the first column will be our first pivot column because $-12 < -4$, and the quotients in the following tableau determine that we pivot at the 4 in the a_{31}-location:

$$\begin{array}{c} \\ y_4 \\ y_5 \\ y_6 \\ \hline u \end{array} \begin{bmatrix} y_1 & y_2 & y_3 & y_4 & y_5 & y_6 & \\ 1 & 1 & -1 & 1 & 0 & 0 & 1 \\ 2 & 0 & 2 & 0 & 1 & 0 & 2 \\ 4 & 1 & -1 & 0 & 0 & 1 & 2 \\ \hline -12 & -4 & 0 & 0 & 0 & 0 & 0 \end{bmatrix} \begin{array}{l} 1/1 = 1 \\ 2/2 = 1 \\ 2/4 = 0.5 \end{array}$$

LINEAR PROGRAMMING

Changing the labels so that y_1 replaces y_6 and multiplying u by the pivot element 4, the pivot has this result:

$$\begin{array}{c} \\ y_4 \\ y_5 \\ y_1 \\ \hline 4u \end{array} \begin{array}{c} \begin{array}{cccccc} y_1 & y_2 & y_3 & y_4 & y_5 & y_6 \end{array} \\ \left[\begin{array}{cccccc|c} 0 & 3 & -3 & 4 & 0 & -1 & 2 \\ 0 & -2 & 10 & 0 & 4 & -2 & 4 \\ 4 & 1 & -1 & 0 & 0 & 1 & 2 \\ \hline 0 & -4 & -12 & 0 & 0 & 12 & 24 \end{array} \right] \end{array}$$

Pivoting at the 10 in the a_{23}-location brings us to

$$\begin{array}{c} \\ y_4 \\ y_3 \\ y_1 \\ \hline 40u \end{array} \begin{array}{c} \begin{array}{cccccc} y_1 & y_2 & y_3 & y_4 & y_5 & y_6 \end{array} \\ \left[\begin{array}{cccccc|c} 0 & 24 & 0 & 40 & 12 & -16 & 32 \\ 0 & -2 & 10 & 0 & 4 & -2 & 4 \\ 40 & 8 & 0 & 0 & 4 & 8 & 24 \\ \hline 0 & -64 & 0 & 0 & 48 & 96 & 288 \end{array} \right] \end{array}$$

At this point, the numbers in the matrix are getting uncomfortably large. Each row represents a linear equation, so we can cut the numbers down to size by dividing each row by a convenient constant. For instance, the numbers in the first row are all multiples of 4; we will divide that row by 4. Similarly, we divide the second row by 2 and the third by 4. We also can divide the last row by 4, so long as we recall that the row represents a linear function of the form

$$40u - 16y_2 + 48y_5 + 96y_6 = 288$$

Therefore, when we divide, we must also divide the label $40u$. Now we have the more manageable tableau

$$\begin{array}{c} \\ y_4 \\ y_3 \\ y_1 \\ \hline 10u \end{array} \begin{array}{c} \begin{array}{cccccc} y_1 & y_2 & y_3 & y_4 & y_5 & y_6 \end{array} \\ \left[\begin{array}{cccccc|c} 0 & 6 & 0 & 10 & 3 & -4 & 8 \\ 0 & -1 & 5 & 0 & 2 & -1 & 2 \\ 10 & 2 & 0 & 0 & 1 & 2 & 6 \\ \hline 0 & -16 & 0 & 0 & 12 & 24 & 72 \end{array} \right] \begin{array}{l} 8/6 = 1.33 \\ \\ 6/2 = 3 \end{array} \end{array}$$

From the quotients on the right, we see that we must pivot at the 6 in the a_{12}-location:

$$\begin{array}{c} \\ y_2 \\ y_3 \\ y_1 \\ \hline 60u \end{array} \begin{array}{c} \begin{array}{cccccc} y_1 & y_2 & y_3 & y_4 & y_5 & y_6 \end{array} \\ \left[\begin{array}{cccccc|c} 0 & 6 & 0 & 10 & 3 & -4 & 8 \\ 0 & 0 & 30 & 10 & 15 & -10 & 20 \\ 60 & 0 & 0 & -20 & 0 & 20 & 20 \\ \hline 0 & 0 & 0 & 160 & 120 & 80 & 560 \end{array} \right] \end{array}$$

Now that all the numbers in the last row are positive or zero, we have completed the simplex algorithm. The answer to the original minimization problem will lie in that row; but first we must express the row in terms of the dependent variable u, rather than in terms of $60u$. We divide the bottom row by 60, expressing the answer in the most convenient form which, in this case, is reduced fractions:

$$u \begin{bmatrix} y_1 & y_2 & y_3 & y_4 & y_5 & y_6 & \\ 0 & 0 & 0 & \frac{8}{3} & 2 & \frac{4}{3} & \frac{28}{3} \end{bmatrix}$$

According to the duality theory that we described at the beginning of this section, the numbers below the slack variables y_4, y_5, and y_6 form the transpose of the solution matrix X in the original minimization problem:

$$X = \begin{bmatrix} x_1 \\ x_2 \\ x_3 \end{bmatrix} = \begin{bmatrix} \frac{8}{3} \\ 2 \\ \frac{4}{3} \end{bmatrix}$$

From the final simplex tableau,

$$\text{Max } u = \frac{560}{60} = \frac{28}{3}$$

so, by the duality theorem, we also know that

$$\text{Min } z = \text{Max } u = \frac{28}{3}$$

It is a good idea to check your work to be sure, first, that the solution

$$x_1 = \frac{8}{3}, \quad x_2 = 2, \quad x_3 = \frac{4}{3}$$

is in the feasible set: this we can do by means of the matrix calculation

$$AX = \begin{bmatrix} 1 & 2 & 4 \\ 1 & 0 & 1 \\ -1 & 2 & -1 \end{bmatrix} \begin{bmatrix} \frac{8}{3} \\ 2 \\ \frac{4}{3} \end{bmatrix} = \begin{bmatrix} 12 \\ 4 \\ 0 \end{bmatrix} \geq \begin{bmatrix} 12 \\ 4 \\ 0 \end{bmatrix} = B$$

We also substitute into the original objective function:

$$z = x_1 + 2x_2 + 2x_3 = \frac{8}{3} + 2(2) + 2\left(\frac{4}{3}\right) = \frac{8}{3} + \frac{12}{3} + \frac{8}{3} = \frac{28}{3}$$

Example 1. A company hires a lazy, but effective, salesperson on the understanding that he will contact at least 100 potential customers a week, by means of personal visits, telephone calls, or personalized letters; that at least ten of the contacts will be by personal visit, and that at least 30 customers will be spoken to each week, either in person or over the telephone. The salesperson decides that he will have to make at least fifteen sales a week in order to support himself from commissions. His experience is that, in general, he can make one sale from two personal visits, from ten telephone calls, or from twenty letters. He estimates that it takes him 1 hour to make a personal visit, 15 minutes for a telephone call, and 6 minutes to dictate a letter. How many personal visits, telephone calls, and letters should he plan so that he can work as few hours as possible?

Solution

Let x_1 be the number of personal visits the salesperson makes, let x_2 be the number of telephone calls, and let x_3 be the number of letters he writes. The conditions of his employment are

$$\begin{aligned} x_1 + x_2 + x_3 &\geq 100 \\ x_1 &\geq 10 \\ x_1 + x_2 &\geq 30 \end{aligned}$$

If he makes x_1 personal visits, since he expects to make a sale in half of them, he will obtain $\frac{1}{2}x_1$ sales in this way. Since one out of ten telephone calls produces a sale, he will make $\frac{1}{10}x_2$ sales from x_2 telephone calls. Similarly, he expects to get $\frac{1}{20}x_3$ sales on the basis of his letters. He needs at least 15 sales in all, so the requirement is

$$\frac{1}{2}x_1 + \frac{1}{10}x_2 + \frac{1}{20}x_3 \geq 15$$

Since the point of our pivoting method is to avoid the use of fractions, we must eliminate the fractions from this inequality. We can easily do so just by multiplying the inequality by 20, which gives the equivalent restriction

$$10x_1 + 2x_2 + x_3 \geq 300$$

We can represent these restrictions on $x_1, x_2,$ and x_3 by the matrix inequality

$$AX \geq B$$

where

$$A = \begin{bmatrix} 1 & 1 & 1 \\ 1 & 0 & 0 \\ 1 & 1 & 0 \\ 10 & 2 & 1 \end{bmatrix} \quad \text{and} \quad B = \begin{bmatrix} 100 \\ 10 \\ 30 \\ 300 \end{bmatrix}$$

It takes the salesperson 1 hour, or 60 minutes, to make a visit; so, it requires $60x_1$ minutes to make x_1 visits. Since a telephone call requires 15 minutes and a letter 6 minutes, the salesperson's total investment of time is

$$\text{Time} = z = 60x_1 + 15x_2 + 6x_3 = CX$$

where

$$C = \begin{bmatrix} 60 & 15 & 6 \end{bmatrix}$$

We note that $C \geq 0$, as required. Forming the transposes

$$A^T = \begin{bmatrix} 1 & 1 & 1 & 10 \\ 1 & 0 & 1 & 2 \\ 1 & 0 & 0 & 1 \end{bmatrix}, \quad B^T = \begin{bmatrix} 100 & 10 & 30 & 300 \end{bmatrix}, \quad \text{and} \quad C^T = \begin{bmatrix} 60 \\ 15 \\ 6 \end{bmatrix}$$

the simplex tableau is

	y_1	y_2	y_3	y_4	y_5	y_6	y_7		
y_5	1	1	1	10	1	0	0	60	$60/10 = 6$
y_6	1	0	1	2	0	1	0	15	$15/2 = 7.5$
y_7	1	0	0	1	0	0	1	6	$6/1 = 6$
u	-100	-10	-30	-300	0	0	0	0	

2.6 CONSEQUENCES OF DUALITY   203

where, this time, the slack variables are y_5, y_6, and y_7. The fourth column is the pivot column. We have a choice of the first or third row as the pivot row; we choose the third so that we can pivot at the 1 in the a_{34}-location. After that pivot we obtain

$$\begin{array}{c} \\ y_5 \\ y_6 \\ y_4 \\ u \end{array} \begin{array}{|ccccccc|c|} y_1 & y_2 & y_3 & y_4 & y_5 & y_6 & y_7 & \\ -9 & 1 & 1 & 0 & 1 & 0 & -10 & 0 \\ -1 & 0 & 1 & 0 & 0 & 1 & -2 & 3 \\ 1 & 0 & 0 & 1 & 0 & 0 & 1 & 6 \\ \hline 200 & -10 & -30 & 0 & 0 & 0 & 300 & 1800 \end{array} \begin{array}{l} 0/1 = 0 \\ 3/1 = 3 \\ \\ \end{array}$$

Then pivoting at the a_{13}-location yields:

$$\begin{array}{c} \\ y_3 \\ y_6 \\ y_4 \\ u \end{array} \begin{array}{|ccccccc|c|} y_1 & y_2 & y_3 & y_4 & y_5 & y_6 & y_7 & \\ -9 & 1 & 1 & 0 & 1 & 0 & -10 & 0 \\ 8 & -1 & 0 & 0 & -1 & 1 & 8 & 3 \\ 1 & 0 & 0 & 1 & 0 & 0 & 1 & 6 \\ \hline -70 & 20 & 0 & 0 & 30 & 0 & 0 & 1800 \end{array} \begin{array}{l} \\ 3/8 = 0.375 \\ 6/1 = 6 \\ \end{array}$$

The final pivot is at the 8 in the a_{21}-location:

$$\begin{array}{c} \\ y_3 \\ y_1 \\ y_4 \\ 8u \end{array} \begin{array}{|ccccccc|c|} y_1 & y_2 & y_3 & y_4 & y_5 & y_6 & y_7 & \\ 0 & -1 & 8 & 0 & -1 & 9 & -8 & 27 \\ 8 & -1 & 0 & 0 & -1 & 1 & 8 & 3 \\ 0 & 1 & 0 & 8 & 1 & -1 & 0 & 45 \\ \hline 0 & 90 & 0 & 0 & 170 & 70 & 560 & 14{,}610 \end{array}$$

The label $8u$ at the left tells us that we must divide the bottom row by 8 to obtain the answer:

$$u \begin{array}{|ccccccc|c|} y_1 & y_2 & y_3 & y_4 & y_5 & y_6 & y_7 & \\ 0 & \frac{90}{8} & 0 & 0 & \frac{170}{8} & \frac{70}{8} & \frac{560}{8} & \frac{14{,}610}{8} \end{array}$$

Duality theory gives us the answer to the original problem: the salesperson should make $x_1 = \frac{170}{8} = 21.25$ visits (on the average), he should make $x_2 = \frac{70}{8} = 8.75$ telephone calls, and should write $x_3 = \frac{560}{8} = 70$ letters each week for a minimal investment of time of $z = \frac{14{,}610}{8}$ minutes, which is a bit less than 30.5 hours. ∎

Linear Programming in General

We have presented the simplex algorithm only for certain types of maximization and minimization problems. A maximization problem,

Maximize: $\quad z = CX$

Subject to: $\quad AX \le B, X \ge 0$, and $B \ge 0$

has a simplex tableau

$$\begin{array}{c} \{x\} \\ z \end{array} \left[\begin{array}{cc|c} A & I & B \\ -C & 0 & 0 \end{array} \right] \begin{array}{c} x_1 \cdots x_N \end{array}$$

with two crucial features: a basis $\{x\}$ for every row but the bottom one, here given by the slack variables, and a last column of nonnegative numbers, so that the corresponding basic solution consists of nonnegative numbers and is therefore feasible. A matrix with these features is said to be in **canonical form**. The simplex algorithm directs us to make a series of pivots that changes the basis and the numbers in the matrix, but at each stage we still have a tableau in canonical form.

In general, an LP-problem is any problem that requires the maximization or the minimization of a linear function, where the values of the independent variables must satisfy a set of linear conditions, either inequalities or equations. For instance, the following is an LP-problem:

Maximize: $z = 2x_1 + 3x_2 - x_3 + x_4$
Subject to:
$$x_1 + 2x_2 - 3x_3 + x_4 \leq -2$$
$$3x_1 \quad\quad + x_3 - 3x_4 \geq 3$$
$$x_2 + 2x_3 + 3x_4 = 1$$
$$x_1 + 4x_2 - x_3 - 2x_4 \geq -4$$
$$x_1, x_2, x_3, x_4 \geq 0$$

The difficulty with an LP-problem of this type is that, even after introducing slack variables to change the inequalities to equations, the problem cannot be put into canonical form immediately so that the simplex algorithm can be applied to it. However, there is a straightforward extension of the simplex algorithm that will solve problems as general as this one.

Another *Farmer's* Problem

To conclude this section, we present a quite different application of the duality theorem of linear programming. We illustrate it with another agricultural problem.

A farmer will plant barley and sugar beets on his 200-acre farm. He has available 2500 acre-inches of water with which to irrigate the crops. An acre of barley requires 5 inches of water and an acre of sugar beets, 15 inches. He estimates that the net profit per acre from each of the crops will be $50 for barley and $80 for sugar beets. How many acres of each crop should he plant to maximize his net profit.

When we write this problem in mathematical form, with the acreage x_1 for barley and x_2 for sugar beets, we obtain

Maximize: Profit $= z = 50x_1 + 80x_2$
Subject to:
$$x_1 + x_2 \leq 200 \quad \text{(land)}$$
$$5x_1 + 15x_2 \leq 2500 \quad \text{(water)}$$
$$x_1, x_2 \geq 0$$

From the matrices

$$A = \begin{bmatrix} 1 & 1 \\ 5 & 15 \end{bmatrix}, \quad B = \begin{bmatrix} 200 \\ 2500 \end{bmatrix}, \quad \text{and} \quad C = \begin{bmatrix} 50 & 80 \end{bmatrix}$$

we have the initial tableau

2.6 CONSEQUENCES OF DUALITY

$$\begin{array}{c} \\ x_3 \\ x_4 \\ z \end{array} \begin{bmatrix} x_1 & x_2 & x_3 & x_4 & \\ 1 & 1 & 1 & 0 & 200 \\ 5 & 15 & 0 & 1 & 2500 \\ \hline -50 & -80 & 0 & 0 & 0 \end{bmatrix}$$

After two pivots, this becomes the tableau

$$\begin{array}{c} \\ x_1 \\ x_2 \\ 150z \end{array} \begin{bmatrix} x_1 & x_2 & x_3 & x_4 & \\ 10 & 0 & 15 & -1 & 500 \\ 0 & 150 & -75 & 15 & 22,500 \\ \hline 0 & 0 & 5250 & 450 & 2,175,000 \end{bmatrix}$$

From this we conclude that the farmer should plant $x_1 = \frac{500}{10} = 50$ acres of barley and $x_2 = \frac{22,500}{150} = 150$ acres of sugar beets. Since the label on the bottom row is $150z$, we divide that row by 150 obtaining:

$$z \begin{bmatrix} x_1 & x_2 & x_3 & x_4 & \\ 0 & 0 & 35 & 3 & 14,500 \end{bmatrix}$$

Thus, the farmer can earn a profit of $14,500.

Marginal Value

So far, we have just used the simplex algorithm; but now we carry our investigation a step further. The farmer's profit is $50 an acre for barley and $80 an acre for sugar beets. The more acreage the farmer plants, the more profit he can make; so, if he had a farm of more than 200 acres, he could have a larger profit. Barley requires 5 inches of water per acre, whereas the more profitable crop, sugar beets, requires 15 inches. Even with the given farm acreage, if the farmer had more water available, he could use more acres for the more profitable crop. We wish to determine how much more profit there would be for the farmer if he could acquire extra land and/or additional water.

In the terminology of economics, the *marginal value* of a resource is the rate at which income increases as the availability of the resource increases. In a linear-programming setting, we suppose that we have a maximization problem and that we know its optimal solution. In a linear inequality of the problem, the **marginal value** of the limitation described by that inequality is the amount we would increase the value of the objective function by making the constant on the right-hand side of the inequality one unit larger and using the corresponding optimal solution. Thus, the marginal value of the land is the additional profit that the farmer could earn if he had one more acre of land, and the same amount, 2500 acre-inches, of water. Similarly, the marginal value of the water is the additional profit from one more acre-inch of water for a 200-acre farm. Of course, we could repeat the preceding simplex algorithm, just changing the first inequality to

$$x_1 + x_2 \leq 201$$

and see what happens to the value of the objective function

$$z = 50x_1 + 80x_2$$

Then we could do the same, replacing the second inequality with

$$5x_1 + 15x_2 \leq 2501$$

Fortunately, all that computation is unnecessary because the marginal values are already available in the final tableau.

To see why the tableau contains this information, we need to use duality. Recall that for a maximization problem,

Maximize: $z = CX$
Subject to: $AX \leq B$ and $X \geq 0$

the objective function of the dual (minimization) problem is $u = B^T Y$, where Y has as many independent variables in it as there are slack variables in the maximization problem. Each independent variable of the dual problem corresponds to an inequality in the system $AX \leq B$, numbered in the same order. For the new *Farmer's* problem we can organize this information in Table 2-17:

Variable of Dual	Equation with Slack Variable	Interpretation
y_1	$x_1 + x_2 + x_3 = 200$	Land
y_2	$5x_1 + 15x_2 + x_4 = 2500$	Water

Table 2-17

Looking at the original matrix B, we see that the objective function of the dual problem is

$$u = 200y_1 + 2500y_2$$

To solve minimization problems, we have been using the fact that, when the bottom row in the final tableau is written to express the objective function z, the numbers below the slack variables of the maximization problem give the solution to the dual minimization problem. In the new *Farmer's* problem we have

$$z \begin{bmatrix} x_1 & x_2 & x_3 & x_4 \\ 0 & 0 & 35 & 3 & | & 14{,}500 \end{bmatrix}$$

thus, the solution to the minimization problem is $y_1 = 35$ and $y_2 = 3$. When we substitute the solutions to the two dual problems into their respective objective functions,

$$z = 50x_1 + 80x_2 = 50(50) + 80(150) = 14{,}500$$
$$u = 200y_1 + 2500y_2 = 200(35) + 2500(3) = 14{,}500$$

we get the same answer, just as the duality theorem says we must.

Now, returning to the subject of marginal value, suppose the farmer had 201 acres of land, instead of 200. This changes the first number in the matrix B, and the objective function of the dual becomes

$$u = 201y_1 + 2500y_2$$

Then, when we substitute the solution into this linear function, the minimum of u rises to

$$u = 201(35) + 2500(3) = 14{,}535$$

But then, by the duality theorem,

$$\text{Max } z = \text{Min } u = 14{,}535$$

so, the farmer's profit increases by the same amount, \$35. Therefore, the marginal value of an acre of land is \$35. In just the same way, if there are 2501 acre-inches of water available,

$$\text{Min } u = 200(35) + 2501(3) = 14{,}503$$

so that the additional acre-inch of water is worth \$3 to the farmer. Thus, the solution $y_1 = 35$, $y_2 = 3$ calculated from the last row of the tableau also gives us the marginal values. Since x_3 is the slack variable for the inequality limiting the availability of land, the number 35 in the last row of that column is the marginal value of an acre of land. Similarly, x_4 is the slack variable for the water inequality; the 3 at the bottom of its column is the marginal value of an acre-inch of water.

Calculating Marginal Values

In the same way, we can apply the duality theorem to determine the marginal values of any maximization problem of the type discussed in the preceding section. For each inequality of the system $AX \leq B$, we have a different slack variable. We suppose that the simplex algorithm has eliminated all negative numbers from the bottom row of the tableau and that we have divided this row so that it represents the objective function z, rather than a multiple of it. Then the number in the bottom row that lies in the column headed by a slack variable is the **marginal value** of the limitation described by the corresponding linear inequality of the maximization problem. That is, the number in that location is the amount the objective function would increase if the limitation were one unit larger.

Example 2. In Example 3 of Section 2.5 we solved the following problem:

An investor has \$10,000 from which she would like to make as much money as possible. She plans to invest some money in stocks, some in bonds, and to put the rest in a savings account. The investor believes she can earn 8% on stocks and 7% on bonds. The savings bank pays 5% interest. To balance her investments, she establishes the following rules. She will limit her investment in stocks to no more than half the amount invested in bonds. Her stock investment must also be less than or equal to the amount in her savings account. Finally, the amount she invests in stocks plus the amount she invests in bonds will be no more than \$8000. How much money should she invest in stocks, how much in bonds, and how much should she put in the bank?

Analyze the marginal values of the restrictions of this problem.

Solution

In the solution, we let x_1 be the amount of money to be invested in stocks, x_2 the amount in bonds, and x_3 the amount in the bank. The restrictions on the investor we listed there were:

$$\begin{aligned} x_1 + x_2 + x_3 &\le 10{,}000 \\ 2x_1 - x_2 &\le 0 \\ x_1 \qquad - x_3 &\le 0 \\ x_1 + x_2 \qquad &\le 8000 \end{aligned}$$

The marginal value analysis will indicate to the investor the financial effect of relaxing each of these restrictions. We introduce the slack variables and give each equation a convenient name to remind ourselves what it represents.

$$\begin{aligned} x_1 + x_2 + x_3 + x_4 \qquad\qquad\qquad &= 10{,}000 \quad \text{(total available)} \\ 2x_1 - x_2 \qquad + x_5 \qquad\qquad &= \qquad 0 \quad \text{(stocks vs. bonds)} \\ x_1 \qquad - x_3 \qquad + x_6 \qquad &= \qquad 0 \quad \text{(stocks vs. savings)} \\ x_1 + x_2 \qquad\qquad\qquad + x_7 &= 8000 \quad \text{(total stocks and bonds)} \end{aligned}$$

The last tableau of Example 3 of Section 2.5 was

	x_1	x_2	x_3	x_4	x_5	x_6	x_7	
x_2	0	8	0	−8	0	−8	16	48,000
x_3	0	0	16	16	0	0	−16	32,000
x_1	16	0	0	16	0	16	−16	32,000
x_5	0	0	0	−6	2	−6	8	4,000
$1600z$	0	0	0	96	0	16	16	1,088,000

Dividing the last row by 1600, so that it represents the objective function z, we obtain

	x_1	x_2	x_3	x_4	x_5	x_6	x_7	
z	0	0	0	0.06	0	0.01	0.01	680

Thus, the marginal value of the total investment of \$10,000 is \$0.06, that is, the first additional dollar invested would yield earnings of 6 cents. Of the two restrictions on the stock investment, the first, in terms of bonds, has zero marginal value and the second, with respect to savings, a marginal value of \$0.01. The final marginal value, \$0.01, on the restriction on the investment in stocks and bonds to \$8000, means that increasing the limit to \$8001 and at the same time keeping the total investment at \$10,000, would only increase the investors earnings by one cent. ∎

We have now given a complete marginal-value analysis of the investor's problem, but we have two further comments on the subject. The solution to that problem in Section 2.5 was

$$x_1 = 2000, \quad x_2 = 6000, \quad \text{and} \quad x_3 = 2000$$

The restriction corresponding to the slack variable x_5, that no more than half the amount invested in bonds be invested in stocks, is more than satisfied, that is, $x_1 = 2000$ is much less than $\frac{1}{2} x_2 = 3000$. The inequality restriction with slack variable x_5 has zero marginal value because this inequality does not, in fact, impose any limitation on the investment decision. Our final comment is that you should not conclude that, because investing an additional dollar will yield 6 cents, that an additional $2000 will give the investor an increase in earnings of $(0.06)(2000) = 120$ dollars, for a total of $680 + 120 = 800$ dollars. In fact, if we do change the total investment restriction to $x_1 + x_2 + x_3 \leq 12{,}000$, but leave all the other restrictions as before, the simplex algorithm finds that the most profitable investment is to use $2,666.67 for stocks, $5,333.33 for bonds, and to leave $4000 in savings, which gives the investor a total earnings of $786.66, not $800.

Although we have restricted our marginal-value analysis to maximization problems, the duality theorem can also be used to analyze marginal value for minimization problems.

Summary

The minimization problem

Minimize: $\qquad z = CX$

Subject to: $\qquad AX \geq B, X \geq 0,$ and $C \geq 0$

can be solved by solving the dual problem

Maximize: $\qquad u = B^T Y$

Subject to: $\qquad A^T Y \leq C^T$ and $Y \geq 0$

by the simplex algorithm. When negative numbers no longer remain in the last row of the tableau of the dual maximization problem, the numbers that lie in that row and below the slack variables form a row matrix that is the transpose of the matrix X that minimizes $z = CX$.

Suppose that, given a maximization problem

Maximize: $\qquad z = CX$

Subject to: $\qquad AX \leq B, X \geq 0,$ and $B \geq 0,$

the simplex algorithm has eliminated all negative numbers from the bottom row of the tableau, and that row has been divided so that it represents the objective function z. Then the number in the bottom row and in a column headed by a slack variable is the **marginal value** of the limitation described by the linear inequality that defines that slack variable. That is, it is the amount the objective function would increase if the constant on the right-hand side of the inequality were one unit larger.

Exercises

In Exercises 1 through 6, solve by the simplex algorithm the minimization problem,

$$\text{Minimize:} \qquad z = CX$$
$$\text{Subject to:} \qquad AX \geq B \text{ and } X \geq 0.$$

1. $X = \begin{bmatrix} x_1 \\ x_2 \end{bmatrix}$, $A = \begin{bmatrix} 6 & 1 \\ 1 & 1 \end{bmatrix}$, $B = \begin{bmatrix} 1 \\ 2 \end{bmatrix}$, and $C = \begin{bmatrix} 8 & 6 \end{bmatrix}$

2. $X = \begin{bmatrix} x_1 \\ x_2 \end{bmatrix}$, $A = \begin{bmatrix} 0 & 1 \\ 2 & 3 \end{bmatrix}$, $B = \begin{bmatrix} 2 \\ 5 \end{bmatrix}$, and $C = \begin{bmatrix} 1 & 3 \end{bmatrix}$

3. $X = \begin{bmatrix} x_1 \\ x_2 \end{bmatrix}$, $A = \begin{bmatrix} 1 & 3 \\ 2 & 1 \\ 0 & 3 \end{bmatrix}$, $B = \begin{bmatrix} 1 \\ 2 \\ 3 \end{bmatrix}$, and $C = \begin{bmatrix} 4 & 2 \end{bmatrix}$

4. $X = \begin{bmatrix} x_1 \\ x_2 \\ x_3 \end{bmatrix}$, $A = \begin{bmatrix} 1 & 0 & 0 \\ 0 & 0 & 4 \\ 1 & 2 & 0 \\ 1 & 1 & 1 \end{bmatrix}$, $B = \begin{bmatrix} 1 \\ 2 \\ 3 \\ 1 \end{bmatrix}$, and $C = \begin{bmatrix} 4 & 2 & 6 \end{bmatrix}$

5. $X = \begin{bmatrix} x_1 \\ x_2 \\ x_3 \end{bmatrix}$, $A = \begin{bmatrix} 1 & 1 & 1 \\ 4 & 0 & 1 \\ 2 & 3 & -1 \end{bmatrix}$, $B = \begin{bmatrix} 10 \\ 8 \\ 0 \end{bmatrix}$, and $C = \begin{bmatrix} 2 & 1 & 4 \end{bmatrix}$

6. $X = \begin{bmatrix} x_1 \\ x_2 \end{bmatrix}$, $A = \begin{bmatrix} 2 & 0 \\ 1 & 1 \\ 1 & 2 \end{bmatrix}$, $B = \begin{bmatrix} 3 \\ 6 \\ 2 \end{bmatrix}$, and $C = \begin{bmatrix} 4 & 2 \end{bmatrix}$

In Exercises 7 through 15, solve by the simplex algorithm.

7. Minimize: $z = 2x_1 + x_2$
   Subject to: $2x_1 \geq x_2$
   $2x_1 + 3x_2 \geq 4$
   $x_1, x_2 \geq 0$

8. Minimize: $z = 5x_1 + 4x_2 + 6x_3$
   Subject to: $2x_1 + x_2 + 3x_3 \geq 2$
   $x_1 + 2x_2 + x_3 \geq 1$
   $x_1, x_2, x_3 \geq 0$

9. Maximize: $z = 3x_1 + 2x_2 + x_3 + 4x_4$
   Subject to: $x_1 + x_2 + x_4 \leq 1$
   $x_1 \leq x_4$
   $x_1 + 2x_2 + 2x_3 \leq 3$
   $x_2 \leq 4$
   $x_2 + x_3 \leq 2$
   $x_1, x_2, x_3, x_4 \geq 0$

10. Maximize: $z = 3x_1 + 2x_2 + 4x_3$
    Subject to: $x_1 + x_2 + 2x_3 \leq 8$
    $x_1 + x_2 \leq 5$
    $2x_2 \leq 1$
    $x_1, x_2, x_3 \geq 0$

2.6 CONSEQUENCES OF DUALITY   211

11. A man plans to plant at least 1000 square feet of his yard with vegetables and perennial flowers. He will devote at least as much space to vegetables as to flowers. The man estimates that growing vegetables requires 2 units of work for each square foot and flowers require 1 unit of work for each square foot. How many square feet should he devote to growing vegetables and how many to growing flowers so as to do the least possible work?

12. A jeweler makes rings, earrings, pins, and necklaces. He wishes to work no more than 40 hours a week. It takes him 2 hours to make a ring, 2 hours to make a pair of earrings, 1 hour to make a pin, and 4 hours to make a necklace. He will make no more pieces of each type of jewelry than he can sell in a week. He estimates that he can sell no more than ten rings, ten pairs of earrings, fifteen pins, and three necklaces in a week. The jeweler charges $50 for a ring, $80 for a pair of earrings, $25 for a pin, and $200 for a necklace.
    (a) How many rings, earrings, pins, and necklaces should he make in order to maximize gross earnings?
    (b) Perform a marginal-value analysis of the jeweler's problem.

13. A woman wishes to design a weekly exercise schedule that will involve jogging, bicycling, and swimming. In order to vary the exercises, she plans to spend at least as many hours bicycling as the total number of hours jogging and swimming. Also, she wishes to swim at least 2 hours a week because she enjoys that activity more than the others. If jogging consumes 600 calories an hour, bicycling uses up 300 calories an hour, and swimming requires 300 calories an hour, and if she wishes to burn up a total of at least 3000 calories a week through exercising, how many hours should she devote to each type of exercise if she wishes to reach this goal in the fewest number of hours?

14. In Exercise 13, in a week in which the woman has only 7 hours to spend on exercising, she realizes that she may not be able to burn up her usual 3000 calories. She still wishes to spend at least as many hours bicycling as the total number of hours jogging and swimming. She also decides that she should not devote more than 5 hours altogether to bicycling and jogging.
    (a) If she wishes to use up as many calories as possible within 7 hours, how much time should she devote to each type of exercise?
    (b) Perform a marginal-value analysis of this form of the exercise problem.

15. A company makes one type of chair and one type of table. The factory consists of a machine shop with 15 employees and an assembly-and-finishing line with 25 employees. Each employee works 8 hours a day. It requires 2 work-hours of machine-shop labor and 2 work-hours of assembly-and-finishing labor to produce a chair, and it takes $\frac{3}{2}$ work-hours of machine-shop labor and 4 work-hours of assembly-and-finishing labor to make a table. The company will manufacture no more than 60 chairs and no more than 45 tables a day because that is the most it can sell to wholesalers. The manufacturer's profit is $75 for each chair and $100 for each table.
    (a) How many chairs and how many tables should the manufacturer produce each day in order to maximize profit?
    (b) Perform a marginal-value analysis of the manufacturer's problem.

2.7 A Farm-Management Problem

In the introduction to this chapter, we described the problem of a farmer in Kern County, California who wanted to plant his 150 acres of land so that it would make as much profit as possible. The major crops in Kern County are cotton, potatoes, alfalfa, sugar beets, and barley. The estimated dollar profit per acre of each crop is given in Table 2-18.

Crop	Profit
Cotton	$207
Potatoes	$200
Alfalfa	$ 86
Sugar Beets	$136
Barley	$ 29

Table 2-18

The Variables and the Objective Function

We write x_1, x_2, x_3, x_4, and x_5, respectively, for the numbers of acres of cotton, potatoes, alfalfa, sugar beets, and barley that the farmer plants. From Table 2-18, the total profit the farmer expects to earn from his farm is calculated by the formula

$$\text{Profit} = z = 207x_1 + 200x_2 + 86x_3 + 136x_4 + 29x_5$$

The farmer's problem is to choose x_1, x_2, x_3, x_4, and x_5, all greater than or equal to zero, to make his profit as large as possible.

Restrictions

There are, of course, restrictions on the possible values of x_1, x_2, x_3, x_4, and x_5 that reflect the circumstances under which the farmer has to operate his farm. For example, since he has a total of just 150 acres in all,

$$x_1 + x_2 + x_3 + x_4 + x_5 \leq 150$$

The other restrictions we stated in the introduction to the chapter were the following: First, the government had limited to 60 acres the amount of cotton he could grow. Thus, we require that

$$x_1 \leq 60$$

In addition, because of his uncertainty with regard to potato prices, he voluntarily limited the number of acres he would devote to potatoes to no more than 50 acres, so that

$$x_2 \leq 50$$

2.7 A FARM-MANAGEMENT PROBLEM

The limited availability of water caused the remaining restrictions. The growing season was divided into three periods, both because each crop requires different amounts of water at different times during its growing season and because the amount of water available depends on the time of year. The needs of each crop for water during each period, in inches per acre, and the amount of water available, in acre-inches, are given in Table 2-19. Thus, for example, during the third growing period, alfalfa requires 11.1 inches of water for each acre from the 730 acre-inches available.

Water Needed for Crop (acre-inches)

Period	Cotton	Potatoes	Alfalfa	Sugar Beets	Barley	Water Available (acre-inches)
1	4.0	13.3	15.8	13.0	6.3	2200
2	16.6	0	22.2	42.7	0	2110
3	7.8	0	11.1	3.3	0	730

Source: James N. Boles, "Linear Programming and Farm Management Analysis," *Journal of Farm Economics* 3 (Feb. 1955).

Table 2-19

If the farmer plants x_1 acres of cotton, x_2 of potatoes, x_3 of alfalfa, x_4 of sugar beets, and x_5 of barley, then his total water needs during the first growing period will be

$$4.0x_1 + 13.3x_2 + 15.8x_3 + 13.0x_4 + 6.3x_5$$

acre-inches of water. Since there are 2200 acre-inches of water available during that period, x_1, x_2, x_3, x_4, and x_5 must be chosen so that

$$4.0x_1 + 13.3x_2 + 15.8x_3 + 13.0x_4 + 6.3x_5 \leq 2200$$

There are 2110 acre-inches of water available during the second period, so from Table 2-19 we have the restriction

$$16.6x_1 + 22.2x_3 + 42.7x_4 \leq 2110$$

Similarly, there is the restriction imposed by the water requirements for the various crops during the third growing period. The table states that 730 acre-inches of irrigation water will be available, so the restriction is

$$7.8x_1 + 11.1x_3 + 3.3x_4 \leq 730$$

Certainly, x_1, x_2, x_3, x_4, and x_5 must be chosen to satisfy all three requirements simultaneously, since, for instance, there is no profit to be had in planting crops that will grow well during the first growing period and then die later for lack of water. Of course, the independent variables that represent acreage must be nonnegative, but as usual, we state this explicitly in the mathematical formulation of the problem.

Mathematical Formulation

The *Farm-Management* problem is actually the maximization problem

Maximize: $z = 207x_1 + 200x_2 + 86x_3 + 136x_4 + 29x_5$

Subject to:
$$\begin{aligned}
x_1 + x_2 + x_3 + x_4 + x_5 &\le 150 \\
x_1 &\le 60 \\
x_2 &\le 50 \\
4.0x_1 + 13.3x_2 + 15.8x_3 + 13.0x_4 + 6.3x_5 &\le 2200 \\
16.6x_1 + 22.2x_3 + 42.7x_4 &\le 2110 \\
7.8x_1 + 11.1x_3 + 3.3x_4 &\le 730 \\
x_1, x_2, x_3, x_4, x_5 &\ge 0
\end{aligned}$$

We express the problem in matrix form in terms of finding the 5×1-matrix of variables

$$X = \begin{bmatrix} x_1 \\ x_2 \\ x_3 \\ x_4 \\ x_5 \end{bmatrix}$$

that maximizes the linear function $z = CX$, where

$$C = \begin{bmatrix} 207 & 200 & 86 & 136 & 29 \end{bmatrix}$$

The matrix X is subject to the restrictions $AX \le B$ and $X \ge 0$, where

$$A = \begin{bmatrix} 1 & 1 & 1 & 1 & 1 \\ 1 & 0 & 0 & 0 & 0 \\ 0 & 1 & 0 & 0 & 0 \\ 4.0 & 13.3 & 15.8 & 13.0 & 6.3 \\ 16.6 & 0 & 22.2 & 42.7 & 0 \\ 7.8 & 0 & 11.1 & 3.3 & 0 \end{bmatrix} \text{ and } B = \begin{bmatrix} 150 \\ 60 \\ 50 \\ 2200 \\ 2110 \\ 730 \end{bmatrix}$$

Since the condition $B \ge 0$ is satisfied, this problem can be solved by the simplex algorithm in the form presented in Section 2.5.

The Solution

Introduce a slack variable for each of the inequalities of the system $AX \le B$, which we label for the marginal-value analysis:

$$\begin{aligned}
x_1 + x_2 + x_3 + x_4 + x_5 + x_6 &= 150 \quad \text{(land)} \\
x_1 + x_7 &= 60 \quad \text{(cotton)} \\
x_2 + x_8 &= 50 \quad \text{(potatoes)} \\
4.0x_1 + 13.3x_2 + 15.8x_3 + 13.0x_4 + 6.3x_5 + x_9 &= 2200 \quad \text{(1st water)} \\
16.6x_1 + 22.2x_3 + 42.7x_4 + x_{10} &= 2110 \quad \text{(2nd water)} \\
7.8x_1 + 11.1x_3 + 3.3x_4 + x_{11} &= 730 \quad \text{(3rd water)}
\end{aligned}$$

2.7 A FARM-MANAGEMENT PROBLEM

The corresponding simplex tableau is shown in Table 2-20.

	x_1	x_2	x_3	x_4	x_5	x_6	x_7	x_8	x_9	x_{10}	x_{11}	
x_6	1	1	1	1	1	1	0	0	0	0	0	150
x_7	1	0	0	0	0	0	1	0	0	0	0	60
x_8	0	1	0	0	0	0	0	1	0	0	0	50
x_9	4.0	13.3	15.8	13.0	6.3	0	0	0	1	0	0	2200
x_{10}	16.6	0	22.2	42.7	0	0	0	0	0	1	0	2110
x_{11}	7.8	0	11.1	3.3	0	0	0	0	0	0	1	730
z	−207	−200	−86	−136	−29	0	0	0	0	0	0	0

Table 2-20

We carry out the steps of the simplex algorithm. The first pivot will take place in the first column and, since $\frac{60}{1} = 60$ is less than all the quotients

$$\frac{150}{1} = 150, \quad \frac{2200}{4} = 550, \quad \frac{2110}{16.6} = 127, \text{ and } \frac{730}{7.8} = 94$$

the first pivot is at the number 1 in the a_{21}-location. Continuing in this manner with the aid of a computer, the form the tableau takes when all the numbers in the last row are either positive or zero is given in Table 2-21 (with all answers rounded to two decimal places of accuracy):

	x_1	x_2	x_3	x_4	x_5	x_6	x_7	x_8	x_9	x_{10}	x_{11}	
x_5	0	0	0	0	1	1	−0.28	−1	0	−0.02	−0.05	4.91
x_1	1	0	0	0	0	0	1	0	0	0	0	60
x_2	0	1	0	0	0	0	0	1	0	0	0	50
x_9	0	0	0	0	0	−6.3	9.06	−7	1	−0.07	−0.66	755.32
x_4	0	0	0	1	0	0	−0.03	0	0	0.07	−0.06	16.34
x_3	0	0	1	0	0	0	−0.69	0	0	−0.01	0.11	18.75
z	0	0	0	0	0	29	135.33	171	0	2.13	0.15	26,397.13

Table 2-21

Reading from the last column of the tableau, we conclude that the farmer will obtain the largest possible profit if he devotes $x_1 = 60$ acres to cotton, $x_2 = 50$ acres to potatoes, $x_3 = 18.75$ acres to alfalfa, $x_4 = 16.34$ acres to sugar beets, and $x_5 = 4.91$ acres to barley. In that case, his total profit will be \$26,379.13.

Marginal Analysis

Looking at the part of the bottom row that lies below the slack variables, we have

$$\begin{array}{cccccc} x_6 & x_7 & x_8 & x_9 & x_{10} & x_{11} \\ 29 & 135.33 & 171 & 0 & 2.13 & 0.15 \end{array}$$

From the equations containing these slack variables, we see that the marginal value to the farmer of another acre of land is $29. The marginal value of another acre of cotton is $135.33. Thus, the farmer would make an additional profit of $135.33 if the government would permit him to plant another acre of cotton. The marginal value of an acre of potatoes is $171.

The marginal value of water during the first growing period is 0. If we compute the amount of water the farmer will actually use in that period:

$$4.0(60) + 13.3(50) + 15.8(18.75) + 13.0(16.34) + 6.3(4.91) = 1444.6$$

that is far less than the 2200 acre-inches available at that time; so, it is reasonable that it will not help the farmer to have more water then. On the other hand, all 2110 acre-inches of water available during the second period will be consumed by the proposed planting schedule; we would expect water to have a positive marginal value, and it does, $2.13, during that period. Similarly, the availability of additional water during the third growing period would increase the farmer's profit, but here the marginal value is only 15 cents for an acre-inch of water.

Exercises

1. Suppose that the government restrictions on the number of acres of cotton were removed, but that all the other information were the same. Show that the farmer still could not plant all 150 acres with cotton, even though that is the most profitable crop.

2. Suppose that the farmer decided to grow only potatoes and barley. If all other information is the same as before, how many acres of each crop should he grow in order to have the greatest possible profit? [*Hint:* Set $x_1 = 0$, $x_3 = 0$, and $x_4 = 0$ in all the inequalities and in the objective function; delete any inequality in which the left-hand side becomes 0.]

3. If the farmer decided to eliminate the restriction of planting no more than 50 acres of potatoes and if all other information remained the same, how many acres of each crop should he grow in order to have the greatest possible profit? Carry all calculations to one decimal place.

4. If the government no longer restricted the farmer's acreage of cotton and if all other information remained the same, how many acres of each crop should he grow in order to have the greatest possible profit? Carry all calculations to one decimal place. (A calculator is advised for doing this problem.)

Review Exercises for Chapter 2

In Exercises 1 and 2, write the dual of the given LP-problem as a problem of maximizing or minimizing a linear function subject to a system of linear inequalities.

REVIEW EXERCISES FOR CHAPTER 2

1. Minimize: $z = 4x_1 + x_2 + x_3 + 2x_4$
   Subject to: $x_1 - x_2 + x_3 + x_4 \geq 5$
   $x_1 + x_2 \geq x_3$
   $x_1, x_2, x_3, x_4 \geq 0$

2. Maximize: $z = x_1 + 3x_3$
   Subject to: $2x_1 + x_2 \leq 3$
   $x_1 \leq x_2$
   $x_1 + x_2 + x_3 \leq 5$
   $x_1 + x_3 \leq 5$
   $x_1, x_2, x_3 \geq 0$

In Exercises 3 and 4, graph the feasible set of the LP-problem, find the coordinates of all extreme points, and solve the problem by the graphical method.

3. Maximize: $z = 2x_1 + 3x_2$
   Subject to: $x_1 + 2x_2 \leq 4$
   $5x_1 + x_2 \leq 10$
   $x_1, x_2 \geq 0$

4. Minimize: $z = \frac{1}{2}x_1 + \frac{1}{5}x_2$
   Subject to: $4x_1 + x_2 \geq 5$
   $x_1 + x_2 \geq 2$
   $x_1 + 8x_2 \geq 4$
   $x_1, x_2 \geq 0$

In Exercises 5 and 6, use the simplex algorithm to solve the maximization problem

Maximize: $z = CX$

Subject to: $AX \leq B$ and $X \geq 0$

for the given matrices.

5. $X = \begin{bmatrix} x_1 \\ x_2 \end{bmatrix}$, $A = \begin{bmatrix} 1 & -1 \\ 0 & 1 \\ -1 & 1 \end{bmatrix}$, $B = \begin{bmatrix} 1 \\ 2 \\ 0 \end{bmatrix}$, and $C = \begin{bmatrix} 3 & 1 \end{bmatrix}$

6. $X = \begin{bmatrix} x_1 \\ x_2 \\ x_3 \end{bmatrix}$, $A = \begin{bmatrix} 1 & 0 & 2 \\ 2 & 1 & 0 \\ 0 & -1 & 2 \end{bmatrix}$, $B = \begin{bmatrix} 1 \\ 1 \\ 3 \end{bmatrix}$, and $C = \begin{bmatrix} 2 & \frac{1}{2} & \frac{1}{2} \end{bmatrix}$

In Exercises 7 through 12, solve by the simplex algorithm.

7. Minimize: $z = CX$
   Subject to: $AX \geq B$ and $X \geq 0$
   where

   $X = \begin{bmatrix} x_1 \\ x_2 \\ x_3 \end{bmatrix}$, $A = \begin{bmatrix} 0 & 1 & 1 \\ 1 & 2 & 0 \\ 0 & 0 & 2 \\ -1 & 0 & 1 \end{bmatrix}$, $B = \begin{bmatrix} 1 \\ 1 \\ 1 \\ 2 \end{bmatrix}$, and $C = \begin{bmatrix} 0 & 1 & 1 \end{bmatrix}$

8. Minimize: $z = 10x_1 + 13x_2$
   Subject to: $x_1 + x_2 \geq 2$
   $x_1 + 2x_2 \geq 1$
   $2x_2 \geq x_1$
   $x_1, x_2 \geq 0$

9. Maximize: $z = x_1 + 2x_2 + x_3$
   Subject to: $x_2 \leq 2x_3$
   $x_1 + x_3 \leq 1$
   $x_2 + 2x_3 \leq 1$
   $x_1, x_2, x_3 \geq 0$

10. Minimize: $z = 4x_1 + 3x_2$
    Subject to: $x_1 + x_2 + x_3 \geq 3$
    $2x_1 + x_2 \geq 2$
    $x_1 + x_3 \geq 2$
    $2x_2 \geq x_3$
    $x_1, x_2, x_3 \geq 0$

11. A convenience food consists of rice and vermicelli (and some seasoning, which we ignore for the purposes of the problem). The manufacturer pays $0.15 a pound for rice and $0.20 a pound for vermicelli. In a box of the food, the total weight of rice and vermicelli must be at least 1 pound. If a box of the food must contain at least $\frac{1}{4}$ pound more vermicelli than rice, how much rice and vermicelli should be used in each box to minimize the manufacturer's cost?

12. A charitable foundation can invest up to $200 million. The foundation will invest in real estate, stocks, and treasury notes. The investment committee decides to invest no more in stocks than half the amount invested in real estate, no more in real estate than the amount invested in stocks plus the amount in treasury notes, and no more than $50 million in treasury notes. The committee estimates that it will earn 12% a year on its investment in real estate, 7% a year from stocks, and 9% a year from treasury notes.
    (a) How much should the foundation invest in real estate, stocks and treasury notes in order to maximize its annual earnings?
    (b) Perform a marginal-value analysis of the foundation's problem.

3
Probability

Probability and the Weather

"Everyone talks about the weather, but no one does anything about it." This is a statement which is so obvious that we regard it as a cliche. Although meteorologists are working on research that they hope *will* lead to some degree of weather control, this goal still lies far in the future. However, meteorologists can and do predict the weather with varying degrees of accuracy. Since forecasts are not certainties, it is natural to express them in a language that describes uncertainty. The language of uncertainty is called "probability." This chapter explains some of the basic principles of the mathematical theory of probability.

Until 1965, precipitation forecasts by the National Weather Service (NWS) described the likelihood of rain or snow in rather general terms. In that year, the NWS began a nationwide program of issuing to the general public "probability of precipitation" forecasts. Instead of a forecast such as "it will rain tomorrow" or "snow is likely by the weekend," we have grown used to hearing "the probability of rain is 20% for tomorrow morning, increasing to 50% by afternoon and evening." Whether this forecast leads you to carry your umbrella to class depends on your own experience of the accuracy of these forecasts in your particular area, but it appears that the NWS is offering you quite a lot of information to help you decide.

The information that the forecasters use to calculate the probability of precipitation comes from many sources, such as weather-satellite pictures and reports from weather observation-stations on the ground. In addition, the weather forecaster can take weather history into account in making the forecast. A forecaster in Seattle in January may well be more likely to predict rain than will a colleague in Phoenix in June—because it usually rains in Seattle in January, but seldom rains in Phoenix in June.

The forecasters in Seattle and Phoenix know the weather histories of their cities because they can look them up in the volumes of *Climatological Data*. This U. S. Government journal has been publishing the daily weather data from every weather-reporting station in the United States since 1914. For instance, the forecaster in Seattle can learn that during the five-year period 1981–85, out of the 5 × 31 = 155 January days, there was measurable precipitation on 96 of those days in Seattle. Instead of stating that it usually rains in Seattle in January, the forecaster can say, more precisely, that the probability of rain on a January day in Seattle, based on the recent climatological record, is $\frac{96}{155}$. On the other hand, during the same 1981–85 period, out of the 5 × 30 = 150 June days, precipitation was recorded in Phoenix on only one of those days. Therefore, it is certainly true that it seldom rains in Phoenix in June. Based on the climatological record, the probability of rain is $\frac{1}{150}$.

Obviously, it is much more likely to rain in Seattle in January than in Phoenix in June, and there is an easy way to compare the probabilities directly. If we divide the numerator by the denominator in each of the fractions $\frac{96}{155}$ and $\frac{1}{150}$, the probabilities become decimal numbers between zero and one. Specifically, letting "p" stand for probability,

$$p(\text{rain in Seattle on a January day}) = \frac{96}{155} = .619 \text{ (approximately)}$$

$$p(\text{rain in Phoenix on a June day}) = \frac{1}{150} = .007 \text{ (approximately)}$$

We see that rain in Seattle in January is far more likely than rain in Phoenix in June.

Probability can also be used to describe the relationship of weather conditions between different locations. For instance, precipitation in New York in January is a fairly common occurrence; the climatological data for the 1980–84 period gives a probability of $\frac{39}{155}$ = .25 (approximately). We would expect similar weather throughout that part of the country, but the extent of the similarity will depend on the particular location. New York lies on the Atlantic seacoast, so the weather in New York should resemble that of other coastal cities more than the weather of cities that lie inland. For instance, we would expect that information about weather in Boston will tell us more about weather in New York than similar information from Pittsburgh. The reason is not primarily that Pittsburgh is somewhat further from New York than Boston is, but rather that Pittsburgh lies west of the Appalachian Mountains. The appropriate probability concepts, together with some computations from the climatological data, can make the relationship of the precipitation patterns between New York and the two other cities, Boston and Pittsburgh, quite precise. We will return to this topic later in the chapter, after we have presented the necessary background information on probability. In the rest of this section, we will describe a more recent development in weather forecasters' effort to make better use of probability in their work. This will introduce a concept of probability not exclusively based on what happened in the past.

Precipitation occurrence lends itself quite well to the use of probabilities. It will either rain or snow—or it won't, and therefore only a simple probability like 20% or 50% can express a forecaster's uncertainty. When the weather forecaster tries to extend the use of probability to temperature forecasting, things get more interesting. In an article in the *Monthly Weather Review* published in 1974, Allan H. Murphy and Robert L. Winkler proposed a solution to the problem of making probability forecasts of temperatures.

At present, temperature forecasts are given either as "point forecasts" such as "the high tomorrow will be 76° (Fahrenheit)" or as "interval forecasts" ("the high tomorrow will be between 73° and 78°"). When temperature forecasts are given in terms of intervals, they suggest the level of uncertainty by the size of the interval. The forecast of "a high between 73° and 75°," which involves a range of only 3 degrees, sounds more certain than "a high in the mid seventies." This is an intuitive interpretation, and to be more precise, the intervals should be accompanied by a probability in order to convey the amount of uncertainty in the forecast. The proposal of Murphy and Winkler is that the NWS give its temperature forecasts in the form of a probability interval called a "credible interval."

To construct a "50% credible interval" for tomorrow's high temperature, the forecaster first chooses a point forecast based on the available weather information and his or her experience. For instance, the forecaster may decide on a point forecast of 73°, meaning that there is a 50% chance of a high temperature of 73° or lower and the same chance that the high will be 73° or higher. Then, in the same way, the forecaster splits the temperatures below 73°, deciding, for instance, that the high is as likely to be 71° or lower as it is to be between 71° and 73°. To put it another way, the forecaster thinks that there is a 25% chance the high temperature will be 71° or lower and also a 25% chance that it will be between 71° and 73°. The same is done with the temperatures above 73°, deciding, for instance, that there is as good a chance that the high temperature will be between 73° and 76° as that it will be 76° or higher. We can describe the forecaster's conclusions by means of Figure 3-1.

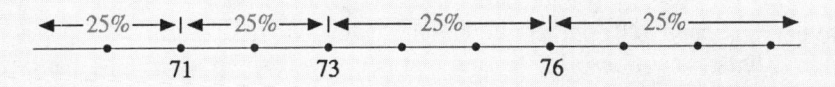

◄—25%—►|◄—25%—►|◄———25%———►|◄———25%———►
　　　　　71　　　　73　　　　　　76

Figure 3-1

> The forecaster would then report the "50% (central) credible interval" as 71° to 76°. In less technical terms, the report to the public would just state: "there is a 50% chance that tomorrow's high temperature will be between 71° and 76°." If the forecaster wants to use a different probability, such as 75%, a similar technique can produce it.
>
> The forecaster's determination of the credible interval was certainly not just based on climatological data. The decision that there is a 50% probability that tomorrow's high temperature will be between 71° and 76°, was based on many sources of information. The climatological data was one element that went into the forecaster's decision, together with satellite pictures of cloud patterns, detailed weather reports from other locations, and the like. Furthermore, in utilizing all this information, the forecaster had to rely on his or her own judgement. Therefore, we would expect a more experienced forecaster to produce more accurate credible-interval forecasts. The role of the forecaster's informed opinion introduces another aspect of probability that we will discuss in the next section.

3.1 Probability and Odds

Probability Based on Experience

Information about probability often comes from experience. Data are collected and made the basis for an **estimate**—that is, a guess or prediction of the value of the probability. It is very likely that an estimate based on sufficient evidence will produce a number close to the true probability. Our first two examples illustrate how we can estimate probability from experience. We begin with a climatological forecast of the probability of rain on an April day in Los Angeles.

Example 1. According to *Climatological Data* for the years 1982 through 1986, during the month of April there was measurable rain (at least $\frac{1}{100}$ of an inch) in Los Angeles on 20 of the days. Use this information to estimate the probability of rain on an April day in Los Angeles.

Solution

There were $5 \cdot 30 = 150$ days in the given period and there was rain on 20 of them; so the estimated probability is $\frac{20}{150} = .13$ (approximately). ∎

Example 2. A study of the traffic at tollbooths on a bridge finds that 3721 cars used the right-hand tollbooth on a certain day, while only 2210 cars used the left-hand booth. Estimate the probability that on that day a car chosen at random that crosses the bridge will pay at the right-hand tollbooth.

Solution

Because 3721 cars used the right-hand tollbooth and 2210 used the one on the left, there were 3721 + 2210 = 5931 cars on the bridge that day. Of the 5931 cars, 3721 used the right-hand tollbooth; so, the probability that a car chosen at random will use the right-hand booth is estimated to be 3721 out of 5931, or $\frac{3721}{5931} = .63$ (approximately).

■

In both examples, there is a number n, the number of observations: the total number of April days in a five-year period in Example 1, and the total number of cars on the bridge on a certain day in Example 2. Also, in each case we know the number of observations in which the outcome we are interested in occurred; call that number k. In Example 1, there were $n = 150$ April days, of which $k = 20$ were rainy. Out of the $n = 5931$ cars on the bridge in Example 2, $k = 3721$ used the right-hand tollbooth. On the basis of this information, we **estimate the probability** (that it will rain on an April day or that a car will use the right-hand tollbooth) to be k divided by n—in symbols,

$$\text{Probability} = \frac{k}{n}$$

It is a good idea to express the answer in decimal form, as we did in both the examples, because decimal numbers are easier to compare with each other than fractions are and we often want to know whether one outcome is more likely than another. For instance, it is not immediately obvious that $\frac{17}{52}$ is larger than $\frac{13}{40}$. But it is easy to compare them in decimal form because $\frac{17}{52}$ is approximately .327, whereas $\frac{13}{40} = .325$.

Probability based on experience is computed by using two numbers: n, the number of observations, and k, the number of "favorable outcomes," the observations we are interested in. Of course, k can never be larger than n, so the fraction $\frac{k}{n}$ can never be larger than 1. It can equal 1, however, if every observation has the desired outcome, in which case we get a probability of $\frac{n}{n} = 1$. The other extreme would occur when no observation produced the favorable outcome. In that case the probability will be $\frac{0}{n} = 0$. To sum up, the probability $\frac{k}{n}$ is always between 0 and 1.

Probability Based on Counting

In contrast to the temperature-forecasting problem and the two preceding examples, there are situations in which probabilities can be predicted accurately—without collecting any data at all—just from the nature of the question being asked. For example, if a coin is flipped, what is the probability that it will come up heads? There are two possibilities, heads and tails; and we assume that the coin is evenly balanced, so that each face is as likely to turn up as the other. Thus, heads should come up half the time. In other words, the probability of heads is $\frac{1}{2}$, or .50.

Example 3. Suppose that a card is drawn at random from a standard deck of cards. What is the probability of drawing an ace?

Solution

There are 52 cards in a standard deck, each with the same chance of being selected in a random draw. Since there are four aces in the deck, the probability of drawing an ace is four chances out of 52. That is, the probability equals $\frac{4}{52}$, or about .077. ∎

Observe that the reasoned-out probabilities were computed in the same way as the estimated ones. The number of ways the favorable event can occur (one way for heads in the coin toss, four ways for the four aces in a deck of cards) was divided by the number of possible outcomes (two when a coin is tossed, 52 when a card is drawn). When equally likely events are studied, this ratio of favorable outcomes to possible outcomes is precisely what is meant by **probability**. Thus, we still have the formula

$$\text{Probability} = \frac{k}{n}$$

where n is the number of possible outcomes and k, the number of favorable ones. Notice that probability must therefore be a number between 0 and 1.

Example 4. Toss two coins at the same time. What is the probability that they will both come up heads?

Solution

Even though they are tossed at the same time, the two coins are separate objects; so we distinguish them by calling one the "first coin" and the other the "second coin." (The problem is the same whether the coins are tossed at the same time or one after the other.) The possible outcomes are listed in Table 3-2. There are four outcomes in all. One of the four outcomes is that both coins turn up heads; so, the probability of this occurring is $\frac{1}{4}$. ∎

First Coin	*Second Coin*
Heads	Heads
Heads	Tails
Tails	Heads
Tails	Tails

Table 3-2

Subjective Probability

In addition to probability based on experience and probability based on reasoning, there is a third kind of probability, that is used in the credible-interval forecasts discussed in the introduction to this chapter. When a forecaster says there is a 50% probability that tomorrow's high temperature will be between 47° and 54°, this probability is certainly not reasoned out like a coin-flipping probability. Even past experience, in the form of climatological data, is just one element among the many that go into the forecaster's decision. Other factors that contribute to the forecast include satellite pictures of cloud patterns, detailed weather reports from other locations, and the like. But above all, the forecaster uses his or her own judgement. So, weather-forecasting probability is an example of the third kind of probability, called **subjective probability**, which is based neither on abstract reasoning nor on a substantial amount of relevant past experience, but rather on instinctive feelings, often with the aid of knowledge and training. Because it is a probability, a subjective-probability estimate is a number between 0 and 1.

Odds

Rather than a decimal probability statement, such as "the probability is .35," subjective-probability estimates are often expressed as "odds." For example, a bettor at a horse race says, "I thought that horse had a good chance to win; but at odds of 5 to 2, it didn't seem like a good bet." The bettor certainly cannot reason out the chances that the horse would win in the same way as he would if coins were being flipped. It is even doubtful that a significant amount of past experience would be terribly useful in making an estimate, because in the horse's previous races, the conditions and competition would have been different. Although the bettor said the horse "had a good chance" to win, the actual probability statement was expressed in the language of odds. To convert odds statements into probability numbers in fraction or decimal form, like those in the previous examples, there is a simple rule.

Odds in horse racing and in many other contexts are given as "odds to fail"; that is, in expressing 5-to-2 odds, the 5 (which we denote by a) refers to failure, and the corresponding probability that the horse will lose (fail to win) is

$$p(\text{Failure}) = \frac{a}{a+b}$$

where b (which is 2 in the example) refers to winning.

Of course, we are usually more interested in the probability that the horse will win. The probability of a successful outcome, given odds of a to b, is, in fraction form,

$$p(\text{Success}) = \frac{b}{a+b}$$

Thus, the bettor will not bet when the odds are "5 to 2" because that would mean that the probability that the horse will win is

$$\frac{2}{5+2} = \frac{2}{7} = .29 \text{ (approximately)}$$

Apparently, the bettor believes, subjectively, that the horse's chance of winning are not *that* good.

Example 5. If a student says she has "only a 10 to 1 chance" of receiving an A in her history course, what is she estimating to be the probability that she will receive an A?

Solution

The phrase "10 to 1 chance" is another way of saying the odds are 10 to 1; therefore, she estimates that the probability she will receive an A is

$$\frac{1}{1+10} = \frac{1}{11} = .09 \text{ (approximately)} \qquad \blacksquare$$

Types of Probability

The reasoned-out sort of probability and the probability estimated on the basis of experience are closely related to one another. To estimate the probability of heads in a single flip of a coin, we can flip the coin many times, record the numbers of heads and tails that result, and divide the number of heads by the total number of flips. If we have flipped the coin enough times, the resulting quotient will be very close to .50. The probability estimated by the experiment of flipping the coin tends toward the probability calculated for equally likely events. This relationship holds in general. That is, if for equally likely outcomes a probability is estimated by an actual experiment, the estimates can be used to predict the ratio of favorable outcomes to possible outcomes, provided the experiment is repeated enough times.

In this chapter we will be concerned with certain basic principles of probability. These principles are most clearly seen in the context of equally likely outcomes, in which the calculation of probability is just a matter of arithmetic. Although the relationship of subjective probability to the other two, more evidently related, types is not entirely clear, we use the word "probability" impartially for all three types and assume these principles always apply. In practice this means we will discuss these principles of probability first in the context of flipped coins, rolled dice, and the like. Then we will apply these principles to problems in which the probability may be someone's subjective opinion or the result of past experience.

Summary

If in the past, a "favorable outcome" happened k times out of n possible occurrences, then the **estimate of the probability** is $\frac{k}{n}$ that a favorable outcome will occur in the future, provided the circumstances do not change. If all possible outcomes of an act are equally likely to happen, then **probability** can be calculated directly: Divide the number of favorable outcomes by the total number of possible outcomes. Probability is always a number between 0 and 1. If **odds** are given as "a to b," that means the probability of failure is $\frac{a}{a+b}$ and of success is $\frac{b}{a+b}$.

Exercises

1. In a marketing survey, a family is given four unmarked tubes of toothpaste, one of which is the product of the company paying for the survey. If the family cannot tell the four brands apart and so chooses one at random, what is the probability that it will choose the sponsoring company's product?

In Exercises 2 through 4, a single die is rolled.

2. What is the probability that the die comes up an odd number?

3. What is the probability that the number showing on the die is less than five?

4. What is the probability that the die will come up with a number at least as large as two?

5. The trains in the kingdom of Oriana are sometimes "very late," that is, more than 24 hours behind schedule. A survey of trains arriving at Oriana City showed that 15 out of 75 were very late. If you travel to Oriana City on a train chosen at random, what is the probability that your train will be very late?

In Exercises 6 through 8, a card is chosen from a standard 52-card deck.[†]

6. What is the probability that the card is a club?

7. What is the probability that the card is either a jack or a queen?

8. If aces are considered to be higher than kings, what is the probability that the card drawn is higher than a nine?

9. During a five-year period, in the last half (15 days) of November, there were a total of 22 days of measurable precipitation (rain or snow) in Minneapolis. Estimate the probability of precipitation on a day in the last half of November in Minneapolis.

[†]See Appendix: The Standard Deck of Cards, p. 296.

228   PROBABILITY

In Exercises 10 through 12, a nursery has a bin of mixed tulip bulbs containing 22 that will produce red tulips, 14 that will yield yellow flowers, and 7 that will produce white tulips. Because all the bulbs look alike, a customer chooses one at random.

10. What is the probability that the bulb chosen is one that will produce a white flower?

11. What is the probability that the bulb will yield either a white or a yellow flower?

12. What is the probability that the flower produced will not be yellow?

Exercises 13 through 16 are concerned with the gambling game of American roulette. The American roulette wheel has 38 sections numbered 0, 00, and 1 through 36. The game amounts to choosing one of these numbers at random.

13. What is the probability that the number 12 will be chosen?

14. In roulette, the numbers 0 and 00 are considered neither odd nor even. What is the probability that the outcome will be an even number?

15. What is the probability that the result of the game is a number greater than 24?

16. What is the probability that the result of the game is an odd number greater than 24?

17. Edgar Allan Poe's *The Gold Bug* contains the solution to a code. A coded message in the story consists of 159 symbols in all, including 33 repetitions of the symbol "8," 26 of ";", and so on. Poe wrote, "Now, in English, the letter which most frequently occurs is e . . . [It] predominates so remarkably that an individual sentence of any length is rarely seen, in which it is not the prevailing character." Thus, the narrator of the story assumes, correctly as it turns out, that the symbol "8" represents the letter e. If the message in Poe's story is typical of English, what is the probability that a letter chosen at random from an English sentence will be an e?

Exercises 18 through 20 are based on a study of a sporting-goods store in which it was found that out of 348 customers who parked in the store's parking lot, 287 made at least one purchase, whereas of 518 customers who did not use that lot, 305 made purchases.

18. Estimate the probability that a customer who uses the parking lot will make at least one purchase.

19. Estimate the probability that a customer who does not use the parking lot will make at least one purchase.

20. Estimate the probability that the next customer who comes into the sporting-goods store will make at least one purchase.

21. The odds against winning a door prize are given as 20 to 1. What is the probability of winning?

Exercises 22 through 25 refer to the odds of losing that are posted on the following tote board for a horse race:

Thundercloud	7 to 5
Caesar III	8 to 1
Birdsong	5 to 3
Whirlwind	6 to 2

22. What is the probability that Birdsong will win?
23. What is the probability that Whirlwind will win?
24. Which horse is the "long shot" (i.e., has the lowest probability of winning the race) and what is that probability?
25. Which horse is the "favorite" (i.e., has the highest probability of winning the race) and what is that probability?

3.2 Counting

The definition of the probability of a certain type of outcome among equally likely events requires us to find two numbers: the total number n of outcomes possible and the number k of favorable outcomes. In the examples and exercises of Section 3.1, these two counting problems were not very difficult. In many important types of probability problems, however, the counting problems are more complicated. To be sure that the numbers are correct, we need a systematic way to count. Counting theory eliminates the need for reasoning out each new type of probability problem from first principles. These counting procedures can turn the calculation of probability in even quite intricate problems into rather routine exercises.

Some Examples of Counting

The business of selling new automobiles is complicated by the fact that the buyer has such a wide choice of styles, colors, and optional extras. The number of possible combinations is large enough so that each new car is virtually a custom product. Let's look at a simplified example that illustrates the problem.

Example 1. Suppose that the buyer has only two types of decisions concerning the new car: the style and the color. Further, assume that there are just three styles available: two-door, four-door, and wagon. And there are five colors: black, gray, blue, green, and red. How many distinct combinations of style and color must the manufacturer be prepared to supply?

Solution

The buyer's choices can be represented by two words, where the first word determines the automobile's style and the second, its color. That is, the buyer must choose two words such as (two-door, red) or (wagon, gray). An expression such as (wagon, gray) containing two words or other symbols, in a specified order, is called an **ordered pair**. Thus, the question is: How many ordered pairs of words, representing style and color, are there when there are three choices of style and five choices of color? One convenient way to represent all the possible pairs is by means of a rectangular diagram like the one shown in Table 3-3. Each asterisk represents the pair determined by its row and column headings; for example, the asterisk in parentheses (*) represents the pair (two-door, blue) and the asterisk in brackets [*] corresponds to (wagon, green). We see that each possible pair is represented by an asterisk, and the number of ordered pairs equals the number of asterisks. In this case, since there are three rows of five asterisks each, we have fifteen pairs. Thus, even with these restricted choices, the manufacturer must produce fifteen different kinds of automobiles.

Style/Color	Black	Gray	Blue	Green	Red
Two-door	*	*	(*)	*	*
Four-door	*	*	*	*	*
Wagon	*	*	*	[*]	*

Table 3-3

We can summarize by writing:

$$(\underbrace{\text{Style}}_{3}, \underbrace{\text{Color}}_{5}) = 15$$

Another way to count the number of different style-color choices is by the **tree diagram** (Figure 3-4), in which we can see each choice of style combined with all the possible colors. If we count the number of branches on the tree, we again see that there are 15 choices in all.

Example 2. Two dice are rolled. How many different outcomes are possible?

Solution

Even though the two dice are rolled at the same time, they are physically distinct; so the possible outcomes will be the same as if they were rolled one at a time. Thus, we refer to them as the "first die" and the "second die" (compare with the "two coins" example in the last section). A possible outcome consists of an ordered pair of the form (first die, second die); for example, (2, 4) represents a two on the first die and a four on the second. The possible outcomes are conveniently described in Table 3-5. Because there are six rows of six asterisks each, there are 36 possible outcomes. Notice the significance of the fact that we consider the *order* in which the dice are

```
                    Style                          Color

                                              ── black
                                              ── grey
                      two-door ←───────────── ── blue
                                              ── green
                                              ── red

                                              ── black
                                              ── grey
        ←──────────── four-door ←──────────── ── blue
                                              ── green
                                              ── red

                                              ── black
                                              ── grey
                      wagon ←───────────────── ── blue
                                              ── green
                                              ── red
```

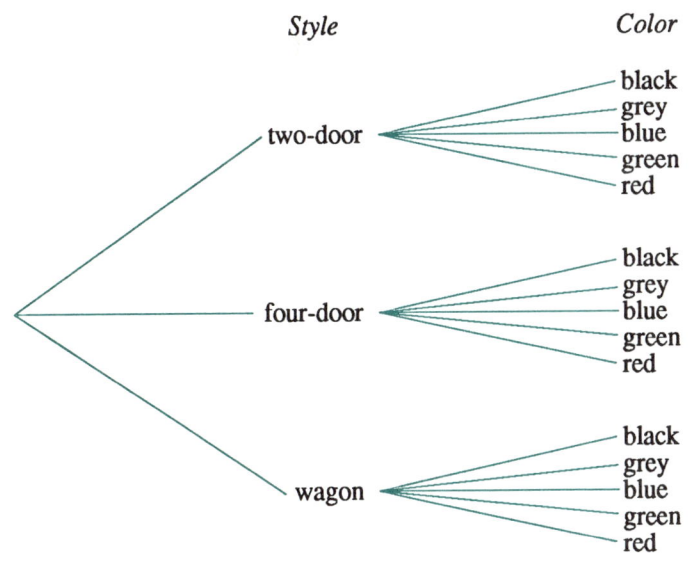

Figure 3-4

thrown: for example, (2, 4) and (4, 2) represent different outcomes, even though the same "face-numbers" are involved. The distinction represents the fact that the two dice are physically distinct, and so the act of first rolling a 2 and then a 4 is different from rolling the 4 first and then the 2.

Again, we can summarize:

(First die, Second die)

$$6 \cdot 6 = 36$$

■

First/Second	1	2	3	4	5	6
1	*	*	*	*	*	*
2	*	*	*	*	*	*
3	*	*	*	*	*	*
4	*	*	*	*	*	*
5	*	*	*	*	*	*
6	*	*	*	*	*	*

Table 3-5

Example 3. Suppose the four aces are separated from the rest of a standard 52-card deck of playing cards. An ace is chosen, and then a second ace is selected from among the three that remain. In how many ways can this be done?

Solution

We are interested in pairs of the form (first ace, second ace). But this situation is rather different from the preceding examples because the choices available for the second ace depend on which ace was chosen first. If you roll a 4 with your first die, the second die can still be *any* number between 1 and 6. But, if your first ace was the ace of spades, the choices for the second ace are restricted to hearts, diamonds, and clubs. Problems of this type can still be represented by a tree diagram (Figure 3-6). Thus, each possible pair (first ace, second ace), corresponds to a branch of the tree. For instance, the branch in boldface represents the pair (diamond, heart). There are twelve distinct branches, so we conclude that there are twelve outcomes in all.

We can still summarize the computation in terms of ordered pairs.

$$(\underbrace{\text{First ace}}_{4}, \underbrace{\text{Second ace}}_{3}) = 12$$

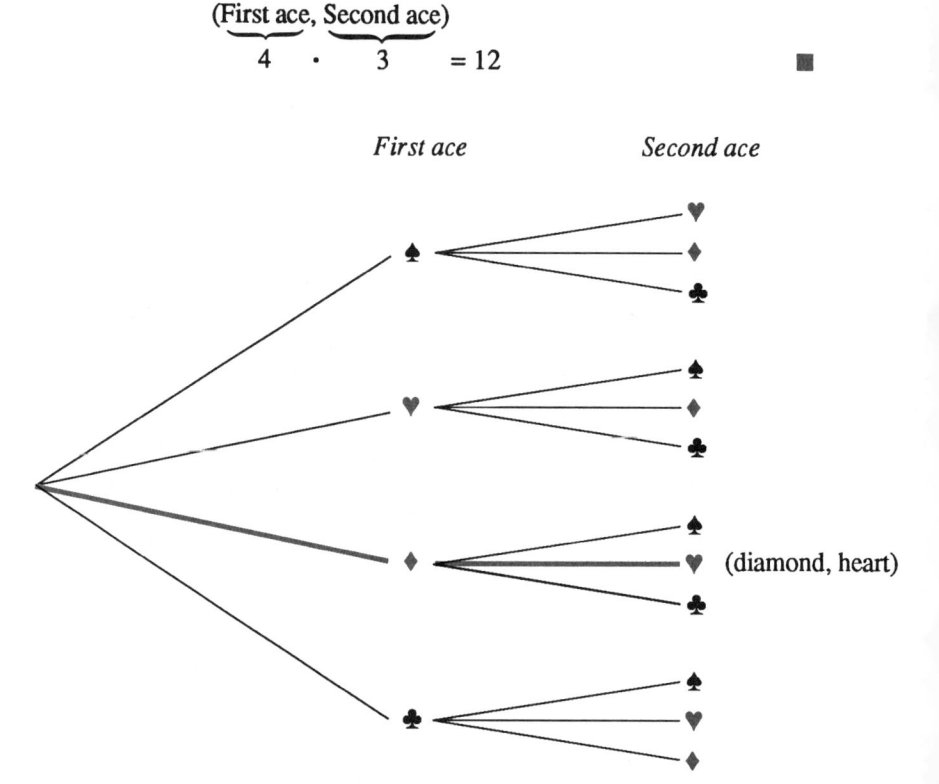

Figure 3-6

A Counting Rule

The automobile buyer in Example 1 had five choices of color no matter which style of car is chosen. In Example 3, the aces remaining for the second choice depend on which ace was chosen first, but the *number* of possibilities was the same—three, no matter which ace was first. Both types of problems can be solved by the same **counting rule**:

If there are n_1 choices for the first member of a pair and n_2 choices for the second, regardless of the first member, then there are $n_1 \cdot n_2$ (n_1 times n_2) ordered pairs in all. In summary form,

$$(\underbrace{\text{First member}}_{n_1}, \underbrace{\text{Second member}}_{n_2}) = n_1 \cdot n_2$$

Example 4. How many two-digit numbers (including 00, 01, etc.) are there? How many of them involve two distinct digits?

Solution

By "digits," we mean the numbers 0, 1, 2, 3, 4, 5, 6, 7, 8, and 9. The two-digit numbers may be thought of as ordered pairs (first digit, second digit) where, for example, the number 25 corresponds to the ordered pair (2, 5). Since there are ten different digits, $n_1 = 10$ and $n_2 = 10$, so there are

$$(\underbrace{\text{First digit}}_{10}, \underbrace{\text{Second digit}}_{10}) = 100$$

two-digit numbers in all. However, if the digits have to be distinct (which means that (1, 4) and (1, 8) are permitted, but (1, 1) is not), after the first digit is selected, the second must be chosen from the remaining nine digits. In this case $n_1 = 10$ but $n_2 = 9$; this time

$$(\underbrace{\text{First digit}}_{10}, \underbrace{\text{Second digit}}_{9}) = 90$$

There are 90 two-digit numbers involving two distinct digits. ∎

A General Counting Rule

A polling organization divides the population into groups, called "strata," according to sex, race (white or nonwhite), and age (18–35, 36–50, over 50). For example, "white male between the ages of 18 and 35" is one stratum. We wish to know how many different strata there are. Notice that we need three categories to determine the stratum: sex, race, and age.

In order to describe more complicated choices, such as this one, we will need the concept of an *ordered k-tuple*. An **ordered k-tuple** is a collection of k symbols in a fixed order—so the ordered pairs we encountered in the previous examples can also be called "ordered 2-tuples." We can represent a stratum as an ordered 3-tuple (triple) of the form (sex, race, age). If the polling organization needed a more refined sampling structure, including educational level, geographical location, income, and so on, the strata could be described by ordered k-tuples, where k is the number of different categories. To count the number of

(Sex, Race, Age)

234   PROBABILITY

strata, we use the tree diagram of Figure 3-7. There are 12 branches, or strata, in this sampling design.

Extending the rule for ordered pairs, the **general counting rule** for ordered k-tuples is:

> If there are n_1 choices for the first member of an ordered k-tuple, n_2 for the second—regardless of the first choice, n_3 for the third—regardless of the first two choices, and so on, up to n_k for the kth (independent of what the previous choices are), then there are
>
> $$n_1 \cdot n_2 \cdot n_3 \ldots n_k$$
>
> such k-tuples in all.

Applying this rule to the number of strata:

$$(\underbrace{\text{Sex}}_{2}, \underbrace{\text{Race}}_{2}, \underbrace{\text{Age}}_{3}) = 12$$

as we saw from the tree.

Example 5. Suppose we allow the car buyer of Example 1 some choice of optional extras on the new car. In addition to choosing the style from among the three possibilities: two-door, four-door, and wagon; and the color from among five choices:

Sex   *Race*   *Age*

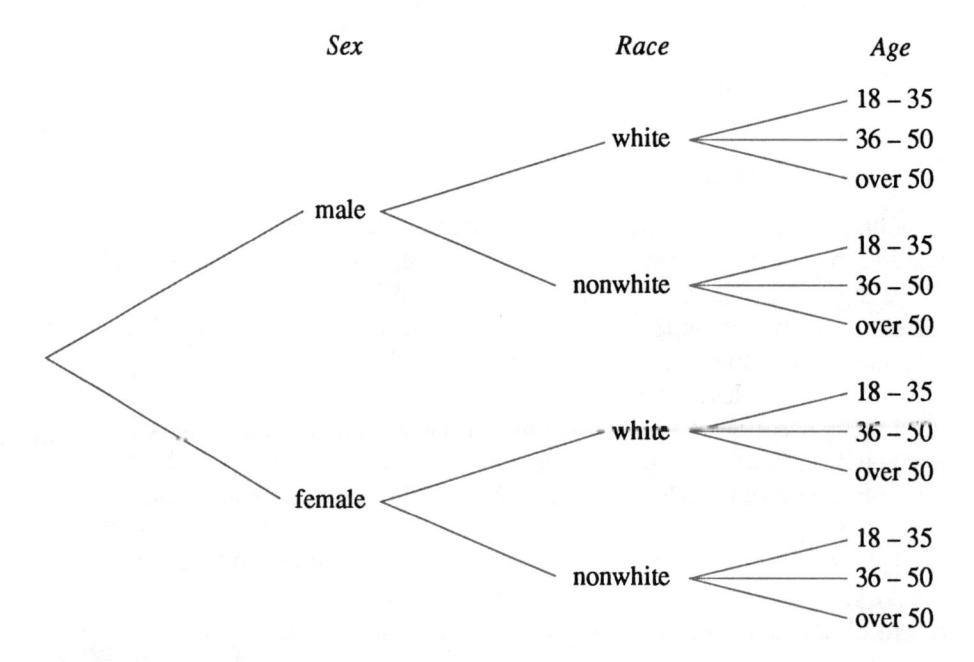

                                18 – 35
                white   36 – 50
                                over 50
        male
                                18 – 35
                nonwhite   36 – 50
                                over 50

                                18 – 35
                white   36 – 50
                                over 50
        female
                                18 – 35
                nonwhite   36 – 50
                                over 50

Figure 3-7

black, gray, blue, green, and red, we will assume that the manufacturer also offers two options: automatic transmission and an air-conditioning unit. The buyer must decide whether to choose the transmission option only, the air-conditioning option only, both options, or neither option. Therefore, the buyer has four possible choices concerning these options. How many distinct combinations of style, color, and optional extras are there?

Solution

We can represent each combination by an ordered triple of the form

(Style, Color, Options)

Applying the general counting rule, we see that there are now

$$\underbrace{3}_{\text{Style}} \cdot \underbrace{5}_{\text{Color}} \cdot \underbrace{4}_{\text{Options}} = 60$$

choices.

Summary

An **ordered k-tuple** is a collection of k symbols in a fixed order. If there are n_1 choices for the first member of an ordered k-tuple, n_2 choices for the second—regardless of the first, n_3 for the third—regardless of the first two, and so on, up to n_k for the kth (independent of what the previous choices are), then there are $n_1 \cdot n_2 \cdot n_3 \ldots n_k$ such k-tuples in all.

Exercises

1. How many ordered triples are there if there are 5 choices for the first symbol in the triple, 7 choices for the second, and 3 choices for the third?

2. In an ordered 4-tuple, there are 6 choices for the first symbol, 5 for the second, 7 for the third, and 3 for the fourth. How many such 4-tuples are there in all?

3. A game is played by rolling a die and then drawing a card from a standard deck. How many different outcomes can this game have?

4. A sport shirt can be purchased in a choice of sizes: small, medium, large, and extra large. It is also available in the colors white, blue, green, gray, and brown. Suppose a store wishes to keep in stock exactly one shirt of this style in each combination of size and color. How many shirts must the store keep in stock?

5. How many different ways are there to fill in an answer sheet if a quiz consists of five true-false questions?

6. How many different ways are there to fill in an answer sheet if a quiz consists of six multiple-choice questions with four possible answers to each question?

7. A family plans to drive from Los Angeles to New York stopping in Chicago to visit relatives. They are considering four different routes between Los Angeles and Chicago and 5 different routes between Chicago and New York. How many routes are being considered in all?

8. A psychologist attempts to teach a monkey to distinguish "words" made up of the symbols O, Π, ∇, and Σ. If words are exactly five symbols long and repetitions of symbols are allowed, how large can the monkey's vocabulary be?

9. A botanist wishes to determine the effects of temperature, humidity, and light on the growth rate of a species of plant. Growth will be tested at constant temperatures of 18, 20, 22, and 25°C; at 50%, 65%, and 80% constant humidity; and under the following lighting schedules; constant light, 12 hours of light alternating with 12 hours of darkness, and 16 hours of light alternating with 8 hours of darkness. If each choice of temperature, humidity, and light schedule constitutes a distinct experiment, how many different experiments must the botanist set up?

10. Scott College requires its students to take a basic course in each of three areas: natural sciences, social sciences, and humanities. It offers six natural-science courses, five social-science courses, and eight humanities courses that satisfy the requirement. In how many different ways can a student fulfill the basic requirements?

11. Suppose the members of the parliament of a country belong to three political parties: Liberal, Moderate, and Conservative. The parliament has 20 Liberal members, 50 Moderates, and 30 Conservatives. If a three-member committee of the parliament is to be made up of one member from each party, how many different such committees can there be?

In Exercises 12 and 13, a three-digit "number" means a sequence of three digits such as 020, 001, or 356.

12. How many three-digit numbers are there in all?

13. How many three-digit numbers are there in which all the digits are distinct?

14. A woman has a choice of six different routes between her home and her work. How many different choices does she have for a round trip if she never uses the same route both going and returning?

In Exercises 15 through 17, three dice are rolled.

15. How many different outcomes are there?

16. In how many different ways can the dice be rolled so that all three dice show odd numbers?

17. In how many ways can the dice be rolled so that the numbers that come up are all different?

In Exercises 18 and 19, a "three-letter word" means a sequence of three letters such as bat, tab, qcc, ccq, and xxx.

18. How many three-letter words are there?

19. How many three-letter words are there that use three different letters?

20. A large corporation uses four-digit numbers for telephone extensions. Dialing 8 or 9 will connect the telephone with an outside line, and 0 calls the operator; so, the first digit of the extension cannot be any of these. How many telephone extension numbers does the company have available?

3.3 Permutations and Factorials

We'll begin this section with an application of the (general) counting rule from the preceding section.

Example 1. In how many ways can the letters of the word "red" be ordered? How many ways for the word "brown"?

Solution

An ordering of the letters of the word "red," such as dre, erd, red, and so on, can be thought of as an ordered triple

(first letter, second letter, third letter)

using each of the three letters r, e, and d exactly once. Choosing any one of the $n_1 = 3$ available letters for the first position, there is a choice of $n_2 = 2$ left for the second; when we get to the third position there is only a single letter remaining; so, $n_3 = 1$. Thus, by the counting rule, there are

$$\text{(first letter, second letter, third letter)} = 3 \cdot 2 \cdot 1 = 6$$

orderings of the word "red." We can also see this by counting the branches of the tree in Figure 3-8.

Similarly, an ordering of the letters of "brown" is an ordered 5-tuple involving each of the letters b, r, o, w, and n exactly once. Now $n_1 = 5$ since any letter can be used in the first position. There are four letters left; so, $n_2 = 4$. Similarly, $n_3 = 3$ and $n_4 = 2$; in the final position there is only a single letter remaining, so, $n_5 = 1$. The counting rule gives

$$\text{(first letter, second letter, third letter, fourth letter, fifth letter)} = 5 \cdot 4 \cdot 3 \cdot 2 \cdot 1 = 120$$

different ways to order the letters of the word "brown." ∎

238   PROBABILITY

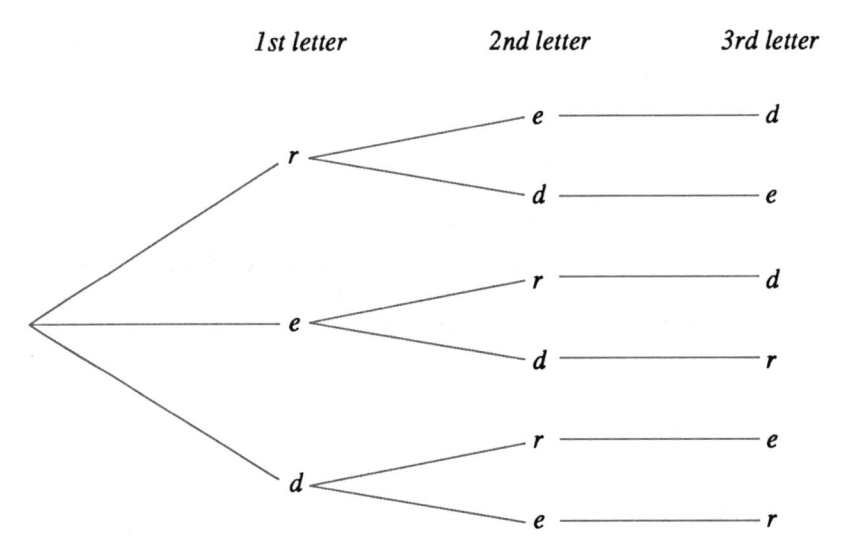

1st letter   *2nd letter*   *3rd letter*

```
         e ———————— d
    r  <
         d ———————— e

         r ———————— d
    e  <
         d ———————— r

         r ———————— e
    d  <
         e ———————— r
```

Figure 3-8

The Number of Orderings

Example 1 illustrates an important consequence of the counting rules. If we are given k distinct symbols and are asked in how many ways they can be ordered, we are being asked to count the number of ordered k-tuples with a different symbol in each position from a list of k symbols. Since all k symbols are available for the first position, $n_1 = k$. At each succeeding position, the choice is reduced by 1 from the list of choices in the preceding position—so, $n_2 = k - 1$, $n_3 = k - 2$, and so on. When we reach the kth position in the ordered k-tuple, there will be just a single choice remaining, and so, $n_k = 1$. By the general counting rule, the number of different orderings must be

$$\underbrace{\text{(first)}}_{k} , \underbrace{\text{second}}_{(k-1)} , \underbrace{\text{third}}_{(k-2)} , \ldots , \underbrace{\text{next to last}}_{2} , \underbrace{\text{last)}}_{1}$$

k-Factorial

A product of the preceding form occurs so often in mathematics that it has a special name, **k factorial**, and a special symbol, $k!$. Thus, by definition,

$$2! = 2 \cdot 1 = 2$$
$$3! = 3 \cdot 2 \cdot 1 = 6$$
$$4! = 4 \cdot 3 \cdot 2 \cdot 1 = 24$$

and so on. In general, if k is a whole number greater than 1,

$$\boxed{k! = k(k-1)(k-2) \ldots 3 \cdot 2 \cdot 1}$$

3.3 PERMUTATIONS AND FACTORIALS

We extend the definition by letting $1! = 1$. In addition, it is convenient to give a meaning to the symbol $0!$ (zero factorial) by simply defining $0! = 1$. (We will later see why it is a good idea to define $0! = 1$.)

The next example illustrates some of the properties of factorials that we will use in this chapter.

Example 2. (a) Express in factorial form:
  i. $7 \cdot 6!$ ii. $8 \cdot 7 \cdot 6!$

  (b) Calculate each of the following:
  i. $\dfrac{7!}{6!}$ ii. $\dfrac{8!}{2!6!}$ iii. $\dfrac{52!}{5!47!}$

Solution

(a) *i.* Since
$$6! = 6 \cdot 5 \cdot 4 \cdot 3 \cdot 2 \cdot 1$$
then
$$7 \cdot 6! = 7 \cdot 6 \cdot 5 \cdot 4 \cdot 3 \cdot 2 \cdot 1$$
which we can recognize as $7!$

  ii. Similarly,
$$8 \cdot 7 \cdot 6! = 8 \cdot 7 \cdot 6 \cdot 5 \cdot 4 \cdot 3 \cdot 2 \cdot 1 = 8!$$

(b) *i.* From part (a), we write $7! = 7 \cdot 6!$, thus
$$\frac{7!}{6!} = \frac{7 \cdot 6!}{6!} = 7$$

  ii. In the same way, we can use the other computation of Part (a) to calculate that
$$\frac{8!}{2!6!} = \frac{8 \cdot 7 \cdot 6!}{2 \cdot 1 \cdot 6!} = \frac{8 \cdot 7}{2 \cdot 1} = 28$$

  iii. These computations become much simpler, even when you use a calculator, if you can recognize common factorial factors. In order to compute the last question, it will help us to look for the factor $47!$ (of the denominator) of the very large number in the numerator: $52!$. Note that
$$52! = 52 \cdot 51 \cdot 50 \cdot 49 \cdot 48 \cdot 47 \cdot 46 \ldots 3 \cdot 2 \cdot 1$$
We can recognize the factors of $47!$ on the right side and express $52!$ as
$$52! = 52 \cdot 51 \cdot 50 \cdot 49 \cdot 48 \cdot 47!$$
Now we see that
$$\frac{52!}{5!47!} = \frac{52 \cdot 51 \cdot 50 \cdot 49 \cdot 48 \cdot 47!}{5!47!} = \frac{52 \cdot 51 \cdot 50 \cdot 49 \cdot 48}{5 \cdot 4 \cdot 3 \cdot 2 \cdot 1}$$

The numerator is still very large, but we can divide 50 by 5, leaving 10. We can also divide 48 by the remaining product of the denominator, that is, 24, obtaining 2, so that the result is
$$52 \cdot 51 \cdot 10 \cdot 49 \cdot 2 = 2{,}598{,}960$$

240 PROBABILITY

In the language of factorials, we have seen that the number of orderings of k (one or more) distinct symbols is equal to $k!$. Now we apply the same reasoning to a slightly different kind of problem, in which we do not use all available symbols.

Example 3. How many slates of officers—president, vice-president, secretary, and treasurer—are possible for an organization of 15 members? Express the answer in terms of factorials.

Solution

We must count 4-tuples of the form

(President, Vice President, Secretary, Treasurer)

Since any member could be president, $n_1 = 15$. This leaves $n_2 = 14$ choices for vice president, $n_3 = 13$ for secretary, and $n_4 = 12$ for treasurer; by the general counting rule there are

(President, Vice president, Secretary, Treasurer)
$$15 \cdot 14 \cdot 13 \cdot 12 = 32{,}760$$

different slates of officers possible in the organization.

The product $15 \cdot 14 \cdot 13 \cdot 12$ looks like the beginning of the definition of $15!$. If we write out the meaning of $15!$ in the form

$$15! = (15 \cdot 14 \cdot 13 \cdot 12) \cdot 11 \cdot 10 \cdot 9 \cdot 8 \cdot 7 \cdot 6 \cdot 5 \cdot 4 \cdot 3 \cdot 2 \cdot 1$$

we can see what has been left out:

$$11 \cdot 10 \cdot 9 \cdot 8 \cdot 7 \cdot 6 \cdot 5 \cdot 4 \cdot 3 \cdot 2 \cdot 1 = 11!$$

Therefore, we can write

$$15! = (15 \cdot 14 \cdot 13 \cdot 12) \cdot 11!$$

Dividing both sides of the equation by $11!$, we can express $15 \cdot 14 \cdot 13 \cdot 12$ in factorial form as

$$15 \cdot 14 \cdot 13 \cdot 12 = \frac{15!}{11!}$$ ∎

The Number of Permutations

You may be wondering what $11!$ has to do with the statement of Example 3. Notice that there are fifteen members, and four of them will become officers; so, the remaining eleven will not. In other words, $11 = 15 - 4$. The same relationship works in general. If there were n members of the organization and k different offices, then there would be $n_1 = n$ choices for the highest office, $n_2 = n - 1$ choices for the next, and so forth. If we try to compute the number of different slates of k officers from an organization with n members, we will come up with a product beginning $n(n-1)(n-2)\ldots$. The choices of the first k officers, starting at n and reducing by 1 each time look like this:

3.3 PERMUTATIONS AND FACTORIALS   241

$$\underbrace{\text{1st officer}}_{n-0} \quad \underbrace{\text{2nd officer}}_{n-1} \quad \underbrace{\text{3rd officer}}_{n-2} \quad \ldots \quad \underbrace{k\text{th officer}}_{n-(k-1)}$$

(The count of k terms ends by subtracting $k-1$, rather than k, because it starts by subtracting 0, rather than 1.) Of course, $n-0$ is just n, and we can write $n-(k-1)$ more simply as $n-k+1$. So, by the counting rule, there are

$$n(n-1)(n-2)\ldots(n-k+1)$$

slates of k officers. As in Example 3, we write $n!$ in this way:

$$n! = [n(n-1)(n-2)\ldots(n-k+1)](n-k)(n-k-1)\ldots(2)(1)$$

The product of the $(n-k)$ numbers to the right of the brackets is a factorial: $(n-k)!$. Therefore,

$$n! = [n(n-1)(n-2)\ldots(n-k+1)] \cdot (n-k)!$$

If we divide both sides of the equation by $(n-k)!$, we can express

$$n(n-1)(n-2)\ldots(n-k+1)$$

in terms of factorials:

$$n(n-1)(n-2)\ldots(n-k+1) = \frac{n!}{(n-k)!}$$

In Example 3, there were $n = 15$ members and there were $k = 4$ officers, so we again have

$$15 \cdot 14 \cdot 13 \cdot 12 = \frac{15!}{(15-4)!} = \frac{15!}{11!}$$

slates of officers. If all n members are to be officers, each with a different title, then a slate of officers is just an ordering of all the members. As we saw at the beginning of this section, there are $n!$ such orderings. Since $n = k$ in this case, then $(n-k)! = 0!$. Remember that we defined $0! = 1$; by doing that, we can still express the number of slates in the form

$$\frac{n!}{(n-k)!}$$

because this time

$$\frac{n!}{(n-n)!} = \frac{n!}{0!} = \frac{n!}{1} = n!$$

A slate of k officers selected from an organization with n members is an example of a permutation.

A **permutation** is an ordered k-tuple made up of k distinct elements from a set containing n elements. We can apply the counting rule just as before:

$$\underbrace{(\text{1st element,}}_{(n-0)} \cdot \underbrace{\text{2nd element,}}_{(n-1)} \cdot \underbrace{\text{3rd element,}}_{(n-2)} \ldots, \underbrace{k\text{th element})}_{(n-k+1)}$$

The total number of permutations of k elements from a set of n elements is denoted by the symbol P_k^n. As we have just seen,

$$P_k^n = n(n-1)(n-2)\ldots(n-k+1) = \frac{n!}{(n-k)!}$$

Example 4. Eight swimmers are in the final of an Olympic event. The first three finishers win gold, silver, and bronze medals, respectively. How many different awards of medals can there be?

Solution

An award of medals is an ordered triple:

(Gold, Silver, Bronze)

so, we are counting permutations. Since there are $n = 8$ swimmers and $k = 3$ medals, then $(n - k)! = (8 - 3)! = 5!$, and thus, the number of possible medal outcomes is

$$P_3^8 = \frac{8!}{5!} = \frac{8 \cdot 7 \cdot 6 \cdot 5!}{5!} = 8 \cdot 7 \cdot 6 = 336$$ ■

Example 5. The city editor of a newspaper has four stories to assign to reporters: a trial, the mayor's speech, the opening of a new library branch, and a fire. There are 7 reporters available and no reporter will be given more than one story to cover. In how many ways can the editor assign the stories?

Solution

The editor's decision can be expressed as an ordered 4-tuple of the form

(Trial, Speech, Library, Fire)

Since there are 7 reporters available, we just compute that there are

$$P_4^7 = \frac{7!}{(7-4)!} = \frac{7 \cdot 6 \cdot 5 \cdot 4 \cdot 3!}{3!} = 7 \cdot 6 \cdot 5 \cdot 4 = 840$$

ways the editor can make the assignments. ■

Summary

The term ***k* factorial** refers to the product

$$k! = k(k-1)(k-2)\ldots 3 \cdot 2 \cdot 1$$

where k is a positive whole number greater than 1; by definition, $1! = 1$ and $0! = 1$. A **permutation** of k elements out of a set containing n elements is an ordered k-tuple of k different elements of the set. The number of permutations of k elements out of a set of n is denoted by P_k^n and is given by the formula

$$P_k^n = n(n-1)(n-2) \ldots (n-k+1) = \frac{n!}{(n-k)!}$$

Exercises

1. Calculate: (a) $5!$ (b) $7!$

2. Calculate: (a) $5 \cdot 4 \cdot 3!$ (b) $10 \cdot 9 \cdot 8 \cdot 7!$

3. Calculate: (a) $\dfrac{10!}{7!}$ (b) $\dfrac{12!}{11!}$ (c) $\dfrac{5!}{2!3!}$ (d) $\dfrac{13!}{4!9!}$

4. Calculate: (a) P_6^8 (b) P_5^{10} (c) P_2^7

5. A golfer has won a total of five trophies from various tournaments. She wishes to line them all up in a row along her mantelpiece. In how many ways can she do this?

6. A college psychology department consists of 12 faculty members. The administration of the department consists of a department chairperson and a vice chairperson. How many different administrations can this department have?

7. The author of a mathematics text has written 11 exercises that she can use for the exercise set in a section of the text. She decides to have just 8 exercises in the set, and she must decide in which order the 8 exercises chosen will appear. In how many ways can she construct this exercise set?

8. A consumer-testing organization buys seven models of microwave ovens and tests all of them, one at a time, for safety. In how many ways can this be done?

9. A social-services office employs eight social workers. Five new cases are added to the responsibilities of the office. Each of the cases will be assigned to a different social worker. In how many ways can this assignment be made?

Exercises 10 and 11 concern tone rows. A tone row is a musical theme consisting of a sequence of notes, each different and each used exactly once in the row. Assume each note is of the same duration.

10. How many tone rows can be constructed out of the notes that spell the name of the composer B-A-C-H (H represents the note "b-flat").

11. The composer Arnold Schoenberg composed tone rows using all 12 notes of the chromatic scale. How many tone rows did he have available?

244 PROBABILITY

12. There are 16 entries in the table-setting competition at the county fair, and first through fourth prizes will be awarded. How many different outcomes can there be?

13. A tourist buys an airplane ticket that permits him to visit any three major European cities from a list, in any order he wishes. The listed cities are Paris, Rome, Frankfurt, Copenhagen, Amsterdam, Madrid, and Athens. An "itinerary" would consist of a decision about the first, second, and third city on his trip. How many itineraries are there?

14. The members of a kindergarten class were asked to list their order of preference in ice-cream flavors from among chocolate, vanilla, strawberry, and cookies-and-cream. Each student was asked to list all four flavors, and no ties were permitted. If each child chose a different ordering, what was the largest possible size of the class?

15. A track coach must fill out the entry blank giving the runners, in their order of competition, for a 400-meter relay race in which each athlete runs 100 meters. If the track team contains 9 sprinters who could be entered in the relay, in how many different ways can the coach fill out the entry blank.

In Exercises 16 and 17, different ceramic coatings are to be applied in layers on samples of two different metal alloys. All available coatings will be used, no coating is to be used on more than one of the alloys, but each alloy must receive at least two coatings. If the same coatings are applied to an alloy in a different order, the effect of the coating may be different—so your answer should take order of application into account.

16. In how many ways can this be done if there are 5 coatings in all?

17. In how many ways can this be done if there are 6 coatings in all?

18. In a chemical experiment, 6 different chemicals will be mixed by adding one at a time to the mixture. There are two chemicals, out of the six, that cannot be added one after the other (in either order) without causing an undesirable reaction. In how many ways can this experiment be performed? [*Hint:* In how many ways can the experiment be performed in which the two reacting chemicals are added one after the other?]

3.4 Combinations

In many counting problems, there is no sense of order. For example, how many different five-card poker hands can be dealt from a standard 52-card deck? To a card player it makes no difference whether the cards are picked up in the order (3♥, 5♣, 3♠, 3♦, 5♦), or (5♦, 3♠, 5♣, 3♥, 3♦), or in some other order; the player will be delighted in any case because she has a very good hand—a "full house." If we thought of a poker hand in the terms of the last section, as an ordered 5-tuple of cards—that is, as a permutation of five cards out of the deck of 52—then we would be looking at it unrealistically because both of these 5-tuples are the same hand. Similarly, if we wish to determine the number of different six-to-three majorities possible in a U.S. Supreme Court decision, there is no sense of ordering among the six justices who make up the majority. Or if a television dealer wants to select two sets to test, out of a shipment of ten, in order to decide whether to accept the entire shipment, the dealer may need to know the number of such samples; but there is no reason to distinguish between the first television set he chooses and the second.

For this very common kind of counting problem we need a formula for the number of distinct sets of k objects we can choose out of a collection of n objects, without imposing an ordering on the k objects. An *unordered* set of this kind is called a **combination** of k elements out of n, and the number of such combinations is denoted by the symbol C_k^n.

Combination problems occur when samples are taken. Suppose that an instructor in a psychology class of 40 students needs three of these students to take part in a demonstration for the class. If the group of three students is chosen all at once without regard to any order of selection, then the three students form a combination of three out of the 40 students. We would like to calculate C_3^{40}, the number of different (unordered) samples of three students possible out of a class of 40 students. We do have a formula for P_3^{40}, the number of permutations (ordered triples) of three members out of 40, namely,

$$P_3^{40} = \frac{40!}{(40-3)!} = \frac{40!}{37!}$$

so it seems reasonable to try to relate C_3^{40} to P_3^{40} in order to take advantage of what we already know. This sampling problem *would* be a permutation problem if each student had a different role in the experiment. To show this, the instructor could give each student a colored flag: one red, one yellow, and the last blue. Then we would represent the sample as an ordered triple, such as

(Red flag, Yellow flag, Blue flag)

and we know from the previous section that there are P_3^{40} of these. In order to relate this "flagged" sample to the undifferentiated one we are really interested in, we will think of taking the flagged sample by means of a two-step process. First, three students are picked as a group by drawing their names from a box containing slips of paper with the names of all the members of the class. The result is shown in Figure 3-9. We show the names this way to emphasize the lack of order

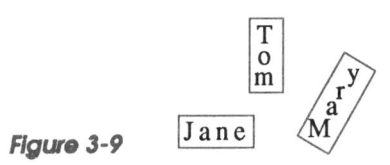

Figure 3-9

among them. As the second step, the instructor gives each of these students different colored flags, for example,

$$\text{Jane – red,} \quad \text{Tom – blue,} \quad \text{Mary – yellow}$$

or, as an ordered triple:

$$(\underbrace{\text{Red student,}}_{\text{Jane}} \underbrace{\text{Yellow student,}}_{\text{Tom}} \underbrace{\text{Blue student}}_{\text{Mary}})$$

So, the choice of three students, each given a different flag, is a permutation obtained by a two-step process. We can represent the process itself as an ordered pair:

$$\text{(First step, Second step)} \leftrightarrow \text{(Choose students, Assign flags)}$$

We will use the counting rule of Section 3.2 to find out how many such ordered pairs there are: If there are n_1 choices for the first member of an ordered pair and n_2 choices for the second, regardless of what the first member is, then there are $n_1 \cdot n_2$ pairs in all. In this example, there are 40 students in the class. Since the symbol for the number of combinations of three students out of the class is C_3^{40}, we write $n_1 = C_3^{40}$ for the number of (unordered) ways we can choose three students, as we did in Figure 3-9.

Once the three students have been selected, we can count the number of ways to assign them flags:

$$(\underbrace{\text{Red flag,}}_{3} \underbrace{\text{Yellow flag,}}_{2} \underbrace{\text{Blue flag}}_{1}) = 3!$$

No matter *which* three students are selected, we get the same answer: there are 3! ways to assign these three students colors, and therefore, $n_2 = 3!$. Thus, by the formula for counting the number of ordered pairs (choose students, assign flags) is

$$P_3^{40} = n_1 \cdot n_2 = \left(C_3^{40}\right)(3!)$$

Using the computational formula

$$P_3^{40} = \frac{40!}{(40-3)!} = \frac{40!}{37!}$$

we write

$$\frac{40!}{37!} = \left(C_3^{40}\right)(3!)$$

Dividing through the equation by 3!,

$$\frac{40!}{3!37!} = C_3^{40}$$

we can now compute:

$$C_3^{40} = \frac{40!}{3!37!} = \frac{40 \cdot 39 \cdot 38 \cdot 37!}{3 \cdot 2 \cdot 1 \cdot 37!} = \frac{40 \cdot 39 \cdot 38}{3 \cdot 2 \cdot 1} = 9880$$

Therefore, there are 9880 different unordered samples (combinations) of three students in a class of 40.

Formula for C_k^n

To obtain a general formula for C_k^n, the number of combinations of k members from a set of size n, we could just repeat the preceding argument for any choices of n and k. An ordered k-tuple is produced through a two-step process: choose a sample of size k from the set of size n, without regard to order; then assign an order to the particular sample you have. We can abbreviate the two- step process as the ordered pair:

Permutation = (Choose the sample, Order the sample)

Again, the formula for the number of ordered pairs tells us that P_k^n, the number of permutations, is of the form $n_1 \cdot n_2$. Furthermore, n_1 is the number of samples, that is, $n_1 = C_k^n$. Once we have the k members of the sample, we need to order them all, and the previous section tells us that there are $k!$ ways to do that, no matter which particular ones have been chosen (which is why we can use the aforementioned formula). Recalling that

$$P_k^n = \frac{n!}{(n-k)!}$$

we have the relationship

$$\frac{n!}{(n-k)!} = P_k^n = n_1 \cdot n_2 = \left(C_k^n\right)(k!)$$

Dividing both sides by $k!$ gives us the general formula:

$$\boxed{C_k^n = \frac{n!}{k!(n-k)!}}$$

Example 1. How many different six-to-three majorities are possible in the U.S. Supreme Court?

Solution

There are nine justices in all, and a six-person majority is a combination of any six of them; that is, it does not matter in which order the justices cast their votes, so there are

$$C_6^9 = \frac{9!}{6!3!}$$

different majorities. For the actual computation of C_6^9, let us write out the numerator:

$$9! = 9 \cdot 8 \cdot 7 \cdot 6 \cdot 5 \cdot 4 \cdot 3 \cdot 2 \cdot 1$$

As in the previous section, you will notice the appearance of $(n - k)! = 3!$. But the larger factor, $k! = 6!$, appears also, that is, $9! = 9 \cdot 8 \cdot 7 \cdot 6!$; if we write $9!$ in this way, the computation is quite manageable because, starting with

$$C_6^9 = \frac{9!}{6!3!} = \frac{9!}{(6!)(3 \cdot 2 \cdot 1)} = \frac{(9 \cdot 8 \cdot 7)(6!)}{(6!)(3 \cdot 2 \cdot 1)}$$

248   PROBABILITY

we then divide both numerator and denominator by 6! to complete the calculation:
$$C_6^9 = \frac{9 \cdot 8 \cdot 7}{3 \cdot 2 \cdot 1} = 84$$

■

This computation illustrates a convenient procedure for calculating C_k^n. First, write out the terms of either $k!$ or $(n-k)!$, whichever is smaller, in the denominator of a fraction. Then, in the numerator, start with n and write numbers in decreasing order until there are as many factors in the numerator as in the denominator. You can calculate C_k^n in this way because, as in the example, the other factors from the formula C_k^n cancel each other.

Example 2. Calculate C_7^{11}.

Solution

In this case $n = 11$ and $k = 7$; so $n - k = 4$. Thus, write out the definition of 4! in the denominator, and the numerator will consist of the four factors beginning with 11. Therefore,
$$C_7^{11} = \frac{11 \cdot 10 \cdot 9 \cdot 8}{4 \cdot 3 \cdot 2 \cdot 1} = 330$$

■

The computation in Example 2 looks just like the computation of C_4^{11}:
$$C_4^{11} = \frac{11!}{4!7!} = \frac{11 \cdot 10 \cdot 9 \cdot 8}{4 \cdot 3 \cdot 2 \cdot 1} = 330$$

and therefore, $C_4^{11} = C_7^{11}$. This equality is not a coincidence. In Example 2, $n = 11$ and $k = 7$—so that $4 = n - k$. In general, if we write out the definition of C_{n-k}^n, here is what we get:
$$C_{n-k}^n = \frac{n!}{(n-k)!(n-(n-k))!} = \frac{n!}{(n-k)!k!} = \frac{n!}{k!(n-k)!} = C_k^n$$

because, in the denominator of the first fraction, we have $n - (n-k) = k$. Thus, for a set of n objects, the number of combinations of size $n - k$ is equal to the number of combinations of size k. In symbols,

$$\boxed{C_{n-k}^n = C_k^n}$$

Example 3. We will choose 5 days at random next April on which to test a new type of rain gauge. How many different samples are there from which to choose?

Solution

There are 30 days in April, and we wish to choose 5 of them. Think of putting cards with the numbers 1 through 30 on them in a hat and drawing out 5 of them. It doesn't matter in which order we take out the cards; that is, if we choose in the order 28, 24, 1, 27, 16, we will regard this as the same as choosing the cards in the order 27, 28, 1, 16, 24. Thus, the answer is given by

$$C_5^{30} = \frac{30 \cdot 29 \cdot 28 \cdot 27 \cdot 26}{5 \cdot 4 \cdot 3 \cdot 2 \cdot 1} = 142{,}506$$

■

Some counting problems require us to combine combinations with other counting techniques, as the next example illustrates.

Example 4. Fifteen people volunteer to participate in a study of the effects of smoking on health. Of this group, 9 are smokers and the other 6 are nonsmokers. If the investigators need a sample of 3 smokers and 3 nonsmokers for the study, how many different samples are possible?

Solution

We will think of a sample as an ordered pair where now, instead of a single symbol in each position in the pair, we have a set:

(Sample of 3 smokers, Sample of 3 nonsmokers)

Since the two groups—smokers and nonsmokers—must be chosen separately, the number of ways we have to choose 3 nonsmokers out of the 6 available does not depend on which 3 smokers were chosen for the first term of the ordered pair. Thus, the number of such ordered pairs is just $n_1 \cdot n_2$, where n_1 is the number of ways of choosing the 3 smokers and n_2 is the same for the nonsmokers. The number of samples of 3 smokers out of 9 available is easy to recognize as a combination problem; that is,

$$n_1 = C_3^9 = \frac{9!}{3!6!} = \frac{9 \cdot 8 \cdot 7}{3 \cdot 2 \cdot 1} = 84$$

and, in just the same way,

$$n_2 = C_3^6 = \frac{6!}{3!3!} = \frac{6 \cdot 5 \cdot 4}{3 \cdot 2 \cdot 1} = 20$$

Therefore, the number of distinct samples of 3 smokers and 3 nonsmokers possible for this experiment is given by

$$n_1 \cdot n_2 = C_3^9 C_3^6 = 84 \cdot 20 = 1680$$

■

Sometimes there is more than one correct way to count, as the final example of this section illustrates.

Example 5. A group of 11 people staying at a vacation cabin have just one car available. All 11 can drive and the car holds 5 people. In how many ways can they choose a driver and four passengers to go for a ride?

Solution

We can think of a choice as an ordered pair

(Driver, 4 passengers)

Any one of the 11 people can be chosen as the driver, so there are 11 possibilities for the first member of the ordered pair. Once the driver has been chosen, there are 10 people available as passengers; no matter which person is the driver, the number of ways to choose the passengers is

$$C_4^{10} = \frac{10!}{4!6!} = \frac{10 \cdot 9 \cdot 8 \cdot 7}{4 \cdot 3 \cdot 2 \cdot 1} = 210$$

Then, as in the last example, the number of ordered pairs is

$$n_1 \cdot n_2 = 11 \cdot C_4^{10} = 11 \cdot 210 = 2310$$

Another way to solve the same problem is to imagine a two-step process for designating the people in the car. First choose 5 people to be in the car, then decide which one of these 5 will do the driving. Thus, we think of the choice as a different ordered pair

(5 people in the car, One of these is driver)

The number of ways to choose the 5 people from all 11 available is

$$C_5^{11} = \frac{11!}{5!6!} = \frac{11 \cdot 10 \cdot 9 \cdot 8 \cdot 7}{5 \cdot 4 \cdot 3 \cdot 2 \cdot 1} = 462$$

Once the 5 are chosen, there are 5 possible drivers; so that, as before, we find that

$$n_1 \cdot n_2 = C_5^{11} \cdot 5 = 462 \cdot 5 = 2310$$ ∎

You will see several more illustrations of these counting methods in the next section, when we apply counting methods to compute probabilities.

Summary

A **combination** of k elements out of a set containing n elements is an unordered subset of k elements out of the set of n. The number of combinations of k elements out of a set of n elements is denoted by C_k^n and is given by the formula

$$C_k^n = \frac{n!}{k!(n-k)!}$$

This number C_k^n has the property that

$$C_k^n = C_{n-k}^n$$

Exercises

1. Calculate: (a) C_2^7 (b) C_6^9
2. Calculate: (a) C_1^3 (b) C_1^5
   (c) What is the value of C_1^n for any whole number n?

In Exercises 3 and 4, a bookstore decides to feature four current best-selling novels in its display window. There are ten novels on the best-seller list.

3. How many choices does the bookstore have if there is a most desirable location in the window, a next most desirable location, a third most desirable one, and a least desirable one?

4. How many choices does the bookstore have if the order in which the books appear in the windows makes no difference?

In Exercises 5 and 6, we suppose that in order to complete the requirements for a major in political science, a student must take four required courses that are all offered during each of the three quarters of the year. Each quarter, the four courses are scheduled so that they meet at four different hours. This makes it possible to take all of them at once.

5. If the student decides to take two of the required courses this quarter, in how many ways can the courses be chosen?

6. If the student decides to take one of the required courses in each of the three quarters this year and wait till next year for the fourth course, in how many ways can this be done?

7. How many triangles can you draw by connecting three of the dots in Figure 3-10?

Figure 3-10

In Exercises 8 and 9, ten children sign up for a playground basketball team. The playground director will choose five of these children as the starting team for a game against another playground team.

252   PROBABILITY

8. In how many ways can this choice be made if these are young children who play without any assignment of individual positions, that is, if a team is formed just by naming the five children who will play?

9. In how many ways can this be done if the playground director assigns each player chosen a distinct playing position (center, power forward, small forward, point guard, or shooting guard)?

10. A pizza restaurant offers a special low price on "our delicious extra-large cheese-and-tomato pizza with your choice of any 3 additional toppings." The 9 toppings available are mushrooms, green peppers, onions, olives, pineapple, anchovies, pepperoni, sausage, and ham. If you choose exactly 3 toppings, how many different choices are there?

11. A town consists of 12 families living north of the railroad tracks and 15 families living south of the tracks. A sampling organization will choose 3 families north of the tracks and 2 families from the southern sector for a survey. In how many ways can this be done?

In Exercises 12 and 13, a coffee store has in stock the following five "regular roast" coffees: Mocha-Java, Mexican, Colombian, Kenyan, and Kona. There are also three "dark roast" coffees: French, Espresso, and Viennese. A customer wishes to make up a pound of coffee by mixing two kinds of "regular roast" coffee with one of the "dark roast" coffees.

12. In how many ways can this be done if the mixture will consist of $\frac{1}{3}$ pound of each kind?

13. In how many ways can this be done if the mixture will consist of $\frac{1}{2}$ pound of one of the three kinds and $\frac{1}{4}$ pound of each of the other two?

14. A friend wants to borrow books from you to take on vacation. He will choose two mysteries and two westerns. If you own 23 mysteries and 14 westerns, how many choices does your friend have?

15. A soccer team consists of 11 players as follows: 3 forwards, 3 midfielders, 4 defenders, and a goalkeeper. Suppose a forward can play any one of the forward positions, but no other position on the team, and the same is true of the other players. If the soccer coach has 6 forwards, 7 midfielders, 6 defenders, and 3 goalkeepers on her squad, how many choices does she have in making up the starting lineup for the next game?

16. A "full-house" hand in 5-card poker consists of 3 cards of one value and 2 of another—for instance, three kings and two 5s. How many different full-house hands are there?

In Exercises 17 and 18, express the answers in terms of the symbols C_k^n, but do not attempt to compute these very large numbers.

17. In the card game "rummy," each of the two players receives 7 cards from a standard 52-card deck. In how many ways can this be done? [*Hint:* For the purpose of

the count, you may assume that first one player is dealt all 7 cards, then the other player receives 7 cards.]

18. Four people are playing poker. Five cards are dealt to each of the four players from the 52-card deck. In how many ways can this be done?

3.5 Computing Probability by Counting

We will now use the counting theory of Sections 3.2 through 3.4 to solve some problems concerning probability with equally likely outcomes. The notation

$$p(\ )$$

will represent the probability that an outcome of the type described within the parentheses will occur. For example, if we toss a coin, we know that $p(\text{heads}) = .50$; if we roll two dice, then $p(\text{sum is 3}) = \frac{2}{36}$; if we draw a card at random from a standard deck, then $p(5\blacklozenge) = \frac{1}{52}$.

In discussing probability, we imagine an experiment such as flipping a coin, dealing a poker hand, or choosing a TV set to test. A certain outcome—for example, heads, a royal flush, finding that the TV set works—is called a **success** and any other outcome is a **failure**. This terminology of successes and failures of experiments is traditional in probability theory; it helps to isolate the mathematical facts being discussed from the context of the particular situation.

If each possible outcome has the same chance of success, then recall that the definition of probability in Section 3.1 tells us that the computation of the probability of success amounts to a counting problem. We will use the notation

$$n(\ )$$

to indicate the number of outcomes of the type described within the parentheses. For example,

$$n(\text{outcomes possible from one die}) = 6$$
$$n(\text{kings in a standard card deck}) = 4$$

and, since there are 100 U. S. senators,

$$n(\text{committees of 4 senators}) = C_4^{100} = \frac{100!}{4!96!} = \frac{100 \cdot 99 \cdot 98 \cdot 97}{4 \cdot 3 \cdot 2 \cdot 1} = 3,921,225$$

The definition of probability in Section 3.1 tells us that

$$p(\text{success}) = \frac{n(\text{successful outcomes})}{n(\text{possible outcomes})}$$

The examples that follow are based on this definition.

254   PROBABILITY

Example 1. What is the probability of being dealt a "royal flush" (ace, king, queen, jack, and ten, all of the same card suit) in poker?

Solution

If the cards are dealt fairly, any hand is as likely to appear as any other. There are 52 cards in the deck and we choose five at a time. Thus,

$$C_5^{52} = \frac{52!}{5!47!} = \frac{52 \cdot 51 \cdot 50 \cdot 49 \cdot 48}{5 \cdot 4 \cdot 3 \cdot 2 \cdot 1} = 2{,}598{,}960$$

Because there are just four royal-flush hands, one for each suit,

$$p(\text{royal flush}) = \frac{4}{2{,}598{,}960} = .0000015 \text{ (approximately)} \blacksquare$$

Example 2. We will choose 5 days at random next April on which to test a new type of rain gauge in Los Angeles. Based on past experience, we assume that there will be only 4 days with measurable rain in Los Angeles next April. Calculate the probability that there will be no rain during the 5 days of the test.

Solution

In Example 3 of the preceding section we observed that choosing 5 days during the month of April is like dealing 5 cards from a deck numbered 1 through 30 (for the days of that month), and that therefore, the number of possible samples of 5 days is given by

$$C_5^{30} = \frac{30 \cdot 29 \cdot 28 \cdot 27 \cdot 26}{5 \cdot 4 \cdot 3 \cdot 2 \cdot 1} = 142{,}506$$

The numerator of the probability computation, that is, $n(\text{successes})$, is the number of samples chosen from the days in which there is no rain, because "success" in computing $p(\text{no rain during test})$ means having no rain during any of the five chosen days. We are told that there will be $30 - 4 = 26$ days without rain. A "success" consists of a sample that comes from those 26 days. Notice that we do not have to know which days are rainless because the *number* of 5-day samples is the same for any 26 days, and it is

$$C_5^{26} = \frac{26 \cdot 25 \cdot 24 \cdot 23 \cdot 22}{5 \cdot 4 \cdot 3 \cdot 2 \cdot 1} = 65{,}780$$

Therefore,

$$p(\text{no rain during test}) = \frac{C_5^{26}}{C_5^{30}} = \frac{65{,}780}{142{,}506} = .46 \text{ (approximately)}$$

so, there is almost 1 chance in 2 that the rain gauge won't have anything to measure. $\blacksquare$

Example 3. A man and a woman eating together in a restaurant play a game to determine who will pay the check. The man takes out a penny, nickel, and dime from his change, the woman—a nickel, dime, quarter, and half-dollar from hers. Each chooses a coin and hides it under a napkin. When the napkin is removed, the man pays if the total value of the two coins is odd, the woman pays if it is even. Calculate the probability that the man will pay the bill.

3.5 COMPUTING PROBABILITY BY COUNTING

```
Man's Coin      Woman's Coin     Total Value

                    5                6
                    10               11
         1
                    25               26
                    50               51

                    5                10
                    10               15
         5
                    25               30
                    50               55

                    5                15
                    10               20
         10
                    25               35
                    50               60
```

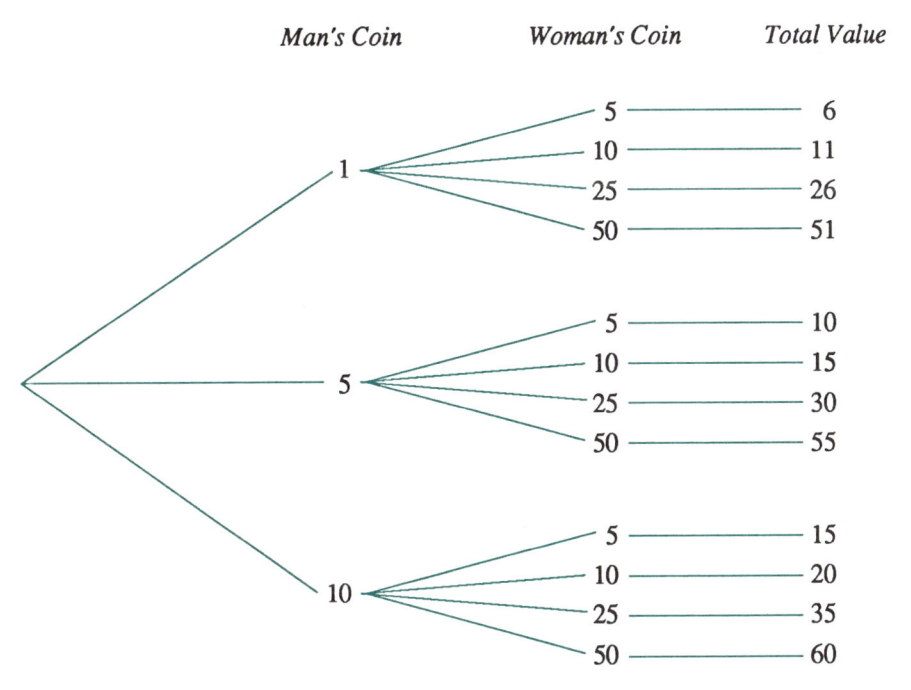

Figure 3-11

Solution

An outcome to the game consists of an ordered pair

$$(\text{Man's coin, Woman's coin})$$

Since the man has 3 coins and the woman 4, there are $3 \cdot 4 = 12$ possible outcomes. To calculate the number of successful outcomes, in which the sum of the value of the coins is odd, it is easiest just to make a tree (Figure 3-11) of these outcomes and see which of the 12 branches add up to an odd value. Since there are 6 outcomes in which the total value is odd, we conclude that this is a fair game because

$$p(\text{man will pay}) = \frac{6}{12} = .50$$

∎

Example 4. A family consists of a father, mother, son, and daughter. What is the probability that all four were born on different days of the week?

Solution

A possible outcome consists of a 4-tuple listing the four (not necessarily different) days of the week in which the four members of the family were born. For instance, the outcome could be

(Father,   Mother,   Son,   Daughter)
(Tuesday, Saturday, Tuesday, Friday)

Any day of the week can appear in any position in the ordered 4-tuple, so there are 7 possibilities for each position. We see that

$$\underbrace{\text{(Father,}}_{7} \cdot \underbrace{\text{Mother,}}_{7} \cdot \underbrace{\text{Son,}}_{7} \cdot \underbrace{\text{Daughter)}}_{7} = 2401$$

On the other hand, a successful outcome is one in which the four days are all different. We can let the father's birthday be any day of the week; then the mother's must be one of the remaining 6. This leaves 5 possibilities for the son and, after that, 4 for the daughter. Thus, the numerator of the probability computation is

$$\underbrace{\text{(Father,}}_{7} \cdot \underbrace{\text{Mother,}}_{6} \cdot \underbrace{\text{Son,}}_{5} \cdot \underbrace{\text{Daughter)}}_{4} = 840$$

Another way to compute the numerator is to notice that we are counting ordered subsets of four of the seven days of the week because, for example, the successful outcomes

(Father,   Mother,   Son,   Daughter)
(Tuesday,   Saturday,   Monday,   Friday)

(Father,   Mother,   Son,   Daughter)
(Saturday, Tuesday,   Monday,   Friday)

are different, even though they involve the same four days of the week. Therefore, we are counting permutations:

$$n(\text{successful outcomes}) = P_4^7 = \frac{7!}{(7-4)!} = 7 \cdot 6 \cdot 5 \cdot 4 = 840$$

The result is

$$p(\text{all born on different days}) = \frac{7 \cdot 6 \cdot 5 \cdot 4}{7 \cdot 7 \cdot 7 \cdot 7} = \frac{840}{2401} = .35 \text{ (approximately)} \quad \blacksquare$$

Summary

$$p(\text{success}) = \frac{n(\text{successful outcomes})}{n(\text{possible outcomes})}$$

where $p(\ )$ indicates probability and $n(\ )$ indicates the number of outcomes.

Exercises

In Exercises 1 and 2, snow falling in Minneapolis will be tested for pollution during the last half (that is, the last 15 days) of November. Six days will be chosen at random for the test.

1. How many different samples are there from which to choose?

2. Based on past experience, we estimate that there will be two days of snow in Minneapolis during the last half of November. Calculate the probability that there will be no measurable snow during the six days of the test.

3. Suppose that in the game described in Example 3, the man's change consists of a penny, nickel, dime, quarter, and half-dollar, whereas the woman has only a nickel and a quarter. The man still pays if the total value of the two coins chosen is odd. Show that the game is no longer fair by computing the probability that the man will pay the check.

In Exercises 4 through 6, the board of directors of a company consists of the chairperson of the board and six other members. The board must select three directors to form an executive committee.

4. In how many ways can the three be chosen from among the seven members of the board?

5. If the board chairperson must be included among the three chosen, how many choices are there?

6. What is the probability that the board chairperson will be included, assuming that it is equally likely that any of the seven members of the board may be chosen for the committee?

7. Calculate the probability of rolling a sum of 10 with two dice.

8. A group of 12 people take a vacation together. There is an opportunity for 4 of them to have a boat ride, and they will choose at random to decide who will go. There are twin sisters among the 12 people. Calculate the probability that both twins will be chosen.

9. Suppose the U. S. Olympic Committee decides to select three of the 50 states at random as the location of special-training facilities in three sports: badminton, gymnastics, and wrestling. Calculate the probability that none of the 6 New England states will be chosen.

10. Thirty mice in a nutrition experiment are divided into three diet groups: 8 mice are fed a high-protein diet, 12 other mice receive the standard diet plus a vitamin supplement, and the remaining 10 form a control group that is fed just the standard diet. If three mice among the 30 are chosen at random for a test, what is the probability that one is chosen from each diet group?

11. A magician invites you to shuffle a deck of cards. She shuffles them some more; then she invites you to turn over the top four cards in order. You turn over A♠, then K♠, then Q♠, and finally J♠. If this were really a randomly shuffled deck, what is the probability you would get exactly this outcome from turning over the top four cards one at a time?

258   PROBABILITY

In Exercises 12 through 14, a handsome prince is informed by his fairy godmother that he may select three wishes from a list of twelve. However, two of the twelve wishes have the undesirable side effect that if he chooses them, he will be instantly and permanently turned into a frog. The prince has no way of knowing which of the twelve wishes on the list have this side effect, so he selects three wishes at random.

12. In how many different ways can the prince choose his wishes?

13. In how many different ways can he choose three that avoid the undesirable side effects?

14. What is the probability that the prince will avoid being turned into a frog?

In Exercises 15 through 22, imagine a card game in which each player receives three cards.

15. How many different three-card hands are possible, using a standard deck?

16. How many hands consisting of three spades are possible?

17. How many different hands consist of three cards of the same suit (flush)?

18. How many different hands consist of three kings?

19. How many different hands consist of three cards of the same value, such as three kings (three of a kind)?

20. What is the probability of drawing a flush?

21. What is the probability of drawing three cards of the same value?

22. If you were writing the rules for this card game (three-card poker), you might make a hand more valuable if the probability of obtaining it is lower. Which then, should be the more valuable hand in three-card poker—a flush or three of a kind?

In Exercises 23 and 24, the casting director of a film will choose actors to play the members of a family consisting of a father, mother, and three children. There are four men being considered for the role of the father, three women for the mother, and 8 child actors for the children's roles.

23. How many choices of cast does the casting director have?

24. Suppose the child of one of the men being considered for the father's role is among those considered for a child's role. If the cast were chosen at random, what is the probability that this father and child would both be chosen?

3.6 Union of Events

A roulette game is an experiment in which a wheel spins and selects one of the possible numbers at random. The player bets on the outcome of the random choice. Although it is possible to bet on a particular number, usually the player

3.6 UNION OF EVENTS

bets on a block of numbers (see Figure 3-12). For example, the player may bet that the number that comes up will be in the first dozen (1–12), the second dozen (13–24), or the last dozen (25–36). In another type of bet, the gambler chooses one column of numbers from among the first column (1, 4, 7, and so on, to 34), the second column (2, 5, 8, and so on, to 35), and the third column (3, 6, 9, and so on, to 36). Roulette players often bet on more than one type of block of numbers in a single play of the game.

An Illustration

Figure 3-12 shows a roulette table where the gambler has placed bets on the first dozen numbers and also on the last column. The next play of the roulette game will be a success from the gambler's point of view if the number chosen turns out to be in the first dozen or the last column, since in either case the gambler will win some money. We wish to compute the probability that the gambler will be successful; that is, we wish to compute p(first dozen or last column). Since we

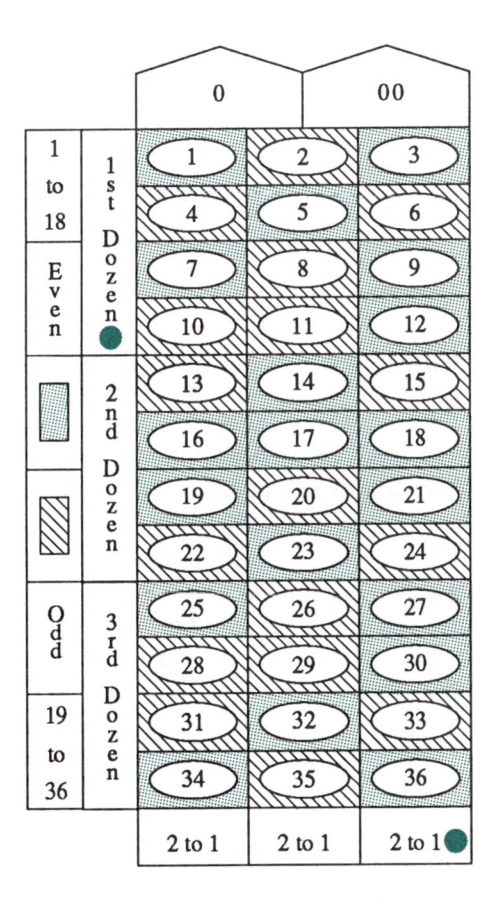

Figure 3-12

260   PROBABILITY

assume that the roulette game is honest, so that any number is as likely to come up as any other, we need to count to determine the probability of winning because

$$p(\text{success}) = \frac{n(\text{successful outcomes})}{n(\text{possible outcomes})}$$

For the roulette problem this means

$$p(\text{first dozen or last column}) = \frac{n(\text{first dozen or last column})}{n(\text{outcomes in roulette})}$$

The numbers 0, 00, 1, 2, 3, ... , 36 yield 38 possible outcomes in roulette; so, the denominator of the fraction is 38. The significant part of the problem is the calculation of the numerator.

We can compute n(first dozen or last column) by listing the outcomes and counting up how many there are. We want to determine how many numbers between 1 and 36 are either less than or equal to 12 (first dozen) or evenly divisible by 3 (since the last column consists of 3, 6, 9, ... , 36). The numbers that satisfy these requirements are

$$1, 2, 3, 4, 5, 6, 7, 8, 9, 10, 11, 12, 15, 18, 21, 24, 27, 30, 33, 36$$

If we count these numbers, we find that n(first dozen or last column) = 20. Therefore,

$$p(\text{first dozen or last column}) = \frac{20}{38}$$

In problems involving large numbers of outcomes, it may not be practical to make such a list. We need a more sophisticated counting procedure.

Another Way To Count

It often happens that there are two classes of successful outcomes; call them A and B. The calculation of $n(A)$ and $n(B)$, the number of outcomes of each class, is simpler than the calculation of $n(A$ or $B)$. Thus, n(first dozen) is certainly 12 because this class consists of the numbers 1–12. Similarly, n(last column) is 12 because the numbers 3, 6, 9, ... , 36 that make up the last column are just the numbers 1–12 each multiplied by 3. In general, it is not possible to calculate $n(A$ or $B)$ knowing only $n(A)$ and $n(B)$. If we try adding $n(A)$ to $n(B)$ in the example, guessing that the number of outcomes in the class *either A or B* is just the number of outcomes in A plus the number in B, we get

$$n(\text{first dozen}) + n(\text{last column}) = 12 + 12 = 24$$

But we know that n(first dozen or last column) is 20. The problem is that a number, such as 6, which is both between 1 and 12 and also a multiple of 3, is counted twice in the sum n(first dozen) + n(last column).

The correct way to relate n(first dozen) + n(last column) to n(first dozen or last column) is to adjust the sum so that no single number is counted twice. The numbers that are counted twice in n(first dozen) + n(last column) are those numbers that are both between 1 and 12 and also multiples of 3, namely, 3, 6, 9, and 12. If we subtract the number of these, which should be written n(first dozen and last column), from the sum, then we are counting each of the numbers 3, 6, 9, and 12 just once. Thus, we have

$$n(\text{first dozen}) + n(\text{last column}) - n(\text{first dozen and last column})$$
$$= 12 + 12 - 4$$
$$= 20$$
$$= n(\text{first dozen or last column})$$

A Counting Rule

The roulette example illustrates a general counting rule. Let us visualize all possible outcomes of some experiment as the points shown in Figure 3-13.

If the experiment consists of a play of roulette, each dot represents one of the numbers on the roulette wheel. If we look at a probability problem as an experiment, then the set of all outcomes of the experiment is called the **sample space**. A subset of the sample space is called an **event**. We assume that the sample space contains at least one outcome.

Suppose there are two classes of successful outcomes for this experiment, class A and class B. Then each of the events A and B is in the sample space of all outcomes. These events are represented in Figure 3-14 by the points within the indicated regions of the Venn diagram.

The computation of the number of outcomes either in event A or in event B amounts to counting the number of points within at least one of the regions. If we count the number $n(A)$ of points in event A and add to it the number $n(B)$ of points in event B, we are counting those points that are in A and not in B once each, and those points that are in B and not in A once each. But points in both A and B are counted twice, once as members of A and once as members of B. Thus, if we subtract from the sum $n(A) + n(B)$ the number $n(A \text{ and } B)$ of points that are both in A and B, then we eliminate the double counting of these elements and we obtain a correct count of the number of successful outcomes. We now have the counting rule:

$$n(A \text{ or } B) = n(A) + n(B) - n(A \text{ and } B)$$

Using the sample space to visualize outcomes suggests a convenient language for stating the counting rule. The sample space is just a set of points and the events

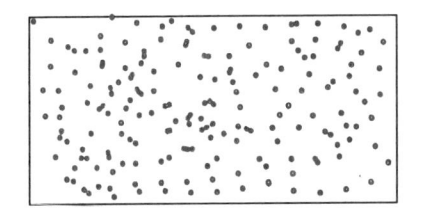

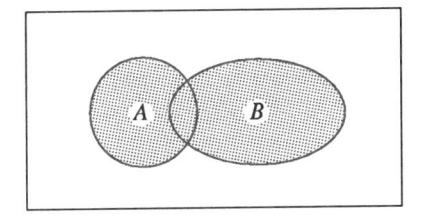

Figure 3-13

Figure 3-14

are subsets. The set of points that are either in the subset A or the subset B or both form a set, the union of A and B, written $A \cup B$. The set of points that are both in A and in B form another subset of the sample space, the intersection of A and B, written $A \cap B$. The counting rule can be expressed in set notation.

> If A and B are sets, then
> $$n(A \cup B) = n(A) + n(B) - n(A \cap B)$$

Example 1. How many numbers from 1 through 1000 are either even or greater than 800?

Solution

It would certainly take a long time to list all the numbers required; so, let A be the set of even numbers from 1 through 1000 and let B be the set of numbers greater than 800, that is, from 801 through 1000. We wish to compute $n(A \cup B)$. Since there are 1000 numbers in all and half are even, $n(A) = 500$. Because 801 can be written in the form $800 + 1$, and $1000 = 800 + 200$, we can see that there are 200 numbers from 801 through 1000; thus, $n(B) = 200$. The set $A \cap B$ consists of the even numbers from 801 through 1000; this set contains half of the numbers in B. Therefore, $n(A \cap B) = 100$. By the counting rule the result is

$$n(A \cup B) = n(A) + n(B) - n(A \cap B) = 500 + 200 - 100 = 600$$ ■

The Probability of a Union of Events

The counting rule leads us to our first important principle of probability. If U is the (nonempty) set of all elements of the sample space (in which each element represents a possible outcome of an experiment), if A and B are two events, and if all outcomes are equally likely, then according to the definition in Section 3.1,

$$p(A \cup B) = \frac{n(A \cup B)}{n(U)}$$

In this equation, $n(U)$ is the (nonzero) number of all possible outcomes. The counting rule tells us that

$$p(A \cup B) = \frac{n(A \cup B)}{n(U)} = \frac{n(A) + n(B) - n(A \cap B)}{n(U)} = \frac{n(A)}{n(U)} + \frac{n(B)}{n(U)} + \frac{n(A \cap B)}{n(U)}$$

By definition,

$$\frac{n(A)}{n(U)}$$

is the probability, $p(A)$, of success for event A, and similarly, for event B. Also, the probability $p(A \cap B)$ that both events will be simultaneously successful is the fraction

$$\frac{n(A \cap B)}{n(U)}$$

We will examine $p(A \cap B)$ more closely in Section 3.9. For now, we restate the preceding equations entirely in terms of probability and, as we will soon see, this will lead us to the general principle of probability:

$$p(A \cup B) = p(A) + p(B) - p(A \cap B)$$

The General Probability Rule for the Union

When we introduced the probability concept in Section 3.1, we identified three types of probability. Subsequently, we have been focusing our attention on outcomes that are equally likely. In this setting probability principles can be deduced from counting rules, as we have just seen in the case of the union of events. Of the other two types of probability, one type is based on experience—for instance, our estimate of the probability of rain in Seattle in January was based on the climatological records for that city. The other kind of probability is a subjective estimate, like the weather forecaster's credible interval estimate of the day's high temperature.

In discussing any type of probability, we will be concerned with a set of possible outcomes, called the sample space U. There is a number $p(A)$, the probability that is associated with a subset A of the sample space; the subset A is still referred to as an event. Thus, the picture we have in mind is again represented as in Figure 3-15.

There is an important difference in how this picture should be interpreted, however. In the case of equally likely outcomes of U, the probability $p(A)$ depended on the "size" of A, that is, the number of outcomes in the subset. For probability in general, we can no longer make this assumption. We just know that a probability $p(A)$, which is a number between 0 and 1, has been assigned to A.

In Section 3.1 we stated that probability is the same whether it is calculated by counting equally likely outcomes, by accumulating experience, or by expressing a subjective opinion. Consequently, since we were able to state the formula for the probability of a union of events entirely in terms of probabilities, rather

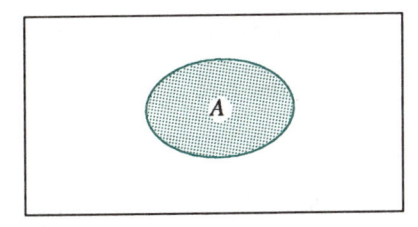

Figure 3-15

than by counting, we can use the general principle

$$p(A \cup B) = p(A) + p(B) - p(A \cap B)$$

whenever we need it in a probability problem, even though we reasoned it out only in the counting setting.

Example 2. Using information from *Climatological Data*, we estimate that on an April day the probability of rain in Los Angeles is .13, the probability of rain in San Diego is .12, and the probability that it will rain in both cities is .08. Calculate the probability that it will rain in at least one of these two cities.

Solution

We let A represent the event (rain in Los Angeles), LA, and B, (rain in San Diego), SD; then by the probability principle and the information given in the statement of the example, we can compute

$$p(A \cup B) = p(\text{LA} \cup \text{SD}) = p(\text{LA}) + p(\text{SD}) - p(\text{LA} \cap \text{SD})$$
$$= .13 + .12 - .08$$
$$= .17$$

We can use the union formula in another way also, as an indirect technique for computing the probability that both A and B will happen, as the following illustrates.

Example 3. An oil company drills two wells. The company estimates that there is a probability of .20 that it will strike oil at the location of the first well and .30 that it will strike oil at the second well. Stockholders are informed that there is a probability of .44 that it will strike oil in at least one of these wells. What is the estimated probability of striking oil at both wells?

Solution

Let $p(A)$ be the probability that the company will strike oil at the first well, that is, $p(A) = .20$; and let $p(B)$ be the probability of striking oil at the second well, so that $p(B) = .30$. The company tells its stockholders that the probability of striking oil at either the first or the second well is $p(A \cup B) = .44$. Substituting these probabilities into the formula

$$p(A \cup B) = p(A) + p(B) - p(A \cap B)$$

we get

$$.44 = .20 + .30 - p(A \cap B)$$

and solving for $p(A \cap B)$, we find that the estimate of the probability of striking oil at both wells is

$$p(A \cap B) = .20 + .30 - .44 = .06$$

Summary

If a probability problem is viewed as an experiment, then the set of all outcomes of the experiment is called the **sample space**. A subset of the sample space is called an **event**. If success in the experiment consists of an outcome in either event A or event B, then

$$p(A \text{ or } B) = p(A) + p(B) - p(A \text{ and } B)$$

which can be written in set notation as

$$p(A \cup B) = p(A) + p(B) - p(A \cap B)$$

Exercises

1. What is the probability of winning in American roulette if bets are placed on the block of numbers (1–18) and (all odd numbers)?

2. What is the probability of winning in American roulette if bets are placed on the second column (2, 5, 8, ..., 35) and (all even numbers)?

3. What is the probability of drawing either a face card (jack, queen, king) or a club from a standard deck of cards?

4. What is the probability of drawing either a red card or an ace from a standard deck of cards?

5. Using information from *Climatological Data*, we estimate that the probability it will freeze (that is, that the low temperature for a 24-hour period will be below 32°F.) on any particular day during the first half of November is .56 in Minneapolis, .47 in Duluth, and there is a probability of .37 that it will freeze in both cities on that day. What is the probability that it will freeze in at least one of the two cities on a particular day in the first half of November?

6. A student needs money to stay in school. He can do so either if he finds a part-time job or if he is awarded a scholarship. He estimates the probability that he will get a job as .75, that he will be awarded a scholarship as .20, and that both will happen as .15. What is the probability that he will be able to stay in school?

7. An assembly line combines components to form a finished product. The probability is .12 that a finished product contains a defective component, .05 that there was an error in assembling the components, and .006 that there is both a defective component and an error in assembly. What is the probability that a finished product either contains a defective component or was assembled incorrectly?

8. At Oak College, 22 faculty members have earned doctorates and 10 have published books. Of the ten who have published books, seven have doctorates. How many members of the faculty have either a doctorate or a published book to their credit?

9. The sales manager of a large corporation estimates that she will be chosen as vice president of a new division if the corporation opens this division and if this year's sales are higher than last year's. She estimates the probability that the corporation will open the new division to be .90, that this year's sales will be higher than last year's to be .70, and that at least one of these will happen to be .98. According to these estimates, what is the probability that she will become the vice president of the new division?

10. A professor gave her students a questionnaire at the beginning of her class and found out that 32 had taken a course in computer science, 61 had taken statistics, and 16 had taken both courses. If, in order to take this professor's course, a student must have taken either computer science or statistics, how many students did the professor have in her class?

11. A precision part is inspected by taking accurate measurements of size and by using rays to detect internal flaws in the metal. The part is defective if either its size is incorrect or if it has internal flaws. The probability that the size is incorrect is .15; the probability of an internal flaw is .01; and the probability that both problems occur is .0015. What is the probability that the part is defective?

12. Suppose that when a car enters a service station, the probability is .08 that it will need oil and .03 that the tires will need air. Suppose the probability is .09 that the car will need either oil or air. What is the probability that a car will need both oil and air?

13. The price of the stock of a ship-building company will rise if the company receives a large number of orders for its ships or if it is granted a low-interest government loan. The president of the company estimates that there is a probability of .40 that the stock will rise in price, because there is a probability of .30 that the company will receive a large number of orders for its ships and of .05 that it will both receive the orders and get the government loan. What does the president estimate to be the probability that his company will be granted the government loan?

14. There is a country with a telephone system so bad that the probability is .55 that a caller will be connected with the wrong number and .35 that a call will be cut off before the answering party can be identified. The probability is .70 that at least one of these will occur. What is the probability that both events will happen, that is, that the caller will be connected to the wrong party, but the line will be cut off before the answering party can be identified?

15. The merger of two companies will take place if two conditions are met: the stockholders of the two companies agree to a stock trade and the government approves the merger. The probability of the merger is estimated to be .60. The probability that at least one of the two conditions, the stock trade or government approval, will be met is estimated to be .90. If the estimated probability that the stockholders will agree to the stock trade is .95, what is the probability of government approval?

3.7 Disjoint Events

An important special case of the probability principle

$$p(A \cup B) = p(A) + p(B) - p(A \cap B)$$

arises when no outcome is in both event A and event B.

Example 1. A gambler wins on the first roll of two dice if the sum of the numbers on the dice adds up to either 7 or 11. What is the probability that the gambler wins on the first roll?

Solution

Remember that the possible outcomes when two dice are rolled may be viewed as ordered pairs (first die, second die), and so, there are 36 outcomes in all. If A represents the outcomes in which the numbers add up to 7 and B those in which the sum is 11, then the probability that the gambler will win on the first roll is $p(A \cup B)$. There is no way that the numbers on the dice can, on the same roll, add up to both 7 and 11, so there is no outcome that is in both events. That is, $n(A \cup B) = 0$. Consequently,

$$p(A \cap B) = \frac{0}{36} = 0$$

and the probability principle reduces in this case to the simple formula

$$p(A \cup B) = p(A) + p(B)$$

The outcomes of type A: (sum of 7) are (1, 6), (2, 5), (3, 4), (4, 3), (5, 2), and (6, 1); so $p(A) = \frac{6}{36}$. The only rolls with a sum of 11 are (5, 6) and (6, 5); so $p(B) = \frac{2}{36}$. Therefore,

$$\begin{aligned} p(\text{win on first roll}) &= p(A \cup B) \\ &= p(A) + p(B) \\ &= \frac{6}{36} + \frac{2}{36} \\ &= \frac{8}{36} \\ &= .22 \text{ approximately} \end{aligned}$$ ∎

Probability of Disjoint Events

We can picture the outcomes of Example 1 as shown in Figure 3-16. This special case arises so often, it is worth stating. In the language of set theory, two sets are said to be **disjoint** if no element is a member of both sets.

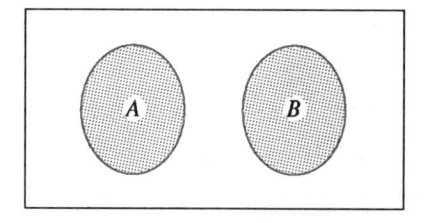

Figure 3-16

> If events A and B are disjoint, then
> $$p(A \cup B) = p(A) + p(B)$$

Example 2. The assistant district manager of a company expects to be promoted to district manager (DM) if the present DM leaves the district office. The DM will leave if she accepts a promotion within the company or if she resigns to take a position with another company. The assistant estimates that the probability that the present DM will be promoted is .25 and the probability that she will resign is .35. What does the assistant estimate to be the probability that the DM will leave?

Solution

Let A be the event (DM accepts a promotion) and let B be the event (DM resigns). Since the DM cannot both accept a promotion and resign, the probability that the DM will leave is given by

$$\begin{aligned} p(\text{DM leaves}) &= p(A \cup B) \\ &= p(A) + p(B) \\ &= .25 + .35 \\ &= .60 \end{aligned}$$ ∎

Probability of Failure

No outcome can be both a success and a failure (that is, not a success) so

$$p(\text{success} \cup \text{failure}) = p(\text{success}) + p(\text{failure})$$

On the other hand, every outcome is either a success or a failure, as we see in Figure 3-17; so

$$p(\text{success} \cup \text{failure}) = p(\text{some outcome}) = 1$$

Thus,

$$p(\text{success}) + p(\text{failure}) = 1$$

and we can relate $p(\text{success})$ to $p(\text{failure})$ by writing either of the following:

3.7 DISJOINT EVENTS

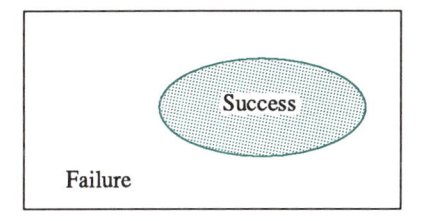

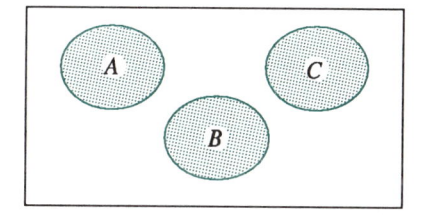

Figure 3-17

Figure 3-18

$$p(\text{success}) = 1 - p(\text{failure})$$
$$p(\text{failure}) = 1 - p(\text{success})$$

Example 3. A television dealer receives ten sets from the manufacturer. The dealer decides to test two of the sets at random. The dealer will send the entire shipment back to the manufacturer if there is anything wrong with either set. If there are, in fact, two faulty sets among the ten, what is the probability that the dealer will reject the shipment?

Solution

The probability that the dealer will succeed—find at least one of the defective sets in the sample—is most easily computed by first determining the probability of failure to find anything wrong. In Section 3.5, we calculated that the probability of failing to find a defective set is

$$p(\text{selecting two good sets}) = \frac{C_2^8}{C_2^{10}} = \frac{28}{45} = .62 \text{ (approximately)}$$

Therefore,

$$p(\text{rejecting the shipment}) = p(\text{success})$$
$$= 1 - p(\text{failure})$$
$$= 1 - .62$$
$$= .38 \text{(approximately)}$$

Many Disjoint Events

If there are three or more ways to succeed in an experiment, we would like to know how to calculate the probability of success, that is, how to find an outcome that is in one of the (successful) events. In other words, we would like a probability principle for $p(A \cup B \cup C)$ and for more general cases. When we know that the various successful events are all disjoint, that is, no outcome belongs to two or more events, the probability principle is quite simple. Figure 3-18 illustrates a sample space with three distinct types of successes, A, B, and C. If you wanted to count the total contents of A, B, and C, you would just count each separately and add them up:

270 PROBABILITY

$$n(A \cup B \cup C) = n(A) + n(B) + n(C)$$

There is nothing to subtract because no outcome is in more than one event; so, no outcome can be counted more than once. More generally, if there are k events—$A_1, A_2, \ldots, A_k$—of successful outcomes for an experiment and if no outcome is in more than one of these events, then

$$n(A_1 \cup A_2 \cup \ldots \cup A_k) = n(A_1) + n(A_2) + \ldots + n(A_k)$$

Since we can count the number of favorable outcomes, there is no difficulty in calculating the probability when all outcomes are equally likely:

$$p(A_1 \cup A_2 \cup \ldots \cup A_k) = \frac{n(A_1) + n(A_2) + \ldots + n(A_k)}{n(U)}$$

$$= \frac{n(A_1)}{n(U)} + \frac{n(A_2)}{n(U)} + \ldots + \frac{n(A_k)}{n(U)}$$

$$= p(A_1) + p(A_2) + \ldots + p(A_k)$$

We state this now as a general principle of probability.

If $A_1, A_2, \ldots, A_k$ are events of a sample space and if no outcome is in more than one event, then

$$p(A_1 \cup A_2 \cup \ldots \cup A_k) = p(A_1) + p(A_2) + \ldots + p(A_k)$$

Example 4. Two dice are rolled. What is the probability that the sum of the numbers on the dice is at least 10?

Solution

Since each die can take on any number between 1 and 6, the sum of the numbers on the dice will be between 2 and 12. The sum will be at least 10 if it is either 10, 11, or 12. Therefore,

$$p(\text{at least } 10) = p(10 \cup 11 \cup 12)$$
$$= p(10) + p(11) + p(12)$$

where, for instance, $p(10)$ means the probability that the sum on the dice is exactly 10. The ways that the dice can add up to 10 are (4, 6), (5, 5), and (6, 4). Therefore, $p(10) = \frac{3}{36}$. There are two outcomes that add up to 11, (5, 6) and (6, 5), and only (6, 6) produces the sum of 12. So, $p(11) = \frac{2}{36}$ and $p(12) = \frac{1}{36}$. Therefore,

$$p(\text{at least } 10) = \frac{3}{36} + \frac{2}{36} + \frac{1}{36} = \frac{6}{36} = \frac{1}{6} \qquad \blacksquare$$

Example 5. Seven individual pieces are selected at random from an assembly line. If the probability is .15 that a piece is defective, then the approximate probability (rounded off to three decimal places) that exactly a given number of pieces in the sample are defective is shown in Table 3-19. What is the probability that at least four of the pieces in the sample are defective?

Number defective	0	1	2	3	4	5	6	7
Probability	.320	.396	.210	.062	.011	.001	.000	.000

Table 3-19

Solution

A sample will contain at least four defective pieces if there are exactly four, exactly five, exactly six, or exactly seven defective pieces. No sample can be of more than one of these types, so

$$p(\text{at least 4 defective}) = p(4 \cup 5 \cup 6 \cup 7)$$
$$= p(4) + p(5) + p(6) + p(7)$$
$$= .011 + .001 + .000 + .000$$
$$= .012$$

■

Some problems can be solved by more than one technique; but often the right choice makes the job easier, as in the following example.

Example 6. What is the probability that a poker hand will contain at least one ace?

Solution

On the model of the previous two examples, we would expect to compute the probability that the hand will contain exactly one ace, the probability of exactly two aces, and so on, for three and four aces. We would then add up these probabilities to solve the problem. Although this approach would work, it is very tedious and there is, fortunately, a much neater way to arrive at the solution. We will use the principle

$$p(\text{success}) = 1 - p(\text{failure})$$

In this example, failure amounts to being dealt a five-card hand in which no ace appears. A hand of that type is just the same as a poker hand drawn from an incomplete deck—namely, the standard 52-card deck with the four aces removed. A poker hand dealt from such a 48-card deck is a combination of five cards out of the 48. According to Section 3.4, there are

$$C_5^{48} = \frac{48!}{5!43!} = \frac{48 \cdot 47 \cdot 46 \cdot 45 \cdot 44}{5 \cdot 4 \cdot 3 \cdot 2 \cdot 1} = 1{,}712{,}304$$

different poker hands with no aces. There are

$$C_5^{52} = \frac{52!}{5!47!} = \frac{52 \cdot 51 \cdot 50 \cdot 49 \cdot 48}{5 \cdot 4 \cdot 3 \cdot 2 \cdot 1} = 2{,}598{,}960$$

poker hands in all, so

$$p(\text{failure}) = p(\text{no aces}) = \frac{1{,}712{,}304}{2{,}598{,}960} = .659 \text{ (approximately)}$$

We conclude that

$$p(\text{at least one ace}) = 1 - p(\text{no aces}) = 1 - .659 = .341 \text{ (approximately)}$$ ∎

Summary

If events A and B are **disjoint**, that is, no outcome is in both event A and event B, then

$$p(A \cup B) = p(A) + p(B)$$

Also,

$$p(\text{success}) = 1 - p(\text{failure}) \quad \text{and} \quad p(\text{failure}) = 1 - p(\text{success})$$

If $A_1, A_2, \ldots, A_k$ are events of a sample space and if there is no outcome that is in more than one of these events, then

$$p(A_1 \cup A_2 \cup \ldots \cup A_k) = p(A_1) + p(A_2) + \ldots + p(A_k)$$

Exercises

1. Households in Lake City are classified by annual income. The probabilities that households are in various categories are given in Table 3-20 (in thousands of dollars). If a family from that city is chosen at random, what is the probability that its annual income will be between $15,000 and $50,000?

Annual income	Under 15	15 to 29	30 to 50	Over 50
Probability	.20	.30	.35	.15

Table 3-20

2. The probability that the high temperature in Los Angeles on an April day will be 80 or more is .21. The probability that the high will be less than 70 is .38. Calculate the probability that the high temperature will be in the 70s.

3. A losing roll in dice on the first roll is a 2, 3, or 12. What is the probability of losing on the first roll?

4. An insurance salesperson will fail to sell a policy on a visit to a customer if she is unable to speak to the customer or if she speaks to the customer, but the customer refuses to buy a policy. The salesperson estimates the probability that she will not be able to speak to the customer to be .30 and the probability that she will speak to the customer, but fail to make the sale to be .15. What is her estimate of the probability that the visit will be a failure?

3.7 DISJOINT EVENTS

5. With respect to a certain gene, 15% of a population is of genotype AA, 80% of genotype Aa, and the remaining 5% of genotype aa. An individual either of genotype AA or of genotype aa is said to homozygous. If an individual is chosen at random from this population, what is the probability that he or she is homozygous with respect to this gene?

6. There are three people running for governor. One of the candidates concludes from the polls that the probability of election for each of her two opponents is .41 and .15, respectively. What does that candidate estimate the probability of her own election to be?

In Exercises 7 through 9, assume there is a country in which 30% of the population is color-blind. If ten individuals are chosen at random, the approximate probabilities $p(k)$ that exactly k of them are color-blind are given in Table 3-21.

k	0	1	2	3	4	5	6	7	8	9	10
$p(k)$	.03	.12	.23	.27	.20	.10	.04	.01	.00	.00	.00

Table 3-21

7. What is the approximate probability that at least five people in the sample will be color-blind?

8. What is the approximate probability that there will be at least one color-blind person in the sample?

9. What is the approximate probability that fewer than three people in the sample will be color-blind?

10. A list of the rulers of England since the Norman Conquest indicates the age of each at death. Of these 39 rulers before Elizabeth II, two died before they turned 30, three died in their thirties (that is from 30 through 39), eight died in their forties, ten in their fifties, ten in their sixties, four in their seventies, and two in their eighties. If a ruler is chosen at random, what is the probability that at death he or she was at least 50 years of age?

11. In a marketing survey for a new brand of shampoo, 12 families are each given four unmarked bottles of shampoo. One of the bottles is the new brand. Each family is asked to choose the shampoo they like best from among the four bottles. If all four are really indistinguishable from each other, so the families choose at random, the probabilities that a certain number of the families will choose the new brand are given in Table 3.22. What is the probability that at least four, but not more than nine, families will choose the new brand?

Number of families	0	1	2	3	4	5	6	7	8	9	10	11	12
Probability	.03	.13	.23	.26	.19	.11	.04	.01	.00	.00	.00	.00	.00

Table 3-22

12. In a test of extrasensory perception, six cards, each from a different standard deck, are placed face down on a table, and a "psychic" is asked to tell the suit of each card. He gets three correct out of the six. If a person guessed each suit at random, the approximate probabilities $p(k)$ of getting exactly k correct out of the six are given in Table 3-23. What is the approximate probability that a person who guesses will do at least as well as the psychic did?

k	0	1	2	3	4	5	6
$p(k)$	.18	.36	.30	.13	.03	.00	.00

Table 3-23

13. A man owns six vacation cabins that he rents out by the week. He supplies television sets at an extra cost. He observes that about $\frac{2}{3}$ of the families who rent cabins from him want televisions sets, so he buys four sets. In a week in which all his cabins are rented, Table 3-24 gives the approximate probabilities $p(k)$ that exactly k of the families will want a television set. What is the approximate probability that he will not have enough sets to satisfy all his customers?

k	0	1	2	3	4	5	6
$p(k)$	.00	.02	.08	.22	.33	.26	.09

Table 3-24

14. Four coins are flipped. What is the probability that at least one comes up heads?

15. Suppose that 55% of the voters in an election for mayor will vote for Smith and the rest will vote for Jones. The approximate probabilities $p(k)$ that out of a random sample of five voters exactly k will vote for Smith are given in Table 3-25. What is the approximate probability that the sample will correctly predict the winner of the election (that is, it will contain a majority of voters who favor Smith)?

k	0	1	2	3	4	5
$p(k)$	.02	.11	.27	.34	.21	.05

Table 3-25

16. Two dice are rolled. (a) What is the probability that both come up odd? (b) What is the probability that at least one comes up even?

In Exercises 17 through 20, a city health department has received complaints that a supermarket is selling ground beef that is excessively fat. A health inspector takes three packages of meat at random from ten such packages to test whether any of them exceed the limit for fat content permitted by state law. In fact, four of the ten packages contain too much fat.

17. In how many ways can the inspector choose the sample of three packages?
18. In how many ways can the inspector choose three packages with legal fat content?
19. What is the probability that the inspector will fail to detect any violations?
20. What is the probability that the inspector will choose at least one package with too high a fat content?

3.8 Conditional Probability

A Betting Problem

Imagine that two gamblers agree to play the following card game. Each player bets $1 and is dealt one card from a standard deck. After looking at the card, the player who was the dealer has the opportunity to raise the value of the bet by $1. If the dealer raises the bet, the other player must meet it. The winning card is the higher of the two dealt (aces high). If both cards are of the same value, the winning card is of the higher suit in the following order (higher suit to the right):

♣   ♦   ♥   ♠

Thus, an ace beats a jack and 6♠ beats 6♦. The player holding the higher card wins both dollars bet or, if the value of the bet has been raised, the entire $4 bet.

Imagine that you are the dealer and that you have dealt yourself the 8♦. Your strategy should be to raise the value of the bet if you think you hold the higher card, but stay at $1 otherwise. You should raise the value of the bet only if the probability is at least .50 that the 8♦ will win the game.

The Wrong Analysis

The cards that are higher than 8♦ are all the aces, kings, queens, jacks, tens, and nines together with the two higher eights: 8♠ and 8♥. Thus, there are 26 cards higher than 8♦. Suppose you reason that since there are 52 cards in all, the probability is $\frac{26}{52}$ = .50 that your opponent holds a higher card; so, there is no harm in raising the value of the bet. If you consistently use this kind of reasoning you are not likely to have a successful gambling career.

The Correct Analysis

The correct way to compute the probability that your opponent holds a higher card makes use of all the information available. The error in the preceding analysis was that you failed to take account of the fact that 8♦ in your hand was from the same deck as the card dealt to your opponent. Thus, your opponent's card must

276   PROBABILITY

be thought of as dealt from a deck of 51 cards, the standard deck with the 8♦ removed. The 26 cards that can beat the 8♦ are still in the deck of 51, so the correct probability is $\frac{26}{51}$, or about .51, that your opponent will hold a higher card. Since your probability of winning with the 8♦ is only about .49, you should not raise the value of the bet.

How can we carry the reasoning of the preceding paragraph over to other probability problems? We will try to visualize the analysis using the language of set theory. Let U be the set of all the cards in the standard 52-card deck. Then U is the sample space for choosing a card at random from a deck. The subset A will consist of all aces, kings, queens, jacks, tens, nines, and the 8♠ and 8♥. In our first analysis of the game, we were visualizing the problem as shown in Figure 3-26. Then we computed the probability as

$$p(A) = \frac{n(A)}{n(U)} = \frac{26}{52} = .50$$

Now let B be the set of outcomes (cards) that are really possible for your opponent when you hold the 8♦. Then B consists of all the cards in the standard deck except 8♦, and we should visualize the situation as shown in Figure 3-27.

Figure 3-26

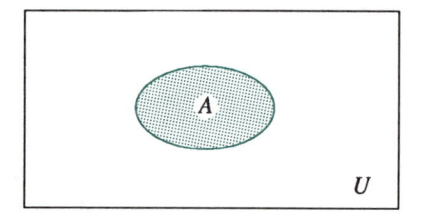

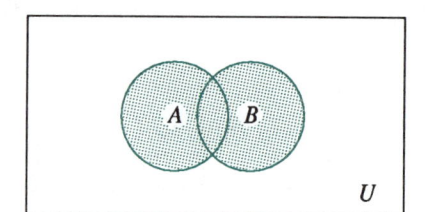

Figure 3-27

Conditional Probability for Equally Likely Outcomes

When we took into account the card you held in your hand, we were computing the **conditional probability** that the outcomes in A will take place—under the conditions imposed by event B. The conditional probability is represented by the symbol $p(A \mid B)$. The condition B tells us that the set of possible outcomes is not really the entire sample space U, but instead, the subset B of U. Thus, the denominator in the calculation of $p(A \mid B)$ is not $n(U)$, the number of all possible outcomes, but rather, $n(B)$, the number of possibilities that actually have a chance of occurring. We might expect that $p(A \mid B)$ would then be $n(A)$, the number of successful outcomes, divided by $n(B)$, the number of outcomes that are really possible. This is correct in the betting example, but, in general, there may be outcomes in A that cannot occur because they are not in B. The probability $p(A)$ is the number of successful outcomes divided by the total number of outcomes while, in general, $p(A \mid B)$ is the number of possible successful outcomes (possible after knowing what is in B) divided by the number of possible outcomes. Thus, the numerator of $p(A \mid B)$ is the number of outcomes that are both successful (in A) and possible (in B), that is, $A \cap B$, the intersection of A and B.

When considering equally likely outcomes, the definition of **conditional probability** is

$$p(A \mid B) = \frac{n(A \cap B)}{n(B)}$$

provided, of course, that $n(B) \neq 0$. In the preceding gambling example, B consists of all cards except for 8♦, whereas A consists of all cards with a higher value than 8♦, that is, 8♥, 8♠, all nines, tens, jacks, queens, kings, and aces. Thus, in this case $A \cap B$ is the same as A because every card in A is a possible outcome for the opponent. We can now use the definition to compute correctly the probability that your opponent will win when you hold the 8♦:

$$p(A \mid B) = \frac{n(A \cap B)}{n(B)} = \frac{26}{51} = .51 \text{ (approximately)}$$

Here is an example in which A is not a subset of B.

Example 1. The accounting office of a company consists of 15 men, of whom 6 are college graduates, and 23 women, of whom 17 are college graduates.
(a) If an employee is chosen at random to represent the office on a planning committee, what is the probability that the person selected will be a college graduate?
(b) If, before the announcement of the representative, the office manager lets slip the fact that a woman was chosen by referring to the representative as "she," what then is the probability that a college graduate will represent the office?

Solution

(a) There are 38 employees in all and 23 of them are college graduates; so, the probability $p(A)$ that an employee chosen at random will be a graduate is

$$p(A) = \frac{n(A)}{n(U)} = \frac{23}{38} = .61 \text{ (approximately)}$$

(b) Now we know that the employee chosen comes from the set B of women employees. Therefore, there are 17 employees who are both college graduates (A) and women (B). The probability that the representative is a college graduate, given that she is a woman, is

$$p(A \mid B) = \frac{n(A \cap B)}{n(B)} = \frac{17}{23} = .74 \text{ (approximately)}$$ ∎

The General Definition

The definition of the conditional probability

$$p(A \mid B) = \frac{n(A \cap B)}{n(B)}$$

that A will occur, knowing the possibilities are limited to B, makes sense for the simple counting probability of equally likely events, provided only that B contains some outcome (since we cannot divide by 0 when $n(B) = 0$). However, we would like a definition of conditional probability that avoids counting, that we can extend to subjective probability and to probability based on experience. Therefore, we must replace this definition of conditional probability with an equivalent one expressed entirely in terms of probability itself. Recall that in the counting context

$$p(B) = \frac{n(B)}{n(U)}$$

so, if we divide the denominator on the right side of

$$p(A \mid B) = \frac{n(A \cap B)}{n(B)}$$

by $n(U)$, the new denominator will be $p(B)$. But in order not to destroy the equality, we have to divide the numerator by $n(U)$ also, with this result

$$p(A \mid B) = \frac{n(A \cap B)}{n(B)} = \frac{\frac{n(A \cap B)}{n(U)}}{\frac{n(B)}{n(U)}} = \frac{p(A \cap B)}{p(B)}$$

We will use this equation to define conditional probability in general.

In an experiment, the **conditional probability** that the outcome will come from a subset A, given that the possible outcomes are restricted to a subset B, is defined to be

$$p(A \mid B) = \frac{p(A \cap B)}{p(B)}$$

provided that $p(B) \neq 0$.

Example 2. The president of an advertising agency estimates the probability that his firm will be selected to design the advertising campaign for a new brand of cosmetics to be .15. He also estimates the probability to be .12 that the new brand will be successful and that his firm will handle its advertising. What does the president estimate to be the probability that the new brand will be successful, if his firm is selected to design the advertising campaign?

Solution

The problem is to calculate the probability $p(A \mid B)$ that the new brand will be successful, given that the agency is selected. Therefore, A is the event that the new brand will be successful, and B is the event that the agency is selected to design the advertising campaign. We are told that $p(A \cap B) = .12$ and that $p(B) = .15$, so we calculate that the president's estimate of success for the new brand of cosmetics is

$$p(A \mid B) = \frac{p(A \cap B)}{p(B)} = \frac{.12}{.15} = .80$$

provided that his firm gets to design the advertising. ∎

Example 3. According to *Climatological Data*, on an April day, the probability of rain in Santa Barbara is .11 and the probability that it will rain both in Los Angeles and in Santa Barbara is .09. Calculate the conditional probability that it will rain in Los Angeles, given that it rains in Santa Barbara.

Solution

Abbreviating Los Angeles by LA and Santa Barbara by SB, we just apply the definition. Then

$$p(\text{rain in LA} \mid \text{rain in SB}) = \frac{p(\text{LA} \cap \text{SB})}{p(\text{SB})} = \frac{.09}{.11} = .82 \text{ (approximately)}$$ ∎

Independent Events

In Example 1, the probability that an outcome from event A would occur changed when we were given additional information B. Once we learned that the office's representative was a woman, the probability that the representative would be a college graduate increased from .61 to .74. You should not get the idea, however, that $p(A \mid B)$ is necessarily different from $p(A)$. Suppose a coin is tossed five times and we wish to calculate the probability that it will come up heads on the fifth flip. A sequence of five tosses may be thought of as an ordered 5-tuple with each position either heads (H) or tails (T). There are $2 \cdot 2 \cdot 2 \cdot 2 \cdot 2 = 32$ possible outcomes. An outcome in which heads occurs on the last toss is represented by a 5-tuple of the form (, , , , H). There are $2 \cdot 2 \cdot 2 \cdot 2 \cdot 1 = 16$ of these outcomes because there is only one possibility for the last position. We conclude that

$$p(\text{fifth toss heads}) = p(A) = \frac{16}{32} = .50$$

Now suppose we were told that the first four tosses came up heads, and we were asked to find the conditional probability that the last flip would come up heads as well. The set B consists of the ordered 5-tuples (H, H, H, H, H) and (H, H, H, H, T). The first of these outcomes is also in A because the last toss is heads. Therefore,

$$p(A \mid B) = \frac{n(A \cap B)}{n(B)} = \frac{1}{2} = .50$$

This calculation indicates that the outcome of the fifth toss of the coin is not affected by the fact that it has just come up heads four times in a row. This is hardly surprising since the coin has no way of remembering what it did in previous flips. (We assume that we are flipping an honest coin.) In technical language, the events A (heads on the fifth toss) and B (heads on the first four tosses) are *independent*.

The definition of independence is the following:

Two events A and B are **independent** if
$$p(A \mid B) = p(A)$$

This definition appears to give more importance to the event A than to B since it says nothing about $p(B)$. However, we can quickly convince ourselves that we could just as well have written the definition in the form

$$p(B \mid A) = p(B)$$

instead. That is why we say "A and B are independent" rather than "A is independent of B" or "B is independent of A." To obtain the second equation from the first, we use the definition of conditional probability to write

$$p(A \mid B) = \frac{p(A \cap B)}{p(B)}$$

and multiply both sides of the equation by $p(B)$, so that the equation becomes

$$p(A \mid B) \cdot p(B) = p(A \cap B)$$

Now if A and B are independent according to the given definition, this tells us that we can replace $p(A \mid B)$ by $p(A)$, since they are equal. (This gives us another equation for independent events that we will find useful in the next section.)

If the events A and B are independent, then
$$p(A) \cdot p(B) = p(A \cap B)$$

To say that an outcome is in $A \cap B$ is the same thing as saying that it is in $B \cap A$, so we always have that $p(A \cap B) = p(B \cap A)$. Therefore, dividing by $p(A)$ and recalling the definition of conditional probability we see that if A and B are independent, then we also have

$$p(B) = \frac{p(A \cap B)}{p(A)} = \frac{p(B \cap A)}{p(A)} = p(B \mid A)$$

provided, of course, that $p(A) \neq 0$. This is what we set out to show.

Example 4. Two dice are rolled and the first die comes up a 3.
(a) What is the probability that the sum of the numbers on the dice is 6?

(b) Is the event (sum equals 6) independent of the event (first die is a 3)?
(c) What is the probability that the roll is a double, i.e., both dice show the same number?
(d) Is the event (double) independent of (first die is a 3)?

Solution

(a) Representing rolls of the dice as ordered pairs, the set B of pairs in which the first die is a 3 consists of the six outcomes (3, 1), (3, 2), (3, 3), (3, 4), (3, 5), and (3, 6). If A is the set of outcomes in which the sum is 6, namely (1, 5), (2, 4), (3, 3), (4, 2), and (5, 1), then only one, (3, 3), is in both A and B. Since the question asks us to find the probability that the sum is 6, given that the first die is a 3, we compute

$$p(A \mid B) = \frac{n(A \cap B)}{n(B)} = \frac{1}{6}$$

(b) To determine whether A (sum equals 6) is independent of B (first die is 3), we calculate

$$p(A) = \frac{n(A)}{n(U)}$$

where U stands for the set of all possible outcomes of the experiment of rolling two dice. We know from Part (a) that there are five pairs that add to 6. The answer, $p(A) = \frac{5}{36}$, is not the same as $p(A \mid B)$ from Part (a); so, the events are not independent. Intuitively, we would expect the probability of getting a sum of 6 to be improved by knowing that the first die is a 3 because this eliminates the possibility that the first die is a 6, in which case it is impossible to get a sum of 6 from two dice.

(c) Now let A stand for the six doubles outcomes (1, 1), (2, 2), (3, 3), (4, 4), (5, 5), and (6, 6). Again, only (3, 3) also belongs to B, the set of outcomes with a 3 on the first die. Thus,

$$p(A \mid B) = \frac{n(A \cap B)}{n(B)} = \frac{1}{6}$$

(d) This time, since we saw in Part (c) that there are six doubles,

$$p(A) = \frac{n(A)}{n(U)} = \frac{6}{36} = \frac{1}{6} = p(A \mid B)$$

Therefore, the events A and B *are* independent by definition. Our intuition tells us that the result makes sense because in order to form doubles, the second die must match the first regardless of which of the six possible numbers turns up on the first die. ∎

Example 5. According to *Climatological Data*, on an April day, the probability of rain in Santa Barbara is .11 and the probability that it will rain both in Los Angeles and in Santa Barbara is .09. The probability that it will rain on an April day in Los Angeles is .13. Determine whether the events (rain in Los Angeles) and (rain in Santa Barbara) are independent.

Solution

The events are not independent. As we showed in Example 3, the conditional probability of rain in Los Angeles on an April day is very high if it rains in the nearby city of Santa Barbara, specifically

$$p(\text{rain in LA} \mid \text{rain in SB}) = \frac{p(\text{LA} \cap \text{SB})}{p(\text{SB})} = \frac{.09}{.11} = .82 \text{ (approximately)}$$

On the other hand, in general, it is not likely to rain in Los Angeles because we were told that $p(\text{rain in LA}) = .13$. ■

Applying Conditional Probability

The way in which we used conditional probability in Examples 3 and 5 suggests an approach to a problem we discussed at the beginning of this chapter. Based on *Climatological Data* for the 1980–84 period, we estimated that the probability of precipitation on a January day in New York to be $\frac{39}{155} = .25$ (approximately). We will assume that precipitation in the North in January is in the form of snow. We reasoned from the general behavior of winter storms that it would be very likely to snow in New York on a day when it snows in Boston. On the other hand, knowing that it is snowing in Pittsburgh on a January day is not as likely to imply snow in New York. The problem is to determine how to use climatological data to describe these relationships more precisely. Conditional probability offers us a very useful tool for this purpose.

To measure the impact of snow in Boston on the likelihood of snow in New York on a January day, we will use the climatological data to compute the conditional probability of snow in New York, given that there is snow that same day in Boston. So, if we let A be the event (snow in New York) and B the event (snow in Boston), we are seeking to compute the conditional probability $p(A \mid B)$, which, by definition is

$$p(A \mid B) = \frac{p(A \cap B)}{p(B)}$$

The type of probability we are discussing is that based on experience. We calculated before that $p(A) = \frac{39}{155} = .25$ (approximately) because during the 155 January days in the 1980–84 period, there was snow in New York on 39 of them. In the same way, we find that Boston reported snow on 43 out of those same 155 days; so, $p(B) = \frac{43}{155}$. To calculate $p(A \cap B)$, we recorded the dates on which snowfall was observed in each city and counted the number of dates that showed snow in both. Again dividing by the total number of January days in this period, we found that $p(A \cap B) = \frac{27}{155}$. Since the denominators are both 155, and therefore divide out, the computation reduces to $p(A \mid B) = \frac{27}{43} = .63$ (approximately). Comparing this result with $p(A) = .25$, we are not at all surprised to learn that the events (snow in New York) and (snow in Boston) are not independent because the probabilities $p(A)$ and $p(A \mid B)$ are not equal.

Referring to *Climatological Data*, there were 69 days of snow in Pittsburgh during the January days of 1980–84, while there was snow in both New York and Pittsburgh on 25 of these days. Therefore, computing just as before, the conditional probability of snow on a January day in New York, given that there was snow that day in Pittsburgh, is $\frac{25}{69} = .36$ (approximately). We note that this is a much smaller number than the conditional probability of .63 under the corresponding condition for Boston. This is again what we would expect; but now we are able to compare the numerical values of the conditional probabilities, instead of just saying that, with respect to snow in January, the similarity of New York to Boston is greater than it is to Pittsburgh.

Summary

The **conditional probability** that event A will occur, given that event B happens, is written $p(A \mid B)$ and is defined by

$$p(A \mid B) = \frac{p(A \cap B)}{p(B)}$$

provided that $p(B) \neq 0$. When all outcomes are equally likely, so that probability depends on counting, then we can use the computational formula

$$p(A \mid B) = \frac{n(A \cap B)}{n(B)}$$

provided that $n(B) \neq 0$. Events A and B are said to be **independent** if

$$p(A \mid B) = p(A)$$

or, equivalently,

$$p(B \mid A) = p(B)$$

or, in another equivalent form,

$$p(A) \cdot p(B) = p(A \cap B)$$

Exercises

1. An economist estimates the probability that interest rates will rise in the next year to be .75 and that both interest rates will rise and the stock market's Dow Jones average will fall during that period to be .30. What does the economist estimate to be the conditional probability that the Dow Jones average will fall next year, if it is assumed that interest rates rise?

2. Two dice are rolled. Calculate the conditional probability that the second die rolled will show a larger number than the first, given that the first die rolled is a 4.

3. A sociologist estimates that the probability that a student selected to take part in an experiment will be a sociology major is .60. She also estimates that the proba-

bility that a student subject will both be a sociology major and will pass a screening test to be .45. What does the sociologist estimate to be the probability that the subject will pass a screening test, given that the subject is a sociology major?

In Exercises 4 and 5, a company offers one management traineeship. The trainee will be selected at random from among 16 candidates. There are six candidates who are employees of the company, two of whom have had at least one year of experience. There are ten candidates who do not work for the company, four of whom have at least one year of experience.

4. What is the probability that the trainee will be an employee of the company?

5. What is the conditional probability that the trainee will be an employee of the company, given that he has at least one year of experience?

6. A publisher estimates that there is a probability of .05 that a new novel will be a book-club selection and .01 that it will both be a book-club selection and be made into a movie. What does she estimate to be the probability that a novel selected by a book club will be made into a movie?

7. On a day in the first half of November, the probability that it will freeze (i. e., the low temperature for a 24-hour period will be below 32°F.) is .56 in Minneapolis and .47 in Duluth, and the probability that it will freeze in both cities is .37. Calculate the conditional probability that it will freeze in Duluth, given that it freezes in Minneapolis.

8. In Exercise 7, are the events (freeze in Duluth) and (freeze in Minneapolis) independent?

In Exercises 9 through 11, imagine that at a food-processing plant, an employee forgot to change the labels in a labeling machine, so that the day's production of 2000 cans of cherries, 1000 cans of peaches, 4000 cans of peas, and 4000 cans of popping corn are all labelled identically as "beans" and cannot otherwise be distinguished because the cans are all the same size. The processor sells the cans at a reduced price and the first customer chooses a can at random.

9. What is the probability that the customer chooses a can of peaches?

10. What is the probability that the customer chooses a can of fruit (peaches or cherries)?

11. The customer picks up a can and shakes it. Since it does not rattle, he knows it is not popping corn. Now, what is the probability that the can contains fruit?

For Exercises 12 and 13, suppose that in King City, 30% of the families with children have no car, 50% have one car, 15% have two cars, and 5% have more than two cars.

12. What is the probability that a family with children has at least two cars?

13. If it is observed that a car belonging to the family is parked in its garage, what then is the probability that the family has at least two cars?

3.8 CONDITIONAL PROBABILITY

14. In Central City, 80% of all households have television sets and 20% have both a television and a stereo. What is the probability that a household with a television selected at random also has a stereo?

In Exercises 15 through 17, a vacation resort consists of 500 apartments with sizes and views as indicated in Table 3-28. Assume that all apartments have the same likelihood of being the next one offered for sale.

	One Bedroom	Two Bedroom
Sea view	240	60
No sea view	160	40

Table 3-28

15. What is the probability that the next apartment for sale has a sea view?
16. If the next apartment offered for sale has two bedrooms, what is the probability that it has a sea view?
17. Are the events (has two bedrooms) and (sea view) independent?

In Exercises 18 through 20, an agricultural experiment station has 120 small plots of land, with types of soil and adequacy of drainage given in Table 3-29.

	Rich Topsoil	Clay Soil
Good drainage	60	20
Poor drainage	10	30

Table 3-29

18. If a plot is chosen at random for an experiment, what is the probability that it will have clay soil?
19. If the plot to be chosen must have good drainage, what is the probability that it will have clay soil?
20. Are the events (clay soil) and (good drainage) independent?
21. An oil company plans to drill two wells. It estimates that there is a probability of .20 that it will strike oil at the first location and of .03 that it will strike oil at both locations. If the oil company assumes that the events of striking oil at the two locations are independent, what is the estimated probability for striking oil at the second location?
22. A manufacturer of model-airplane kits finds that 3% of its kits contain a broken part, 7% of the boxes in which the parts are packed are defective, and in 1% of its production the box is damaged and there is a broken part in the kit. Are the events (part broken) and (box damaged) independent, or should the manufacturer suspect that there is a connection between its two problems?

286 PROBABILITY

23. An automobile manufacturer estimates the probability that the government will make emission standards for engines tougher at .60, that the cost of steel will go up at .40, and that both will happen at .24. Is the manufacturer assuming that the events (tougher emission standards) and (higher steel prices) are independent?

In Exercises 24 and 25, suppose the price of stock of an oil company will rise substantially this year if either its current oil explorations uncover a large amount of oil or the government permits a rise in domestic oil prices. The president of the company estimates there is a probability of .25 that the company will find a large amount of oil, of .60 that the government will permit oil prices to rise, and of .09 that both will happen.

24. What does the president of the company estimate to be the probability that the government will permit oil prices to rise if the company finds a large amount of oil?

25. Is the president assuming that a rise in the government ceiling on the price of oil is independent of whether the company finds large amounts of oil?

In Exercises 26 through 28, two dice are rolled.

26. What is the probability that the sum is greater than or equal to 4 and less than or equal to 8?

27. If the first die is a 3, what is the probability that the sum is greater than or equal to 4 and less than or equal to 8?

28. Are the events (first die a 3) and (sum is greater than or equal to 4 and less than or equal to 8) independent?

3.9 Intersection of Events

One of the most popular card games in gambling casinos is known as "blackjack" or "21." We suppose that at the start of the game, each player is dealt two cards from a standard deck. A very desirable opening hand consists of two aces. We can compute the probability of getting such a hand using the counting procedure of Section 3.4. The number of two-card hands of two aces is

$$C_2^4 = \frac{4!}{2!2!} = 6$$

because these hands must come from just the four aces. The total number of two-card hands is

$$C_2^{52} = \frac{52!}{2!50!} = \frac{52 \cdot 52}{2 \cdot 1} = 1326$$

We conclude that

$$p(\text{two aces}) = \frac{6}{1326} = \frac{1}{221} = .0045 \text{ (approximately)}$$

The Probability of A and B

We can solve the same problem in a different way, using a principle that is very useful in analyzing probability. Essentially, this principle is nothing more than a way of rewriting the definition of conditional probability, which we know to be

$$p(A \mid B) = \frac{p(A \cap B)}{p(B)}$$

Multiplying both sides of the equation by $p(B)$, we obtain

$$p(A \cap B) = p(B) \cdot p(A \mid B)$$

Thus, the probability that both A and B will take place is the probability that B will happen times the conditional probability of the event A, given B.

There is another way of expressing $p(A \cap B)$, again based on the definition of conditional probability. This formula is often more convenient than the one we just found for computing the probability of an intersection; but we will need both expressions in the next chapter. Recall that we noticed earlier, when discussing independent events, that $p(A \cap B) = p(B \cap A)$ because both sides of the equation express the probability of the same subset of the sample space. Just as in that discussion, we can write the definition of the conditional probability $p(B \mid A)$ in the form

$$p(B \mid A) = \frac{p(B \cap A)}{p(A)} = \frac{p(A \cap B)}{p(A)}$$

Multiplying both sides of this equation, this time by $p(A)$, gives us the other formula for $p(A \cap B)$, which is

$$p(A \cap B) = p(A) \cdot p(B \mid A)$$

Let us apply this formula to compute the probability of being dealt two aces in a blackjack hand of two cards. Think of receiving the two aces as an intersection of events: A (ace on the first deal) and B (ace on the second deal). Then we can use

$$p(\text{two aces}) = p(A \cap B) = p(A) \cdot p(B \mid A)$$

We see right away that $p(A) = \frac{4}{52}$. Now the remaining symbol $p(B \mid A)$ means the probability of being dealt an ace on the second deal, after one ace has already been dealt. Since there are three aces left in this 51-card deck, $p(B \mid A)$ is $\frac{3}{51}$. Thus, we compute

$$p(\text{two aces}) = p(A) \cdot p(B \mid A) = \frac{4}{52} \cdot \frac{3}{51} = \frac{12}{2652} = \frac{1}{221}$$

Example 1. On an April day, the probability of rain in Santa Barbara (SB) is .11 while the conditional probability of rain in San Diego (SD), given that it does rain in Santa Barbara, is .64. Calculate the probability that it will rain in both cities.

Solution

We can easily use the formula directly, even without renaming the events as A and B, to compute

$$p(SB \cap SD) = p(SB) \cdot p(SD \mid SB) = (.11)(.64) = .07$$ ■

Example 2. Many automobiles have dual braking systems. That is, if the main system fails, there is a second, less effective, system that attempts to stop the car. Suppose the probability that the main brake system will fail is .001. The probability that, if the main system fails, the backup system will be unable to stop the car is .02. What is the probability that when the brakes are applied the car will not stop?

Solution

If we let A represent the event (the main brake system fails), then we know that $p(A) = .001$. Then let B be (the backup system fails). The probability that the backup system will fail, under the condition that the main system fails, $p(B \mid A)$, is stated as .02. (We do not know $p(B)$, the probability of failure of the backup system by itself. It is probably lower than .02 because that given figure takes into account the likelihood that whatever damaged the main braking system was major enough to take the backup system with it. In any event, we are not interested in how the backup system would behave if it were the only system on the car, because it isn't.) By the formula for computing probability for the intersection of events,

$$\begin{aligned}
p(\text{car cannot stop}) &= p(\text{both brake systems fail}) \\
&= p(A \cap B) \\
&= p(A) \cdot p(B \mid A) \\
&= (.001)(.02) \\
&= .00002
\end{aligned}$$

To put it another way, with the main braking system alone, the chances are 999 in 1000 that when the driver steps on the brake pedal the car will stop; but the dual system brings the chances up to 99,998 out of 100,000. ■

Independent Events

At the end of the previous section, we discussed independent events, i.e., those in which the conditional probability was the same as the probability without any condition. More precisely, event A is independent of event B if $p(A \mid B) = p(A)$. If A is independent of B, then we may substitute $p(A)$ for the conditional probability $p(A \mid B)$ in the formula $p(A \cap B) = p(A \mid B) \cdot p(B)$. We again obtain the very useful formula,

3.9 INTERSECTION OF EVENTS

$$p(A \cap B) = p(A) \cdot p(B) \quad \text{if } A \text{ and } B \text{ are independent.}$$

Example 3. On any day in April, the probability of rain in Los Angeles is .13, of rain in Santa Barbara is .11 and of rain in both cities is .09. Are the events (rain in Los Angeles) and (rain in Santa Barbara) independent?

Solution

We could solve this problem by the method of the preceding section—for instance, by computing the conditional probability that there will be rain in Los Angeles, given that there is rain in Santa Barbara, and comparing it to the given probability of rain in Los Angeles. But it is even easier to notice that if the events (rain in Los Angeles) and (rain in Santa Barbara) were independent, then the probability of rain in both cities would be

$$p(\text{LA}) \cdot p(\text{SB}) = (.13)(.11) = .01$$

by the preceding formula, instead of the given probability of .09. Therefore, the events are not independent. ∎

More Than Two Events

The formula $p(A \cap B) = p(A) \cdot p(B)$ for independent events extends to the case of more than two events.

If the events $A_1, A_2, \ldots, A_k$ are independent,[†] then

$$p(A_1 \cap A_2 \cap \ldots \cap A_k) = p(A_1) \cdot p(A_2) \ldots p(A_k)$$

Example 4. A market-research firm chooses four people from among the shoppers in a supermarket. Each shopper is asked to try three samples of potato chips, one of which is prepared by a new process, and to choose the one he or she likes the best. Suppose the shoppers can't really tell any difference, so they each choose a chip at random. Assume that the decisions of the four shoppers are independent. What is the probability that all four will choose the new potato chip?

Solution

Since each shopper chooses a potato chip at random, the probability of each one choosing the new chip is $\frac{1}{3}$. Because the choices are independent,

---

[†]The definition of independence for more than two events requires the corresponding formula for each subcollection of the events $A_1, A_2, \ldots, A_k$; but the stated property is the only part of the definition that we will use.

$p(\text{all choose the new chip}) = \frac{1}{3} \cdot \frac{1}{3} \cdot \frac{1}{3} \cdot \frac{1}{3} = \left(\frac{1}{3}\right)^4 = .012 \text{ (approximately)}$

Thus, there is a chance (though, fortunately a small one) that the new chips will seem a great success even though the shoppers really don't taste any difference at all. ∎

We must be careful when working with probabilities not to confuse the concepts "disjoint" and "independent."

If events A and B are disjoint, then

$$p(A \cup B) = p(A) + p(B)$$

If events A and B are independent, then

$$p(A \cap B) = p(A) \cdot p(B)$$

In fact, except in one extreme case, disjoint events can *never* be independent. If A and B are disjoint, we have the situation pictured in Figure 3-30; so, "disjoint" certainly implies that $p(A \cap B) = 0$. On the other hand, if A and B are independent, then we would have $0 = p(A \cap B) = p(A) \cdot p(B)$, which implies that either $p(A) = 0$ or $p(B) = 0$, that is, that at least one of these events is impossible. For the same reason, unless one of the events is impossible, independent events are never disjoint.

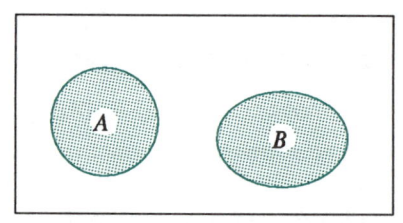

Figure 3-30

Summary

For any events A and B,

$$p(A \cap B) = p(B) \cdot p(A \mid B)$$

$$p(A \cap B) = p(A) \cdot p(B \mid A)$$

If A and B are independent, then

$$p(A \cap B) = p(A) \cdot p(B)$$

If events $A_1, A_2, \ldots, A_k$ are independent, then

$$p(A_1 \cap A_2 \cap \ldots \cap A_k) = p(A_1) \cdot p(A_2) \ldots p(A_k)$$

Exercises

1. According to *Climatological Data*, on a day in the first half of November, the probability of a freeze in Redwood Falls, Minnesota is .40, whereas the conditional probability of a freeze in Minneapolis, given that it freezes in Redwood Falls, is .87. Calculate the probability that it will freeze in both cities.

2. The president of an electronics company submits a bid for a large government contract. She estimates the probability that her firm will be awarded the contract to be .20. She further estimates the conditional probability that her company will expand by purchasing an electronics supplier, given that her company gets the government contract, to be .90. What does she estimate to be the probability that her company will be awarded the government contract and will also purchase the electronics supplier?

3. A mayor, running for reelection, hires a public-opinion firm to survey the city's voters. The firm estimates the probability that a voter, chosen at random, voted for the mayor in the last election to be .40. The probability that a voter will vote for the mayor in the upcoming election, given that he or she did so in the previous election is estimated to be .80. What is the probability that a voter will have voted for the mayor in both elections?

4. A player draws a card from a standard deck, then draws a second card from those that remain. Calculate the probability that the player draws an ace as the first card and a king as the second card.

5. A life-insurance agent calls families who have just had babies. His experience is that 15% of those he calls agree to make an appointment with him. Of those who agree to see him, one in four buys insurance from him. If he calls a family with a new baby, what is the probability that they will make an appointment and buy insurance from him?

6. A psychologist tests two rats in a maze. A rat who goes through the maze successfully will reach food. One rat has run through the maze before and the psychologist estimates the probability that this rat will reach the food to be .95. The other rat has never been in a maze, and the psychologist estimates the probability that this rat will reach food to be only .20. Assuming that the behavior of the two rats is independent, what is the probability that both will reach food?

In Exercises 7 through 11, a spinning arrow is set up on a circular card. Half of the card is colored white and the other half is divided into equal parts, colored red, blue, and green. The arrow is spun twice.

7. What is the probability of the arrow stopping at red both times?

8. What is the probability of getting red on the first spin and either red or green on the second?

9. What is the probability of spinning white the first time and green the second?

10. What is the probability of getting white on one spin and green on the other? (Notice that there are two ways that this can happen.)

292   PROBABILITY

11. What is the probability of getting either red or blue on the first spin and a color other than green on the second?

In Exercises 12 and 13, a real-estate agent believes that he will be able to arrange the sale of a house either if he gives an effective sales talk or if the seller decides to lower the asking price for the house. He estimates that there is a probability of .65 that his sales talk will be effective and a probability of .30 that the seller will lower the price. Assume that the effectiveness of the agent's talk and the asking price are independent.

12. What is the probability that the real-estate agent will give an effective sales talk and the seller will also lower the asking price for the house?

13. What is the probability that the real-estate agent will be able to arrange the sale of the house?

14. We will use a baseball player's batting average as an estimate of the probability that he will get a hit his next time at bat. Suppose the first two batters in a team's lineup have batting averages of .315 and .290, respectively. Assuming independence, what is the probability that they will both get hits to start the game?

15. A soft-drink bottler conducts a marketing survey in which families are given five unmarked bottles of cola drinks and asked to rank them for flavor in order of preference, with no ties permitted. The five different colas include both the bottler's product and that of the leading competitor. If a family thinks that all five colas taste alike and so ranks them in a random order, what is the probability that they will rank the bottler's product as the first choice and the leading competitor's as the last?

16. Suppose that 70% of the workers on an assembly line are women and that 20% of the women in the line are left-handed. If someone is chosen at random from the line, what is the probability that the person is a left-handed woman?

17. In an experiment conducted by the Stanford Research Institute, the psychic Uri Geller correctly called the uppermost face of a die that had been shaken inside a steel box on eight successive occasions. If Geller was guessing at random, what is the probability that this would have happened?

18. In an attempt to stay in business, a financially troubled company takes three actions. It markets a new model of its main product, it applies to its bank for a loan, and it attempts to sell a money-losing subsidiary. The company can stay in business if at least one of these actions is successful. Otherwise, the company will have to cease operation. It estimates the probability that the new model will sell successfully at .35, that it will get a loan at .20, and that it will sell the subsidiary at .45. If these actions are independent, what is the probability that the company will stay in business? (*Hint:* What is the probability that all three actions will be failures?)

19. A manufacturer of electronic parts must be sure that its product is dependable; so, it hires two inspectors, with the second inspector checking everything the first inspector approves. Suppose the first inspector finds 98% of the defective units

and the second inspector catches 75% of the defective units approved by the first inspector. What is the probability that a defective unit will get by both inspectors?

In Exercises 20 through 22, three people each flip a coin to see who will pay the bill after lunch in a restaurant. If two coins come up the same way (for example, two heads and one tail) then the person whose coin fails to match the others pays. If all three coins come up the same, then the game is repeated until exactly two agree.

20. What is the probability that the person who pays the bill will be chosen on the first play of the game?

21. What is the probability that they will need exactly two plays of the game to decide who pays?

22. What is the probability they will have to play exactly three times to decide who pays?

23. A TV-production company makes pilot films for three proposed TV series: a comedy, a drama, and an adventure series. The production company will have to fire some of its employees unless at least one the the series is purchased by a TV network. The company estimates the probability of sale to a network to be .60 for the comedy, .50 for the drama, and .20 for the adventure series. Assuming the events of selling the three series are independent, calculate the probability that the production company will have to fire some of its employees.

Review Exercises for Chapter 3

1. How many different 5-digit numbers (that is, numbers between 10,000 and 99,999) are there in which no digit appears more than once?

2. In a certain species of animals, only individuals of genotype aa (with respect to a particular gene) can be albino. The probability that an animal of genotype aa will be albino is .40. The probability is .15 that an individual of this species will be of genotype aa. Picking an animal of this species at random, what is the probability that it will be albino?

3. An organization with 100 members, of whom 6 are officers, plans to elect delegates to attend a convention. There are to be two delegates; one must be an officer and the other cannot be an officer. In addition, an alternate delegate, either an officer or not, will be elected and will attend if one of the regular delegates is unable to do so. How many different outcomes can this election have?

4. According to *Climatological Data*, on any day in the first half of November, the probability of a freeze in Redwood Falls, Minnesota is .46, of a freeze in Duluth is .47, and of a freeze in both cities is .39. Are the events (freeze in Duluth) and (freeze in Redwood Falls) independent?

5. A new system of watering fruit trees is tried out on ten trees. If there is a probability of .40 that a tree's yield of fruit will increase significantly as a result of the new system, the approximate probabilities that exactly k of the ten trees will produce higher yields are given in Table 3-31. What is the approximate probability that at least five of the trees will produce higher yields of fruit?

k	0	1	2	3	4	5	6	7	8	9	10
$p(k)$	.01	.04	.12	.22	.25	.20	.11	.04	.01	.00	.00

Table 3-31

6. The senate of the State of Nevada has 20 members. Suppose there are 13 Democrats and 7 Republicans in the senate. If a four-person committee is selected at random, what is the probability that it will consist of two senators of each party?

7. A motel has 120 rooms of which 80 are equipped with kitchenettes. The other 40 are not, but they have correspondingly lower room rates. On a typical night, 60 of the rooms with kitchenettes and 30 of the rooms without are rented.
    (a) Are the events (equipped with kitchenettes) and (room rented) independent?
    (b) Does this evidence suggest that the public prefers its motel rooms to be equipped with kitchenettes, even if it has to pay higher room rates for this feature?

8. An architect will receive a substantial bonus from her firm either if her design for a large building is chosen by the owners as the one to be constructed or if she wins an award for a building she designed previously. She estimates the probability that her design for the new building will be chosen as .30, that the building she designed previously will win an award as .05, and that both will happen as .015. What does she estimate to be the probability that she will receive a bonus?

9. An expert wine taster is asked to judge a wine competition at a county fair. There are 11 entries and the taster is to select a first, second, and third prize winner, with no ties allowed. How many different outcomes can there be?

10. A polling place has 5 voting machines. Voters coming to the polling place are equally likely to choose any one of the machines. Suppose each voter's choice is independent of any other's. If one of the machines is defective, what is the probability that at least one of the next 4 voters who come to the polling place will choose the defective machine?

11. Historical records indicate the odds (to lose) that a first-term incumbent U.S. congressmember will be reelected are 9 to 11. Calculate the probability that a first-term incumbent will be reelected.

12. According to *Climatological Data*, the probability that the high temperature in Duluth on a given day in November will be above 45°F is .26, while the probability that the high will be below 32°F is .23. What is the probability that the high will be between 32° and 45°?

13. Public-health officials estimate that 15% of a population suffer from a particular disease. They also estimate the probability to be .35 that a person chosen at

random from that population is free of the disease and will also test positive for the presence of the disease in a screening test. Calculate the probability that the test will alarm a healthy person unnecessarily, that is, calculate the conditional probability that the test will record a positive reading, given that the person is actually free of the disease.

14. Military planners estimate that the probability a missile will be destroyed on the ground by an opposing land-based missile to be .30, that it will be destroyed in the air by a laser device to be .25, and that it will be untouched by missiles or lasers but nevertheless malfunction, to be .30. If the missile will reach its destination unless one of these things happens, what is the probability that a missile will reach its destination?

15. A "tone row" is a musical theme consisting of a sequence of notes each different and each used exactly once in the row. A tone row is to be constructed from the notes B-A-C-H. Furthermore, exactly one note in the tone row will be a half-note in duration whereas each of the other three will be a quarter-note. How many such tone rows are possible?

16. Two people walk into a department store. One wants to buy a hammer and the other, a tennis racquet. Suppose the probability is .40 of being served promptly in the hardware department and .60 in the sporting-goods department. If we are told that the probability is .76 that at least one of the two will be served promptly, what is the probability that they will both be served promptly?

In Exercises 17 and 18, an expedition leader and six other mountain climbers must choose three of their number to make the final attempt at the summit, while the other four remain behind at the base camp.

17. In how many ways can the three be chosen from among the seven climbers?

18. If the expedition leader must be included among the three chosen, how many choices are there?

Exercises 19 through 21 concern a children's party at which each child plays three games. In the first game, a child rolls two dice and wins if both numbers are the same. In the second, a child picks one card from a standard deck and wins if a face card (king, queen, jack) appears. In the third, a child flips two coins and wins if they are both heads or both tails.

19. What is the probability that a child will win all three games?

20. What is the probability that a child will win the dice game and lose the others?

21. What is the probability that a child will win exactly one game?

Exercises 22 through 24 use the following information: Two evenly matched teams are in a basketball playoff, that is, we assume that each team has exactly .50 probability of winning any given game. The winner of the playoff is the first team to win four games. The probability is .125 that the playoff will last only four games, .25 that it will last exactly five games, .3125 that it will last exactly six games, and .3125 that it will require all seven games.

296   PROBABILITY

22. What is the probability that the playoff will last at least five games?

23. What is the probability that the playoff will require fewer than seven games?

24. What is the probability that the playoff will require fewer than four games?

In Exercises 25 through 28, a family consisting of a father, mother, and child is chosen at random and is asked on what day of the week each of them was born.

25. What is the probability that all three were born on different days, given that the father was born on a Monday?

26. Is the event (all were born on different days) independent of (father was born on a Monday)?

27. What is the probability that all were born on different days, given that the father was born on a Monday and the mother on a Tuesday?

28. Is the event (all were born on different days) independent of (father was born on a Monday and mother on a Tuesday)?

29. According to *Climatological Data*, on a November day the probability of precipitation in Minneapolis is .30, while in Duluth it is .22; the probability that there will be precipitation in both cities is .14. Calculate the probability of precipitation in at least one of the two cities.

30. An instructor lists 10 topics from which he will choose 5 for the final examination. A student has prepared only 7 of the topics adequately. Calculate the probability that all the topics that appear on the examination will be ones that the student has prepared adequately.

Appendix: The Standard Deck of Cards

A standard deck of playing cards consists of 52 different cards. Each card is identified by its value and its suit. There are four cards of each value and thirteen different values:

2, 3, 4, 5, 6, 7, 8, 9, 10, Jack, Queen, King, Ace

The values are ordered from left to right. Thus, an eight is higher than a three, a jack is higher than an eight, and a king is higher than a jack.

The face cards are the jacks, queens, and kings, so there are twelve face cards in all. Each of the four cards of a fixed value is of a different suit. The suits are:

clubs (♣),   diamonds (♦),   hearts (♥),   spades (♠)

Spades and clubs are printed with black ink and hearts and diamonds are printed with red ink; so, half the cards (26) in the standard deck are black and the other half are red. Thus, a red seven means either the seven of hearts (7♥) or the seven of diamonds (7♦).

4

More Probability

Drug Testing

News stories telling of new proposals for drug testing of athletes, pilots, bus drivers, railway engineers, government officials, or of testing as a condition of employment for any job that effects the public safety have appeared frequently in recent years. One of the first incidents that brought drug testing of athletes to the public consciousness happened at the Munich Olympics in 1972 when U.S. swimmer Rick Demont was forced to return his gold medal after such a test. Demont was penalized because his sinus medicine contained a drug on the proscribed list. A similar case arose in July 1987 when Scott Young, an American hockey player, jeopardized his chances of making the American Olympic hockey team because he took Sine-Aid, a nonprescription, over-the-counter decongestant and, as a result, tested positive for ephedrine, just as Demont had 15 years earlier. In 1988 the Canadian sprinter Ben Johnson tested positive for steroids and lost his gold medal and world record for the 100-yard dash. While this was headline news all over the world, the controversy that resulted was not about the propriety of the test, but about whether other athletes were also using steroids in a less obvious way—yet getting away with it.

Testing of Olympic athletes is relatively noncontroversial because the tests are administered to only a few people and those people are doing something very special. The tests are perceived as a way of keeping the competitions fair and also of protecting young people from coaches and trainers who want to win so badly that they disregard the long-term effects performance drugs can have on the health of the athletes. However, drug testing as a condition of employment for broad categories of jobs, such as airline pilots, police officers, and heavy-equipment operators, is much more controversial.

Another important issue that must be considered is the accuracy of the tests. If society decides that, for example, airline pilots should undergo periodic testing for drugs, how accurate are those tests? What are the chances that a person who is not using and has never used barbiturates, amphetamines, cocaine, or marijuana will test positive? Or, on the other side, what are the chances that a person who is using one of these drugs will test negative and perhaps cause an accident?

In urine drug screening, like all tests, not every result can be guaranteed to be correct. These analytical situations present many pitfalls which are described by their own special vocabulary. A "false-positive" result of a test means finding a drug that isn't there. A "false-negative" is the failure of the test to find a drug that is there. A "misidentification" means finding a drug that is there, but identifying it as the wrong drug. In theory, none of these things have to happen, but they all do.

From 1972 through 1981 the Centers for Disease Control in conjunction with the National Institute on Drug Abuse conducted a proficiency testing (PT) program for drugs-of-abuse screening laboratories. The original program involved mailing 40 samples per year to each lab and if 80% of the samples were correct, the lab was rated as satisfactory. However, since these samples were clearly identified as coming from the PT program, early in the program claims were made that at some laboratories these samples were not subjected to the same testing procedures as the lab's routine patient samples. So a "blind test" was devised in which PT samples were submitted along with regular ones, from various physician's offices and drug-treatment facilities, and not identified as coming from the PT program. Although the percentage of drugs detected by laboratories in the two studies ranged from 76% to 100% (average, 98%) on mailed PT samples, the percentage on blind PT samples for the same laboratories testing identical samples ranged from 11% to 100% (average, 69%). Additional CDC blind studies of six laboratories in 1978 provided similar results. The percentage of drugs detected ranged from 37% to to 74% (average, 61%).

As these figures indicate, laboratory tests are often not sensitive enough. There is a high rate of false-negatives; evidence of drug abuse is present in the urine sample, but the laboratory fails to find it. In general, the rate of false-positives is much lower. But even a quite low false-positive rate can lead to serious problems in the mass testing of populations with a low prevalence of drug use, such as in our airline-pilot example. We will use the probability theory we discussed in Chapter 3, along with some new topics in probability, to analyze this false-positive problem.

4.1 Partitions

A Gambling Problem

Two gamblers agree to play the following game. After placing a bet, each player will draw a card from a standard deck, show it to the other player and return it to the deck. The higher card (with aces above kings) wins the money. The first player draws a seven and puts it back in the deck. But before the second player can draw her card, the first says that he is afraid she will cheat, and to protect himself he demands that she draw two cards from the deck and use the second card. The second player does not intend to cheat. She knows the probability is $\frac{28}{52} = .5385$ (approximately) that she will beat the seven on her first draw (because there are 28 cards from eight to ace), but she wonders whether her opponent's demand might be a trick to reduce the probability that she will win. Thus, the second player wishes to know the probability that she will draw an eight or higher on her second card.

The new feature of this question is the fact that the probability of drawing an eight or higher after the first card is removed depends on what the first card is—but we do not know what card this is. After the first card is removed, 51 cards remain. If the first card is an eight or higher, then only 27 cards at least as high as an eight remain in the deck; so, the probability of getting at least an eight on the second card drops to $\frac{27}{51} = .5294$ (approximately). On the other hand, if the first card is lower than an eight, then all 28 winning cards remain, and the second player has a probability of $\frac{28}{51} = .5490$ (approximately) that she will win. Neither calculation answers the player's question about the probability of winning because she needs the answer *before* she knows what the first card is.

Analysis of the Problem

In order to analyze the gambler's problem, we will use A to represent the winning outcomes. That is, A is the set of ordered pairs

(any card, eight or higher)

where the two cards in the pair are different because the first is kept out of the deck when the second is drawn. We need a convenient computational formula for the probability $p(A)$. We will separate the successful outcomes, the pairs in A, into two types: those in which the first card of the pair is an eight or higher and those in which it is not. The way we will do this is first to divide the entire sample space U of all the possible outcomes (pairs of distinct cards), successful or not, in the same way. That is, let B_1 indicate that the first card in a pair of distinct cards is an eight or higher and the second card has any value. Let B_2 indicate that the first card is a seven or lower, also with any second card. We divide A into outcomes of the type

first card   second card
(eight or higher, eight or higher)

and those of the type

first card   second card
(seven or lower, eight or higher)

The first type is both in A, because the second card is at least an eight, and in B_1 because the first card is also at least an eight. Thus, the first type of successful outcome is described by $A \cap B_1$. Similarly, the other type of successful outcome is described by $A \cap B_2$. Since every outcome in A is either in $A \cap B_1$ or $A \cap B_2$, it must be true that

$$p(A) = p((A \cap B_1) \cup (A \cap B_2)).$$

We can see from the definition and from Figure 4-1 that no outcome can be both of type B_1 and of type B_2. So every successful outcome is either in event $(A \cap B_1)$ or event $(A \cap B_2)$ and no successful outcome can belong to both events, that is, these events are disjoint sets. Therefore, the principle of probability in Section 3.7 (written there as $p(A \cup B) = p(A) + p(B)$) applies, and we have

$$p(A) = p((A \cap B_1) \cup (A \cap B_2))$$
$$= p(A \cap B_1) + p(A \cap B_2)$$

In Section 3.9, we found the following formula for computing the probability of the intersection of events:

$$p(A \cap B) = p(B) \cdot p(A \mid B)$$

For the card problem,

$$p(A \cap B_1) = p(B_1) \cdot p(A \mid B_1)$$

and similarly,

$$p(A \cap B_2) = p(B_2) \cdot p(A \mid B_2)$$

Substituting into the preceding equation for $p(A)$, we obtain the formula we need:

$$p(A) = p(B_1) \cdot p(A \mid B_1) + p(B_2) \cdot p(A \mid B_2)$$

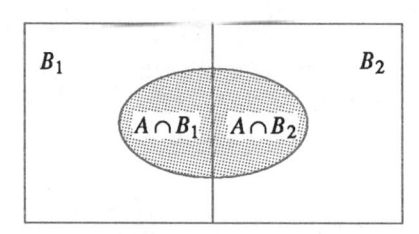

Figure 4-1

Solving the Problem

If we apply the formula to the gambler's problem, $p(B_1)$ just represents the probability of drawing an eight or higher on the first card, which is $\frac{28}{52}$, whereas the probability of drawing a card lower than an eight is $p(B_2) = \frac{24}{52}$. The symbol $p(A \mid B_1)$ means the probability of drawing an eight or higher on the second card when the first card is at least as high as an eight. We calculated before that $p(A \mid B_1) = \frac{27}{51}$. We also calculated that $p(A \mid B_2)$, the probability of drawing an eight or higher on the second draw when the first produced a card below an eight, is $\frac{28}{51}$. By the formula,

$$p(\text{win with 2nd card drawn}) = p(A) = p(B_1) \cdot p(A \mid B_1) + p(B_2) \cdot p(A \mid B_2)$$

$$= \frac{28}{52} \cdot \frac{27}{51} + \frac{24}{52} \cdot \frac{28}{51} = \frac{28 \cdot 27 + 24 \cdot 28}{52 \cdot 51}$$

$$= \frac{28(27 + 24)}{52 \cdot 51} = \frac{28 \cdot 51}{52 \cdot 51} = \frac{28}{52}$$

which is the same as under the original rules of the game, that is, drawing a single card. So the gambler might just as well agree to use the second card to settle the bet, since it will make her opponent feel more secure.

Further Analysis

Figure 4-1 is the basis for the formula

$$p(A) = p(B_1) \cdot p(A \mid B_1) + p(B_2) \cdot p(A \mid B_2)$$

The successful outcomes A form a subset of the sample space U of all outcomes (pairs of distinct cards). Furthermore, B_1 and B_2 are also subsets of outcomes. Those in B_1 are pairs in which the first member is at least as high as eight; all the other pairs are in B_2. Thus, U is divided into subsets B_1 and B_2, and no outcome is in both subsets.

The probability formula can also be pictured by means of a probability tree, as illustrated in Figure 4-2. The left-hand side of the tree represents U separated

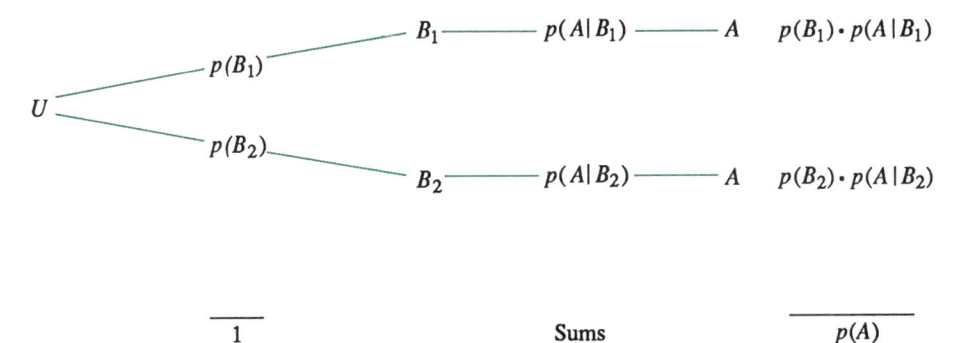

Figure 4-2

into subsets B_1 and B_2. The sum 1 reminds us that every outcome in U is either in B_1 or B_2; so

$$p(B_1) + p(B_2) = 1$$

The calculation of $p(A)$ is shown to the right of the tree. Probabilities on the same branch of the tree are multiplied; then the products are added to compute $p(A)$. The probability tree for the gambling problem that introduced this section illustrates the procedure, as shown in Figure 4-3.

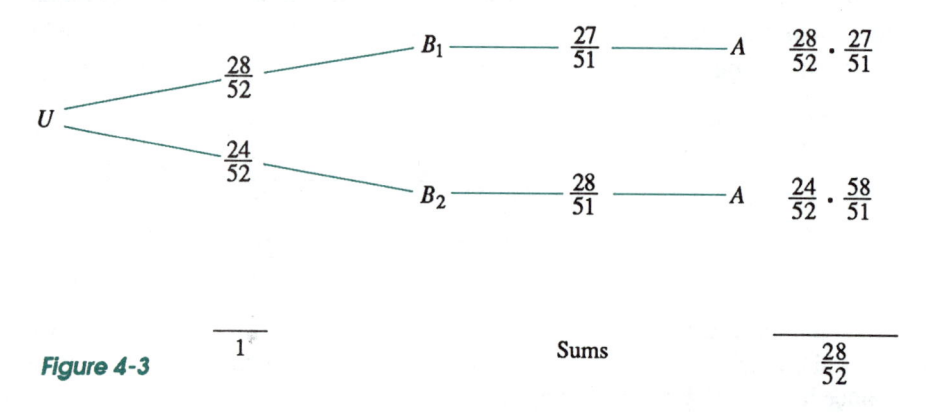

Figure 4-3

$$U \begin{cases} \frac{28}{52} \to B_1 \to \frac{27}{51} \to A \qquad \frac{28}{52} \cdot \frac{27}{51} \\ \frac{24}{52} \to B_2 \to \frac{28}{51} \to A \qquad \frac{24}{52} \cdot \frac{58}{51} \end{cases}$$

Sums: 1 ... $\frac{28}{52}$

Example 1. Suppose that 3% of the members of a population use an illegal drug. A screening test is 95% accurate in the sense that the rate of false-positives is 5% and the rate of false-negatives is the same. If a person is chosen at random from the population, calculate the probability that he or she will test positive.

Solution

We wish to calculate $p(A)$, the probability that a person will test positive. The probability of a positive result depends on whether or not the person actually uses the drug—with a 95% chance of a positive result among drug users, but only a 5% chance among nonusers. To express this information in terms of probabilities, we partition the population U into B_1, the drug users, and B_2, the nonusers. Since we are told that $p(B_1) = .03$, we have $p(B_2) = .97$. We also know that $p(A \mid B_1) = .95$ and $p(A \mid B_2) = .05$. The probability tree is shown in Figure 4-4, from which it is seen that the probability that a person chosen at random from the entire population will test positive is .077. ∎

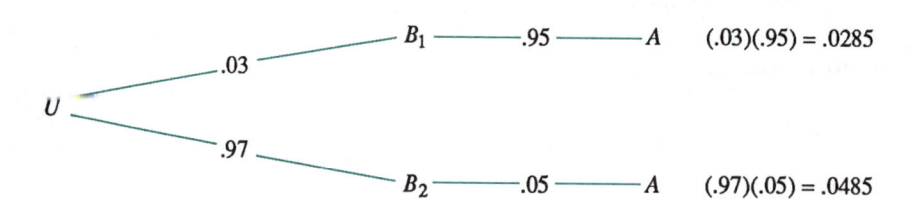

Figure 4-4

$$U \begin{cases} .03 \to B_1 \to .95 \to A \qquad (.03)(.95) = .0285 \\ .97 \to B_2 \to .05 \to A \qquad (.97)(.05) = .0485 \end{cases}$$

Sums: 1 ... $.077$

Partitions

The division of U, the set of all outcomes, into the subsets B_1 and B_2 is a special case of a procedure that frequently occurs in studies of probability. A **partition** of the sample space U is a collection of subsets of U in which each outcome of U belongs to exactly one of the subsets (see Figure 4-5).

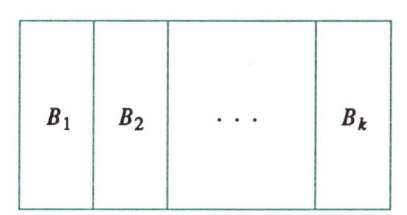

Figure 4-5

General Probability Formula

If $B_1, B_2, \ldots, B_k$ is a partition of a sample space U of outcomes, and if A is viewed as the set of successful outcomes of the experiment, then we visualize the situation as shown in Figure 4-6. An argument exactly like the one we used for the gambler's problem, just involving more symbols, shows that:

> If $B_1, B_2, \ldots, B_k$ is a partition of a sample space U and if A is any subset of U, then
> $$p(A) = p(B_1) \cdot p(A \mid B_1) + p(B_2) \cdot p(A \mid B_2) + \ldots + p(B_k) \cdot p(A \mid B_k)$$

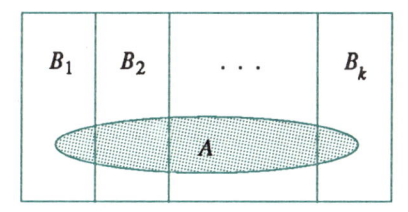

Figure 4-6

Again, a probability tree, such as Figure 4-7, will help us remember the formula.

Example 2. A manufacturer has three assembly lines to produce the same product. Line 1, which uses old equipment, is so slow that it produces only 5% of the total output, and it is so inaccurate that 2% of the units produced are defective. (Line 1 has a "defect rate" of .02.) Assembly line 2 produces 50% of the total output with a defect rate of .005, whereas line 3 produces 45% of the output with a defect rate of only .003. What is the manufacturer's overall defect rate for this product, i.e., what is the probability that a unit chosen at random will be defective?

304   MORE PROBABILITY

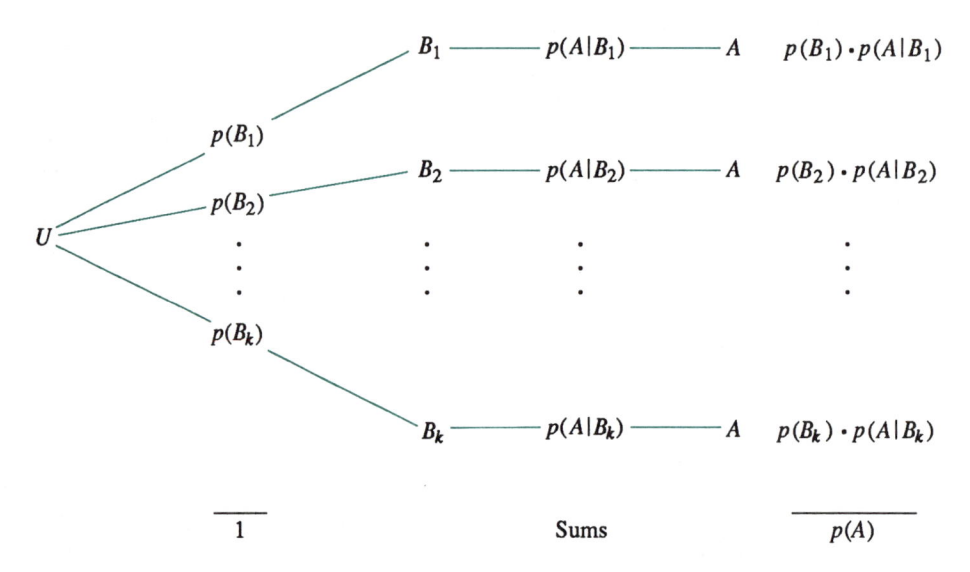

```
                  B₁ ──── p(A|B₁) ──── A     p(B₁) · p(A|B₁)
          p(B₁)
                  B₂ ──── p(A|B₂) ──── A     p(B₂) · p(A|B₂)
        p(B₂)
   U                ⋮           ⋮              ⋮
        p(Bₖ)
                  Bₖ ──── p(A|Bₖ) ──── A     p(Bₖ) · p(A|Bₖ)

         1                    Sums                 p(A)
```

Figure 4-7

Solution

The outcomes are units of the product and we partition the units into subsets B_1, B_2, and B_3, according to which assembly line they come from. If a unit is chosen at random, then $p(B_1)$, the probability it comes from line 1, is .05. Likewise, $p(B_2) = .50$ and $p(B_3) = .45$, according to the statement of the problem. We wish to know $p(A)$, the probability that a unit is defective. The "defect rate .02 for line 1" is the probability that a unit is defective under the condition that it comes from line 1. Thus, we can write $p(A \mid B_1) = .02$. In the same way, $p(A \mid B_2) = .005$ and $p(A \mid B_3) = .003$. The probability tree is shown in Figure 4-8. From the tree we see that the defect rate for the entire factory is .00485. ∎

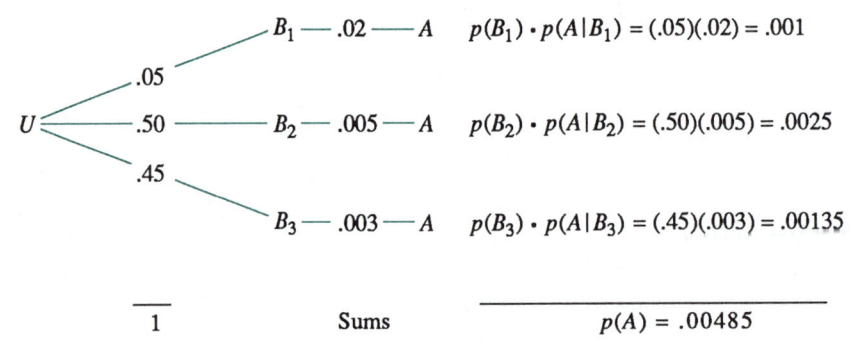

```
                 B₁ ── .02 ── A    p(B₁) · p(A|B₁) = (.05)(.02) = .001
          .05
   U ──── .50 ──── B₂ ── .005 ── A    p(B₂) · p(A|B₂) = (.50)(.005) = .0025
          .45
                 B₃ ── .003 ── A    p(B₃) · p(A|B₃) = (.45)(.003) = .00135

          1               Sums                     p(A) = .00485
```

Figure 4-8

Summary

Suppose that a sample space U is partitioned into subsets B_1 and B_2, that is, U is the union of B_1 and B_2, where B_1 and B_2 are disjoint. Thus, every element of U is in either B_1 or B_2. Then, for any subset A of U,

$$p(A) = p(B_1) \cdot p(A \mid B_1) + p(B_2) \cdot p(A \mid B_2)$$

In general, a **partition** of a sample space U is a collection of subsets such that each outcome in U is in exactly one of the subsets. If $B_1, B_2, \ldots, B_k$ forms a partition of a sample space U and if A is any subset of U, then

$$p(A) = p(B_1) \cdot p(A \mid B_1) + p(B_2) \cdot (A \mid B_2) + \ldots + p(B_k) \cdot p(A \mid B_k)$$

Exercises

1. A sample space U is partitioned into subsets B_1, B_2, B_3, and B_4, with $p(B_1) = .10$, $p(B_2) = .55$, $p(B_3) = .05$, and $p(B_4) = .30$. Given the conditional probabilities $p(A \mid B_1) = .50$, $p(A \mid B_2) = .80$, $p(A \mid B_3) = .10$, and $p(A \mid B_4) = 1$, calculate $p(A)$.

2. In a certain population, 15% of the men have difficulty hearing, whereas only 12% of the women have this problem. Assuming that there are as many men as women in the population, what is the probability that a person chosen at random from the population will have difficulty hearing?

3. In a test for a particular disease, the probability of a positive result is .80 for a carrier of the disease, but there is also a probability of .15 that the test will be positive even though the subject does not carry the disease. If 5% of the population are carriers of this disease, what is the probability that a person chosen at random will have a positive test?

4. A sales representative for a manufacturer visits a department store. The representative estimates that the probability is .40 that the store's head buyer will be in her office and .60 that she will be out of town, causing the representative to see an assistant buyer instead. The representative also estimates the probability of a sale to the store on this visit is .50 if the head buyer is in, but only .30 if the assistant buyer must be seen. What does the representative estimate to be the probability of a sale to the department store on this visit?

5. An insurance company knows that 20% of the drivers covered by its policies are under 25 years of age and 35% are over 60. The company estimates that the accident rate for its drivers under 25 is .012; that is, there is a probability of .012 that a driver in this age group will have an accident during a given year. The accident rate for its over-60 drivers is .009, and for all other drivers (ages 25 to 60) is .005. Estimate the probability that a driver covered by this company's policies will have an accident in a given year.

306   MORE PROBABILITY

6. The human blood types are A, B, AB, and O. Suppose that, in some population, 35% of the population are of type A, 42% of type B, 18% of type AB, and 5% of type O. Imagine that for some blood disease, an individual's susceptibility depends on his or her blood type. The probability of getting the disease is .001 for type A- and type B-individuals, .0005 for those of type AB, and .005 for those of type O. What is the probability that an individual chosen at random from this population will contract the disease?

7. A factory uses three shifts, with 1000 employees during the day shift (8 A.M. to 4 P.M.), 500 on the swing shift (4 P.M. to midnight), and 300 on the graveyard shift (midnight to 8 A.M.). The absentee rate, that is, the probability that an employee will fail to show up for work on a given day, is .02 for the day shift, .05 for the swing shift, and .07 for the graveyard shift. What is the absentee rate for the factory as a whole?

8. A Democrat and a Republican are running for mayor. The voters of the city are 50% Democrats, 30% Republicans, and 20% independents. If a poll indicates that 60% of the Democrats will vote for the Democratic candidate and the rest for the Republican, that 90% of the Republicans will vote for their party's candidate and the rest for the Democrat, and that the independents are evenly divided between the candidates, who does the poll indicate will win the election?

9. A psychology experiment involves 20 human subjects. Five of the subjects rest for an hour before participating in the experiment, seven others exercise for an hour before the experiment, and the remaining eight do mathematics problems for an hour before the experiment. The psychologist running the experiment estimates the probability that a subject will succeed in the experiment to be $\frac{3}{5}$ if the subject has rested, $\frac{1}{2}$ if the subject has exercised, and $\frac{1}{10}$ if the subject has been doing mathematics problems. If a subject is chosen at random, what is the probability that he or she will succeed in the experiment?

10. With respect to a particular gene, biologists classify individuals into three distinct genotypes: AA, Aa, and aa. Suppose that 30% of some population are of type AA, 60% are of type Aa, and 10% are of type aa. Further, suppose that genotypes with respect to this gene influence the probability that an infant will reach maturity as follows: 70% of type AA individuals, 60% of type Aa, and 30% of type aa. What is the probability that an infant chosen at random from this population will reach maturity?

4.2 Bayes' Theorem

A Fundamental Relationship

In Section 3.9, the definition of conditional probability led us to two different formulas for the probability, $p(A \cap B)$, that the result of an experiment will be both of type A and of type B. The formulas were

$$p(A \cap B) = p(B) \cdot p(A \mid B)$$
$$p(A \cap B) = p(A) \cdot p(B \mid A)$$

Since both of the products $p(B) \cdot p(A \mid B)$ and $p(A) \cdot p(B \mid A)$ are equal to $p(A \cap B)$, we conclude that they equal each other:

$$p(A) \cdot p(B \mid A) = p(B) \cdot p(A \mid B)$$

Dividing both sides of the equation by $p(A)$ gives a fundamental relationship between the two conditional probabilities $p(B \mid A)$ and $p(A \mid B)$:

$$p(B \mid A) = \frac{p(B) \cdot p(A \mid B)}{p(A)}, \text{ providing that } p(A) \neq 0$$

Example 1. An accounting-office staff consists of 20 men, 8 of whom were college graduates, and 30 women, of whom 18 were college graduates. If being named to the planning committee requires a college degree, what is the probability that the representative is a woman?

Solution

We will let A represent the set of workers with college degrees and B, the set of women workers. In this example we must calculate $p(B \mid A)$, the probability that the representative chosen is a woman, under the condition that the representative is a college graduate. We can see that $p(A) = \frac{26}{50}$ and $p(B) = \frac{30}{50}$. We also need to use a formula from Chapter 3 to see that

$$p(A \mid B) = \frac{n(A \cap B)}{n(B)} = \frac{18}{30}$$

Of course, we could compute $p(B \mid A)$ in the same way; but instead, let's practice using the new formula to compute

$$p(B \mid A) = \frac{p(B) \cdot p(A \mid B)}{p(A)} = \frac{\frac{30}{50} \cdot \frac{18}{30}}{\frac{26}{50}} = \frac{18}{26} = .69 \text{ (approximately)} \blacksquare$$

Example 2. Suppose that two dice are rolled.

(a) What is the probability, $p(A \mid B)$, that the sum of the numbers is 6, given that the first die is a 2.

(b) What does $p(B \mid A)$ mean?

(c) Compute $p(B \mid A)$.

Solution

(a) We solved this problem in Chapter 3. The second die can have 6 different values. But since the first die is a 2, the second die must be a 4 in order for the sum to be 6; so, the answer is $p(A \mid B) = \frac{1}{6}$.

(b) From the way in which the question of Part (a) is asked, we know that A is the class of outcomes in which the sum is 6, whereas B is the class of outcomes in which the first die is a 2. Therefore, the symbol $p(B \mid A)$ denotes the probability that the first die is a 2, given that the sum of the two dice is 6.

(c) In Part (a) we calculated that $p(A \mid B) = \frac{1}{6}$. There are 5 ways two dice can have a sum of 6; so, $p(A) = \frac{5}{36}$. Since B consists of pairs in which the first member is a 2, but the second can be any number on the die,

$$p(B) = \frac{6}{36} = \frac{1}{6}$$

By the formula relating $p(B \mid A)$ with $p(A \mid B)$,

$$p(B \mid A) = \frac{p(B) \cdot p(A \mid B)}{p(A)} = \frac{\frac{1}{6} \cdot \frac{1}{6}}{\frac{5}{6}} = \frac{1}{5} = .20 \qquad \blacksquare$$

Partitions

A very common use of the formula

$$p(B \mid A) = \frac{p(B) \cdot p(A \mid B)}{p(A)}$$

is in partition questions of the following kind: Given a partition $B_1, B_2, \ldots, B_k$ of a sample space and knowing that the outcome of an experiment was successful in the sense of belonging to some set A, what is the probability $p(B_i \mid A)$ that the outcome belongs to some specific set B_i of the partition? The formula that answers the question is

$$p(B_i \mid A) = \frac{p(B_i) \cdot p(A \mid B_i)}{p(A)}$$

which is just the preceding formula with B_i in place of B.

Example 3. In order to illustrate the use of the formula to determine the probability that a successful outcome belongs to a particular set in the partition, we recall Example 2 of the preceding section. In that example, a manufacturer has three assembly lines—1, 2, and 3—producing the same product. Line 1 produces only 5% of the total output, and it has a defect rate of .02. Assembly line 2 produces 50% of the total output, with a defect rate of .005; and line 3 produces 45% of the total output, with a defect rate of .003. We now wish to know the probability that a defective unit comes from line 1. This will tell us how much of the defective production of the entire factory is due to the outmoded assembly line 1.

Solution

We partition the factory's output into sets B_1, B_2, and B_3 according to which line produced the unit. The example states that $p(B_1) = .05$, $p(B_2) = .50$, and $p(B_3) = .45$. For A, the set of defective units, the example also gives the defect rates $p(A \mid B_1) = .02$, $p(A \mid B_2) = .005$, and $p(A \mid B_3) = .003$. The manufacturer wants to know the probability $p(B_1 \mid A)$ that a defective unit came from line 1. The solution to the example in the preceding section showed that $p(A) = .00485$. Therefore, by the formula,

$$p(B_1 \mid A) = \frac{p(B_1) \cdot p(A \mid B_1)}{p(A)} = \frac{(.05)(.02)}{.00485} = .21 \text{ (approximately)}$$

Thus, line 1, which produces only 5% of the total output, is responsible for more than 20% of the defective units produced by the entire factory. ■

Notice that the first two examples of this section, where the probability was based on counting, could have been done by the methods of Chapter 3. In the example we just completed, however, in which the probabilities were based on the factory's experience, we needed to use the new formula in order to compute $p(B_1 \mid A)$.

Bayes' Theorem

We have been breaking the problem of computing $p(B_i \mid A)$ into two parts. First we use a formula from the previous section to calculate

$$p(A) = p(B_1) \cdot p(A \mid B_1) + p(B_2) \cdot p(A \mid B_2) + \ldots + p(B_k) \cdot p(A \mid B_k)$$

and then we substitute the result into

$$p(B_i \mid A) = \frac{p(B_i) \cdot p(A \mid B_i)}{p(A)}$$

If, instead, we substitute the formula for $p(A)$ for the denominator of the formula for $p(B_i \mid A)$, we have a one-step formula, called *Bayes' Theorem*, for calculating $p(B_i \mid A)$.

Bayes' Theorem: If $B_1, B_2, \ldots, B_k$ form a partition of a sample space U and if A is any subset of U, then

$$p(B_i \mid A) = \frac{p(B_i) \cdot p(A \mid B_i)}{p(B_1) \cdot p(A \mid B_1) + p(B_2) \cdot p(A \mid B_2) + \ldots + p(B_k) \cdot p(A \mid B_k)}$$

MORE PROBABILITY

Notice that the numerator of the formula is identical to one of the terms in the denominator. From the point of view of a probability tree, the numerator is the probability along one "branch" of the tree, whereas the denominator is the sum of the probabilities along all the branches. Specifically, the numerator comes from the branch of the same set B_i in the partition as the probability we are trying to compute. (See Figure 4-9, in which $i = 1$ and $k = 3$.) The next example illustrates how the probability tree helps us apply Bayes' Theorem.

Example 4. Ecologists band individuals in a small population of animals of an endangered species to study their habits. Suppose 10% of the population is banded with plastic tags and 30% with metal tags. There is a 50% probability that a plastic tag will fall off an animal, but the metal tags have only a 5% probability of falling off. If an animal from that population is found without a tag, what is the probability that it once had a plastic tag?

Solution

Let A be the set of animals without tags. Partition the animal population into B_1, those animals that were originally given plastic tags; B_2, those that were given metal tags; and B_3, those animals that were never tagged. Since we are told that $p(B_1) = .10$ and $p(B_2) = .30$, we see that $p(B_3) = .60$. According to the problem, the probability that an animal once wore a plastic tag and no longer has it, that is $p(A \mid B_1)$, is .50. Also, $p(A \mid B_2) = .05$, and of course, $p(A \mid B_3) = 1.00$ because an animal that never had a tag can hardly acquire one by itself. The probability we are seeking, that an animal without a tag once wore a plastic one, is $p(B_1 \mid A)$. The probability tree is shown in Figure 4-9. We highlighted the B_1 branch because that branch gives us the numerator in Bayes' Theorem. The denominator is computed in the tree just as in the preceding section; so, we conclude that the probability that an animal without a tag once wore a plastic tag is

$$p(B_1 \mid A) = \frac{.05}{.665} = .075 \text{ (approximately)}$$ ■

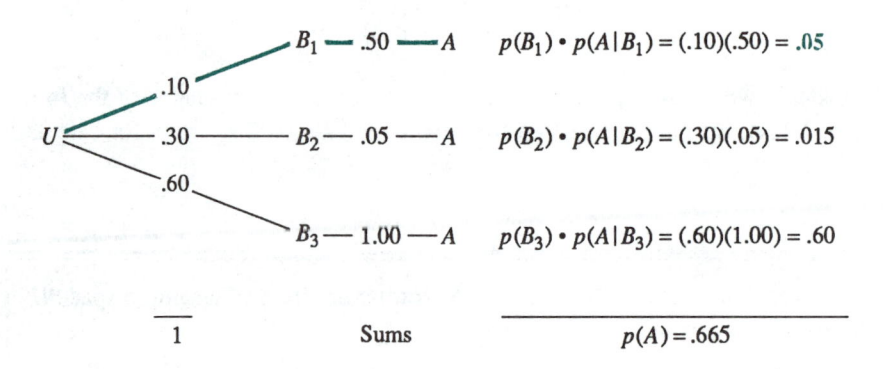

```
                    B₁ — .50 — A    p(B₁) · p(A|B₁) = (.10)(.50) = .05
            .10
    U ———— .30 ———— B₂ — .05 — A    p(B₂) · p(A|B₂) = (.30)(.05) = .015
            .60
                    B₃ — 1.00 — A   p(B₃) · p(A|B₃) = (.60)(1.00) = .60

    1            Sums                       p(A) = .665
```

Figure 4-9

Drug Screening

When we apply Bayes' Theorem to the topic of drug screening, it has something rather surprising to teach us, as you will see. Recalling Example 1 (and the information in the introduction to the chapter), we suppose that a "95% accurate" drug-screening test is applied to a population of which 3% are users of the drug. Since the rate of false-positives is 5%, for a person who is not a user of the drug, there is a 5% risk that the test will be positive anyway. If a person uses the drug, there is a probability of .95 that the test will recognize this fact. Now suppose you are an employer, and it is reported to you that one of your employees tested positive. With such an accurate test, you should feel very confident that the employee is a user of the drug, perhaps 95% or more certain. Right?

Wrong! To see just how wrong, let's first recall the notation of Example 1. We denoted by A the event "test positive" and we partitioned the population U into B_1, the drug users, and B_2, the nonusers. The accuracy of the test tells us that $p(A \mid B_1) = .95$ and $p(A \mid B_2) = .05$. Furthermore, $p(B_1) = .03$, so that $p(B_2) = .97$. If the employee tested positive, that means we know the condition, namely A, of a conditional probability. As an employer, you want to know if that person is really a drug user; so, the probability you need to evaluate is $p(B_1 \mid A)$. From the probability tree shown in Figure 4-10, we see that

$$p(B_1 \mid A) = \frac{.0285}{.077} = .37 \text{ (approximately)}$$

You are not so certain that the employee is a drug user, after all. The probability is only a little more than one out of three. Or, to put it more forcefully, there are almost two chances in three that the employee reported to you as a drug user was falsely accused of breaking the law.

You can see from this analysis how important it is that people understand the probabilistic implications of information from drug-screening tests. The problem becomes even more acute if the accuracy of the test is less than the relatively

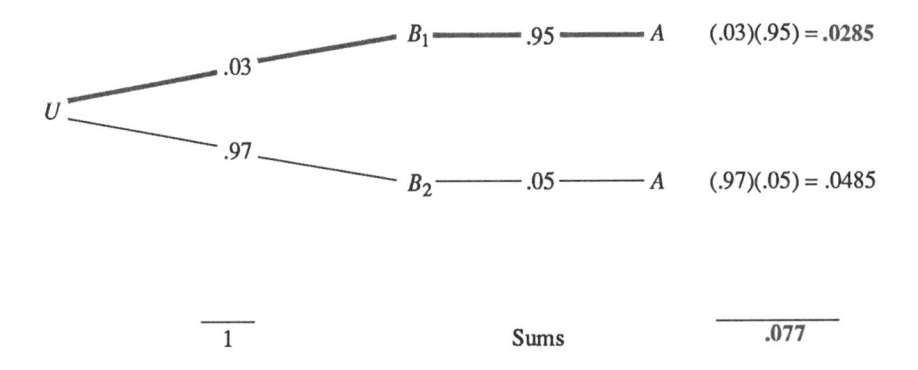

Figure 4-10

312   MORE PROBABILITY

ideal level that is suggested by the Center for Disease Control, but which is seldom actually attained. The Center did not classify the performance of a laboratory as unsatisfactory unless its rate of false-positives or false-negatives exceeded 25%. Suppose, for instance, the accuracy rate in both respects was 20%, and you still were testing a population with 3% drug users. In that case, you can show (see Exercise 12) that most of the people who test positive are actually not users of the drug, but instead, are false-positives.

Summary

The conditional probabilities $p(B \mid A)$ and $p(A \mid B)$ are related by the formula

$$p(B \mid A) = \frac{p(B) \cdot p(A \mid B)}{p(A)}, \text{ provided that } p(A) \neq 0$$

Let $B_1, B_2, \ldots, B_k$ be a partition of a sample space U and let A be a subset of U. Then **Bayes' Theorem** states that

$$p(B_i \mid A) = \frac{p(B_i) \cdot p(A \mid B_i)}{p(B_1) \cdot p(A \mid B_1) + p(B_2) \cdot p(A \mid B_2) + \ldots + p(B_k) \cdot p(A \mid B_k)}$$

Exercises

1. Suppose that A and B are subsets of a sample space U. If $p(A) = .14$, $p(B) = .35$, and $p(A \mid B) = .20$, what is $p(B \mid A)$?

2. A sample space is partitioned into subsets B_1, B_2, and B_3, with $p(B_1) = \frac{2}{5}$, $p(B_2) = \frac{1}{5}$, and $p(B_3) = \frac{2}{5}$. Given that $p(A \mid B_1) = \frac{1}{2}$, $p(A \mid B_2) = \frac{3}{4}$, and $p(A \mid B_3) = \frac{1}{4}$, use Bayes' Theorem to calculate $p(B_2 \mid A)$.

3. In a certain population, 10% of the men and 2% of the women are color-blind. Assuming that there are as many men as women in the population, if a person chosen at random is color-blind, what is the probability that it is a man?

For Exercises 4 and 5, Table 4-11 states accident rates (probabilities that a driver will have an accident in a given year) for drivers of various ages as well as the percentage of drivers in each age group covered by the policies of an insurance company.

Table 4-11

Age	Accident Rate	Percentage of Total
Under 25	.015	20%
25–35	.006	35%
36–50	.004	30%
Over 50	.011	15%

4. What percentage of accidents by drivers covered by the company's policies involve drivers under 25? That is, if a driver covered by the company is involved in an accident, what is the probability that the driver is under 25?

5. What percentage of accidents covered by the company involve drivers between the ages of 36 and 50?

6. In a test for a certain disease, the probability of a positive result is .90 for a carrier of the disease, but there is also a probability of .20 that a person who does not have the disease will have a positive reaction to the test. If 5% of the population carries the disease, what is the probability that a person who has a positive reaction to the test really has the disease?

7. A Democrat and a Republican are running for governor. The voters of the state are 40% Democrats, 40% Republicans, and 20% independents. Suppose a poll indicates that 70% of the Democrats will vote for the candidate of their party and 30% against her. Furthermore, 80% of the Republicans will vote for their party's candidate and the rest will vote for the Democrat. The independents support the Democratic candidate 60% to 40%. How much of the Democratic candidate's support is coming from her own party; that is, if a person who will vote for the Democratic candidate is chosen at random, what is the probability that he or she will be a Democrat?

8. Suppose that in some country, 25% of the people have parents who are divorced. Suppose also that there is a probability of .30 that a person will be divorced if his or her parents were divorced and a probability of .20 if they were not. If a person is divorced, what is the probability that his or her parents were also divorced?

9. A manufacturer of sports cars buys its speedometers from companies that specialize in their manufacture. One small company produces speedometers with a high degree of precision; only one unit in 1000 is defective (that is, the defect rate is .001). However, this company can supply only 80% of the car maker's requirements. The other speedometers needed must be bought from a large company whose mass-production methods give rise to a defect rate of .05. What is the probability that a defective speedometer comes from the large speedometer manufacturer?

10. Table 4-12 lists the number of employees in each shift at a factory and the absentee rate for each shift (the probability that any particular employee in that shift will fail to show up for work on any given day). What is the probability that an absent employee, chosen at random, was supposed to be working during the graveyard shift?

Table 4-12

Shift	Number of Employees	Absentee Rate
Day (8 A.M. to 4 P.M.)	500	.01
Swing (4 P.M. to Midnight)	500	.03
Graveyard (Midnight to 8 A.M.)	300	.07

314   MORE PROBABILITY

11. Wheat fields are classified as poor, fair, or good on the basis of the quality of the soil. In a particular area, 25% of the fields have poor soil, 45% fair soil, and 30% good soil. A new variety of wheat is used on all the fields in the area; this variety will reportedly increase the yield per acre on 80% of all fields with poor soil, on 40% of fields rated fair, and on 5% of fields with good soil. How much of the improvement in wheat yields from the new variety is the result of increasing the yield of fields with poor soil? That is, if a field from the area whose yield per acre has been improved is chosen at random, what is the probability that it has poor soil?

12. A population in which 3% are users of a drug is screened by a test that is 80% accurate in the sense that the probability of a false-positive is .20 and the probability of a false-negative is also .20. If a person tests positive, calculate the probability that he or she is, in fact, not a user of the drug.

4.3 Random Variables and Probability Distributions

In Chapter 3, we discussed probability problems such as the following:

1. (*Three Dice*)
   Three dice are rolled. What is the probability of getting no more than two 6s?

2. (*Picnic Committee*)
   Five men and four women employees volunteer to serve on a committee to plan the office picnic. The four employees who will actually serve on the committee are chosen at random from among the volunteers. What is the probability that exactly three of those chosen will be men?

3. (*Fruit Trees*)
   A new system of grafting fruit trees is tested on eight trees. The approximate probability that exactly k of the trees will produce higher yields than the average for such trees is given in Table 4-13. What is the probability that at least four of the trees will produce higher-than-average yields?

k	0	1	2	3	4	5	6	7	8
$p(k)$	.05	.17	.28	.27	.16	.06	.00	.00	.00

Table 4-13

Each of these problems asks for the probability of a particular outcome; but these problems also have another important characteristic in common. Problem 1 includes the phrase "no more than two 6s," Problem 2 concerns "exactly three ... men," and Problem 3 asks about "at least four of the trees." Suppose we think of each problem as describing an experiment: rolling dice, choosing employees, measuring the yields of trees. With each possible outcome of the experiment, we

4.3 RANDOM VARIABLES AND PROBABILITY DISTRIBUTIONS

associate a *number*; the number of 6s, the number of men, the number of trees with higher-than-average yield. The problem in each case is to calculate the probability that this number will take on a particular sort of value: no more than two, exactly three, at least four.

Random Variables

A **random variable** X is a rule that associates a number with each outcome of a sample space. The random variable in Problem 1 represents the number of 6s because the problem asks a question concerning the number of 6s. For instance, if the outcome of rolling the three dice is (3, 6, 2), then there is one 6; and so, we write $X(3, 6, 2) = 1$. Similarly, $X(1, 2, 1) = 0$ because no 6 appears, and $X(6, 6, 5) = 2$.

Example 1. In Problem 2 *(Picnic Committee)*, we represent the five men employees by the symbols M_1, M_2, M_3, M_4, M_5 and the four women employees by W_1, W_2, W_3, W_4. What is the random variable X in that problem? Evaluate

$$X(M_1, W_3, M_4, M_2) \quad \text{and} \quad X(W_4, W_2, M_5, W_3)$$

Solution

The question is: What is the probability that exactly three of those chosen will be men? Therefore, the random variable X indicates the number of men in the committee of four. Thus,

$$X(M_1, W_3, M_4, M_2) = 3 \quad \text{and} \quad X(W_4, W_2, M_5, W_3) = 1 \quad \blacksquare$$

Example 2. We represent the eight trees in Problem 3 by an ordered 8-tuple, where in each position we will write a "Y" (*yes*) if the tree has higher-than-average yield and an "N," otherwise. What is the random variable X in this problem? If trees numbered 2, 7, and 8 are the ones with higher-than-average yields, what is the value of X in this case; that is, what is $X(N, Y, N, N, N, N, Y, Y)$? If no tree has a higher-than-average yield, what then is the value of X, that is, what is $X(N, N, N, N, N, N, N, N)$?

Solution

The random variable X represents the number of trees with higher-than-average yield. (Look back at the question in Problem 3.) Thus,

$$X(N, Y, N, N, N, N, Y, Y) = 3 \quad \text{and} \quad X(N, N, N, N, N, N, N, N) = 0 \quad \blacksquare$$

The Probability of 6s

We return once again to Problem 1. If three dice are rolled, there are $6 \cdot 6 \cdot 6 = 216$ possible outcomes. The random variable X in this problem counts the number of 6s in an outcome. Since there can be either zero, one, two, or three 6s showing on the three dice, these are the possible values that X can take. If an outcome contains no 6s, then each of the three dice must show one of the numbers 1, 2, 3, 4, or 5, and there are $5 \cdot 5 \cdot 5 = 125$ such outcomes. Because all outcomes are equally likely, the probability that no 6s will come up when three dice are rolled is $\frac{125}{216}$. We write the probability that X will take on the value 0 as $p(X = 0)$ or, more briefly, $p(0)$. This probability is then

$$p(0) = \frac{125}{216} = .579 \text{ (approximately)}$$

At the other extreme,

$$p(3) = \frac{1}{216} = .005 \text{ (approximately)}$$

because there is only one outcome, (6, 6, 6), for which the value of the random variable is 3.

One way the three dice can come up with just one 6 is for the 6 to appear on the first die and not on the others. This type of outcome can be described by the ordered triple:

$$\underbrace{\text{First die}}_{(6,} \quad \underbrace{\text{Second die}}_{\text{not a 6,}} \quad \underbrace{\text{Third die}}_{\text{not a 6)}}$$

Since there are 5 numbers other than a 6 possible on a die, the number of such outcomes is calculated in the usual way:

$$\begin{array}{ccc} (6, & \text{Not a 6,} & \text{Not a 6)} \\ \downarrow & \downarrow & \downarrow \\ 1 & \cdot \quad 5 \quad \cdot & 5 \quad = 25 \end{array}$$

But there is a distinctly different way to get exactly one 6—namely on the second die:

$$\underbrace{\text{First die}}_{(\text{not a 6,}} \quad \underbrace{\text{Second die}}_{6,} \quad \underbrace{\text{Third die}}_{\text{not a 6)}}$$

and, for the same reason, there are 25 outcomes of this type. The remaining pattern of this type

$$\underbrace{\text{First die}}_{(\text{not a 6,}} \quad \underbrace{\text{Second die}}_{\text{not a 6,}} \quad \underbrace{\text{Third die}}_{6)}$$

produces another 25 such outcomes. Therefore, there are $25 + 25 + 25 = 75$ ways to roll three dice and have exactly one of them show a 6. Recalling that there are altogether 216 outcomes from rolling three dice, we conclude that

$$p(1) = \frac{75}{216} = .347 \text{ (approximately)}$$

We could use a similar counting argument for $p(2)$, but it is not necessary to do that. We already know that out of the 216 possible outcomes from rolling three dice, 125 of them have no 6s, 75 have one 6, and 1 has three 6s. That accounts for $125 + 75 + 1 = 201$ of the outcomes; the remaining 15 outcomes must be the ones having exactly two 6s, since that is the only alternative left. This tells us that

4.3 RANDOM VARIABLES AND PROBABILITY DISTRIBUTIONS

$$p(2) = \frac{15}{216} = .069 \text{ (approximately)}$$

and we summarize all the calculations of the probabilities of the number of 6s in Table 4-14.

Table 4-14

k	0	1	2	3
p(k)	.579	.347	.069	.005

Probability Distributions

For every outcome of a sample space U, a random variable X takes on a value k. The table for $p(k)$ assigns a probability to each possible value and the probabilities in the table must add up to one. The table describes how the total probability, one, is distributed among the possible k. For example, from Table 4-14 we see that there is a better-than-even chance that the three dice will show no sixes, since $p(0) = .579$, whereas $p(3) = .005$ tells us that it is quite unlikely that all three dice will be sixes. A listing of the values k together with the corresponding probabilities $p(k)$ that the random variable will assume is called a **probability distribution**. Notice that the probability distribution for Problem 3 (*Fruit Trees*) is given as part of that problem (in Table 4-13).

Histograms

A table for $p(k)$ contains complete information about the probability distribution. However, when k takes on whole-number values, it is convenient to draw a picture, a **histogram**, that allows us to see just how the probability is distributed. Figure 4-15 is the histogram of the probability distribution for Problem 1 (*Three*

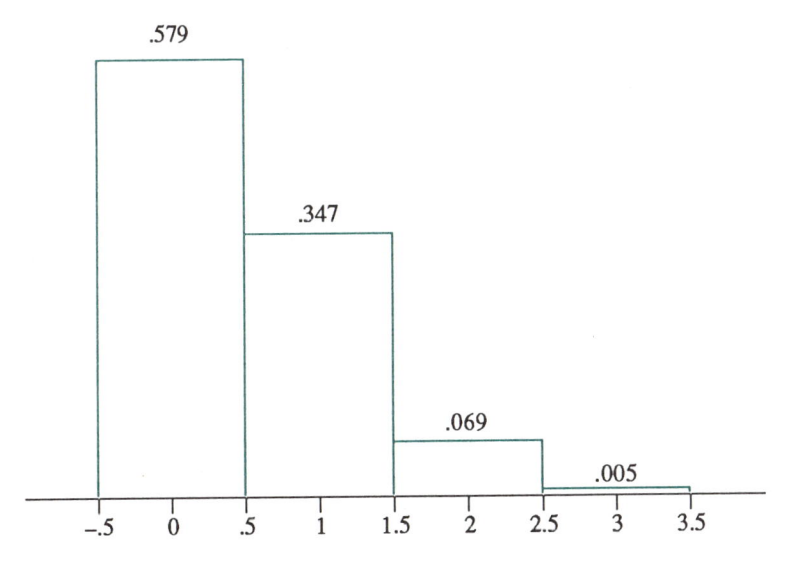

Figure 4-15

Dice) that we calculated previously. In the histogram, the probability $p(k)$ is represented by a rectangle whose base, centered at k, is of length one and whose height is $p(k)$; thus, the probability $p(k)$ equals the area of the rectangle. Notice, for instance, that the rectangle corresponding to $k = 0$ is very large, whereas the one for $k = 3$ is very small. We will take advantage of this correspondence between probability and area later in the chapter. Histograms need not be limited to random variables whose values k are whole numbers. Any probability distribution can be pictured so that area is proportional to probability, but the whole-number case is the only one we will discuss.

Example 3. Draw the histogram of the probability distribution of Problem 3 (*Fruit Trees*).

Solution

Since the possible values of k in this problem go from 0 to 8, we again start at −.5 and note the points .5, 1.5, 2.5, and so on, midway between the indicated whole numbers. We extend to 8.5, and over each interval we construct a rectangle, the height of which is given in Table 4-13, with the result shown in Figure 4-16. ∎

Example 4. Determine the probability distribution for the random variable of Problem 2 (*Picnic Committee*), draw its histogram, and solve the problem.

Solution

There are five men and four women volunteers in the office. An outcome is an unordered subset of four of these nine employees, so there are

$$C_4^9 = \frac{9 \cdot 8 \cdot 7 \cdot 6}{4 \cdot 3 \cdot 2 \cdot 1} = 126$$

possible outcomes. All 126 outcomes are equally likely. Just one outcome, the one consisting of the four women, has no men in it ($X = 0$), so

$$p(0) = \frac{1}{126} = .01 \text{ (approximately)}$$

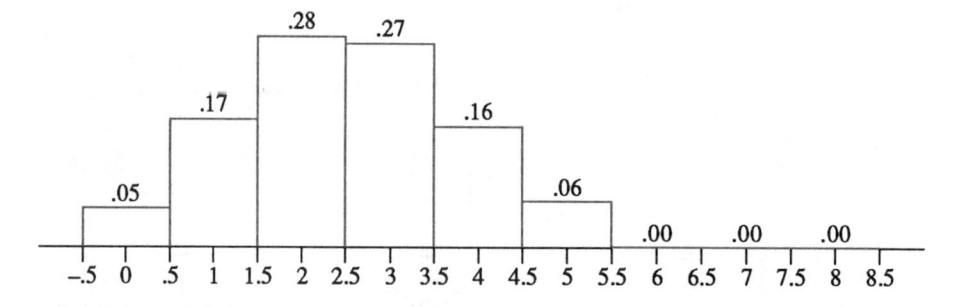

Figure 4-16

4.3 RANDOM VARIABLES AND PROBABILITY DISTRIBUTIONS

A subset for which $X = 1$ would consist of one man and three women, so we are counting ordered pairs of the form

(One man, Three women)

There are five men and four women in all, so there are

$$C_1^5 \cdot C_3^4 = 5 \cdot 4 = 20$$

such ordered pairs (corresponding to committees with one man and three women) and therefore,

$$p(1) = \frac{20}{126} = .16 \text{ (approximately)}$$

Similarly,

$$p(2) = \frac{C_2^5 \cdot C_2^4}{126} = \frac{10 \cdot 6}{126} = \frac{60}{126} = .47 \text{ (approximately)}$$

$$p(3) = \frac{C_3^5 \cdot C_1^4}{126} = \frac{10 \cdot 4}{126} = \frac{40}{126} = .32 \text{ (approximately)}$$

Finally, there are $C_4^5 = 5$ ways to choose a committee of four men, so

$$p(4) = \frac{5}{126} = .04 \text{ (approximately)}$$

The probability distribution is presented in Table 4-17 and the histogram is shown in Figure 4-18. The answer to Problem 2 is $p(3) = .32$. ∎

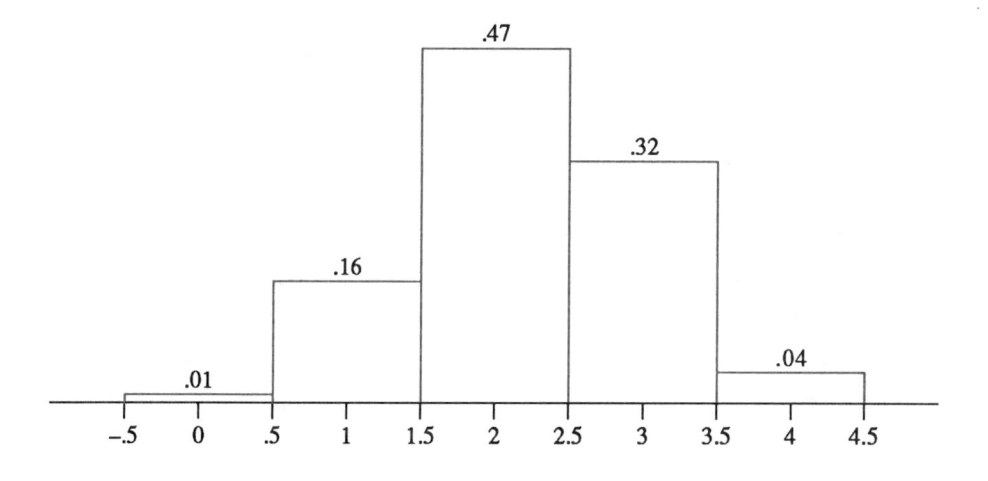

Table 4-17

k	0	1	2	3	4
p(k)	.01	.16	.47	.32	.04

Figure 4-18

Summary

A **random variable** X is a rule that assigns a number to each outcome of a sample space. Given a sample space U and a random variable X defined on the sample space, a **probability distribution** gives the probability $p(k)$ for each possible value k of X, where $p(k)$ is the probability that an outcome chosen at random from U will be one for which the value of X is k. Thus, we can also write $p(k)$ as $p(X = k)$. If the random variable X takes whole-number values k, then a **histogram** of the probability distribution can be constructed by erecting, for each possible value k, a rectangle of height $p(k)$ over an interval of length one that is centered at k.

Exercises

In Exercises 1 through 6, do not attempt to solve the stated probability problem. Instead, answer Questions (a) and (b) in each case; that is, describe the random variable X in words and calculate X for the given outcome.

1. Five cards are drawn at random from a standard deck. What is the probability of drawing fewer than two aces?
    (a) What is the random variable X?
    (b) What is $X(2\spadesuit, A\diamondsuit, 9\diamondsuit, 3\clubsuit, A\clubsuit)$, where $2\spadesuit$ means the 2 of spades, $A\diamondsuit$ means the ace of diamonds, and so on?

2. There are two tollbooths at the approach to a bridge, one on the left and the other on the right. It is estimated that the probability is $\frac{2}{5}$ that a car will pay the toll at the left booth and $\frac{3}{5}$ that it will pay at the right. Of the next seven cars on the bridge, what is the probability that exactly six will pay at the right-hand tollbooth?
    (a) What is the random variable X?
    (b) If L means that a car pays at the left-hand booth and R that it pays at the right, what is $X(R, R, L, L, R, L, R)$?

3. A city health-department inspector selects 5 packages of hamburger at random out of 20 in a supermarket meat counter to test for fat content. If 4 of the 20 packages have fat content that exceeds the legal limit, what is the probability that the inspector will select only legal packages?
    (a) What is the random variable X?
    (b) If L represents a package with legal fat content and F a package with illegal fat content, what is $X(L, L, L, L, L)$?

4. Each time a sharpshooter fires, the probability that she will hit a target is $\frac{9}{10}$. What is the probability that she will hit the target in at least four out of five shots?
    (a) What is the random variable X?
    (b) What is $X(\text{miss, miss, miss, hit, miss})$?

5. An inspector chooses ten units at random from an assembly line and tests them. If there is a probability of .08 that a unit will fail the test, what is the probability that at most two units will fail the test?
   (a) What is the random variable X?
   (b) If a unit that passes the test is represented by P and one that fails by F, what is $X(P, P, P, P, F, F, P, P, P, P)$?

6. Two evenly matched teams are in a basketball playoff. That is, we assume the probability is $\frac{1}{2}$ that either team will win each game. The winner of the playoff is the first team to win four games. What is the probability that the playoff will last at least five games?
   (a) What is the random variable X?
   (b) If the teams are called C and L and the results of the games are L, L, C, L, C, L, what is the corresponding value of X?

In Exercises 7 through 10, find the probability distribution $p(k)$ for the random variable and draw the histogram.

7. In Central City, 30% of the families own no car, 50% have one car, 15% have two cars, and 5% have three cars. The random variable X is the number of cars owned by a family.

8. Three coins are flipped, and the random variable X is the number of heads.

9. Two dice are rolled, and the random variable is the sum of the numbers on the dice.

10. In a marketing survey, a family is asked to test three brands of frozen fruit juice—A, B, and C. They rate each brand as acceptable or not acceptable. The probability is $\frac{1}{2}$ that a family will find brand A acceptable, $\frac{1}{3}$ that brand B will be acceptable, and $\frac{1}{4}$ that brand C will be acceptable. The decisions on the three brands are assumed to be independent. The random variable X is the number of brands a family rates as acceptable.

4.4 Expected Value and Variance

Roulette versus Keno

A gambler who plays at a casino that offers both American roulette and a game called "keno" wants to decide which game to play. If he understands probability, he knows that he can expect to lose money at both games. But if he views his losses as the cost of the entertainment provided by playing the game, he still might wonder which game, roulette or keno, is the cheaper form of entertainment.

Suppose that when he plays roulette, the gambler always bets $1 that the winning number will be odd (one of the numbers 1, 3, 5, and so on, to 35). If the winning number is not one of these (when it is 0, 00, 2, 4, through 36), then he loses $1. But if the number chosen is odd, he makes $1 on the $1 bet. Consequently, out of the 38 possible outcomes in roulette, 18 outcomes will cause the gambler to win $1 and the remaining 20 outcomes will cause him to lose $1.

In the game of keno, the gambler selects a number from 1 through 80. Twenty numbers from 1 through 80 are chosen at random. If the gambler bets $1 and his number is one of the 20 selected, he wins $2. If his number is not selected, he loses his $1 bet.

In order to compare the games, the gambler can think of each play of a game as a single unit of entertainment and ask how much a single play of roulette costs compared to a single game of keno. Of course, the outcome of a single game of roulette is either a win of $1 or a loss of $1, whereas the outcome from keno is either a win of $2 or a loss of $1. But what the gambler really wants to know is the cost per game over the long run.

Expected Value of a Game

The measure of the cost of a game that we will use is called the *expected value* of the game. To calculate the expected value of any gambling game, we divide up the possible outcomes according to the amount of the payoff, grouping together all outcomes with the same payoff. The payoff will be negative if the outcomes result in a loss. We determine the probabilities that the result will be an outcome in each group. The expected value of the game is calculated by multiplying the value of any outcome in a group by the probability that the result of the game will come from that group and adding up the amounts calculated for all such groups.

For example, the gambler's strategy in roulette was to bet $1 on the odd numbers. We have seen that if one of these numbers is selected, then the gambler wins $1. Since there are 18 winning numbers, the probability of winning $1 is $\frac{18}{38} = .47$ (approximately). The 20 remaining numbers produce a loss of $1 if they are selected. A $1 loss is expressed as a payoff of –$1. The probability is $\frac{20}{38} = .53$ that this payoff will occur. By the rule we just stated for calculating expected value,

$$\begin{aligned}\text{Expected value} &= (\$1)(\text{Probability of a \$1 payoff}) \\ &\quad + (-\$1)(\text{Probability of a }-\$1\text{ payoff}) \\ &= (\$1)(.47) + (-\$1)(.53) \\ &= -\$0.06\end{aligned}$$

In the keno game, with 20 winning numbers out of 80, the probability of a $2 payoff is the probability $\frac{20}{80} = .25$ of winning. The probability of a loss is $\frac{60}{80} = .75$, and the amount lost is $1; again, this payoff is –$1. Therefore,

$$\text{Expected value} = (\$2)(.25) + (-\$1)(.75) = \$0.50 - \$0.75 = -\$0.25$$

Thus, we see that keno at $1 a game is more expensive than the proposed roulette strategy. The gambler who wants as many plays as possible for his money would be better off betting on roulette than on keno.

Example 1. A player bets $4, rolls a single die, and gets back $1.10 times the number on the die. What is the expected value of this game to the player?

Solution

The player who rolls a 1 on the die pays in $4 and receives only (1)($1.10) = $1.10; the payoff for this outcome is

$$\$1.10 - \$4 = -\$2.90$$

The probability of rolling a 1 is $\frac{1}{6}$. For a roll of 2, the payoff is

$$2(\$1.10) - \$4 = \$2.20 - \$4 = -\$1.80$$

and again the probability is $\frac{1}{6}$. For the remaining outcomes, the payoffs are

$$\$3.30 - \$4 = -\$0.70 \qquad \$4.40 - \$4 = \$0.40$$
$$\$5.50 - \$4 = \$1.50 \qquad \$6.60 - \$4 = \$2.60$$

The probability is $\frac{1}{6}$ for each outcome, so

$$\text{Expected value} = (-\$2.90)\left(\frac{1}{6}\right) + (-\$1.80)\left(\frac{1}{6}\right) + (-\$0.70)\left(\frac{1}{6}\right) + (\$0.40)\left(\frac{1}{6}\right)$$
$$+ (\$1.50)\left(\frac{1}{6}\right) + (\$2.60)\left(\frac{1}{6}\right) = (-\$0.90)\left(\frac{1}{6}\right) = -\$0.15 \qquad ∎$$

The General Definition

The concept of expected value arises in many circumstances that have nothing to do with gambling; so, we have to define it in a more abstract way. To see how this can be done, notice that there is a random variable X, the payoff, associated with any gambling problem. For instance, in the roulette problem that began this section, an outcome of 17, an odd number, yields $1; so, $X(17) = 1$. If the outcome is 16, the gambler loses $1; so, $X(16) = -1$. The random variable in this problem can take on two different values, 1 and -1. We have seen that the corresponding probability distribution is given in Table 4-19.

There are two symbols commonly used for expected value. If the random variable is X, then $E(X)$ is often used. However, the more compact symbol μ (Greek "mu") is convenient if it will appear in a formula. For the roulette problem, we saw that

$$\mu = E(X) = 1 \cdot p(1) + (-1) \cdot p(-1)$$

Now suppose that we have a sample space U and a random variable X defined on the outcomes of U. The random variable X can take on several different values that we write as k_1, k_2, and so on, to k_n. (For the random variable in the roulette problem, the possible values were $k_1 = 1$ and $k_2 = -1$.) We write the general probability distribution as in Table 4-20.

k	1	-1
$p(k)$	.47	.53

Table 4-19

k	k_1	k_2	...	k_n
$p(k)$	$p(k_1)$	$p(k_2)$	...	$p(k_n)$

Table 4-20

324 MORE PROBABILITY

> The **expected value** $\mu = E(X)$ of the random variable X is
>
> $$\mu = E(X) = k_1 p(k_1) + k_2 p(k_2) + \ldots + k_n p(k_n)$$

Expected value is not limited to gambling, as the next example illustrates.

Example 2. A manufacturer's representative calls on four retail stores one day in order to sell the manufacturer's products. The representative estimates the probability that none of the stores will place an order to be .05, that exactly one store will place an order to be .25, that exactly two will place orders to be .35, that three will order to be .20, and that all four stores will order to be .15. What is the expected value of the number of orders the representative will receive that day?

Solution

The random variable X represents the number of stores that place orders, so X can take on the values 0, 1, 2, 3, and 4. We are told in the statement of the problem that the probability distribution for the random variable X, the number of stores placing an order, is given in Table 4-21. Therefore, the expected value is

$$\begin{aligned} E(X) &= 0 \cdot p(0) + 1 \cdot p(1) + 2 \cdot p(2) + 3 \cdot p(3) + 4 \cdot p(4) \\ &= 0(.05) + 1(.25) + 2(.35) + 3(.20) + 4(.15) \\ &= 2.15 \text{ (orders)} \end{aligned}$$ ■

Table 4-21

k	0	1	2	3	4
$p(k)$	.05	.25	.35	.20	.15

Picturing Expected Value

The expected value of −$0.15 for the betting game of Example 1 can be interpreted to mean that if the player plays the game many times, the estimated cost (loss) per game is 15 cents. The 15-cent loss represents the estimated average over many plays. Similarly, the expected value of 2.15 orders in Example 2 tells us that if the probability distribution of orders stays the same over a period of time, the estimate of the average number of orders received per day is 2.15.

Another way to understand expected value is in relation to the histogram of the probability distribution. For instance, suppose that we consider the probability distribution of the keno game. Thus,

$$p(-1) = .75 \quad \text{and} \quad p(2) = .25$$

so that k ranges over the whole numbers from −1 to 2, as shown in Table 4-22. Then we can draw the corresponding histogram as Figure 4-23. The expected value $\mu = -.25$ is the "point of balance" for the histogram. That is, if we built a

4.4 EXPECTED VALUE AND VARIANCE

Table 4-22

k	−1	0	1	2
p(k)	.25	0	0	.75

model of the histogram out of wood or some such material, we could balance the model exactly at the point −.25 because the smaller rectangle representing $p(2)$ is just far enough away from −.25 to balance the larger, but closer, rectangle representing $p(-1)$.

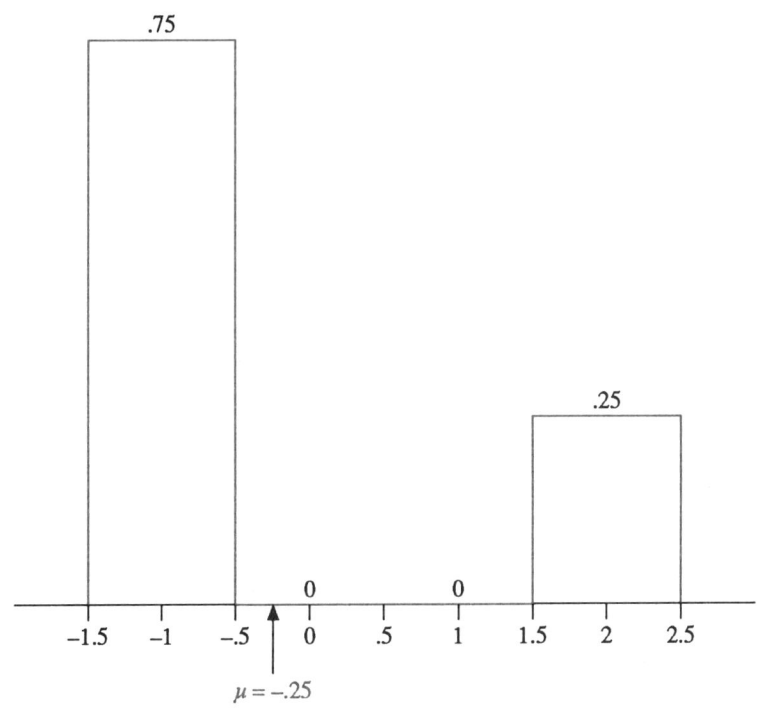

Figure 4-23

Variance

The expected value of the random variable whose probability distribution is given in Table 4-24 is

$$\mu = E(X) = (-1) \cdot p(1) + 0 \cdot p(0) + 1 \cdot p(1)$$
$$= (-1)(.375) + (0)(.50) + (1)(.125)$$
$$= -.375 + .125$$
$$= -.25$$

just as in the keno game of Figure 4-23; its histogram, Figure 4-25, balances at the point $\mu = -.25$. If you compare the histograms of the two figures, you will

Table 4-24

k	−1	0	1
p(k)	.375	.50	.125

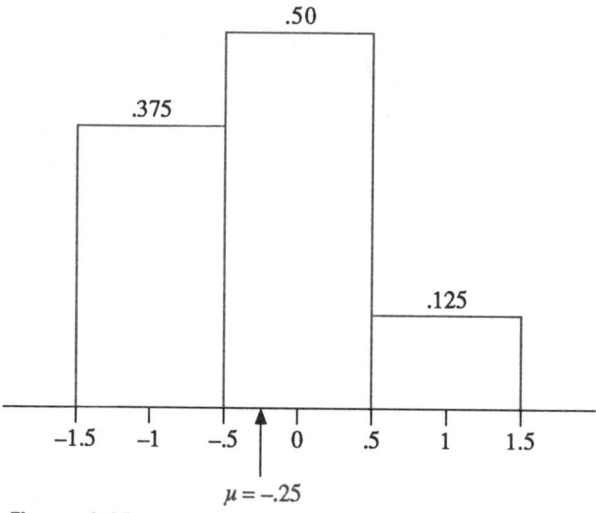

Figure 4-25

notice that although they both balance at the same point, Figure 4-23 is more spread out than Figure 4-25. If the latter figure represents a gambling game with payoffs of "lose \$1" ($k = -1$), "no payment" ($k = 0$), and "win \$1" ($k = 1$), there is less variation in the outcomes than there is in the keno game. We do not always need all the information about a random variable contained in a probability distribution. But in general, the expected value by itself is not enough. In addition, we need to describe how the histogram spreads out around its balance point.

A number commonly used to measure the variability of the probability distribution is the *variance* of a random variable X. It is represented by the symbol Var(X) or, more compactly, by σ^2 (Greek "sigma," squared).

For the general probability distribution given by Table 4-26, with expected value $\mu = E(X)$, the **variance** is defined by

$$\sigma^2 = \text{Var}(X) = (k_1 - \mu)^2 \cdot p(k_1) + (k_2 - \mu)^2 \cdot p(k_2) + \ldots + (k_n - \mu)^2 \cdot p(k_n)$$

---

k	k_1	k_2	$\ldots$	k_n
$p(k)$	$p(k_1)$	$p(k_2)$	$\ldots$	$p(k_n)$

Table 4-26

The variance measures how the possible values k of the random variable X disperses from the expected value $\mu = E(X)$. The expressions are squared; thus, the differences are always positive. Notice that the definition resembles the definition of the expected value. Like the expected value, the variance is a sort of average: here, the average of the squares of the amounts that the random variable differs from μ.

Computing the Variance

The definition shows us how the variance measures the spread of the probability distribution about the expected value, but this is not the most convenient way to write it. To find a simpler formula, we will look once more at the probability distribution of the keno problem: $p(-1) = .75$ and $p(2) = .25$. We calculated that the expected value is

$$\mu = (-1) \cdot p(-1) + 2 \cdot p(2) = (-1)(.75) + (2)(.25) = -.25$$

We will write μ, rather than $-.25$, in the formula for σ^2 because then it will be easier to find the general formula. From the definition of the variance,

$$\sigma^2 = ((-1) - \mu)^2 \cdot p(-1) + (2 - \mu)^2 \cdot p(2) = ((-1) - \mu)^2(.75) + (2 - \mu)^2(.25)$$

By elementary algebra, we have the computation

$$(k - \mu)^2 = (k - \mu)(k - \mu) = k^2 - 2k\mu + \mu^2$$

In particular,

$$((-1) - \mu)^2 = (-1)^2 - 2 \cdot (-1) \cdot \mu + \mu^2$$

and

$$(2 - \mu)^2 = 2^2 - 2 \cdot 2 \cdot \mu + \mu^2$$

When we substitute and then multiply out the expressions, we have

$$\sigma^2 = [(-1)^2 - 2 \cdot (-1) \cdot \mu + \mu^2](.75) + [2^2 - 2 \cdot 2 \cdot \mu + \mu^2](.25)$$
$$= (-1)^2(.75) - 2(-1)\mu(.75) + \mu^2(.75) + (2)^2(.25) - 2(2)\mu(.25) + \mu^2(.25)$$

Then we collect the six summands into three groups. The first group consists only of numbers, the second group consists of expressions containing μ, and the last group consists of expressions with μ^2.

$$\sigma^2 = [(-1)^2(.75) + (2)^2(.25)] + [(-2)(-1)\mu(.75) + (-2)(2)\mu(.25)]$$
$$+ [\mu^2(.75) + \mu^2(.25)]$$

When we look at the first group,

$$(-1)^2(.75) + (2)^2(.25)$$

we notice that it looks just like the formula for $\mu = E(X)$ used previously, except that we squared the "k" numbers, -1 and 2. Thus, we will call this expression $E(X^2)$. The formula for μ is actually contained in the next group, as we find by factoring the common term -2μ and recalling that $\mu = -.25$:

$$(-2)(-1)\mu(.75) + (-2)(2)\mu(.25) = -2\mu[(-1)(.75) + (2)(.25)]$$
$$= -2\mu(-.25)$$
$$= -2\mu(\mu)$$
$$= -2\mu^2$$

When we factor the common term out of the last group, this is the result:

$$\mu^2(.75) + \mu^2(.25) = \mu^2[(.75) + (.25)] = \mu^2 \cdot 1 = \mu^2$$

328   MORE PROBABILITY

Now when we substitute these expressions into the three groups, we have the formula
$$\sigma^2 = E(X^2) - 2\mu^2 + \mu^2 = E(X^2) - \mu^2$$

The same algebraic simplification works for the general probability distribution given in Table 4-27, showing that in general,

$$\boxed{\sigma^2 = E(X^2) - \mu^2}$$

This means that we should compute both
$$\mu = k_1 p(k_1) + k_2 p(k_2) + \ldots + k_n p(k_n)$$
and
$$E(X^2) = k_1^2 p(k_1) + k_2^2 p(k_2) + \ldots + k_n^2 p(k_n)$$

and then substitute them into this general formula to calculate σ^2.

For the keno example, in which $p(-1) = .75$ and $p(2) = .25$, we have already seen that $\mu = -.25$. We compute
$$E(X^2) = (-1)^2(.75) + 2^2(.25) = 1(.75) + 4(.25) = 1.75$$
and thus,
$$\sigma^2 = E(X^2) - \mu^2 = 1.75 - (-.25)^2 = 1.75 - .0625 = 1.6875$$

When we compare this to the distribution of Table 4-28 (see Table 4-24), which also had expected value $-.25$, we find that
$$E(X^2) = (-1)^2(.375) + 0^2(.50) + 1^2(.125) = .375 + .125 = .50$$

Thus, the variance of this probability distribution is
$$\sigma^2 = E(X^2) - \mu^2 = .50 - (-.25)^2 = .50 - .0625 = .4375$$

which is much smaller, just as Figures 4-23 and 4-25 suggested.

k	k_1	k_2	...	k_n
$p(k)$	$p(k_1)$	$p(k_2)$	...	$p(k_n)$

Table 4-27

k	-1	0	1
$p(k)$	.375	.50	.125

Table 4-28

Summary

For a sample space U and a random variable X defined on U with probability distribution given in Table 4-29, the **expected value of X** is
$$\mu = E(X) = k_1 p(k_1) + k_2 p(k_2) + \ldots + k_n p(k_n)$$
and the **variance** is

4.4 EXPECTED VALUE AND VARIANCE

$$\sigma^2 = \text{Var}(X) = (k_1 - \mu)^2 \cdot p(k_1) + (k_2 - \mu)^2 \cdot p(k_2) + \ldots + (k_n - \mu)^2 \cdot p(k_n)$$

which can also be computed by the formula

$$\sigma^2 = E(X^2) - \mu^2$$

where

$$E(X^2) = k_1^2 p(k_1) + k_2^2 p(k_2) + \ldots + k_n^2 p(k_n)$$

k	k_1	k_2	$\ldots$	k_n
$p(k)$	$p(k_1)$	$p(k_2)$	$\ldots$	$p(k_n)$

Table 4-29

Exercises

1. The probability function $p(k)$ for a random variable X is given by $p(0) = .10$, $p(1) = .15, p(2) = .20, p(3) = .25, p(4) = .20$, and $p(5) = .10$. Calculate μ and σ^2.

2. Two coins are flipped. The random variable X represents the number of heads. Calculate μ and σ^2.

3. If eight items are chosen at random from an assembly line, the approximate probability that exactly k of them are defective is given in Table 4-30. If X represents the number of defective items in the sample, find μ and σ^2.

k	0	1	2	3	4	5	6	7	8
$p(k)$	.60	.32	.07	.01	.00	.00	.00	.00	.00

Table 4-30

4. The European roulette wheel consists of the numbers 0, 1, 2, and so on, to 36. A gambler wins $1 for his bet of $1 on the odd numbers 1, 3, 5, and so on, to 35 and loses $1 if the number that comes up is one of 0, 2, 4, and so on, to 36. What is the expected value of this game to the gambler? (Remember that European roulette uses only 0; it does not use 00.)

5. In a chess tournament, a player receives 1 point for a win, $\frac{1}{2}$ point for a draw, and 0 for a loss. Based on her past performance, a player estimates there is no chance she will earn fewer than 3, out of 5 possible points in an upcoming tournament. The probability that she will receive each possible score k between 3 and 5 points is given in Table 4-31. What is the expected value of the number of points she will earn in the tournament?

k	3	$3\frac{1}{2}$	4	$4\frac{1}{2}$	5
$p(k)$	.15	.20	.30	.25	.10

Table 4-31

6. Gamblers choose numbers from 000 to 999 for a lottery and bet $1 per number. The lottery selects one such number at random and pays $450 to the winner, including the winner's own $1. Calculate the expected value of the gambler's gain (or loss) in the lottery and the variance of the probability distribution.

7. A customer in an automobile showroom is trying to decide whether to buy a car, and, if so, whether to buy the simplest model of sedan or else a station wagon with many optional extras. If the customer chooses the sedan, the dealer's profit is $400, but for the wagon the dealer's profit is $800. Of course, if the customer fails to buy at all, the dealer has 0 profit. The dealer estimates the probability the customer will buy the sedan to be .40 and the probability the customer will buy the wagon to be .10. Calculate the expected value of the dealer's profit and the variance of the probability distribution.

8. Suppose the probability that a litter of a species of animal will consist of just one offspring is .08, the probability of two offspring is .27, of three is .39, of four is .22, and of five is .04. Calculate the expected value of litter size for this species and the variance of the probability distribution.

9. In "6-spot keno," a gambler pays $1 and selects six numbers from 1 through 80. Twenty numbers from 1 through 80 are chosen at random as the winning numbers. If he has selected two or fewer of the winning numbers, he loses the $1 bet. If the gambler has chosen three winning numbers, he gets back his $1, so that the payoff is 0. If the gambler gets four winners, he makes $2 (he is paid $3, but that includes the $1 he paid in). If the gambler has five winners, he makes $89, and if he has all six winning numbers, he makes $1799. The approximate probability of choosing exactly three winners is .1298, for four winners it is .0285, for five winners, .0031, and for six winners, .0001. The probability of choosing two or fewer winners, and thus losing the $1, is therefore .8385. What is the expected value of "6-spot keno" to the gambler, calculated to the nearest cent?

10. An insurance company charges a woman a premium of $150 for an accident-insurance policy for the coming year. If the policy holder has a major accident during the year, the company will pay her $5000, and for a minor accident it will pay $1000. The company does not return the $150 premium. Assume the policy holder is such a careful driver that the probability is essentially 0 that she will have more than one accident in a year. If the probability is .08 that she will have one minor accident, .005 that she will have one major one, and thus .915 that she will have no accident at all, what is the expected value of the insurance policy to the company?

11. A gambler chooses a number between 000 and 999 for a lottery and bets $1. The lottery selects one such number at random. If the gambler has chosen the winning number, she receives $250, which includes her own dollar. If the first two digits of the gambler's number are the same as the winning number (for example, if the winning number is 134 and she holds 137), then she receives $25, again including her $1. What is the expected value of the lottery to the gambler?

12. A company is deciding whether to bring out a new product. If the product is moderately successful, the dividend per share of stock in the company will increase by $0.50 the next year. If the product is very successful, the dividend per share

will go up by $1. On the other hand, if the product is somewhat unsuccessful, the dividend per share the next year will decline by $0.75, and if the product is a complete failure, the dividend will go down $1.50 a share. The company estimates the probability that the product will be moderately successful to be .50 and the probability that it will be very successful to be .10. The estimated probability is .20 that the product will be somewhat unsuccessful and .20 that it will be a complete failure. What is the expected value of the change in the dividend per share of the company's stock the next year if it brings out the new product?

4.5 Binomial Experiments

Suppose that a baseball player has a batting average of .300. If he has four (official) chances to bat in a single game, what is the probability that he will get exactly one hit? (We will make the simplifying assumption that the player has the same probability, $.3 = \frac{3}{10}$, of getting a hit each time he comes to the plate.) We think of the baseball player's four times at bat as an experiment consisting of four parts. Each part of the experiment occurs under identical conditions, and there is the same probability, $\frac{3}{10}$, of success for each part. Thus, the overall experiment consists of a single experiment repeated four times. If getting a hit is considered to be a success in a time at bat and making an out is a failure, then we are asking for the probability $p(1)$ of exactly one success out of four tries.

One way in which the overall experiment can succeed—that is, the player gets one hit out of four official times at bat—is for the player to get a hit his first time at bat and then fail to get a hit the other times. The symbol S will represent success in a trial and F, failure. Then this case is described by (S, F, F, F). Since the hit can occur in any of the four times at bat, we see that each of the following distinct patterns of success and failure in four tries also satisfies the requirement for one hit in four times at bat: $(F, S, F, F), (F, F, S, F), (F, F, F, S)$. Therefore,

$$p(1) = p(S, F, F, F) + p(F, S, F, F) + p(F, F, S, F) + p(F, F, F, S)$$

What is the probability that the pattern will be the (S, F, F, F) case? We assume that each time at bat is independent of the others. According to Section 3.9,

$$p(S, F, F, F) = p(S) \cdot p(F) \cdot p(F) \cdot p(F)$$

The batting average tells us that $p(S)$, the probability of getting a hit, is $\frac{3}{10} = .3$. The probability $p(F)$ of failing to get a hit is then .7 and we conclude that

$$p(S, F, F, F) = (.3)(.7)(.7)(.7) = (.3)(.7)^3$$

By the same reasoning,

$$p(F, S, F, F) = p(F) \cdot p(S) \cdot p(F) \cdot p(F) = (.7)(.3)(.7)(.7) = (.3)(.7)^3$$

and similarly,

$$p(F, F, S, F) = p(F, F, F, S) = (.3)(.7)^3$$

Since the probability of each of the four patterns is the same,

$$p(1) = 4(.30)(.70)^3 = .4116$$

More generally, suppose that the baseball player has a probability p of getting a hit each time at bat. In our notation, this means that $p(S) = p$, and therefore, $p(F) = 1 - p$. Since the preceding argument didn't depend on the number $\frac{3}{10}$, we may repeat it replacing .3 by p and .7 by $1 - p$ to conclude that, for any probability p,

$$p(1) = 4p(1-p)^3$$

Example 1. A die is rolled three times. What is the probability that 4 will come up exactly twice?

Solution

We consider success to be rolling a 4; we wish to compute $p(2)$, where $p = p(S)$ is the probability of rolling a 4, namely $\frac{1}{6}$. There are three patterns of two successes and one failure, which are (S, S, F), (S, F, S), and (F, S, S). For example, we have

$$p(S, S, F) = p(S) \cdot p(S) \cdot p(F) = \frac{1}{6} \cdot \frac{1}{6} \cdot \frac{5}{6}$$

Clearly, we obtain the same fractions from each pattern, it follows that

$$p(2) = 3 \cdot \left(\frac{1}{6}\right)^2 \cdot \frac{5}{6} = \frac{15}{216} = .07 \text{ (approximately)}$$ ∎

The Probability of k Successes

A sequence of n independent, identical repetitions of a trial that has two possible outcomes (success or failure), and the same probability p of success each time, is called a **binomial experiment**. Now we are ready to find a formula for $p(k)$, which denotes the probability of exactly k successes. The probability of any particular pattern of k successes out of n trials is the probability p of success multiplied by itself k times, times the probability $1 - p$ of failure multiplied by itself the number of times the experiment fails. There are n trials in all and k of them are successes, so the remaining $n - k$ fail. In symbols, the probability of any particular pattern of k successes in n trials is therefore $p^k \cdot (1 - p)^{n-k}$. Since the probability $p(k)$ is just the sum of the probabilities of all possible distinct patterns, and the probability of each pattern is $p^k \cdot (1 - p)^{n-k}$, then

$$p(k) = (\text{number of patterns}) \cdot p^k \cdot (1 - p)^{n-k}$$

The Number of Patterns

To use the formula for calculating $p(k)$, we still have to determine the number of possible patterns of k successes in n trials. Think of the n repetitions of the action as a deck consisting of n cards with a different number, 1, 2, and so on, to n, printed on each card. If we are dealt a "card hand" containing k cards, then that is the same thing as choosing k numbers from 1 to n. The k numbers on the cards in the hand will be the turns in which success occurs. For example, if we are dealt the card numbered 3, then the third time the act takes place, it will be a success. As in poker, it does not matter whether our "card hand" received the 3 as the first card dealt, the second card, or whatever. The number of k card hands from a deck of n cards is

$$C_k^n = \frac{n!}{k!(n-k)!}$$

so, there are C_k^n different patterns of k successes in n trials.

The Formula

We have now solved the problem of computing $p(k)$.

> The probability of k successes in a binomial experiment with n trials in which each trial has probability p of success is
>
> $$p(k) = C_k^n \cdot p^k \cdot (1-p)^{n-k} = \frac{n!}{k!(n-k)!} \cdot p^k \cdot (1-p)^{n-k}$$

Example 2. If a coin is weighted so that the probability of it coming up heads is $\frac{2}{3}$, what is the probability that it will come up heads exactly twice out of four tosses?

Solution

There are $n = 4$ tosses of the weighted coin and we wish to know the probability of $k = 2$ successes, where the probability p of success (heads) on each toss is $\frac{2}{3}$. By the preceding formula,

$$p(2) = C_2^4 \left(\frac{2}{3}\right)^2 \left(\frac{1}{3}\right)^2$$

where

$$C_2^4 = \frac{4!}{2!2!} = 6$$

Thus,

$$p(2) = 6\left(\frac{2}{3}\right)^2 \left(\frac{1}{3}\right)^2 = \frac{24}{81} = .30 \text{ (approximately)} \blacksquare$$

Example 3. The probability that any unit chosen at random from a production line will pass inspection is .90. At five randomly chosen times during the day, a unit is selected at random from the production line, inspected, and returned to the line. What is the probability that at least four out of the five times the unit will pass inspection?

Solution

If exactly four of the times the inspection takes place the unit passes (and the other time it fails), the experiment will satisfy the stated conditions. In addition, the inspection may be passed on all five occasions. The probability of at least four passes out of five is the sum of the probability of exactly four passes out of five, $p(4)$, plus the probability, $p(5)$, of five passed inspections. Applying the formula with $p = .9$ gives

$$p(4) = C_4^5(.9)^4(.1)^1 = 5(.9)^4(.1) = .33 \text{ (approximately)}$$

and

$$p(5) = C_5^5(.9)^5(.1)^0 = .59 \text{ (approximately)}$$

Therefore,

$$p(\text{at least four passes}) = p(4) + p(5) = .33 + .59 = .92 \text{ (approximately)} \quad \blacksquare$$

We solved Example 3 just by using the formula for $p(k)$. However, we can see the solution more clearly if we draw the histogram for the corresponding probability distribution. From the formulas for $p(k)$ when $n = 5$ and $p = .9$, that is, from

$$p(0) = C_0^5(.9)^0(.1)^5 = .00001, \text{ or approximately, } .00$$

and so on (including the computations of $p(4)$ and $p(5)$ we made in the example), the probability distribution is given in Table 4-32. This is shown in the histogram of Figure 4-33. The shaded area of the figure represents geometrically what we calculated in Example 3, the probability that at least 4 units will pass inspection.

Table 4-32

k	0	1	2	3	4	5
$p(k)$	.00	.00	.01	.07	.33	.59

Example 4. A die is rolled five times. Calculate the probability that a number smaller than 3 appears at least once.

Solution

We have $n = 5$ repeated trials. The numbers on a die smaller than 3 are 1 and 2, so the probability of success is

$$p = \frac{2}{6} = \frac{1}{3}$$

Since $k = 1$ success out of five satisfies the condition of "at least one" success, as do $k = 2, 3, 4,$ and 5, we could compute

$$p(\text{number less than 3 at least once}) = p(1) + p(2) + p(3) + p(4) + p(5)$$

However, there is a much easier solution to the problem. Out of five trials, there certainly must be between zero and five successes. Thus,

$$p(\text{something happening}) = p(0) + p(1) + p(2) + p(3) + p(4) + p(5) = 1$$

that is,

$$p(0) + p(\text{number less than 3 at least once}) = 1$$

4.5 BINOMIAL EXPERIMENTS

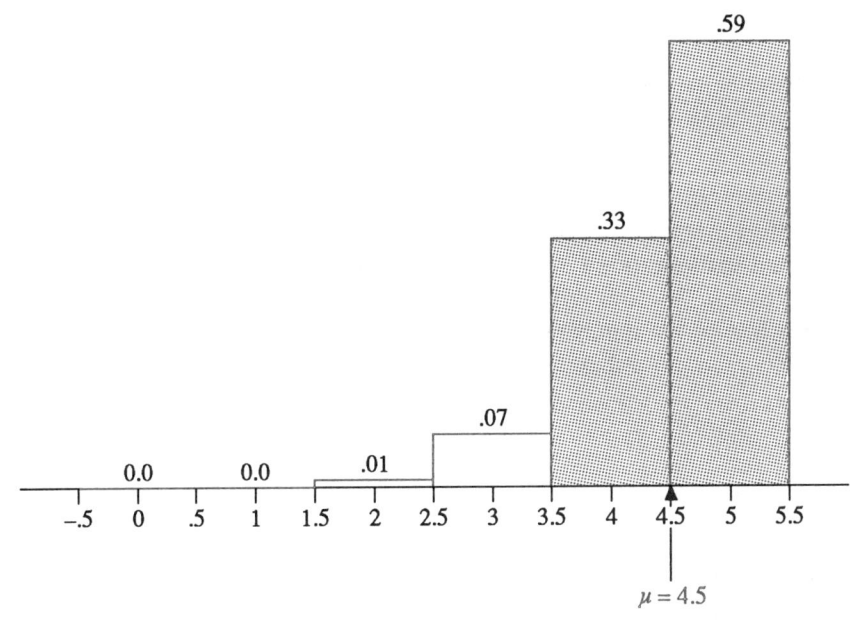

Figure 4-33

Therefore,
$$p(\text{number less than 3 at least once}) = 1 - p(0)$$
The area corresponding to $p(0)$ is indicated in Figure 4-34. We calculate

$$p(0) = C_0^5 \left(\frac{1}{3}\right)^0 \left(\frac{2}{3}\right)^5 = 1 \cdot \left(\frac{2}{3}\right)^5 = \frac{32}{243} = .13 \text{ (approximately)}$$

Consequently,

$$p(\text{number less than 3 at least once}) = 1 - .13 = .87 \text{ (approximately)} \quad \blacksquare$$

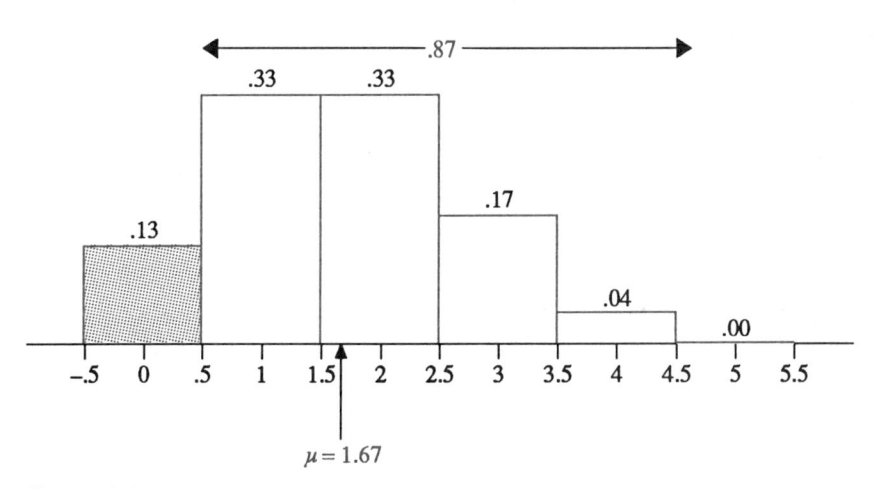

Figure 4-34

The Binomial Distribution

For a binomial experiment with n trials and probability p of success on each trial, the probability distribution with $k = 0, 1, \ldots, n$ and with $p(k)$ computed by the formula

$$p(k) = C_k^n \cdot p^k \cdot (1-p)^{n-k}$$

is called the **binomial probability distribution**. We will need to find the expected value μ and variance σ^2 of this probability distribution, as we did for probability distributions in Section 4.4. Since the calculation of the probabilities $p(k)$ depended just on the numbers n and p, it will not surprise us that μ and σ^2 also depend on n and p. Furthermore, the relationship is given by simple formulas, as we will see.

Expected Value

We will compute the expected value μ for the case of $n = 2$. This special case will illustrate what happens in general. We will not specify p, so we can obtain a formula that is valid for any given probability of success in each trial. From the formulas for $p(k)$ we have

$$p(0) = C_0^2 \cdot p^0 (1-p)^2 = (1-p)^2$$

since $C_0^2 = 1$ and also, $p^0 = 1$. For the binomial probability distribution with $n = 2$ and any p, we also need $p(1)$ and $p(2)$, which are obtained in a similar manner (see Exercise 16). The results are given in Table 4-35. Substituting the formulas for $p(k)$ into the definition, we see that

$$\begin{aligned}
\mu &= E(X) \\
&= 0 \cdot p(0) + 1 \cdot p(1) + 2 \cdot p(2) \\
&= 0 \cdot [(1-p)^2] + 1 \cdot 2p(1-p) + 2 \cdot p^2 \\
&= 0 + 2p(1-p) + 2p^2 \\
&= (2p - 2p^2) + 2p^2 \\
&= 2p
\end{aligned}$$

So, in the case of $n = 2$, the expected value of the binomial distribution is $\mu = 2p$, no matter what the probability p is. The calculation of μ for the binomial distribution with any number n of trials is similar to the case of $n = 2$, except that the algebraic manipulation is more complicated. The general result has the same pattern:

$$\mu = np$$

Although we did not do the algebraic work for the general formula, we can see that the result $\mu = np$ agrees not only with the case of $n = 2$, which we did work out, but also with a common-sense interpretation of the expected value of a

Table 4-35

k	0	1	2
$p(k)$	$(1-p)^2$	$2p(1-p)$	p^2

4.5 BINOMIAL EXPERIMENTS

binomial distribution. For instance, if we flip an *honest* coin (where p, the probability of heads, is .50) $n = 10$ times, then our best guess for the number of heads we expect is 5, which is $np = (10)(.50)$. As another example, suppose units in a production line have the probability .90 of passing inspection, as in Example 3, and we chose 100 units at random from the production line. We would expect 90 out of the 100 units to pass inspection. We do not mean that exactly 90 must be satisfactory, but rather, that this is the most sensible estimate, based on the overall quality of the units. Noticing that $90 = 100(.90) = np$, we find that our intuitive understanding of expectation agrees with the formula.

Variance

It is difficult to give an intuitive explanation of variance that can help us interpret its meaning for the binomial distribution. However, we can use the computational formula

$$\sigma^2 = E(X^2) - \mu^2$$

we worked out earlier to calculate the variance when $n = 2$ and when the probability is p. From Table 4-35 we calculate that

$$E(X^2) = 0^2 \cdot (1-p)^2 + 1^2 \cdot 2p \cdot (1-p) + 2^2 \cdot p^2$$
$$= 2p(1-p) + 4p^2$$

We showed that $\mu = 2p$, so that

$$\sigma^2 = E(X^2) - \mu^2$$
$$= [2p(1-p) + 4p^2] - [2p]^2$$
$$= 2p(1-p) + 4p^2 - 4p^2$$
$$= 2p(1-p)$$

Just as we noticed that $\mu = 2p$ is of the form $\mu = np$ with $n = 2$, so too, this computation gives us $\sigma^2 = np(1-p)$ when $n = 2$.

> The general formulas for the expected value and variance of the binomial probability distribution for n trials with probability p of success on each trial are
>
> $$\mu = np \quad \text{and} \quad \sigma^2 = np(1-p)$$

Example 5

(a) (*Compare with Example 3.*) The probability that any unit chosen at random from a production line will pass inspection is .90. At five randomly chosen times during the day, a unit is selected at random from the production line, inspected, and returned to the line. Calculate the expected value of the num-

338   MORE PROBABILITY

ber of units that will pass inspection, and find the variance of this binomial distribution.

(b) *(Compare with Example 4.)* A die is rolled five times. Calculate the expected value of the number of times a number smaller than 3 appears, and find the variance of this binomial distribution.

Solution

(a)
$$\mu = np = 5(.9) = 4.5 \quad \text{(See Figure 4-33.)}$$
$$\sigma^2 = np(1-p) = 5(.9)(.1) = (4.5)(.1) = .45$$

(b) Recalling from Example 4 that $p = \frac{1}{3}$, we have

$$\mu = np = 5 \cdot \frac{1}{3} = 1.67 \text{ (approximately)} \quad \text{(See Figure 4-34.)}$$
$$\sigma^2 = np(1-p) = 5 \cdot \frac{1}{3} \cdot \frac{2}{3} = \frac{10}{9} = 1.11 \text{ (approximately)}$$ ∎

Summary

A sequence of n independent, identical repetitions of a trial that has two possible outcomes (success or failure), and the same probability p of success each time, is called a **binomial experiment**. The probability of exactly k successes in the experiment is

$$p(k) = C_k^n \cdot p^k \cdot (1-p)^{n-k}$$

where

$$C_k^n = \frac{n!}{k!(n-k)!}$$

For a binomial experiment with n trials and probability p of success on each trial, the probability distribution for $k = 0, 1, \ldots, n$ and $p(k)$ computed by the formula is called the **binomial probability distribution**. The expected value and variance of the binomial probability distribution for n trials with probability p of success on each trial can be computed by the formulas $\mu = np$ and $\sigma^2 = np(1-p)$.

Exercises

1. If there are eight binomial experiments and the probability of success on each one is $\frac{1}{2}$, calculate $p(5)$.

2. If there are five binomial experiments and if the probability of success on each one is $\frac{2}{3}$:

(a) Calculate the expected value μ and variance σ^2.
(b) What is the probability that there will be at least four successes out of the five?

In Exercises 3 and 4, suppose that there are three binomial experiments and that the probability of success on each is .1.

3. (a) Calculate $p(0), p(1), p(2)$, and $p(3)$.
   (b) Draw the histogram for this binomial probability distribution.
   (c) Calculate the expected value μ from the definition.
   (d) Use Parts (a) and (c) to calculate the variance σ^2.

4. Use the results of Exercise 3(a) to calculate the probability of at least two successes in three trials.

5. An insurance agent sells a policy at 15% of his appointments. If, one day, the agent has five appointments, what is the probability that he will make exactly two sales?

6. In an election for mayor in which 60% of the voters prefer White and 40% prefer Green, the individuals in a random sample of seven voters are asked their preferences. What is the probability that the sample will predict the outcome of the election correctly, that is, that a majority of the sample voters (at least four) will prefer White?

7. Suppose that 80% of all drivers on an interstate use their seat belts.
   (a) What is the probability that, if there were accidents involving six cars on the interstate, more than four of the drivers would be wearing seat belts?
   (b) Suppose there were accidents involving six cars on the interstate. Calculate the expected value of the number of drivers wearing seat belts.

8. What is the probability that a family with six children will have three boys and three girls, assuming the probability that a child will be a boy is precisely .50?

For Exercises 9 and 10, suppose there is a country in which the probability that an infant will live to the age of 60 is .65.

9. If six infants are chosen at random in that country, what is the probability that at least two of them will reach the age of 60?

10. Calculate the expected value of the number, out of six infants from that country, that will live to the age of 60.

11. A sales representative will call on six customers in one day. The representative estimates the probability of a sale to any particular customer to be $\frac{1}{2}$. What is the probability that the representative will make at least two sales that day?

Exercises 12 and 13 concern a test of extrasensory perception in which five cards, each from a different deck, are placed face down on a table and a psychic is asked the value (6, queen, and so on) of each card. The psychic gets two right out of 5.

12. What is the probability that a person guessing at random will choose incorrect values for all five cards?

13. What is the probability that a person guessing at random would do at least as well as the psychic did?

14. A woman owns six vacation cabins that she rents out by the week. She supplies television sets at an extra cost to renters requesting them. Her experience is that $\frac{1}{2}$ of the time a television is requested. She owns four sets. In a week in which all the cabins are rented, what is the probability that she will have enough sets to supply all who want them?

15. A cardboard-box factory tests five of its boxes, chosen at random, to determine whether they will withstand a certain amount of pressure. If 15% of the boxes from the factory would fail the test, what is the probability that no more than one of the five boxes in the sample will fail the test?

16. Show that for a binomial distribution with $n = 2$ trials and probability p of success on each trial that $p(1) = 2p(1-p)$ and that $p(2) = p^2$.

17. For a binomial distribution with $n = 3$ trials and probability p of success on each trial, perform the following calculations.
    (a) Determine formulas for $p(k)$ in the probability distribution, for $k = 0, 1, 2, 3$.
    (b) Use Part (a) to show that the expected value is $\mu = 3p$.
    (c) Use Parts (a) and (b) to show that $\sigma^2 = 3p(1-p)$.

4.6 The Normal Distribution

A Histogram for Coin Flipping

If we flip an honest coin $n = 10$ times, the probability distribution for the random variable X = "number of heads" is shown in the histogram of Figure 4-36.

Thus, for instance, the probability of exactly 8 heads out of 10 flips is .04, whereas the expected number of heads, that is $\mu = 5$, which is the most likely result, has only a probability of .25. Notice the shape of this histogram: it rises to a

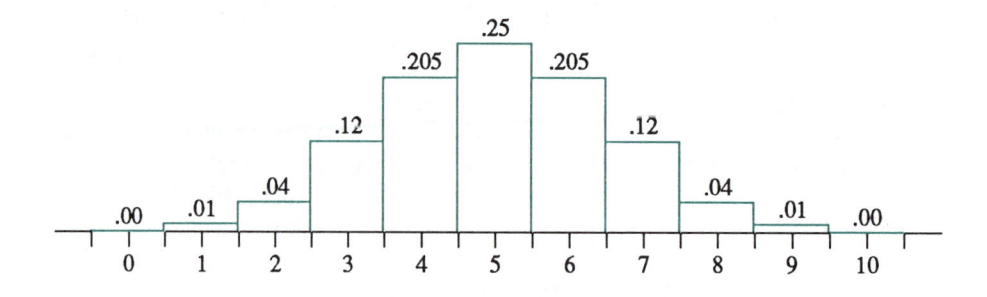

Figure 4-36

4.6 THE NORMAL DISTRIBUTION

peak in the middle and tapers off on both sides. The histogram of a binomial distribution of n trials with probability p of success on each trial generally takes approximately this shape. See, for instance, Figure 4-34 of Section 4.5, where $n = 5$ and $p = \frac{1}{3}$. This is not always the case with binomial distributions, as we saw in Figure 4-33 of Section 4.5; but the shape is characteristic of most binomial distributions, especially when n is large. Its recognition will lead us to a useful tool for working with binomial distributions. For the present, we will focus on the shape itself and on some of its other uses.

A Common Histogram Shape

We can summarize numerical information by means of a histogram construction. For instance, suppose we measure the heights, to the nearest inch, of the 950 boys at Turtle Creek High School. We find that 151 of them are 68 inches tall. Over the interval from 67.5 to 68.5 on the horizontal axis we construct a rectangle of height $\frac{151}{950}$. There are 137 boys 69 inches tall; over the interval from 68.5 to 69.5 we construct a rectangle of height $\frac{137}{950}$. We do a similar construction for 67 inches, and so on, to obtain Figure 4-37. The rectangles still represent probabilities because, for instance, if we choose a boy at random from this school, the probability that his height will be 68 inches is $\frac{151}{950} = .16$ (approximately).

Histograms of data often take the shape of those we observed for binomial distributions: rising to a peak in the middle and tapering off on the sides. The histogram of boys' heights is of this shape because most boys are of about average height, with relatively few either very short or very tall. A histogram of scores on a nationally administered college test, with numerical scores represented by intervals along a horizontal axis and rectangle heights determined by the fraction of students who achieve the score, would have the same shape. Most students would receive approximately average scores, though a few would do very well and a few very poorly.

There is also a mathematical reason why this shape occurs so often. The *Central Limit Theorem* states that histograms based on large collections of num-

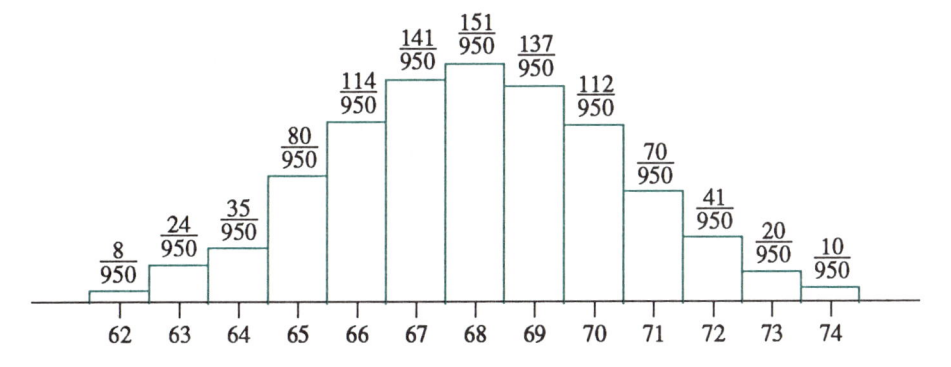

Figure 4-37

bers will have approximately this shape if those numbers represent averages. Many histograms of data describe averages. For instance, a boy's height can be thought of as the average of the many genetic and environmental factors that effect height. Similarly, performance on an examination depends on many things: intelligence, study habits, concentration, and so on.

Continuous Random Variables

Smoothing out a histogram that peaks in the middle and tapers off at the extremes produces a shape like that shown in Figure 4-38. This shape is called a **normal curve** or **normal distribution**. The word "distribution" that we encountered earlier in this chapter is appropriate because this curve can be thought of as the "histogram" of a random variable X. However, unlike the discrete random variables X we studied in previous sections, those in which X that can take only a finite number of values k, this random variable is continuous; it takes values along the entire number line.

The histogram of the probability distribution of a discrete random variable is a collection of rectangles, as we have seen, whereas for a continuous random variable the corresponding information is described by a smooth curve. However, the two types of random variables share many features. Information about the probability distribution of a continuous random variable X is summarized by the expected value $\mu = E(X)$, the balance point of the curve, and by the variance σ^2 that measures how spread out the curve is. Furthermore, probability is still described in terms of areas determined by the curve, though we no longer talk in terms of $p(k)$, the probability that X takes on a specific value k. Rather, we are concerned with the probability that the continuous random variable X takes on values over an interval. The probability that X takes a value between a and b is given by the area between the curve and the horizontal axis that lies above the interval from a to b. (See, for instance, Figure 4-40, in which a is 65 and b is 71; the figure tells us that the probability that X takes a value in that interval is .34 + .34 = .68.)

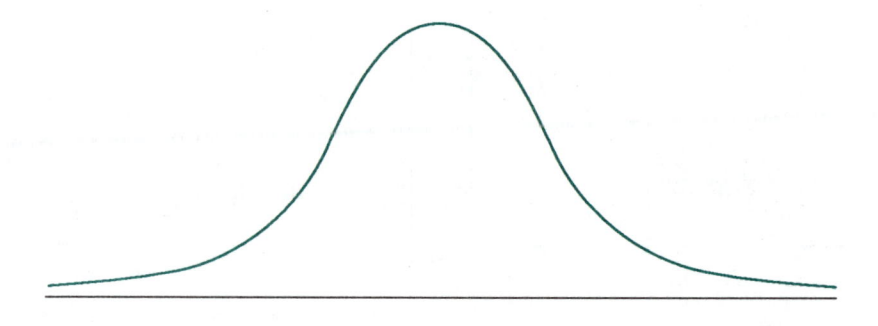

Figure 4-38

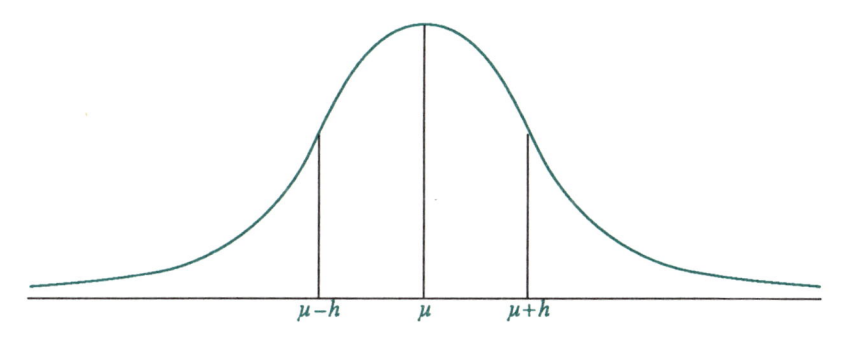

Figure 4-39

Properties of the Normal Distribution

For a normal distribution the curve reaches its high point above the expected value, μ, which is also called the **mean** of the normal distribution. A normal distribution is symmetric about its mean. Suppose that h is any positive number and points h units to the right and left of the mean are marked off on the horizontal axis, as in Figure 4-39. The shape of the curve in the regions lying to the left of the mean is a mirror image of what lies to the right. Thus, the area under the curve and above the interval between $\mu - h$ and μ is exactly the same as the area under the curve between μ and $\mu + h$.

For example, if the heights of the male high-school students are approximated by a normal distribution with mean $\mu = 68$ inches, then 68 inches should be the most common height: more boys are 68 inches tall than any other height. (It will also approximate the average height of all the boys in the school.) Furthermore, if the probability that a boy chosen at random will be between 65 and 68 inches tall is .34, then the probability is also .34 that such a boy will be between 68 and 71 inches because $65 = 68 - 3$ and $71 = 68 + 3$. To put it in more concrete terms, if 34% of all the boys in that school are between 65 and 68 inches tall, then (approximately) 34% should be expected to be between 68 and 71 inches in height (see Figure 4-40).

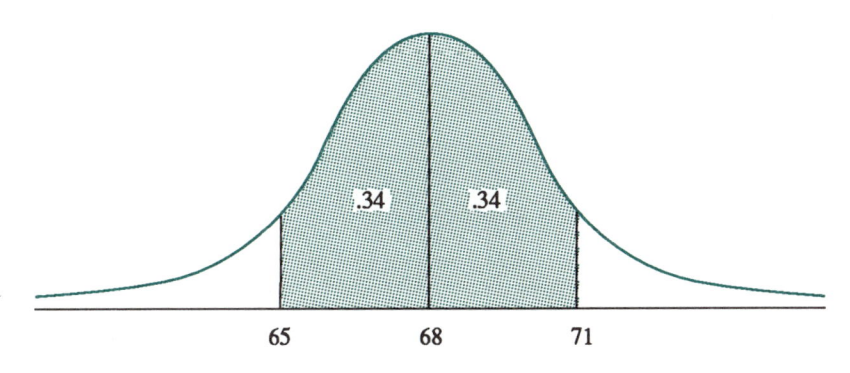

Figure 4-40

The Standard Deviation

The two normal distributions shown in Figure 4-41 are quite different, even though they have the same mean. In the first distribution the curve represents measurements clustered around the mean, whereas in the second, the measurements are much more spread out. We have discussed a measure of the spread of a histogram: the variance, σ^2, of the probability distribution.

The difference between the two normal distributions in Figure 4-41 lies in fact that in Part (a) the distribution has a small variance, whereas the one in Part (b) has a much larger variance.

(a)

(b)

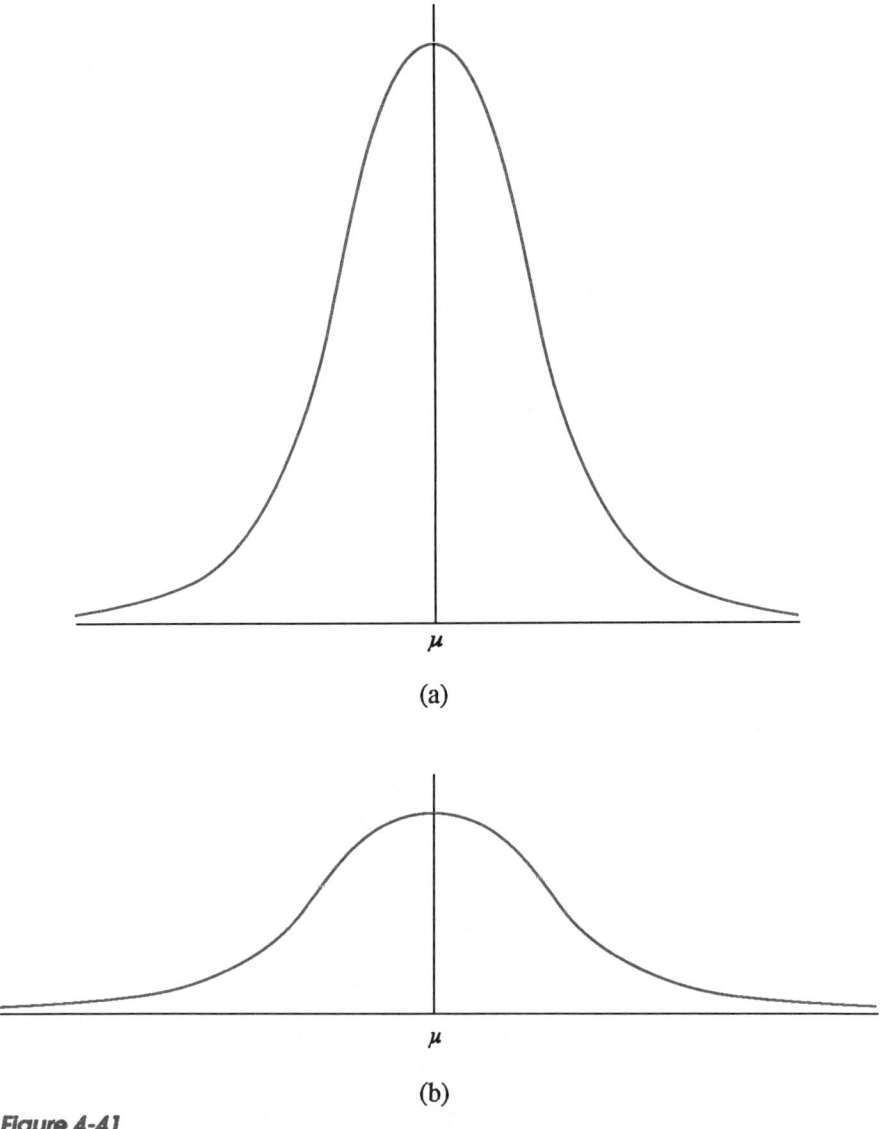

Figure 4-41

4.6 THE NORMAL DISTRIBUTION

The spread of a normal distribution is usually described by the (positive) square root of the variance, rather than by the variance itself. The square root of the variance is called the **standard deviation** and is represented by the symbol σ. We use the standard deviation in order to calculate areas under the normal curve by means of Table A on page 427. We can then interpret these areas as probabilities, as we shall see.

We use Table A to find the area under a normal curve between its mean μ and $\mu + h$ for some number h. Table A presents such areas in terms of z, where z is related to h by the equation

$$z = \frac{h}{\sigma}$$

Multiplying by σ, we see that

$$h = z\sigma$$

so, the area we wish to find lies under the normal curve between μ and $\mu + z\sigma$. For example, if the normal distribution has standard deviation $\sigma = 6$ and we wish to consider the portion of the distribution between μ and $\mu + 9$, then we compute

$$z = \frac{h}{\sigma} = \frac{9}{6} = 1.5$$

so, $\mu + 9 = \mu + 1.5\sigma$. According to Table A, the area under a normal curve between the mean μ and $\mu + 1.5\sigma$ is .4332. We found this number in the table by thinking of 1.5 = 1.50 as 1.5 + .00, looking across the row labeled 1.5 and stopping at the first column, headed by .00. Table A tells us that in a normal distribution about 43% of the area under the curve lies between μ and $\mu + 1.5\sigma$, as shown in Figure 4-42.

Using Table A

Before we present applications of the normal distribution, we need more practice in the use of Table A. This table gives the area of the region under a normal-distribution curve above the interval between μ and $\mu + z\sigma$. More briefly, we will refer to this area as between μ and $\mu + z\sigma$. This area depends only on the value of z.

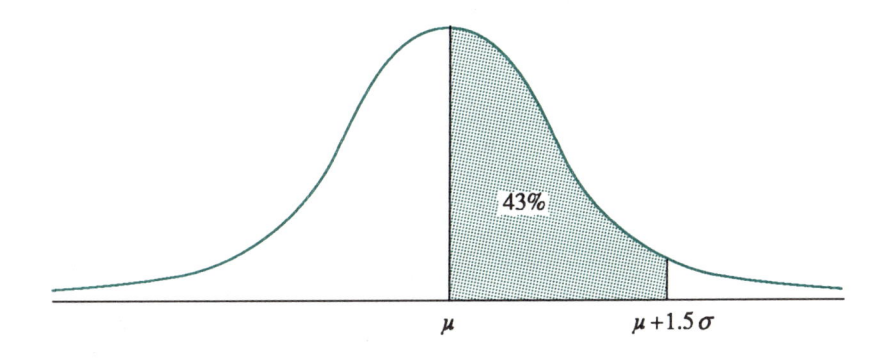

Figure 4-42

346   MORE PROBABILITY

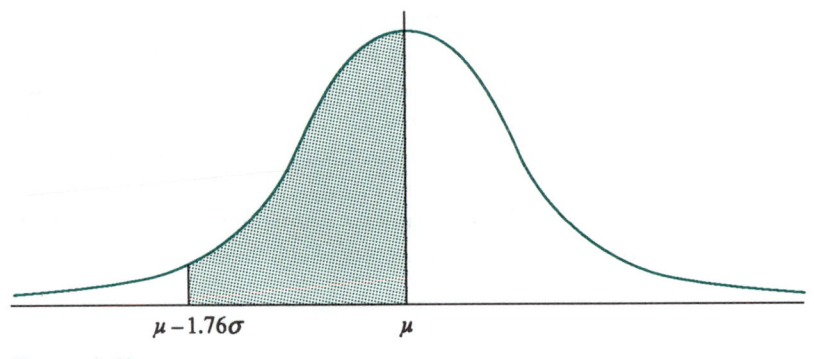

Figure with normal curve shaded between $\mu - 1.76\sigma$ and μ

Figure 4-43

Example 1. Find the area under a normal-distribution curve between $\mu - 1.76\sigma$ and μ.

Solution

The area is shown in Figure 4-43. By the symmetry property of a normal distribution, the area between $\mu - 1.76\sigma = \mu + (-1.76)\sigma$ and μ equals the area from μ to $\mu + 1.76\sigma$. To use Table A, we think of the number $z = 1.76$ in the form $1.76 = 1.7 + .06$ and locate $z = 1.7$ along the left side. We look across that row until we find the number in the column headed .06. That number is .4608; so, the area between $\mu - 1.76\sigma$ and μ is .4608. This means that 46.08% of the area below the curve lies over this interval. ∎

Example 2. Find the area under a normal-distribution curve to the right of $\mu - .43\sigma$.

Solution

The area looks like that in Figure 4-44. The area between $\mu - .43\sigma$ and μ equals the area between μ and $\mu + .43\sigma$. Writing $.43 = 0.4 + .03$, we find the row that begins with 0.4, and then look along this row to the column headed by .03, where we learn that the area is .1664. Because of the symmetry of the curve about the mean, half the area under the curve, that is, .5000, lies to the right of μ. Therefore, the area to the right of $\mu - .43\sigma$ consists of the .1664 between $\mu - .43\sigma$ and μ together with the .5000 to the right of μ, for a total area of $.1664 + .5000 = .6664$. ∎

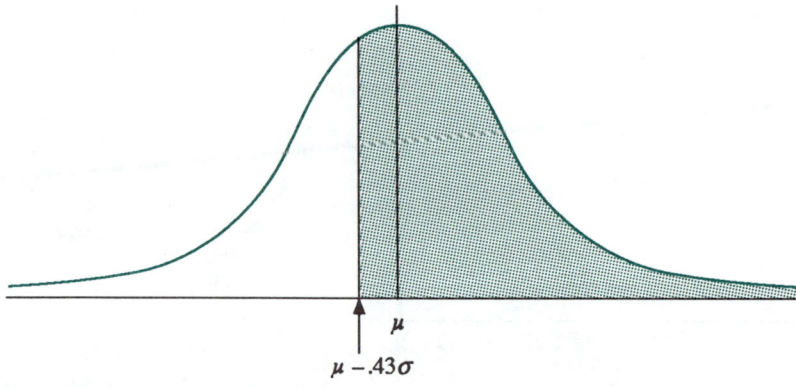

Figure with normal curve shaded to the right of $\mu - .43\sigma$

Figure 4-44

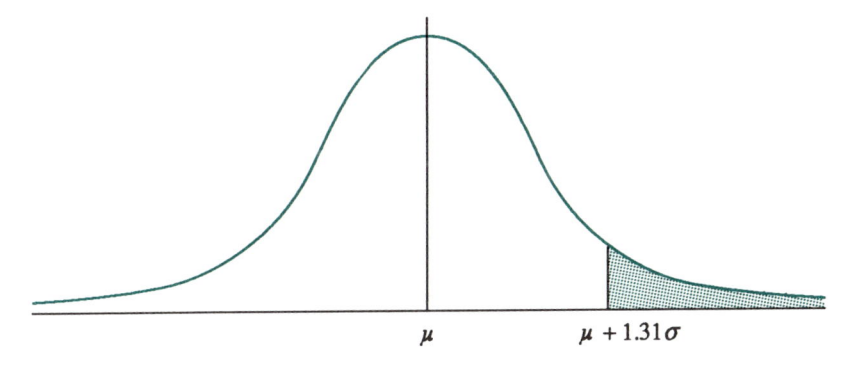

Figure 4-45

Example 3. Find the area under a normal-distribution curve to the right of $\mu + 1.31\sigma$.

Solution

We wish to find the area of the shaded region in Figure 4-45. By symmetry, half the area of the distribution (.5000) lies to the right of μ. According to Table A, the area between μ and $\mu + 1.31\sigma$ is .4049. Subtracting that area from all the area to the right of μ, we find that the area remaining is .5000 − .4049 = .0951. ∎

Applications of Table A

Now we suppose we have a large collection of data described by a histogram whose shape can be approximated by a normal distribution with mean μ and standard deviation σ. We can apply Table A to estimate the percentage of the data that lies over an interval between μ and another value, which we will call x. In order to use Table A, we must express x in the form $\mu + z\sigma$. We solve $x = \mu + z\sigma$ for z as follows. Subtracting μ from both sides, we have

$$z\sigma = x - \mu$$

Dividing by σ, which is not zero, the result is

$$\boxed{z = \frac{x - \mu}{\sigma}}$$

The area given in Table A that corresponds to this value of z is the percentage of the data between μ and x. The next two examples illustrate how we apply Table A to such data.

Example 4. Suppose the histogram of the operating lives of a type of flashlight battery can be approximated by a normal distribution with mean 5 hours and standard deviation 1.4 hours. What percentage of all batteries will last for at least 4 hours of operation?

348   MORE PROBABILITY

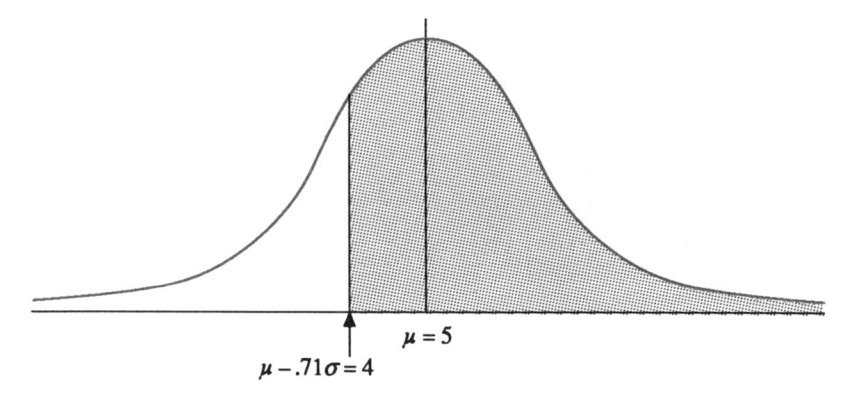

$\mu - .71\sigma = 4 \qquad \mu = 5$

Figure 4-46

Solution

We picture the histogram of operating lives as a normal curve centered at $\mu = 5$. We are trying to determine the percentage of measurements (battery lives) indicated by the shaded region in Figure 4-46. By the symmetry about the mean of a normal distribution, half of all measurements are to the right of the mean; so, in this case we know that 50% of all the battery lives are greater than 5 hours. To determine the percentage of the batteries that last between 4 and 5 hours, we use the preceding formula with $x = 4$.

$$z = \frac{x - \mu}{\sigma} = \frac{4 - 5}{1.4} = -.71 \text{ (approximately)}$$

This tells us that $4 = \mu - .71\sigma$. Referring to Table A, the area in question is .2611. Adding the area to the right of μ (compare with Example 2) the total is $.2611 + .5000 = .7611$. Therefore, about 76% of the batteries last at least 4 hours. ∎

Example 5. A study for a fruit-juice producer estimates that the amount of liquid consumed by a sample group of people in one day can be approximated by a normal distribution with mean 40.5 ounces and standard deviation 3 ounces. What percentage of this group drinks less than 35 ounces of liquid a day?

Solution

Figure 4-47 shows the area of concern. To calculate the area under the normal curve between $x = 35$ and $\mu = 40.5$, we find

$$z = \frac{x - \mu}{\sigma} = \frac{35 - 40.5}{3} = \frac{-5.5}{3} = -1.83 \text{ (approximately)}$$

From Table A, the area between $x = \mu - 1.83\sigma$ and μ is .4664. By the symmetry of the normal distribution, half the people drink less than the mean of 40.5 ounces. Since the area between 35 and 40.5 ounces is .4664, the area to the left of $x = 35$ is $.5000 - .4664 = .0336$. In other words, about 3% of the people in this group drink less than 35 ounces of liquid a day. ∎

$\mu - 1.83\sigma = 35 \qquad \mu = 40.5$

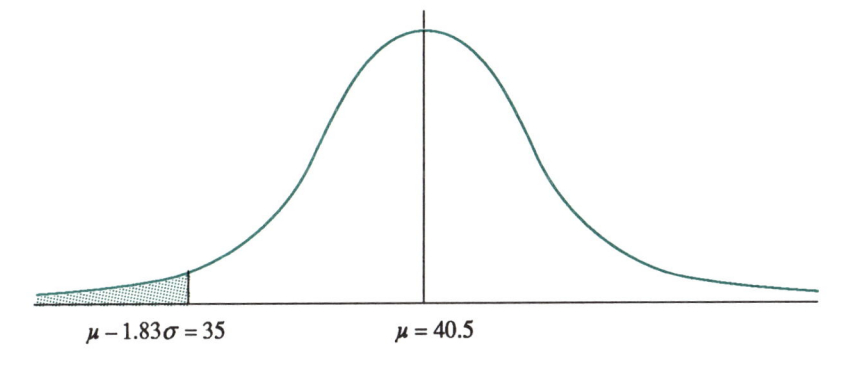

Figure 4-47

Summary

The **normal distribution** is a curve with the shape shown in Figure 4-48. For any normal distribution, the curve of the histogram reaches its high point above the expected value μ, which is also called the **mean**. The area under the curve of Figure 4-48 from μ to $\mu + h$, for any number h, is the same as that between μ and $\mu - h$. The (positive) square root of the variance of a probability distribution is called the **standard deviation**, and is denoted by the symbol σ. The probability that a normally distributed random variable with mean μ and standard deviation σ will take a value between μ and $\mu + z\sigma$ is the value found in Table A on page 427. Given a large collection of data described by a histogram whose shape can be approximated by a normal distribution with mean μ and standard deviation σ, to find the percentage of the data between the mean and some number x, calculate

$$z = \frac{x - \mu}{\sigma}$$

Then the value of the area corresponding to z (or to the absolute value of z if z is negative) in Table A is the required percentage.

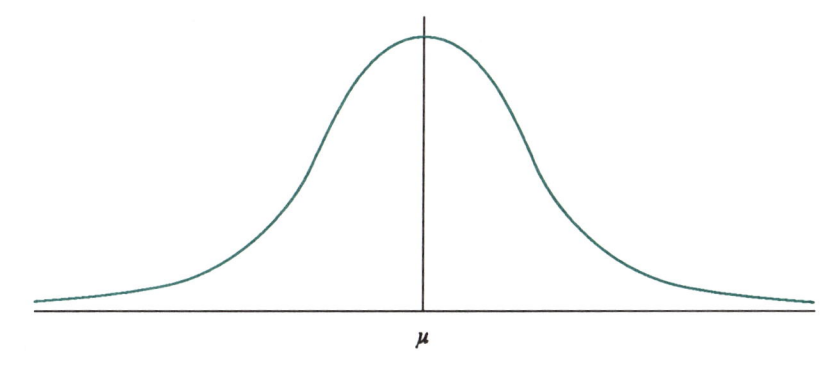

μ

Figure 4-48

Exercises

In Exercises 1–8, use Table A to find the area under the normal curve corresponding to the indicated interval.

1. Between $\mu - 2.13\sigma$ and μ
2. Between μ and $\mu + 1.37\sigma$
3. To the left of $\mu + .47\sigma$
4. To the right of $\mu - 1.51\sigma$
5. To the left of $\mu - 2.89\sigma$
6. To the right of $\mu + 1.42\sigma$
7. To the right of $\mu - .78\sigma$
8. To the left of $\mu + 2.84\sigma$

9. The durations of long-distance telephone calls between two cities are normally distributed with mean 6 minutes and standard deviation 1.5 minutes. What percentage of all calls are between 5 and 6 minutes long?

10. The distribution of high-school teachers' salaries is normally distributed with mean $30,000 a year and standard deviation $4000. What percentage of teachers earn at least $24,000 a year?

11. Suppose that the life expectancy of American women can be described by a normal distribution with mean 75 years and standard deviation 7 years. What percentage of American women live at least 80 years?

12. The mean temperature in Fairbanks, Alaska, in January is –11°F. Suppose the temperature in Fairbanks in January is normally distributed with standard deviation 14°F. During what percentage of the month of January is the temperature in Fairbanks below –30°F.?

13. Suppose that the heights of 10-year-old American boys form a normal distribution with mean 54 inches and standard deviation 3 inches. What percentage of 10-year-old American boys are less than 60 inches tall?

14. If the life of a television-picture tube is normally distributed with mean 7 years and standard deviation .85 year and if the manufacturer guarantees to replace the tube if it burns out within the first five years, what percentage of picture tubes must be replaced under the guarantee?

15. The mean discharge of water from the mouth of the Missouri River is 70,000 cubic feet per second. Suppose that the rate of discharge is normally distributed with standard deviation 6000 cubic feet per second. What percentage of the time does the Missouri River discharge at the rate of at least 65,000 cubic feet per second?

16. A box of dry breakfast cereal holds 272 cubic inches. The machine that fills the boxes discharges amounts of breakfast cereal that form a normal distribution with mean 270 cubic inches and standard deviation 1.2 cubic inches. What percentage of the time does the machine put so much breakfast cereal in the box that it overflows?

4.7 Normal Approximation for Binomial Distributions

Acceptance Sampling

A wholesale hardware dealer orders a large shipment of nails. The dealer is concerned with the quality of the shipment. In the production of nails a certain percentage will be defective; some will be bent, some will have damaged heads or will have missing points. The hardware dealer does not want to accept a shipment with too large a percentage of defectives. However, it would be impractical to examine individually each of the tens of thousands of nails in the shipment. Instead, the dealer can use a sampling procedure.

Acceptance sampling is a common sampling procedure. The customer selects a number n of units at random for testing. A number, called the **acceptance number** and represented by the symbol a, is decided upon before the testing procedure. If, out of the n units, a or fewer are defective, then the customer will accept the entire shipment. If more than a units in the sample are defective, then the customer will reject the entire shipment.

Acceptance sampling involves two kinds of risks. One risk is that the customer will reject a shipment when, in fact, it is of acceptable quality. It is possible that the sample the customer selects will just happen to contain many defective units. For instance, suppose the hardware dealer chooses 200 nails at random and will accept the shipment if 12 or fewer are defective. In other words, the dealer's acceptance number is $a = 12$. Since 12 is 6% of 200, the dealer would accept a shipment in which 6% of the nails have defects. Now suppose that, in fact, only 4% of the tens of thousands of nails shipped are defective. It is still possible that 20 of the 200 nails chosen for the sample will be defective. Since 20 is larger than the acceptance number 12, this entire shipment would be rejected because of the unfortunate sample, even though the shipment actually satisfies the customer's requirements.

The other risk is that the customer will happen to choose a sample with an uncharacteristically small number of defectives, and therefore accept a shipment that contains too many defective units. For instance, for a large shipment in which 10% of the units are defective, the customer may happen to choose a random sample of size 200 containing only 9 defective units. In that case, the entire shipment will be accepted, even though it is unsatisfactory.

Businesses use acceptance sampling because it works *most of the time*. The shipments the customer accepts are usually satisfactory and the rejected shipments usually do contain too many defective units. Probability theory plays an important role in acceptance sampling because it permits the customer to measure the risks we have described. Accurate information about these risks permits the customer to design an appropriate acceptance-sampling plan.

Binomial-Experiment Interpretation

We can describe acceptance-sampling risks by means of the binomial probability distribution. Suppose the customer chooses a unit for the acceptance sample, for

352 MORE PROBABILITY

instance, one nail from among tens of thousands, and inspects it to determine whether or not it is defective. Each such choice is thought of as a trial in a binomial experiment with n trials, where n is the sample size. (For the hardware dealer, $n = 200$.) The actual percentage of defective units determines the probability of success (finding a defective unit) on each trial. Thus, if 6% of the units in the entire shipment are defective, the probability of success on each trial is $p = .06$.

Strictly speaking, the trials are not identical because once a unit is chosen for the acceptance sample, it is removed from the shipment; so, the shipment is slightly different when the next unit is chosen. However, the total shipment is so large that the removal of a few units does not appreciably change the probability of choosing a defective unit. Thus, the binomial probability distribution gives an accurate mathematical description of acceptance sampling. Consequently, to evaluate the risks of acceptance sampling, we view an acceptance sample of n units as a binomial experiment with n trials. The percentage of defective units in the entire shipment determines the probability p that each trial will be successful in choosing a defective unit. The probability that the customer will accept the shipment is the probability that the n trials will produce a (the customer's chosen acceptance number) or fewer successes. The probability that the customer will reject the shipment is the probability of more than a successes.

We found a formula for calculating the binomial probability distribution in Section 4.5. However, when the number of trials n is large, as often happens in acceptance-sampling applications, instead of using the formula, we can make probability calculations by taking advantage of the approximately normal shape of the histogram of the binomial probability distribution. This procedure is much easier than using the formula for $p(k)$. Furthermore, from this geometric approach we get a better understanding of acceptance sampling and other applications of the binomial distribution.

Normal Approximation

Figure 4-49 shows the histogram of the binomial distribution for $n = 25$ and $p = .3$ with a normal distribution drawn over it. (We omitted the portion of the histogram

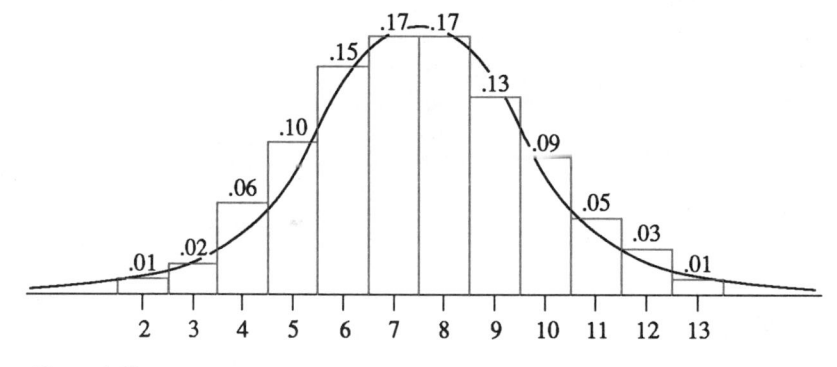

Figure 4-49

4.7 NORMAL APPROXIMATION FOR BINOMIAL DISTRIBUTIONS

from $k = 14$ to $k = 25$ because the probabilities there are all almost zero.) Notice that the shapes are very much alike. If n is large, the histogram of the binomial distribution will usually look like a normal distribution. This resemblance permits us to approximate a binomial distribution by a normal distribution. We saw in Section 4.6 that a single short table (Table A) contains all the information we require about normal distributions; that same table will permit us to compute binomial probabilities, as well.

Not all binomial distributions look like normal distributions. The binomial distribution for given n and p can be approximated very well by a normal distribution provided that both np and $n(1 - p)$ are at least 5. In the example of Figure 4-49, we have $np = 25(.3) = 7.5$ and $n(1 - p) = 25(.7) = 17.5$. On the other hand, Figure 4-33 of Section 4.5, the histogram of the binomial distribution with $n = 5$ and $p = .9$, and thus with $n(1 - p) = 5(.1) = .5$, did not resemble a normal distribution.

Once we conclude that the histogram of a binomial distribution can be approximated by a normal distribution, we have to be more specific about the characteristics of the normal distribution that we will use. As we observed in the preceding section, we have to specify the mean μ and standard deviation σ of a normal distribution in order to calculate z for Table A. Recall that the mean of a normal distribution is just another name for its expected value. In Section 4.5, we saw that the expected value $\mu = E(X)$ for the probability distribution for n binomial experiments with probability of success p on each trial was calculated by the formula $E(X) = np$. The approximating normal distribution has the same mean: $\mu = np$. We also discussed the fact that the variance of such a binomial distribution was calculated by the formula $\sigma^2 = np(1 - p)$. Therefore, for the standard deviation of the binomial distribution, that is, for the square root of the variance, we have the formula

$$\sigma = \sqrt{np(1 - p)}$$

We use the same formula for the standard deviation of the approximating normal distribution. Thus, the normal distribution of Figure 4-49 that approximates the binomial distribution with $n = 25$ and $p = .3$ has mean

$$\mu = np = 25(.3) = 7.5$$

and standard deviation

$$\sigma = \sqrt{np(1 - p)} = \sqrt{(7.5)(.7)} = \sqrt{5.25} = 2.3 \text{ (approximately)}$$

Example 1. Calculate the approximate probability of 9 or more successes in $n = 25$ trials in a binomial experiment with probability of success $p = .3$ on each trial.

Solution

In Figure 4-50 we have shaded the rectangles representing the (nonzero) probability we wish to calculate. The shaded rectangle furthest to the left represents the proba-

354   MORE PROBABILITY

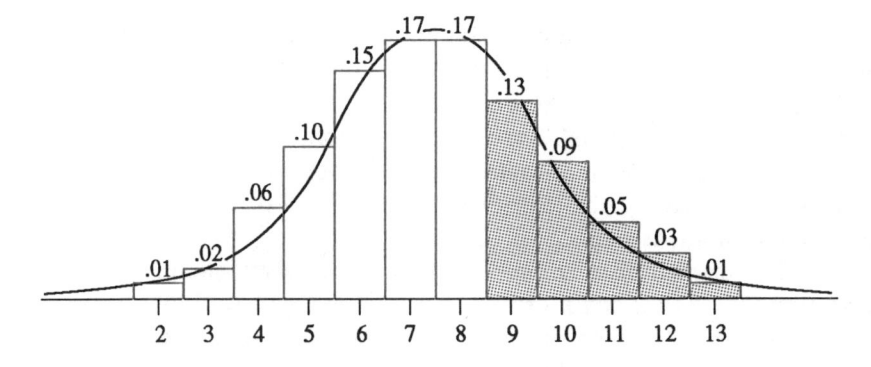

Figure 4-50

bility of exactly 9 successes. Recall from the construction of the histogram of the binomial probability distribution that the base of this rectangle is one unit long, and is centered at 9. Thus, it goes from halfway between 8 and 9, that is, from 8.5, to 9.5, which is halfway between 9 and 10. Therefore, as we show in Figure 4-51, we wish to calculate the area to the right of 8.5 (*not* 9) in the approximating normal distribution. We calculated that for the binomial distribution with $n = 25$ and $p = .3$, the mean μ and standard deviation σ of the approximating normal distribution are

$$\mu = 7.5 \quad \text{and} \quad \sigma = 2.3 \text{ (approximately)}$$

We will calculate the area of the shaded portion of Figure 4-51 from Table A, just as in Section 4.6. That is, we express $x = 8.5$ in terms of z by the formula

$$z = \frac{x - \mu}{\sigma} = \frac{8.5 - 7.5}{2.3} = \frac{1}{2.3} = .43 \text{ (approximately)}$$

and apply Table A. According to this table, the area between the mean and 8.5 is .1664; to the right of 8.5 in Figure 4-51, the area is .5000 − .1664 = .3336. Therefore,

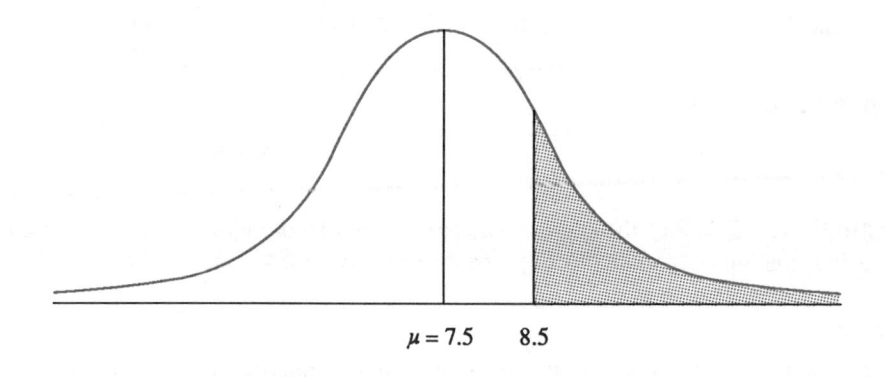

Figure 4-51

4.7 NORMAL APPROXIMATION FOR BINOMIAL DISTRIBUTIONS

the approximate probability of 9 or more successes in $n = 25$ binomial trials with probability of success $p = .3$ on each trial is .3336.

Let us compare this approximation with the probabilities shown for the rectangles in Figure 4-49, which were calculated from the exact formula

$$p(k) = C_k^n \cdot p^k \cdot (1-p)^{n-k}$$

of Section 4.5. The probabilities in the figure add up to .99 because the probability of 14 through 25 successes is .01. Therefore, by the method of Section 4.5,

$$\begin{aligned} p(9 \text{ or more successes}) &= p(9) + p(10) + p(11) + p(12) + p(13) \\ &\quad + p(14 \text{ through } 25) \\ &= .13 + .09 + .05 + .03 + .01 + .01 \\ &= .32 \end{aligned}$$

∎

In our next example we apply the normal approximation to an acceptance-sampling problem.

Example 2. A wholesale hardware dealer receives a large shipment of nails. The wholesaler tests 200 nails and will accept the shipment if 12 or fewer are defective.

(a) If 4% of the nails are defective, what is the probability that the wholesaler will reject the shipment?

(b) If 10% of the nails are defective, what is the probability that the shipment will be accepted?

Solution

(a) We may describe this acceptance-sampling problem by means of a binomial distribution with $n = 200$ and $p = .04$. Since $np = 200(.04) = 8$ is larger than 5, and certainly, $n(1 - p) = 200(.96) = 192$ is very large, we can approximate by a normal distribution. The mean of the normal distribution is

$$\mu = np = 8$$

and its standard deviation is

$$\sigma = \sqrt{np(1-p)} = \sqrt{8(.96)} = \sqrt{7.68} = 2.77 \text{ (approximately)}$$

Since the wholesaler will reject the shipment if there are 13 or more defective nails in the sample, we wish to approximate the sum $p(13) + p(14) + \ldots + p(200)$. The portion of the histogram of this binomial distribution about $k = 13$ is indicated in Figure 4-52, with the regions whose corresponding probabilities we wish to compute shaded. We see that the approximation should be of the region under the normal distribution curve to the right of 12.5 (and *not* 13), as indicated in Figure 4-53. As in the preceding section, we compute

$$z = \frac{x - \mu}{\sigma} = \frac{12.5 - 8}{2.77} = 1.62 \text{ (approximately)}$$

and apply Table A. According to Table A, the area under the normal curve to the right of 12.5, as in Figure 4-53, is $.5000 - .4474 = .0526$. Therefore, we estimate

356 MORE PROBABILITY

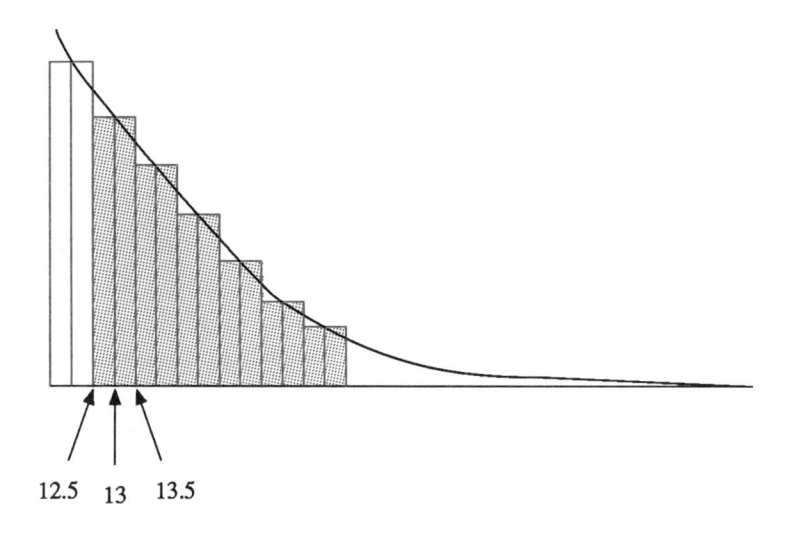

12.5 13 13.5

Figure 4-52

the risk that the wholesaler will reject a shipment containing 4% defectives, under the acceptance-sampling plan with $n = 200$ and $a = 12$, as .0526.

(b) This time, the binomial distribution has $n = 200$ and $p = .10$. The wholesaler will accept the shipment if there are 12 or fewer defective nails. Thus, we wish to approximate the histogram of the binomial distribution to the left of 12.5, since the interval from 11.5 to 12.5 contains 12. See Figure 4-54.

It is easy to check that the approximation is permitted, and we will omit these verifications from now on. The approximating normal distribution has mean

$$\mu = np = 200(.10) = 20$$

and standard deviation

$$\sigma = \sqrt{np(1-p)} = \sqrt{20(.90)} = \sqrt{18} = 4.24 \text{ (approximately)}$$

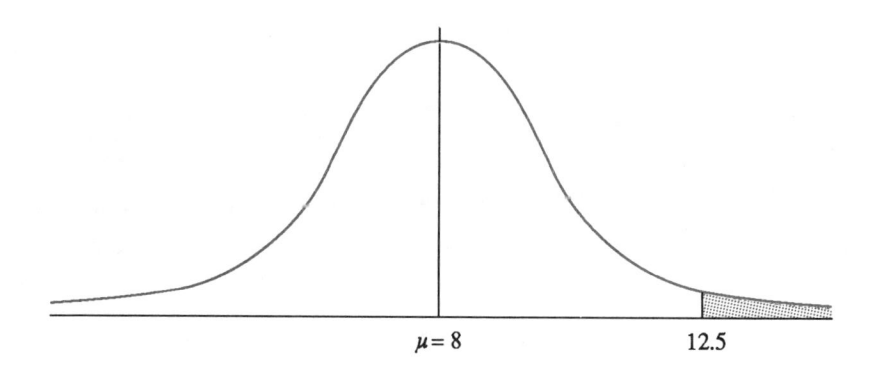

$\mu = 8$ 12.5

Figure 4-53

4.7 NORMAL APPROXIMATION FOR BINOMIAL DISTRIBUTIONS

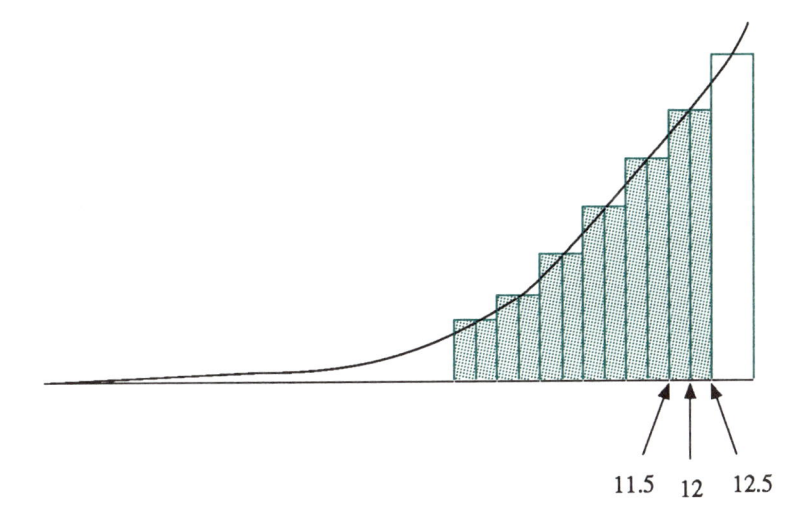

11.5 12 12.5

Figure 4-54

To find the area for the normal distribution to the left of 12.5, as in Figure 4-55, we calculate

$$z = \frac{x - \mu}{\sigma} = \frac{12.5 - 20}{4.24} = -1.77 \text{ (approximately)}$$

According to Table A that area is .5000 − .4616 = .0384. Therefore, we estimate the risk that the acceptance-sampling plan with $n = 200$ and $a = 12$ will accept a shipment containing 10% defective nails to be .0384. ∎

The normal approximation to the binomial distribution has many uses besides acceptance sampling.

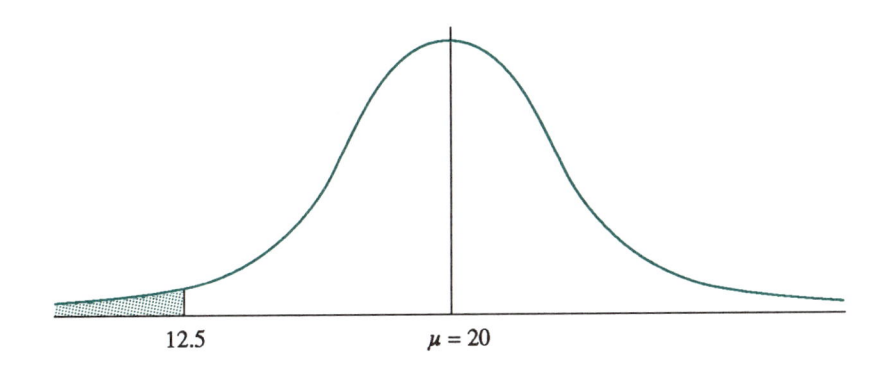

12.5 $\mu = 20$

Figure 4-55

358 MORE PROBABILITY

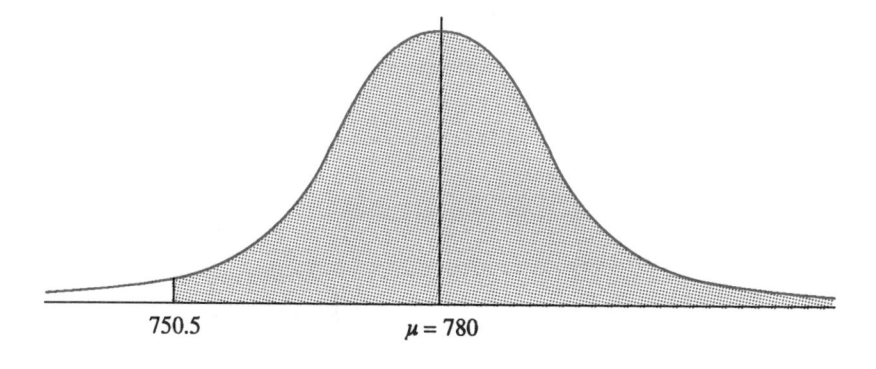

750.5 $\mu = 780$

Figure 4-56

Example 3. The Gallup polling organization usually takes a sample of 1500 voters for a nationwide survey. Suppose that 52% of all voters favor a certain presidential candidate. What is the probability that the sample will correctly predict the winner of the popular vote by containing at least 751 voters who favor that same candidate?

Solution

For the binomial distribution with $n = 1500$ and $p = .52$, we wish to approximate the sum

$$p(751) + p(752) + \ldots + p(1500)$$

In this example, $p(k)$ means the probability that exactly k voters in the sample favor the candidate. The mean is

$$\mu = np = 1500(.52) = 780$$

and the standard deviation is

$$\sigma = \sqrt{np(1-p)} = \sqrt{780(.48)} = \sqrt{374.4} = 19.35 \text{ (approximately)}$$

The portion of the binomial histogram about 751 begins at 750.5; we must find the shaded area in Figure 4-56. We calculate

$$z = \frac{x - \mu}{\sigma} = \frac{750.5 - 780}{19.35} = -1.52 \text{ (approximately)}$$

so by Table A, the area under the normal curve between 750.5 and $\mu = 780$ is .4357. Therefore, the area to the right of 750.5 is .4357 + .5000 = .9357. The approximate probability that the sample of $n = 1500$ voters will correctly predict the outcome of the popular vote is .9357. ∎

Summary

Acceptance sampling of a shipment depends on randomly selecting a number n of units to be tested and choosing an **acceptance number** a. If, out of the n units, a or fewer are defective, then the customer will accept the entire ship-

4.7 NORMAL APPROXIMATION FOR BINOMIAL DISTRIBUTIONS

ment. If more than a units in the sample are defective, then the customer will reject the entire shipment. An acceptance sample of n units is a binomial experiment with n trials. The percentage of defective units in the entire shipment determines the probability p that each trial will be successful in finding a defective unit. For a binomial distribution with n trials and probability p of success in each trial, if both np and $n(1 - p)$ are at least 5, then the binomial distribution can be approximated by the normal distribution with mean

$$\mu = np$$

and standard deviation

$$\sigma = \sqrt{np(1 - p)}$$

Exercises

In Exercises 1 through 4, use the normal distribution to calculate the approximate probability of the given number of successes, out of n trials, in a binomial experiment with probability p of success in each trial, as indicated.

1. For $n = 50$ and $p = .40$, calculate the probability of 25 or more successes.

2. For $n = 90$ and $p = .20$, calculate the probability of 15 or more successes.

3. For $n = 40$ and $p = .15$, calculate the probability of 4 or fewer successes.

4. For $n = 75$ and $p = .65$, calculate the probability of 55 or more successes.

In Exercises 5 through 7, a roofer receives a shipment of wooden shingles from a supplier. The roofer examines 60 shingles at random and will reject the entire shipment if 13 or more are so poorly cut that they cannot be used without further trimming. Otherwise, when 12 or fewer shingles in the sample have defects, the roofer will accept the entire shipment. In other words, the acceptance-sampling plan has $n = 60$ and $a = 12$.

5. Suppose that 15% of the shingles are poorly cut. What is the approximate probability that the roofer will reject the shipment?

6. If 25% of the shingles are poorly cut, what is the approximate probability that the shipment will be accepted?

7. If 40% of the shingles are poorly cut, what is the approximate probability that the roofer will accept the shipment?

In Exercises 8 through 10, the assembly department of a manufacturer of precision optical equipment receives the day's supply of brackets from the fabrication department. One-hundred brackets are chosen at random and the entire supply is returned to the fabrication department if 8 or more of the 100 fail the inspection. In other words, the acceptance-sampling plan has $n = 100$ and $a = 7$.

8. If 5% of the brackets are defective, what is the approximate probability that the assembly department will reject the supply?

9. If 10% of the brackets are defective, what is the approximate probability that the supply will be accepted?

10. If 15% of the brackets are defective, what is the approximate probability that the assembly department will accept the supply?

11. A market-research organization asks 48 people to try a new soft drink. If 25% of all people would think the soft drink is too sweet, what is the approximate probability that 18 or more out of the 48 people in the sample will consider the drink too sweet?

12. A sample of 90 graduating seniors at a university are asked whether, given the choice again, they would choose the same university. If 80% of all graduating seniors at that university would choose the same one again, what is the approximate probability that at least 80 out of the 90 in the sample would choose that university again?

In Exercises 13 and 14, a fast-food chain orders a shipment of drinking cups. The chain will sample 300 cups at random and accept the shipment if 19 or fewer of the cups are defective. Otherwise, it will return the entire shipment to the manufacturer.

13. If 5% of the cups are defective, what is the approximate probability that the chain will reject the shipment?

14. If 10% of the cups are defective, what is the approximate probability that the chain will accept the shipment?

In Exercises 15 and 16, a food-processing plant receives a shipment of tomatoes from a farm cooperative. The processing plant selects 75 tomatoes at random and examines them to determine whether they meet standards of size and condition. The plant will accept the shipment if 10 or fewer tomatoes fail to meet the standards and will reject the shipment otherwise.

15. If 10% of the tomatoes fail to meet the standards, what is the approximate probability that the processing plant will reject the shipment?

16. If 25% of the tomatoes fail to meet the standards, what is the approximate probability that the shipment will be accepted?

17. Two-hundred paper clips are chosen at random from the output of a machine that forms the clips. The machine will be stopped and adjusted if 16 or more of the 200 are defective. If 10% of the paper clips produced by the machine are defective, what is the approximate probability the machine will be stopped?

18. An opinion-polling organization asks 500 adults whether they favor stronger gun-control laws. If 45% of all adults favor such laws, what is the approximate probability that at least 250 people in the sample favor stronger gun-control laws?

19. A sample of 300 television viewers are asked if they can remember the name of the sponsor of a certain television program. If 60% of all television viewers can

recall the name of the sponsor, what is the approximate probability that 200 or fewer viewers in the sample will remember the name?

20. In a survey of consumer sensitivity to package size, milk is made available both in half-gallon and in 1.5-liter containers, at the same cost per ounce. If 50% of all shoppers would choose the 1.5-liter container, what is the approximate probability that 34 or fewer, out of a sample of 80 shoppers, would choose the 1.5-liter container?

Review Exercises for Chapter 4

In Exercises 1 and 2, a manufacturer of plastic bags chooses 60 bags at random from the production line and tests each to determine how much weight it will hold before breaking. If 10 or more bags in the sample break carrying less than the required weight, then the production line will be stopped and the production process adjusted.

1. If 12% of all bags would break carrying less than the required weight, what is the probability that the production line will be stopped?

2. If 25% of all bags would break carrying less than the required weight, what is the probability that production will continue without stopping?

In Exercises 3 through 6, a sample of four men chosen at random are asked to try a new type of razor and to state whether they prefer it to the type they are presently using. Suppose the probability is .10 that each man will prefer the new type and that the men's preferences are independent. The random variable X represents the number of men who prefer the new razor.

3. If a "yes" response means the man prefers the new razor, what is X(no, yes, no, no)?

4. Calculate the probability distribution $p(k)$ for the random variable X and draw the corresponding histogram.

5. Calculate the expected value of the number of men out of a sample of four who prefer the new type of razor.

6. Calculate the variance of the probability distribution of Exercise 4.

7. An automobile manufacturer makes three models. The economy model accounts for 45% of its production, 35% of the production is of the luxury model, and the remaining 20% is of the sports model. If the probability that an automobile will require repairs during the warranty period is .15 for the economy model, .20 for the luxury model, and .35 for the sports model, calculate the repair rate for the manufacturer, that is, the probability that one of its cars, chosen at random, will need repairs during the warranty period.

8. A salesperson who will call on three customers one day estimates the probability of selling her product to none of the three to be .06, the probability of a sale to one customer to be .29, to two customers, .44, and to all three customers, .21. If

X represents the number of customers who buy the salesperson's product, find the expected value, $E(X)$.

9. What is the probability of getting at least eight answers correct on an examination with ten true-false questions, if the student has a probability of .80 of choosing the correct answer to each question?

10. Testing for a certain communicable disease has a probability of .05 of false-positives and of .30 of false-negatives. The test is given only to individuals in a high-risk population in which 20% of the individuals are infected with the disease. If a person tests negative, what is the probability that this person nevertheless has the disease?

In Exercises 11 and 12, a garage owns three recreational vehicles which it rents by the week to vacationers. For a particular week, the garage estimates the probability that each of the vehicles will be rented to be $\frac{2}{3}$; the renting of any one vehicle is independent of the renting of the others.

11. Calculate the probability distribution of the random variable "number of recreational vehicles rented" and draw its histogram.

12. If a vehicle is rented for the week, the garage makes a profit of $100. There is no profit if the vehicle is not rented. Calculate the expected value of the profit for that week.

13. Students entering a college chemistry course are given a test of their mathematical background. Students who pass the test have a probability of .85 of passing the chemistry course, whereas students who fail the mathematics test have only a .35 probability of passing chemistry. If 20% of the students in the chemistry course fail the mathematics test, what will the pass rate for this course be, that is, what is the probability that a student in the course will pass it?

14. According to a survey of judges in criminal courts, if a jury finds a defendant guilty, the probability is .85 that the judge thinks this is the correct verdict. In the case of innocent verdicts, the probability is .65 that the judge agrees with the verdict, and thus .35 that the judge believes the defendant was really guilty. In cases where there is a mistrial because the jury was unable to reach a unanimous verdict, there is a probability of .70 that the judge thinks the defendant was guilty. In 45% of all trials, the jury brings in a guilty verdict, in 30% a verdict of innocent, and in the remaining 25% there is a mistrial. If the judge in a case thinks the defendant is guilty, calculate the probability that the jury will find the defendant guilty.

15. Suppose the probability that a person would like cherry peanut-butter ice cream is .20. A sample of 10 people are asked to try cherry peanut-butter flavored ice cream and decide whether or not they like it. Calculate the probability that at least 2 out of the 10 will like the ice cream.

16. The tax department of a country divides taxpayers according to their income: 25% are considered to have low income, 65% to have medium income, and 10% to have high income. The department estimates that the probability of cheating on a tax return is .20 for low-income taxpayers, .30 for those of medium-income,

and .55 for high-income taxpayers. Calculate the estimated cheating rate, that is, the probability that a taxpayer chosen at random is one that cheated on his or her return.

In Exercises 17 through 19, a manufacturer's representative calls on three retail stores. Based on his records and past experience, he estimates that the probability he will receive an order from the first store is .80, from the second is .50, and from the third is .70. The decisions on whether to place an order are independent of one another. Let X be the random variable: the number of orders he receives, that is, the number of stores that decide to order from him.

17. Calculate the probability distribution for this random variable and make a histogram of it.

18. Calculate the expected value of the number of orders he will receive.

19. Calculate the variance of the probability distribution of Exercise 17.

20. The purity of a drug in capsule form is checked by testing 200 capsules chosen at random. If 3% of all capsules would fail the test, what is the probability that ten or more of the capsules in the sample will fail?

21. A restaurant owner's records show that 15% of the parties dining at her restaurant consist of people eating alone, 60% are parties of moderate size (two to four people), and the remaining 25% consists of larger parties. The records also indicate that the probability that a person eating alone will order wine is .40, that the probability rises to .65 for parties of moderate size, and to .70 for larger parties. If a party at that restaurant orders wine, what is the probability the wine is for a person eating alone?

22. Suppose the mean IQ is 100, the standard deviation of IQ's is 12, and IQ's are normally distributed. What percentage of the population has an IQ of 80 or higher?

5

Markov Chains

Brand-Share Prediction

When a company develops a new product, it is often test-marketed in one area of the country before it is distributed nationally. The company must determine as quickly as possible whether the product is going to be successful so that, if it is, the company can give it nationwide distribution before a rival firm brings out a similar product. If, on the other hand, the product is a failure, the company can stop production and hold its losses to a minimum.

An example of this kind of test-marketing is illustrated by a study written by Benjamin Lipstein for the *Harvard Business Review*.[†] The article shows how predictions about a product were made from a study in the Chicago area. Fictitious names are used, but the information is based on an actual marketing situation.

When a product Lipstein calls "Electra margarine" was test-marketed from November, 1958 to May, 1959, its brand share (the percentage of the available market that the new brand received) rose to 15% in mid-December, but then slipped in mid-January to 4%, a percentage that remained relatively stable throughout the 6-month trial period. At the same time, the percentage of repeat buyers of Electra showed a steady increase, while the average repeat rate for competitive brands dropped.

[†]Benjamin Lipstein, "Tests for Test Marketing," *Harvard Business Review* 76 (March–April, 1961), pp. 365–369.

To understand what this means in terms of Electra's eventual success or failure in the marketplace, it helps to look at what was happening to all margarine sales. Before Electra was introduced, brand-loyal margarine buyers (those who devoted $\frac{3}{4}$ or more of their purchases of margarine in any 6-week period to one brand) made up 70% of the market. During the first 6 weeks of Electra's introduction, brand-loyal buyers declined to 65%, primarily because some of them were trying Electra. The number of buyers of competitive brands decreased further (to 58%) with the advent of the Electra advertising campaign and coupon distribution. Then the percentage of repeat buyers of all margarines increased because repeat buyers of Electra margarine were included in the total. The repeat rate for Electra at the end of 4 months was 2% above the market average, but this did not necessarily guarantee a satisfactory brand share for Electra because the whole market was in turmoil.

As the market began to stabilize toward the end of the 6-month trial period, the rate of new buyers had leveled off, but the repeat-buyer rate continued to climb. In other words, Electra seemed to have claimed a share of the market and it acquired some loyal customers. The marketing managers needed to know what Electra's eventual brand share was going to be. We will show you how such a prediction can be made, using a combination of probability theory and matrix arithmetic in a "Markov chain process."

5.1 Matrices and Probability

Instead of immediately trying to analyze all aspects of the margarine marketing problem, as presented in the introductory material, we begin with a simplified example. We divide margarine purchases into two classes: Electra or any other brand. We suppose that 14% of all margarine buyers have purchased Electra within a certain time period. Studies of buyer behavior indicate that 23% of Electra purchasers will choose the same brand the next time they buy margarine. Ten percent of the people who bought another brand the previous time will switch to Electra. What percentage of all margarine buyers will select Electra for their margarine in the next time period? For simplicity, we will assume that each buyer makes one margarine purchase in each time period.

We may solve this problem by the partition technique of probability theory that was the subject of the first section of Chapter 4. Let B_1 be the set of all purchasers of Electra margarine in the given time period and let B_2 be the set of margarine buyers who chose another brand at that time. If a margarine purchaser is chosen at random, we are assuming that the probability is $p(B_1) = .14$ that this person bought Electra margarine and therefore, $p(B_2) = .86$ that the choice was some other brand. Let A_1 represent the event that a person buying margarine will select Electra in the next period. We use this notation because the B's indicate "before" and the A's "after." According to Chapter 4, the probability $p(A_1)$ that a random purchaser in the next time period will select Electra can be calculated by means of the formula

$$p(A_1) = p(B_1) \cdot p(A_1 \mid B_1) + p(B_2) \cdot p(A_1 \mid B_2)$$

The conditional probability $p(A_1 \mid B_1)$ is the probability that an Electra buyer in the original time period will buy Electra again in the next period, so we have been told that $p(A_1 \mid B_1) = .23$. Similarly, $p(A_1 \mid B_2)$ represents the probability that a purchaser of another brand will switch to Electra in the next time period, and thus we know that $p(A_1 \mid B_2) = .10$. Substituting these values into the formula, we find that

$$p(A_1) = (.14)(.23) + (.86)(.10) = .12 \text{ (approximately)}$$

Electra's share of the margarine market will drop from 14% in the original time period to 12% in the next period.

Another Partition

The partition B_1, B_2 divided all margarine buyers at the "before" time period into those who purchased Electra and those who bought another brand. The event A_1 represented the Electra buyers in the next, or "after" time period, so there is another way to partition the margarine buyers: those in A_1, who purchased Electra brand in the later period, and the rest, A_2, who chose another brand. We now have two partitions of the sample space U of all margarine buyers. The partition B_1 (Electra brand), B_2 (other brands) refers to the original time period while the other partition into A_1 (Electra brand) and A_2 (other brands) concerns the buyers in the succeeding time period. We emphasize that although B_1 and A_1 concern buyers in the same "state"—purchasers of Electra brand—they are generally different subsets of U.

The Transition Matrix

In the computation

$$p(A_1) = p(B_1) \cdot p(A_1 \mid B_1) + p(B_2) \cdot p(A_1 \mid B_2) = (.14)(.23) + (.86)(.10) = .12$$

the conditional probabilities concern both partitions. For instance, $p(A_1 \mid B_1) = .23$ indicates that, if an individual is an Electra purchaser ("state 1") in the "before" time period, event B_1, the conditional probability of buying Electra in the "after" time period, event A_1 (being in state 1 again), is .23. The formula also uses the information that $p(A_1 \mid B_2) = .10$, where $p(A_1 \mid B_2)$ is the probability that a purchaser of a rival brand in the original time period, that is, a member of B_2, will select Electra in the later time period, and so belong to the set A_1.

Of the margarine buyers in B_1, those who purchased Electra in the first time period, although 23% chose Electra again, the remaining 77% must have chosen some other brand since we are assuming that each buyer makes a purchase in each time period. Thus, these buyers belong to the set A_2 in the partition of the later period, and we see that $p(A_2 \mid B_1) = .77$. In the same way, since 90% of the members of B_2 did not switch to Electra in the succeeding time period, $p(A_2 \mid B_2) = .90$.

We summarize all this information in Table 5-1. This table has a special structure with respect to the two partitions B_1, B_2 and A_1, A_2 of the sample space U of all margarine buyers. This is indicated in Table 5-2.

		Original Period	
		Electra (B_1)	Other Brands (B_2)
Next Period	Electra (A_1)	.23	.10
	Other Brands (A_2)	.77	.90

Table 5-1

		Before	
		B_1	B_2
After	A_1	$p(A_1 \mid B_1)$	$p(A_1 \mid B_2)$
	A_2	$p(A_2 \mid B_1)$	$p(A_2 \mid B_2)$

Table 5-2

We will view these tables as matrices, just as we did the input-output tables of Chapter 1. We let

$$P = \begin{bmatrix} p(A_1 \mid B_1) & p(A_1 \mid B_2) \\ p(A_2 \mid B_1) & p(A_2 \mid B_2) \end{bmatrix} = \begin{bmatrix} .23 & .10 \\ .77 & .90 \end{bmatrix}$$

The matrix P is called a **transition matrix** because the conditional probability $p(A_i \mid B_j)$ in the ith row and jth column of the matrix P represents the probability of a transition from state j in the original "before" time period to state i in the succeeding "after" period. Thus, for example, $p(A_2 \mid B_1) = .77$ is the probability that a margarine purchaser in state 1 (Electra) in the original time period will make the transition to state 2 (Other Brand) in the next time period. Even a probability like $p(A_1 \mid B_1)$, that an Electra purchaser will stay with Electra, is a transitional probability in the sense that it tells us the probability that no transition out of state 1 will take place.

The numbers in each column of the transition matrix will always add up to 1, because, by our assumption, a buyer of Electra in the previous period must make a purchase and thus either repeats the choice of Electra or changes to some other brand. So,

$$p(A_1 \mid B_1) + p(A_2 \mid B_1) = 1$$

Probability Formulas Using Matrix Multiplication

We calculated $p(A_1)$, the probability that a margarine buyer will purchase Electra in the later time period, by the formula

$$p(A_1) = p(B_1) \cdot p(A_1 \mid B_1) + p(B_2) \cdot p(A_1 \mid B_2) = (.14)(.23) + (.86)(.10) = .12$$

In order to relate this formula to matrix multiplication, we need to reverse the order in which we multiply, that is, we rewrite the formula as

$$p(A_1) = p(A_1 \mid B_1) \cdot p(B_1) + p(A_1 \mid B_2) \cdot p(B_2)$$

This gives us the same result, namely,

$$p(A_1) = (.23)(.14) + (.10)(.86) = .12$$

In the same way, if instead of A_1, we wish to compute the probability of the event A_2 that buyers choose a brand other than Electra in the later time period, the partition probability formula of Chapter 4, with A_2 replacing A throughout, can be written as

$$p(A_2) = p(A_2 \mid B_1) \cdot p(B_1) + p(A_2 \mid B_2) \cdot p(B_2)$$

The advantage of writing the probability formulas in this way is that we can express both $p(A_1)$ and $p(A_2)$ by means of a single matrix multiplication, as in Section 1.2:

$$\begin{bmatrix} p(A_1) \\ p(A_2) \end{bmatrix} = \begin{bmatrix} p(A_1 \mid B_1) & p(A_1 \mid B_2) \\ p(A_2 \mid B_1) & p(A_2 \mid B_2) \end{bmatrix} \begin{bmatrix} p(B_1) \\ p(B_2) \end{bmatrix} = \begin{bmatrix} p(A_1 \mid B_1) \cdot p(B_1) + p(A_1 \mid B_2) \cdot p(B_2) \\ p(A_2 \mid B_1) \cdot p(B_1) + p(A_2 \mid B_2) \cdot p(B_2) \end{bmatrix}$$

In even more compact notation, if we let

$$S^{(0)} = \begin{bmatrix} p(B_1) \\ p(B_2) \end{bmatrix} \quad \text{and} \quad S^{(1)} = \begin{bmatrix} p(A_1) \\ p(A_2) \end{bmatrix}$$

and let P be the transition matrix as before, the formula that takes us from the "before" matrix $S^{(0)}$ to the "after" matrix $S^{(1)}$ is: $S^{(1)} = PS^{(0)}$, i. e., the product of matrix P by matrix $S^{(0)}$.

Comparison with Another Brand

We will next use two partitions of the sample space U to analyze not only how Electra margarine will perform from one buying period to the next, but also how this behavior will compare to that of its major competitor, "Marigold." More detailed results from the study of margarine buyer behavior are summarized in Table 5-3.

The meaning of the first column is that there is a probability of .23 that a purchaser of Electra in the first time period will choose the same brand in the next

		Original Period		
		Electra (B_1)	Marigold (B_2)	Others (B_3)
Next Period	Electra (A_1)	.23	.05	.12
	Marigold (A_2)	.04	.25	.15
	Others (A_3)	.73	.70	.73

Table 5-3

period, a probability of .04 that the buyer will switch to Marigold, and therefore, a probability of $1 - (.23 + .04) = .73$ that some brand other than Electra or Marigold will be purchased. Similarly, the second column shows the expected buying behavior of Marigold purchasers; for instance, the probability that a Marigold purchaser will switch to Electra is .05. Similarly, the third column describes the behavior of the buyers of the remaining brands.

Now we partition the margarine buyers into three classes according to their purchases in the original time period: Electra buyers (B_1), Marigold buyers (B_2), and purchasers of all other brands (B_3). Suppose that in the original time period Electra is purchased by 14% of margarine buyers and Marigold by 20%. In terms of the partition, we have $p(B_1) = .14$ and $p(B_2) = .20$, so that $p(B_3) = .66$. We would like to know the distribution of Electra buyers, Marigold buyers, and purchasers of other brands in the next buying period.

The partition B_1, B_2, B_3, of the sample space U of all margarine buyers classifies the members of U according to a fixed criterion, namely, margarine purchased in a given time period. Each member of U is considered with respect to brand purchase and is placed in an appropriate subset determined by the partition as a result. Formally, we say that each outcome can be in exactly one of three possible states: Electra purchaser, Marigold purchaser, or purchaser of another brand, and that the partition B_1, B_2, B_3 is determined by which outcomes (purchasers) are in which state (which brand they choose). We continue to assume that each buyer makes one margarine choice in each time period.

We would like to know the probability that in the next time period a margarine purchaser will buy Electra, Marigold, or some other brand. Just as in the original time period, we can examine each outcome to determine which state it is in during the next time period. This examination produces a new partition A_1, A_2, A_3 of the sample space U of all purchasers, where A_1 denotes those who choose Electra during the next time period, A_2, those who choose Marigold, and A_3, those who select one of the other brands. The subscript 1 denotes Electra, 2 denotes Marigold, and 3 denotes other brands for both partitions. The problem, now, is to calculate $p(A_1)$, $p(A_2)$, and $p(A_3)$ from the available information.

The Transition Matrix

With respect to the partitions B_1, B_2, B_3 and A_1, A_2, A_3, the numbers in Table 5-3 represent all possible conditional probabilities. For example, the given probability of .23 that an Electra purchaser in the given time period will again buy Electra in the next period is the conditional probability that an outcome will be in subset A_1 in the next time period under the condition that it was in subset B_1 in the original period. In symbols, the table informs us that

$$p(A_1 \mid B_1) = .23$$

The probability that a Marigold purchaser will switch to Electra is the probability that an outcome will be in subset A_1, given that it was originally in B_2. Thus, the table states that

$$p(A_1 \mid B_2) = .05$$

("state 2 before, state 1 after"). In the same way, we read from the table that the probability a purchaser of a brand other than Electra or Marigold will choose Marigold in the next period is

$$p(A_2 \mid B_3) = .15$$

We write Table 5-3 as a transition matrix:

$$P = \begin{bmatrix} p(A_1\mid B_1) & p(A_1\mid B_2) & p(A_1\mid B_3) \\ p(A_2\mid B_1) & p(A_2\mid B_2) & p(A_2\mid B_3) \\ p(A_3\mid B_1) & p(A_3\mid B_2) & p(A_3\mid B_3) \end{bmatrix} = \begin{bmatrix} .23 & .05 & .12 \\ .04 & .25 & .15 \\ .73 & .70 & .73 \end{bmatrix}$$

Notice that again the numbers in each column of the transition matrix add up to 1, because, for instance, a buyer of Electra in the preceding period must either repeat that choice, change to Marigold, or change to some other brand. Thus,

$$p(A_1 \mid B_1) + p(A_2 \mid B_1) + p(A_3 \mid B_1) = 1$$

Matrix Versions of the Probability Formulas

The partition probability formula of Chapter 4 for $p(A_1)$, the probability that a margarine buyer will select Electra in the next time period, can be written as

$$p(A_1) = p(A_1 \mid B_1) \cdot p(B_1) + p(A_1 \mid B_2) \cdot p(B_2) + p(A_1 \mid B_3) \cdot p(B_3)$$

Similarly, for Marigold,

$$p(A_2) = p(A_2 \mid B_1) \cdot p(B_1) + p(A_2 \mid B_2) \cdot p(B_2) + p(A_2 \mid B_3) \cdot p(B_3)$$

and for all other brands,

$$p(A_3) = p(A_3 \mid B_1) \cdot p(B_1) + p(A_3 \mid B_2) \cdot p(B_2) + p(A_3 \mid B_3) \cdot p(B_3)$$

In terms of matrices, the three equations for the probabilities $p(A_i)$ can be replaced by the single matrix equation

$$\begin{bmatrix} p(A_1) \\ p(A_2) \\ p(A_3) \end{bmatrix} = \begin{bmatrix} p(A_1\mid B_1) & p(A_1\mid B_2) & p(A_1\mid B_3) \\ p(A_2\mid B_1) & p(A_2\mid B_2) & p(A_2\mid B_3) \\ p(A_3\mid B_1) & p(A_3\mid B_2) & p(A_3\mid B_3) \end{bmatrix} \begin{bmatrix} p(B_1) \\ p(B_2) \\ p(B_3) \end{bmatrix}$$

If we now let

$$S^{(0)} = \begin{bmatrix} p(B_1) \\ p(B_2) \\ p(B_3) \end{bmatrix} \quad \text{and} \quad S^{(1)} = \begin{bmatrix} p(A_1) \\ p(A_2) \\ p(A_3) \end{bmatrix}$$

and let P be the transition matrix, then we have the same formula to take us from the "before" matrix $S^{(0)}$ to the "after" matrix $S^{(1)}$, that is, $S^{(1)} = PS^{(0)}$.

In the margarine marketing problem, we were given that

$$P = \begin{bmatrix} .23 & .05 & .12 \\ .04 & .25 & .15 \\ .73 & .70 & .73 \end{bmatrix} \text{ and } S^{(0)} = \begin{bmatrix} .14 \\ .20 \\ .66 \end{bmatrix}$$

Therefore, multiplying the matrices and rounding the answers, we have

$$S^{(1)} = PS^{(0)} = \begin{bmatrix} .23 & .05 & .12 \\ .04 & .25 & .15 \\ .73 & .70 & .73 \end{bmatrix} \begin{bmatrix} .14 \\ .20 \\ .66 \end{bmatrix} = \begin{bmatrix} .121 \\ .155 \\ .724 \end{bmatrix} \text{ (approximately)}$$

Thus, the calculation predicts that Electra's share of the market will drop from 14% to 12% in the next purchasing period and that, at the same time, Marigold's share will drop from 20% to about 15.5%.

General Formulation

We now turn to a general class of probability problem, of which the brand-share problem we have been studying is a special case. We consider a sample space U and a fixed set of states $s_1, s_2, \ldots, s_k$ that determine a partition of U. That is, we assume that each element of U is in exactly one state and that we have a way of deciding which state it is in. We suppose we are to perform an experiment that will result in placing each member of U in one of the states $s_1, s_2, \ldots, s_k$, possibly the state it is in already or perhaps another one. We are given a $k \times k$-transition matrix P of numbers p_{ij}, where p_{ij} is the probability that an element of U in state s_j when the experiment begins will be in state s_i at the conclusion of the experiment. Thus, if we let $B_1, B_2, \ldots, B_k$ be the partition of U determined by the states when the experiment begins and if $A_1, A_2, \ldots, A_k$ is the corresponding partition afterward, then the p_{ij} are the conditional probabilities $p_{ij} = p(A_i \mid B_j)$. The numbers in each column of the transition matrix will always add up to 1 because an element of U in state s_j before the experiment must end up in one of the states $s_1, s_2, \ldots, s_k$ after the experiment; this implies that

$$p_{1j} + p_{2j} + \ldots + p_{kj} = 1$$

for each choice of j.

In the next section, we will want to use matrices of probabilities to look further into the future; not just from "before" to "after," but several periods ahead. In order to write down what we are doing, it will be necessary to give a number to each time period, and to modify the notation a bit so that there is room for the period number. The "before" period we numbered as period 0, and the "after" period as 1, that is, the first period after the beginning one. We introduced this notation when we called the column probability matrices $S^{(0)}$ and $S^{(1)}$. So, for instance, instead of writing $p(B_2)$ for the probability of being in state 2 at the beginning, that is, in time period 0, we will write $p(s_2)^{(0)}$, which we read as the "probability of state 2 in the 0 time period." Similarly, the probability of being in state 3 in the next ("after") period becomes $p(s_3)^{(1)}$, instead of $p(A_3)$. Now the matrices $S^{(0)}$ and $S^{(1)}$ have this general form:

$$S^{(0)} = \begin{bmatrix} p(s_1)^{(0)} \\ p(s_2)^{(0)} \\ \vdots \\ p(s_k)^{(0)} \end{bmatrix} \quad S^{(1)} = \begin{bmatrix} p(s_1)^{(1)} \\ p(s_2)^{(1)} \\ \vdots \\ p(s_k)^{(1)} \end{bmatrix}$$

The matrices $S^{(0)}$ and $S^{(1)}$ represent exactly what they did before we changed the notation, so we still have the formula

$$\boxed{S^{(1)} = PS^{(0)}}$$

Example 1. In an automobile-leasing company's study of its customers who lease a new car every year, it is found that 30% of those who leased sedans the preceding year changed to station wagons, whereas 40% of those who had leased wagons changed to sedans. The rest of the customers repeated the type of vehicle they had before. If 20% of the cars leased to customers last year were wagons, what percentage of customers will lease wagons this year?

Solution

We describe the possible states as s_1, leasing a sedan, and s_2, leasing a station wagon. We are given the data represented by Table 5-4. Thus, the transition matrix is given by

$$P = \begin{bmatrix} .70 & .40 \\ .30 & .60 \end{bmatrix}$$

Last year, 20% of the customers leased wagons, that is, were in state s_2; thus, 80% chose sedans (state s_1) last year. Therefore,

$$S^{(0)} = \begin{bmatrix} p(s_1)^{(0)} \\ p(s_2)^{(0)} \end{bmatrix} = \begin{bmatrix} .80 \\ .20 \end{bmatrix}$$

The prediction is that

$$S^{(1)} = PS^{(0)} = \begin{bmatrix} .70 & .40 \\ .30 & .60 \end{bmatrix} \begin{bmatrix} .80 \\ .20 \end{bmatrix} = \begin{bmatrix} .54 \\ .36 \end{bmatrix} = \begin{bmatrix} p(s_1)^{(1)} \\ p(s_2)^{(1)} \end{bmatrix}$$

Next year, 36% of the customers will lease wagons. ■

		Previous Year	
		Sedan (s_1)	Wagon (s_2)
Present	Sedan (s_1)	.70	.40
Year	Wagon (s_2)	.30	.60

Table 5-4

Example 2. Suppose the probability that a girl whose mother graduated from college will also graduate from college is .70, that she will graduate from high school but not from college is .25, and that she will not finish high school is .05. If her mother finished high school but not college, then the probability that the daughter will graduate from college is .50, will graduate from high school but not college is .40, and will not finish high school is .10. If the mother did not graduate from high school, the probabilities are .25 that the daughter will finish college, .55 that she will graduate from high school but not college, and .20 that she will not finish high school. If 30% of the mothers' generation graduated from college, 50% graduated from high school but not college, and 20% did not finish high school, what percentage of the daughters will graduate from college? What percentage will graduate from high school but not college? What percentage will not graduate from high school?

Solution

From the data of the problem, we have Table 5-5. The sample space U consists of women, and the possible states are levels of education: s_1 = college graduate, s_2 = high-school but not college graduate, and s_3 = not a high-school graduate. The "0" time period indicates the mothers' generation and the "1" time period, the daughters'. Thus, for example, $p(s_1)^{(0)}$ is the probability that a mother graduated from college and $p(s_1)^{(1)}$ is the probability that a daughter will graduate from college. From Table 5-5 we see that the transition matrix is given by

$$P = \begin{bmatrix} .70 & .50 & .25 \\ .25 & .40 & .55 \\ .05 & .10 & .20 \end{bmatrix}$$

The distribution by education of the mothers' generation tells us that

$$S^{(0)} = \begin{bmatrix} p(s_1)^{(0)} \\ p(s_2)^{(0)} \\ p(s_3)^{(0)} \end{bmatrix} = \begin{bmatrix} .30 \\ .50 \\ .20 \end{bmatrix}$$

and we calculate that

$$S^{(1)} = PS^{(0)} = \begin{bmatrix} .70 & .50 & .25 \\ .25 & .40 & .55 \\ .05 & .10 & .20 \end{bmatrix} \begin{bmatrix} .30 \\ .50 \\ .20 \end{bmatrix} = \begin{bmatrix} .510 \\ .385 \\ .105 \end{bmatrix} = \begin{bmatrix} p(s_1)^{(1)} \\ p(s_2)^{(1)} \\ p(s_3)^{(1)} \end{bmatrix}$$

Therefore, in the daughters' generation, 51% will graduate from college, 38.5% will graduate from high school, and 10.5% will fail to graduate from high school. ∎

		Mother's Eduction		
		College Grad	*High-School Grad*	*No Grad*
Daughter's Education	*College Grad*	.70	.50	.25
	High-School Grad	.25	.40	.55
	No Grad	.05	.10	.20

Table 5-5

Summary

Suppose that each element of a sample space U can be in exactly one of a set of **states** $s_1, s_2, \ldots, s_k$. An experiment is performed, and for each i and j between 1 and k, there is a given probability p_{ij} that an element of U in state j before the experiment begins will be transformed to state i as a result of the experiment. The $k \times k$-matrix P with p_{ij} in the ith row and the jth column is the **transition matrix**. The numbers in each column of P add up to 1. Let $p(s_i)^{(0)}$ be the probability that an element of U is in state s_i before the experiment and let $p(s_i)^{(1)}$ be the probability that an element is in state s_i after the experiment. We define $k \times 1$-matrices $S^{(0)}$ and $S^{(1)}$ by letting the number in the ith row of $S^{(0)}$ be $p(s_i)^{(0)}$ and of $S^{(1)}$ be $p(s_i)^{(1)}$; then $S^{(1)} = PS^{(0)}$.

Exercises

1. A sampling of opinion with regard to a coming election indicates that 60% of those who voted for the Democratic candidate in the last election will again vote for the Democrat, whereas 40% will vote for the Republican opponent. On the other hand, 80% of those who voted for the Republican in the last election will again support the Republican, but 20% will vote for the Democrat. If the Democrat won the last election with 55% of the vote against 45% for the Republican, who would we expect to win the present election? What percentage of the vote will the winner receive?

2. A chess player notices that her performance in the second game of a tournament is very much influenced by the outcome of the first game. The probabilities that she will win, lose, or draw the second game, given the outcome of the first, are presented in Table 5-6. In the first games of previous tournaments she has won 60% of the time, lost 10%, and drawn 30%. Estimate the probability that in the next tournament she enters she will win the second game she plays.

		First Game		
		Win	Lose	Draw
Second Game	Win	.50	.20	.40
	Lose	.10	.30	.20
	Draw	.40	.50	.40

Table 5-6

3. A door-to-door salesperson finds that, although regular customers will buy something several times a year, only 35% make a purchase on two successive visits. On the other hand, if a regular customer did not buy anything during the preceding visit, there is a 55% probability of a purchase the next time. If during the last round of visits, the salesperson sold something to 40% of the customers, what percentage of the present round of visits will produce sales?

4. The weather in Lake City is classified as fair, cloudy (without rain), or rainy. An investigation of past records indicates the probability that a fair day will be followed by another fair day is .60, by a cloudy day, .25, and by a rainy day, .15. A cloudy day has a .30 probability of being followed by a fair day, .40 by a cloudy day, and .30 by a rainy day. A rainy day has a .50 chance of being followed by a fair day, .20 by a cloudy day, and .30 by another day of rain. If the forecast for tomorrow's weather is cloudy with a 50% chance of rain (and thus, a 50% chance of cloudiness without rain), what should the Lake City forecast for the day after tomorrow be?

5. Suppose the probability that a person will be divorced is .35 if his or her parents were divorced and .20 otherwise. If 25% of the people in the preceding generation were divorced, what is the probability that a person of the present generation will be divorced?

6. The records of previous classes in a two-semester Western Civilization course indicates the probability that a student who receives a certain grade (A, B, C, or D) in the first semester will receive a grade in the second semester as indicated in Table 5-7. Among the students who are going on to the second semester this year, 20% received A's, 40% B's, 35% C's, and 5% D's the first semester. What is the expected distribution of grades for the second semester of the course?

		First Semester			
		A	B	C	D
	A	.80	.40	.20	.05
Second	B	.15	.40	.30	.10
Semester	C	.05	.15	.30	.40
	D	0	.05	.20	.45

Table 5-7

7. Suppose that in a study of inherited size in an animal species, it is found that, of the male offspring of large males, 50% are also large, 40% are of medium size, and 10% are small. If the father is of medium size, then 20% of the male offspring are large, 60% medium, and 20% small. If the father is small, 5% of his male offspring are large, 70% medium, and 25% small. If in the present adult-male population, 30% are large, 60% medium, and 10% small, what will be the distribution of sizes among males of the next generation?

8. In a midwestern farming area, wheat fields are classified as poor, fair, good, or excellent in yield per acre. A new variety of wheat is introduced and used in all the fields in the area. As a result of using the new variety, it is expected that a field classified as poor has a 50% chance of remaining in the poor classification, a 40% chance of improving its yield to fair, and a 10% chance of improving to the good classification. For a field rated fair, there is a 5% risk that the new variety will cause the field to drop to poor, a 45% chance that it will retain its fair classification, a 40% chance that it will advance to good, and a 10% chance that its yield will move it up to the excellent classification. If a field is classified as

good, there is a 5% chance that the introduction of the new variety will cause a drop in yield to fair, a 55% chance that the field will continue to have a good yield, and a 40% chance that it will be classified as excellent. A field rated as excellent runs a 5% risk of dropping to good and there is a 95% probability that it will still have an excellent yield with the new variety of wheat. If at present, 15% of all wheat fields in the area are rated as poor, 30% as fair, 40% as good, and 15% as excellent, what will the distribution of classifications be after the introduction of the new variety of wheat?

5.2 Markov Chain Processes

A company that markets a new product, such as the Electra brand margarine we have been using as an example, is not primarily interested in the product's immediate popularity. Rather, the company is concerned with whether the product will continually command a significant share of the market.

In the preceding section we saw that if the company could estimate the probability that a margarine buyer would repeat the brand purchased or switch to another brand, then given the product's share of the market in a base period, it could predict its share in the next period. Now we will see that, if the market has settled down sufficiently so that repeating or switching brands is relatively stable, then it is possible to make a long-range prediction. We again assume that each buyer makes one purchase of margarine in each time period.

To illustrate the meaning of stability we refer to Table 5-3 on page 368. We see that the probability is .23 that an Electra purchaser will buy Electra in the next period, .04 of Electra buyers will switch from Electra to Marigold, .05 of these buyers will switch brands in the opposite direction, and so on. To make long-term predictions we must assume that the probabilities in this table stay the same throughout the prediction period. Furthermore, this behavior is assumed to be independent of the number of buyers who choose each brand. For example, we assume that whether Electra has a large share of the market or a small one, 23% of its buyers in one period will repeat their purchase in the next period.

Changing the Time Period

Recall from the brand-share example that, in the original time period, Electra had been the choice of 14% of all margarine buyers surveyed, that Marigold had been the choice of 20%, leaving 66% for other brands. This is described by the matrix

$$S^{(0)} = \begin{bmatrix} .14 \\ .20 \\ .66 \end{bmatrix}$$

From Table 5-3 we obtain the transition matrix

$$P = \begin{bmatrix} .23 & .05 & .12 \\ .04 & .25 & .15 \\ .73 & .70 & .73 \end{bmatrix}$$

As in the preceding section, we calculate that the market shares in the next time period will be

$$S^{(1)} = PS^{(0)} = \begin{bmatrix} .1214 \\ .1546 \\ .7240 \end{bmatrix}$$

If we think of the "1" time period as a new base for prediction, then the next time period should be numbered "2," and the market shares for the various brands in period 2 will be represented by the matrix

$$S^{(2)} = \begin{bmatrix} p(s_1)^{(2)} \\ p(s_2)^{(2)} \\ p(s_3)^{(2)} \end{bmatrix}$$

where $p(s_1)^{(2)}$ is the probability that a margarine buyer chooses Electra in time period 2, $p(s_2)^{(2)}$ is the probability for Marigold in period 2, and $p(s_3)^{(2)}$ represents the probability of another choice in that time period. Our assumption is that buyer behavior does not change over time. This means that we should use the same transition matrix P to calculate market shares in period 2 as we used in period 1. The situation is exactly as before, except for the base time period (1 instead of 0); thus,

$$S^{(2)} = PS^{(1)}$$

For this particular example,

$$S^{(2)} = \begin{bmatrix} .23 & .05 & .12 \\ .04 & .25 & .15 \\ .73 & .70 & .73 \end{bmatrix} \begin{bmatrix} .1214 \\ .1546 \\ .7240 \end{bmatrix} = \begin{bmatrix} .1225 \\ .1521 \\ .7254 \end{bmatrix}$$

We can see that Electra will continue to hold about 12% of the market in period 2, but Marigold's share will drop slightly from period 1 to period 2.

A Direct Calculation

We could have computed the matrix $S^{(2)}$ another way, without first calculating $S^{(1)}$. We used the two equations

$$S^{(1)} = PS^{(0)}$$
$$S^{(2)} = PS^{(1)}$$

If we substitute the right-hand side of the equation for $S^{(1)}$ in the one for $S^{(2)}$, then

$$S^{(2)} = P(S^{(1)}) = P(PS^{(0)}) = P^2 S^{(0)}$$

Here, $S^{(2)}$, with 2 in parenthesis, indicates the *second time period*, whereas P^2 means PP (that is, P *squared*). Thus, we could first calculate that

$$P^2 = \begin{bmatrix} .23 & .05 & .12 \\ .04 & .25 & .15 \\ .73 & .70 & .73 \end{bmatrix} \begin{bmatrix} .23 & .05 & .12 \\ .04 & .25 & .15 \\ .73 & .70 & .73 \end{bmatrix} = \begin{bmatrix} .1425 & .1080 & .1227 \\ .1287 & .1695 & .1518 \\ .7288 & .7225 & .7255 \end{bmatrix}$$

and then

$$S^{(2)} = \begin{bmatrix} .1425 & .1080 & .1227 \\ .1287 & .1695 & .1518 \\ .7288 & .7225 & .7255 \end{bmatrix} \begin{bmatrix} .14 \\ .20 \\ .66 \end{bmatrix} = \begin{bmatrix} .1225 \\ .1521 \\ .7254 \end{bmatrix}$$

Similarly, if we want to calculate $S^{(3)}$, then because

$$S^{(3)} = PS^{(2)}$$

we obtain

$$S^{(3)} = PS^{(2)} = P(P^2 S^{(0)}) = P^3 S^{(0)}$$

Markov Chain Processes

The long-range prediction of brand share illustrates an important type of probability problem. There is a sample space U and a fixed set of states $s_1, s_2, \ldots, s_k$, as in the preceding section. We still have an experiment that puts each element of U into exactly one of the states $s_1, s_2, \ldots, s_k$. The new aspect of the situation is that we now suppose that the experiment is repeated several times, each time under identical circumstances. This means that the probability p_{ij}, that an element of U will change from state s_j to state s_i as a result of the experiment is the same each time the experiment takes place. The transition matrix P of the probabilities p_{ij} therefore remains the same each time. A sequence of experiments of this type is called a **Markov chain process**. We still write $p(s_i)^{(0)}$ for the probability that an element of U is in state s_i at the beginning of the process and $S^{(0)}$ for the corresponding $k \times 1$-matrix. Let $p(s_i)^{(m)}$ stand for the probability that a member of U is in state s_i after the experiment has been repeated m times (corresponding to the mth time period). For $S^{(m)}$, the $k \times 1$-matrix with $p(s_i)^{(m)}$ in the ith row, we have the equation

$$\boxed{S^{(m)} = P^m S^{(0)}}$$

in which P^m is the mth power of the matrix P, that is, the product of m factors of P.

Example 1. In an animal species, the percentage of male offspring of large males that are also large is 50%, that are medium sized is 40%, and that are small is 10%. If the father is of medium size, then 20% of his male offspring will be large, 60% will be medium sized and 20% will be small. If the father is small, the corresponding percentages are 5% large, 70% medium, and 25% small. If in the present adult-male population, 20% are large, 70% are medium sized, and 10% are small, and if the prob-

5.2 MARKOV CHAIN PROCESSES

abilities of male offspring of various sizes do not change from generation to generation, what will be the distribution of sizes of males in the fourth generation?

Solution

We express the given information in terms of the matrices

$$P = \begin{bmatrix} .50 & .20 & .05 \\ .40 & .60 & .70 \\ .10 & .20 & .25 \end{bmatrix} \text{ and } S^{(0)} = \begin{bmatrix} .20 \\ .70 \\ .10 \end{bmatrix}$$

where $S^{(0)}$ describes the present adult-male population. We want to know $S^{(4)}$, which, according to the general theory, is calculated by the formula

$$S^{(4)} = P^4 S^{(0)}$$

To two decimal places of accuracy,

$$P^4 = \begin{bmatrix} .26 & .24 & .24 \\ .56 & .57 & .57 \\ .18 & .19 & .19 \end{bmatrix}$$

and therefore,

$$S^{(4)} = \begin{bmatrix} .26 & .24 & .24 \\ .56 & .57 & .57 \\ .18 & .19 & .19 \end{bmatrix} \begin{bmatrix} .20 \\ .70 \\ .10 \end{bmatrix} = \begin{bmatrix} .24 \\ .57 \\ .19 \end{bmatrix}$$

Thus, four generations from now, 24% of the males will be large, 57% will be of medium size, and 19% will be small. ∎

Equilibrium

Next, let us return to the margarine brand-share problem and calculate one more time period, using the same transition matrix

$$P = \begin{bmatrix} .23 & .05 & .12 \\ .04 & .25 & .15 \\ .73 & .70 & .73 \end{bmatrix}$$

we used before. The next time period is number 3, and since we previously showed that

$$S^{(2)} = \begin{bmatrix} .12 \\ .15 \\ .73 \end{bmatrix}$$

we obtain

$$S^{(3)} = PS^{(2)} = \begin{bmatrix} .23 & .05 & .12 \\ .04 & .25 & .15 \\ .73 & .70 & .73 \end{bmatrix} \begin{bmatrix} .12 \\ .15 \\ .73 \end{bmatrix} = \begin{bmatrix} .12 \\ .15 \\ .73 \end{bmatrix}$$

where, once again, we round to two decimal places. Thus, to this degree of accuracy, there is no change in market penetration for Electra and Marigold brands from period 2 to period 3. We might therefore suspect that the margarine market has settled down. Electra can expect to continue to receive 12% of all margarine sales and Marigold 15%. We can demonstrate this by supposing that at some future period m

$$S^{(m)} = \begin{bmatrix} .12 \\ .15 \\ .73 \end{bmatrix}$$

Then matrix multiplication shows us that when we round to two decimal places,

$$S^{(m+1)} = PS^{(m)} = \begin{bmatrix} .23 & .05 & .12 \\ .04 & .25 & .15 \\ .73 & .70 & .73 \end{bmatrix} \begin{bmatrix} .12 \\ .15 \\ .73 \end{bmatrix} = \begin{bmatrix} .12 \\ .15 \\ .73 \end{bmatrix}$$

This implies that the market shares for Electra and Marigold will remain constant. The conclusion, of course, assumes that the customers' behavior with regard to brand loyalty does not change. In fact, the margarine market does reach equilibrium and Electra receives a 12% brand share.

Thus, it may happen that in a Markov chain process, after a certain number of repetitions of the experiment, there is no further change in the probabilities that an element of U will be in each of the possible states. In symbols, we say that for some m,

$$S^{(m)} = S^{(m+1)} = S^{(m+2)} = \ldots$$

In this case we say the Markov chain process has reached **equilibrium**. Since Markov chain processes are used most often to find out how something will turn out in the long run—for instance, what share of the market a product can hope to command on a (more-or-less) permanent basis, the question of whether the process reaches equilibrium is crucial. (Strictly speaking, $S^{(m)}$, $S^{(m+1)}$, and so on will usually change very slightly as they "approach equilibrium," and they never quite reach this state. However, because we calculate the numbers (in these matrices) only to a fixed number of decimal places, the matrices will appear to be equal from some point on.)

Approaching Equilibrium

It can be demonstrated that for every Markov chain process there is a matrix S for which $PS = S$, where P is the transition matrix of the process. However, it may not be possible to reach it by calculating $S^{(0)}$, $S^{(1)} = PS^{(0)}$, $S^{(2)} = P^2 S^{(0)}$, and so on, for a given matrix $S^{(0)}$. For instance, if

$$P = \begin{bmatrix} 0 & 1 \\ 1 & 0 \end{bmatrix}$$

and if it happens that we start with

$$S^{(0)} = \begin{bmatrix} \frac{1}{2} \\ \frac{1}{2} \end{bmatrix}$$

then it is easy to see that

$$S^{(1)} = PS^{(0)} = \begin{bmatrix} 0 & 1 \\ 1 & 0 \end{bmatrix} \begin{bmatrix} \frac{1}{2} \\ \frac{1}{2} \end{bmatrix} = \begin{bmatrix} \frac{1}{2} \\ \frac{1}{2} \end{bmatrix}$$

so that this initial matrix $S^{(0)}$ gives equilibrium for this particular Markov chain. On the other hand, if we begin with the matrix

$$S^{(0)} = \begin{bmatrix} \frac{1}{4} \\ \frac{3}{4} \end{bmatrix}$$

then

$$S^{(1)} = \begin{bmatrix} \frac{3}{4} \\ \frac{1}{4} \end{bmatrix}, \quad S^{(2)} = \begin{bmatrix} \frac{1}{4} \\ \frac{3}{4} \end{bmatrix}, \quad S^{(3)} = \begin{bmatrix} \frac{3}{4} \\ \frac{1}{4} \end{bmatrix}, \quad \text{and} \quad S^{(4)} = \begin{bmatrix} \frac{1}{4} \\ \frac{3}{4} \end{bmatrix}$$

and so on. Thus, for this choice of $S^{(0)}$, the $S^{(m)}$ jump back and forth, depending on whether m is even or odd, and the process never reaches equilibrium.

Powers of the Transition Matrix

We know that $S^{(m)} = P^m S^{(0)}$; if we want to understand how $S^{(m)}$, $S^{(m+1)}$, $S^{(m+2)}$, and so on, behave, it makes sense to try to understand what happens when we continue to multiply P by itself to form P^m, P^{m+1}, P^{m+2}, and so on. Let us experiment with the animal-size transition matrix of Example 1, which was

$$P = \begin{bmatrix} .50 & .20 & .05 \\ .40 & .60 & .70 \\ .10 & .20 & .25 \end{bmatrix}$$

In that example, we obtained

$$P^4 = \begin{bmatrix} .26 & .24 & .24 \\ .56 & .57 & .57 \\ .18 & .19 & .19 \end{bmatrix}$$

Continued calculation will establish that, to two decimal places,

$$P^6 = \begin{bmatrix} .25 & .25 & .25 \\ .57 & .57 & .57 \\ .18 & .18 & .18 \end{bmatrix}$$

which is a matrix with an unusual structure: all the columns are the same. Such a matrix behaves in an unusual way. Suppose we start with a matrix M of the form

$$M = \begin{bmatrix} a \\ b \\ c \end{bmatrix}$$

where $a + b + c = 1$. When we calculate

$$P^6 M = \begin{bmatrix} .25 & .25 & .25 \\ .57 & .57 & .57 \\ .18 & .18 & .18 \end{bmatrix} \begin{bmatrix} a \\ b \\ c \end{bmatrix} = \begin{bmatrix} .25a + .25b + .25c \\ .57a + .57b + .57c \\ .18a + .18b + .18c \end{bmatrix} = \begin{bmatrix} .25(a+b+c) \\ .57(a+b+c) \\ .18(a+b+c) \end{bmatrix}$$

$$= \begin{bmatrix} .25(1) \\ .57(1) \\ .18(1) \end{bmatrix} = \begin{bmatrix} .25 \\ .57 \\ .18 \end{bmatrix}$$

we see that the repeated column of P^6 is duplicated. Then, if we calculate $P^7 = P^6 P$, we obtain

$$P^7 = P^6 P = \begin{bmatrix} .25 & .25 & .25 \\ .57 & .57 & .57 \\ .18 & .18 & .18 \end{bmatrix} \begin{bmatrix} .50 & .20 & .05 \\ .40 & .60 & .70 \\ .10 & .20 & .25 \end{bmatrix} = \begin{bmatrix} .25 & .25 & .25 \\ .57 & .57 & .57 \\ .18 & .18 & .18 \end{bmatrix} = P^6$$

because, in the matrix multiplication, each column of P is multiplied by all of P^6 and the numbers in each column of P add up to one. The same thing will happen if we multiply by P again; therefore, $P^6 = P^7 = P^8$, and so on. If we start with any initial matrix $S^{(0)}$, since $S^{(m)} = P^m S^{(0)}$, then $S^{(6)} = S^{(7)} = S^{(8)}$, and so on. Thus, no matter which matrix $S^{(0)}$ we start with, we will reach equilibrium at $S^{(6)}$. Furthermore, we know what the equilibrium is. The numbers in $S^{(0)}$ add up to 1; thus, $S^{(6)} = P^6 S^{(0)}$ looks just like the columns of P^6. Therefore, the equilibrium distribution of animal sizes is given by

$$S^{(6)} = \begin{bmatrix} .25 \\ .57 \\ .18 \end{bmatrix}$$

That is, 25% of the males will be large, 57% will be medium, and 18% will be small; these percentages will remain constant.

Positive Transition Matrices

To summarize, we have a sample space U, states $s_1, s_2, \ldots, s_k$, and a transition matrix P of probabilities p_{ij} that an element of U changes from state j to state i. We have seen that if we have a transition matrix P in which the columns of some power P^m are all the same (to some degree of accuracy), then starting with any $S^{(0)}$, the Markov process reaches equilibrium at $S^{(m)} = P^m S^{(0)}$ and the equilibrium matrix $S^{(m)}$ is identical with the columns of P^m. What was so special about the transition matrix

$$P = \begin{bmatrix} .50 & .20 & .05 \\ .40 & .60 & .70 \\ .10 & .20 & .25 \end{bmatrix}$$

of the animal-size example that caused P^6 to have such an unusual structure? The answer is that all the numbers in this transition matrix are positive. That is, the conditional probability p_{ij} of a change from any state j to state i is greater than zero. Unlike the transition matrix

$$P = \begin{bmatrix} 0 & 1 \\ 1 & 0 \end{bmatrix}$$

which, as we saw, behaves so differently, the animal-size transition matrix contains no zeros. We call a matrix **positive** if all the numbers in it are positive. When we form the powers P, P^2, P^3 of any positive transition matrix, with each multiplication the numbers in each row come to resemble each other more and more. For instance, in the animal-size example

$$P = \begin{bmatrix} .50 & .20 & .05 \\ .40 & .60 & .70 \\ .10 & .20 & .25 \end{bmatrix}, \quad P^2 = \begin{bmatrix} .33 & .223 & .18 \\ .51 & .58 & .61 \\ .16 & .19 & .21 \end{bmatrix}, \quad \text{and} \quad P^3 = \begin{bmatrix} .28 & .24 & .22 \\ .55 & .57 & .59 \\ .17 & .19 & .19 \end{bmatrix}$$

$$P^4 = \begin{bmatrix} .26 & .24 & .24 \\ .56 & .57 & .57 \\ .18 & .19 & .19 \end{bmatrix}, \quad P^5 = \begin{bmatrix} .25 & .25 & .24 \\ .57 & .57 & .57 \\ .18 & .18 & .19 \end{bmatrix}, \quad \text{and} \quad P^6 = \begin{bmatrix} .25 & .25 & .25 \\ .57 & .57 & .57 \\ .18 & .18 & .18 \end{bmatrix}$$

To see what is happening, use the symbol z for the smallest number in the transition matrix P. Then the largest number in P must be no bigger than $1 - z$ because the numbers in the column containing it add up to one. Now take any row of P; then the difference between the largest and smallest numbers in that row must be no greater than the largest number in the entire matrix minus the smallest number. In other words, this difference must be no greater than

$$(1 - z) - z = 1 - 2z$$

If we carefully analyzed the matrix multiplication of the positive transition matrix P with itself, we would find that in P^2 the difference between the largest and smallest numbers in any row could be no greater than $(1 - 2z)^2$, where z still means the smallest number in the original matrix P itself. Generally, in P^m the corresponding difference is no greater than $(1 - 2z)^m$. The reason that, for large m, each row of P^m consists of a single (repeated) number, and thus, the columns of P^m are identical, is that the row differences must be less than $(1 - 2z)^m$, which gets very small as m becomes large. For instance, if the smallest number in P is $z = .1$,

then
$$(1 - 2(.1))^{20} = (.8)^{20} = .01$$
so that the difference between any numbers in the same row of P^{20} must be less than .01. In fact, they would all look the same when the answer is rounded to two decimal places.

Regular Markov Chains

In order to have the sort of equilibrium behavior we saw in the animal-size example, we need not insist that the transition matrix be positive. For instance, the transition matrix

$$P = \begin{bmatrix} 0 & .4 \\ 1 & .6 \end{bmatrix}$$

is not positive, but

$$P^2 = \begin{bmatrix} .4 & .24 \\ .6 & .76 \end{bmatrix}$$

is positive; thus, if we calculate the powers of P^2, after several multiplications the columns of the matrices we obtain will come to resemble one another. For instance, for $P^4 = (P^2)^2$ we already have,

$$P^4 = \begin{bmatrix} .30 & .28 \\ .70 & .72 \end{bmatrix}$$

to two places of accuracy. A transition matrix P of a Markov chain process is said to be **regular** if some power P^r is positive. In this case we say that the process itself is **regular**. As we have seen, if we continue to multiply a regular transition matrix by itself to form powers, eventually we will obtain a matrix P^m in which all the columns are identical and equal to the equilibrium matrix, which is $S^{(m)} = P^m S^{(0)}$, for any choice of $S^{(0)}$.

Identifying Regular Transition Matrices

If a transition matrix P is regular, the locations of the zeros in P, if there are any, will "fill in" with positive numbers as we form the powers P^2, P^3 and so on, as happened at P^2 in the 2 × 2-example just given. If for any 2 × 2-transition matrix P there are still zeros in P^2, then P cannot be regular and you do not have to continue to multiply (see Exercises 16 through 18). In the 3 × 3-case, it can be shown that if P^5 contains zeros, then P is not regular. In general, if there are zeros in the $[(n - 1)^2 + 1]$th power of an $n \times n$-transition matrix, then the matrix is not regular. Thus, we can decide on the regularity of a transition matrix by taking a predetermined power of it.

5.2 MARKOV CHAIN PROCESSES

Example 2. Show that the following transition matrix P is regular.

$$P = \begin{bmatrix} 0.90 & 0.30 & 0.50 \\ 0.10 & 0.60 & 0 \\ 0 & 0.10 & 0.50 \end{bmatrix}$$

Solution

Since P itself contains zeros, we calculate

$$P^2 = \begin{bmatrix} .84 & .50 & .70 \\ .15 & .39 & .05 \\ .01 & .11 & .25 \end{bmatrix}$$

which has no zeroes; this shows that P is regular. ∎

Example 3. Show that if a transition matrix is of the form

$$P = \begin{bmatrix} 1 & a \\ 0 & b \end{bmatrix}$$

where a and b are positive numbers, then it cannot be regular.

Solution

Notice that $p_{21} = 0$, where p_{21} is the entry in the second row and first column of P. The corresponding entry is still zero in P^2:

$$P^2 = \begin{bmatrix} 1 & a \\ 0 & b \end{bmatrix} \begin{bmatrix} 1 & a \\ 0 & b \end{bmatrix} = \begin{bmatrix} 1 & a+ab \\ 0 & b^2 \end{bmatrix}$$

The reason is that to obtain the number in the second row and first column of the product, we multiply the second row of P by the first column. Thus, we are calculating

$$\begin{bmatrix} 0 & b \end{bmatrix} \begin{bmatrix} 1 \\ 0 \end{bmatrix} = 0 \cdot 1 + b \cdot 0 = 0$$

When we go on to $P(P^2) = P^3$, we notice that the first column of P^2 is identical with the first column of P; we make exactly the same calculation to fill the p_{21}-location of P^3. No matter how many times we repeat this, the calculation for the p_{21}-location will never change; so, every power P^m will have $p_{21} = 0$. Therefore, P is not regular. ∎

Once we know that a Markov chain process with transition matrix P is regular (because some P^r is positive), we can reach the equilibrium of the process by taking the powers P^r, $(P^r)^2 = P^{2r}$, P^{3r}, and so on, until the columns become identical. There is, however, a more efficient technique for determining the equilibrium $S^{(m)}$ of a regular Markov process; this is based on the Gauss-Jordan method of Chapter 1. We will describe it in the next section.

Summary

A **Markov chain process** is a sequence of identical experiments of the following type: There is a sample space U and a set of states $s_1, s_2, \ldots, s_k$. The effect of the experiment is to place each element of U in one of the states. The probability p_{ij} that an element in state s_j at the beginning of the experiment will be in a state s_i at the conclusion is the same each time the experiment is performed. The $k \times k$-matrix P of the p_{ij} is the **transition matrix** of the Markov chain process.

Let $p(s_i)^{(0)}$ be the probability that an element of U is in state s_i before the first performance of the experiment in a Markov chain process and let $p(s_i)^{(m)}$ be the probability that an element is in state s_i after the experiment has taken place m times. Define $S^{(0)}$ to be the $k \times 1$-matrix of the $p(s_i)^{(0)}$ and $S^{(m)}$ to be the $k \times 1$-matrix of the $p(s_i)^{(m)}$. Then $S^{(m)} = P^m S^{(0)}$, where P^m is the product of m copies of the transition matrix P. A Markov chain process reaches **equilibrium** if $S^{(m)} = S^{(m+1)}$ for some m. A Markov chain process is **regular** if its transition matrix is **regular**, that is, if for some r, the power P^r is a **positive** matrix (all the numbers in it are positive). If a transition matrix P is regular, then for large enough m, the columns of P^m are all the same (to the prescribed level of accuracy); starting with any $S^{(0)}$, the Markov process reaches equilibrium at $S^{(m)} = P^m S^{(0)}$ and the equilibrium matrix $S^{(m)}$ is identical with the columns of P^m.

Exercises

In Exercises 1 through 4, calculate $S^{(1)}$, $S^{(2)}$, and $S^{(3)}$ for the Markov chain process with the given $S^{(0)}$ and P.

1. $S^{(0)} = \begin{bmatrix} .30 \\ .70 \end{bmatrix}$ and $P = \begin{bmatrix} .25 & .05 \\ .75 & .95 \end{bmatrix}$

2. $S^{(0)} = \begin{bmatrix} .10 \\ .80 \\ .10 \end{bmatrix}$ and $P = \begin{bmatrix} 0 & 0.10 & 0.60 \\ 0.20 & 0 & 0 \\ 0.80 & 0.90 & 0.40 \end{bmatrix}$

3. $S^{(0)} = \begin{bmatrix} 0 \\ 1 \\ 0 \end{bmatrix}$ and $P = \begin{bmatrix} 0.20 & 0 & 0.40 \\ 0.20 & 0.40 & 0.20 \\ 0.60 & 0.60 & 0.40 \end{bmatrix}$

4. $S^{(0)} = \begin{bmatrix} .30 \\ .30 \\ .40 \end{bmatrix}$ and $P = \begin{bmatrix} 1 & .70 & 0 \\ 0 & .10 & 0 \\ 0 & .20 & 1 \end{bmatrix}$

5. Suppose that in some population of animals there is a genetic defect carried only by females. The probability is .80 that the female offspring of a defective female will also be defective. A female whose female parent does not have the defect still has a probability of .10 of possessing the defect through mutation. If, in some generation, 25% of the females are defective, what proportion of females in the next three generations will be defective?

6. The weather in Central City is classified as fair, cloudy (without rain), and rainy. An investigation of weather records indicates the probability that a fair day will be followed by another fair day is .50, by a cloudy day, .30, and by a rainy day, .20. A cloudy day has a .40 probability of being followed by a fair day, a .40 probability by a cloudy day, and a .20 probability by a rainy day. A rainy day has a .45 chance of being followed by a fair day, a .25 probability by a cloudy day, and a .30 probability by another day of rain. If it is raining today, that is, if

$$S^{(0)} = \begin{bmatrix} 0 \\ 0 \\ 1 \end{bmatrix}$$

what is the probability of fair weather, cloudy weather, and rain in each of the next three days?

In Exercises 7 and 8, calculate $S^{(3)}$ for the given $S^{(0)}$ and P, without also calculating $S^{(1)}$ and $S^{(2)}$, but rather by calculating P^3.

7. $S^{(0)} = \begin{bmatrix} .22 \\ .78 \end{bmatrix}$ and $P = \begin{bmatrix} .20 & .40 \\ .80 & .60 \end{bmatrix}$

8. $S^{(0)} = \begin{bmatrix} .10 \\ .20 \\ .70 \end{bmatrix}$ and $P = \begin{bmatrix} 0 & 0 & 0.30 \\ 0.50 & 0.80 & 0.70 \\ 0.50 & 0.20 & 0 \end{bmatrix}$

9. A study of coffee-buying habits indicates that about 70% of the people who bought Sunrise coffee in a given buying period would again choose this brand in the next period, whereas 20% of those who chose another brand in the given period would switch to Sunrise in the next period. If, at the time of the study, 34% of the coffee buyers chose Sunrise, calculate P^3 to find Sunrise's share of the market three time periods after the initial one.

10. Cedar Grove College's records indicate that for each class—freshman, sophomore, and junior—each year 10% of its humanities majors change to a major in the social sciences, whereas the rest remain in the humanities. Records also show that 10% of social-science majors in each class change to majors in the humanities, 10% change to the natural sciences, and the rest remain in the social sciences. Natural-science majors, however, do not change out of that area of studies. If at

that college 50% of the freshman class choose majors in the humanities, 30% in the social sciences, and 20% in the natural sciences, calculate the proportion of majors in each area for these same students in their senior year without doing the same for the intervening years (that is, calculate P^3 directly).

In Exercises 11 and 12, verify that the given transition matrix is regular.

11. $\begin{bmatrix} 0 & 0.20 & 0 \\ 0.10 & 0 & 0.30 \\ 0.90 & 0.80 & 0.70 \end{bmatrix}$

12. $\begin{bmatrix} 0.50 & 0.10 & 0 & 0 \\ 0 & 0 & 0.20 & 0 \\ 0 & 0.90 & 0 & 0.10 \\ 0.50 & 0 & 0.80 & 0.90 \end{bmatrix}$

In Exercises 13 and 14, verify that the given Markov chain process reaches equilibrium at the indicated values if numbers are rounded to two decimal places.

13. $S^{(0)} = \begin{bmatrix} 1 \\ 0 \end{bmatrix}$ and $P = \begin{bmatrix} .15 & .20 \\ .85 & .80 \end{bmatrix}$; Equilibrium $= \begin{bmatrix} .19 \\ .81 \end{bmatrix}$

14. $S^{(0)} = \begin{bmatrix} 0.33 \\ 0.34 \\ 0.33 \end{bmatrix}$ and $P = \begin{bmatrix} 0.10 & 0.30 & 0.70 \\ 0 & 0.20 & 0 \\ 0.90 & 0.50 & 0.30 \end{bmatrix}$; Equilibrium $= \begin{bmatrix} 0.44 \\ 0 \\ 0.56 \end{bmatrix}$

15. Suppose there is a society in which all women are classified in the opinion of the community as either noble or common on the basis of both birth and accomplishments. Suppose also that the probability that the daughter of a noblewoman will be a noble is .70, but the probability of attaining nobility drops to just .10 for the daughter of a common woman. Show that, in the long run, the society will stabilize with 25% of all women classified as noble.

16. Find the four 2 × 2-transition matrices containing two zeros and show that none of them are regular.

17. For positive numbers a and b, which of the following transition matrices are regular?

$$P_1 = \begin{bmatrix} 0 & a \\ 1 & b \end{bmatrix}, \quad P_2 = \begin{bmatrix} a & 0 \\ b & 1 \end{bmatrix}, \quad \text{and} \quad P_3 = \begin{bmatrix} a & 1 \\ b & 0 \end{bmatrix}$$

18. Suppose that for a 2 × 2-transition matrix P, the matrix P^2 contains a zero. Use Exercises 16 and 17 to explain why P cannot be regular.

5.3 Equilibrium

A Markov chain process reaches equilibrium after m repetitions of the experiment if

$$S^{(m)} = S^{(m+1)} = S^{(m+2)} = \dots$$

The probability that an element of the sample space U will be in each of the states $s_1, s_2, \dots, s_k$ does not change as a result of further repetitions of the experiment.

From the preceding section we know that once we reach the stage m for which $S^{(m)} = S^{(m+1)}$, we have reached equilibrium and need not check that

$$S^{(m+1)} = S^{(m+2)} = S^{(m+3)} = \dots$$

To determine the equilibrium, we need to find the matrix $S^{(m)}$ for which $S^{(m+1)} = S^{(m)}$. Since $PS^{(m)} = S^{(m+1)}$, we see that the matrix equation we must solve is

$$PS^{(m)} = S^{(m)}$$

A Sales Problem

We will illustrate the procedure for finding the matrix $S^{(m)}$ by analyzing the following problem. A salesperson for a manufacturer of hand tools calls on hardware stores at regular intervals. The hardware stores do not need to order tools from the salesperson on each visit because they may have enough stock on hand to last until the next visit. However, each store will order fairly often because it does not want to tie up money and display space with a large inventory of hand tools. The salesperson's experience is that 20% of the stores that ordered tools during the preceding visit will order again on the next round, whereas 70% of the stores that did not order the preceding time will place an order. If this process has reached equilibrium, at what percentage of the stores can the salesperson expect to receive an order for tools? We summarize the probabilities in Table 5-8.

		Previous Visit	
		Order	No Order
Present	Order	.20	.70
Visit	No Order	.80	.30

Table 5-8

The transition matrix for this Markov chain process is therefore

$$P = \begin{bmatrix} .20 & .70 \\ .80 & .30 \end{bmatrix}$$

The Matrix Problem

On the mth visit we will have a matrix

$$S^{(m)} = \begin{bmatrix} p(s_1)^{(m)} \\ p(s_2)^{(m)} \end{bmatrix}$$

in which $p(s_1)^{(m)}$ is the probability a store will order on the mth visit and $p(s_2)^{(m)}$ is the probability it will not. The salesperson wants to find the equilibrium matrix, that is, the matrix $S^{(m)}$ for which $PS^{(m)} = S^{(m)}$. Thinking in matrix terms, we are trying to solve a matrix equation for a 2×1-matrix; so, we can simplify the notation by replacing $S^{(m)}$ by a matrix of unknowns

$$X = \begin{bmatrix} x \\ y \end{bmatrix}$$

and attempting to solve the matrix equation

$$PX = X$$

where P is the transition matrix. Subtracting X from both sides of the equation gives

$$PX - X = O$$

where O denotes the 2×1-matrix of zeros. As in Chapter 1, we can use the property of the 2×2-identity matrix I, that $IX = X$, and the distributive law to rewrite the equation as

$$PX - X = PX - IX = (P - I)X = O$$

There is an obvious solution to the equation, namely

$$X = 0 = \begin{bmatrix} 0 \\ 0 \end{bmatrix}$$

because $(P - I)O = O$. But $X = O$ is a useless solution because the numbers in $S^{(m)}$ that X replaces in the equation represent the probability a store will place an order, and the probability it will not. Certainly these probabilities must add up to one. Thus, the matrix X must not only be a solution of the equation $PX = X$, that is $(P - I)X = O$, but the numbers in X must also add up to one:

$$x + y = 1$$

In order to combine the matrix equation $(P - I)X = O$ and the ordinary equation $x + y = 1$ into a single problem, we recall that

$$P = \begin{bmatrix} .20 & .70 \\ .80 & .30 \end{bmatrix}$$

and we write out the matrix equation $(P - I)X = O$ as

$$(P - I)X = \left(\begin{bmatrix} .20 & .70 \\ .80 & .30 \end{bmatrix} - \begin{bmatrix} 1 & 0 \\ 0 & 1 \end{bmatrix} \right) \begin{bmatrix} x \\ y \end{bmatrix} = \begin{bmatrix} -.80 & .70 \\ .80 & -.70 \end{bmatrix} \begin{bmatrix} x \\ y \end{bmatrix}$$

$$= \begin{bmatrix} (-.80)x + .70y \\ .80x + (-.70)y \end{bmatrix} = \begin{bmatrix} 0 \\ 0 \end{bmatrix}$$

So, we see that the matrix equation $(P - I)X = O$ represents two ordinary equations,

$$(-.80)x + .70y = 0$$
$$.80x + (-.70)y = 0$$

Now we combine these equations with $x + y = 1$:

$$x + y = 1$$
$$(-.80)x + .70y = 0$$
$$.80x + (-.70)y = 0$$

We can write this system of linear equations in matrix form. If we let

$$A = \begin{bmatrix} 1 & 1 \\ -0.80 & 0.70 \\ 0.80 & -0.70 \end{bmatrix}$$

then AX represents the left-hand sides of the equations. Representing the right-hand sides by

$$B = \begin{bmatrix} 1 \\ 0 \\ 0 \end{bmatrix}$$

the preceding system of three linear equations becomes the single matrix equation $AX = B$.

In conclusion, we have replaced the two conditions on X, namely, $x + y = 1$ and $PX = X$, by a single matrix equation $AX = B$, where A is of the form

$$A = \begin{bmatrix} 1 & 1 \\ P - I & \end{bmatrix}, \quad \text{i.e.,} \quad A = \begin{bmatrix} 1 & 1 \\ -0.80 & 0.70 \\ 0.80 & -0.70 \end{bmatrix}$$

(I is the 2×2-identity matrix) and

$$B = \begin{bmatrix} 1 \\ 0 \\ 0 \end{bmatrix}$$

The General Equilibrium Problem

We can use the same approach to find the equilibrium $PS^{(m)} = S^{(m)}$ for a Markov chain process with any number of states $s_1, s_2, \ldots, s_k$. If we replace $S^{(m)}$ by a matrix

$$X = \begin{bmatrix} x_1 \\ x_2 \\ \vdots \\ x_k \end{bmatrix}$$

of unknowns, the numbers in X are probabilities that must add up to 1. Therefore, we must require

and solve $(P - I)X = O$. These conditions can be represented by the single matrix equation $AX = B$, where

$$A = \begin{bmatrix} 1 & 1 & \cdots & 1 \\ & P-I & & \end{bmatrix}$$

(the first row consists of k 1s and I is the $k \times k$-identity matrix) and

$$B = \begin{bmatrix} 1 \\ 0 \\ \vdots \\ 0 \end{bmatrix}$$

The equilibrium problem has been reduced to solving a matrix equation of the form $AX = B$, where X is a $k \times 1$-matrix of unknowns. The matrix A is a $(k+1) \times k$-matrix since P is $k \times k$ and we have added one row: the top row of 1s. Therefore, B is a $(k+1) \times 1$-matrix.

Applying the Gauss-Jordan Method

We studied matrix equations $AX = B$ of this type in Section 1.7. We applied the Gauss-Jordan method to the augmented matrix

$$[A \mid B]$$

to produce a matrix

$$[A' \mid B']$$

where the columns of A' were as much like those of an identity matrix as possible, so that it was easy to find the solutions of $A'X = B'$ that are also the solutions of $AX = B$. In Chapter 1 we encountered examples of matrix equations of the form $AX = B$ that had an infinite number of solutions and others that had no solution at all. However, if the transition matrix P is regular, there is a single solution that is the equilibrium matrix for the Markov chain process. As we saw in the last section, this matrix is identical to the columns of P^m for m sufficiently large.

We will apply the Gauss-Jordan method to the tool salesperson's problem described earlier in this section, in which

$$A = \begin{bmatrix} 1 & 1 \\ -0.80 & 0.70 \\ 0.80 & -0.70 \end{bmatrix}, \quad X = \begin{bmatrix} x \\ y \end{bmatrix}, \quad \text{and} \quad B = \begin{bmatrix} 1 \\ 0 \\ 0 \end{bmatrix}$$

where $x = p(s_1)^{(m)}$ is the probability that the store will place an order at the mth visit, and $y = p(s_2)^{(m)}$ is the probability that it will not. We express the numbers of the augmented matrix in fraction form for greater accuracy in calculation.

$$[A|B] = \begin{bmatrix} \overset{x}{1} & \overset{y}{1} & | & 1 \\ -0.80 & 0.70 & | & 0 \\ 0.80 & -0.70 & | & 0 \end{bmatrix} = \begin{bmatrix} \overset{x}{1} & \overset{y}{1} & | & 1 \\ -\frac{8}{10} & \frac{7}{10} & | & 0 \\ \frac{8}{10} & -\frac{7}{10} & | & 0 \end{bmatrix}$$

Since we already have a 1 in the upper right-hand corner, we next produce a 0 below it by adding $\frac{8}{10}$ times the first row to the second row:

$$\begin{bmatrix} 1 & 1 & | & 1 \\ 0 & \frac{15}{10} & | & \frac{8}{10} \\ \frac{8}{10} & -\frac{7}{10} & | & 0 \end{bmatrix}$$

Then we add $-\frac{8}{10}$ times the first row to the third row.

$$\begin{bmatrix} 1 & 1 & | & 1 \\ 0 & \frac{15}{10} & | & \frac{8}{10} \\ 0 & -\frac{15}{10} & | & -\frac{8}{10} \end{bmatrix}$$

Now that the first column is that of an identity matrix, we obtain a 1 in the main-diagonal location of the second column by dividing the second row by the number presently there, $\frac{15}{10}$, which is the same as multiplying the row by $\frac{10}{15}$:

$$\begin{bmatrix} 1 & 1 & | & 1 \\ 0 & 1 & | & \frac{8}{15} \\ 0 & -\frac{15}{10} & | & -\frac{8}{10} \end{bmatrix}$$

Subtracting the second row from the first changes the number at the top of the second column to a zero:

$$\begin{bmatrix} 1 & 0 & | & \frac{7}{15} \\ 0 & 1 & | & \frac{8}{15} \\ 0 & -\frac{15}{10} & | & -\frac{8}{10} \end{bmatrix}$$

We complete the second column by adding $\frac{15}{10}$ times the second row to the third row:

394 MARKOV CHAINS

$$\begin{bmatrix} & x & y & \\ 1 & 0 & \bigg| & \frac{7}{15} \\ 0 & 1 & \bigg| & \frac{8}{15} \\ 0 & 0 & \bigg| & 0 \end{bmatrix}$$

We can eliminate the last row, which tells us only that $0 = 0$, and we easily see that the solution of $AX = B$ is $x = \frac{7}{15}$ and $y = \frac{8}{15}$. This means the salesperson can expect to receive an order from $\frac{7}{15}$, or about 47%, of the hardware stores visited on each round.

A Simplification

The last step of the Gauss-Jordan method turned the bottom row of the matrix into a row of zeros, so we could eliminate it. The technique was thus telling us that the information in that row was already contained in the rows above it. That is hardly surprising when we look back at the beginning of the process:

$$\begin{bmatrix} & x & y & \\ 1 & 1 & \bigg| & 1 \\ -0.80 & 0.70 & \bigg| & 0 \\ 0.80 & -0.70 & \bigg| & 0 \end{bmatrix}$$

The numbers in the last row are just the same as those in the second row, but with pluses and minuses reversed. To look at it another way, the second and third rows stand for the equations

$$-.80x + .70y = 0$$
$$.80x - .70y = 0$$

and clearly, the last row can be obtained from the second by multiplying through by -1. So we can see from the start that the last row will eventually reduce to the statement $0 = 0$ and will be eliminated because it contains exactly the same information as the row above it. We could have dropped that last row to start with, and saved quite a bit of arithmetic.

In fact, for a transition matrix P of any size, the last row of the augmented matrix for the equilibrium problem $AX = B$ will eventually reduce to a row of zeros. Thus, we remove that row, and in the remaining augmented matrix we have a square matrix to the left of the vertical line. Applying the Gauss-Jordan method, we get an identity matrix to the left of the line and, therefore, the equilibrium probabilities to the right. We illustrate this simplification in the next example.

Example 1. An automobile-leasing company's study indicates that 60% of its regular customers who lease a two-door sedan in one time period will lease the same type of car the next time, 30% will change to a four-door sedan, and the remaining 10% will switch to a station wagon. Of the customers who lease a four-door sedan, 20% will change to a two-door sedan the next time, 60% will choose a four-door sedan again, and 20% will change to a station wagon. Of the customers who leased a station wagon in the preceding period, none will switch to a two-door sedan, 50% will change to a four-door sedan, and the other 50% will again choose a station wagon. Suppose these buying patterns remain constant over a sufficiently long period so that the demand for each type of car is the same from period to period. When the leasing company orders new cars from the manufacturer, what should be the percentages of two-door sedans, four-door sedans, and station wagons?

Solution

The matrix of unknowns is

$$X = \begin{bmatrix} x \\ y \\ z \end{bmatrix}$$

where x refers to two-door sedans, y to four-door sedans, and z to station wagons. From the probability information in the example, we obtain Table 5-9.

		Previous Choice		
		2-door Sedan	4-door Sedan	Wagon
Next Choice	2-door Sedan	.60	.20	0
	4-door Sedan	.30	.60	.50
	Wagon	.10	.20	.50

Table 5-9

Therefore, the transition matrix is

$$P = \begin{bmatrix} 0.60 & 0.20 & 0 \\ 0.30 & 0.60 & 0.50 \\ 0.10 & 0.20 & 0.50 \end{bmatrix}$$

Although we do not need the information to solve the problem, the reader might wish to verify that P is regular, because P^2 contains no zeros. The equilibrium problem is of the form $AX = B$, where

$$A = \begin{bmatrix} 1 & 1 & 1 \\ P-I \end{bmatrix} = \begin{bmatrix} 1 & 1 & 1 \\ -0.40 & 0.20 & 0 \\ 0.30 & -0.40 & 0.50 \\ 0.10 & 0.20 & -0.50 \end{bmatrix} \quad \text{and} \quad B = \begin{bmatrix} 1 \\ 0 \\ 0 \\ 0 \end{bmatrix}$$

The augmented matrix for $AX = B$ is given by

$$\begin{array}{c}xyz\\\left[\begin{array}{ccc|c}1 & 1 & 1 & 1\\-0.40 & 0.20 & 0 & 0\\0.30 & -0.40 & 0.50 & 0\\0.10 & 0.20 & -0.50 & 0\end{array}\right]\end{array}$$

We will delete the last row, as we discussed. (In this case, notice that when you add the second and third row; you get the negative of the fourth row. It is therefore not difficult to see that the fourth row contains no new information.) Writing the numbers in fraction form, we have

$$\begin{array}{c}\phantom{-\frac{0}{0}}x\phantom{\frac{0}{0}}y\phantom{\frac{0}{0}}z\\\left[\begin{array}{ccc|c}1 & 1 & 1 & 1\\-\frac{2}{5} & \frac{1}{5} & 0 & 0\\\frac{3}{10} & -\frac{2}{5} & \frac{1}{2} & 0\end{array}\right]\end{array}$$

In the next two steps we add multiples of the first row to the other rows to convert the first column to that of the 3×3-identity matrix:

$$\left[\begin{array}{ccc|c}1 & 1 & 1 & 1\\0 & \frac{3}{5} & \frac{2}{5} & \frac{2}{5}\\\frac{3}{10} & -\frac{2}{5} & \frac{1}{2} & 0\end{array}\right]$$

$$\left[\begin{array}{ccc|c}1 & 1 & 1 & 1\\0 & \frac{3}{5} & \frac{2}{5} & \frac{2}{5}\\0 & -\frac{7}{10} & \frac{1}{5} & -\frac{3}{10}\end{array}\right]$$

Next we multiply the second row by $\frac{5}{3}$ to obtain a 1 in the diagonal location of the second column:

$$\left[\begin{array}{ccc|c}1 & 1 & 1 & 1\\0 & 1 & \frac{2}{3} & \frac{2}{3}\\0 & -\frac{7}{10} & \frac{1}{5} & -\frac{3}{10}\end{array}\right]$$

After adding multiples of the second row to the other two rows, the second column, also, is that of the identity matrix.

$$\begin{bmatrix} 1 & 0 & \frac{1}{3} & | & \frac{1}{3} \\ 0 & 1 & \frac{2}{3} & | & \frac{2}{3} \\ 0 & 0 & \frac{2}{3} & | & \frac{1}{6} \end{bmatrix}$$

The last steps give us:

$$\begin{bmatrix} 1 & 0 & \frac{1}{3} & | & \frac{1}{3} \\ 0 & 1 & \frac{2}{3} & | & \frac{2}{3} \\ 0 & 0 & 1 & | & \frac{1}{4} \end{bmatrix}$$

$$\begin{array}{ccc} x & y & z \end{array}$$
$$\begin{bmatrix} 1 & 0 & 0 & | & \frac{1}{4} \\ 0 & 1 & 0 & | & \frac{1}{2} \\ 0 & 0 & 1 & | & \frac{1}{4} \end{bmatrix}$$

Thus, the solution of the equation is

$$X = \begin{bmatrix} \frac{1}{4} \\ \frac{1}{2} \\ \frac{1}{4} \end{bmatrix}$$

which means that 25% of the cars ordered should be two-door sedans, 50% should be four-door sedans, and the remaining 25% should be station wagons. ∎

Summary

A Markov chain process with $k \times k$-transition matrix P reaches equilibrium after m repetitions of the experiment if $PS^{(m)} = S^{(m)}$. The equilibrium matrix $S^{(m)}$ can be calculated by solving a matrix equation of the form $AX = B$ for a $k \times 1$-matrix X of unknowns, where A is the $(k + 1) \times k$-matrix

$$A = \begin{bmatrix} 1 & 1 & \cdots & 1 \\ & P - I & \end{bmatrix}$$

in which the first row consists of k 1s, I the $k \times k$-identity matrix, and B is the $(k+1) \times 1$-matrix

$$B = \begin{bmatrix} 1 \\ 0 \\ \vdots \\ 0 \end{bmatrix}$$

The matrix equation $AX = B$ is solved by the Gauss-Jordan method of Chapter 1.

Exercises

In Exercises 1 through 4, the given matrix is the transition matrix of a Markov chain process. Find the equilibrium matrix for the process.

1. $\begin{bmatrix} .60 & .40 \\ .40 & .60 \end{bmatrix}$

2. $\begin{bmatrix} 0.20 & 0.80 & 0.40 \\ 0.80 & 0 & 0.20 \\ 0 & 0.20 & 0.40 \end{bmatrix}$

3. $\begin{bmatrix} .35 & .95 \\ .65 & .05 \end{bmatrix}$

4. $\begin{bmatrix} 0.10 & 0.10 & 0.70 \\ 0 & 0.50 & 0.20 \\ 0.90 & 0.40 & 0.10 \end{bmatrix}$

5. In a study of voting patterns in some country, it is observed that 90% of the daughters of women who vote in most elections also vote regularly, but only 75% of the daughters of women who do not usually vote are themselves regular voters. If this pattern were to continue, eventually what percentage of the women in that country would vote in most elections?

6. Suppose that the occupations available to men in a South Pacific agricultural society are those of farmer, craftsman, and leader (i.e., ruler or priest). Studies indicate the probability that the son of a man who has a particular occupation will follow each of the available occupations is given in Table 5-10. The proportion of men in each occupation stays the same from generation to generation. What percentage of men are farmers, craftsmen, and leaders in that society?

		Father's Occupation		
		Farmer	*Craftsman*	*Leader*
Son's Occupation	*Farmer*	.90	.70	.30
	Craftsman	.10	.20	.20
	Leader	0	.10	.50

Table 5-10

7. The females in an animal population are classified as aggressive and passive. It is observed that 70% of the female offspring of aggressive females are themselves aggressive, whereas 50% of the offspring of passive females are aggressive. In

the long run, assuming these percentages remain constant, what proportion of the females in this animal population will be aggressive?

8. A study of coffee buying indicates that the probability of brand change in the next time period for Sunrise brand, Meadowlark brand, and all other brands is as given in Table 5-11. If this table remains valid in the future, so that the market for coffee becomes stable, what percentage of the coffee market will belong to Sunrise brand, to Meadowlark brand, and to all other brands?

		Present Period		
		Sunrise	Meadow Lark	Others
Next Period	Sunrise	.75	.10	.10
	Meadow Lark	.05	.65	.05
	Others	.20	.25	.85

Table 5-11

9. A chess player's probability of winning, drawing, or losing a game is influenced by her performance in the previous game to the extent that, if she wins a game, in the next game the probability is .50 that she will win, .40 that she will draw, and .10 that she will lose. If she draws, the probabilities become .20 for a win, .50 for another draw, and .30 for a loss in the next game. If she loses, the probability is .30 that she will win her next game, .40 that she will draw, and .30 that she will lose again. What are the probabilities that she will win, draw, or lose when equilibrium is attained?

10. In Gotham City, the probability of precipitation (rain or snow) on the next day is .25 if the present day has precipitation and .10 if it does not. What percentage of days have precipitation?

5.4 Absorbing Markov Chains

Imagine a brand of some product so superior that no person who tries the brand ever switches from it. If we represent a purchase of this brand as state 1 of a Markov chain process, then we are saying that $p(A_1 \mid B_1) = 1$ (once in state 1, the purchaser is certain to stay there), whereas $p(A_i \mid B_1) = 0$ for i not 1 (the purchaser in state 1 will never change). Thus, the first column of the transition matrix looks like this:

$$\begin{bmatrix} 1 \\ 0 \\ \vdots \\ 0 \end{bmatrix}$$

Table 5-12 presents the probabilities for a market consisting of three brands of a product, the first of which is the type just described.

		Original Period	
	Brand 1	Brand 2	Brand 3
Next Period Brand 1	1	.20	.30
Brand 2	0	.70	.50
Brand 3	0	.10	.20

Table 5-12

There is a chance that a purchaser of brand 2 will change to brand 1, specifically, $p(A_1 \mid B_2) = .20$, and also that a brand 3 user will switch to brand 1, $(p(A_1 \mid B_3) = .30)$; but no purchaser of brand 1 will ever change. You might expect that, eventually, everyone will buy only brand 1, and you would be correct. For example, if initially the three brands share the market equally, that is, if

$$S^{(0)} = \begin{bmatrix} \frac{1}{3} \\ \frac{1}{3} \\ \frac{1}{3} \end{bmatrix}$$

then we can compute the situation after 5 time periods. If for the transition matrix

$$P = \begin{bmatrix} 1 & .20 & .30 \\ 0 & .70 & .50 \\ 0 & .10 & .20 \end{bmatrix}$$

we calculate the 5th power P^5, then

$$S^{(5)} = P^5 S^{(0)} = \begin{bmatrix} 1 & .6946 & .7391 \\ 0 & .2608 & .2228 \\ 0 & .0446 & .0381 \end{bmatrix} \begin{bmatrix} \frac{1}{3} \\ \frac{1}{3} \\ \frac{1}{3} \end{bmatrix} = \begin{bmatrix} .81 \\ .16 \\ .03 \end{bmatrix}$$

Thus, brand 1 has already grown from $\frac{1}{3}$ to more than $\frac{4}{5}$ of the market. Multiplying again by P^5 to compute

$$S^{(10)} = P^{10} S^{(0)} = P^5 (P^5 S^{(0)})$$

we can easily see how dominant brand 1 has become by the 10th time period:

$$S^{(10)} = \begin{bmatrix} .94 \\ .05 \\ .01 \end{bmatrix}$$

Absorbing States

A state of a Markov chain that behaves like brand 1 does in this example is called an *absorbing state*: once in that state, the probability of remaining in that state is 1. In symbols, a state j is an **absorbing state** if $p_{jj} = p(A_j \mid B_j) = 1$. Therefore, $p_{ij} = 0$ for all i not equal to j since we recall that the sum of each column must be 1. We can recognize an absorbing state in the transition matrix because the corresponding column will look like a column of an identity matrix: 1 on the main diagonal and 0 elsewhere.

The preceding transition matrix comes from what is called an "absorbing" Markov chain process, but we must be careful in describing what we mean by such a process. The probability table for a Markov chain presented in Table 5-13 also has state 1 absorbing.

		Original Period		
		State 1	State 2	State 3
Next Period	State 1	1	0	0
	State 2	0	0.5	0.7
	State 3	0	0.5	0.3

Table 5-13

However, state 1 cannot capture everything. In fact, there is no way to get from state 2 to state 1 because $p(A_1 \mid B_2) = 0$, or from state 3 to state 1 since $p(A_1 \mid B_3) = 0$. Whatever share of the market brand 1 has initially, it will never change. For instance, if we begin with

$$S^{(0)} = \begin{bmatrix} .50 \\ .10 \\ .40 \end{bmatrix}$$

then since the transition matrix is

$$\begin{bmatrix} 1 & 0 & 0 \\ 0 & 0.50 & 0.70 \\ 0 & 0.50 & 0.30 \end{bmatrix}$$

in the next time period we will have

$$S^{(1)} = PS^{(0)} = \begin{bmatrix} 1 & 0 & 0 \\ 0 & 0.50 & 0.70 \\ 0 & 0.50 & 0.30 \end{bmatrix} \begin{bmatrix} 0.50 \\ 0.10 \\ 0.40 \end{bmatrix} = \begin{bmatrix} 0.50 \\ 0.33 \\ 0.17 \end{bmatrix}$$

So, although $S^{(1)}$ is quite different from $S^{(0)}$, the top number in the matrix is still exactly .50. It will never change, no matter how many times we multiply the result by P. The reason is that the top number in $S^{(m+1)}$ will always be computed by multiplying $S^{(m)}$ on the left by the top row of P,

$$\begin{bmatrix} 1 & 0 & 0 \end{bmatrix}$$

which is the first row of an identity matrix.

Absorbing Markov Chain Processes

In an "absorbing Markov chain," we must certainly have at least one absorbing state (a column with a 1 in the diagonal location and 0s elsewhere). Thus, the matrix corresponding to Table 5-12 has this property:

$$\begin{bmatrix} 1 & .20 & .30 \\ 0 & .70 & .50 \\ 0 & .10 & .20 \end{bmatrix}$$

But, in addition, we require that "everything eventually gets absorbed," that is, we need to know that if something starts in any state, it can eventually reach an absorbing state. The word "eventually" permits us to include Markov chains like the one described in Table 5-14.

		Original Period		
		State 1	State 2	State 3
Next Period	State 1	1	0	.10
	State 2	0	.50	.70
	State 3	0	.50	.20

Table 5-14

In Table 5-14, the only absorbing state is state 1, and we can go directly from state 3 to state 1 because the table tells us that $p(A_1 \mid B_3) = .10$. But we can also see from looking at the table that there is no chance to go directly from state 2 to state 1. Nevertheless, we want to call this Markov chain process "absorbing" because, no matter what $S^{(0)}$ is, we really will end up with

$$S^{(m)} = \begin{bmatrix} 1 \\ 0 \\ 0 \end{bmatrix}$$

when m is large enough. The reason is that, as we can see from Table 5-14, it *is* possible to get from state 2 to state 3 because $p(A_3 \mid B_2) = .50$. So, we have a two-step process that goes from state 2 to state 3 and then on to state 1, the absorbing state. Eventually, everything ends up in state 1, no matter where it starts.

It would be awkward to try to determine in the same way whether or not every state eventually leads to an absorbing state, especially in a large transition matrix. Fortunately, the following definition of an "absorbing Markov chain" can be routinely checked, if necessary, with the aid of a computer.

An absorbing Markov chain process is a Markov chain

1. with at least one absorbing state and
2. with the property that for any nonabsorbing state j there is an absorbing state i and a positive integer m such that in P^m (the mth power of the transition matrix P of the chain) we have $p_{ij} > 0$.

5.4 ABSORBING MARKOV CHAINS

We can check part 2 of the definition quite easily if we put the transition matrix in the right form. We will also use this form in the next section, for another purpose. All we have to do to obtain this form is to order the states of the process so that the absorbing states come first. For instance, suppose we are given Table 5-15 of conditional probabilities.

		Original Period			
		State 1	State 2	State 3	State 4
Next Period	State 1	0	0	.10	0
	State 2	.40	1	.70	0
	State 3	.50	0	.20	0
	State 4	.10	0	0	1

Table 5-15

It is easy to recognize the absorbing states 2 and 4. We will reorder the states so that these two become the first ones in the new ordering. It will be more convenient to work with the transition matrix rather than the table. We will also label the rows and columns of the matrix with the numbers of the corresponding states of the probability table:

$$\begin{array}{c} \\ 1 \\ 2 \\ 3 \\ 4 \end{array} \begin{array}{cccc} 1 & 2 & 3 & 4 \\ \left[\begin{array}{cccc} 0 & 0 & 0.10 & 0 \\ 0.40 & 1 & 0.70 & 0 \\ 0.50 & 0 & 0.20 & 0 \\ 0.10 & 0 & 0 & 1 \end{array}\right] \end{array}$$

We will first interchange states 1 and 2. This means that we must interchange both the first two columns and the first two rows since the ordering of the states certainly determines both the rows and columns. To avoid errors, it is a good idea to do this in two steps—first interchanging the columns, including the column labels:

$$\begin{array}{c} \\ 1 \\ 2 \\ 3 \\ 4 \end{array} \begin{array}{cccc} 2 & 1 & 3 & 4 \\ \left[\begin{array}{cccc} 0 & 0 & 0.10 & 0 \\ 1 & 0.40 & 0.70 & 0 \\ 0 & 0.50 & 0.20 & 0 \\ 0 & 0.10 & 0 & 1 \end{array}\right] \end{array}$$

—then interchanging the first two rows:

$$\begin{array}{c} \\ 2 \\ 1 \\ 3 \\ 4 \end{array} \begin{array}{cccc} 2 & 1 & 3 & 4 \\ \left[\begin{array}{cccc} 1 & 0.40 & 0.70 & 0 \\ 0 & 0 & 0.10 & 0 \\ 0 & 0.50 & 0.20 & 0 \\ 0 & 0.10 & 0 & 1 \end{array}\right] \end{array}$$

Since we want the other absorbing state, presently state 4, to be next, we interchange the second and fourth columns of this new matrix, that is, we interchange state 1 and state 4

$$\begin{array}{c} \\ 2 \\ 1 \\ 3 \\ 4 \end{array} \begin{array}{cccc} 2 & 4 & 3 & 1 \\ \left[\begin{array}{cccc} 1 & 0 & 0.70 & 0.40 \\ 0 & 0 & 0.10 & 0 \\ 0 & 0 & 0.20 & 0.50 \\ 0 & 1 & 0 & 0.10 \end{array}\right] \end{array}$$

and then the second and fourth rows:

$$\begin{array}{c} \\ 2 \\ 4 \\ 3 \\ 1 \end{array} \begin{array}{cccc} 2 & 4 & 3 & 1 \\ \left[\begin{array}{cccc} 1 & 0 & 0.70 & 0.40 \\ 0 & 1 & 0 & 0.10 \\ 0 & 0 & 0.20 & 0.50 \\ 0 & 0 & 0.10 & 0 \end{array}\right] \end{array}$$

The new transition matrix represents the Markov chain of Table 5-15; all we did was to change the order in which the four states appear so that the absorbing states now come first.

R-Q Form

To describe the matrix we just obtained, think of the first two columns, for the absorbing states, in terms of two standard kinds of matrices—a 2 × 2-identity matrix I placed on top of a 2 × 2-zero matrix O:

$$\begin{array}{c} \\ 2 \\ 4 \\ 3 \\ 1 \end{array} \begin{array}{cc} 2 & 4 \\ \left[\begin{array}{cc} 1 & 0 \\ 0 & 1 \\ 0 & 0 \\ 0 & 0 \end{array}\right] \end{array} = \left[\begin{array}{c} I \\ O \end{array}\right]$$

The rest of the transition matrix also divides up in a natural way. The matrix that lies to the right of the identity matrix is called R; in this case

$$R = \begin{array}{c} 2 \\ 4 \end{array} \begin{array}{cc} 3 & 1 \\ \left[\begin{array}{cc} 0.70 & 0.40 \\ 0 & 0.10 \end{array}\right] \end{array}$$

Below R, to the right of the O matrix, is a matrix we call Q. In this example,

$$Q = \begin{array}{c} 3 \\ 1 \end{array} \begin{array}{cc} 3 & 1 \\ \left[\begin{array}{cc} 0.20 & 0.50 \\ 0.10 & 0 \end{array}\right] \end{array}$$

5.4 ABSORBING MARKOV CHAINS

Thus, we have put the transition matrix into the form

$$\begin{array}{c} \\ 2 \\ 4 \\ 3 \\ 1 \end{array} \begin{array}{cccc} 2 & 4 & 3 & 1 \\ \left[\begin{array}{cc|cc} 1 & 0 & 0.70 & 0.40 \\ 0 & 1 & 0 & 0.10 \\ \hline 0 & 0 & 0.20 & 0.50 \\ 0 & 0 & 0.10 & 0 \end{array}\right] \end{array} = \left[\begin{array}{cc} I & R \\ 0 & Q \end{array}\right]$$

We will call this form of the transition matrix the **R-Q form**.

Example 1. Put the following transition matrices into R-Q form:

(a)
$$\begin{array}{c} \\ 1 \\ 2 \\ 3 \end{array} \begin{array}{ccc} 1 & 2 & 3 \\ \left[\begin{array}{ccc} 1 & 0 & 0.10 \\ 0 & 0.50 & 0.70 \\ 0 & 0.50 & 0.20 \end{array}\right] \end{array}$$

(b)
$$\begin{array}{c} \\ 1 \\ 2 \\ 3 \\ 4 \end{array} \begin{array}{cccc} 1 & 2 & 3 & 4 \\ \left[\begin{array}{cccc} 0.10 & 0.80 & 0 & 0 \\ 0.50 & 0 & 0 & 0.50 \\ 0 & 0.20 & 1 & 0.50 \\ 0.40 & 0 & 0 & 0 \end{array}\right] \end{array}$$

(c)
$$\begin{array}{c} \\ 1 \\ 2 \\ 3 \end{array} \begin{array}{ccc} 1 & 2 & 3 \\ \left[\begin{array}{ccc} 0.10 & 0 & 0 \\ 0.60 & 1 & 0 \\ 0.30 & 0 & 1 \end{array}\right] \end{array}$$

Solution

(a) The only absorbing state is state 1. The identity matrix is 1×1 and the zero matrix is 2×1. We divide up the matrix as follows:

$$\begin{array}{c} \\ 1 \\ 2 \\ 3 \end{array} \begin{array}{ccc} 1 & 2 & 3 \\ \left[\begin{array}{c|cc} 1 & 0 & 0.10 \\ \hline 0 & 0.50 & 0.70 \\ 0 & 0.50 & 0.20 \end{array}\right] \end{array}$$

We have the R-Q form with

$$R = 1\begin{array}{c} \\ \end{array} \begin{array}{cc} 2 & 3 \\ \left[\begin{array}{cc} 0 & .10 \end{array}\right] \end{array} \quad \text{and} \quad Q = \begin{array}{c} \\ 2 \\ 3 \end{array} \begin{array}{cc} 2 & 3 \\ \left[\begin{array}{cc} .50 & .70 \\ .50 & .20 \end{array}\right] \end{array}$$

(b) The single absorbing state is represented by the third column, so to make it the first state, we interchange the first and third columns:

$$\begin{array}{c} \\ 1 \\ 2 \\ 3 \\ 4 \end{array} \begin{array}{cccc} 3 & 2 & 1 & 4 \\ \left[\begin{array}{cccc} 0 & 0.80 & 0.10 & 0 \\ 0 & 0 & 0.50 & 0.50 \\ 1 & 0.20 & 0 & 0.50 \\ 0 & 0 & 0.40 & 0 \end{array}\right] \end{array}$$

then the first and third rows:

$$\begin{array}{c} \\ 3 \\ 2 \\ 1 \\ 4 \end{array} \begin{array}{cccc} 3 & 2 & 1 & 4 \\ \left[\begin{array}{cccc} 1 & 0.20 & 0 & 0.50 \\ 0 & 0 & 0.50 & 0.50 \\ 0 & 0.80 & 0.10 & 0 \\ 0 & 0 & 0.40 & 0 \end{array}\right] \end{array}$$

Again, the identity matrix is 1×1 and the $R\text{-}Q$ form is

$$\begin{array}{c} \\ 3 \\ 2 \\ 1 \\ 4 \end{array} \begin{array}{cccc} 3 & 2 & 1 & 4 \\ \left[\begin{array}{c|ccc} 1 & 0.20 & 0 & 0.50 \\ \hline 0 & 0 & 0.50 & 0.50 \\ 0 & 0.80 & 0.10 & 0 \\ 0 & 0 & 0.40 & 0 \end{array}\right] \end{array} = \left[\begin{array}{cc} I & R \\ 0 & Q \end{array}\right]$$

(c) The absorbing states are represented by the second and third columns; so, we first interchange the first and second columns, then the first and second rows:

$$\begin{array}{c} \\ 1 \\ 2 \\ 3 \end{array} \begin{array}{ccc} 2 & 1 & 3 \\ \left[\begin{array}{ccc} 0 & .10 & 0 \\ 1 & .60 & 0 \\ 0 & .30 & 1 \end{array}\right] \end{array} \text{ and } \begin{array}{c} \\ 2 \\ 1 \\ 3 \end{array} \begin{array}{ccc} 2 & 1 & 3 \\ \left[\begin{array}{ccc} 1 & .60 & 0 \\ 0 & .10 & 0 \\ 0 & .30 & 1 \end{array}\right] \end{array}$$

Then we do the same with the second and third columns and rows.

$$\begin{array}{c} \\ 2 \\ 1 \\ 3 \end{array} \begin{array}{ccc} 2 & 3 & 1 \\ \left[\begin{array}{ccc} 1 & 0 & .60 \\ 0 & 0 & .10 \\ 0 & 1 & .30 \end{array}\right] \end{array} \text{ and } \begin{array}{c} \\ 2 \\ 3 \\ 1 \end{array} \begin{array}{ccc} 2 & 3 & 1 \\ \left[\begin{array}{ccc} 1 & 0 & .60 \\ 0 & 1 & .30 \\ 0 & 0 & .10 \end{array}\right] \end{array}$$

Notice that although we had several interchanges to make in this example, we did not combine steps because that can lead to errors. In the last of these matrices, we have the 2×2-identity matrix in the upper-left corner and a 1×2-zero matrix below it; this gives the $R\text{-}Q$ form:

$$\begin{array}{c} \\ 2 \\ 3 \\ 1 \end{array} \begin{array}{ccc} 2 & 3 & 1 \\ \left[\begin{array}{cc|c} 1 & 0 & .60 \\ 0 & 1 & .30 \\ \hline 0 & 0 & .10 \end{array}\right] \end{array} = \left[\begin{array}{cc} I & R \\ 0 & Q \end{array}\right] \qquad \blacksquare$$

Absorbing and Nonabsorbing States

The size of the identity matrix in the $R\text{-}Q$ form depends on the number of absorbing states in the Markov chain. If there are k absorbing states, then after we interchange rows and columns to make those states the first ones, a $k \times k$-identity matrix appears in the upper-left corner of the $R\text{-}Q$ form. For instance, in Part (c) of Example 1, there were two absorbing states, 2 and 3, and the $R\text{-}Q$ form has a

2 × 2-identity matrix in the upper-left corner. Like the identity matrix, the matrix Q is a square matrix, but its size is determined by the number of nonabsorbing states. If there are n states in all and if k of them are absorbing, then Q will be $(n-k) \times (n-k)$. In Part (b) of Example 1, only state 3 is absorbing, so the remaining three states are nonabsorbing; we note that in the R-Q form, the matrix Q is 3×3. As we have seen, the zero matrix and the matrix R are not necessarily square.

Identifying Absorbing Markov Chains

Any transition matrix with absorbing states can be put in R-Q form, but that does not make it the transition matrix of an absorbing Markov chain. However, putting the transition matrix in R-Q form permits us to determine whether or not it is the transition matrix of an absorbing Markov chain. Notice what happens when we multiply the R-Q form of the matrix of Part (a) of Example 1 by itself:

$$\begin{array}{c} \begin{array}{ccc}1 & 2 & 3\end{array} \\ \begin{array}{c}1\\2\\3\end{array}\begin{bmatrix} 1 & 0 & 0.10 \\ 0 & 0.50 & 0.70 \\ 0 & 0.50 & 0.20 \end{bmatrix}\end{array} \begin{array}{c} \begin{array}{ccc}1 & 2 & 3\end{array} \\ \begin{bmatrix} 1 & 0 & 0.10 \\ 0 & 0.50 & 0.70 \\ 0 & 0.50 & 0.20 \end{bmatrix}\end{array} = \begin{array}{c} \begin{array}{ccc}1 & 2 & 3\end{array} \\ \begin{bmatrix} 1 & 0.05 & 0.12 \\ 0 & 0.60 & 0.49 \\ 0 & 0.35 & 0.39 \end{bmatrix}\end{array}$$

The product is still in R-Q form. But although in the transition matrix P, the R-matrix

$$\begin{array}{c}\begin{array}{cc}2 & 3\end{array}\\1\begin{bmatrix}0 & .10\end{bmatrix}\end{array}$$

contains a zero entry, for the R-matrix of P^2 we have

$$\begin{array}{c}\begin{array}{cc}2 & 3\end{array}\\1\begin{bmatrix}.05 & .12\end{bmatrix}\end{array}$$

with all entries nonzero. This is our test of condition 2 of the definition of an absorbing Markov chain: "for any nonabsorbing state j there is an absorbing state i and a positive integer m such that in P^m we have $p_{ij} > 0$." We put P into R-Q form and keep multiplying it by itself. Each time we multiply, the product will remain in R-Q form. If we obtain a power P^m for which the R matrix of the R-Q form is a positive matrix, then the Markov chain with transition matrix P is absorbing. Thus, our computation of P^2 in Part (a) of Example 1 shows that P is the transition matrix of an absorbing Markov chain.

Example 2. Determine the power P^m of the transition matrix in R-Q form that shows the following are transition matrices of absorbing Markov chains:

(a) $\begin{array}{c}\begin{array}{cccc}1 & 2 & 3 & 4\end{array}\\\begin{array}{c}1\\2\\3\\4\end{array}\begin{bmatrix}0 & 0 & 0.10 & 0\\0.40 & 1 & 0.70 & 0\\0.50 & 0 & 0.20 & 0\\0.10 & 0 & 0 & 1\end{bmatrix}\end{array}$ (b) $\begin{array}{c}\begin{array}{cccc}1 & 2 & 3 & 4\end{array}\\\begin{array}{c}1\\2\\3\\4\end{array}\begin{bmatrix}0.10 & 0.80 & 0 & 0\\0.50 & 0 & 0 & 0.50\\0 & 0.20 & 1 & 0.50\\0.40 & 0 & 0 & 0\end{bmatrix}\end{array}$

(c)
$$\begin{array}{c} \\ 1 \\ 2 \\ 3 \end{array} \begin{array}{ccc} 1 & 2 & 3 \\ \begin{bmatrix} 0.10 & 0 & 0 \\ 0.60 & 1 & 0 \\ 0.30 & 0 & 1 \end{bmatrix} \end{array}$$

Solution

(a) Just before Example 1, we showed that the R-Q form of this transition matrix is

$$\begin{array}{c} \\ 2 \\ 4 \\ 3 \\ 1 \end{array} \begin{array}{cccc} 2 & 4 & 3 & 1 \\ \begin{bmatrix} 1 & 0 & 0.70 & 0.40 \\ 0 & 1 & 0 & 0.10 \\ 0 & 0 & 0.20 & 0.50 \\ 0 & 0 & 0.10 & 0 \end{bmatrix} \end{array}$$

Since the matrix

$$R = \begin{array}{c} 2 \\ 4 \end{array} \begin{bmatrix} 0.70 & 0.40 \\ 0 & 0.10 \end{bmatrix}$$

contains a zero, we compute

$$\begin{array}{c} 2 \\ 4 \\ 3 \\ 1 \end{array} \begin{bmatrix} 1 & 0 & 0.70 & 0.40 \\ 0 & 1 & 0 & 0.10 \\ 0 & 0 & 0.20 & 0.50 \\ 0 & 0 & 0.10 & 0 \end{bmatrix} \begin{bmatrix} 1 & 0 & 0.70 & 0.40 \\ 0 & 1 & 0 & 0.10 \\ 0 & 0 & 0.20 & 0.50 \\ 0 & 0 & 0.10 & 0 \end{bmatrix} = \begin{bmatrix} 1 & 0 & 0.88 & 0.75 \\ 0 & 1 & 0.01 & 0.10 \\ 0 & 0 & 0.09 & 0.10 \\ 0 & 0 & 0.02 & 0.05 \end{bmatrix}$$

Now

$$R = \begin{array}{c} 2 \\ 4 \end{array} \begin{bmatrix} .88 & .75 \\ .01 & .10 \end{bmatrix}$$

is a positive matrix, and we conclude that the Markov chain process is absorbing.

(b) By Part (b) of Example 1, the R-Q form is

$$\begin{array}{c} 3 \\ 2 \\ 1 \\ 4 \end{array} \begin{bmatrix} 1 & 0.20 & 0 & 0.50 \\ 0 & 0 & 0.50 & 0.50 \\ 0 & 0.80 & 0.10 & 0 \\ 0 & 0 & 0.40 & 0 \end{bmatrix}$$

There is a zero in

$$R = 3 \begin{bmatrix} .20 & 0 & .50 \end{bmatrix}$$

Again, it is sufficient to calculate

$$\begin{array}{c} \begin{array}{cccc} 3 & 2 & 1 & 4 \end{array} \\ \begin{array}{c} 3 \\ 2 \\ 1 \\ 4 \end{array}\left[\begin{array}{cccc} 1 & 0.20 & 0 & 0.50 \\ 0 & 0 & 0.50 & 0.50 \\ 0 & 0.80 & 0.10 & 0 \\ 0 & 0 & 0.40 & 0 \end{array}\right] \end{array} \begin{array}{c} \begin{array}{cccc} 3 & 2 & 1 & 4 \end{array} \\ \left[\begin{array}{cccc} 1 & 0.20 & 0 & 0.50 \\ 0 & 0 & 0.50 & 0.50 \\ 0 & 0.80 & 0.10 & 0 \\ 0 & 0 & 0.40 & 0 \end{array}\right] \end{array}$$

$$= \begin{array}{c} \begin{array}{cccc} 3 & 2 & 1 & 4 \end{array} \\ \left[\begin{array}{cccc} 1 & 0.20 & 0.30 & 0.60 \\ 0 & 0.40 & 0.25 & 0 \\ 0 & 0.08 & 0.41 & 0.40 \\ 0 & 0.32 & 0.04 & 0 \end{array}\right] \end{array}$$

because then

$$R = 3\begin{array}{c} \begin{array}{ccc} 2 & 1 & 4 \end{array} \\ \left[.20 \quad .30 \quad .60 \right] \end{array}$$

is a positive matrix.

(c) By Part (c) of Example 1, the R-Q form is

$$\begin{array}{c} \begin{array}{ccc} 2 & 3 & 1 \end{array} \\ \begin{array}{c} 2 \\ 3 \\ 1 \end{array}\left[\begin{array}{ccc} 1 & 0 & .60 \\ 0 & 1 & .30 \\ 0 & 0 & .10 \end{array}\right] \end{array}$$

and since

$$R = \left[\begin{array}{c} .60 \\ .30 \end{array}\right]$$

in this case, we already know that this is the transition matrix of an absorbing Markov chain. ∎

This procedure for identifying absorbing Markov chains is similar to the definition of a regular Markov chain, given in Section 5.2, that P^m must be a positive matrix for some mth power of the transition matrix P. We also quoted the rule: If P is $n \times n$ and P^m contains zeros when $m = (n - 1)^2 + 1$, then it is not necessary to calculate higher powers of P because the Markov chain cannot be regular. There is also a similar rule that limits the test of whether a Markov chain with absorbing states is an absorbing Markov chain. That is, if the transition matrix P of such a Markov chain is put in R-Q form and if there are zeros in the R-matrix of P^m for m large enough (again depending on the size of P), then the Markov chain is not absorbing.

Long-term Behavior of Absorbing Markov Chains

An absorbing Markov chain with a single absorbing state behaves as in the computation at the beginning of this section. For m large enough, $S^{(m)}$ will look like a column of an identity matrix, with the 1 in the row of the absorbing state. That state eventually absorbs everything.

The long-term behavior of an absorbing Markov chain process is more interesting when there is more than one absorbing state. It is still true that eventually everything ends up in an absorbing state, but these states must "share the market" in some way. The following is an example of this type of process, but with more believable absorbing states than that wonderful brand above that, once tried, is never abandoned.

An Inspection Problem

In the final inspection of a complex product, the inspector classifies a unit into one of three categories: "pass" (the unit is satisfactory for sale), "fail" (the unit is of such poor quality that it must be destroyed), and "repairable" (the unit is sent back for improvements and will be inspected again). After the units are repaired, at the next inspection 70% pass, 10% fail, and 20% require further repairs. These percentages stay the same (at least approximately) no matter how many times the unit is repaired. We wish to know the probability that a unit returned for repairs will eventually pass inspection. (If that probability is too low, the management of the company may decide to eliminate the repair process altogether and destroy any unit that cannot pass inspection the first time.)

To construct the transition matrix for this Markov chain process, we label the three states: 1 for pass, 2 for fail, and 3 for repair. The first two are absorbing—once a unit passes or fails inspection, it stays that way; however, in the next round of inspections, a unit in the "repair" state 3 can pass to any of the three states, with the probabilities from the preceding paragraph: $p(A_1 \mid B_3) = p_{13} = .70$, $p_{23} = .10$, and $p_{33} = .20$. Thus, the transition matrix is

$$\begin{array}{c} \\ 1 \\ 2 \\ 3 \end{array} \begin{array}{ccc} 1 & 2 & 3 \end{array} \\ \begin{bmatrix} 1 & 0 & .70 \\ 0 & 1 & .20 \\ 0 & 0 & .10 \end{bmatrix}$$

We are only interested in units that have been repaired, that is, that are in state 3, so we begin with

$$S^{(0)} = \begin{array}{c} 1 \\ 2 \\ 3 \end{array} \begin{bmatrix} 0 \\ 0 \\ 1 \end{bmatrix}$$

We compute $S^{(1)}$, $S^{(2)}$, and so on, to get some idea of what will happen in the long run:

$$S^{(1)} = PS^{(0)} = \begin{array}{c} 1 \\ 2 \\ 3 \end{array} \begin{array}{ccc} 1 & 2 & 3 \end{array} \\ \begin{bmatrix} 1 & 0 & .70 \\ 0 & 1 & .20 \\ 0 & 0 & .10 \end{bmatrix} \begin{bmatrix} 0 \\ 0 \\ 1 \end{bmatrix} = \begin{bmatrix} .70 \\ .20 \\ .10 \end{bmatrix}$$

$$S^{(2)} = PS^{(1)} = \begin{matrix} 1 \\ 2 \\ 3 \end{matrix} \begin{bmatrix} 1 & 0 & .70 \\ 0 & 1 & .20 \\ 0 & 0 & .10 \end{bmatrix} \begin{bmatrix} .70 \\ .20 \\ .10 \end{bmatrix} = \begin{bmatrix} .77 \\ .22 \\ .01 \end{bmatrix}$$

After a few more steps we find that

$$S^{(m)} = \begin{matrix} 1 \\ 2 \\ 3 \end{matrix} \begin{bmatrix} 0.78 \\ 0.22 \\ 0 \end{bmatrix}$$

About 78% of the repairable units eventually pass inspection.

Summary

A state j of a Markov chain process is an **absorbing state** if $p_{jj} = p(A_j \mid B_j) = 1$ and $p_{ij} = 0$ for all i not equal to j. A Markov chain process is **absorbing** if it has at least one absorbing state and for any nonabsorbing state j, there is an absorbing state i and a positive integer m such that in P^m (where P is the transition matrix of the chain), each $p_{ij} > 0$. The order of states can be modified by interchanging rows and columns of the transition matrix so that the absorbing states come first. Then the transition matrix has the **R-Q form**

$$P = \begin{bmatrix} I & R \\ O & Q \end{bmatrix}$$

The powers of P remain in R-Q form. If the R-matrix of some power P^m is a positive matrix, then the Markov chain is absorbing. If an absorbing Markov chain has a single absorbing state, then for m large enough, $S^{(m)}$ will look like a column of an identity matrix, with the 1 in the row of the absorbing state. If there is more than one absorbing state, the probability of reaching a particular absorbing state depends on the structure of the transition matrix.

Exercises

In Exercises 1 through 10, the given matrix is the transition matrix of an absorbing Markov chain. For each matrix:

(a) interchange rows and columns, if necessary, so that the absorbing states are placed before the nonabsorbing states;
(b) identify the matrices R and Q in the resulting R-Q form;
(c) find the smallest power m so that the R-matrix in P^m contains no zeros.

1.
$$\begin{array}{c}123\\ \begin{array}{c}1\\2\\3\end{array}\left[\begin{array}{ccc}1 & 0 & .20\\ 0 & 1 & .60\\ 0 & 0 & .20\end{array}\right]\end{array}$$

2.
$$\begin{array}{c}1234\\ \begin{array}{c}1\\2\\3\\4\end{array}\left[\begin{array}{cccc}1 & 0.70 & 0.20 & 0.50\\ 0 & 0 & 0.80 & 0.10\\ 0 & 0.10 & 0 & 0.10\\ 0 & 0.20 & 0 & 0.30\end{array}\right]\end{array}$$

3.
$$\begin{array}{c}123\\ \begin{array}{c}1\\2\\3\end{array}\left[\begin{array}{ccc}.30 & 0 & .60\\ .20 & 1 & .30\\ .50 & 0 & .10\end{array}\right]\end{array}$$

4.
$$\begin{array}{c}1234\\ \begin{array}{c}1\\2\\3\\4\end{array}\left[\begin{array}{cccc}1 & 0 & 0 & 0.90\\ 0 & 1 & 0.40 & 0.10\\ 0 & 0 & 0.50 & 0\\ 0 & 0 & 0.10 & 0\end{array}\right]\end{array}$$

5.
$$\begin{array}{c}123\\ \begin{array}{c}1\\2\\3\end{array}\left[\begin{array}{ccc}.20 & 0 & 0\\ .70 & 1 & 0\\ .10 & 0 & 1\end{array}\right]\end{array}$$

6.
$$\begin{array}{c}1234\\ \begin{array}{c}1\\2\\3\\4\end{array}\left[\begin{array}{cccc}.70 & 0 & 0 & 0.50\\ .10 & 0.20 & 0 & 0\\ .10 & 0.60 & 1 & 0\\ .10 & 0.20 & 0 & 0.50\end{array}\right]\end{array}$$

7.
$$\begin{array}{c}123\\ \begin{array}{c}1\\2\\3\end{array}\left[\begin{array}{ccc}0.60 & 0.30 & 0\\ 0.40 & 0.50 & 0\\ 0 & 0.20 & 1\end{array}\right]\end{array}$$

8.
$$\begin{array}{c}1234\\ \begin{array}{c}1\\2\\3\\4\end{array}\left[\begin{array}{cccc}1 & .60 & 0 & 0\\ 0 & .10 & 0 & 0\\ 0 & .20 & 1 & 0\\ 0 & .10 & 0 & 1\end{array}\right]\end{array}$$

9.
$$\begin{array}{c}1234\\ \begin{array}{c}1\\2\\3\\4\end{array}\left[\begin{array}{cccc}0.10 & 0 & 0 & 0.30\\ 0.90 & 1 & 0 & 0\\ 0 & 0 & 0.60 & 0\\ 0 & 0 & 0.40 & 0.70\end{array}\right]\end{array}$$

10.
$$\begin{array}{c}1234\\ \begin{array}{c}1\\2\\3\\4\end{array}\left[\begin{array}{cccc}0 & 0 & 0 & 0.30\\ 0.80 & 1 & 0 & 0.30\\ 0.10 & 0 & 1 & 0.20\\ 0.10 & 0 & 0 & 0.20\end{array}\right]\end{array}$$

5.5 The Fundamental Matrix

In the inspection problem at the end of the preceding section the transition matrix

$$P=\begin{array}{c}123\\ \begin{array}{c}1\\2\\3\end{array}\left[\begin{array}{ccc}1 & 0 & .70\\ 0 & 1 & .20\\ 0 & 0 & .10\end{array}\right]\end{array}$$

corresponds to three states: state 1 means a unit passes inspection, state 2 means a unit fails inspection, and state 3 means a unit is sent back for repairs before reinspection. Starting with a unit that is sent back for repairs, that is, with

$$S^{(0)} = \begin{matrix} 1 \\ 2 \\ 3 \end{matrix} \begin{bmatrix} 0 \\ 0 \\ 1 \end{bmatrix}$$

we want to know the probability that the unit will eventually pass inspection.

Eventual Probabilities for Three States

We can use what we know about matrices and probability to find a simple formula that will tell us how, eventually, the absorbing states 1 and 2 will share the probabilities. At present, we will concentrate on transition matrices like the one for the inspection problem in which there are three states, the first two of which are absorbing. The transition matrix has the form

$$\begin{matrix} & 1 & 2 & 3 \end{matrix}$$
$$\begin{matrix} 1 \\ 2 \\ 3 \end{matrix} \begin{bmatrix} 1 & 0 & p_{13} \\ 0 & 1 & p_{23} \\ 0 & 0 & p_{33} \end{bmatrix}$$

We want to calculate the probability that a unit in state 3 (sent back for repairs) will eventually end up in the absorbing state 1 (passes inspection). We will represent this probability by the symbol e_{13}. We will also compute the probability e_{23} that a unit in state 3 will eventually be in state 2, that is, will fail inspection.

The probability e_{13} is a conditional probability because we can write

$$e_{13} = p(\text{eventually in state 1} \mid \text{presently in state 3})$$

How can a unit in state 3 go to state 1? There are two different ways this can happen: (a) go to state 1 in the next time period or (b) go to (the same) state 3 in the next time period and eventually get to state 1. There is no third possibility, since if the unit were to go to state 2 in the next time period (fail inspection) it could never get out of that absorbing state, and so never get to state 1. We emphasized "or" to remind you of the formula for the probability of disjoint events that tells us that

$$e_{13} = p(a) + p(b)$$

Here, $p(a)$ is the probability p_{13} from the transition matrix. In alternative (b), we emphasized "and" to remind you of the formula

$$p(A \cap B) = p(B) \cdot p(A \mid B)$$

Therefore,

$p(b) = p(\text{in state 3 next period and go eventually to state 1})$
$= p(\text{in state 3 next period}) \cdot p(\text{eventually in state 1} \mid \text{presently in state 3})$
$= p_{33} e_{13}$

Substituting $p_{33}e_{13}$ for $p(b)$ in the formula $e_{13} = p(a) + p(b)$, we obtain

$$e_{13} = p_{13} + p_{33} e_{13}$$

We can solve this equation for e_{13} by elementary algebra:

$$e_{13} - p_{33}e_{13} = p_{13}$$
$$e_{13}(1 - p_{33}) = p_{13}$$
$$e_{13} = \frac{p_{13}}{1 - p_{33}}$$

The Inspection Example

In the inspection example,

$$P = \begin{array}{c} \\ 1 \\ 2 \\ 3 \end{array} \begin{array}{ccc} 1 & 2 & 3 \\ \left[\begin{array}{ccc} 1 & 0 & .70 \\ 0 & 1 & .20 \\ 0 & 0 & .10 \end{array}\right] \end{array}$$

We calculate that

$$p(\text{repaired item will eventually pass}) = e_{13} = \frac{p_{13}}{1 - p_{33}} = \frac{.70}{1 - .10}$$
$$= \frac{.70}{.90} = .78 \text{ (approximately)}$$

which is what the matrix computations $S^{(1)} = PS^{(0)}$, $S^{(2)} = PS^{(1)}$, ..., in Section 5.4, indicated would happen.

A Matrix Equation

The same argument also gives us the following, similar formula for the probability that a repaired item will eventually fail inspection:

$$e_{23} = p_{23} + p_{33}e_{23}$$

We may express these formulas for e_{13} and e_{23} as a single matrix equation. The transition matrix P is already in the R-Q form discussed in the preceding section:

$$\begin{array}{c} \\ 1 \\ 2 \\ 3 \end{array} \begin{array}{ccc} 1 & 2 & 3 \\ \left[\begin{array}{cc|c} 1 & 0 & p_{13} \\ 0 & 1 & p_{23} \\ \hline 0 & 0 & p_{33} \end{array}\right] \end{array} = \left[\begin{array}{cc} I & R \\ O & Q \end{array}\right]$$

We form the matrix of the eventual probabilities

$$E = \begin{array}{c} \\ 1 \\ 2 \end{array} \begin{array}{c} 3 \\ \left[\begin{array}{c} e_{13} \\ e_{23} \end{array}\right] \end{array}$$

Notice that the symbols in E have the same subscripts as those in the R-matrix. That is, in e_{ij} the subscript i represents only absorbing states and j only nonabsorbing states.

5.5 THE FUNDAMENTAL MATRIX

In order to express the probability formulas in terms of matrix multiplication, we rewrite $p_{33}e_{13}$ as $e_{13}p_{33}$. The formulas

$$e_{13} = p_{13} + e_{13}p_{33}$$
$$e_{23} = p_{23} + e_{23}p_{33}$$

can be written as the single matrix equation

$$E = \begin{bmatrix} e_{13} \\ e_{23} \end{bmatrix} = \begin{bmatrix} p_{13} + e_{13}p_{33} \\ p_{23} + e_{23}p_{33} \end{bmatrix} = \begin{bmatrix} p_{13} \\ p_{23} \end{bmatrix} + \begin{bmatrix} e_{13} \\ e_{23} \end{bmatrix} \begin{bmatrix} p_{33} \end{bmatrix} = R + EQ$$

In the inspection example,

$$P = \begin{matrix} & \begin{matrix} 1 & 2 & 3 \end{matrix} \\ \begin{matrix} 1 \\ 2 \\ 3 \end{matrix} & \begin{bmatrix} 1 & 0 & .70 \\ 0 & 1 & .20 \\ 0 & 0 & .10 \end{bmatrix} \end{matrix}$$

and the equation $E = R + EQ$ looks like this:

$$E = \begin{bmatrix} e_{13} \\ e_{23} \end{bmatrix} = \begin{bmatrix} p_{13} + e_{13}p_{33} \\ p_{23} + e_{23}p_{33} \end{bmatrix} = \begin{bmatrix} .70 + e_{13}(.10) \\ .20 + e_{23}(.10) \end{bmatrix} = \begin{bmatrix} .70 \\ .20 \end{bmatrix} + \begin{bmatrix} e_{13} \\ e_{23} \end{bmatrix} \begin{bmatrix} .10 \end{bmatrix} = R + EQ$$

The General Case

Turning to the general case, we suppose we have an absorbing Markov chain process, as in the preceding section. By interchanging rows and columns we can always put its transition matrix into the R-Q form

$$P = \begin{bmatrix} I & R \\ O & Q \end{bmatrix}$$

We define $E = [e_{ij}]$, where e_{ij} is the conditional probability

$e_{ij} = p$(go eventually to absorbing state i | start in nonabsorbing state j)

Then E will be the same size as the matrix R in the R-Q form. If we repeat the steps of the inspection example in the general case, we arrive at the same matrix equation

$$E = R + EQ$$

We must solve for E to obtain a matrix formula for the eventual probabilities e_{ij}. We subtract

$$E - EQ = R$$

Then we write $E = EI$, for the identity matrix I of the correct size, and we substitute

$$EI - EQ = R$$

416 MARKOV CHAINS

By the distributive law, we have
$$E(I - Q) = R$$

If $I - Q$ is nonsingular, we can multiply both sides of the equation on the right by the inverse of $I - Q$ to obtain
$$E(I - Q)(I - Q)^{-1} = R(I - Q)^{-1}$$

Our final result is then

$$\boxed{E = R(I - Q)^{-1}}$$

The Fundamental Matrix

In an absorbing Markov chain process the matrix $I - Q$ is always nonsingular, so the matrix $F = (I - Q)^{-1}$, called the **fundamental matrix** of the process, always exists; and we can calculate the eventual probabilities: $E = RF$. The fundamental matrix contains further information. We will use the symbol f_{ij} for the numbers in $F = (I - Q)^{-1}$. The i, j-pairs are the same as in Q, that is, i and j are both nonabsorbing states. Choose some nonabsorbing state j and add up all the f_{ij} in that jth column. The sum is the expected value, in the sense of Chapter 4, of the number of steps of the Markov chain process it will take for a unit in nonabsorbing state j to attain one of the absorbing states. Thus, for instance, in the inspection example, in which $Q = [p_{33}] = [.10]$,

$$(I - Q)^{-1} = ([1] - [.10])^{-1} = [.90]^{-1} = \left[\frac{1}{.90}\right] = [1.11] \text{ (approximately)}$$

This calculation tells us that, "on the average," it will take a repairable item slightly more than one repair either to be corrected (reach state 1) or be destroyed (reach state 2).

Example 1. Given a Markov chain process with transition matrix

$$P = \begin{array}{c} \\ 1 \\ 2 \\ 3 \\ 4 \end{array} \begin{array}{c} \begin{array}{cccc} 1 & 2 & 3 & 4 \end{array} \\ \left[\begin{array}{cccc} 1 & 0 & 0.20 & 0 \\ 0 & 1 & 0.10 & 0.40 \\ 0 & 0 & 0.40 & 0.10 \\ 0 & 0 & 0.30 & 0.50 \end{array}\right] \end{array}$$

(a) Calculate the expected number of steps of the process it will take for a unit initially in each of the nonabsorbing states 3 or 4 to be absorbed into one of the absorbing states 1 or 2.
(b) Calculate the probability that a unit initially in state 4 will eventually be absorbed into state 1.

Solution

Since P is already in R-Q form, we see that

$$I - Q = \begin{bmatrix} 1 & 0 \\ 0 & 1 \end{bmatrix} - \begin{bmatrix} .40 & .10 \\ .30 & .50 \end{bmatrix} = \begin{matrix} & 3 & 4 \\ 1 \\ 2 \end{matrix} \begin{bmatrix} .60 & -.10 \\ -.30 & .50 \end{bmatrix}$$

We will find the inverse of this 2 × 2-matrix using the method of Section 1.4. The determinant of $I - Q$ is given by

$$\Delta = (.60)(.50) - (-.10)(-.30) = .27$$

Therefore, to two decimal places, the fundamental matrix is

$$F = (I - Q)^{-1} = \begin{bmatrix} \frac{.50}{.27} & \frac{.10}{.27} \\ \frac{.30}{.27} & \frac{.60}{.27} \end{bmatrix} = \begin{matrix} & 3 & 4 \\ 1 \\ 2 \end{matrix} \begin{bmatrix} 1.85 & 0.37 \\ 1.11 & 2.22 \end{bmatrix}$$

(a) The first column of the fundamental matrix corresponds to state 3, so we calculate the expected value to be $1.85 + 1.11 = 2.96$. Thus, starting in state 3, the expected number of steps to absorption is almost 3; in the expected-value sense, absorption is quicker from state 4, namely $.37 + 2.22 = 2.59$ steps.

(b) Since the absorbing states are 1 and 2 and the nonabsorbing states 3 and 4,

$$E = \begin{bmatrix} e_{13} & e_{14} \\ e_{23} & e_{24} \end{bmatrix}$$

and we wish to find e_{14}, the number in the upper right-hand corner of the matrix. We calculate

$$E = RF = \begin{bmatrix} .20 & 0 \\ .10 & 0.40 \end{bmatrix} \begin{bmatrix} 1.85 & 0.37 \\ 1.11 & 2.22 \end{bmatrix} = \begin{bmatrix} 0.37 & 0.07 \\ 0.63 & 0.93 \end{bmatrix}$$

so, $e_{14} = .07$. This tell us that even though it is impossible to go directly from state 4 to state 1 in this Markov chain process because $p_{14} = 0$, there is a positive probability, namely .07, of getting there indirectly. ■

Example 2. Two players, call them H and T, are each given $2 at the start of a game. A coin is flipped and if the result is heads, player T must give $1 to H; if the result is tails, H pays $1 to T. The game continues until one player, H or T, runs out of money.

  (a) Show that if the coin is fair (probability of heads is .50), then the game is fair in the sense that H has a probability of .50 of winning all of T's money and the same is true for T.
  (b) Calculate H's chances of winning if, instead of being equally divided at the start of the game, H starts with $3 and T with only $1.

Solution

We represent this game as a Markov chain process with states corresponding to the amounts of money that H can have: $0, $1, $2, $3, $4. The states corresponding to $0 (T wins) and $4 (H wins) are absorbing because the game ends when it reaches either of these states. Table 5-16 describes the conditional probabilities that H will have

418   MARKOV CHAINS

		Before Coin Flip			
	$0	$1	$2	$3	$4
After Coin Flip $0	1	.50	0	0	0
$1	0	0	.50	0	0
$2	0	.50	0	.50	0
$3	0	0	.50	0	0
$4	0	0	0	.50	1

Table 5-16

various amounts of money after a coin flip, given the amount, denoted by the column heading, that H had before the coin was flipped. For instance, if H had $1 and the coin comes up heads, T will pay her a dollar, making H's total then $2. But if the coin comes up tails, then H pays a dollar to T and has $0 left. Since the probability that H will win a dollar is .50 and also .50 that she will lose a dollar, the column headed "$1" consists of .50 in the rows for $0 and $2, and 0 elsewhere. Similarly, if H has $2, she can either win another dollar for a total of $3, or drop to $1; each possible outcome has probability .50, as we see in the column headed "$2."

The transition matrix is

$$\begin{array}{c} \\ 0 \\ 1 \\ 2 \\ 3 \\ 4 \end{array} \begin{array}{c} 0 1 2 3 4 \end{array} \\ \begin{bmatrix} 1 & 0.50 & 0 & 0 & 0 \\ 0 & 0 & 0.50 & 0 & 0 \\ 0 & 0.50 & 0 & 0.50 & 0 \\ 0 & 0 & 0.50 & 0 & 0 \\ 0 & 0 & 0 & 0.50 & 1 \end{bmatrix}$$

The rows and columns are labeled to indicate the amount of money held by H in each state. In order to put the transition matrix into R-Q form, we will interchange the $1 state with the $4 state, so that the absorbing states $0 and $4 come first. As in the preceding section, we first interchange columns:

$$\begin{array}{c} 0 \\ 1 \\ 2 \\ 3 \\ 4 \end{array} \begin{array}{ccccc} 0 & 4 & 2 & 3 & 1 \end{array} \\ \begin{bmatrix} 1 & 0 & 0 & 0 & 0.50 \\ 0 & 0 & 0.50 & 0 & 0 \\ 0 & 0 & 0 & 0.50 & 0.50 \\ 0 & 0 & 0.50 & 0 & 0 \\ 0 & 1 & 0 & 0.50 & 0 \end{bmatrix}$$

We then complete the process by interchanging rows to obtain the transition matrix P in R-Q form:

$$\begin{array}{c} 0 \\ 4 \\ 2 \\ 3 \\ 1 \end{array} \begin{bmatrix} 1 & 0 & 0 & 0 & 0.50 \\ 0 & 1 & 0 & 0.50 & 0 \\ \hline 0 & 0 & 0 & 0.50 & 0.50 \\ 0 & 0 & 0.50 & 0 & 0 \\ 0 & 0 & 0.50 & 0 & 0 \end{bmatrix} = \begin{bmatrix} I & R \\ 0 & Q \end{bmatrix}$$

5.5 THE FUNDAMENTAL MATRIX

In order to analyze the problem by the methods of this section, we must have an absorbing Markov chain. In the R-Q form

$$R = \begin{array}{c} \\ 0 \\ 4 \end{array} \begin{array}{c} 2 \quad\ 3 \quad\ 1 \\ \begin{bmatrix} 0 & 0 & 0.50 \\ 0 & 0.50 & 0 \end{bmatrix} \end{array}$$

there are zeros. Therefore, as we discussed in the preceding section, we must calculate powers of the transition matrix to determine whether the Markov chain process is absorbing. When we multiply the transition matrix P by itself, the resulting matrix is still in R-Q form:

$$\begin{array}{c} \\ 0 \\ 4 \\ 2 \\ 3 \\ 1 \end{array} \begin{array}{c} 0 \quad 4 \quad\ 2 \quad\ \ 3 \quad\ \ 1 \\ \left[\begin{array}{cc|ccc} 1 & 0 & 0.25 & 0 & 0.50 \\ 0 & 1 & 0.25 & 0.50 & 0 \\ \hline 0 & 0 & 0.50 & 0 & 0 \\ 0 & 0 & 0 & 0.25 & 0.25 \\ 0 & 0 & 0 & 0.25 & 0.25 \end{array} \right] \end{array}$$

There are still zeros in the R-matrix of P^2, though fewer than before; thus, we multiply by the transition matrix once more.

$$\begin{array}{c} \\ 0 \\ 4 \\ 2 \\ 3 \\ 1 \end{array} \begin{array}{c} 0 \quad 4 \quad\ 2 \quad\ \ 3 \quad\ \ 1 \\ \left[\begin{array}{cc|ccc} 1 & 0 & 0.25 & 0.125 & 0.625 \\ 0 & 1 & 0.25 & 0.625 & 0.125 \\ \hline 0 & 0 & 0 & 0.25 & 0.25 \\ 0 & 0 & 0.25 & 0 & 0 \\ 0 & 0 & 0.25 & 0 & 0 \end{array} \right] \end{array}$$

Now the R-matrix of P^3 is positive, so we know that this Markov chain process is absorbing.

We wish to compute the probability of H winning, that is, ending up in the $4 state. In Part (a) she starts with $2. In Part (b) she starts with $3. Thus, for Part (a), we need to compute e_{42}, the probability that H will eventually reach state 4, given that she starts in state 2. For Part (b) we compute e_{43}. To do this, we must find $E = RF$, where $F = (I - Q)^{-1}$ is the fundamental matrix for

$$I - Q = \begin{bmatrix} 1 & 0 & 0 \\ 0 & 1 & 0 \\ 0 & 0 & 1 \end{bmatrix} - \begin{bmatrix} 0 & 0.50 & 0.50 \\ 0.50 & 0 & 0 \\ 0.50 & 0 & 0 \end{bmatrix} = \begin{bmatrix} 1 & -0.50 & -0.50 \\ -0.50 & 1 & 0 \\ -0.50 & 0 & 1 \end{bmatrix}$$

We use the Gauss-Jordan method to find the inverse of this matrix. It is a little easier to work with fractions rather than decimals, so we write the augmented matrix as

$$\begin{bmatrix} 1 & -\frac{1}{2} & -\frac{1}{2} & 1 & 0 & 0 \\ -\frac{1}{2} & 1 & 0 & 0 & 1 & 0 \\ -\frac{1}{2} & 0 & 1 & 0 & 0 & 1 \end{bmatrix}$$

We add $\frac{1}{2}$ times the first row to each of the second and third rows:

$$\begin{bmatrix} 1 & -\frac{1}{2} & -\frac{1}{2} & 1 & 0 & 0 \\ 0 & \frac{3}{4} & -\frac{1}{4} & \frac{1}{2} & 1 & 0 \\ 0 & -\frac{1}{4} & \frac{3}{4} & \frac{1}{2} & 0 & 1 \end{bmatrix}$$

Next, we divide the second row by $\frac{3}{4}$.

$$\begin{bmatrix} 1 & -\frac{1}{2} & -\frac{1}{2} & 1 & 0 & 0 \\ 0 & 1 & -\frac{1}{3} & \frac{2}{3} & \frac{4}{3} & 0 \\ 0 & -\frac{1}{4} & \frac{3}{4} & \frac{1}{2} & 0 & 1 \end{bmatrix}$$

To convert the second column into the corresponding column of the identity matrix, we add $\frac{1}{2}$ times the second row to the first row and $\frac{1}{4}$ times the second row to the third:

$$\begin{bmatrix} 1 & 0 & -\frac{2}{3} & \frac{4}{3} & \frac{2}{3} & 0 \\ 0 & 1 & -\frac{1}{3} & \frac{2}{3} & \frac{4}{3} & 0 \\ 0 & 0 & \frac{2}{3} & \frac{2}{3} & \frac{1}{3} & 1 \end{bmatrix}$$

Multiplying the third row by $\frac{3}{2}$, we obtain

$$\begin{bmatrix} 1 & 0 & -\frac{2}{3} & \frac{4}{3} & \frac{2}{3} & 0 \\ 0 & 1 & -\frac{1}{3} & \frac{2}{3} & \frac{4}{3} & 0 \\ 0 & 0 & 1 & 1 & \frac{1}{2} & \frac{3}{2} \end{bmatrix}$$

Finally, we add $\frac{2}{3}$ times the third row to the first row and $\frac{1}{3}$ times the third row to the second:

$$\begin{bmatrix} 1 & 0 & 0 & 2 & 1 & 1 \\ 0 & 1 & 0 & 1 & \frac{3}{2} & \frac{1}{2} \\ 0 & 0 & 1 & 1 & \frac{1}{2} & \frac{3}{2} \end{bmatrix}$$

We write the inverse in decimal form

$$F = (I - Q)^{-1} = \begin{bmatrix} 2 & 1 & 1 \\ 1 & 1.50 & 0.50 \\ 1 & 0.50 & 1.50 \end{bmatrix}$$

Remember that in E the rows correspond to the absorbing states $0 and $4, and the columns correspond to the nonabsorbing states $2, $3, and $1. Thus,

$$E = \begin{bmatrix} e_{02} & e_{03} & e_{01} \\ e_{42} & e_{43} & e_{41} \end{bmatrix} = RF = \begin{bmatrix} 0 & 0 & 0.50 \\ 0 & 0.50 & 0 \end{bmatrix} \begin{bmatrix} 2 & 1 & 1 \\ 1 & 1.50 & 0.50 \\ 1 & 0.50 & 1.50 \end{bmatrix}$$

$$= \begin{bmatrix} 0.50 & 0.25 & 0.75 \\ 0.50 & 0.75 & 0.25 \end{bmatrix}$$

We see that in Part (a), where H starts out with $2, her probability of winning is given by $e_{42} = .5$; thus, the game is fair. In Part (b), H starts with $3, and the probability is $e_{43} = .75$ that she will win—so H has a substantial advantage. ∎

Example 2 illustrates the fact that in casino gambling, since the individual player has much less money than the "house," the casino has a great advantage.

Summary

If we write the transition matrix for any absorbing Markov chain process with the absorbing states listed first, it has the *R-Q* form

$$P = \begin{bmatrix} I & R \\ O & Q \end{bmatrix}$$

Here, the rows of R are the absorbing states and the columns—the nonabsorbing states; also the rows and columns of Q correspond to the nonabsorbing states. We define E to be the matrix of conditional probabilities, given by

$e_{ij} = p$(eventually in absorbing state i | start in nonabsorbing state j)

Then $E = RF$, where $F = (I - Q)^{-1}$ is called the **fundamental matrix** of the process. If we choose a nonabsorbing state j and add up all the numbers in the corresponding column of F, the sum is the expected value of the number of steps of the Markov chain process for a unit in the nonabsorbing state j to reach some absorbing state.

Exercises

1. Given the transition matrix of an absorbing Markov chain process

$$\begin{array}{c} \\ 1 \\ 2 \\ 3 \end{array} \begin{bmatrix} 1 & 2 & 3 \\ 1 & 0 & .10 \\ 0 & 1 & .50 \\ 0 & 0 & .40 \end{bmatrix}$$

   (a) Calculate the expected value of the number of time periods until a unit in state 3 is absorbed.
   (b) Calculate the probability that a unit in state 3 will be absorbed into state 1.

2. Given the transition matrix of an absorbing Markov chain process

$$\begin{array}{c} \\ 1 \\ 2 \\ 3 \end{array} \begin{bmatrix} 1 & 2 & 3 \\ .40 & 0 & 0 \\ .20 & 1 & 0.50 \\ .40 & 0 & 0.50 \end{bmatrix}$$

   Calculate the expected value of the number of time periods required for a unit in state 1 to be absorbed.

3. Given the transition matrix of an absorbing Markov chain process

$$\begin{array}{c} \\ 1 \\ 2 \\ 3 \\ 4 \end{array} \begin{bmatrix} 1 & 2 & 3 & 4 \\ 0.50 & 0 & 0 & 0 \\ 0 & 1 & 0 & 0.30 \\ 0.20 & 0 & 1 & 0.20 \\ 0.30 & 0 & 0 & 0.50 \end{bmatrix}$$

   calculate the probability that a unit in state 1 will eventually be absorbed into state 3.

4. Given the transition matrix of an absorbing Markov chain process

$$\begin{array}{c} \\ 1 \\ 2 \\ 3 \\ 4 \end{array} \begin{bmatrix} 1 & 2 & 3 & 4 \\ 1 & \frac{1}{2} & \frac{1}{3} & \frac{1}{3} \\ 0 & 0 & \frac{1}{3} & 0 \\ 0 & 0 & \frac{1}{3} & \frac{1}{3} \\ 0 & \frac{1}{2} & 0 & \frac{1}{3} \end{bmatrix}$$

   calculate the expected value of the number of time periods it will take for a unit in state 4 to reach state 1.

5. In Example 2, calculate the expected value of the number of coin flips until the game ends:
   (a) when each player starts with $2;
   (b) when player H starts with $3 and T with $1.

6. In a learning experiment, a subject's response is classified as follows: knew the correct answer (C), guessed and got the right answer (R), guessed and got the wrong answer (W). The subject is told whether the guess is right or not, but not what the correct answer is if the guess is not correct. In a model of the learning process, we suppose that the transition matrix is the same at every repetition of the experiment and that it is

$$\begin{array}{c} \\ C \\ R \\ W \end{array} \begin{array}{c} C \quad R \quad W \\ \begin{bmatrix} 1 & \frac{1}{3} & \frac{1}{4} \\ 0 & \frac{2}{15} & \frac{3}{20} \\ 0 & \frac{8}{15} & \frac{3}{5} \end{bmatrix} \end{array}$$

Calculate the expected value of the number of wrong answers a subject will give before knowing the correct one.

7. A bank categorizes short-term loans as paid off (P), continuing payment (C), or defaulted (D), with transition matrix

$$\begin{array}{c} \\ P \\ C \\ D \end{array} \begin{array}{c} P \quad C \quad D \\ \begin{bmatrix} 1 & .13 & 0 \\ 0 & .85 & 0 \\ 0 & .02 & 1 \end{bmatrix} \end{array}$$

Calculate the probability that a short-term loan will eventually be paid off.

8. Members of the Islandian Merchant Marine are classified as seamen on active duty (E), officers on active duty (O), retired (R), and dismissed (D). The transition matrix for a given year is

$$\begin{array}{c} \\ E \\ O \\ R \\ D \end{array} \begin{array}{c} E \quad\quad O \quad\quad R \quad D \\ \begin{bmatrix} 0.85 & 0 & 0 & 0 \\ 0.04 & 0.955 & 0 & 0 \\ 0.10 & 0.040 & 1 & 0 \\ 0.01 & 0.005 & 0 & 1 \end{bmatrix} \end{array}$$

(a) Calculate the expected number of years an officer will serve (before retiring or being dismissed).
(b) What is the probability that a seaman will eventually be dismissed?

9. A nutrient in a river can be located in the soil of the river bottom (S), in the vegetation growing in the river (V), in an animal in the river (A), or it can pass out of the river as the soil is carried as silt to the ocean (O), from which it cannot return to the river. We assume there is no chance the nutrient will leave the river in vegetation or in an animal. The transition matrix for each period of time is

$$\begin{array}{c} \begin{array}{cccc} S & V & A & O \end{array} \\ \begin{array}{c} S \\ V \\ A \\ O \end{array} \left[\begin{array}{cccc} 0.60 & 0.20 & 0.30 & 0 \\ 0.30 & 0.60 & 0 & 0 \\ 0 & 0.20 & 0.70 & 0 \\ 0.10 & 0 & 0 & 1 \end{array} \right] \end{array}$$

(a) Show that this is the transition matrix of an absorbing Markov chain process.

(b) If a unit of the nutrient is in the soil, calculate the expected number of time periods it will remain in the river before it is carried to the ocean.

10. Suppose in Example 2 that the coin being flipped is weighted so that the probability of heads is $\frac{2}{3}$.

(a) Calculate H's chances of winning this game if both players start with $2.

(b) Calculate H's chances of winning this game if she starts with $3, as opposed to $1 for her opponent.

Review Exercises for Chapter 5

1. In a message sent in code in the English alphabet, the probability that a vowel will be followed by another vowel is found to be .15, and the probability that a consonant will be followed by another consonant is .40.

(a) Write the transition matrix for the states "vowel" and "consonant."

(b) Estimate the percentage of vowels in the messages in this code by computing the equilibrium distribution of vowels and consonants.

2. Given the transition matrix of an absorbing Markov chain process,

$$\begin{array}{c} \begin{array}{cccc} 1 & 2 & 3 & 4 \end{array} \\ \begin{array}{c} 1 \\ 2 \\ 3 \\ 4 \end{array} \left[\begin{array}{cccc} 1 & \frac{1}{2} & \frac{3}{4} & \frac{1}{2} \\ 0 & 0 & 0 & \frac{1}{2} \\ 0 & \frac{1}{2} & 0 & 0 \\ 0 & 0 & \frac{1}{4} & 0 \end{array} \right] \end{array}$$

calculate the expected number of time periods for a unit in state 2 to get to state 1.

3. The president of a chain of fast-food restaurants classifies the profitability of each restaurant as excellent, good, fair, or poor. The president's estimates of the probabilities that a restaurant will change classification from one month to the next are given in Table 5-17. If in a particular month, 10% of the restaurants in the chain are classified as excellent, 30% as good, 40% as fair, and 20% as poor, what is the estimate for the next month?

		Present Month			
		Excellent	Good	Fair	Poor
Next Month	Excellent	.80	.20	0	0
	Good	.10	.70	.10	0
	Fair	.10	.10	.80	.10
	Poor	0	0	.10	.90

Table 5-17

4. Suppose that the female of some animal species is known from her physical characteristics to be of type aa with respect to some gene, but that the identity of her mate is unknown. If she mates with a male of genotype AA, the offspring will all be of type Aa. If she mates with a type-Aa male, then according to Mendel's law, the probability that the offspring will be of type Aa is .50 and of type aa, .50. If she mates with a male who is also of type aa, then all the offspring will be aa, as well. Suppose that 30% of the males of this species are of genotype AA, 50% of type Aa, and 20% of type aa, and suppose further, that mating is independent of this gene.
   (a) What is the probability that the first offspring of this female will be of genotype AA?
   (b) What is the probability that the first offspring will be of genotype Aa?
   (c) What is the probability that the first offspring will be of genotype aa?

5. Consumers of soft drinks are divided according to their favorite beverage into cola drinkers, citrus drinkers, and drinkers of other flavors. Suppose that drinking preferences during the year change according to Table 5-18. If this table remains valid in the future, what will be the eventual (that is, equilibrium) percentage of drinkers who prefer cola?

		Start of Year		
		Cola	Citrus	Other
End of Year	Cola	.80	.20	.20
	Citrus	.20	.70	.20
	Other	0	.10	.60

Table 5-18

6. In an animal population, the probability that a female offspring of a long-haired female will be long-haired is .80, whereas the offspring of a female that is not long-haired has a probability of .10 of being long-haired. At the present time, 20% of the females in this population are long-haired. Three generations into the future, what percentage of the females will be long-haired?

7. Students at a university are classified as lower-division (L), upper-division (U), graduated (G), and failed (F). We assume that a failed student never returns to the university. The transition matrix for each term at the university is

$$\begin{array}{c c} & \begin{array}{cccc} L & U & G & F \end{array} \\ \begin{array}{c} L \\ U \\ G \\ F \end{array} & \left[\begin{array}{cccc} 0.50 & 0 & 0 & 0 \\ 0.40 & 0.50 & 0 & 0 \\ 0 & 0.45 & 1 & 0 \\ 0.10 & 0.05 & 0 & 1 \end{array} \right] \end{array}$$

Calculate the probability that a lower-division student will eventually graduate from this university.

8. A new chocolate bar—Tiger brand—is introduced into a market dominated by a single brand—Lion. The probability that a purchaser of Tiger, Lion, or some other brand will change brands with the next purchase is given in Table 5-19. Show that if these probabilities stay the same, then eventually $\frac{1}{15}$ of all purchasers will choose Tiger and $\frac{8}{15}$ will choose Lion.

		Present Purchase		
		Tiger Brand	Lion Brand	Other
Next Purchase	Tiger Brand	$\frac{7}{10}$	0	$\frac{1}{20}$
	Lion Brand	$\frac{1}{5}$	$\frac{9}{10}$	$\frac{1}{10}$
	Other	$\frac{1}{10}$	$\frac{1}{10}$	$\frac{17}{20}$

Table 5-19

6

Game Theory

Political Campaigns

Traditionally, political scientists have based their work on historical studies and philosophical theories. In recent years, however, they have also made use of statistics and other mathematically based methods. A good example of this mathematical approach to political science is found in *Game Theory and Politics* by Stephen J. Brams. In his book Brams shows that the mathematical theory of games (or game theory) is particularly useful in analyzing political campaigns. Brams uses game theory to compare the ideal presidential campaign strategies of the candidates of the two major political parties under the present electoral-college system with the strategies that would be used if the president were elected by direct popular vote. In making his analysis Brams assumes that the two candidates have approximately the same amount of committed support at the start of the official campaign. So, the campaigning that begins about the first of September for the November election is really a contest to win over a majority of the uncommitted voters. The twenty to forty percent of the electorate who are still undecided at that point are more than sufficient to determine the outcome of most presidential elections.

Brams assumes that the candidates have approximately equal resources of time and money and that, generally, the more resources a candidate spends in a state compared to those spent by the opposing candidate, the more uncommitted votes he or she will receive. So, the

428   GAME THEORY

> major problem for each candidate is how to allocate total resources to win over as many undecided voters as possible without alienating any committed voters.
>
> Brams' major conclusion is that under the electoral-college system, the best strategy for both candidates is to spend virtually all available time and money in the small number of large states that have many electoral votes. On the other hand, if the president were elected by direct popular vote, although the candidates would still find it most profitable to devote much of their resources to the larger states, the contrast would not be as extreme.
>
> This chapter presents a limited, but useful, portion of game theory. Therefore, we will not attempt as sophisticated an analysis of campaign resource allocation as the one used in Brams' book. We will, however, develop a simple model that illustrates the effectiveness of the game-theoretic approach to political campaigning.

6.1 Matrix Games

Imagine a contest in which each of the two contestants, we'll call them Ruth and Carl, sits out of sight of the other. Each has two buttons, one silver and one gold, in front of them. Both contestants can see the display shown in Table 6-1.

		Carl	
		Silver	Gold
Ruth	Silver	−$10	$10
	Gold	0	$5

Table 6-1

Each contestant will push one button, gold or silver. The display shows the players the possible results. For example, if both push their gold buttons, then Carl must pay $5 to Ruth. The negative entry in the display indicates a payment of $10 from Ruth to Carl when they both push silver buttons. If Ruth pushes gold and Carl pushes silver, the 0 indicates that no payment is made by either player. Thus, the numbers in the display describe the outcomes from Ruth's point of view.

At a signal, the contestants push the button of their choice; when both buttons have been pushed, the outcome registers on the display and the payoff is made. In this way neither player knows what the other's choice is until his or her own choice has been recorded.

Looking at the display from Carl's perspective, we realize that there is no reason for him to push the gold button. No matter which button Ruth pushes, if Carl pushes the gold button, he will lose money. Thus, we may assume that Carl will push the silver button. Ruth can also see that Carl has no reason to push the gold button. Therefore, Ruth concludes that the choice of the silver button would cost her $10 since the outcome will then be silver-silver. So, Ruth will push the gold button, giving the combination Ruth-gold and Carl-silver, which is represented by 0 on the display. Consequently, no money changes hands.

Thus, as a spectator sport, watching this game would be somewhat less exciting than watching grass grow. We described this game not to furnish entertainment, but as a simple example to introduce the type of games we will analyze in this chapter.

Matrix Games

The first thing to notice about the game is that it is a **two-person game**: there are just two players in the game, Ruth and Carl. Next we observe that each contestant has a number of clear-cut choices; they are: push the silver button, push the gold button. In general, we will not restrict the number of such choices, nor will we require that the number or types of choices available to each of the two players be the same. We demand only that all the players' choices, called **pure strategies** in game theory, be distinct and be unambiguously described in the game.

In the example, the consequences of every possible combination of pure-strategy choices by the two players were shown in the display. The general form of such a display is given in Table 6-2.

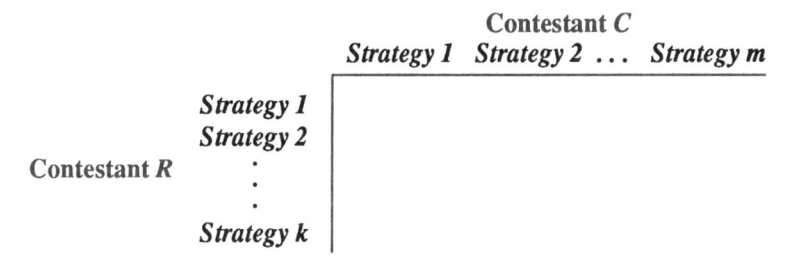

	Contestant C			
	Strategy 1	*Strategy 2*	...	*Strategy m*
Strategy 1				
Strategy 2				
Contestant R ⋮				
Strategy k				

Table 6-2

We recognize the display as a matrix, just as in Chapter 1. It is a $k \times m$-matrix, where k, the number of rows, equals the number of pure strategies for contestant R, and m, the number of columns, is the number of pure strategies for C. The outcome of play is thought of in terms of gains and losses for the players (though not necessarily as amounts of money), so the outcomes are called **payoffs** and the matrix is the **payoff matrix** G of the game. A positive payoff g_{ij} at the location corresponding to a choice of the strategies of row i by R and column j by C indicates a transfer of g_{ij} units from contestant C to contestant R. A negative entry means that the transfer goes in the opposite direction, so the payoff matrix takes the point of view of the player who chooses the rows. Notice that according to the rules, for any payoff the gain to the winner equals the loss to the loser. A game with this feature is called a **zero-sum game**.

430 GAME THEORY

An important feature of the type of game that we are describing is that both players have complete information on the consequences of their own and their opponent's strategy choices, as specified by the payoff matrix. In terms of our example, it is crucial that both contestants be able to see the display. We summarize in the following definition.

> A two-person, zero-sum game in which the payoffs are specified by a matrix, visible to both players, and the pure strategies consist of a choice of row for one contestant and a choice of column for the other is called a **matrix game** because the payoff matrix G contains the entire mathematical content of the game.

In the game described at the start of this section, Ruth and Carl sat out of sight of each other, and neither knew the other's strategy choice until both had made their own choices. The purpose of these rules was to ensure that there was no communication between the contestants.

The final feature of game theory that is illustrated in our simple example is the assumed rationality of the contestants. Carl rejected the strategy "push the gold button" because the payoff matrix showed no opportunity for gain in that column. Ruth was not tempted to push the silver button in the hope that Carl would choose the gold button, resulting in her winning $10. Rather, Ruth assumed that Carl would ignore the gold button as unprofitable, and therefore she would lose $10 if she pushed the silver button. Thus, not only are both contestants assumed to be rational, but, in addition they each expect their opponent to be rational. Game theorists realize that, in real-life competition, motives and behavior are not that simple. But mathematics is not well-adapted to the analysis of irrational actions, so game theory assumes the players to be rational.

We now describe some competitive situations by means of matrix games.

Example 1. A football quarterback must decide whether to call a running play or a pass. As the play begins, the defending linebackers must decide whether it is a running play or a pass. A gain of yardage for one team is a loss for the other, and vice versa. Suppose that, if the quarterback calls a running play, the expected gain is only 2 yards, provided that the opposing linebackers defend against a running play. But the expected gain is 8 yards if the linebackers think the quarterback will attempt a pass. If the quarterback calls a pass but the linebackers expect a running play, then the expected outcome will be a 15-yard gain for the offense. However, if the linebackers defend against a pass, the expectation is that the offense will suffer a 6-yard loss. Determine the payoff matrix for this game.

Solution

We let the quarterback's choices be represented by the rows of the payoff matrix and the linebackers' strategy be given by the columns. Then the form of the matrix is shown in Table 6-3.

6.1 MATRIX GAMES

		Linebackers (Defense)	
		Run	*Pass*
Quarterback (Offense)	*Run*		
	Pass		

Table 6-3

Since one team's gain of yardage is the other's loss, this is a zero-sum game. In a matrix game, payoffs to the row player are positive. Therefore, the payoffs represent gains and losses for the offense. From the given data, the payoff matrix is indicated in Table 6-4. ∎

		Linebackers	
		Run	*Pass*
Quarterback	*Run*	2	8
	Pass	15	−6

Table 6-4

Our next example is suggested by the discussion of strategies for the distribution of resources by the political candidates in the introduction to this chapter.

Example 2. Rond and Cali are the two candidates for president of their country, Islandia, and the election will be won by the candidate who captures the majority of the popular vote. For purposes of campaigning, Islandia is divided into three regions: East, Central, and West; each region has different social and economic characteristics, which can be expected to affect voting preferences. The candidates must decide how to distribute their campaign resources to win undecided votes. (For simplicity, we always assume that undecided voters will vote for one of the candidates.) If each candidate chooses to put all available resources into a single region, a political scientist predicts the behavior of the undecided voters as follows. He concludes that if Rond and Cali choose to put their resources in the same region, they will each receive the same number of undecided votes. If Rond puts all his resources in the East, he will receive the same number of undecided votes as will Cali if she puts all her resources in the Central region, but one million more of these votes than Cali will if she puts all her resources into the West. If Rond chooses the Central region, then he will obtain one million more undecided votes if Cali chooses to concentrate on the East, but if Cali chooses the West, then she will come out ahead by two million undecided votes. If Rond chooses the West, then he will lose two million votes to Cali if she chooses to concentrate on the East, but will gain one million votes if she chooses the Central region. Determine the payoff matrix for this game.

Solution

The pure strategies for each candidate consist of using all resources for a single region. Rond's choices are represented by rows and Cali's by columns. The form of the matrix is then given in Table 6-5.

432 GAME THEORY

		Cali	
	East	Central	West
Rond East			
Central			
West			

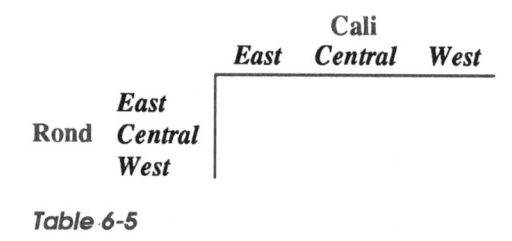

Table 6-5

The entries in the payoff matrix will represent millions of undecided voters, with a positive number indicating Rond's advantage over Cali and a negative number the opposite. Since there is no advantage if both candidates choose to concentrate on the same region, all the diagonal entries are 0. The other entries, according to the political scientist's estimates are represented in Table 6-6. ∎

		Cali		
		East	Central	West
Rond	East	0	0	1
	Central	1	0	−2
	West	−2	1	0

Table 6-6

Of course, the candidates for president of Islandia need not restrict themselves to strategies that put all their resources into a single region. However, by formulating the game in terms of these pure strategies, we will be able to use the concept of *mixed strategy* that we will introduce in Section 6.3 to describe more sensible strategies for the candidates. Then, when we have the necessary tools available in Section 6.5, we will determine the best strategies for the two candidates and we will predict the corresponding outcome of the distribution of the undecided vote in the campaign for President of Islandia.

Our final example illustrates a "game against nature." In such a game, one player is viewed as competing with circumstances beyond his or her control.

Example 3. In planning a backpacking trip, a hiker has to decide whether or not to take a heavy jacket. If the weather is warm, she estimates the unpleasantness of carrying the extra weight of the jacket will cause her to lose 5 points on her personal scale of happiness. If the weather is cool, the greater comfort from wearing the jacket part of the time will just balance the bother of carrying it the rest of the time, with no effect on total happiness. However, if the weather is cold, the satisfaction of having made the correct decision together with the greater comfort will add 7 points to her happiness. Alternatively, if it is warm, leaving the jacket behind will give the hiker 5 points of happiness. If it is cool there will be no happiness effect if she does not have the jacket. But in cold weather the shivering hiker will give up 10 points of happiness if she has no jacket. Determine the payoff matrix of this game.

Solution

The hiker is certainly one contestant in this game. Her opponent is the weather. We are only interested in the hiker's gains and losses in happiness, so the convenient approach is to treat this, and all other such games against nature, as zero-sum games. The backpacker's gain is the weather's loss, while her loss in happiness is interpreted as a victory for the weather. The hiker's options are to take the jacket or not, which we denote by *Yes* and *No*. The weather has the choice of being warm, cool, or cold. Therefore, the payoff matrix is as shown in Table 6-7. ∎

$$
\begin{array}{c c}
 & \text{Weather} \\
 & \begin{array}{c c c} \text{Warm} & \text{Cool} & \text{Cold} \end{array} \\
\text{Hiker} \begin{array}{c} \text{Yes} \\ \text{No} \end{array} & \left| \begin{array}{c c c} -5 & 0 & 7 \\ 5 & 0 & -10 \end{array} \right.
\end{array}
$$

Table 6-7

Bimatrix Games

Although the mathematical content of the present chapter will be limited to the subject of matrix games, we do not want to give the impression that game theory limits itself to games of this type. In another important class of two-person games, one player chooses a row of a matrix and the other a column, as in a matrix game, but now each player has his or her own payoff matrix. The matrix describes payments into or out of a "bank," which is distinct from the two players. We still suppose Ruth chooses a row as pure strategy and Carl chooses a column. Suppose the payoffs for the players are given by the matrices

$$
\text{For Ruth} \qquad \text{For Carl}
$$
$$
\begin{bmatrix} 3 & -1 \\ 4 & 0 \end{bmatrix} \qquad \begin{bmatrix} 1 & 4 \\ -1 & 0 \end{bmatrix}
$$

If Ruth chooses row 1 and Carl chooses column 1, then Ruth's matrix indicates that she will receive 3 dollars from the bank and from Carl's matrix we learn that he will receive one dollar from the bank. If Ruth chooses row 1 but Carl chooses column 2, then the entry of −1 in the upper right-hand corner of Ruth's matrix means that Ruth must pay one dollar to the bank, whereas the corresponding entry in Carl's matrix indicates a payment to Carl of 4 dollars from the bank. On the other hand, the choices of row 2 and column 1 produce a gain of 4 dollars from the bank to Ruth and a payment of a dollar to the bank by Carl.

Instead of two payoff matrices, one for each player, it is convenient to express the payoffs to the two players by means of a single matrix in which each entry is an ordered pair of numbers, rather than a single number. The first number of the pair indicates the payoff to the row player from the bank (or to the bank if the number is negative) and the second number is the payoff to the column player.

A matrix in which the entries are ordered pairs is called a **bimatrix**. The payoff bimatrix for the **bimatrix game** we are discussing is

$$\begin{bmatrix} 3,1 & -1,4 \\ 4,-1 & 0,0 \end{bmatrix}$$

In contrast to a matrix game, in which the players compete against each other (since a gain for one player requires the equal loss to the other), players in a bimatrix game can cooperate with each other to obtain as much as possible from the bank. Thus, in the present game, if the players agree on the cooperative strategy "row 1, column 1," then they obtain a total of 4 dollars from the bank. Any other combination of strategies will give them, as a team, at most three dollars, though one player can do better individually. In order to make the row 1, column 1-strategy attractive to Carl, who receives only one dollar compared to Ruth's three, Ruth can offer Carl a **side-payment** of one dollar, so each will end up with two dollars. Bimatrix games illustrate the fact that game theory can analyze cooperation as well as competition.

n-person Games

Games with more than two players, called ***n*-person games**, introduce another important aspect of game theory, the formation of **coalitions** in which some of the players agree to cooperate among each other to the advantage of the group and, generally, the disadvantage of those outside the coalition. We may illustrate the coalition concept by a three-person game in which the players—Amy, Bill, and Craig—must each give one of the other players one dollar. Suppose that Amy and Bill form a coalition and agree to give each other a dollar. Since Craig cannot keep his dollar, he must give it to either Amy or Bill. As part of their agreement, Amy and Bill can arrange a side-payment in which the player who receives Craig's dollar splits it with the other member of the coalition. In this way, by forming a coalition, Amy and Bill guarantee themselves a profit of 50 cents each, whereas Craig must lose his dollar.

Game theory has developed tools to analyze games other than the types we have mentioned. Some of these games have **chance moves** in which part of the play of the game is determined by chance—for instance, by the outcome of tossing a coin or rolling a die. Thus, a wide variety of situations involving competition and cooperation can be modeled and studied by means of games.

Summary

A **matrix game** is described by its **payoff matrix** G with entries g_{ij}, the **payoffs** of the game. In a **two-person game**, the pure strategies for one player, R, are given by the rows of the payoff matrix and those of the other, C, by the col-

> umns. A positive payoff g_{ij} at the location corresponding to a choice of row i by R and of column j by C indicates a transfer of g_{ij} units from contestant C to contestant R. A negative entry indicates the transfer goes in the opposite direction. In a **zero-sum game**, the gain by the winner equals the loss to the loser. The payoff matrix is visible to both players, the players cannot communicate with each other, and the game is played only once.

Exercises

Exercises 1 through 10 describe competitive situations that can be expressed as matrix games. Determine the payoff matrix for each of these games.

1. A band of smugglers can choose between two routes to bring a shipment of gold into the kingdom of Gondor without paying duty on it: over a mountain pass or along a river. The Gondor customs department has enough officers to watch only one route. If the smugglers and the customs officers choose the same route, the smugglers will be caught and the customs department will receive $2 million from impounding the gold, reselling it, and fining the smugglers. On the other hand, if the customs officers choose one route and the smugglers the other, then the smugglers gain $1 million from the sale of the duty-free gold.

2. Mr. Allen and Mr. Baker are the two candidates for Mayor of River City. On the ballot there will also be a referendum for a local sales tax to finance redevelopment of downtown River City. A poll indicates that among the undecided voters, if Mr. Allen supports the sales tax, he will obtain 12,000 more votes than Mr. Baker if Baker also supports the tax. Allen will obtain 2000 more votes than Baker if Baker is neutral on the tax, and Allen will obtain 2000 fewer votes than Baker if Baker opposes the sales tax. If Mr. Allen opposes the tax, neither candidate will have an advantage over the other, provided that Mr. Baker supports the sales tax. Allen will get 8000 fewer votes than Baker if Baker is neutral on the sales tax and 2000 fewer votes than Baker if Baker also opposes the sales tax.

3. An investor wants to invest a small inheritance in the stock market. If she buys conservative stocks, then at the end of the year the expected value of her investment will be $8000 if market prices decline, $10,500 if the market holds steady, and $11,500 if the stock market averages increase. However, if she chooses to invest in speculative stocks, then the expected value of her investment will be $3000 at the end of the year if the stock market declines, $9500 if the market is steady, and $16,000 if market prices increase.

4. In a military-training exercise, the soldiers are divided into two groups: the green army and the blue army. The green army must send reinforcements to a threatened city by one of three routes: over a mountain pass, through a river valley, or over a plain. The opposing blue army has only enough soldiers to set up an

ambush at one location. If the blue army fails to intercept these reinforcements, the green army is awarded points depending on the route used, and an equal number of points is taken from the blue army. Since the mountain-pass route is fastest, it is worth 20 points; the valley route is worth 15 points, and the plain 5 points. If, however, the blue-army ambush is set up on the route the green army chooses, blue receives 30 points for the mountain pass, 20 points for the river valley, and 10 points for the plain.

5. A developer buys a large parcel of land in a suburban area and must decide what to build on it. A major highway may be constructed near this parcel. If she builds a housing development and the highway is not located nearby, she will make a 40% profit on her investment, but if the highway is constructed nearby, the houses will be less desirable and she will make only 10% on her money. If she builds a shopping center and there is no highway nearby, more housing will be built for more shoppers, resulting in a 60% profit. But if the highway is close by, the area will become industrial and the shopping center will fail, with a 40% loss to the developer. If she builds an office complex on the land and there is no highway, office space will be hard to rent and the developer will suffer a 30% loss. But with the highway close by, the office complex will be successful and she will make a 75% profit. The final possibility, an industrial development, will fail if there is no highway nearby, and the developer will lose 50% of her investment. But if the highway comes close to her industrial development, she will make a 100% profit.

6. The Miller family is planning to have a picnic, but it looks as though it may rain. If they go on the picnic and it does not rain, the pleasure resulting from the picnic will be worth 5 points, but if they go and it rains on their picnic, their displeasure will lose them 5 points. If they stay home and it does rain, the Millers gain 2 points from the satisfaction of having made the correct decision. If they stay home and it fails to rain, their disappointment costs the Millers 2 points.

7. I hide either a nickel, a dime, or a quarter, and you must guess which coin I choose. If you guess correctly, you win the coin; if you guess wrong, you must give me an amount of money equal to the "difference" between the amount of your guess and the coin I hid. (In computing the payments, it does not matter which coin has more value, so if I hid the dime and you guessed a quarter or if I hid a quarter and you guessed a dime, you must pay me 15 cents in either case.)

8. Two airlines share a commuter route between two large cities. Since the number of travelers is essentially constant, a gain in business (measured in average percentage of seats filled) for one is a corresponding loss for the other. Both airlines are considering two innovations: redesigning the interiors of their planes to make them more attractive and serving complimentary snacks. They can adopt both innovations, either, or neither. Each innovation is worth a 5% increase to the airline that introduces it, provided the other airline does not do so. Thus, for example, if one airline chooses both innovations and the other airline chooses just one, the first airline gains 5%. However, if one airline redesigns interiors and the other airline serves snacks, the result is no change in passenger business.

9. There is the threat of a flu epidemic in which the flu can be of three types—A, B, and C. Four antibiotics are available; call them I, II, III, and IV. If a person is infected by type-A flu, then the probability of a speedy cure is .80 if antibiotic I is administered, .30 if II is used; the other two antibiotics are ineffective against this type of flu, so the probability of a speedy cure is 0 in these cases. For type-B flu, antibiotics I and III are ineffective, antibiotic II has a .90 probability, and antibiotic IV has a .50 probability of effecting a speedy cure against type B. Type-C flu can be cured speedily with probability .60 if antibiotic I is used, with probability .20 if either II or IV is used, whereas if antibiotic III is administered, the probability of a speedy cure is .90.

10. Two television networks, the Red and the Blue, must decide on programming for the same three time slots. Each network has a sports show (S), an entertainment show (E), and a cultural show (C) to schedule. The sports and entertainment programs are equally popular and so do not affect the overall ratings of the networks. Scheduling the cultural show at the time of the other network's entertainment show will cause the network with the cultural show to lose 3 points in the ratings to the other network. Running the cultural program against the sports show loses the network running it 5 points to its competitor. For example, suppose the Red network runs the programs in the order SEC, while the Blue network chooses the order ECS. The networks break even on the first time slot (S against E), then the Red network gains 3 points over the Blue network in the next time slot by running E against C, but the Red network loses 5 points during the last time slot by running C against S. So the total effect of the schedules SEC for Red and ECS for Blue is that the Red network loses 2 points to the Blue in the ratings. [*Hint*: Each network has a choice of 6 (that is, 3!) different strategies.]

6.2 Strictly Determined Games

All the information for a matrix game is contained in an ordinary $k \times m$-matrix, the payoff matrix with entries g_{ij}. For example, the rules of the game with payoff matrix

$$\begin{bmatrix} 1 & 3 & -1 & 0 & -1 \\ 5 & -2 & -1 & 2 & -2 \\ 4 & -1 & -2 & 0 & 0 \\ 3 & 2 & 1 & 2 & 3 \end{bmatrix}$$

are that player R has a choice of four strategies, one for each of the 4 rows, and player C has one strategy corresponding to each of the 5 columns. The payoff for the choices of row i, column j is g_{ij} units: from C to R, if g_{ij} is positive, and from R to C, if g_{ij} is negative. We will again give our players names when it is convenient to do so: Ruth for player R and Carl for player C.

Analysis of the Game

We begin our analysis of this game by studying Ruth's alternatives. If Ruth chooses strategy (row) 1, then a number of different payoffs are possible, depending on what Carl decides to do. For example, if Carl were to choose strategy (column) 2, then Ruth would receive $3 from Carl; a choice of strategy 5 by Carl would result in the loss to Ruth of $1. By the conditions of the game described in Section 6.1, Ruth has no communication with her opponent about the game, and so, she will assume the worst. That is, whatever strategy she chooses, she must assume that Carl's strategy will be the most undesirable one from her point of view. So if Ruth chooses row 1, the worst possible payoff is −1, which appears both in the third and fifth columns. Ruth must pay $1 to Carl. The payoff of −1 is called Ruth's **security level** for her row-1 strategy because she knows that nothing worse can happen if she makes that choice. We will keep a record of this security level by printing the −1s in row 1 in bold-face.

In general, for each row of a payoff matrix, we indicate the security level for Ruth, the row player, by printing the minimum number (which may appear more than once) in that row in boldface. The preceding matrix becomes

$$\begin{bmatrix} 1 & 3 & \mathbf{-1} & 0 & \mathbf{-1} \\ 5 & \mathbf{-2} & -1 & 2 & \mathbf{-2} \\ 4 & -1 & \mathbf{-2} & 0 & 0 \\ 3 & 2 & \mathbf{1} & 2 & 3 \end{bmatrix}$$

because the security levels for rows 2 and 3 are both −2, and the security level for row 4 is 1. It seems clear which strategy is best for Ruth in this case. If she chooses row 4, she will win at least one dollar, no matter what Carl chooses to do. Since game theory assumes that contestants make rational choices, the row player will choose the strategy with the maximum security level. This will be true even if all the security levels are negative. For negative numbers, the maximum means the least negative number (−2 is preferable to −5), so the row player will want to lose the least possible. The choice of the row with the maximum security level is called the **optimum pure strategy** for player R.

Optimum Pure Column Strategy

Finding an optimum pure strategy for Carl, whose pure strategies are the columns, involves the same sort of reasoning. Carl examines strategy (column) 1 and, since he has no information on what Ruth intends to do, he looks at the matrix to discover what the worst possible outcome would be if he selects column 1. The worst payoff for Carl in column 1 is the $5 loss that results if Ruth chooses row 2. Thus, $5 is the security level for player C associated with strategy 1. In general, the **security level** for the column player C is the *maximum* number in the column. Carl records the security level of strategy 1 by putting an asterisk at the payoff 5. He continues this analysis of the payoff matrix column by column, marking the security level by means of an asterisk to the right of the maximum number in the column, which may appear more than once. For the given matrix this results in

$$\begin{bmatrix} 1 & 3^* & -1 & 0 & -1 \\ 5^* & -2 & -1 & 2^* & -2 \\ 4 & -1 & -2 & 0 & 0 \\ 3 & 2 & 1^* & 2^* & 3^* \end{bmatrix}$$

The **optimum pure strategy** for the column player C is the strategy (column) with the *minimum* security level. The minimum is to player C's advantage because the larger the number, the more player R wins. In this example, then, column 3 is the optimum pure strategy for Carl because its security level, 1, is less than the other security levels, which are 2, 3, and 5. If Carl chooses column 3, no matter what strategy Ruth adopts, Carl's loss will be no more than $1; the selection of any other strategy by Carl might result in a greater loss.

Each player, Ruth and Carl, has complete information about the game, so they can identify not only their own optimum pure strategy, but their opponent's optimum pure strategy as well. Will this information matter to the players? Ruth can, for instance, see that Carl's optimum strategy is to select column 3. But certainly, that information reinforces Ruth's choice of row 4 since any other row choice will lead to a loss if Carl does choose column 3. In the same way, Carl can observe that row 4 is Ruth's optimum choice, and thus that the selection of any column other than 3 would produce a greater loss. So, Carl has no temptation to stray from the optimum pure strategy either. Thus, we see that in this example, the knowledge of the opponent's optimum pure strategy has the effect of strengthening each player's resolve to employ his or her own optimum strategy.

Saddle Point and Value

A game, like the one in our example, in which the optimum pure strategy for each player remains the most desirable choice when we assume that the opponent will play the opposing optimum pure strategy, is called a **strictly determined game**.

> A matrix game is strictly determined if there is an entry in the payoff matrix that is simultaneously the minimum of its row and the maximum of its column. Such an entry is called a **saddle point** of the game. In a strictly determined game, that is, in a game with a saddle point, the value of the saddle point is called the **value** of the game.

The row containing the saddle point is the optimum pure strategy for player R, while the column in which the saddle point occurs is the optimum pure strategy for player C. The identification of the value of the game and the corresponding optimum pure strategies is called a **solution** of a strictly determined game.

In a payoff matrix, if we indicate both the minimum of each row in boldface and the maximum of each column by an asterisk, then a saddle point appears as a boldface number with an asterisk, as happens in our example at the g_{43}-location. (For emphasis, we highlight this in color in the following matrix.)

$$\begin{bmatrix} 1 & 3* & -1 & 0 & -1 \\ 5* & -2 & -1 & 2* & -2 \\ 4 & -1 & -2 & 0 & 0 \\ 3 & 2 & 1* & 2* & 3* \end{bmatrix}$$

Example 1. Solve the matrix game with payoff matrix

$$\begin{bmatrix} 5 & -10 & -10 \\ -5 & 5 & -10 \\ -5 & 0 & -5 \\ -15 & -15 & -5 \\ 0 & 5 & -10 \\ -5 & 5 & -5 \end{bmatrix}$$

Solution

Indicating the minimum numbers of each row in boldface and the maximum numbers of each column by asterisks gives us

$$\begin{bmatrix} 5* & -10 & -10 \\ -5 & 5* & -10 \\ -5 & 0 & -5* \\ -15 & -15 & -5* \\ 0 & 5* & -10 \\ -5 & 5* & -5* \end{bmatrix}$$

Since the number -5 in the third and sixth rows of the third column is in boldface and has an asterisk, for an optimum pure strategy player R may choose either row 3 or row 6, while player C must choose column 3. The value of the game is -5. ∎

Several Saddle Points

As Example 1 shows, there may be more than one number in the payoff matrix that is both in boldface and has an asterisk. Thus, there is more than one solution of this game. If the game matrix has two saddle points in the same column, as in

6.2 STRICTLY DETERMINED GAMES

Example 1, they must have the same value since a saddle point is the maximum of its column. Similarly, two saddle points in the same row must have the same value, the minimum of the row.

Suppose that a payoff matrix has two saddle points that are in different rows and columns, as indicated by w and v in the following schematically drawn matrix:

$$\begin{bmatrix} \cdot & \cdot & \cdot & \cdot & \cdot & \cdot \\ \cdot & \cdot & a & \cdot & w^* & \cdot \\ \cdot & \cdot & \cdot & \cdot & \cdot & \cdot \\ \cdot & \cdot & v^* & \cdot & b & \cdot \\ \cdot & \cdot & \cdot & \cdot & \cdot & \cdot \end{bmatrix}$$

Because w is the minimum of its row, $w \leq a$, furthermore, v is the maximum of its column, so $a \leq v$, and therefore $w \leq v$. But on the other hand $w \geq b$, whereas $b \geq v$; so $w \geq v$ as well, which tells us that $w = v$ (and both equal a and b as well).

We have shown that no matter how many saddle points the matrix of a strictly determined game has, they all have the same value. Thus, when we define the value of a strictly determined game to be the value of a saddle point, we do not have to specify which saddle point we have in mind. Such a game has many solutions because it has many optimal pure strategies; but it has just a single value.

Example 2. Happy Toy Inc. must choose one of three possible marketing plans for the dolls it will sell next Christmas. Plan I is to sell last year's dolls at low prices. Plan II is to make moderate changes in design, but spend money on an advertising campaign, financed by slightly higher prices. Plan III is to redesign the dolls and charge considerably higher prices. If the national economy is strong next Christmas, the public will prefer newer looking dolls regardless of price, so in that case plan I would cause the price of the manufacturer's stock to drop 10 points ($10 a share), plan II would result in a rise of 7 points, and plan III would result in a rise of 20 points. If the economy is weak at Christmas time, price will be important; so the cheaper dolls of plan I would sell well, bringing the stock price up 15 points. Plan II would produce a 10-point rise in stock value and plan III would produce a 30-point loss. Finally, if the economy is steady, that is, neither particularly strong nor weak, plan I would lead to an unchanged stock price, while plans II and III would both bring the stock value up 7 points. Use game theory to determine which of the plans Happy Toy should adopt.

Solution

We will represent Happy Toy's plans by the 3 rows of a matrix. The opposing player is then the national economy, which has strategies *strong*, *weak*, and *steady*, represented by the columns of the matrix. The payoff values are in terms of changes in stock price, as described in the statement of the example. Thus, the table of payoffs is given in Table 6.8.

		Strong	Economy Weak	Steady
Happy Toy	Plan I	−10	15	0
	Plan II	7	10	7
	Plan III	20	−30	7

Table 6-8

If we indicate the row minima and column maxima in the usual way, the payoff matrix looks like this:

$$\begin{bmatrix} -10 & 15^* & 0 \\ 7 & 10 & 7^* \\ 20^* & -30 & 7^* \end{bmatrix}$$

The saddle point at the g_{23}-location indicates that the value of this game is 7. Therefore, the optimal pure strategy for the manufacturer is plan II: make moderate changes in design, advertise, and raise prices slightly. This would result in a 7-point rise in Happy Toy stock. ∎

Nonstrictly Determined Games

If we analyze the matrix game with payoff matrix

$$\begin{bmatrix} 3 & -2 \\ -1 & 4 \end{bmatrix}$$

by indicating the row minima and column maxima of the payoff matrix in the usual way, we obtain

$$\begin{bmatrix} 3^* & -2 \\ -1 & 4^* \end{bmatrix}$$

Here, no boldface entry has an asterisk, which means there is no saddle point. We again give the row and column players the names Ruth and Carl. Row 2 would seem a better choice than row 1 for Ruth because its security level, −1, is greater than the security level −2 of row 1. For Carl, the security level of column 1, which is 3, is less than the security level, 4, of column 2; so it would seem that Carl should choose column 1. If Ruth does choose row 2 and Carl column 1, the payoff is a 1-unit gain for Carl. But now, if Ruth believes that Carl *will* choose the column with the minimum security level, Ruth is strongly tempted to abandon the choice of row 2 and select row 1 because then, instead of a 1-unit loss, she would obtain a 3-unit gain. On the other hand, Carl will be well aware of Ruth's temptation and so, in defense, would consider choosing column 2 instead of column 1. If Carl chooses column 2 and Ruth does give in to the temptation of choosing row 1, then Carl is ahead 2 units. But then again, perhaps Ruth will not go with row 1 after all—in which case Carl will lose 4 units.

This discussion shows that the matrix game

$$\begin{bmatrix} 3 & -2 \\ -1 & 4 \end{bmatrix}$$

represents a highly unstable situation. Such a game is called a **nonstrictly determined** matrix game and is characterized by the absence of a saddle point. In contrast to a strictly determined game, a game of this sort can have no solution in the sense of mutually agreeable optimum pure strategies for both players. Thus, in Section 6.3 we will present a definition of "solution" that makes sense for nonstrictly determined games.

Summary

In a matrix game, let player R be the competitor whose strategies are represented by the rows of the payoff matrix and let player C be the opponent, whose strategies are represented by the columns. The **security level** of each row for player R is the minimum of that row. An **optimum pure strategy** for player R is a row with the maximum security level. For each column, the **security level** for player C is the maximum of that column; an **optimum pure strategy** for player C is a column with the minimum security level. A **strictly determined game** is a matrix game in which there appears an entry, called a **saddle point**, that is both the minimum of its row and the maximum of its column. The value of a saddle point is called the **value** of the strictly determined game. A row containing a saddle point is an optimum pure strategy for player R and a column containing a saddle point is an optimum pure strategy for player C. In a strictly determined game there is no advantage for a player choosing a strategy other than an optimum pure strategy—assuming that the opposing player chooses an optimum pure strategy. The identification of a saddle point that gives the value of the game, together with the corresponding optimum pure strategies is a **solution** of a strictly determined game. Many matrix games fail to have solutions in this sense because they are not strictly determined.

Exercises

Exercises 1 through 8 present the payoff matrices of various strictly determined games. Solve these games.

1. $\begin{bmatrix} -1 & 1 & 0 \\ 0 & 2 & -1 \\ 1 & 1 & 3 \end{bmatrix}$

2. $\begin{bmatrix} 7 & 5 & 4 & 6 \\ 1 & 2 & 0 & 2 \end{bmatrix}$

3. $\begin{bmatrix} -2 & 4 & -5 \\ -1 & -3 & -4 \\ -1 & 1 & -1 \\ -2 & -1 & -3 \end{bmatrix}$

4. $\begin{bmatrix} 0.12 & 0.76 & 0 & -0.31 & 0 \\ -0.01 & 0 & 0.05 & -0.40 & -0.01 \\ -0.22 & -0.15 & -0.12 & -0.30 & -0.01 \end{bmatrix}$

5. $\begin{bmatrix} -1 & 1 \\ 3 & 2 \\ -1 & -1 \\ 0 & 2 \end{bmatrix}$

6. $\begin{bmatrix} 0 & \frac{1}{2} & 0 & 0 \\ -\frac{1}{2} & 0 & -1 & -1 \\ -\frac{1}{4} & 1 & 0 & 1 \\ 0 & 1 & 0 & 0 \end{bmatrix}$

7. $\begin{bmatrix} -1 & -2 & -2 & -2 \\ -2 & -1 & -3 & -4 \\ -1 & -\frac{1}{2} & -2 & -2 \end{bmatrix}$

8. $\begin{bmatrix} -\frac{1}{4} & \frac{2}{3} & -\frac{1}{5} \\ -\frac{1}{4} & -\frac{2}{3} & \frac{2}{3} \\ -\frac{1}{5} & \frac{2}{5} & \frac{3}{4} \end{bmatrix}$

9. Imagine a game in which player R can choose one of the numbers $-1, 0$, or 1 and player C can also choose one of the same three numbers. The payoff to R is calculated by squaring the number chosen by player R and then subtracting the number chosen by player C. What are the optimum strategies for players R and C, and what is the value of the game?

10. Jeff will buy either a conservative or a speculative stock. He estimates that if the Dow-Jones industrial average (which uses a selection of stocks to measure the stock market as a whole) goes up over a 12-month period, then the value of the conservative stock will increase by 10%, while the value of the speculative stock will increase 30%. If the Dow-Jones average remains constant over a 12-month period, then the value of the conservative stock will remain constant, but the value of the speculative stock will decline by 5%. If the Dow-Jones average decreases over a 12-month period, then the value of the conservative stock will go down by 10% and the value of the speculative stock will decrease by 20%. Solve the game to determine whether Jeff should buy the conservative stock or the speculative stock.

11. In a movie script bank robbers ride out of town at nightfall. They will go either to their hideout in the hills or across the border into Mexico. The sheriff and his men ride after them but, since it soon becomes dark, they cannot see the robbers' trail. If the sheriff rides to the hideout and the robbers have gone there, he will capture them and thus increase by .10 the probability that he will be reelected. However, if he rides to the hideout and the robbers have gone to Mexico, the probability of his reelection drops by .10. If the sheriff heads for the border while

the robbers go to their hideout, there is a chance he will realize his mistake in time to come back and capture some of them, thus raising his reelection probability by .05. Finally, if both the robbers and the sheriff head for the border, the outcome will not influence the sheriff's reelection chances at all. The script writer wants the sheriff to have the best chance of reelection. Solve this game to determine whether in the script the sheriff should head for the hideout or for the Mexican border.

12. A farmer must decide whether to plant potatoes, sugar beets, or barley. If there is little rain during the growing season, the potato crop will earn the farmer $25 an acre, the sugar-beet crop will fail, causing the farmer to lose $50 an acre, and the barley crop will bring in $25 an acre. In a year of moderate rain, the potatoes will earn $200 an acre, the sugar beets will bring in $100 an acre, and the barley will be worth $50 an acre. In a year of heavy rain, the potato crop will earn $20 an acre, the sugar beets will earn $150 an acre, and the barley crop will earn $25 an acre. Solve the game to determine which crop the farmer should plant.

13. The Jackson family is trying to decide whether to take its vacation at the seashore, in the mountains, or at home. The Jacksons estimate the pleasure they will receive from the vacation on a scale from –3 (displeasure) to +3 (pleasure), depending on whether the weather is warm, cool, cold, or rainy during the vacation. If the Jacksons go to the seashore, their estimate on the pleasure scale is +2 if the weather is warm, 0 if it is cool, –2 if it is cold, and –3 if it is rainy. If they go to the mountains, warm weather will produce pleasure worth +2, cool weather, +3, cold weather, +1, and rainy weather, –1. If the Jacksons stay home, they estimate +2 points if the weather is warm and +1 pleasure point for each of the other three types of weather. Solve the game to determine which vacation the Jacksons should take.

14. An automobile company plans to open a new manufacturing plant and must decide whether to build its standard-sized car, its compact car, or its subcompact at this plant. The United States government is considering an annual tax on automobiles by size. The tax will be very high in the case of standard-sized cars and quite substantial even for compact cars, but does not apply to subcompacts. If the government does not institute the tax, the plant would normally run at 70% capacity if it manufactures standard-sized cars and 80% capacity if either compact or subcompacts are built there. However, if the tax law passes, the plant will run at only 30% of capacity if it manufactures the standard model, at 50% of capacity if it builds the compact, but at 90% of capacity if it makes subcompacts. Solve the game to determine which model the manufacturer should build at this plant.

15. A defense attorney must advise a client accused of attempted burglary how to plead. Because of careless handling of the evidence by the arresting officer, there is a chance that the judge will dismiss the case on technical grounds. For this reason, the district attorney has agreed to accept a guilty plea to the lesser charge of breaking and entering. If the client pleads innocent and the judge is harsh, the trial for attempted burglary will take place; in this case the defense attorney believes her client will be found guilty and will be sentenced to 5 years in prison.

However, if the judge is lenient, the case will be dismissed and the client will go free. On the other hand, if the client pleads guilty to the breaking-and-entering charge, a harsh judge will give a 1-year prison sentence and a lenient judge, a 3-month ($\frac{1}{4}$-year) sentence. Solve the game to determine how the defense attorney should advise her client to plead.

6.3 Mixed Strategies

At the end of Section 6.2, we saw that the game with payoff matrix

$$\begin{bmatrix} 3 & -2 \\ -1 & 4 \end{bmatrix}$$

is not strictly determined, so we cannot solve the game by the method of that section. We still want to find optimum strategies so that if Ruth believes Carl will play his optimum strategy, then any strategy other than her own optimum one will be less desirable, and similarly for Carl. Before finding the optimum strategies for this game, we need to extend the concept of strategy beyond the *pure strategies* we have used so far.

The pure strategies for Ruth in a matrix game with two rows are: choose row 1 or choose row 2. Now we will permit Ruth to use a strategy such as: flip a coin; choose row 1 if the coin comes up heads and row 2 if it comes up tails. Ruth still has complete charge of designing the strategy, but her design includes an element of chance: the outcome of the coin. Assuming that the coin is evenly balanced, there is a probability of $\frac{1}{2}$ that row 1 will be chosen and $\frac{1}{2}$ that row 2 will be chosen. Thus, Ruth's strategy really amounts to assigning probabilities of $\frac{1}{2}$ to the selection of each row.

Carl may also use probability to decide which column to choose. For instance, he could design the strategy: roll a die and choose column 1 if the die shows one or two dots, and column 2 otherwise. In probability terms, Carl's strategy assigns a probability of $\frac{2}{6} = \frac{1}{3}$ to column 1 and a probability of $\frac{2}{3}$ to column 2.

Pure and Mixed Strategies

The strategy with probability p_1 for row 1 and probability p_2 for row 2 is a **mixed strategy** in a game with two rows. The pure strategies of the preceding section can be viewed as special types of mixed strategies. For instance, the pure strategy "choose row 1" is the same as the mixed strategy "probability 1 for row 1 and probability 0 for row 2." In general, for a $k \times m$-matrix game, a **mixed strategy** for player R assigns probabilities p_1 for selecting row 1, p_2 for row 2, and so on, to

p_k for the last row. The only restriction is that, since the player must choose exactly one row, the numbers p_1 through p_k must add up to one.

Ruth can carry out the mixed strategy $p_1 = \frac{1}{2}$ and $p_2 = \frac{1}{2}$ by flipping an honest coin, but how could she implement $p_1 = .17$ and $p_2 = .83$. One way to do this would be to use a spinner which rotates about a circle as shown in Figure 6-9. The circular region is divided into two smaller regions—one occupying 17% of the area and the other, the remaining 83%. Out of Carl's sight, Ruth spins the spinner and chooses row 1 if the spinner stops in the smaller region and row 2 otherwise. A player with more than two pure strategies to choose from would divide the circular region into several smaller regions.

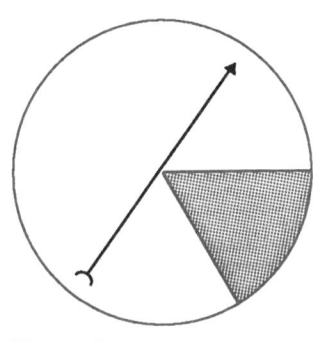

Figure 6-9

Payoff

In the game we have been using, with payoff matrix

$$\begin{bmatrix} 3 & -2 \\ -1 & 4 \end{bmatrix}$$

if Ruth adopts the pure strategy "choose row 2" and Carl chooses the pure strategy "choose column 1," we know that the payoff will be a $1 gain for Carl. But suppose that Ruth uses the coin-flipping strategy with probability $\frac{1}{2}$ for row 1 and probability $\frac{1}{2}$ for row 2, while Carl uses the dice-rolling strategy with probability $\frac{1}{3}$ for column 1 and probability $\frac{2}{3}$ for column 2. What do we now mean by the payoff?

The definition of payoff for mixed strategies depends on the theory of probability. There are only four possible outcomes for a play of a 2 × 2-game:

(row 1, column 1) (row 1, column 2)
(row 2, column 1) (row 2, column 2)

We can calculate the probability that each of the four possibilities will occur as follows: Ruth flips a coin to select a row and Carl rolls a die in order to choose a column. Neither outcome depends on the other; in fact, neither player knows what the other has done until the play is recorded. In the language of Chapter 3, the selection of the row and the selection of the column are independent events. We wish to calculate the probability that both events will occur. Thus, when we write $p(\text{row } i, \text{column } j)$ we mean $p(\text{row } i \text{ and column } j)$. In this case, as we recall from Section 3.9, the probability that both will take place is the product of the two probabilities. That is,

$$p(\text{row } i, \text{column } j) = p(\text{row } i \text{ and column } j) = p(\text{row } i) \cdot p(\text{column } j)$$

For example, since the probability is $\frac{1}{2}$ that Ruth will choose row 1 and $\frac{1}{3}$ that Carl will choose column 1, then

$$p(\text{row } 1, \text{column } 1) = p(\text{row } 1) \cdot p(\text{column } 1) = \frac{1}{2} \cdot \frac{1}{3} = \frac{1}{6}$$

If Ruth chooses row 1 and Carl column 1, the payoff matrix indicates that the payoff will be 3; so we may also write

$$p(\text{payoff is } 3) = \frac{1}{6}$$

or, more briefly,

$$p(3) = \frac{1}{6}$$

Similarly,

$$p(\text{row } 1, \text{column } 2) = \frac{1}{2} \cdot \frac{2}{3} = \frac{1}{3}$$

and since the payoff is -3, we have

$$p(-3) = \frac{1}{3}$$

Also,

$$p(\text{row } 2, \text{column } 1) = \frac{1}{6} \quad \text{and} \quad p(\text{row } 2, \text{column } 2) = \frac{1}{3}$$

or, in terms of payoffs,

$$p(-1) = \frac{1}{6} \quad \text{and} \quad p(4) = \frac{1}{3}$$

We will use the expected-value concept to define the payoff, so we briefly review what this means. Suppose we have a list of n types of outcomes, values (monetary or otherwise) of $k_1, k_2, \ldots, k_n$ corresponding to each type, and the probabilities $p(k_1), p(k_2), \ldots, p(k_n)$ that the outcome will be of each possible type. In Section 4.4, the **expected value** E was calculated by multiplying the value of each type of outcome times the probability that the type of outcome will take place and adding up the results, that is,

$$E = k_1 p(k_1) + k_2 p(k_2) + \ldots + k_n p(k_n)$$

The values in a matrix game are the numbers in the payoff matrix. For the game with payoff matrix

$$\begin{bmatrix} 3 & -2 \\ -1 & 4 \end{bmatrix}$$

if Ruth uses the coin strategy and Carl the dice strategy, according to the probabilities we have calculated, the expected value is

$$E = (3)p(3) + (-2)p(-2) + (-1)p(-1) + (4)p(4)$$
$$= (3)\left(\frac{1}{6}\right) + (-2)\left(\frac{1}{3}\right) + (-1)\left(\frac{1}{6}\right) + (4)\left(\frac{1}{3}\right)$$
$$= 1$$

The expected value will be the payoff for these strategies.

Suppose both players choose pure strategies—for instance, Ruth chooses row 2 and Carl chooses column 1. Viewing Ruth's choice as a mixed strategy, the probabilities are $p(\text{row } 1) = 0$ and $p(\text{row } 2) = 1$; similarly, for Carl's choice, $p(\text{column } 1) = 1$ and $p(\text{column } 2) = 0$. Then, using independence to calculate the probabilities, the expected value is given by

$$E = (3)p(3) + (-2)p(-2) + (-1)p(-1) + (4)p(4)$$
$$= (3)(0) + (-2)(0) + (-1)(1) + (4)(0)$$
$$= -1$$

This is the number in the second row and first column of the payoff matrix. The example illustrates the fact that if we call the expected value the payoff, then that terminology will agree with the definition for pure strategies that we used in the previous sections.

Payoffs for General Matrix Games

A matrix game is specified by a $k \times m$-matrix G, the payoff matrix, which gives, for each row i and column j, the payoff g_{ij} that occurs if player R chooses row i and player C chooses column j. Player R has a choice of k different pure strategies (rows of the matrix).

> **A mixed strategy** for player R consists of the numbers $p_1, p_2, \ldots, p_k$, where p_i is the probability that player R will select the ith row. Thus, a mixed strategy is an ordered k-tuple of numbers $P = (p_1, p_2, \ldots, p_k)$, all of which are greater than or equal to zero. Since exactly one row must be selected, the probabilities must add up to one; that is,
> $$p_1 + p_2 + \ldots + p_k = 1$$

Similarly, player C has a choice of m distinct columns.

> An ordered m-tuple $Q = (q_1, q_2, \ldots, q_m)$ of numbers greater than or equal to zero with the property that
>
> $$q_1 + q_2 + \ldots + q_m = 1$$
>
> is called a **mixed strategy** for player C.

If player R selects mixed strategy P and player C chooses mixed strategy Q, the payoff for these strategies, which we denote by $E(P, Q)$, is calculated as an expected value. The probability that player R will choose row i is p_i, and the probability that player C will choose column j is q_j. Since we are assuming that the events "row choice" and "column choice" are independent by the rules of the game,

$$p(\text{row } i, \text{column } j) = p_i q_j$$

according to Section 3.9. The value of the outcome "row i, column j" is the number g_{ij} from the game matrix G. Expected value is defined by multiplying the value of each outcome by its probability and then adding up all the products. Consequently, the **payoff** $E(P, Q)$ for the strategies P and Q is

$$\begin{aligned} E(P, Q) &= g_{11} p(g_{11}) + g_{12} p(g_{12}) + \ldots + g_{1m} p(g_{1m}) + g_{21} p(g_{21}) + \ldots + g_{km} p(g_{km}) \\ &= g_{11}(p_1 q_1) + g_{12}(p_1 q_2) + \ldots + g_{1m}(p_1 q_m) + g_{21}(p_2 q_1) + \ldots + g_{km}(p_k q_m) \end{aligned}$$

It is convenient to state this formula for the payoff in matrix terms. We write the strategy P, an ordered k-tuple, as a row matrix

$$P = \begin{bmatrix} p_1 & p_2 & \cdots & p_k \end{bmatrix}$$

and the ordered m-tuple Q as the column matrix

$$Q = \begin{bmatrix} q_1 \\ q_2 \\ \vdots \\ q_m \end{bmatrix}$$

Then, because we can write

$$E(P, Q) = g_{11}(p_1 q_1) + g_{12}(p_1 q_2) + \ldots + g_{1m}(p_1 q_m) + g_{21}(p_2 q_1) + \ldots + g_{km}(p_k q_m)$$

in the form

$$E(P, Q) = p_1 g_{11} q_1 + p_1 g_{12} q_2 + \ldots + p_1 g_{1m} q_m + p_2 g_{21} q_1 + \ldots + p_k g_{km} q_m$$

we see that the payoff can be computed as a product of matrices:

$$E(P, Q) = PGQ$$

Example 1. In the game with payoff matrix

$$G = \begin{bmatrix} 0.15 & -0.31 & 0 & -0.55 \\ 0.76 & 0.02 & -0.62 & -0.10 \end{bmatrix}$$

player R's strategy is: $p(\text{row } 1) = .25$ and $p(\text{row } 2) = .75$. For player C, all columns are equally likely to be chosen. What is the payoff for these strategies?

Solution

Since each of the four columns is equally likely to be chosen, $p(\text{column } j) = .25$ for each j. Thus, in matrix form,

$$P = \begin{bmatrix} .25 & .75 \end{bmatrix} \quad \text{and} \quad Q = \begin{bmatrix} .25 \\ .25 \\ .25 \\ .25 \end{bmatrix}$$

and the payoff is given by

$$E(P, Q) = PGQ = \begin{bmatrix} .25 & .75 \end{bmatrix} \begin{bmatrix} 0.15 & -0.31 & 0 & -0.55 \\ 0.76 & 0.02 & -0.62 & -0.10 \end{bmatrix} \begin{bmatrix} .25 \\ .25 \\ .25 \\ .25 \end{bmatrix}$$

$$= \begin{bmatrix} .6075 & -.0625 & -.4650 & -.2125 \end{bmatrix} \begin{bmatrix} .25 \\ .25 \\ .25 \\ .25 \end{bmatrix}$$

$$= -.033125 \qquad \blacksquare$$

Security Level

Now we suppose that the row player, Ruth, has chosen a particular mixed strategy, which we will represent by $\hat{P}$, rather than by P, to remind ourselves that Ruth's strategy will stay fixed throughout this discussion. There are many mixed strategies available to the column player, Carl, and the payoff $E(\hat{P}, Q)$ will depend on which strategy Q Carl selects. Since the game matrix G is written from Ruth's point of view, with positive numbers indicating a profit for Ruth and negative numbers a loss, Ruth would prefer Carl to choose a strategy Q that makes $E(\hat{P}, Q)$

452 GAME THEORY

the maximum possible value. But Ruth has no information on what Carl intends to do; as in the preceding section, Ruth had better assume the worst. Among all the mixed strategies Q available to Carl, there is a strategy, call it $Q^\#$, which from Ruth's viewpoint is the worst response Carl could make to $\hat{P}$. That is, by the definition of $Q^\#$,

$$E(\hat{P}, Q^\#) \leq E(\hat{P}, Q) \text{ for any } Q$$

The payoff $E(\hat{P}, Q^\#)$, written $\underline{v}(\hat{P})$, is called the **security level** for Ruth's mixed strategy $\hat{P}$.

Example 2. Suppose that in the game with payoff matrix

$$G = \begin{bmatrix} -1 & 0 & 1 \\ 0 & -1 & 0 \\ -1 & 0 & 1 \end{bmatrix}$$

player R chooses the mixed strategy

$$\hat{P}: p(\text{row } 1) = p(\text{row } 2) = p(\text{row } 3) = \frac{1}{3}$$

Which of the following strategies Q for player C is least desirable from player R's point of view?

(a) $p(\text{column } 1) = p(\text{column } 2) = \frac{1}{2}$, $p(\text{column } 3) = 0$

(b) $p(\text{column } 1) = p(\text{column } 2) = p(\text{column } 3) = \frac{1}{3}$

(c) $p(\text{column } 1) = \frac{1}{2}$, $p(\text{column } 2) = \frac{2}{5}$, $p(\text{column } 3) = \frac{1}{10}$

(d) a pure strategy

Solution

Since $E(\hat{P}, Q) = \hat{P}GQ$, we calculate

$$\hat{P}G = \begin{bmatrix} \frac{1}{3} & \frac{1}{3} & \frac{1}{3} \end{bmatrix} \begin{bmatrix} -1 & 0 & 1 \\ 0 & -1 & 0 \\ -1 & 0 & 1 \end{bmatrix} = \begin{bmatrix} -\frac{2}{3} & -\frac{1}{3} & \frac{2}{3} \end{bmatrix}$$

and use $\hat{P}G$ over and over as we vary Q. Thus, for Part (a),

$$E(\hat{P}, Q) = \hat{P}GQ = \begin{bmatrix} -\frac{2}{3} & -\frac{1}{3} & \frac{2}{3} \end{bmatrix} \begin{bmatrix} \frac{1}{2} \\ \frac{1}{2} \\ 0 \end{bmatrix} = -\frac{1}{2}$$

For Part (b),

$$E(\hat{P}, Q) = \hat{P}GQ = \begin{bmatrix} -\frac{2}{3} & -\frac{1}{3} & \frac{2}{3} \end{bmatrix} \begin{bmatrix} \frac{1}{3} \\ \frac{1}{3} \\ \frac{1}{3} \end{bmatrix} = -\frac{1}{9}$$

For Part (c),

$$E(\hat{P}, Q) = \hat{P}GQ = \begin{bmatrix} -\frac{2}{3} & -\frac{1}{3} & \frac{2}{3} \end{bmatrix} \begin{bmatrix} \frac{1}{2} \\ \frac{2}{5} \\ \frac{1}{10} \end{bmatrix} = -\frac{6}{15}$$

The pure strategy for player C, "choose column 1" is written in matrix form as

$$Q = \begin{bmatrix} 1 \\ 0 \\ 0 \end{bmatrix}$$

and similarly, we obtain the matrices

$$Q = \begin{bmatrix} 0 \\ 1 \\ 0 \end{bmatrix} \quad \text{and} \quad Q = \begin{bmatrix} 0 \\ 0 \\ 1 \end{bmatrix}$$

for the other pure strategies. Thus, for Part (d),

$$E(\hat{P}, Q) = \begin{bmatrix} -\frac{2}{3} & -\frac{1}{3} & \frac{2}{3} \end{bmatrix} \begin{bmatrix} 1 \\ 0 \\ 0 \end{bmatrix} = -\frac{2}{3}$$

$$E(\hat{P}, Q) = \begin{bmatrix} -\frac{2}{3} & -\frac{1}{3} & \frac{2}{3} \end{bmatrix} \begin{bmatrix} 0 \\ 1 \\ 0 \end{bmatrix} = -\frac{1}{3}$$

$$E(\hat{P}, Q) = \begin{bmatrix} -\frac{2}{3} & -\frac{1}{3} & \frac{2}{3} \end{bmatrix} \begin{bmatrix} 0 \\ 0 \\ 1 \end{bmatrix} = \frac{2}{3}$$

Therefore, from player R's point of view, the pure strategy for player C, "choose column 1," is the worst that player C can choose against $\hat{P}$ among those given because in that case $E(\hat{P}, Q) = -\frac{2}{3}$, the minimum payoff among the ones we computed. ∎

Computing the Security Level

In Example 2, among the strategies for player C evaluated, the one that was the most undesirable from player R's point of view was a pure strategy: choose column 1. In fact, that pure strategy would have produced the minimum payoff when compared to any strategy for player C. It can be shown that for any choice of strategy $\hat{P}$ for player R, the least desirable response from player C, which we called $Q^{\#}$, is always a *pure* strategy. Since we demonstrated in Part (d) of Example 2 that column 1 was the most undesirable response among the pure strategies for player C, it is in fact the most undesirable compared to all the mixed strategies as well.

You may have noticed that in Part (d), when we multiplied $\hat{P}G$ by each of the matrices Q for the pure column strategies, we just obtained one of the entries in $\hat{P}G$. The pure strategies Q are the columns of an identity matrix. In Chapter 1, we saw that multiplying a matrix by an identity matrix (of the appropriate size) just copies the original matrix; in fact, each column of the identity matrix copies the corresponding column of the original matrix. Each column of $\hat{P}G$ is a single number. Since the security level for the row strategy $\hat{P}$ is the minimum of the products $\hat{P}GQ$ when Q is the matrix of a pure column strategy, the security level $\hat{P}$ is the minimum entry of $\hat{P}G$.

Part (d) of Example 2 shows us that the security level for the strategy

$$\hat{P} = \begin{bmatrix} \frac{1}{3} & \frac{1}{3} & \frac{1}{3} \end{bmatrix}$$

is the minimum entry in

$$\hat{P}G = \begin{bmatrix} -\frac{2}{3} & -\frac{1}{3} & \frac{2}{3} \end{bmatrix}$$

that is, $\underline{v}(\hat{P}G) = -\frac{2}{3}$.

If the column player C chooses a particular strategy $\hat{Q}$, then there is a worst possible response from player R; this is a strategy $P^{\#}$ for which

$$E(P, \hat{Q}) \leq E(P^{\#}, \hat{Q})$$

compared to any other mixed strategy P for player R. Define

$$E(P^{\#}, \hat{Q}) = \overline{v}(\hat{Q})$$

to be the **security level** for player C associated with the mixed strategy $\hat{Q}$.

6.3 MIXED STRATEGIES

Again, we do not have to calculate $E(P, \hat{Q})$ for all mixed strategies P to determine the security level $\bar{v}(\hat{Q})$ of $\hat{Q}$ because it can be shown that the worst response $P^{\#}$ will be a *pure* strategy for player R. Therefore, we need only determine the maximum of the $E(P, \hat{Q})$ among the pure strategies P.

Example 3. Calculate the security level of the strategy for player C, where

$$\hat{Q}: p(\text{column } 1) = \frac{2}{5}, \quad p(\text{column } 2) = \frac{1}{2}, \quad \text{and} \quad p(\text{column } 3) = \frac{1}{10}$$

in the game with payoff matrix

$$G = \begin{bmatrix} -1 & 2 & 1 \\ 0 & 0 & -1 \end{bmatrix}$$

Solution

We write $\hat{Q}$ as the matrix

$$\hat{Q} = \begin{bmatrix} \frac{2}{5} \\ \frac{1}{2} \\ \frac{1}{10} \end{bmatrix}$$

Then we need to calculate $E(P, \hat{Q}) = PG\hat{Q}$ for the pure strategies P for player R. We first find $G\hat{Q}$ since we can use it more than once (just as we found $\hat{P}G$ in Example 2).

$$G\hat{Q} = \begin{bmatrix} -1 & 2 & 1 \\ 0 & 0 & -1 \end{bmatrix} \begin{bmatrix} \frac{2}{5} \\ \frac{1}{2} \\ \frac{1}{10} \end{bmatrix} = \begin{bmatrix} \frac{7}{10} \\ -\frac{1}{10} \end{bmatrix}$$

Now for the two pure strategies for player R, we have

$$E(P, \hat{Q}) = \begin{bmatrix} 1 & 0 \end{bmatrix} \begin{bmatrix} \frac{7}{10} \\ -\frac{1}{10} \end{bmatrix} = \frac{7}{10}$$

and

$$E(P, \hat{Q}) = \begin{bmatrix} 0 & 1 \end{bmatrix} \begin{bmatrix} \frac{7}{10} \\ -\frac{1}{10} \end{bmatrix} = -\frac{1}{10}$$

456 GAME THEORY

Because $\frac{7}{10}$ is certainly greater than $-\frac{1}{10}$, we conclude that the security level of the strategy $\hat{Q}$ is $\overline{v}(\hat{Q}) = \frac{7}{10}$. ∎

In Example 2, multiplication of $\hat{P}G$ by the matrix Q of each pure column strategy selected one entry from $\hat{P}G$. Similarly, Example 3 illustrates that fact that multiplying $G\hat{Q}$ by the matrix P of each pure row strategy, which is a row of an identity matrix, selects an entry of $G(\hat{Q})$. Thus, in order to determine the security level $\overline{v}(\hat{Q})$ of $\hat{Q}$, we just find the maximum entry in $G\hat{Q}$.

Example 4. Given the game with matrix

$$G = \begin{bmatrix} 1 & 0 & -1 \\ 1 & 2 & -1 \\ -1 & -2 & 0 \end{bmatrix}$$

calculate the security levels of the strategies

(a) $\hat{P} = \begin{bmatrix} \frac{1}{6} & \frac{1}{6} & \frac{2}{3} \end{bmatrix}$

(b) $\hat{Q} = \begin{bmatrix} \frac{3}{4} \\ \frac{1}{4} \\ 0 \end{bmatrix}$

Solution

(a) The security level for the strategy $\hat{P}$ is the minimum entry in

$$\hat{P}G = \begin{bmatrix} \frac{1}{6} & \frac{1}{6} & \frac{2}{3} \end{bmatrix} \begin{bmatrix} 1 & 0 & -1 \\ 1 & 2 & -1 \\ -1 & -2 & 0 \end{bmatrix} = \begin{bmatrix} -\frac{1}{3} & -1 & -\frac{1}{3} \end{bmatrix}$$

that is, -1.

(b) Now we calculate

$$G\hat{Q} = \begin{bmatrix} 1 & 0 & -1 \\ 1 & 2 & -1 \\ -1 & -2 & 0 \end{bmatrix} \begin{bmatrix} \frac{3}{4} \\ \frac{1}{4} \\ 0 \end{bmatrix} = \begin{bmatrix} \frac{3}{4} \\ \frac{5}{4} \\ -\frac{5}{4} \end{bmatrix}$$

The security level of the strategy $\hat{Q}$ for player C is the maximum entry in $G\hat{Q}$, that is, $\frac{5}{4}$. ∎

Optimum Strategies

If Ruth chooses strategy $\hat{P}$, then the worst that can happen is that Carl will choose strategy $Q^{\#}$, in which case the payoff will be the security level $E(\hat{P}, Q^{\#}) = \underline{v}(\hat{P})$. Ruth might, however, choose a different mixed strategy $\hat{\hat{P}}$. In that case, there is a worst possible outcome, some (pure) strategy $Q^{\#\#}$ that Carl could choose. Now Ruth can compare $E(\hat{\hat{P}}, Q^{\#\#}) = \underline{v}(\hat{\hat{P}})$, the security level of mixed strategy $\hat{\hat{P}}$, with the security level $\underline{v}(\hat{P})$ of $\hat{P}$. Suppose that $\underline{v}(\hat{\hat{P}})$ is greater than $\underline{v}(\hat{P})$. This means the worst that can happen to Ruth if strategy $\hat{\hat{P}}$ is chosen is not as bad as the worst that can occur if she chooses $\hat{P}$. So Ruth will prefer strategy $\hat{\hat{P}}$ to $\hat{P}$.

As Ruth considers each possible mixed strategy $\hat{P}$, there is a security level $\underline{v}(\hat{P})$ associated with it. The value of $\underline{v}(\hat{P})$ changes from strategy to strategy and there is a maximum security level, which we call $\underline{v}$. The maximum of the $\underline{v}(\hat{P})$ is the security level for some particular strategy, which we write as P^*, that is, $\underline{v} = \underline{v}(P^*)$. Thus, the worst that can happen to Ruth when she uses strategy P^* is better than the worst that can happen to her if she were to use any other strategy. The mixed strategy P^* is called an **optimum strategy** for player R.

We have also seen that for each strategy $\hat{Q}$ for player C, there is a corresponding security level $\overline{v}(\hat{Q})$. Among all those security levels there is one, $\overline{v} = \overline{v}(Q^*)$, which is the least harmful from player C's point of view, that is, the minimum of all the security levels. The corresponding mixed strategy Q^* is called the **optimum strategy** for player C.

A key fact in the theory of matrix games is that

$\underline{v}$ and $\overline{v}$ are, in fact, the same number, called the **value** v of the game. It is the payoff when both players use their optimum strategies, that is,

$$E(P^*, Q^*) = \underline{v} = \overline{v} = v$$

Neither player is tempted to depart from his or her optimum strategy because if the opposing player employs the optimum strategy, then a less desirable payoff than $v = \underline{v} = \bar{v}$ will result. Thus, every matrix game has a **solution**: stable optimum mixed strategies P^* and Q^* for the two players and a corresponding value v for the game. In Section 6.4, we will show that $\underline{v} = \bar{v}$ in any matrix game with a two-by-two payoff matrix and we will describe how to find the solution for any such matrix game. Then, in Section 6.5, we will find that any matrix game can be solved by the simplex algorithm of linear programming. Also, from that discussion we will see that the equation $\underline{v} = \bar{v}$ is a consequence of the Duality Theorem of Linear Programming, as described in Section 2.3.

Strictly Determined Games

How does the concept of optimum mixed strategies relate to our earlier discussion of the solution of a strictly determined game? We discuss only pure strategies for the moment. Then the payoff $E(P, Q)$ is an entry in the payoff matrix G. The reason is that P is represented by a row of an identity matrix and Q, by a column of a, perhaps different, identity matrix. Therefore, GQ will then be a copy of the jth column of the matrix G. And $E(P, Q) = PGQ = P(GQ)$ is the ith row of the matrix GQ, that is, the entry g_{ij}, in row i and column j, of G.

If player R chooses a pure strategy $\hat{P}$ and compares $E(\hat{P}, Q)$ for all pure strategies available to player C to determine the security level of $\hat{P}$, then player R is just comparing all the numbers in the matrix $\hat{P}G$, which is a row of the matrix G. Therefore, $\underline{v}(\hat{P})$ is the minimum number of that row. Then $\underline{v}$ is found by comparing the minima of the various rows and choosing the maximum of these numbers. The optimum pure strategy P^* for player R is the selection of the row in which $\underline{v}$ lies. Similarly, for player C the maximum number in each column is the security level for the corresponding pure strategy, $\bar{v}$ is the minimum of these security levels, and the optimum pure strategy Q^* for player B is the column in which $\bar{v}$ lies.

A strictly determined game is one with a saddle point, an entry that is both the minimum of its row and the maximum of its column. This number will satisfy the conditions for both $\bar{v}$ and for $\underline{v}$. Thus, the solution of a strictly determined game is the pair of optimum pure strategies (P^*, Q^*) given by the row and column of the saddle point. This is the same solution as in Section 6.2.

In finding optimum strategies for a strictly determined game, we considered only pure strategies even though, in general, optimum strategies are chosen from among all possible mixed strategies. However, it can be demonstrated that in a strictly determined game, there is a pure strategy for each player that is preferable to any mixed strategy; so in seeking the optimum in this case, it is sufficient to examine only pure strategies.

Summary

In a matrix game in which player R has a choice of k rows and player C has a choice of m columns, let G be the $k \times m$-payoff matrix of the game. A **mixed strategy** for player R is a $1 \times k$-matrix P with entries that are greater than or equal to zero and that add up to one. The number in the ith column of the matrix represents the probability that player R will choose row i of the payoff matrix. A mixed strategy for player C is an $m \times 1$-matrix Q with entries that are all greater than or equal to zero and that add up to one. The entry in the jth row represents the probability that player C will choose column j. The **payoff** if player R uses mixed strategy P and player C uses mixed strategy Q, written $E(P, Q)$, is defined to be

$$E(P, Q) = PGQ$$

Let $\hat{P}$ be a mixed strategy for player R, and suppose that $Q^\#$ is a mixed strategy for player C with the property that for all Q,

$$E(\hat{P}, Q^\#) \leq E(\hat{P}, Q)$$

Then $E(\hat{P}, Q^\#) = \underline{v}(\hat{P})$ is the **security level** for strategy $\hat{P}$; it is the minimum entry of the matrix $\hat{P}G$. A mixed strategy P^* for player R for which the security level $\underline{v} = \underline{v}(P^*)$ is the maximum is called an **optimum strategy** for player R. If $\hat{Q}$ is a mixed strategy for player C and $P^\#$ is a mixed strategy for player R for which

$$E(P, \hat{Q}) \leq E(P^\#, \hat{Q}) = \overline{v}(\hat{Q})$$

no matter which mixed strategy P player R adopts, then $\overline{v}(\hat{Q})$ is the **security level** of mixed strategy $\hat{Q}$; it is the maximum entry of the matrix $G\hat{Q}$. A mixed strategy Q^* for player C is an **optimum strategy** if the corresponding security level $\overline{v} = \overline{v}(Q^*)$ is the minimum. It can be shown that

$$E(P^*, Q^*) = \overline{v} = \underline{v} = v$$

and the number v is called the **value** of the game. A **solution** of a matrix game is a pair (P^*, Q^*) of optimum strategies for players R and C, respectively, together with the value v of the game. If the game is strictly determined, this solution is the pure strategies given by the row and column that contain a saddle point and the value of the saddle point.

Exercises

In Exercises 1 through 5, calculate the payoff for the mixed strategies P and Q in the game with the payoff matrix G.

1.
$$G = \begin{bmatrix} 1 & -2 \\ -1 & 2 \end{bmatrix}, \quad P = \begin{bmatrix} \frac{1}{2} & \frac{1}{2} \end{bmatrix}, \quad Q = \begin{bmatrix} \frac{1}{3} \\ \frac{2}{3} \end{bmatrix}$$

2.
$$G = \begin{bmatrix} 1 & 0 & 0 & -1 \\ 1 & 1 & -1 & 0 \\ 1 & -1 & 2 & 0 \end{bmatrix}, \quad P = \begin{bmatrix} \frac{1}{2} & \frac{1}{4} & \frac{1}{4} \end{bmatrix}, \quad Q = \begin{bmatrix} 0 \\ \frac{1}{8} \\ \frac{1}{8} \\ \frac{3}{4} \end{bmatrix}$$

3.
$$G = \begin{bmatrix} -1 & 2 & 0 \\ 1 & 3 & -1 \end{bmatrix}, \quad P = [.2 \quad .8], \quad Q = \begin{bmatrix} .7 \\ .1 \\ .2 \end{bmatrix}$$

4.
$$G = \begin{bmatrix} 0.1 & 0.2 & 0 \\ 0.4 & 0.3 & 0.1 \\ 0 & 0 & -0.4 \end{bmatrix}, \quad P = [.7 \quad .1 \quad .2], \quad Q = \begin{bmatrix} .5 \\ .4 \\ .1 \end{bmatrix}$$

5.
$$G = \begin{bmatrix} 1 & 0 & -1 \\ \frac{1}{2} & \frac{1}{2} & -1 \\ -\frac{1}{2} & 2 & -\frac{1}{2} \\ 1 & -1 & 2 \end{bmatrix}, \quad P = \begin{bmatrix} \frac{1}{5} & \frac{2}{5} & \frac{1}{10} & \frac{3}{10} \end{bmatrix}, \quad Q = \begin{bmatrix} \frac{1}{5} \\ \frac{4}{5} \\ 0 \end{bmatrix}$$

In Exercises 6 through 10, calculate the security levels of the given strategies, for the games with the given payoff matrices.

6. $G = \begin{bmatrix} 2 & 1 & 0 & -1 \\ 0 & 2 & -1 & 0 \end{bmatrix}$

(a) $P = \begin{bmatrix} \frac{1}{6} & \frac{5}{6} \end{bmatrix}$
(b) $Q = \begin{bmatrix} \frac{1}{3} \\ \frac{1}{3} \\ \frac{1}{6} \\ \frac{1}{6} \end{bmatrix}$

7. $G = \begin{bmatrix} -2 & 0 \\ 0 & 1 \end{bmatrix}$

(a) $P = \begin{bmatrix} \frac{1}{3} & \frac{2}{3} \end{bmatrix}$ (b) $P = \begin{bmatrix} 1 & 0 \end{bmatrix}$ (c) $Q = \begin{bmatrix} \frac{1}{3} \\ \frac{2}{3} \end{bmatrix}$ (d) $Q = \begin{bmatrix} 1 \\ 0 \end{bmatrix}$

8. $G = \begin{bmatrix} -1 & 2 \\ 0 & 0 \\ 1 & -1 \end{bmatrix}$

(a) $Q = \begin{bmatrix} \frac{1}{2} \\ \frac{1}{2} \end{bmatrix}$ (b) $Q = \begin{bmatrix} \frac{9}{10} \\ \frac{1}{10} \end{bmatrix}$ (c) $Q = \begin{bmatrix} \frac{1}{10} \\ \frac{9}{10} \end{bmatrix}$

9. $G = \begin{bmatrix} 2 & 1 & 1 & -4 \\ -3 & -2 & -1 & 3 \end{bmatrix}$

(a) $P = \begin{bmatrix} 0 & 1 \end{bmatrix}$ (b) $P = \begin{bmatrix} \frac{1}{2} & \frac{1}{2} \end{bmatrix}$ (c) $P = \begin{bmatrix} \frac{2}{3} & \frac{1}{3} \end{bmatrix}$

10. $G = \begin{bmatrix} 0 & 1 & 1 \\ -1 & 0 & 1 \\ -1 & -1 & 0 \end{bmatrix}$

(a) $P = \begin{bmatrix} .3 & .3 & .4 \end{bmatrix}$ (b) $P = \begin{bmatrix} .5 & 0 & .5 \end{bmatrix}$

(c) $Q = \begin{bmatrix} .3 \\ .4 \\ .3 \end{bmatrix}$ (d) $Q = \begin{bmatrix} .2 \\ .7 \\ .1 \end{bmatrix}$ (e) $Q = \begin{bmatrix} 0 \\ 0.6 \\ 0.4 \end{bmatrix}$

6.4 Two-by-Two Games

In Section 6.2, we found the solution, that is, the optimum strategies for both players and the value for any strictly determined game. Now we will find the solution for any nonstrictly determined game whose payoff matrix is two-by-two, so that each player has a choice of just two pure strategies. To see how this is done, we use the matrix discussed in the two preceding sections,

$$G = \begin{bmatrix} 3 & -2 \\ -1 & 4 \end{bmatrix}$$

A mixed strategy for the row player, Ruth, is represented by a matrix

$$P = \begin{bmatrix} p_1 & p_2 \end{bmatrix}$$

where $p_1 + p_2 = 1$. To simplify the notation, we abbreviate p_1 as p, which is a probability and is thus a number between 0 and 1. We call the corresponding strategy P_p. Then $p + p_2 = 1$, so that $p_2 = 1 - p$. We write

$$P_p = \begin{bmatrix} p & 1-p \end{bmatrix}$$

Security Level of P_p

To determine the security level of this strategy, recall from Section 6.3 that we calculate

$$\begin{aligned} P_p G &= \begin{bmatrix} p & 1-p \end{bmatrix} \begin{bmatrix} 3 & -2 \\ -1 & 4 \end{bmatrix} \\ &= \begin{bmatrix} (p)(3) + (1-p)(-1) & (p)(-2) + (1-p)(4) \end{bmatrix} \\ &= \begin{bmatrix} 4p - 1 & -6p + 4 \end{bmatrix} \end{aligned}$$

The security level $\underline{v}(P_p)$ of P_p is the minimum of the numbers $4p - 1$ and $-6p + 4$. Whether the minimum is $4p - 1$ or $-6p + 4$ depends on the value of p. For instance, when $p = 0$ then

$$P_0 G = \begin{bmatrix} 4(0) - 1 & -6(0) + 4 \end{bmatrix} = \begin{bmatrix} -1 & 4 \end{bmatrix}$$

so that $\underline{v}(P_0) = -1$, which is of the form $4p - 1$; but when $p = 1$, then

$$P_1 G = \begin{bmatrix} 4(1) - 1 & -6(1) + 4 \end{bmatrix} = \begin{bmatrix} 3 & -2 \end{bmatrix}$$

and thus, $\underline{v}(P_1) = -2$, that is, $6p + 4$. Clearly the choice of $p = 0$ would be better for Ruth than $p = 1$, since the corresponding security level -1 is greater than -2. What we wish to find is the value of p for which $\underline{v}(P_p)$ will be a maximum, since for this number p, the strategy P_p will be Ruth's optimum strategy.

Notice that if Ruth uses $p = 0$, then the payoff will be the security level $\underline{v}(P_0)$ if Carl chooses column 1, whereas if Ruth lets $p = 1$, then the security level $\underline{v}(P_1)$ is the payoff if Carl chooses column 2. We will now have Ruth choose a value of p for which the payoff is the security level no matter which column Carl chooses. Since the payoff will be either $4p - 1$ or $-6p + 4$, this can only happen if she chooses the value of p for which these are the same, that is, for which

$$4p - 1 = -6p + 4$$

It is easy to find the value of p that will accomplish this by rewriting the equation as $10p = 5$ and concluding that $p = \frac{1}{2}$. We let P^* be the corresponding strategy for Ruth, that is,

$$P^* = \begin{bmatrix} \frac{1}{2} & \frac{1}{2} \end{bmatrix}$$

Since

$$P^*G = \begin{bmatrix} \frac{1}{2} & \frac{1}{2} \end{bmatrix} \begin{bmatrix} 3 & -2 \\ -1 & 4 \end{bmatrix} = \begin{bmatrix} 1 & 1 \end{bmatrix}$$

the security level of P^* is $\underline{v}(P^*) = 1$. Certainly a choice of $p = \frac{1}{2}$ is far more desirable for Ruth than either $p = 0$ or $p = 1$. We will see that it is the most desirable choice of all, that is, $\underline{v}(P^*) = 1$ is the maximum security level for all possible choices of p, and therefore P^* is the optimum strategy for Ruth. For now, we observe that we have shown that Ruth can be sure of a security level of at least 1 because the strategy P^* has that security level. We will soon see that she cannot do better.

Equalizing Strategies

Recall that a mixed strategy for Carl can be represented by the matrix

$$Q = \begin{bmatrix} q_1 \\ q_2 \end{bmatrix}$$

where $q_1 + q_2 = 1$. If we write q_1 as just q, then

$$Q = \begin{bmatrix} q \\ 1 - q \end{bmatrix}$$

where q can take any value between 0 and 1. If Ruth chooses the strategy,

$$P^* = \begin{bmatrix} \frac{1}{2} & \frac{1}{2} \end{bmatrix}$$

then the payoff is given by

$$E(P^*, Q) = P^*GQ = \begin{bmatrix} 1 & 1 \end{bmatrix} \begin{bmatrix} q \\ 1 - q \end{bmatrix} = q + (1 - q) = 1$$

The payoff equals 1 for *any* value of q that Carl chooses. In other words, by using the strategy P^*, Ruth has determined the payoff, no matter what strategy Carl uses. A strategy for one player in which the payoff is the same no matter what strategy the opponent chooses is called an **equalizing strategy**, since all payoffs are equal. Thus, P^* is an equalizing strategy for Ruth.

But Carl also has an equalizing strategy. For any choice of q, we have the strategy

$$Q_q = \begin{bmatrix} q \\ 1 - q \end{bmatrix}$$

and by calculating

$$GQ_q = \begin{bmatrix} 3 & -2 \\ -1 & 4 \end{bmatrix} \begin{bmatrix} q \\ 1-q \end{bmatrix} = \begin{bmatrix} 3q + (-2)(1-q) \\ (-1)q + 4(1-q) \end{bmatrix} = \begin{bmatrix} 5q - 2 \\ -5q + 4 \end{bmatrix}$$

we see that the security level $\overline{v}(Q_q)$ is the minimum of the numbers $5q - 2$ and $-5q + 4$. If we set these numbers equal, that is, if

$$5q - 2 = -5q + 4$$

then $10q = 6$, so that $q = \frac{3}{5}$. Substituting $q = \frac{3}{5}$ into either side of the equation produces a value of 1; therefore, the corresponding security level for Carl is 1. Calling this strategy Q^*, that is, letting

$$Q^* = \begin{bmatrix} \frac{3}{5} \\ \frac{2}{5} \end{bmatrix}$$

for any choice P_p of strategy for Ruth,

$$E(P_p, Q^*) = P_p GQ^* = \begin{bmatrix} p & 1-p \end{bmatrix} \begin{bmatrix} 3 & -2 \\ -1 & 4 \end{bmatrix} \begin{bmatrix} \frac{3}{5} \\ \frac{2}{5} \end{bmatrix}$$

$$= \begin{bmatrix} p & 1-p \end{bmatrix} \begin{bmatrix} 1 \\ 1 \end{bmatrix} = 1$$

Thus, Q^* is an equalizing strategy for Carl; the payoff is 1 no matter what Ruth does.

Now we can see that $\underline{v}(P^*) = 1$ is indeed the maximum security level for Ruth. No matter what value of p she chooses, Carl can always use his strategy Q^* to guarantee that the payoff is 1, so that there is no way Ruth can do better. On the other hand, no choice of q can produce a more favorable security level for Carl than $\overline{v}(Q^*) = 1$, since Ruth can force all payoffs to have that value by using her strategy P^*. Thus, $\overline{v}(Q^*) = 1$ is the minimum security level for Carl. Now we know that the optimum strategies in this game are

$$P^* = \begin{bmatrix} \frac{1}{2} & \frac{1}{2} \end{bmatrix} \qquad Q^* = \begin{bmatrix} \frac{3}{5} \\ \frac{2}{5} \end{bmatrix}$$

and that the value of the game is $\underline{v} = \overline{v} = 1$. Consequently, we have solved the game.

The General Two-by-Two Game

The payoff matrix for any two-by-two game is of the form

6.4 TWO-BY-TWO GAMES

$$G = \begin{bmatrix} a & b \\ c & d \end{bmatrix}$$

If there is a saddle point, we can solve the game as we did in Section 6.2: optimum strategies for the two players are given by any row and the corresponding column in which a saddle point occurs; the value of the game is the value of the saddle point. If there is no saddle point, we can analyze this game just as we did the special case

$$G = \begin{bmatrix} 3 & -2 \\ -1 & 4 \end{bmatrix}$$

to find the (mixed) optimum strategies for the players and the value of the game, as we shall now demonstrate.

Assume that we have a payoff matrix

$$G = \begin{bmatrix} a & b \\ c & d \end{bmatrix}$$

that does not have a saddle point. Suppose the entry a is larger than or equal to b, that is $a \geq b$, so that $a - b$ is nonnegative. Because b is the minimum of its row, it cannot be the maximum of its column; otherwise, it would be a saddle point. Thus, d is the maximum of the second column, and since d cannot therefore also be the minimum of its row, d must be (strictly) larger than c, that is, $d - c > 0$. If $a \geq b$, so that $a - b$ is nonnegative, then we have shown that

$$(a - b) + (d - c) > 0$$

We will now see why it is important to know this fact. We calculate that

$$P_p G = \begin{bmatrix} p & 1-p \end{bmatrix} \begin{bmatrix} a & b \\ c & d \end{bmatrix}$$
$$= \begin{bmatrix} pa + (1-p)c & pb + (1-p)d \end{bmatrix}$$
$$= \begin{bmatrix} (a-c)p + c & (b-d)p + d \end{bmatrix}$$

We wish to find the equalizing strategy for player R, so we solve

$$(a - c)p + c = (b - d)p + d$$
$$[(a - c) - (b - d)]p = d - c$$

Now we observe that

$$(a - c) - (b - d) = (a - b) + (d - c) > 0$$

Because this number is not zero, we can divide both sides of the preceding equation by it to obtain

$$p^* = \frac{d - c}{(a - b) + (d - c)}$$

Both the numerator and the denominator are nonnegative and the numerator, a summand of the denominator, is no larger than the denominator. Thus, we are assured that p^* lies between 0 and 1.

It is easy to see that

$$1 - p^* = \frac{a-b}{(a-b)+(d-c)}$$

because this fraction together with the one for p^* sum to one. Thus when G has no saddle point and $a \geq b$, the equalizing strategy for player R is given by the matrix

$$P^* = \left[\frac{d-c}{(a-b)+(d-c)} \quad \frac{a-b}{(a-b)+(d-c)} \right]$$

The same formula holds when $a \leq b$, provided that G has no saddle point. (See Exercises 13 and 14.)

To find the security level $\underline{v}(P^*)$ of P^*, we can calculate

$$P^*G = \left[\frac{d-c}{(a-b)+(d-c)} \quad \frac{a-b}{(a-b)+(d-c)} \right] \begin{bmatrix} a & b \\ c & d \end{bmatrix}$$

The two entries of the matrix P^*G must be equal, so it is sufficient to calculate one of them. The matrix P^* multiplied by the first column of G gives

$$\underline{v}(P^*) = \frac{d-c}{(a-b)+(d-c)} \cdot a + \frac{a-b}{(a-b)+(d-c)} \cdot c$$

$$= \frac{(d-c)a}{(a-b)+(d-c)} + \frac{(a-b)c}{(a-b)+(d-c)}$$

$$= \frac{(d-c)a + (a-b)c}{(a-b)+(d-c)}$$

$$= \frac{da - ca + ac - bc}{(a-b)+(d-c)}$$

After simplifying the numerator, we find that

$$\underline{v}(P^*) = \frac{ad - bc}{(a-b)+(d-c)}$$

Equalizing Column Strategy

To find the equalizing strategy for player C, we compute

$$GQ_q = \begin{bmatrix} a & b \\ c & d \end{bmatrix} \begin{bmatrix} q \\ 1-q \end{bmatrix} = \begin{bmatrix} (a-b)q + b \\ (c-d)q + d \end{bmatrix}$$

Setting the two entries equal to each other, we obtain

$$(a-b)q + b = (c-d)q + d$$

and thus,
$$[(a-b)-(c-d)]q = d-b$$

When we write $(a-b)-(c-d)$ in the form $(a-b)+(d-c)$, we see that we are dividing by the same term as before. Thus,

$$q^* = \frac{d-b}{(a-b)+(d-c)}$$

and the equalizing strategy for player C is the matrix

$$Q^* = \begin{bmatrix} \frac{d-b}{(a-b)+(d-c)} \\ \frac{a-c}{(a-b)+(d-c)} \end{bmatrix}$$

Computing one of the entries of GQ^* (see Exercise 15), shows us that the security level of this strategy is

$$\overline{v}(Q^*) = \frac{ad-bc}{(a-b)+(d-c)}$$

and that consequently, $\overline{v} = \underline{v}$. Since player R can guarantee a security level $\underline{v}$ by using the strategy P^*, but cannot do better if player C uses strategy Q^*, we see that P^* is the optimum strategy for player R and, similarly, Q^* is the optimum for player C.

Solution of a Two-by-Two Game

To summarize,

if the payoff matrix

$$G = \begin{bmatrix} a & b \\ c & d \end{bmatrix}$$

has a saddle point, then the optimum solutions are pure strategies: P^* is to choose a row in which a saddle point occurs and Q^* should choose a column containing a saddle point; the value of the game is the value of any saddle point. If G has no saddle point, then the optimum solutions of the game are the equalizing strategies

$$P^* = \begin{bmatrix} \frac{d-c}{(a-b)+(d-c)} & \frac{a-b}{(a-b)+(d-c)} \end{bmatrix} \quad \text{and} \quad Q^* = \begin{bmatrix} \frac{d-b}{(a-b)+(d-c)} \\ \frac{a-c}{(a-b)+(d-c)} \end{bmatrix}$$

and the value of the game is

$$v = \frac{ad-bc}{(a-b)+(d-c)}$$

Example 1. Solve the game with payoff matrix

$$G = \begin{bmatrix} 0 & -3 \\ -5 & 6 \end{bmatrix}$$

Solution

Since the row minima are negative numbers (–5 and –3) and the column maxima are nonnegative numbers (0 and 6), there is no saddle point. We have

$$\begin{bmatrix} a & b \\ c & d \end{bmatrix} = \begin{bmatrix} 0 & -3 \\ -5 & 6 \end{bmatrix}$$

and therefore,

$$(a-b) + (d-c) = (0-(-3)) + (6-(-5)) = 3 + 11 = 14$$

The optimum solution for player R is given by

$$P^* = \begin{bmatrix} \dfrac{d-c}{(a-b)+(d-c)} & \dfrac{a-b}{(a-b)+(d-c)} \end{bmatrix} = \begin{bmatrix} \dfrac{11}{14} & \dfrac{3}{14} \end{bmatrix}$$

For player C, the optimum is

$$Q^* = \begin{bmatrix} \dfrac{d-b}{(a-b)+(d-c)} \\ \dfrac{a-c}{(a-b)+(d-c)} \end{bmatrix} = \begin{bmatrix} \dfrac{6-(-3)}{14} \\ \dfrac{0-(-5)}{14} \end{bmatrix} = \begin{bmatrix} \dfrac{9}{14} \\ \dfrac{5}{14} \end{bmatrix}$$

The value of this game is

$$v = \frac{ad-bc}{(a-b)+(d-c)} = \frac{(0)(6)-(-3)(-5)}{14} = -\frac{15}{14} \qquad \blacksquare$$

We now apply our solution technique to a campaign-resources problem in which each candidate has just two pure strategies.

Example 2. Mr. Green and Ms. White are running for mayor of Central City. The candidates must decide how to distribute their time between personal contacts (attending neighborhood receptions, greeting workers outside factories, and so on) and formal speeches (at political rallies or in debates). Analysis of the abilities of the two candidates indicates that, among undecided voters (who, we assume, will vote for one of the two candidates) the outcome will be as follows: if Mr. Green devotes all his time to personal contacts, he will lose 20,000 votes to Ms. White if she also devotes all her time to personal contacts, but will gain 30,000 votes from Ms. White if she spends all her time on formal speeches. If Mr. Green spends all his time on formal speeches, he will gain 10,000 votes from Ms. White if she spends all her time on personal contacts but will lose 40,000 votes to her if she also devotes all her time to formal speeches. Use game theory to determine how the candidates should distribute their time and to predict the outcome of the undecided vote.

6.4 TWO-BY-TWO GAMES

Solution

Since the information in the problem was given in terms of Mr. Green's gains and losses, his choices will constitute the rows of the table of payoffs:

		Ms. White Personal Contact	Formal Speeches
Mr. Green	Personal Contact	−20,000	30,000
	Formal Speeches	10,000	−40,000

Table 6-10

We can simplify the arithmetic by using 10,000 voters as a unit and thus writing the payoff matrix as

$$G = \begin{bmatrix} -2 & 3 \\ 1 & -4 \end{bmatrix}$$

Since the row minima are negative and the column maxima positive, there is no saddle point. Therefore, the optimum strategies are the equalizing ones. It is convenient to calculate first

$$(a-b) + (d-c) = ((-2) - 3) + ((-4) - 1) = -10$$

which is the denominator of the fractions in the formulas for P^* and Q^*. Thus,

$$P^* = \begin{bmatrix} \dfrac{d-c}{(a-b)+(d-c)} & \dfrac{a-b}{(a-b)+(d-c)} \end{bmatrix}$$

$$= \begin{bmatrix} \dfrac{(-4)-1}{-10} & \dfrac{(-2)-3}{-10} \end{bmatrix} = \begin{bmatrix} \dfrac{1}{2} & \dfrac{1}{2} \end{bmatrix}$$

and

$$Q^* = \begin{bmatrix} \dfrac{d-b}{(a-b)+(d-c)} \\ \dfrac{a-c}{(a-b)+(d-c)} \end{bmatrix} = \begin{bmatrix} \dfrac{(-4)-3}{-10} \\ \dfrac{(-2)-1}{-10} \end{bmatrix} = \begin{bmatrix} \dfrac{7}{10} \\ \dfrac{3}{10} \end{bmatrix}$$

The interpretation of these mixed strategies is that Mr. Green should divide his time evenly between personal contacts with voters and formal speeches, whereas Ms. White should spend 70% of her time on personal contacts and the remaining 30% on formal speeches. To predict the outcome of the undecided vote, we calculate the value of the game to be

$$v = \frac{ad - bc}{(a-b)+(d-c)} = \frac{(-2)(-4) - (3)(1)}{-10} = \frac{5}{-10} = -\frac{1}{2}$$

Recall that the units are 10,000 voters and that Mr. Green is represented by the rows. Thus, a negative value is in Ms. White's favor. We conclude that Ms. White will receive $\frac{1}{2}(10,000) = 5000$ more undecided votes than Mr. Green will. ∎

Strictly Determined Two-by-Two Games

In the solution of Examples 1 and 2, we first checked that the game matrices had no saddle points before going on to find the equalizing strategies. To see what happens when we apply the formulas for equalizing strategies to a game that has a saddle point, we use the "silver-gold" game of Section 6.1. This game has the payoff matrix

$$G = \begin{bmatrix} -10 & 10 \\ 0 & 5 \end{bmatrix}$$

Calculating $(a - b) + (d - c) = -15$, we then have

$$P^* = \begin{bmatrix} \dfrac{d-c}{(a-b)+(d-c)} & \dfrac{a-b}{(a-b)+(d-c)} \end{bmatrix}$$

$$= \begin{bmatrix} \dfrac{5-0}{-15} & \dfrac{(-10)-10}{-15} \end{bmatrix}$$

$$= \begin{bmatrix} -\dfrac{1}{3} & \dfrac{4}{3} \end{bmatrix}$$

This does not represent a strategy for player R because neither $-\frac{1}{3}$ nor $\frac{4}{3}$ can be a probability. On the other hand, there *is* an equalizing strategy for player C because

$$Q^* = \begin{bmatrix} \dfrac{d-b}{(a-b)+(d-c)} \\ \dfrac{a-c}{(a-b)+(d-c)} \end{bmatrix} = \begin{bmatrix} \dfrac{5-10}{-15} \\ \dfrac{-10-0}{-15} \end{bmatrix} = \begin{bmatrix} \dfrac{1}{3} \\ \dfrac{2}{3} \end{bmatrix}$$

and $\frac{1}{3}$ and $\frac{2}{3}$ are positive numbers that add to one. Furthermore,

$$GQ^* = \begin{bmatrix} -10 & 10 \\ 0 & 5 \end{bmatrix} \begin{bmatrix} \dfrac{1}{3} \\ \dfrac{2}{3} \end{bmatrix} = \begin{bmatrix} \dfrac{10}{3} \\ \dfrac{10}{3} \end{bmatrix}$$

so that the security level of Q^* is $\frac{10}{3}$. However, this equalizing strategy is certainly not optimal for player C since the pure strategy: "choose column 1" has a much more favorable security level, 0.

As this example indicates, if the payoff matrix G has a saddle point, then there may not be equalizing strategies for the two players, and even when there are equalizing strategies, they may not be optimum. Thus, to solve a two-by-two game, we should first check the payoff matrix for saddle points; we then use the formulas for equalizing strategies only if there is no saddle point.

Summary

If the payoff matrix of a two-by-two game

$$G = \begin{bmatrix} a & b \\ c & d \end{bmatrix}$$

has a saddle point, then the optimum solutions are pure strategies: P^* is to choose a row in which a saddle point occurs and Q^* is to choose a column containing a saddle point; the value of the game is the value of any saddle point. If G has no saddle point, then the optimum solutions of the game are the equalizing strategies

$$P^* = \begin{bmatrix} \dfrac{d-c}{(a-b)+(d-c)} & \dfrac{a-b}{(a-b)+(d-c)} \end{bmatrix}$$

and

$$Q^* = \begin{bmatrix} \dfrac{d-b}{(a-b)+(d-c)} \\ \dfrac{a-c}{(a-b)+(d-c)} \end{bmatrix}$$

and the value of the game is

$$v = \frac{ad-bc}{(a-b)+(d-c)}$$

Exercises

In Exercises 1 through 4, solve the game with the given payoff matrix.

1. $\begin{bmatrix} 2 & 0 \\ -1 & 1 \end{bmatrix}$ 2. $\begin{bmatrix} -3 & 1 \\ 5 & -1 \end{bmatrix}$ 3. $\begin{bmatrix} -1 & -3 \\ 1 & 2 \end{bmatrix}$ 4. $\begin{bmatrix} \frac{1}{2} & 0 \\ -\frac{1}{2} & 1 \end{bmatrix}$

5. Mrs. Adams and Miss Butler are the candidates of their respective parties for governor of the state. Each must decide the fraction of their newspaper advertising budgets to be spent on the state's two major papers, the *Sun* and the *Review*. If Mrs. Adams spends all of her budget on the *Sun*, she will gain 1% of the undecided vote over Miss Butler, provided that Miss Butler also spends all of her budget on the *Sun*; she will lose 2% of this vote to Miss Butler if Miss Butler spends all her budget on the *Review*. If Mrs. Adams spends all of her budget on the *Review*, she will lose 1% of the undecided vote to Miss Butler if Miss Butler spends all her budget on the *Sun*, and will gain 3% of this vote from Miss

Butler if Miss Butler also spends all her budget on the *Review*. Determine the optimum fraction of each candidate's advertising budget to be spent on the two newspapers and predict the resulting outcome of the undecided vote.

6. A developer owns a parcel of land near the proposed site of a new rapid-transit station, and she must decide whether to build a shopping mall or an office complex on this land. If she builds the shopping mall, she will receive an 80% return on her investment provided that the station is built, but only a 30% return if it is not. If she builds the office complex, she will receive a 40% return on her investment, whether or not the station is built. Determine the developer's optimum strategy and predict the return on her investment.

7. When a penalty kick is awarded in soccer, the kicker has a shot at the goal, which is then defended only by the goalkeeper. The kicker will aim at either the right or left half of the goal and, simultaneously, the goalkeeper will dive either right or left in an effort to deflect the ball. Suppose that if a kicker aims at the right half of the goal and the goalkeeper dives in that direction, the probability is .20 that a goal will be scored, whereas if the goalkeeper dives to the left, the probability of a score is .80. On the other hand, if the kicker aims to the left and the goalkeeper dives to the right, the probability of a goal is .90, but if the goalkeeper dives to the left, then the probability of scoring is only .10. Determine the optimum strategies for both the kicker and the goalkeeper, and calculate the probability that a goal will be scored.

8. Ruth and Carl play the game called two-finger Morra. They simultaneously show either one or two fingers each. If the sum of the number of fingers shown is even, then Carl pays that number of dollars to Ruth, if it is odd, then Ruth pays the sum to Carl. Solve the two-finger Morra game.

9. The Spicy Chicken and Happy Hen fast-food chains each plan to open a new restaurant in the vicinity of River City. In terms of their portion of the fast-food chicken market in the River City area, the choice of a downtown location for Spicy Chicken will be a 3% gain over Happy Hen if Happy Hen also chooses a downtown location, but a 1% loss if Happy Hen chooses a suburban location. If Spicy Chicken decides on a suburban location, it will have a sale loss of 2% to Happy Hen provided that Happy Hen chooses a downtown location, and a of loss of 3% if Happy Hen also chooses a suburban location. Assuming that both chains decide on their locations independently, what are their optimum location strategies?

10. The General Umbrella Company makes both umbrellas to protect against rain and beach umbrellas for protection from the sun. If General Umbrella makes only rain umbrellas for next summer's market, it will earn $3 million if the summer is rainy, but only $1 million if it is sunny. If General Umbrella makes only beach umbrellas, then it will make $1 million in a rainy summer and $4 million in one that is sunny. Determine the fraction of each kind of umbrella, rain and beach, that General Umbrella should make for next summer's market, and predict its sales earnings.

6.4 TWO-BY-TWO GAMES

11. Two all-news cable-television networks, NewsTV and AllNews, broadcast both general news and sports news. A media consultant estimates that if NewsTV broadcasts only general news, it will lose 5 rating points to AllNews provided that AllNews also broadcasts only general news, but it will gain 7 points if AllNews broadcasts only sports news. In the event that NewsTV broadcasts only sports news, then it would neither gain nor lose rating points to AllNews if AllNews broadcasts only general news, but it would lose 2 points if AllNews also broadcast only sports news. Determine the optimum fractions of general news and sports news for NewsTV and AllNews and estimate the net effect on the ratings.

12. A fishing boat must decide whether to fish close to or far from shore. The value of the catch will depend on the presence or absence of predatory fish since many fish will stay close to shore if there are predators present, but otherwise, will swim far from shore. If the boat fishes close to shore, then the value of the catch will be $8000 in the presence of predators but $2000 if there are no predators. If the boat fishes far from shore, then the values will be $1000 if there are predators present, and $10,000 otherwise. Determine an optimum fishing strategy for the boat and calculate the expected value of the catch.

In Exercises 13 through 17, suppose that the matrix

$$G = \begin{bmatrix} a & b \\ c & d \end{bmatrix}$$

has no saddle point.

13. Show that if $a \leq b$, then $(a - b) + (d - c) < 0$.

14. Show that if $a \leq b$, then

$$p^* = \frac{d - c}{(a - b) + (d - c)}$$

lies between 0 and 1.

15. Calculate GQ^*, where

$$Q^* = \begin{bmatrix} \dfrac{d - b}{(a - b) + (d - c)} \\ \dfrac{a - c}{(a - b) + (d - c)} \end{bmatrix}$$

16. Show that if $a \geq b$, then

$$q^* = \frac{d - b}{(a - b) + (d - c)}$$

is a number between 0 and 1.

17. Show that if $a \leq b$, then

$$q^* = \frac{d - b}{(a - b) + (d - c)}$$

is a number between 0 and 1.

6.5 Games and Linear Programming

We next present a technique for solving any matrix game. Given the payoff matrix G of the game, we will be able to calculate the optimum mixed strategies P^* and Q^* for players R and C, respectively.

A Positive Matrix Game

We illustrate the method by means of the game with the 2-by-3 payoff matrix

$$G = \begin{bmatrix} 1 & 4 \\ 3 & 1 \\ 2 & 2 \end{bmatrix}$$

This matrix is too large to treat by our previous methods.

Notice that all the entries of the matrix are positive. Every payoff is a gain for the row player R. Recall that a matrix in which every entry is positive is called a **positive matrix**. We will begin by discussing games with positive payoff matrices; afterward, we will extend the method to all matrix games.

First we discuss the game from the point of view of player R. In Section 6.3 we saw that for a matrix with three rows, a mixed strategy for player R can be represented as a matrix

$$P = \begin{bmatrix} p_1 & p_2 & p_3 \end{bmatrix}$$

Here, the probabilities p_1, p_2, and p_3 are, of course, nonnegative, that is,

$$p_1, p_2, p_3 \geq 0$$

and they add up to one:

$$p_1 + p_2 + p_3 = 1$$

The security level corresponding to the strategy P is the minimum entry of the matrix PG. In the present example, since

$$PG = \begin{bmatrix} p_1 & p_2 & p_3 \end{bmatrix} \begin{bmatrix} 1 & 4 \\ 3 & 1 \\ 2 & 2 \end{bmatrix}$$

$$= \begin{bmatrix} p_1 + 3p_2 + 2p_3 & 4p_1 + p_2 + 2p_3 \end{bmatrix}$$

the security level, s, of P is the minimum of the two numbers

$$p_1 + 3p_2 + 2p_3 \quad \text{and} \quad 4p_1 + p_2 + 2p_3$$

The p_i cannot be negative, so the expressions $p_1 + 3p_2 + 2p_3$ and $4p_1 + p_2 + 2p_3$ are nonnegative; since s equals one of these numbers, we conclude that $s \geq 0$. One reason for choosing our matrix G to be positive and for restricting our present discussion to games with positive matrices is to obtain the property that $s \geq 0$.

6.5 GAMES AND LINEAR PROGRAMMING

Furthermore, since s is the minimum of the two numbers, it cannot be greater than either of them; that is, we have the inequalities

$$p_1 + 3p_2 + 2p_3 \geq s \quad \text{and} \quad 4p_1 + p_2 + 2p_3 \geq s$$

which can also be written as

$$s - p_1 - 3p_2 - 2p_3 \leq 0$$

and

$$s - 4p_1 - p_2 - 2p_3 \leq 0$$

Linear-Programming Problems

An optimum strategy

$$P^* = \begin{bmatrix} p_1 & p_2 & p_3 \end{bmatrix}$$

for player R is a strategy whose security level s is a maximum. To find an optimum strategy, we must find numbers s, p_1, p_2, and p_3 that solve the following problem:

Maximize: s

Subject to:
$$p_1 + p_2 + p_3 \leq 1$$
$$s - p_1 - 3p_2 - 2p_3 \leq 0$$
$$s - 4p_1 - p_2 - 2p_3 \leq 0$$
$$s, p_1, p_2, p_3 \geq 0$$

This problem is a maximization problem of linear programming, as discussed in Chapter 2. It is of the form

Maximize: CX

Subject to: $AX \leq B$ and $X \geq 0$

where the matrices are

$$X = \begin{bmatrix} s \\ p_1 \\ p_2 \\ p_3 \end{bmatrix}, \quad C = \begin{bmatrix} 1 & 0 & 0 & 0 \end{bmatrix}, \quad A = \begin{bmatrix} 0 & 1 & 1 & 1 \\ 1 & -1 & -3 & -2 \\ 1 & -4 & -1 & -2 \end{bmatrix}, \quad \text{and} \quad B = \begin{bmatrix} 1 \\ 0 \\ 0 \end{bmatrix}$$

In order to obtain a maximization problem, we replaced the equation

$$p_1 + p_2 + p_3 = 1$$

by the inequality

$$p_1 + p_2 + p_3 \leq 1$$

However, it can be shown that if s, p_1, p_2, p_3 are numbers satisfying the conditions of the maximization problem with

$$p_1 + p_2 + p_3 < 1$$

476 GAME THEORY

then s *cannot* be the maximum. Thus, the solution s, p_1, p_2, p_3 to the maximization problem must have the additional property that

$$p_1 + p_2 + p_3 = 1$$

Consequently, $P^* = [p_1, p_2, p_3]$ will be an optimum strategy for player R.

Now let us look at this same example from the point of view of the other player, who chooses a column of the matrix. For a mixed strategy

$$Q = \begin{bmatrix} q_1 \\ q_2 \end{bmatrix}$$

for player C, we of course require $q_1, q_2 \geq 0$ and $q_1 + q_2 = 1$. Since

$$GQ = \begin{bmatrix} 1 & 4 \\ 3 & 1 \\ 2 & 2 \end{bmatrix} \begin{bmatrix} q_1 \\ q_2 \end{bmatrix} = \begin{bmatrix} q_1 + 4q_2 \\ 3q_1 + q_2 \\ 2q_1 + 2q_2 \end{bmatrix}$$

the security level, which we will call t, of the strategy Q, is the maximum of the three entries of the last matrix. In particular,

$$q_1 + 4q_2 \leq t, \quad 3q_1 + q_2 \leq t, \quad \text{and} \quad 2q_1 + 2q_2 \leq t$$

Player C seeks to minimize the security level t. Therefore, the problem of finding the optimum strategy Q^* is a minimization problem of linear programming:

Minimize: t
Subject to:
$$q_1 + q_2 \geq 1$$
$$t - q_1 - 4q_2 \geq 0$$
$$t - 3q_1 - q_2 \geq 0$$
$$t - 2q_1 - 2q_2 \geq 0$$
$$t, q_1, q_2 \geq 0$$

Duality

Looking back to the LP-problem whose solution is the optimum strategy P^* for player R, we see that the minimization problem is the dual LP-problem, in the sense of Section 2.3. That is, it is of the form

Minimize: $B^T Y$
Subject to: $A^T Y \geq C^T$ and $Y \geq 0$

where A, B, and C are as in the maximization problem and

$$Y = \begin{bmatrix} t \\ q_1 \\ q_2 \end{bmatrix}$$

6.5 GAMES AND LINEAR PROGRAMMING   477

Therefore, the Duality Theorem tells us that the maximum value of the objective function, s, of the maximization problem is equal to the minimum value of the objective function, t, of the dual minimization problem. In Section 6.3, we defined $\underline{v}$ to be the maximum of the security levels s for player R and $\bar{v}$ to be the minimum of the security levels t for player C. Thus the equation $\underline{v} = \bar{v}$ is a consequence of the Duality Theorem of linear programming.

Duality theory also tells us that if we can solve one of these problems by the simplex algorithm, then the other, dual, problem is solved as well. Rather than apply the simplex algorithm to these problems as we have presented them, however, we will first undertake a significant simplification, which will make the computations easier. Recalling that $q_1 + q_2 = 1$ and that

$$q_1 + 4q_2 \leq t, \quad 3q_1 + q_2 \leq t, \quad \text{and} \quad 2q_1 + 2q_2 \leq t$$

we now write the problem of finding the optimum strategy Q^* for player C in this form:

Minimize: t

Subject to:
$$q_1 + q_2 = 1$$
$$q_1 + 4q_2 \leq t$$
$$3q_1 + q_2 \leq t$$
$$2q_1 + 2q_2 \leq t$$
$$t, q_1, q_2 \geq 0$$

Since at least one of the probabilities q_1 and q_2 must be greater than 0, none of the expressions $q_1 + 4q_2$, $3q_1 + q_2$, and $2q_1 + 2q_2$ can be equal to zero. Thus, we know that the same is true of t, that is, $t > 0$. (This is where we make full use of the fact that the matrix G is positive.) When both sides of an inequality are divided by a positive number, the direction of the inequality does not change. Therefore, we can rewrite the restrictions of the problem as

$$\frac{q_1}{t} + \frac{q_2}{t} = \frac{1}{t}$$

$$\frac{q_1}{t} + 4\frac{q_2}{t} \leq 1$$

$$3\frac{q_1}{t} + \frac{q_2}{t} \leq 1$$

$$2\frac{q_1}{t} + 2\frac{q_2}{t} \leq 1$$

We rename the variables:

$$x_1 = \frac{q_1}{t} \quad \text{and} \quad x_2 = \frac{q_2}{t}$$

so that the first restriction is that

$$x_1 + x_2 = \frac{1}{t}$$

478 GAME THEORY

Keeping in mind that we know that t must be greater than zero, the smaller t becomes, the greater is the value of $x_1 + x_2$. Therefore, the objective "Minimize t" is the same as the objective "Maximize $x_1 + x_2$." This observation allows us to eliminate the variable t from the problem altogether when we substitute x_1 and x_2 into the inequalities. The minimization problem of finding the optimum strategy Q^* for player C is equivalent to the following problem:

$$\begin{aligned}
\text{Maximize:} \quad & x_1 + x_2 \\
\text{Subject to:} \quad & x_1 + 4x_2 \leq 1 \\
& 3x_1 + x_2 \leq 1 \\
& 2x_1 + 2x_2 \leq 1 \\
& x_1, x_2 \geq 0
\end{aligned}$$

This is a maximization problem of the type

$$\text{Maximize:} \quad z = CX$$
$$\text{Subject to:} \quad AX \leq B, X \geq 0, \text{ and also } B \geq 0$$

because we can now see that

$$X = \begin{bmatrix} x_1 \\ x_2 \end{bmatrix}, \quad C = \begin{bmatrix} 1 & 1 \end{bmatrix}, \quad A = \begin{bmatrix} 1 & 4 \\ 3 & 1 \\ 2 & 2 \end{bmatrix}, \quad \text{and} \quad B = \begin{bmatrix} 1 \\ 1 \\ 1 \end{bmatrix}$$

Notice that B and C have very special forms; all their entries are 1s. Furthermore the matrix A is the payoff matrix G of the game.

The Simplex Tableau

Our discussion of the example would apply to any matrix game with a positive payoff matrix.

To calculate the optimum solution $Q^* = (q_1, q_2, \ldots, q_m)$ of a game with a positive $k \times m$-payoff matrix G, we solve the maximization problem

$$\text{Maximize:} \quad z = CX$$
$$\text{Subject to:} \quad AX \leq B, X \geq 0, \text{ and also } B \geq 0$$

where

$$X = \begin{bmatrix} x_1 \\ x_2 \\ \vdots \\ x_m \end{bmatrix}$$

Here, $A = G$, and B and C are matrices whose entries are all 1s.

6.5 GAMES AND LINEAR PROGRAMMING

We solve this LP-problem by using the simplex algorithm of Section 2.5. In this problem, the simplex tableau has a rather special form:

$$\begin{array}{c} \phantom{\{x\}} \\ \{x\} \\ z \end{array} \begin{array}{c} \overbrace{}^{x_1 \quad \cdots \quad x_N} \\ \left[\begin{array}{cc|c} G & I & B \\ -C & 0 & 0 \end{array} \right] \end{array}$$

because G is the payoff matrix of the game, and the matrices B and C consist entirely of 1s.

Solving the Example

Thus, the simplex tableau of the example is

$$\begin{array}{c} x_3 \\ x_4 \\ x_5 \\ z \end{array} \left[\begin{array}{ccccc|c} x_1 & x_2 & x_3 & x_4 & x_5 & \\ 1 & 4 & 1 & 0 & 0 & 1 \\ 3 & 1 & 0 & 1 & 0 & 1 \\ 2 & 2 & 0 & 0 & 1 & 1 \\ -1 & -1 & 0 & 0 & 0 & 0 \end{array} \right]$$

We pivot in the a_{21}-location, with the result given by

$$\begin{array}{c} x_3 \\ x_1 \\ x_5 \\ 3z \end{array} \left[\begin{array}{ccccc|c} x_1 & x_2 & x_3 & x_4 & x_5 & \\ 0 & 11 & 3 & -1 & 0 & 2 \\ 3 & 1 & 0 & 1 & 0 & 1 \\ 0 & 4 & 0 & -2 & 3 & 1 \\ 0 & -2 & 0 & 1 & 0 & 1 \end{array} \right]$$

The next pivot at the a_{12}-location finishes the problem:

$$\begin{array}{c} x_2 \\ x_1 \\ x_5 \\ 33z \end{array} \left[\begin{array}{ccccc|c} x_1 & x_2 & x_3 & x_4 & x_5 & \\ 0 & 11 & 3 & -1 & 0 & 2 \\ 33 & 0 & -3 & 12 & 0 & 9 \\ 0 & 0 & -12 & -18 & 33 & 3 \\ 0 & 0 & 6 & 9 & 0 & 15 \end{array} \right]$$

We conclude that the solution of this LP-problem is

$$x_1 = \frac{9}{33} = \frac{3}{11}, \quad x_2 = \frac{2}{11}, \quad \text{and} \quad z = \frac{15}{33} = \frac{5}{11} = x_1 + x_2$$

We will use this solution to obtain the optimum strategy Q^*. In the example, we used the equation

$$x_1 + x_2 = \frac{1}{t}$$

and in the general case we have

$$x_1 + x_2 + \ldots + x_m = \frac{1}{t}$$

When the x's are chosen so that the objective function is a maximum, then t, the security level for player C, is the minimum security level and thus equals the value v of the game. Therefore, we may write

$$x_1 + x_2 + \ldots + x_m = \frac{1}{v}$$

Thus, when we solve the LP-problem and add the numbers in the solution, we know the value v of the game because it is the reciprocal of that sum. In the example,

$$z = x_1 + x_2 = \frac{5}{11}$$

and therefore, the value of the game is

$$v = \frac{1}{z} = \frac{11}{5}$$

which is also the security level for the optimum strategy Q^*. We defined the x's in terms of the probabilities by means of the equations

$$x_1 = \frac{q_1}{t}, \quad x_2 = \frac{q_2}{t}, \quad \text{and so on}$$

where t now represents the value v of the game. For instance,

$$x_1 = \frac{q_1}{v}$$

which implies that

$$q_1 = vx_1$$

In general, we have

$$\boxed{Q^* = (q_1, q_2, \ldots, q_m) = (vx_1, vx_2, \ldots, vx_m)}$$

For the example,

$$Q^* = (vx_1, vx_2) = \left(\frac{11}{5} \cdot \frac{3}{11}, \frac{11}{5} \cdot \frac{2}{11}\right) = \left(\frac{3}{5}, \frac{2}{5}\right)$$

By duality theory, the solution of the dual LP-problem lies below the slack variables in the final tableau, that is

$$y_1 = \frac{6}{33} = \frac{2}{11}, \quad y_2 = \frac{9}{33} = \frac{3}{11}, \quad \text{and} \quad y_3 = 0$$

An argument similar to the one we gave, now from player R's point of view, shows that the same relationship holds between the y's and the probabilities of the optimal solution for player R:

$$\boxed{P^* = (p_1, p_2, \ldots, p_k) = (vy_1, vy_2, \ldots, vy_k)}$$

In the example, then

$$P^* = (vy_1, vy_2, vy_3) = \left(\frac{11}{5} \cdot \frac{2}{11}, \frac{11}{5} \cdot \frac{3}{11}, \frac{11}{5} \cdot 0\right) = \left(\frac{2}{5}, \frac{3}{5}, 0\right)$$

Nonpositive Matrices

To extend the linear-programming technique we have developed to games for which the payoff matrix is not positive, we will use the example

$$G = \begin{bmatrix} 3 & -2 \\ -1 & 4 \end{bmatrix}$$

of the preceding section. We will add a number h to each entry of G to form a new matrix G^+, with h any positive number large enough so that G^+ is a positive matrix. For our illustration, we will let $h = 5$, so that

$$G^+ = \begin{bmatrix} 3+5 & -2+5 \\ -1+5 & 4+5 \end{bmatrix} = \begin{bmatrix} 8 & 3 \\ 4 & 9 \end{bmatrix}$$

Notice that if we let

$$H = \begin{bmatrix} 5 & 5 \\ 5 & 5 \end{bmatrix}$$

then we can describe the relationship between the original payoff matrix G and the positive matrix G^+ by the matrix equation

$$G^+ = G + H$$

Now let P and Q be any strategies for the players in a two-by-two game. If we represent P by a row matrix and Q by a column matrix, then the payoff from these strategies in the game with payoff matrix G is

$$E(P, Q) = PGQ$$

If the payoff matrix is G^+, we call the payoff for the strategies in this game $E^+(P, Q)$, where

$$E^+(P, Q) = PG^+Q$$

Writing G^+ as $G + H$, we use the distributive law of Section 1.2 twice to establish a simple relationship between $E(P, Q)$ and $E^+(P, Q)$:

$$E^+(P, Q) = PG^+Q = P(G + H)Q = (PG + PH)Q = PGQ + PHQ = E(P, Q) + PHQ$$

When we calculate PHQ for the matrix H of the example, we find that

$$PHQ = \begin{bmatrix} p & 1-p \end{bmatrix} \begin{bmatrix} 5 & 5 \\ 5 & 5 \end{bmatrix} \begin{bmatrix} q \\ 1-q \end{bmatrix} = \begin{bmatrix} p & 1-p \end{bmatrix} \begin{bmatrix} 5q + 5(1-q) \\ 5q + 5(1-q) \end{bmatrix}$$

$$= \begin{bmatrix} p & 1-p \end{bmatrix} \begin{bmatrix} 5q + 5 - 5q \\ 5q + 5 - 5q \end{bmatrix} = \begin{bmatrix} p & 1-p \end{bmatrix} \begin{bmatrix} 5 \\ 5 \end{bmatrix} = [p \cdot 5 + (1-p) \cdot 5] = [5] = 5$$

so that the value of PHQ is precisely the number 5, that we added to each entry of G to form G^+.

The fact that $PHQ = h$ follows from the structure of the matrices, P and Q, whose entries add to 1, and of H, in which all entries are equal. (See Exercise 21.) In general, suppose we let $G^+ = G + H$ for any payoff matrix G and any matrix H all of whose entries are the same number, h. For any strategies P and Q we have

$$E^+(P, Q) = E(P, Q) + h$$

Since all payoffs change by h from the game with payoff matrix G to the game with payoff matrix G^+, the security levels of all strategies change by exactly the same amount. Thus, suppose that P^* is the optimum strategy for player R in the game with matrix G. Because its security level $\underline{v}(P^*)$ is the maximum possible, in the game with matrix G^+ the security level of P^*, that is $\underline{v}(P^*) + h$, will still be the maximum, so P^* remains the optimum strategy. In exactly the same way, since all security levels for player C change by h, the optimum strategy Q^* of the game with matrix G is also the optimum strategy in the game with matrix G^+, but its security level has changed to $\overline{v}(Q^*) + h$.

Thus, the game with positive payoff matrix

$$G^+ = \begin{bmatrix} 8 & 3 \\ 4 & 9 \end{bmatrix}$$

has the same optimum strategies P^* and Q^* for the two players as the original game with payoff matrix

$$G = \begin{bmatrix} 3 & -2 \\ -1 & 4 \end{bmatrix}$$

However, the value of the game with payoff matrix G^+ is $h = 5$ units greater than that of the original game. Therefore, to solve the original game by linear programming, which we will do as Example 1, we use the positive matrix G^+ to find the optimum strategies and then subtract $h = 5$ from the value we obtain from the linear-programming solution in order to determine the value of the original game.

In general, to solve any matrix game with payoff matrix G, we let H be a matrix of the same size all of whose entries equal h, a positive number large enough so that $G^+ = G + H$ is a positive matrix. We solve the game with payoff matrix G^+ by the simplex method of linear programming, obtaining optimum strategies P^* and Q^* and the value v^+. The strategies P^* and Q^* are also the optimum strategies in the game with payoff matrix G and the value of that game is $v = v^+ - h$.

Example 1. Solve the game with payoff matrix

$$G = \begin{bmatrix} 3 & -2 \\ -1 & 4 \end{bmatrix}$$

by the simplex method of linear programming.

Solution

Using the positive matrix

$$G^+ = \begin{bmatrix} 8 & 3 \\ 4 & 9 \end{bmatrix}$$

we form the simplex tableau

$$\begin{array}{c|cccc|c} & x_1 & x_2 & x_3 & x_4 & \\ \hline x_3 & 8 & 3 & 1 & 0 & 1 \\ x_4 & 4 & 9 & 0 & 1 & 1 \\ \hline z & -1 & -1 & 0 & 0 & 0 \end{array}$$

and pivot in the a_{11}-location

$$\begin{array}{c|cccc|c} & x_1 & x_2 & x_3 & x_4 & \\ \hline x_1 & 8 & 3 & 1 & 0 & 1 \\ x_4 & 0 & 60 & -4 & 8 & 4 \\ \hline 8z & 0 & -5 & 1 & 0 & 1 \end{array}$$

When we pivot at the a_{22}-location, we obtain

$$\begin{array}{c|cccc|c} & x_1 & x_2 & x_3 & x_4 & \\ \hline x_1 & 480 & 0 & 72 & -24 & 48 \\ x_2 & 0 & 60 & -4 & 8 & 4 \\ \hline 480z & 0 & 0 & 40 & 40 & 80 \end{array}$$

The solution of the LP-problem is

$$x_1 = \frac{48}{480} = \frac{1}{10}, \quad x_2 = \frac{4}{60} = \frac{1}{15}, \quad z = \frac{80}{480} = \frac{1}{6}$$

The optimum strategy for player C is thus

$$Q^* = \left(6 \cdot \frac{1}{10}, \ 6 \cdot \frac{1}{15}\right) = \left(\frac{3}{5}, \frac{2}{5}\right)$$

The solution of the dual problem is

$$y_1 = \frac{40}{480} = \frac{1}{12}, \quad y_2 = \frac{40}{480} = \frac{1}{12}$$

and therefore the optimum strategy for player R is

$$P^* = \left(6 \cdot \frac{1}{12}, \ 6 \cdot \frac{1}{12}\right) = \left(\frac{1}{2}, \frac{1}{2}\right)$$

Since the value of the game with payoff matrix G^+ was $v^+ = 6$, the value of the given game is $v = v^+ - h = 6 - 5 = 1$. ∎

The solution in Example 1 is the same as the one we obtained in Section 6.4 by using the formulas for equalizing strategies in two-by-two games. The linear-programming method involves somewhat more work, but it can be used for matrices of any size. In particular, we now have the tools to solve the campaign-resource-distribution problem that we formulated as Example 2 of Section 6.1.

Example 2. Rond and Cali are the two candidates for president of their country, Islandia, and the election will be won by the candidate who captures the majority of the popular vote. For purposes of campaigning, Islandia is divided into three regions: East, Central, and West; each region has different social and economic characteristics, which can be expected to affect voting preferences. The candidates must decide how to distribute their campaign resources to win undecided votes. (We always assume that undecided voters will vote for one of the candidates.) If each candidate chooses to put all available resources into a single region, a political scientist predicts the behavior of the undecided voters as follows. He concludes that if Rond and Cali choose to put their resources in the same region, they will each receive the same number of undecided votes. If Rond puts all his resources in the East, he will receive the same number of undecided votes as will Cali if she puts all her resources in the Central region, but one million more of these votes than Cali will if she puts all her resources into the West. If Rond chooses the Central region, then he will obtain one million more undecided votes if Cali chooses to concentrate on the East; but if Cali chooses the West, then she will come out ahead by two million undecided votes. If Rond chooses the West, then he will lose two million votes to Cali if she chooses to concentrate on the East, but will gain one million votes if she chooses the Central region.

Use game theory to determine how the candidates should distribute their campaign funds and predict the outcome of the undecided vote.

Solution

In the solution to Example 2 of Section 6.1, we used the following table, in which the entries represented millions of undecided voters, with a positive number indicating Rond's advantage over Cali and a negative number the opposite.

		Cali		
		East	Central	West
Rond	East	0	0	1
	Central	1	0	−2
	West	−2	1	0

Table 6-11

Therefore, the payoff matrix is

$$G = \begin{bmatrix} 0 & 0 & 1 \\ 1 & 0 & -2 \\ -2 & 1 & 0 \end{bmatrix}$$

Since this is not a positive matrix, we must add a positive constant h to each entry. To make the arithmetic as simple as possible, we use the smallest whole number that

6.5 GAMES AND LINEAR PROGRAMMING

will make every entry positive, that is $h = 3$. Thus,

$$G^+ = \begin{bmatrix} 3 & 3 & 4 \\ 4 & 3 & 1 \\ 1 & 4 & 3 \end{bmatrix}$$

Starting with the simplex tableau

$$\begin{array}{c|cccccc|c} & x_1 & x_2 & x_3 & x_4 & x_5 & x_6 & \\ \hline x_4 & 3 & 3 & 4 & 1 & 0 & 0 & 1 \\ x_5 & 4 & 3 & 1 & 0 & 1 & 0 & 1 \\ x_6 & 1 & 4 & 3 & 0 & 0 & 1 & 1 \\ \hline z & -1 & -1 & -1 & 0 & 0 & 0 & 0 \end{array}$$

we pivot in the a_{21}-location

$$\begin{array}{c|cccccc|c} & x_1 & x_2 & x_3 & x_4 & x_5 & x_6 & \\ \hline x_4 & 0 & 3 & 13 & 4 & -3 & 0 & 1 \\ x_1 & 4 & 3 & 1 & 0 & 1 & 0 & 1 \\ x_6 & 0 & 13 & 11 & 0 & -1 & 4 & 3 \\ \hline 4z & 0 & -1 & -3 & 0 & 1 & 0 & 1 \end{array}$$

and then at the a_{13}-location

$$\begin{array}{c|cccccc|c} & x_1 & x_2 & x_3 & x_4 & x_5 & x_6 & \\ \hline x_3 & 0 & 3 & 13 & 4 & -3 & 0 & 1 \\ x_1 & 52 & 36 & 0 & -4 & 16 & 0 & 12 \\ x_6 & 0 & 136 & 0 & -44 & 20 & 52 & 28 \\ \hline 52z & 0 & -4 & 0 & 12 & 4 & 0 & 16 \end{array}$$

The numbers in the tableau are becoming uncomfortably large, so we divide each row except the first one by 4, to obtain the equivalent tableau

$$\begin{array}{c|cccccc|c} & x_1 & x_2 & x_3 & x_4 & x_5 & x_6 & \\ \hline x_3 & 0 & 3 & 13 & 4 & -3 & 0 & 1 \\ x_1 & 13 & 9 & 0 & -1 & 4 & 0 & 3 \\ x_6 & 0 & 34 & 0 & -11 & 5 & 13 & 7 \\ \hline 13z & 0 & -1 & 0 & 3 & 1 & 0 & 4 \end{array}$$

Even with this step, the final pivot, at the a_{32}-location, is most easily carried out with the aid of a calculator

$$\begin{array}{c|cccccc|c} & x_1 & x_2 & x_3 & x_4 & x_5 & x_6 & \\ \hline x_3 & 0 & 0 & 442 & 169 & 117 & -39 & 13 \\ x_1 & 442 & 0 & 0 & 65 & 91 & 117 & 39 \\ x_2 & 0 & 34 & 0 & -11 & 5 & 13 & 7 \\ \hline 442z & 0 & 0 & 0 & 91 & 39 & 13 & 143 \end{array}$$

Thus, the solution of the linear-programming problem is

$$x_1 = \frac{39}{442} = \frac{3}{34}, \quad x_2 = \frac{7}{34}, \quad x_3 = \frac{13}{442} = \frac{1}{34}, \quad z = \frac{143}{442} = \frac{11}{34}$$

and therefore, multiplying by the value $\frac{1}{z} = \frac{34}{11}$, the optimum strategy for Cali is

$$Q^* = \left(\frac{34}{11} \cdot \frac{3}{34}, \frac{34}{11} \cdot \frac{7}{34}, \frac{34}{11} \cdot \frac{1}{34}\right) = \left(\frac{3}{11}, \frac{7}{11}, \frac{1}{11}\right)$$

The solution of the dual problem is

$$y_1 = \frac{91}{442} = \frac{7}{34}, \quad y_2 = \frac{39}{442} = \frac{3}{34}, \quad y_3 = \frac{13}{442} = \frac{1}{34}$$

Again multiplying by $\frac{34}{11}$, we find the optimum strategy for Rond:

$$P^* = \left(\frac{7}{11}, \frac{3}{11}, \frac{1}{11}\right)$$

Noting that the *approximate* values of these fractions are

$$\frac{7}{11} = .64, \quad \frac{3}{11} = .27, \quad \text{and} \quad \frac{1}{11} = .09$$

the political scientist's advice to the candidates would be as follows. Rond should spend 64% of his campaign funds on the Eastern region of Islandia, 27% on the Central, and 9% on the Western region. Cali should spend 27% of her campaign funds on the Eastern region, 64% on the Central, and 9% on the Western region.

In order to predict the outcome of the undecided vote, we note that the value of the game with payoff matrix G^+ was $v^+ = \frac{34}{11}$. Since we added $h = 3$ to each entry in the original payoff matrix G, the value of the original game is

$$v = v^+ - 3 = \frac{34}{11} - 3 = \frac{1}{11}$$

The units in the matrix represented one million votes; since the value v is positive, the result favors Rond, who will receive $\frac{1}{11}$ million, that is, a little less than 91,000, more of the undecided votes than Cali will, if both make optimum use of their campaign funds. ∎

Summary

A matrix in which every number is positive is called a **positive matrix**. To solve a game with a positive $k \times m$-payoff matrix G, use the simplex algorithm to solve the LP-problem whose simplex tableau is

$$\begin{array}{c} \\ \{x\} \\ z \end{array} \begin{array}{c} \begin{array}{ccc} x_1 & \cdots & x_N \end{array} \\ \left[\begin{array}{c|c|c} G & I & B \\ \hline -C & 0 & 0 \end{array}\right] \end{array}$$

6.5 GAMES AND LINEAR PROGRAMMING 487

where the matrices B and C consist entirely of 1s. Let $x_1, x_2, \ldots, x_m$ be the solution of the LP-problem and let z be the minimum value of the objective function. The optimum strategy for player C is

$$Q^* = (q_1, q_2, \ldots, q_m) = (vx_1, vx_2, \ldots, vx_m)$$

where $v = \frac{1}{z}$ is the value of the game. Let $y_1, y_2, \ldots, y_k$ be the solution of the dual LP-problem. Then the optimum strategy for player R is

$$P^* = (p_1, p_2, \ldots, p_k) = (vy_1, vy_2, \ldots, vy_k)$$

To solve any matrix game with payoff matrix G, solve the game with payoff matrix $G^+ = G + H$, where all the entries of H equal h, a positive number large enough so that G^+ is a positive matrix. This yields optimum strategies P^* and Q^* and the value v^+. The optimum strategies of the game with matrix G are therefore P^* and Q^* and the value of this game is $v^+ - h$.

Exercises

In each of Exercises 1–12, if the game with the given payoff matrix is strictly determined, find the optimum pure strategies and the value of the game. If the game is not strictly determined, use the simplex algorithm of linear programming to solve it.

1. $\begin{bmatrix} 3 & 1 \\ 1 & 2 \end{bmatrix}$
2. $\begin{bmatrix} 0 & 2 \\ 2 & -1 \end{bmatrix}$
3. $\begin{bmatrix} 1 & 2 & 1 \\ 2 & 1 & 2 \end{bmatrix}$

4. $\begin{bmatrix} 0 & 0 & -1 & 0 \\ -1 & 0 & 0 & 0 \end{bmatrix}$
5. $\begin{bmatrix} 1 & -3 \\ -1 & 0 \end{bmatrix}$
6. $\begin{bmatrix} 2 & -1 \\ 0 & 1 \\ -1 & -2 \end{bmatrix}$

7. $\begin{bmatrix} 1 & 2 & 1 & 1 \\ 1 & 1 & 1 & 2 \\ 2 & 1 & 1 & 2 \end{bmatrix}$
8. $\begin{bmatrix} 0 & 1 & -1 \\ -1 & 0 & 2 \end{bmatrix}$
9. $\begin{bmatrix} -2 & 2 & 3 \\ -1 & 0 & -1 \\ -2 & -3 & 3 \end{bmatrix}$

10. $\begin{bmatrix} 1 & 0 & 1 & 1 \\ 0 & 1 & 1 & 0 \end{bmatrix}$
11. $\begin{bmatrix} \frac{1}{2} & 0 \\ -\frac{1}{2} & \frac{1}{2} \\ 0 & 0 \end{bmatrix}$
12. $\begin{bmatrix} 1 & 0 & -\frac{1}{2} \\ 1 & 1 & 0 \end{bmatrix}$

In each of Exercises 13–20, if the game is strictly determined, find the optimum pure strategies and the value of the game. If the game is not strictly determined, use the simplex algorithm of linear programming to solve it.

13. Amy and Bill each hold a playing card and, unseen by the other player, each may place the card either face up or face down under a cloth. When the cloths are removed, if both cards are face up or both face down, then Amy wins $1. If one

card is face up and the other face down, then Bill wins $1. Calculate the optimum strategies for Amy and Bill and the value of the game.

14. In planning a backpacking trip into the wilderness, a hiker has to decide whether to take along a heavy jacket. If the weather is warm, she estimates the unpleasantness of carrying the extra weight of the jacket will cause her to lose 1 point on her personal scale of happiness resulting from the trip. If the weather is cool, the greater comfort from wearing the jacket part of the time will just balance the bother of carrying it the rest of the time, with no effect on total happiness. However, if the weather is cold, the satisfaction of having made the correct decision together with the greater comfort will add 2 points to her happiness. Alternatively, if it is warm, leaving the jacket behind will please the hiker 1 point worth, and if it is cool, leaving the jacket will have no effect on her happiness. But in cold weather the shivering hiker will give up 3 points of happiness if she has no jacket. The hiker will use a spinner on a card to decide whether or not to take the jacket. She draws a circle with a center where the spinner rotates and the circle divided into two regions: take the jacket, do not take the jacket. What is the optimum design for the spinner, that is, what fraction of the area of the circle indicates that she should take the jacket? Also, by finding the value of the game, determine whether she is likely to enjoy the trip.

15. Jeff has a penny and a nickel and Matt has a penny, a nickel, and a dime. Each chooses a coin. If the sum of the values of the two coins chosen is an even number of cents, then Matt gives Jeff the coin which Matt chose. For example, if Jeff chose a penny and Matt chose a nickel, then the sum ($.06) is even, so Matt gives Jeff the nickel. If the sum is an odd number of cents, then Jeff gives Matt the coin Jeff chose. Calculate the optimum strategy for each player and the value of the game.

16. An investor has $10,000 to invest in the stock market. If he buys conservative stocks, then at the end of the year the expected value of the investment will be $8000 if the market prices decline, $10,000 if the market holds steady, and $12,000 if that year the stock market averages increase. However, if he invests in speculative stocks, then the expected amount of the investment is $3000 at the end of the year if the stock market declines, $9000 in a year of steady market values, and $16,000 if average stock prices increase. What is his optimum investment strategy? That is, how much money should he invest in conservative stocks and how much money should he invest in speculative stocks?

17. A quarterback must decide whether to call a running play or a pass. As the play begins, the opposing linebackers must decide whether it is a running play or a pass. A gain of yardage for the offense is a loss for the defense. Suppose that when the quarterback calls a running play, the average gain is only 2 yards if the linebackers expect a running play, but the average gain is 4 yards if the linebackers think the quarterback is about to pass. If the quarterback calls for a pass and the linebackers expect a running play, then, on the average, the outcome is an 8-yard gain for the offense. If the linebackers expect the quarterback's pass, the average result will be that the quarterback is thrown for a 3-yard loss. Calculate the optimum strategies for the offense and for the defense, and determine the expected outcome of a play by computing the value of this game.

18. A farmer must decide how much acreage to allot to potatoes, sugar beets, and barley. If there is little rain during the growing season, the potato crop will earn the farmer $20 an acre, the sugar-beet crop will bring in $40 an acre, and the barley crop will be worth $50 an acre. In a year of moderate rain, the potatoes will earn $200 an acre, the sugar beets will bring in $100 an acre, and the barley will be worth $50 an acre. In a year of heavy rain, the potato crop will earn $20 an acre, the sugar beets will bring in $150 an acre, and the barley crop will be worth $20 an acre. What is an optimum planting strategy for the farmer (which can be interpreted as the fraction of land devoted to each crop)?

19. In the children's game "paper, rock, scissors," each of the two players chooses one of these three words. The rules of the game are given by the chant "paper covers rock, rock breaks scissors, scissors cut paper." In other words if a player chooses paper and the other chooses rock, the player who chooses paper wins. Similarly, rock beats scissors and scissors beats paper. If both players choose the same word, the payoff is 0. Otherwise, the payoff to the winning player is 1 unit. Calculate the optimum strategy for each player and the value of the game.

20. In a military exercise, the soldiers are divided into two groups, the green army and the blue army. The green army must send reinforcements to a threatened city. There are three routes available and the green-army commander must choose one: over a mountain pass, through a river valley, or across a plain. The opposing blue army has just enough solders available to set up an ambush at one location. If the blue army fails to intercept the green-army reinforcements, the green army is awarded points and an equal number of points are taken from the blue army. Points are awarded to the green army depending on which route is used. The green army receives 1 point for choosing the mountain-pass route—provided the blue-army ambush was set up on another route. Under the same circumstances, the valley route and the plains route are each worth $\frac{1}{2}$ point to the green army. However, if the blue-army ambush is set up on either the mountain-pass or the river-valley route and if the green-army commander chooses the same route, then the blue army receives 1 point and 1 point is taken from the green army. On the other hand, if the ambush is set up on the plain by the blue army and if the green army chooses that route, the chances of a successful ambush in open country are so uncertain that neither army is awarded any points. Determine the optimum strategies for the commanders and calculate the value of the game to find out which side is likely to benefit.

21. Given the matrices

$$P = \begin{bmatrix} p_1 & p_2 & p_3 & p_4 \end{bmatrix} \quad \text{and} \quad Q = \begin{bmatrix} q_1 \\ q_2 \\ q_3 \end{bmatrix}$$

where $p_1 + p_2 + p_3 + p_4 = 1$ and $q_1 + q_2 + q_3 = 1$, let

$$H = \begin{bmatrix} h & h & h \\ h & h & h \\ h & h & h \\ h & h & h \end{bmatrix}$$

where h is any number. Show that $PHQ = h$.

6.6 Sensitivity Analysis of Games

The game with 2×3-payoff matrix

$$\begin{bmatrix} 1 & 2 & 5 \\ 3 & 4 & 4 \end{bmatrix}$$

has a saddle point at $g_{21} = 3$; so, as in Section 6.2, the optimum strategies for the players are for player R to choose row 2 and player C to choose column 1. Now suppose that the game is expanded to permit an additional pure strategy for player R, and that the payoffs are determined by this 3×3-matrix, in which the first two rows are the same as before:

$$\begin{bmatrix} 1 & 2 & 5 \\ 3 & 4 & 4 \\ 2 & 6 & 5 \end{bmatrix}$$

The payoff $g_{21} = 3$ is still the minimum of its row and the maximum of its column, so the optimum strategies are exactly as before and the value of the game is still 3. As this example illustrates, the expansion of the game by adding another row will not necessarily change the solution.

On the other hand, if we add

$$\begin{bmatrix} 7 & 6 & 5 \end{bmatrix}$$

as a third row to the original 2×3-game matrix, so that the payoffs are given by

$$\begin{bmatrix} 1 & 2 & 5 \\ 3 & 4 & 4 \\ 7 & 6 & 5 \end{bmatrix}$$

there is no longer a saddle point at g_{21} because 3 is no longer the maximum of its column. This game does have a saddle point, however, at $g_{33} = 5$. Thus, the solution of the game has changed entirely: player R should now choose row 3, the row we added, and player C should now choose column 3. The value of this game is 5.

Parametrized Games

The payoff matrices for the two different 3×3 games, that is,

$$\begin{bmatrix} 1 & 2 & 5 \\ 3 & 4 & 4 \\ 2 & 6 & 5 \end{bmatrix} \quad \text{and} \quad \begin{bmatrix} 1 & 2 & 5 \\ 3 & 4 & 4 \\ 7 & 6 & 5 \end{bmatrix}$$

are identical except at the g_{31}-location, yet they have entirely different solutions. These two games are part of a "family" of games with payoff matrices

6.6 SENSITIVITY ANALYSIS OF GAMES

$$\begin{bmatrix} 1 & 2 & 5 \\ 3 & 4 & 4 \\ t & 6 & 5 \end{bmatrix}$$

where t can be any number. (In the preceding discussion we considered $t = 2$ and $t = 7$.)

> A family of matrix games, identical except for a single payoff, which is allowed to take on any value, is called a **parametrized matrix game** and the undetermined payoff is called the **parameter** of the game.

In previous sections, the solution to a matrix game consisted of the optimum strategies P^* and Q^* for the players and the value v of the game. For each value of t, the **solution** of a parametrized game indicates the corresponding optimum strategies P_t^* and Q_t^*, as well as the value v_t.

To solve the given parametrized game for all values of t, and not just for $t = 2$ and $t = 7$, as in Section 6.2 we first find the row minima and column maxima for those rows and columns that do not contain the parameter t. As usual, we indicate row minima with boldface print and column maxima with asterisks:

$$\begin{bmatrix} 1 & 2 & 5^* \\ \mathbf{3} & 4 & 4 \\ t & 6^* & 5^* \end{bmatrix}$$

The column maximum in the first column depends on which is larger, 3 or t. If 3 is larger than t, then $g_{21} = 3$ is the maximum of its column as well as the minimum of its row, and therefore is still a saddle point. This is why the solution did not change from that of the original 2 × 3-game when $t = 2$. In general, then, if $t \leq 3$, the solution is "row 2, column 1," and the value is $v_t = 3$.

Let us then investigate what happens when $t \geq 3$, so that t is the maximum of the first column. The minimum value in the third row depends on the value of t. If t is not greater than 5, then t is the row minimum. But if $t \geq 5$, then $g_{33} = 5$ is the minimum of its row as well as the maximum of its column, and thus is a saddle point. We have seen this solution for the case $t = 7$. To complete our analysis, we suppose that t is between 3 and 5. Because it is not greater than 5, it follows that $g_{31} = t$ is the minimum in its row; but since it is at least 3, it is also the maximum of the first column. Thus, in this case we have a saddle point at $g_{31} = t$. The optimum strategies are given by the third row and first column, and the value of the game is now $v_t = t$.

We can summarize the solution to the parametrized matrix game with payoff matrix

492 GAME THEORY

$$\begin{bmatrix} 1 & 2 & 5 \\ 3 & 4 & 4 \\ t & 6 & 5 \end{bmatrix}$$

by means of Table 6-12:

Parameter	P_t^*	Q_t^*	v_t
$t \leq 3$	row 2	column 1	3
$3 \leq t \leq 5$	row 3	column 1	t
$t \geq 5$	row 3	column 3	5

Table 6-12

According to Table 6-12, the game has two saddle points at the values $t = 3$ and $t = 5$.

The most significant feature of the parametrized game just analyzed is that, although the solution depended on the parameter t, the game is of the same type for all t, that is, strictly determined. Therefore, the solution of the game depended on finding the location of the saddle point for all values of t. Example 1 demonstrates that in general, for parametrized games, the existence of a saddle point depends on the value of t.

Example 1. Determine the values of t for which the parametrized matrix game with payoff matrix

$$\begin{bmatrix} 1 & 2 \\ 4 & t \end{bmatrix}$$

has a saddle point, and solve the game for those values.

Solution

The minimum of the first row and maximum of the first column are indicated in the usual way:

$$\begin{bmatrix} 1 & 2 \\ 4^* & t \end{bmatrix}$$

If $t \geq 4$, then $g_{21} = 4$ is the minimum of its row as well as the maximum of its column, and is thus a saddle point. If $t \leq 4$, then t is the minimum of its row, but in order to be the maximum of its column, t must be at least 2. Thus, if $2 \leq t \leq 4$, then $g_{22} = t$ is a saddle point. However, if $t \leq 2$, then 1 and t are the row minima, whereas 2 and 4 are the column maxima. Consequently, there is no saddle point. We summarize this in Table 6-13. ∎

Parameter	P_t^*	Q_t^*	v_t
$t \leq 2$	no saddle point		
$2 \leq t \leq 4$	row 2	column 2	t
$t \geq 4$	row 2	column 1	4

Table 6-13

6.6 SENSITIVITY ANALYSIS OF GAMES

Since the game of Example 1 is 2×2, we may use formulas from Section 6.4 to solve the parametrized game for all values of t. Thus, if $t \leq 2$ in Example 1, then the value of the game is calculated to be

$$v_t = \frac{ad - bc}{(a-b)+(d-c)} = \frac{(1)(t)-(2)(4)}{(1-2)+(t-4)} = \frac{t-8}{t-5}$$

There are similar formulas for the optimal strategies P_t^* and Q_t^*. (See Exercises 5 and 6.)

A Nonstrictly Determined Game

As Example 1 of Section 6.5, we solved the game with payoff matrix

$$G = \begin{bmatrix} 3 & -2 \\ -1 & 4 \end{bmatrix}$$

by means of linear programming. We now suppose that the game is modified by adding a row containing a parameter, as follows:

$$G_t = \begin{bmatrix} 3 & -2 \\ -1 & 4 \\ t & -4 \end{bmatrix}$$

From the parametrized games we have discussed previously, we might expect that for some values of the parameter t, the solution would remain the same as the one we found in Section 6.5; but for other values the solution would be entirely different. We will show that this happens and determine the values of the parameter for which the solution stays the same as it was previously. We will do this by using an important technique from linear programming called *sensitivity analysis*.

Before we discuss the parametrized game, we briefly review how the game with payoff matrix G was solved in Section 6.5. First we added a constant $h = 5$ to each payoff to produce the positive matrix

$$G^+ = \begin{bmatrix} 8 & 3 \\ 4 & 9 \end{bmatrix}$$

Then we formed the simplex tableau

$$\begin{array}{c} \{x\} \\ z \end{array} \left[\begin{array}{c|c|c} \begin{array}{ccc} x_1 & \cdots & x_N \end{array} & & \\ \hline G^+ & I & B \\ \hline -C & 0 & 0 \end{array} \right] = \begin{array}{c} x_3 \\ x_4 \\ z \end{array} \left[\begin{array}{cccc|c} \overset{x_1}{8} & \overset{x_2}{3} & \overset{x_3}{1} & \overset{x_4}{0} & 1 \\ 4 & 9 & 0 & 1 & 1 \\ \hline -1 & -1 & 0 & 0 & 0 \end{array} \right]$$

We solved this LP-problem by the simplex algorithm, with the final tableau

494   GAME THEORY

$$\begin{array}{c} \begin{array}{cccc} x_1 & x_2 & x_3 & x_4 \end{array} \\ \begin{array}{c} x_1 \\ x_2 \\ 480z \end{array} \left[\begin{array}{cccc|c} 480 & 0 & 72 & -24 & 48 \\ 0 & 60 & -4 & 8 & 4 \\ 0 & 0 & 40 & 40 & 80 \end{array} \right] \end{array}$$

Since $z = \frac{80}{480} = \frac{1}{6}$, the value of the positive game was $v^+ = 6$ and therefore the solution to the original game was

$$P^* = \left(6 \cdot \frac{40}{480},\ 6 \cdot \frac{40}{480}\right) = \left(\frac{1}{2},\ \frac{1}{2}\right)$$

$$Q^* = \left(6 \cdot \frac{48}{480},\ 6 \cdot \frac{4}{60}\right) = \left(\frac{3}{5},\ \frac{2}{5}\right)$$

$$v = v^+ - h = 6 - 5 = 1$$

Returning to the parametrized game, we observe that if $t \leq -1$, then the row player would never choose the third row of the payoff matrix

$$G_t = \begin{bmatrix} 3 & -2 \\ -1 & 4 \\ t & -4 \end{bmatrix}$$

because no matter which column the opponent chooses, the payoff will be better in either of the other rows. In fact, even for a value of t as great as 3, the first row produces payoffs at least as desirable for the row player as row 3 does, no matter which column is chosen. We say that row 1 *dominates* row 3 if $t \leq 3$, so that row 3 would never be chosen. The solution of the modified game must therefore be the same as the solution of the original game, which would now be written as

$$P^* = \left(\frac{1}{2},\ \frac{1}{2},\ 0\right), \quad Q^* = \left(\frac{3}{5},\ \frac{2}{5}\right), \quad v = 1$$

If we add 5 to each element of G_t, with $t > 3$, then the matrix

$$G_t^+ = \begin{bmatrix} 8 & 3 \\ 4 & 9 \\ t+5 & 1 \end{bmatrix}$$

is positive and we can solve the game by the simplex algorithm, starting with the tableau

$$\begin{array}{c} \begin{array}{ccccc} x_1 & x_2 & x_3 & x_4 & x_5 \end{array} \\ \begin{array}{c} x_3 \\ x_4 \\ x_5 \\ z \end{array} \left[\begin{array}{ccccc|c} 8 & 3 & 1 & 0 & 0 & 1 \\ 4 & 9 & 0 & 1 & 0 & 1 \\ t+5 & 1 & 0 & 0 & 1 & 1 \\ -1 & -1 & 0 & 0 & 0 & 0 \end{array} \right] \end{array}$$

The solution would depend on the value of the parameter t. In this way, we can determine the values of t for which the solution is the same as for the original game.

Sensitivity Analysis

There is a technique of linear programming that offers a simpler and more efficient way to determine these values of t. Instead of carrying out the simplex algorithm starting with a new tableau, we can take advantage of the final tableau of the smaller game. This approach comes from a part of linear programming called "sensitivity analysis" that we will describe briefly at the end of this section. But first, we will discuss how sensitivity analysis deals with the parametrized game.

Instead of expanding the initial tableau of the simplex algorithm, we add the row with slack variable x_5 to the *final* tableau, including an additional column to accommodate this slack variable:

$$\begin{array}{c} \\ x_1 \\ x_2 \\ x_5 \\ 480z \end{array} \begin{bmatrix} x_1 & x_2 & x_3 & x_4 & x_5 & \\ 480 & 0 & 72 & -24 & 0 & 48 \\ 0 & 60 & -4 & 8 & 0 & 4 \\ t+5 & 1 & 0 & 0 & 1 & 1 \\ \hline 0 & 0 & 40 & 40 & 0 & 80 \end{bmatrix}$$

We must perform some additional arithmetic on the tableau so, before going further, we divide the rows by constants to reduce the size of the numbers. Thus, we divide the first row by 24, the second by 4, and the fourth by 40:

$$\begin{array}{c} \\ x_1 \\ x_2 \\ x_5 \\ 12z \end{array} \begin{bmatrix} x_1 & x_2 & x_3 & x_4 & x_5 & \\ 20 & 0 & 3 & -1 & 0 & 2 \\ 0 & 15 & -1 & 2 & 0 & 1 \\ t+5 & 1 & 0 & 0 & 1 & 1 \\ \hline 0 & 0 & 1 & 1 & 0 & 2 \end{bmatrix}$$

Although we have written the variables x_1, x_2, and x_5 to the left of the tableau as though this formed a basis for the matrix, it is not a basis in the sense of Section 2.4. For instance, the variable x_1 appears twice in the first column: in the first row, but also in the third row, the one we added to the final tableau of the original game. We can correct this defect by pivoting at the $a_{11} = 20$ location, as in Section 2.4. To begin the pivot, we replace $t + 5$ by 0, so that the first column has the required form. In place of 1 in the a_{32}-location, we have $20 \cdot 1 - 0 \cdot (t + 5) = 20$. Continuing this process, the parameter t appears in several locations. For instance, at the a_{33}-location, 0 is replaced by $20 \cdot 0 - 3 \cdot (t + 5) = -3t - 15$. We obtain

$$\begin{array}{c} \\ x_1 \\ x_2 \\ x_5 \\ 12z \end{array} \begin{bmatrix} x_1 & x_2 & x_3 & x_4 & x_5 & \\ 20 & 0 & 3 & -1 & 0 & 2 \\ 0 & 15 & -1 & 2 & 0 & 1 \\ 0 & 20 & -3t-15 & t+5 & 20 & -2t+10 \\ \hline 0 & 0 & 1 & 1 & 0 & 2 \end{bmatrix}$$

If we followed the rules of Section 2.4, each of the numbers in the second and fourth rows would be multiplied by the pivot element, 20. But we could then divide these rows by 20, so instead we do not change them at all. The rows that need not change when we pivot are those for which the entry in the pivot column is 0.

We still do not have a true basis because the variable x_2 appears twice in the second column. That can be corrected in the same way, by pivoting at the a_{22}-location. Thus, at the a_{33}-location we change $-3t - 15$ to

$$15(-3t - 15) - (-1)20 = -45t - 205$$

and so on:

$$\begin{array}{c} \\ x_1 \\ x_2 \\ x_5 \\ 12z \end{array} \begin{array}{c} x_1 \quad x_2 \quad x_3 \quad x_4 \quad x_5 \\ \left[\begin{array}{ccccc|c} 20 & 0 & 3 & -1 & 0 & 2 \\ 0 & 15 & -1 & 2 & 0 & 1 \\ 0 & 0 & -45t-205 & 15t+35 & 300 & -30t+130 \\ \hline 0 & 0 & 1 & 1 & 0 & 2 \end{array} \right] \end{array}$$

We can understand the meaning of this last tableau better if we choose some specific values of the parameter t. We first choose $t = 4$, with the result

$$\begin{array}{c} \\ x_1 \\ x_2 \\ x_5 \\ 12z \end{array} \begin{array}{c} x_1 \quad x_2 \quad x_3 \quad x_4 \quad x_5 \\ \left[\begin{array}{ccccc|c} 20 & 0 & 3 & -1 & 0 & 2 \\ 0 & 15 & -1 & 2 & 0 & 1 \\ 0 & 0 & -385 & 95 & 300 & 10 \\ \hline 0 & 0 & 1 & 1 & 0 & 2 \end{array} \right] \end{array}$$

Since the last row contains no negative numbers, we are at the end of the simplex algorithm. We observe that $z = \frac{2}{12} = \frac{1}{6}$. The solution of the dual problem is given by the elements in the last row beneath the slack variables, which are x_3, x_4, and x_5 in this tableau, and thus

$$P^* = \left(6 \cdot \frac{1}{12},\ 6 \cdot \frac{1}{12},\ 6 \cdot 0\right) = \left(\frac{1}{2},\ \frac{1}{2},\ 0\right)$$

The optimal strategy is $Q^* = \left(\frac{3}{5}, \frac{2}{5}\right)$, just as in the original game, so we see that for the parameter value $t = 4$, the solution does not change when the new row is added to the game.

However, if we choose $t = 5$, this is the final tableau:

$$\begin{array}{c} \\ x_1 \\ x_2 \\ x_5 \\ 12z \end{array} \begin{array}{c} x_1 \quad x_2 \quad x_3 \quad x_4 \quad x_5 \\ \left[\begin{array}{ccccc|c} 20 & 0 & 3 & -1 & 0 & 2 \\ 0 & 15 & -1 & 2 & 0 & 1 \\ 0 & 0 & -430 & 110 & 300 & -20 \\ \hline 0 & 0 & 1 & 1 & 0 & 2 \end{array} \right] \end{array}$$

The -20 in the last column tells us that the pivots have not put the tableau in canonical form (see Section 2.6). For the purposes of game theory, this means that the solution of the parametrized game is not determined by the original, smaller game when $t = 5$, as it was for $t = 4$. If we want to find out what the solution is when $t = 5$, it is not necessary to start the simplex algorithm from the beginning.

Instead, we could take advantage of the fact that there are still no negative numbers in the last row of the tableau, though there is a negative number in the last column. A variation of the simplex algorithm, called the "dual simplex algorithm" may be used in this setting. Although we will not discuss this algorithm, if we were to use it, the algorithm would require us to make just one more pivot. We would find that for the game with payoff matrix

$$G_5 = \begin{bmatrix} 3 & -2 \\ -1 & 4 \\ 5 & -4 \end{bmatrix}$$

the solution is entirely different:

$$P_5^* = \left(0, \frac{9}{14}, \frac{5}{14}\right), \quad Q_5^* = \left(\frac{4}{7}, \frac{3}{7}\right), \quad \text{and} \quad v_5 = \frac{8}{7}$$

From these examples, we see that whether or not the original solution is still valid for the expanded game depends on the expression $-30t + 130$ in the last row. If $-30t + 130 \geq 0$, then the solution has not changed. To find the corresponding values of t, we add $-30t$ to both sides of the inequality to obtain $130 \geq 30t$. We divide both sides of the inequality by the positive number 30, so the direction of the inequality does not change, and we find the answer to be

$$t \leq \frac{130}{30} = \frac{13}{3}$$

If $t > \frac{13}{3}$, then the last column will contain a negative number and the dual simplex algorithm could be used to calculate the new solution. On the other hand, if we only want to find the values of the parameter t for which the original solution is still valid, it is not necessary even to carry out the final pivot completely, as the next example illustrates.

Example 2.

(a) For the game with payoff matrix

$$G = \begin{bmatrix} 1 & -3 & 2 \\ 2 & 4 & -1 \end{bmatrix}$$

if $h = 4$ is added to every payoff to make the matrix positive and if the corresponding LP-problem is solved by the simplex algorithm, then the final tableau is

	x_1	x_2	x_3	x_4	x_5	
x_3	34	0	45	8	-1	7
x_2	7	15	0	-1	2	1
$9z$	2	0	0	1	1	2

Write the solution to the game with payoff matrix G.

(b) The game with matrix G is made a parametrized game by adding a row as follows:

498   GAME THEORY

$$G_t = \begin{bmatrix} 1 & -3 & 2 \\ 2 & 4 & -1 \\ -2 & t & 1 \end{bmatrix}$$

Show that if $t \leq -3$, then the solution of the game is the same as that of Part (a).

(c) Determine all values of the parameter t for which the solution of the game with payoff matrix G_t is the same as that of Part (a).

Solution

(a) Since $z = \frac{2}{9}$, we know that $v^+ = \frac{9}{2}$. Therefore, the solution is

$$P^* = \left(\frac{9}{2} \cdot \frac{1}{9}, \frac{9}{2} \cdot \frac{1}{9}\right) = \left(\frac{1}{2}, \frac{1}{2}\right)$$

$$Q^* = \left(\frac{9}{2} \cdot 0, \frac{9}{2} \cdot \frac{1}{15}, \frac{9}{2} \cdot \frac{7}{45}\right) = \left(0, \frac{3}{10}, \frac{7}{10}\right)$$

$$v = \frac{9}{2} - 4 = \frac{1}{2}$$

(b) If $t \leq -3$, then row 3 is dominated by row 1 in the sense that, no matter which column player C chooses, the payoff for player R is better for row 1 than for row 3. Therefore, no optimal strategy for player R would involve row 3. We conclude that the additional row will have no effect on the game and that the solution will be the same as for the original game.

(c) By Part (b) we may assume that $t > -3$, so that $t + 4 > 0$. Therefore, when we add $h = 4$ to the entries of G_t, the matrix is positive. We introduce the new row into the final tableau of Part (a) by means of an additional slack variable x_6:

	x_1	x_2	x_3	x_4	x_5	x_6	
x_3	34	0	45	8	-1	0	7
x_2	7	15	0	-1	2	0	1
x_6	2	$t+4$	5	0	0	1	1
$9z$	2	0	0	1	1	0	2

We pivot at 15 in the a_{22}-location, to obtain

	x_1	x_2	x_3	x_4	x_5	x_6	
x_3	34	0	45	8	-1	0	7
x_2	7	15	0	-1	2	0	1
x_6	$-7t+2$	0	75	$t+4$	$-2t-8$	15	$-t+11$
$9z$	2	0	0	1	1	0	2

Instead of carrying out a complete pivot at 45 in the a_{13}-location, we just calculate the change the pivot would make in the present entry, $-t + 11$, at the end of the third row. We find that it would become

$$45(-t + 11) - 7 \cdot 75 = -45t - 30$$

and we need to know when $-45t - 30 \geq 0$. Adding $45t$ to both sides of the inequality, we have $-30 > 45t$; and we may divide by the positive number 45, so that $t \leq -\frac{30}{45} = -\frac{2}{3}$. We conclude that if $t \leq -\frac{2}{3}$, then the optimal strategy for player R does not involve row 3 and therefore the solution is the same as that of Part (a). ∎

Dominance

To use linear programming to analyze the parametrized game, we had to know that the matrix became positive when we added h to every entry. Therefore, it was necessary to restrict the values of the parameter t so that $t + h$ would be positive. We accomplished this in the examples by restricting the parametrized games so that if $t + h$ is not positive, the added row was *dominated* by another row.

> In general, row i of a payoff matrix is **dominated** by row i' if, in each column, the payoff to player R is at least as good in row i' as it is in row i. In symbols, row i' dominates row i if $g_{i'j} \geq g_{ij}$ for all j.

If a row is dominated by another row, then the dominated row will not be used in an optimal strategy for player R.

Sensitivity Analysis in Linear Programming

Sensitivity analysis is the name given to a collection of techniques of linear programming that are carried out after an LP-problem has been solved by the simplex algorithm. We described a part of this subject in Section 2.6 when we found marginal values.

In this section, the part of sensitivity analysis we have employed is the technique that is used when an additional constraint is added to an LP-problem after the solution has been reached. When we apply linear programming to game theory, the addition of a new row to the payoff matrix creates an additional constraint in the corresponding LP-problem. This sensitivity-analysis technique is often helpful in complex applications when a linear-programming formulation of a problem produces an impractical solution and an investigation reveals that the source of the difficulty is that there are restrictions that had not been included in the formulation. Rather than start the simplex algorithm all over again, which can take considerable computer time for a large problem, the new constraints are incorporated into the final tableau that has already been calculated. A relatively small number of pivots then puts the problem into a form where it may be solved quite efficiently by the dual simplex algorithm.

GAME THEORY

Other aspects of sensitivity analysis deal with problems in which the numbers used are only estimates. For instance, in the *Farm-Management* problem of Section 2.7, it is impossible to say with certainty that the farmer's profit per acre of cotton will be exactly $207 when the crop is harvested some time in the future. Sensitivity analysis investigates how the planting schedule for the most profitable use of the farmer's land would be affected by greater or lesser profit per acre for that crop. In other words, the analysis would determine how sensitive the solution to the problem is to estimates of the numbers used in it, and that is how this collection of techniques got its name.

Summary

A family of matrix games, identical except for a single payoff, which is allowed to take on any value, is called a **parametrized matrix game** and the undetermined payoff is called the **parameter** of the game. The **solution** of a parametrized game states, for each value of t, the corresponding optimum strategies P_t^* and Q_t^*, and the value v_t. Row i of a payoff matrix is **dominated** by row i' if, in each column, the payoff to player R is at least as good in row i' as it is in row i: that is, if $g_{i'j} \geq g_{ij}$ for all j. If a row is dominated by another row, then the dominated row will not be used in an optimal strategy for player R. If a parametrized row is added to the payoff matrix of a game that has been solved by the simplex algorithm of linear programming, a technique from sensitivity analysis can be used to determine the values of the parameter for which the expanded game has the same solutions as the original game. A row corresponding to the new row of the payoff matrix is added to the final simplex tableau. After pivoting so that each basis variable appears in just one row of the column headed by that variable, the values of t for which the solution is unchanged are those for which the number in the added row and last column is nonnegative.

Exercises

In Exercises 1 through 4, determine the values of t for which the parametrized matrix game with the given payoff matrix has a saddle point and solve the game for these values.

1. $\begin{bmatrix} -1 & 2 & 1 \\ t & 3 & 4 \end{bmatrix}$

2. $\begin{bmatrix} 2 & -1 \\ 2 & 1 \\ 4 & t \end{bmatrix}$

3. $\begin{bmatrix} 2 & 1 & 3 \\ -1 & t & -2 \end{bmatrix}$

4. $\begin{bmatrix} 0 & 2 & 4 \\ -2 & -1 & 3 \\ t & 1 & 2 \end{bmatrix}$

6.6 SENSITIVITY ANALYSIS OF GAMES

In Exercises 5 and 6, solve the parametrized matrix game for all values of the parameter t.

5. $\begin{bmatrix} 2 & -1 \\ t & 3 \end{bmatrix}$

6. $\begin{bmatrix} 5 & -1 \\ 0 & t \end{bmatrix}$

7. (a) For the game with payoff matrix

$$G = \begin{bmatrix} 2 & 3 & 1 \\ 3 & 1 & 4 \end{bmatrix}$$

the corresponding LP-problem is solved by the simplex algorithm. The final tableau is

	x_1	x_2	x_3	x_4	x_5	
x_2	5	11	0	4	-1	3
x_3	7	0	11	-1	3	2
$11z$	1	0	0	3	2	5

Write the solution of the game with payoff matrix G.

(b) The game with matrix G is made a parametrized game by adding a row as follows:

$$G_t = \begin{bmatrix} 2 & 3 & 1 \\ 3 & 1 & 4 \\ 1 & 1 & t \end{bmatrix}$$

Show that row 3 is dominated by row 1 if $t \leq 1$.

(c) Determine all values of the parameter t for which the solution of the game with payoff matrix G_t is the same as that of Part (a).

8. (a) For the game with payoff matrix

$$G = \begin{bmatrix} -1 & -2 & 1 \\ 2 & 0 & -1 \end{bmatrix}$$

if $h = 3$ is added to every payoff to make the matrix positive and if the corresponding LP-problem is solved by the simplex algorithm, then the final tableau is

	x_1	x_2	x_3	x_4	x_5	
x_3	1	0	10	3	-1	2
x_2	8	5	0	-1	2	1
$10z$	7	0	0	1	3	4

Write the solution of the game with payoff matrix G.

(b) The game with matrix G is made a parametrized game by adding a row as follows:

$$G_t = \begin{bmatrix} -1 & -2 & 1 \\ 2 & 0 & -1 \\ -1 & t & 0 \end{bmatrix}$$

502 GAME THEORY

Show that row 3 is dominated by row 1 if $t \leq -2$.

(c) Determine all values of the parameter t for which the solution of the game with payoff matrix G_t is the same as that of Part (a).

9. (a) For the game with payoff matrix

$$G = \begin{bmatrix} -4 & 2 & -3 & 1 \\ 1 & -1 & 0 & -3 \end{bmatrix}$$

if $h = 5$ is added to every payoff to make the matrix positive and if the corresponding LP-problem is solved by the simplex algorithm, then the final tableau is

	x_1	x_2	x_3	x_4	x_5	x_6	
x_4	-7	27	0	26	5	2	3
x_3	17	5	13	0	-1	3	2
$26z$	0	11	0	0	3	4	7

Write the solution of the game with payoff matrix G.

(b) The game with matrix G is made a parametrized game by adding a row as follows:

$$G_t = \begin{bmatrix} -4 & 2 & -3 & 1 \\ 1 & -1 & 0 & -3 \\ 0 & -1 & -1 & t \end{bmatrix}$$

Show that row 3 is dominated by row 2 if $t \leq -3$.

(c) Determine all values of the parameter t for which the solution of the game with payoff matrix G_t is the same as that of Part (a).

10. (a) For the game with payoff matrix

$$G = \begin{bmatrix} 1 & 2 \\ 0 & 1 \\ 3 & 0 \end{bmatrix}$$

if $h = 1$ is added to every payoff to make the matrix positive and if the corresponding LP-problem is solved by the simplex algorithm, then the final tableau is

	x_1	x_2	x_3	x_4	x_5	
x_2	0	5	2	0	-1	1
x_4	0	0	-7	10	1	4
x_1	10	0	-1	0	3	2
$10z$	0	0	3	0	1	4

Write the solution of the game with payoff matrix G.

(b) The game with matrix G is made a parametrized game by adding a row as follows:

$$G_t = \begin{bmatrix} 1 & 2 \\ 0 & 1 \\ 3 & 0 \\ t & 2 \end{bmatrix}$$

Show that row 4 is dominated by row 1 if $t \leq 1$.

(c) Determine all values of the parameter t for which the solution of the game with payoff matrix G_t is the same as that of Part (a).

Review Exercises for Chapter 6

In Exercises 1 through 6, solve the game with the given payoff matrix.

1. $\begin{bmatrix} -4 & -1 \\ 1 & -2 \end{bmatrix}$
2. $\begin{bmatrix} 1 & 3 \\ 2 & 2 \end{bmatrix}$
3. $\begin{bmatrix} -1 & 0 & -2 & 2 \\ 1 & -2 & 0 & 1 \end{bmatrix}$

4. $\begin{bmatrix} 0 & 1 \\ -1 & 0 \\ 2 & -1 \end{bmatrix}$
5. $\begin{bmatrix} -2 & -3 & 1 \\ 2 & -1 & -1 \\ -3 & -2 & 0 \end{bmatrix}$
6. $\begin{bmatrix} 2 & 1 & 1 \\ 1 & 2 & 1 \\ 1 & 2 & 2 \end{bmatrix}$

In Exercises 7 and 8, calculate the security level of the given strategy for the game with the given payoff matrix G.

7.
$$G = \begin{bmatrix} -1 & 4 & 0 & 1 \\ 2 & 1 & 2 & 1 \\ 1 & 2 & -1 & 1 \end{bmatrix} \quad \text{and} \quad Q = \begin{bmatrix} \frac{1}{4} \\ \frac{5}{8} \\ 0 \\ \frac{1}{8} \end{bmatrix}$$

8.
$$G = \begin{bmatrix} 2 & 1 \\ -\frac{1}{2} & 0 \\ 1 & \frac{1}{2} \end{bmatrix} \quad \text{and} \quad P = \begin{bmatrix} \frac{1}{6} & \frac{1}{2} & \frac{1}{3} \end{bmatrix}$$

9. The Major Manufacturing Company makes three models of pop-up toaster: deluxe, regular, and economy. Major Manufacturing's planners estimate the value of each model to the company in the following year by assigning profitability points as follows: If the economy is strong, the deluxe model will be worth 3 points to the company, the regular model 2 points, and the economy model 1 point. If the economy is weak, the deluxe model is valued at 0 points, the regular model is

worth 2 points, and the economy model gets 5 points. What fraction of its production should Major Manufacturing devote to each model of toaster?

10. Smugglers have a choice of two routes to bring a shipment of jewels into a country without paying duty on them: over a mountain pass or along a river. The customs department of the country has only enough officers to watch one route. If the smugglers and the customs officers choose the same route, the smugglers will be caught and the customs department will gain a total of $500,000 by impounding the jewels and reselling them. On the other hand, if the customs officers choose one route and the smugglers the other, then the smugglers make a profit of $250,000 through the sale of the duty-free jewels. Determine optimum strategies for both the smugglers and the customs officers.

11. Mr. Allen and Mr. Baker are the two candidates for Mayor of Greenwood. On the ballot there will also be a referendum for a local sales tax to finance the redevelopment of the downtown area. A poll indicates that among the undecided voters, if Mr. Allen supports the sales tax, he will gain 12,000 votes provided that Baker also supports the tax, he will gain 2000 votes if Baker is neutral on the tax, and he will lose 2000 votes to Baker if Baker opposes the sales tax. Assuming Mr. Allen is neutral on the sales tax, he will gain 3000 votes over Baker if Baker supports the sales tax, he will lose 5000 votes to Baker if Baker is also neutral on the tax, and he will lose 4000 votes to Baker if Baker opposes the tax. And if Mr. Allen opposes the tax, neither candidate will have an advantage over the other provided that Baker supports the sales tax, Allen will lose 8000 votes to Baker if Baker is neutral on the issue, and he will lose 2000 votes to Baker if Baker also opposes the sales tax. Determine the optimum strategy for each candidate and the outcome of the undecided vote if both candidates use their optimum strategies.

12. The town of Centerville has two pizza restaurants, Antonio's and Maria's. A marketing consultant tells Antonio that if he cuts his prices, then he will get 80% of Centerville's pizza business if Maria charges full price, but only 40% if Maria also cuts prices. If Antonio charges full price, then his restaurant will get 50% of Centerville's pizza business provided that Maria also charges full price, but only 30% if Maria cuts prices. Determine the optimum strategies for Antonio and Maria and the percentage of the pizza business of Centerville that Antonio will have if both use their optimum strategies.

13. Three distinct strains of flu, types A, B, and C, are threatening the population of Valley City. Two types of vaccine, I and II, are available to inoculate the people of Valley City. If a person is inoculated with only type-I vaccine, the probability of contracting type-A flu is $\frac{1}{10}$, the probability of type B is $\frac{2}{10}$, and the probability of type C is $\frac{3}{10}$. If a person is inoculated with only type-II vaccine, the probability of contracting type-A flu is $\frac{2}{10}$, the probability of type B is $\frac{1}{10}$, and the probability of type C is $\frac{1}{10}$. Use game theory to determine the optimum mixture of the two types of vaccine that should be used. (*Hint*: Flu is the row player.)

14. In a basketball game Jill has the ball within shooting distance of the basket and she must decide whether to shoot or pass to a teammate. Alice, the opposing player guarding Jill, must decide whether to concentrate on defending against a shot or against a pass. If Jill shoots and Alice defends against the shot, there is a probability of .30 that Jill will score a basket, but if Alice defends against a pass, the probability of a basket is .70. If Jill passes to a teammate and Alice is concentrating on defending against a shot, the probability is .90 that Jill's pass will result in a basket whereas if Alice defends against the pass, the probability that the pass will result in a score is .60. Determine the optimum strategies for Jill and Alice.

15. Solve the parametrized game for all values of the parameter t, where the payoff matrix is given by
$$G_t = \begin{bmatrix} -1 & 1 & 3 \\ 0 & 0 & 2 \\ t & 1 & 3 \end{bmatrix}$$

16. (a) For the game with payoff matrix
$$G = \begin{bmatrix} 1 & 1 & 0 \\ 2 & 0 & 1 \end{bmatrix}$$
if $h = 1$ is added to every payoff to make the matrix positive and if the corresponding LP-problem is solved by the simplex algorithm, then the final tableau is

$$\begin{array}{c|ccccc|c} & x_1 & x_2 & x_3 & x_4 & x_5 & \\ \hline x_2 & 1 & 3 & 0 & 2 & -1 & 1 \\ x_3 & 4 & 0 & 3 & -1 & 2 & 1 \\ \hline 3z & 2 & 0 & 0 & 1 & 1 & 2 \end{array}$$

Write the solution of the game with payoff matrix G.

(b) The game with matrix G is made a parametrized game by adding a row as follows:
$$G_t = \begin{bmatrix} 1 & 1 & 0 \\ 2 & 0 & 1 \\ 0 & 0 & t \end{bmatrix}$$
Show that row 3 is dominated by row 2 if $t \leq 1$.

(c) Determine all values of the parameter t for which the solution of the game with payoff matrix G_t is the same as that of Part (a).

7

Graphs and Networks

A Trading Network

For many years, anthropologists have studied the trading network known as the Kula ring. This is a system of gift exchanges carried on among East Papua Melanesian tribal groups that inhabit islands to the southeast of Papua, New Guinea. The Kula gift-exchange system developed among these tribal societies as a way of creating friendly political alliances that, in turn, encouraged intertribal trade.

The tribal groups are mutually dependent on each other's agricultural and manufactured products. For instance, the Trobriand Island group produces yams, taro, sweet potatoes, bananas, and coconuts for trade and also manufactures wooden bowls, combs, and baskets. The Amphlett Islands and Tubetube groups export pots. Northwest Dobu provides sago, Gawa makes canoes, Woodlark produces Kula armshells and canoes, and so on.

In a study published in 1983, Bernard Grofman and Janet Landa[†] analyzed this trade, using graph theory as a tool to investigate the underlying market structure of the Kula ring. They found that although it would seem most convenient for an island to trade only with its nearest neighbors, the variety of products involved and the complexity

---

[†]Bernard Grofman and Janet Landa, "The Development of Trading Networks among Spacially Separated Trades as a Process of Proto-Coalition Formation: The Kula Trade," *Social Networks* 5 (1983), pp. 347-365.

of the trading relationships made a more elaborate trading network necessary.

In this chapter, we will not attempt to reproduce the content of the Grofman-Landa paper, but we will demonstrate how graphs can be used to model such trading relationships. We will also develop matrix techniques that will permit us to study some of the aspects of the structure of this trading system.

7.1 Graphs

Figure 7-1 is a map of the Trobriand Island area of the Pacific Ocean. The lines on the map present an example of a system of trade routes among some of these islands. If we replace the islands by dots to simplify the picture and clarify the relationship among the trade routes, we obtain the graph of the trade-route information shown in Figure 7-2.

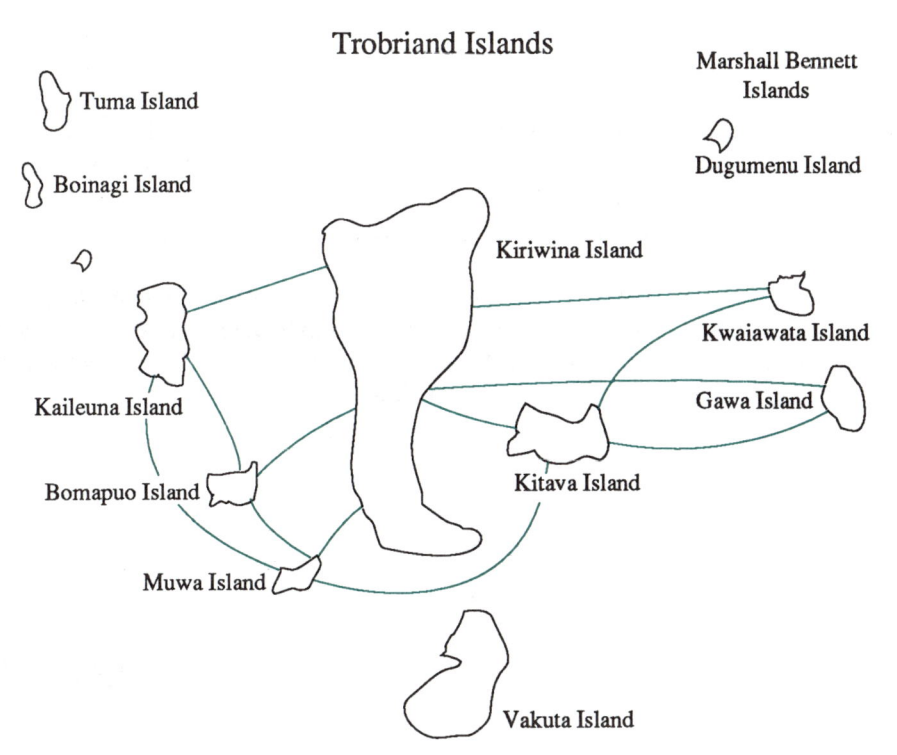

Figure 7-1

508 GRAPHS AND NETWORKS

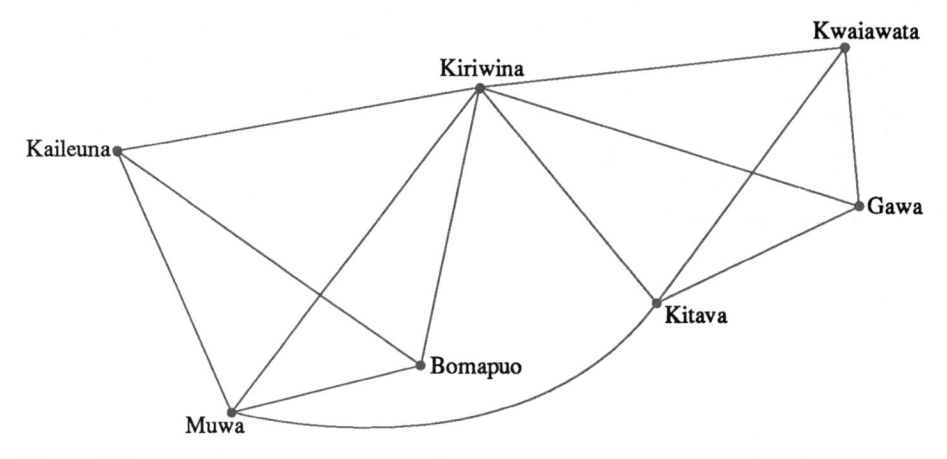

Figure 7-2

Thus, when we use the word "graph" in this chapter we do not mean it in the sense of Section 1.6, where we graphed linear functions. Rather, a graph consists of a set of points (to be called *vertices*) together with lines, straight or curved, (to be called *edges*) that join the points. Figure 7-3 illustrates some general features of graphs.

Vertices and Edges

> A **graph** consists of a set V, the **vertices** of the graph, and a set E, the **edges** of the graph. Each edge e is associated with a set of two (not necessarily different) vertices, v and v'.

We write either $e = \{v, v'\}$ or $e = \{v', v\}$ since the order in which we list the vertices does not matter. An edge $e = \{v, v'\}$ is said to be **incident** on v and v' and we visualize e as a line connecting these vertices, as in Figure 7-3. If $e = \{v, v'\}$, we also say that the vertices v and v' are **incident** on e.

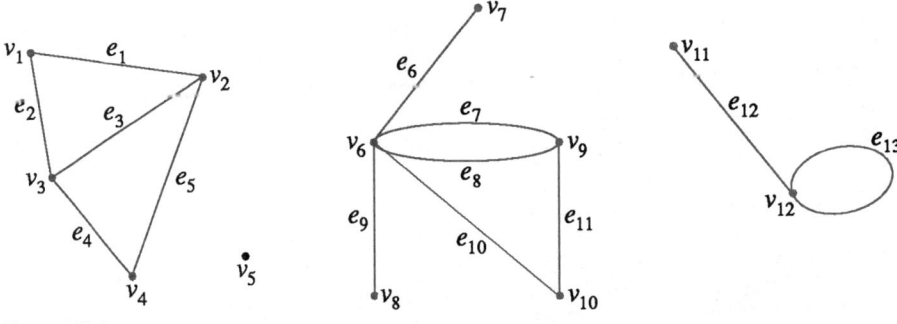

Figure 7-3

An edge $e = \{v, v\}$ that connects a vertex with itself is called a **loop**. Thus, in Figure 7-3, $e_{13} = \{v_{12}, v_{12}\}$ is a loop. Two edges may connect the same two vertices, in which case they are called **parallel edges**, as is illustrated by the edges e_7 and e_8 incident on v_6 and v_9 in Figure 7-3. Not every vertex of a graph must have edges incident on it; for instance, in Figure 7-3, there is no edge incident on vertex v_5.

Example 1. For the graph given in Figure 7-4:
  (a) Determine the vertex set, V.
  (b) Determine the edge set, E, indicating the vertices incident on each edge.
  (c) List the loops.
  (d) Find any parallel edges.

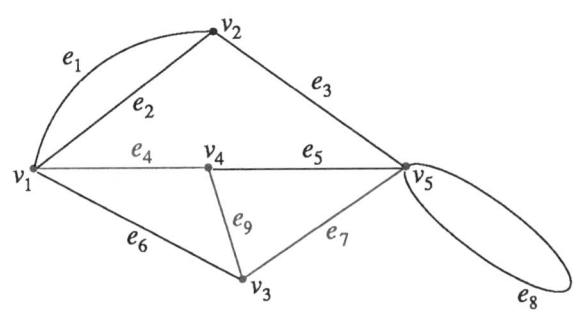

Figure 7-4

Solution

(a) The vertex set $V = \{v_1, v_2, v_3, v_4, v_5\}$.
(b) The edge set $E = \{e_1, e_2, e_3, e_4, e_5, e_6, e_7, e_8, e_9\}$; here $e_1 = \{v_1, v_2\}$, $e_2 = \{v_1, v_2\}$, $e_3 = \{v_2, v_5\}$, $e_4 = \{v_1, v_4\}$, $e_5 = \{v_4, v_5\}$, $e_6 = \{v_1, v_3\}$, $e_7 = \{v_3, v_5\}$, $e_8 = \{v_5, v_5\}$, and $e_9 = \{v_3, v_4\}$.
(c) There is one loop, e_8.
(d) The parallel edges are e_1 and e_2. ∎

Connected Graphs

The graph of Figure 7-4 differs from that of Figure 7-3 in an important way. In Figure 7-3, the graph is *disconnected*, that is, it consists of several pieces, whereas the graph of Figure 7-4 is *connected*. The precise definition of a connected graph requires the concept of a path between two vertices.

> A **path** between the vertices v and v' is a sequence of edges, written as $e_1 e_2 \ldots e_r$, which is such that v is incident on e_1, v' is incident on e_r, and successive edges e_i and e_{i+1} meet at a vertex, that is, they are incident on the same vertex.

510   GRAPHS AND NETWORKS

For instance, in Figure 7-4, $e_4e_9e_7$ is a path between v_1 and v_5; here e_4 and e_9 meet at v_4, and e_9 and e_7 meet at v_3. Notice that there can be several paths between the same two vertices. In Figure 7-4, e_2e_3 is also a path between v_1 and v_5.

> A graph is **connected** if there is a path between any two of its vertices, and it is **disconnected** otherwise.

For example, in the disconnected graph of Figure 7-3, there is no path between the vertices v_{10} and v_{11}.

We have seen that given a graph we can identify the vertex set V, the edge set E, and we can determine which vertices are incident on each edge. Reversing the process, if we have a vertex set V and an edge set E, and if we know which vertices are incident on each edge, we can then draw the graph, as Example 2 illustrates.

Example 2. Draw a graph with vertex set $V = \{v_1, v_2, v_3, v_4, v_5, v_6, v_7\}$ and edge set $E = \{e_1, e_2, e_3, e_4, e_5, e_6, e_7, e_8\}$, given that $e_1 = \{v_1, v_2\}$, $e_2 = \{v_1, v_3\}$, $e_3 = \{v_1, v_4\}$, $e_4 = \{v_2, v_4\}$, $e_5 = \{v_3, v_4\}$, $e_6 = \{v_3, v_5\}$, $e_7 = \{v_6, v_6\}$, and $e_8 = \{v_6, v_7\}$.

Solution

We represent the vertices by seven dots and then connect the dots according to the information about the incidence relations. We may allow the edges to cross each other, provided that we understand that the point of crossing does not represent a vertex when there is no label on it. The graph is shown in Figure 7-5.   ■

We could have drawn Figure 7-5 in such a way that the edges met only at vertices, rather than crossed each other, but there is no need to do so. Further-

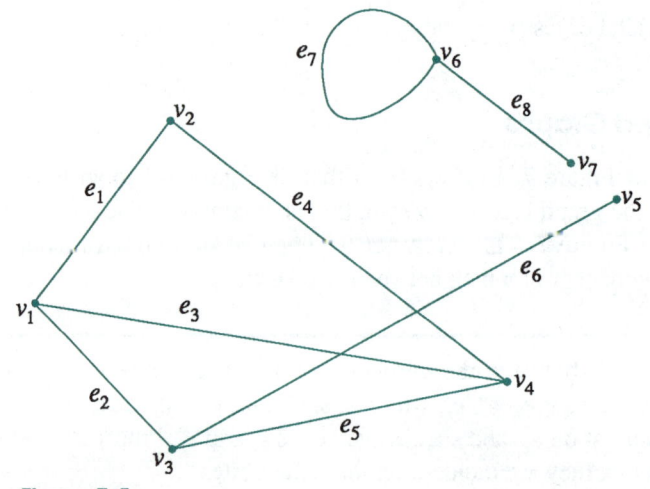

Figure 7-5

more, there are graphs, called "nonplanar graphs," which cannot be drawn without permitting the edges to cross.

Graphs may be used to describe social relationships, organizational structures, and interpersonal communication, as Example 3 indicates.

Example 3. Draw a graph that indicates which of the following boards of directors have members in common:

United Health Drive: West, Goldberg, Lopez, Manning, Johnson
Youth Charity Board: Johnson, Sanchez, Friedman, West, Chang
Third National Bank: West, Manning, Colman, Richards, Strauss, Izumi
American Savings Bank: Goldberg, Johnson, Alsop, Jones, Davis
Friends of Cats: Richards, Sanchez, Crosley, Williams, Anderson
Save the Planet: Johnson, Colman, Alsop, Cohen

Solution

The vertices of the graph represent the organizations; we use the first initials of each organization. Thus, $V = \{U, Y, T, A, F, S\}$. We connect two vertices by an edge if there is a person on the board of both organizations represented by these vertices. Directors of the United Health Drive sit on the boards of Youth Charity Drive (West and Johnson), Third National Bank (West), American Savings (Goldberg and Johnson), Save the Planet (Johnson). Therefore, the graph will require edges $e_1 = \{U, Y\}$, $e_2 = \{U, T\}$, $e_3 = \{U, A\}$, and $e_4 = \{U, S\}$. Continuing in the same way, the remaining edges are: $e_5 = \{Y, T\}$, $e_6 = \{Y, A\}$, $e_7 = \{Y, F\}$, $e_8 = \{Y, S\}$, $e_9 = \{T, F\}$, and $e_{10} = \{A, S\}$. We then draw the graph in Figure 7-6. ∎

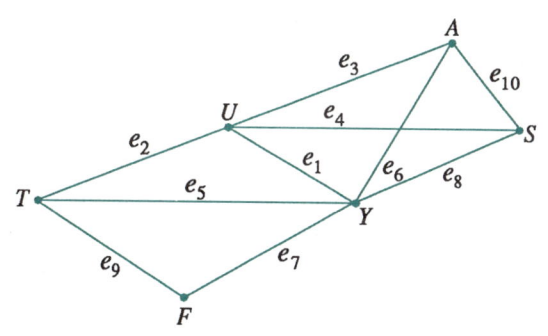

Figure 7-6

A Snowplow Problem

One natural use for graphs is to describe a network of roads. Figure 7-7 shows the street system of the village of Hillsboro.

The street system is represented by the graph in Figure 7-8, where each vertex in $V = \{v_1, v_2, v_3, v_4, v_5, v_6, v_7, v_8, v_9, v_{10}\}$ corresponds to a street intersection and the edges represent the streets connecting these intersections.

512 GRAPHS AND NETWORKS

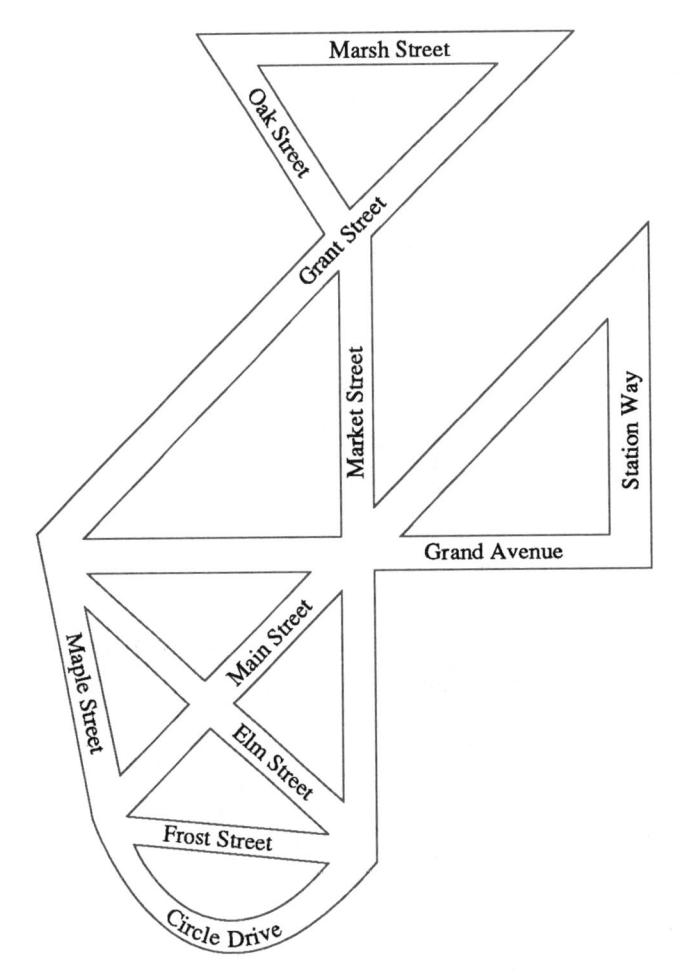

Figure 7-7

Hillsboro owns one snow plow, which is stored near the intersection of Main Street and Station Way, represented by the vertex v_6. The Hillsboro maintenance department wants to find an efficient route for the plow to clear the streets of snow, that is, a route that starts and ends at v_6 and travels along each street exactly once.

> A route through a graph that traverses each edge exactly once is called an **Euler circuit**.

This is named for the Swiss mathematician, Leonard Euler. We next turn to the problem that motivated Euler's interest in this subject. Later in the section, we will apply Euler's method to construct Hillsboro's snow-plow route.

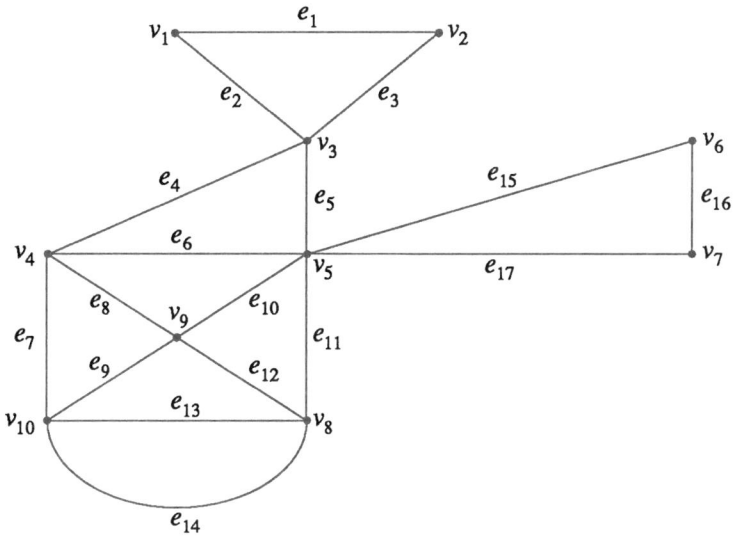

Figure 7-8

The Königsberg Bridge Problem

The problem that Euler solved in 1736 is called the *Königsberg Bridge Problem*. The town of Königsberg (now Kaliningrad in Lithuania) is crossed by the Pregel River, and there are two islands in the river at Königsberg. In the 18th century, the islands were connected to each other by one bridge, the larger island was connected to each bank by two bridges, and the smaller island was connected to each bank by one bridge—for a total of seven bridges. See Figure 7-9.

The people of Königsberg, who enjoyed Sunday walks around their town, wondered if it would be possible to begin at one of the banks or islands, walk across each bridge exactly once, and return to the starting point. The problem is easier to see if we turn it into the graph of Figure 7-10. Here, A is the smaller island, B and C are the river banks and D is the larger island.

Can we start at any one of the vertices A, B, C, or D and, tracing each edge exactly once, return to the starting vertex? Euler answered the question by solving

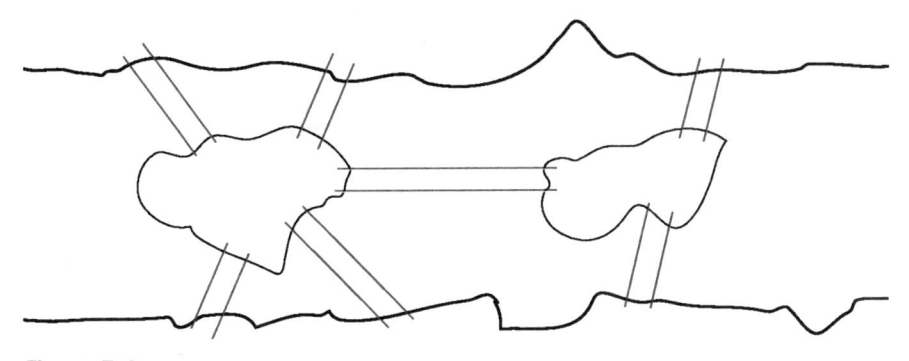

Figure 7-9

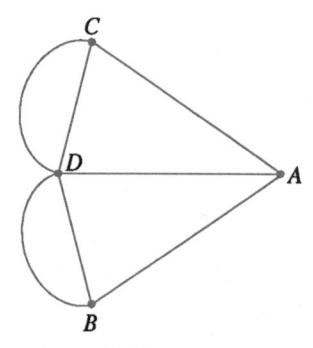

Figure 7-10

a general problem that is called the *Euler Circuit Problem*. A **circuit** is a path in which no edge appears more than once and that starts and ends at the same vertex v. Euler's solution depends on the concept of the **degree** of a vertex v, which is the number of edges incident on v. If there is a loop at the vertex, it raises the degree by 2. In Figure 7-10 we can see that vertex A has 3 edges incident on it and therefore is of degree 3. The vertices B and C are also of degree 3 and D is of degree 5. Euler determined which graphs have an **Euler circuit**, that is, a circuit that includes every edge in the graph. He showed that every connected graph in which each vertex has even degree has an Euler circuit. Furthermore, no other graph may have an Euler circuit. In particular, there is no Euler circuit in the Königsberg bridge graph, so any walk would have to cross some bridge twice.

To see why a graph with an Euler circuit must have only vertices of even order, suppose we trace such a circuit. Since the circuit must go along every edge and since the graph is connected, it will go through every vertex at least once. Each time we arrive at a vertex v we must also leave it, and since each edge is used only once, the edges incident on v must occur in pairs. The edges incident on the starting vertex occur in pairs because the edge on which we leave the starting vertex is paired with the one on which we return to that vertex for the last time. Therefore, every vertex has even degree. See Figure 7-11.

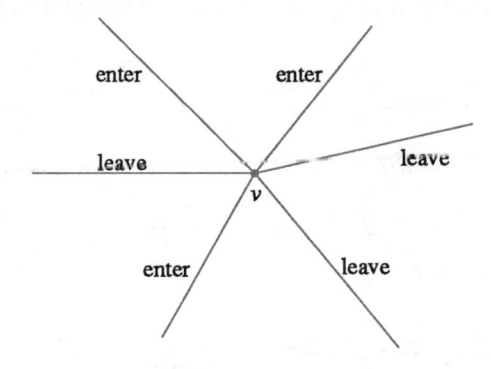

Figure 7-11

Constructing Euler Circuits

Euler showed that, starting with any connected graph in which every vertex is of even degree, it is possible to construct an Euler circuit starting at any vertex. We will illustrate a method for constructing Euler circuits using the graph of Figure 7-12. We observe that this graph is connected, vertices v_1, v_2, and v_8 are of degree 2, and the rest of the vertices are of degree 4; so, the graph has the required properties.

We can construct our Euler circuit at any vertex; we choose v_1. We will first find any circuit that starts and ends at v_1 (called, a **circuit at** v_1). An easy circuit to see is the one determined by the rectangle at the top of the figure, that is, $e_1 e_3 e_4 e_2$. It is not an Euler circuit for the graph since we have used only four of the 13 available edges. The next step in the construction of the Euler circuit is to choose a vertex that is in the circuit we just found, but that is incident on an edge that we have not yet used. To make it easier to find such a circuit, in Figure 7-13 we removed the circuit at v_1 that we just constructed, leaving the vertices v_3 and v_4, and printing them in boldface because they had edges in the part of the graph just removed. Notice that all the other vertices in the graph of Figure 7-13 still have even degree. Choosing one of the boldface vertices, v_3, we construct a circuit at that vertex in the graph of Figure 7-13: $e_5 e_{12} e_{13} e_{10} e_6$. Removing those edges from the graph of Figure 7-13 to form the graph of Figure 7-14, it is easy to see that the re-maining edges form a circuit, which we write in the order $e_7 e_9 e_{11} e_8$, so that it is a circuit at the boldface vertex v_4 that was used in a circuit constructed previously. The three circuits we constructed, starting at v_1, v_3, and v_4 respectively, have now used each of the 13 edges of the graph exactly once.

To construct the Euler circuit, we just combine these three circuits into one. In the first circuit, at v_1, the vertex v_3 was the meeting point of e_4 and e_2. If we insert between e_4 and e_2 the circuit we constructed before at v_3, we obtain a longer circuit at v_1: $e_1 e_3 e_4 e_5 e_{12} e_{13} e_{10} e_6 e_2$. We still have not used all the edges, but we can perform the preceding construction once more. We have the circuit at v_4 re-

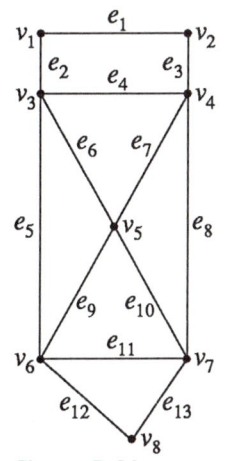

Figure 7-12 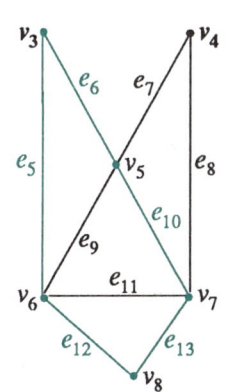 *Figure 7-13* 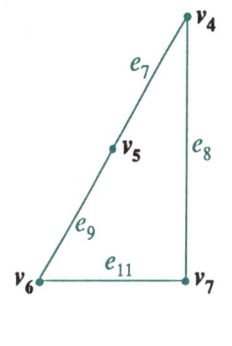 *Figure 7-14*

maining. We locate v_4 in the circuit just constructed as the vertex where e_3 and e_4 meet. Then we insert the final circuit between e_3 and e_4 to obtain:

$$e_1 e_3 e_7 e_9 e_{11} e_8 e_4 e_5 e_{12} e_{13} e_{10} e_6 e_2$$

It is easy to see that each of the 13 edges is listed just once in this circuit, and tracing the circuit in Figure 7-12 convinces us that each edge in the list has a vertex in common with the next edge. Thus, we have obtained the required Euler circuit.

We now summarize the rules that we followed in constructing an Euler circuit. We assume we have a connected graph in which each vertex is of even degree. For any vertex we wish, we construct a circuit at that vertex. If we have not used all the edges in the graph, we choose a vertex in the circuit just constructed that is incident on an edge we have not yet used. The connectedness of the graph guarantees that such a vertex must exist. We construct a circuit at that vertex, without using any of the edges previously chosen. If there are still edges in the graph that we have not used, we choose a vertex in one of the circuits constructed previously that has an unused edge incident on it and construct another circuit according to the same rules. We continue until all the edges of the graph have been accounted for in some circuit. Now viewing the circuits in the order that they were constructed, from the second on, we see that each is at a vertex at the point where two edges of the preceding circuit meet. We insert this circuit into the preceding one at that point. Once all the circuits have been inserted, we have the required Euler circuit.

Now that we know how to build Euler circuits, we can design a snow-plowing plan for the village of Hillsboro.

Example 4. Trace an Euler circuit for the graph in Figure 7-15(a), the street map of Hillsboro, starting and ending at vertex v_6.

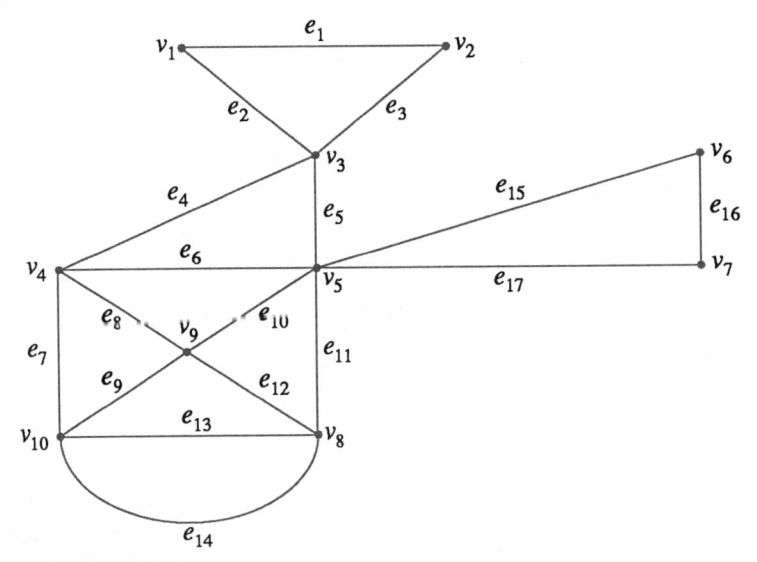

Figure 7-15(a)

Solution

An obvious circuit at v_6 is $e_{15}e_{17}e_{16}$, as shown in Figure 7-15(b).

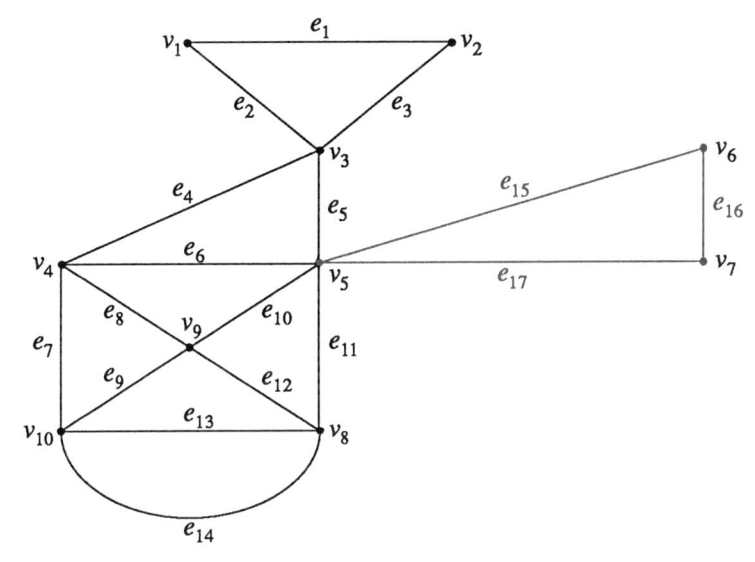

Figure 7-15(b)

Removing that circuit to produce the graph of Figure 7-16, we see that the only boldface vertex in that circuit, that is, the only vertex incident to the circuit just removed, is v_5, and we form the circuit $e_5e_2e_1e_3e_4e_6$ at v_5. Removing this circuit to produce Figure 7-17, we choose another circuit at v_5, namely $e_{10}e_9e_{14}e_{11}$. The remaining edges can be viewed as a circuit, as we see in Figure 7-18, and we view it as based at v_4, that is, $e_7e_{13}e_{12}e_8$. We have now assigned every edge to a circuit, so we

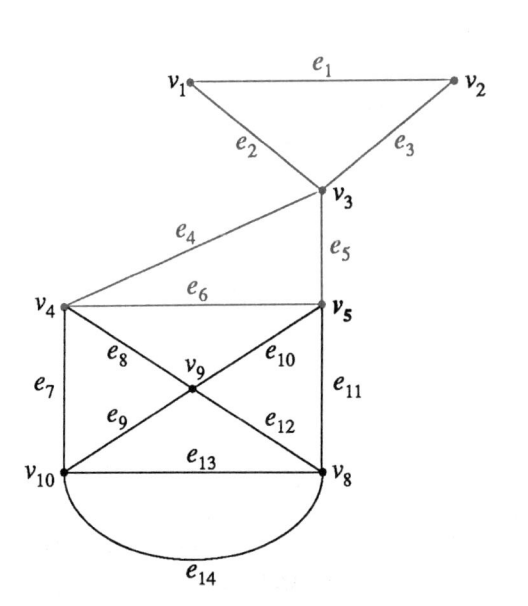

Figure 7-16

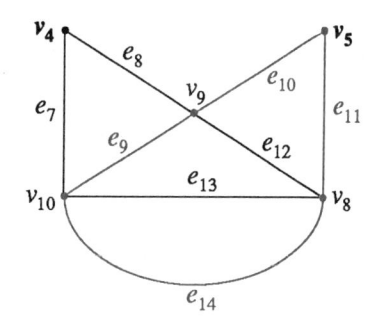

Figure 7-17

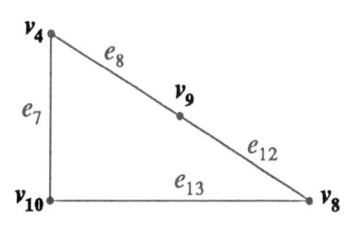

Figure 7-18

insert the second circuit constructed between e_{15} and e_{17}, the edges that meet at v_5 in the first circuit, to obtain $e_{15}e_5e_2e_1e_3e_4e_6e_{17}e_{16}$. The next circuit is also at v_5, which we locate as the meeting point of e_6 and e_{17} in the circuit just constructed; so we form there: $e_{15}e_5e_2e_1e_3e_4e_6e_{10}e_9e_{14}e_{11}e_{17}e_{16}$. (We could have inserted $e_{10}e_9e_{14}e_{11}$ between e_{15} and e_5, instead.) The final circuit is at v_4, which is the meeting point of e_4 and e_6; to obtain the required Euler circuit, we form the circuit

$$e_{15}e_5e_2e_1e_3e_4e_7e_{13}e_{12}e_8e_6e_{10}e_9e_{14}e_{11}e_{17}e_{16}$$

■

We can now refer back to Figure 7-7, which tells us the snow-plow routing in terms of the streets of Hillsboro. If the plow follows the route:

Main – Market – Oak – Marsh – Grant – Maple – Frost – Elm– Grand – Main – Circle – Market – Grand – Station

it will go the length of each of the town's streets exactly once and return to its original location.

Other Applications of Graphs

Graphs have many other uses. For instance, anthropologists find graphs useful when studying tribal relationships. In most societies, the lives of the people are strongly influenced by kinship and one of the most important aspects of kinship is consanguinity. Two people are called consanguineal or blood relatives if they share a common ancestor. Siblings (common parent), cousins (common grandparent), and a boy and his father's sister are instances of consanguineal relatives. This information can be modeled by a graph in which each vertex represents an inhabitant of the village and in which an edge connects two inhabitants if they are consanguineal.

Graphs can be good tools for working out transportation and communication problems, particularly those that must satisfy special conditions, such as moving a house, while avoiding low underpasses and steep grades. Other uses include finding the most efficient way for a sales representative to cover his or her territory and the routing of municipal trash-collecting trucks in a large city.

Summary

A **graph** consists of a set V, the **vertices** of the graph, and a set E, the **edges** of the graph. Each edge $e = \{v, v'\}$ is associated with a set of two (not necessarily different) vertices that are **incident** on e, and e is said to be **incident** on v and v'. An edge $e = \{v, v\}$ that connects a vertex with itself is called a **loop**. Two edges incident on the same two vertices are called **parallel edges**. In a graph

G, a **path** between vertices v and v' is a sequence of edges; we write this as $e_1 e_2 \ldots e_r$, so that v is incident on e_1, v' is incident on e_r, and successive edges $e_i e_{i+1}$ meet at a vertex—that is, they are incident on the same vertex. A graph is **connected** if there is a path between any two of its vertices, and it is **disconnected** otherwise. A **circuit** at v is a path that starts and ends at the same vertex v and in which no edge appears more than once. The **degree** of a vertex v is the number of edges incident on v; if there is a loop at the vertex, it raises the degree by 2. The graphs that have an **Euler circuit**, that is, a circuit that includes every edge in the graph, are precisely those connected graphs in which each vertex has even degree. To construct an Euler circuit, construct a circuit at any vertex. If it does not use all the edges in the graph, choose a vertex in the circuit just constructed that is incident on an unused edge. Construct a circuit at that vertex without using any of the edges previously chosen. Continue to construct circuits in this way until all the edges of the graph are in some circuit. Taking the circuits in the order that they were constructed, from the second on, each is at a vertex where two edges of the preceding circuit meet. Insert the circuit into the previous one at that point until all the circuits have been inserted.

Exercises

In Exercises 1 through 4, draw each graph from the given sets of vertices V and edges E.

1. $V = \{v_1, v_2, v_3, v_4\}$; $E = \{e_1, e_2, e_3, e_4, e_5\}$, $e_1 = \{v_1, v_4\}$, $e_2 = \{v_4, v_3\}$, $e_3 = \{v_3, v_2\}$, $e_4 = \{v_3, v_1\}$, and $e_5 = \{v_3, v_1\}$.

2. $V = \{v_1, v_2, v_3, v_4, v_5\}$; $E = \{e_1, e_2, e_3, e_4, e_5, e_6\}$, $e_1 = \{v_1, v_2\}$, $e_2 = \{v_2, v_2\}$, $e_3 = \{v_2, v_3\}$, $e_4 = \{v_3, v_4\}$, $e_5 = \{v_4, v_5\}$, and $e_6 = \{v_5, v_1\}$.

3. $V = \{v_1, v_2, v_3, v_4, v_5\}$; $E = \{e_1, e_2, e_3, e_4, e_5, e_6, e_7\}$, $e_1 = \{v_1, v_2\}$, $e_2 = \{v_1, v_3\}$, $e_3 = \{v_1, v_4\}$, $e_4 = \{v_3, v_5\}$, $e_5 = \{v_5, v_4\}$, $e_6 = \{v_5, v_2\}$, and $e_7 = \{v_2, v_5\}$.

4. $V = \{v_1, v_2, v_3, v_4, v_5\}$; $E = \{e_1, e_2, e_3, e_4\}$, $e_1 = \{v_1, v_2\}$, $e_2 = \{v_1, v_3\}$, $e_3 = \{v_1, v_4\}$, and $e_4 = \{v_1, v_5\}$.

In Exercises 5 through 8, perform the following analysis of the given graph:
 (a) Determine the vertex set V.
 (b) Determine the edge set E.
 (c) List any loops or parallel edges.
 (d) State whether or not the graph is connected.
 (e) Determine the degree of each vertex.

520   GRAPHS AND NETWORKS

5.

Figure 7-19

6.

Figure 7-20

7.

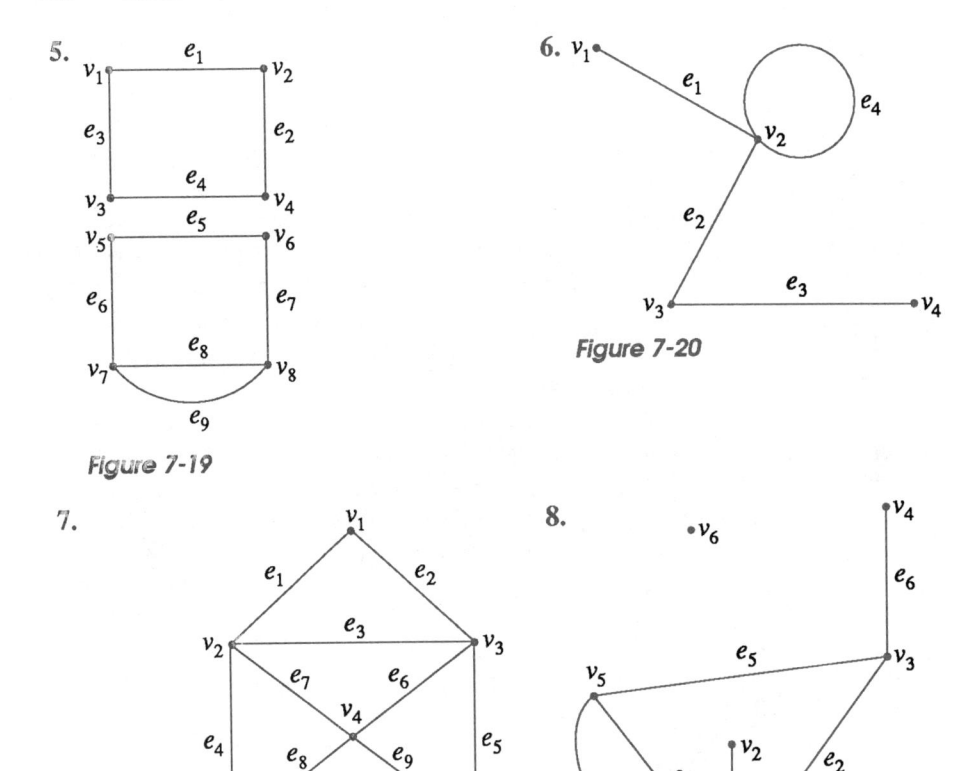

Figure 7-21

8.

Figure 7-22

9. Represent the following playmate relationships observed among a group of children by a graph in which the vertices represent the children and in which an edge is drawn between children who play with each other.

>Jennifer is observed playing with Max, Wendy, and Lisa
>David plays with Max and Lisa
>Wendy plays with Lisa

10. Represent the following relationships by a graph in which the vertices represent the people and in which an edge is drawn between vertices when the corresponding people are together on one or more committees.

>John is on the building committee with Lisa and Mark
>Kathy is on the planning committee with Mark, Sue, and Tom
>Lisa is on the beautification committee with John, Mark, and Tom
>Mark is on the finance committee with John, Kathy, and Lisa
>Sue is on the program committee with Kathy
>Tom is on the hospitality committee with Kathy and Lisa

11. In a modified round-robin soccer tournament each of the eight teams opposed three others. Represent each team by a vertex and join two vertices whenever corresponding teams opposed each other.

   Lions played Bears, Pirates, and Raiders
   Tigers played Wolves, Chargers, and Giants
   Bears played Lions, Raiders, and Chargers
   Wolves played Tigers, Pirates, and Giants
   Pirates played Lions, Wolves, and Chargers
   Raiders played Lions, Bears, and Giants
   Chargers played Tigers, Bears, and Pirates
   Giants played Tigers, Wolves, and Raiders

In Exercises 12 through 15, determine whether there is an Euler circuit for the graph and, if there is, find one.

12.

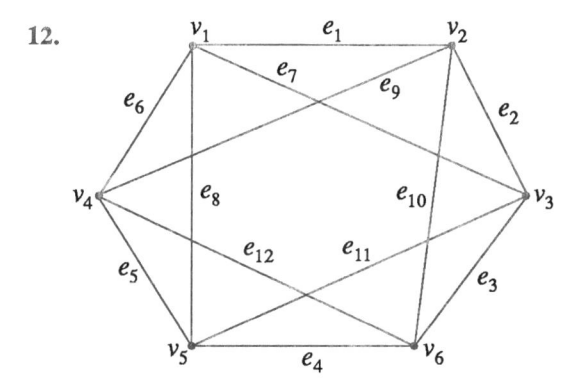

Figure 7-23

13.

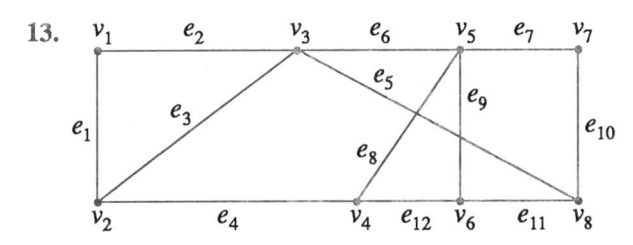

Figure 7-24

14.

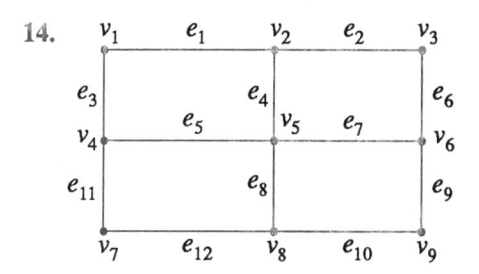

Figure 7-25

15.

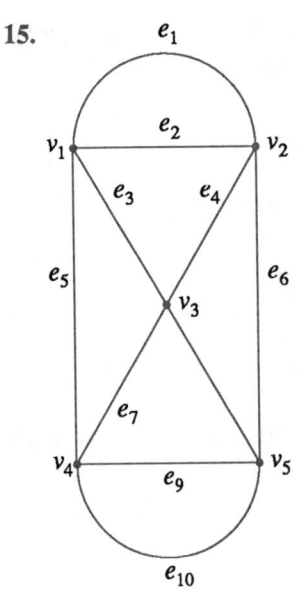

Figure 7-26

16. A street sweeper must clean both sides of the street in the area shown in Figure 7-27.
    (a) Represent this figure by a graph in which a vertex denotes an intersection and an edge denotes one side of a street.
    (b) Determine a route for sweeping the street so that the street sweeper does not drive down the same side of any street more than once.

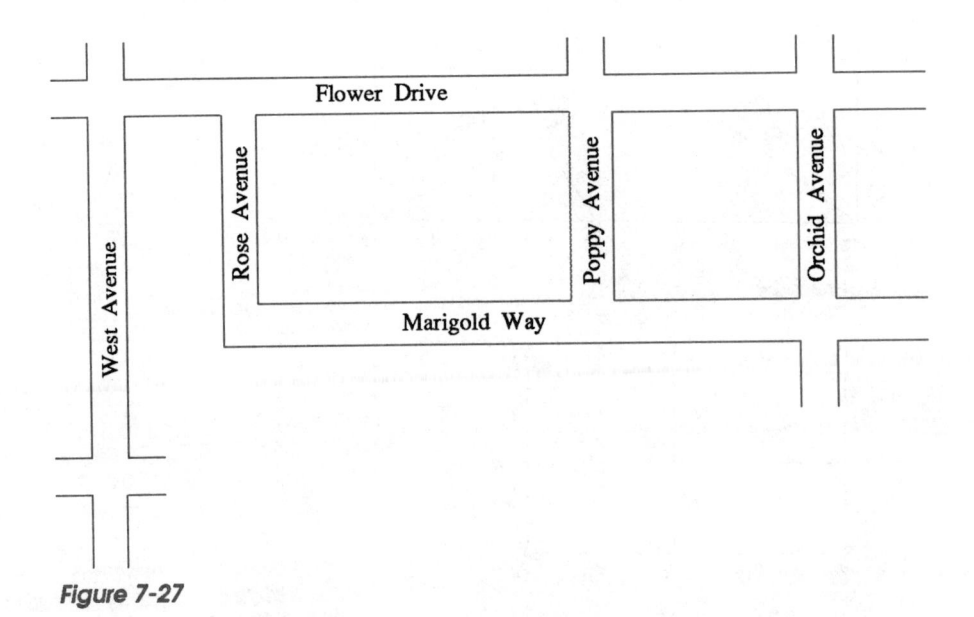

Figure 7-27

17. General Manufacturing Company employs a security guard to patrol its factory at night. The factory plan is shown in Figure 7-28. The guard is required to walk the length of every corridor in the building, as well as to inspect each room.

   (a) Represent the factory plan by a graph with a vertex for each of the designated areas and an edge for each corridor: Assembly Area, Storage Area, Main Office, Guard's Quarters, Tool Shop, and Production Area.
   (b) Find a path that starts and ends in the guard's quarters and passes through each corridor exactly once.

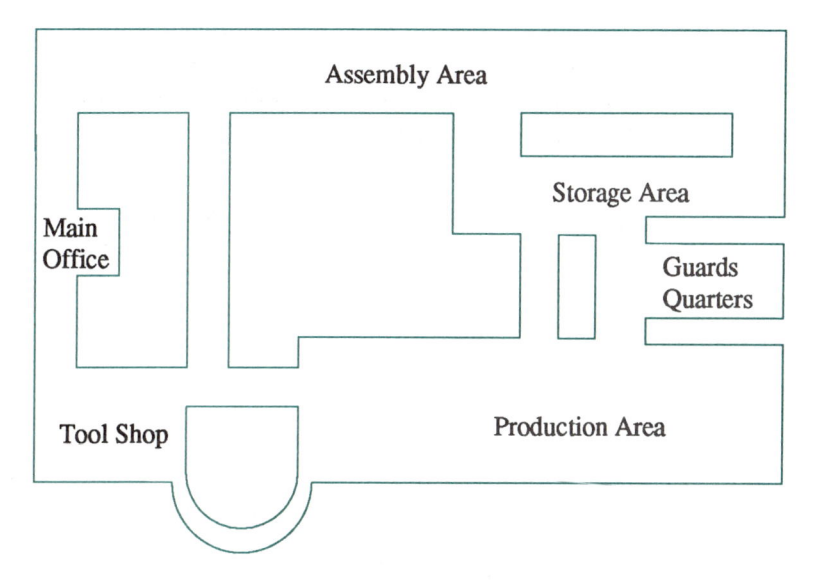

Figure 7-28

7.2 Digraphs

In Section 7.1, we discussed properties of graphs in which an edge $e = \{v, v'\}$ connected two vertices. Thinking of the graph as an abstract version of a street map, each edge in the graph served as a two-way street. However, if a city is looking for the most efficient route for garbage trucks to follow in a district that has one-way streets, the planners must take into account the direction of the traffic flow on these streets. For problems of this kind, another type of graph is needed.

Vertices and Arcs

> A **directed graph** or, more briefly, a **digraph** is a set $V = \{v_1, v_2, \ldots, v_k\}$, the **vertices**, together with a set A of *ordered* pairs of distinct vertices, called **arcs** (or **directed edges**). We denote an arc by $a = (v, v')$ and say that a is an arc **from** v **to** v'; here, v is the **initial vertex** and v' is the **terminal vertex** of the arc.

As in graphs, we represent digraphs by drawing dots for each vertex and by connecting the dots corresponding to v and v' with a line, representing the arc (v, v'). However, we also place an arrowhead on the line heading from v towards v'. Notice that for arcs, unlike edges, $a_1 = (v, v')$ is not the same as $a_2 = (v', v)$.

Example 1. Digraphs are used to study food webs in ecology. In Figure 7-29, each arc indicates who eats whom. Find the vertex set V and the arc set A for this digraph.

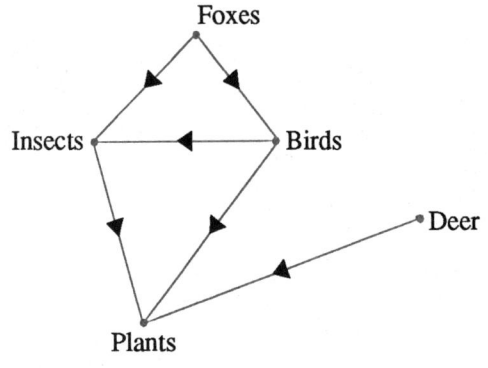

Figure 7-29

Solution

Using the first letter of each species to identify a vertex of the digraph, we see that

$$V = \{F, I, B, P, D\} \text{ and } A = \{a_1, a_2, a_3, a_4, a_5, a_6\}$$

where

$$a_1 = (F, I), a_2 = (F, B), a_3 = (B, I), a_4 = (I, P), a_5 = (B, P), \text{ and } a_6 = (D, P) \blacksquare$$

Example 2. Draw a digraph illustrating the chain of command in a research group consisting of a professor (P), who obtained a grant and who gives orders to the assistant professor (AP), who is actually directing the research. She, in turn, gives orders to two postdoctoral research scholars, (PD1) and (PD2), who supervise the two different types of research taking place. Each postdoctoral scholar supervises two graduate students—PD1 supervises (GSa) and (GSb) and PD2 supervises (GSc) and (GSd). The secretary (S) takes orders only from the professor and the assistant professor. Two undergraduates (UG1) and (UG2) have been hired; one to assist graduate students GSa and GSb and the other to work for the other two graduate students.

Solution

The professor gives direct orders to the assistant professor and the secretary, so we have arcs $a_1 = (P, AP)$ and $a_2 = (P, S)$. The arcs $a_3 = (AP, S)$, $a_4 = (AP, PD1)$ and $a_5 = (AP, PD2)$ indicate the direct supervision by the assistant professor. The chain of command from the postdoctoral scholars to the graduate students is given by $a_6 = (PD1, GSa)$, $a_7 = (PD1, GSb)$, $a_8 = (PD2, GSc)$, and $a_9 = (PD2, GSd)$. Finally, to indicate the authority of the graduate students over the undergraduates, we have $a_{10} = (GSa, UG1)$, $a_{11} = (GSb, UG1)$, $a_{12} = (GSc, UG2)$, and $a_{13} = (GSd, UG2)$. The graph is shown in Figure 7-30. (We do not label the arcs in the figure because we will not make use of them.) ∎

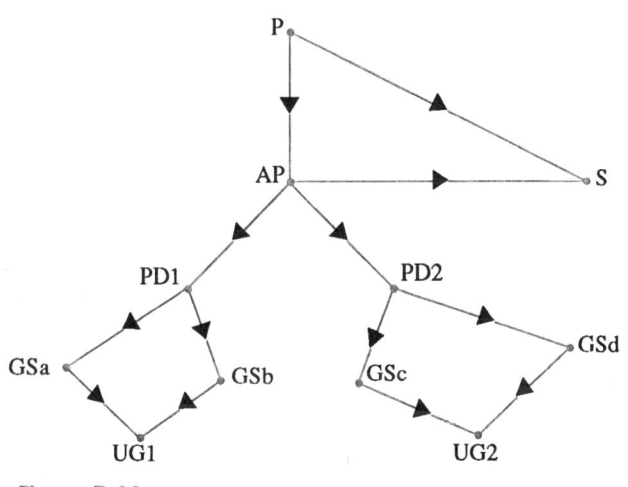

Figure 7-30

Status

> A (**directed**) **path of length** r in a digraph is a sequence of arcs $a_1 a_2 \ldots a_r$ with the property that the terminal vertex of each arc is the initial vertex of the next one; that is, if $a_i = (v, v')$, then $a_{i+1} = (v', w)$ for some vertex w. If there is a path from the vertex u to the vertex v, then v is said to be **reachable** from u. If v is reachable from u, there may be several paths from u to v; the (directed) **distance** from u to v is the minimum of the lengths of all paths from u to v. (If v is not reachable from u, then the distance from u to v is not defined.)

We emphasize that this is a *directed* distance since, unlike the more familiar concept of distance, the distance from u to v may not equal the distance from v to u. It can happen that v is reachable from u—but u is not reachable from v, as we observe in Figure 7-29, in which P (plants) is reachable from D (deer), but not vice versa. The **status** of a vertex is defined as the sum of the distances from u to

526 GRAPHS AND NETWORKS

each vertex reachable from u. Sociological studies of organizations calculate the importance of an individual in an organization by this precise formulation of status, as the next example illustrates.

Example 3. Calculate the status of the following individuals in the research group of Example 2, which is pictured in Figure 7-30.
- (a) The assistant professor;
- (b) The secretary;
- (c) Postdoctoral scholar 2.

Solution

(a) From AP, there are paths of length 1 to each of the two postdoctoral scholars and to the secretary, of length 2 to each of the four graduate students, and of length 3 to each of the undergraduates. The status of the assistant professor is therefore

$$1 + 1 + 1 + 2 + 2 + 2 + 2 + 3 + 3 = 17$$

(b) There are no arcs for which the secretary is the initial vertex, so the status of this individual is 0.

(c) Paths of length 1 from postdoctoral scholar 2 reach two of the graduate students and the distance from postdoctoral scholar 2 to undergraduate student 2 is 2. So, PD2's status is consequently $1 + 1 + 2 = 4$. ∎

Tournaments

A digraph can be used to show what happens in a "round robin" competition in a sport or game in which each player competes once against every other participant in the competition. Each player is represented by a vertex of the digraph, and whenever player v defeats player v', the arc (v, v') appears in the digraph. The digraph of the completed round-robin competition has the property that between any pair of its vertices there is exactly one arc.

A digraph with the property that there is exactly one arc between any two vertices is called a **tournament**. The **score** of a vertex is the number of arcs of which it is the initial vertex. In the digraph of a round-robin competition, the score of the vertex counts the number of competitors defeated by the corresponding player. The **score sequence** of the digraph lists the scores of all the vertices.

Example 4. Draw a tournament and give the score sequence for the following round-robin soccer competition.

> The Bears defeat the Tigers and the Wolves
> The Chargers defeat the Bears, Tigers, and Wolves
> The Tigers defeat the Wolves and the Lions
> The Wolves do not defeat anyone
> The Lions defeat the Bears, Chargers, and Wolves

Solution

The vertex set can be identified with the first initial of each team name:

$$V = \{C, T, B, L, W\}$$

The tournament is given in Figure 7-31.

The score for the Bears is 2, as we see by the two arcs (B, T) and (B, W) that correspond to the Bears' victories over the Tigers and Wolves. Similarly, we can complete the score sequence: 3 for the Chargers, 2 for the Tigers, 0 for the Wolves, and 3 for the Lions. Writing the score sequence in the same order as listed in the vertex set $\{C, T, B, L, W\}$, we have $(3, 2, 2, 3, 0)$. ∎

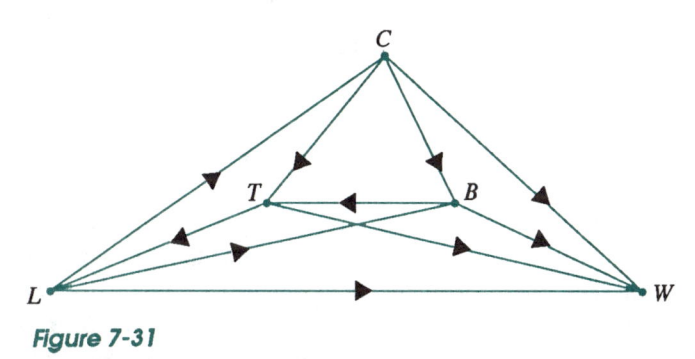

Figure 7-31

Transitive and Cyclic Triples

In a competition, if A defeats B and B defeats C, we might expect that A should defeat C.

> A set $\{v, v', v''\}$ of three vertices in a tournament for which the arcs between them are (v, v'), (v', v''), and (v, v'') is called a **transitive triple**.

For instance, in Figure 7-31, the set of vertices $\{B, T, W\}$ forms a transitive triple since the Bears beat the Tigers, the Tigers beat the Wolves, and, as would be expected, the Bears also beat the Wolves. However, in any competition, upsets can occur. In the tournament represented by Figure 7-31, although the Lions defeated the Chargers and the Chargers defeated the Tigers, the Tigers beat the Lions.

> A set of three vertices that does *not* form a transitive triple, like $\{L, C, T\}$ in Figure 7-31, is called a **cyclic triple**.

Example 5. For the tournament represented by Figure 7-31, how many transitive triples and how many cyclic triples are there?

Solution

We first determine the total number of triples, either transitive or cyclic, in the tournament. Now, in a tournament each vertex is connected with each other vertex by an arc; thus, there is either a transitive or cyclic triple corresponding to each set of three vertices. There are 5 vertices (teams) in this tournament, and each triple, either transitive or cyclic, corresponds to an unordered set of three vertices. Therefore, in the language of Section 3.4, the number of such triples is a combination of 3 objects out of a set of 5. According to a formula from that section, there are

$$C_3^5 = \frac{5!}{3!2!} = \frac{5 \cdot 4 \cdot 3}{3 \cdot 2 \cdot 1} = 10$$

such triples in all. If we look carefully at Figure 7-31, we can find two cyclic triples: $\{B, L, T\}$ and $\{C, L, T\}$. Therefore the remaining 8 triples are transitive.

Transitive triples are used as a measure of consistency in psychological testing. In Example 6, which follows, a preference test is given in which Ralph, the subject, is asked to choose which leisure activity he prefers from a group of eight. The experimenter presents the activities two at a time and all pairs of activities are included. The digraph that corresponds to the subject's choices is a tournament. In this case, the fewer the number of cyclic triples, the greater the consistency of the subject. However, even in Figure 7-31, a tournament with 5 vertices, it was not easy to be sure we have counted the number of cyclic triples accurately. It would certainly be very difficult to draw the tournament of the eight leisure activities and then count the number of cyclic triples by looking for them since there are

$$C_3^8 = \frac{8 \cdot 7 \cdot 6}{3 \cdot 2 \cdot 1} = 56$$

triples in this tournament. Fortunately, there is another formula that tells us how to find the number of cyclic triples from information that is easier to obtain: the scores of their vertices.

Counting Cyclic Triples

The first step in developing this formula is to find the total number of triples. We have already seen how to do this, by using the number of combinations. Letting m be the number of vertices in the digraph, the total number of triples is seen to be

$$C_3^m = \frac{m(m-1)(m-2)}{3 \cdot 2 \cdot 1} = \frac{1}{6} m(m-1)(m-2)$$

If v is the initial vertex of two arcs, $a_1 = (v, v')$, and $a_2 = (v, v'')$, then there has to be a transitive triple corresponding to the set $\{v, v', v''\}$ because we have exactly one of the situations pictured in Figure 7-32.

The score of a vertex v in a tournament was defined to be the number of arcs with initial vertex v; so the arcs a_1 and a_2 contribute 2 to the score of v. On the other hand, since there is a transitive triple for every two arcs with initial vertex v, we see that if the score of v is s, then we have located

Figure 7-32

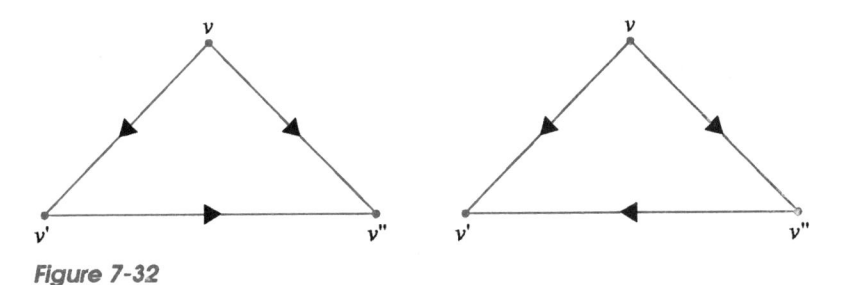

$$C_2^s = \frac{s(s-1)}{2 \cdot 1} = \frac{s(s-1)}{2}$$

transitive triples. As Figure 7-32 further illustrates, there can be only one vertex in a transitive triple that is the initial vertex of two arcs; so each transitive triple depends on a pair of arcs that contribute to the score of just one vertex. Therefore:

> If a tournament with vertices $v_1, v_2, \ldots, v_m$ has score sequence $s_1, s_2, \ldots, s_m$, then the total number of transitive triples is given by
> $$T = \frac{s_1(s_1-1)}{2} + \frac{s_2(s_2-1)}{2} + \ldots + \frac{s_m(s_m-1)}{2}$$
> $$= \frac{1}{2}[s_1(s_1-1) + s_2(s_2-1) + \ldots + s_m(s_m-1)]$$

Of course, if a vertex has score $s = 1$, the single vertex cannot contribute any transitive triples; but then $s(s-1) = 1 \cdot 0 = 0$—so it does not count, anyway. Similarly, if $s = 0$, then the contribution is $(0)(-1) = 0$.

Once we count the transitive triples, since all the other triples are cyclic:

> The number of cyclic triples in a tournament is
> $$C_3^m - T = \frac{1}{6}m(m-1)(m-2) - \frac{1}{2}[s_1(s_1-1) + s_2(s_2-1) + \ldots + s_m(s_m-1)]$$

To apply this formula to the tournament of Figure 7-31, recall from Example 4 that the score sequence for this tournament is (2, 3, 2, 0, 3). The number of transitive triples in this tournament is

$$\frac{1}{2}[(2)(1) + (3)(2) + (2)(1) + (0)(-1) + (3)(2)] = \frac{1}{2}[2 + 6 + 2 + 6] = 8$$

and there are $C_3^5 - T = 10 - 8 = 2$ cyclic triples, as we observed previously.

Example 6. Ralph lists his preferences among pairs of leisure activities—with scores in parentheses—

reading over jogging, swimming, painting, knitting (4)
jogging over painting, knitting (2)
gardening over jogging, reading, travel, painting, knitting (5)
knitting over no activity (0)
cooking over painting, travel, reading, jogging, gardening, knitting (6)
painting over knitting (1)
travel over reading, jogging, knitting, painting (4)
swimming over jogging, gardening, knitting, cooking, painting, travel (6)

Find the number of cyclic triples in this tournament to evaluate the consistency of Ralph's answers.

Solution

We have already observed that there are $C_3^8 = 56$ triples that are either transitive or cyclic. Reading off the score sequence from the given information, we have

$$(4, 2, 5, 0, 6, 1, 4, 6)$$

and therefore

$$T = \frac{1}{2}[(4)(3) + (2)(1) + (5)(4) + (0)(-1) + (6)(5) + (1)(0) + (4)(3) + (6)(5)]$$

$$= \frac{1}{2}[12 + 2 + 20 + 30 + 12 + 30]$$

$$= 53$$

Thus, there are only $56 - 53 = 3$ cyclic triples in this tournament, which suggests that Ralph's answers are quite consistent. ∎

Summary

A **digraph** is a set $V = \{v_1, v_2, \ldots, v_k\}$, the **vertices**, together with a set A of ordered pairs of distinct vertices called **arcs**. Denote an arc by $a = (v, v')$ and say that a is an arc from v, the **initial vertex**, to v', the **terminal vertex**. A **(directed) path** of length r is a sequence of arcs $a_1 a_2 \ldots a_r$ with the property that the terminal vertex of each arc is the initial vertex of the next one. If there is a directed path from the vertex u to the vertex v, then v is said to be **reachable** from u. If v is reachable from u, the **(directed) distance** from u to v is the minimum of the lengths of all directed paths from u to v; otherwise, the distance from u to v is not defined. The **status** of a vertex is defined as the sum of the distances from u to each vertex reachable from u. A digraph with the property that between any pair of its vertices there is exactly one arc is called a **tournament**. The **score** of a vertex is the number of arcs of which the vertex is the initial vertex and the **score sequence** of the digraph lists the scores of all the vertices. An unordered set of three vertices in a tournament in which one vertex is the initial vertex of two arcs for which the other two vertices are terminal vertices is called a **transitive triple**. A set of three vertices in a tournament that does not form a transitive triple is called a **cyclic triple**. For a tournament with score sequence $s_1, s_2, \ldots, s_m$, the total number of transitive triples is

$$T = \frac{1}{2}[s_1(s_1 - 1) + s_2(s_2 - 1) + \ldots + s_m(s_m - 1)]$$

and the number of cyclic triples is

$$\frac{1}{6}(m)(m-1)(m-2) - T$$

Exercises

In Exercises 1 through 4, compute the status of each vertex in the given digraph.

1.

Figure 7-33

2.

Figure 7-34

3.

Figure 7-35

4.

Figure 7-36

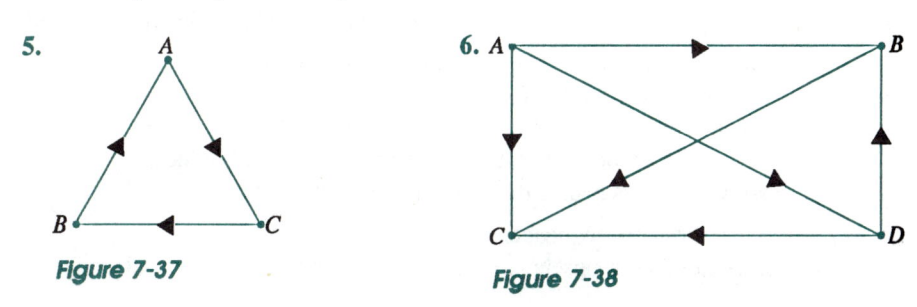

In Exercises 5 and 6, find the score sequence, the number of transitive triples, and the number of cyclic triples for the pictured tournament.

5.

Figure 7-37

6.

Figure 7-38

7. Construct a digraph showing the following relationships.

>Alice likes Bob
>Bob likes Alice and Evelyn
>Carl likes David
>David likes Carl and Fran
>Evelyn likes Bob
>Fran likes Alice, David and Evelyn
>Greta likes Harold
>Harold likes Greta

8. At the Jameson Advertising Agency, the president supervises three vice presidents, who each have different areas of responsibility. The first vice president supervises the head copywriter and the head graphic artist. The second vice president supervises the sales manager and the public-relations officer. The third vice president supervises the chief accountant. Represent this organizational structure by a digraph and compute the status of each member.

9. In a sociological study of street gangs the following interactions determining a chain of command have been observed.

>Ricky gives orders to Matt and Jeff
>Jeff gives orders to Bill, Al, and Sam
>Matt gives orders to Duke, Hank, and Tom
>Bill gives orders to Al
>Duke gives orders to Tom

Determine the digraph and compute the status of each gang member.

10. As part of a market-research study, the researcher uses the method of paired comparisons to find out which low-calorie frozen entree her subjects prefer. She finds that:

Weight Watchers is preferred over Lean Cuisine, Lucky's Lean, and Divinely Slim
Lean Cuisine is preferred over Slim Gourmet, Lucky's Lean, and Divinely Slim
Slim Gourmet is preferred over Weight Watchers, Lucky's Lean, and Divinely Slim
Divinely Slim is preferred over Lucky's Lean

(a) Determine the score sequence.
(b) Examine the consistency of the responses by determining the number of cyclic triples in the tournament.

11. In a series of dual meets between YMCA swim teams, the following are the results:

>Santa Monica beat Westchester and Pacific Palisades
>Culver City beat Santa Monica
>Westchester beat Culver City and Pacific Palisades
>Beverly Hills beat Santa Monica, Culver City, and Westchester
>Pacific Palisades beat Culver City and Beverly Hills

(a) Determine the score sequence.
(b) Count the number of upsets in the tournament by counting the cyclic triples.

12. A scientist studies a group of chimpanzees consisting of Joe, an old male; three mature females: Betty, Sue, and Debby; two immature males: Harry and Frank; and Kathy, an immature female. The dominance relationships observed are:

> Joe dominates all the other animals in the group
> Betty dominates Sue and the three immature animals
> Sue dominates Debby, Harry, and Kathy
> Debbie dominates Betty, Harry, Frank, and Kathy
> Frank dominates Sue and Kathy
> Harry dominates Frank
> Kathy dominates Harry

(a) Determine the score sequence.
(b) Count the number of cyclic triples in this digraph.

7.3 Digraph Matrices

The digraph of Figure 7-39 describes a communications network. Each arc represents a channel available for the one-directional flow of information between the locations indicated by the vertices. For instance, v_1 can transmit directly to v_2 and v_6, but v_1 can only receive directly from v_6. We can describe the digraph either by using this figure or, as in the preceding section, by listing the vertices

$$V = \{v_1, v_2, v_3, v_4, v_5, v_6\}$$

and the arcs

$$A = \{a_1, a_2, a_3, a_4, a_5, a_6, a_7, a_8\}$$

where

$a_1 = (v_1, v_2)$, $a_2 = (v_1, v_6)$, $a_3 = (v_2, v_3)$, $a_4 = (v_2, v_4)$, $a_5 = (v_3, v_5)$, $a_6 = (v_4, v_3)$, $a_7 = (v_5, v_6)$, and $a_8 = (v_6, v_1)$.

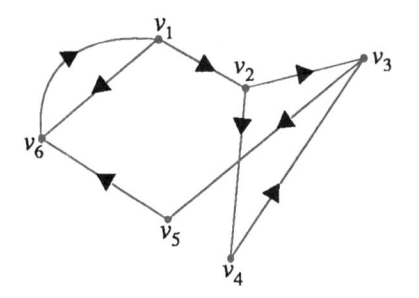

Figure 7-39

The Adjacency Matrix

Another way the information in Figure 7-39 may be presented is through its *adjacency matrix*.

534   GRAPHS AND NETWORKS

> Given a digraph with vertex set $V = \{v_1, v_2, \ldots, v_k\}$, the **adjacency matrix** of the digraph is the $k \times k$-matrix M with a 1 in the a_{ij}-location if the arc $a = (v_i, v_j)$ is in the digraph or otherwise with a 0 there.

Thus, the matrix M contains as many 1s as there are arcs in A. For the communications digraph of Figure 7-39, we see that there are two arcs with initial vertex v_1, namely, $a_1 = (v_1, v_2)$ and $a_2 = (v_1, v_6)$; so in the first row of M we have a 1 in the a_{12}- and a_{16}-locations. Therefore, that row is

$$\begin{array}{c c c c c c c} & v_1 & v_2 & v_3 & v_4 & v_5 & v_6 \\ v_1 & 0 & 1 & 0 & 0 & 0 & 1 \end{array}$$

Continuing in this manner, we find that the adjacency matrix of the digraph of Figure 7-39 is

$$M = \begin{array}{c} \\ v_1 \\ v_2 \\ v_3 \\ v_4 \\ v_5 \\ v_6 \end{array} \begin{array}{c} \begin{array}{cccccc} v_1 & v_2 & v_3 & v_4 & v_5 & v_6 \end{array} \\ \left[\begin{array}{cccccc} 0 & 1 & 0 & 0 & 0 & 1 \\ 0 & 0 & 1 & 1 & 0 & 0 \\ 0 & 0 & 0 & 0 & 1 & 0 \\ 0 & 0 & 1 & 0 & 0 & 0 \\ 0 & 0 & 0 & 0 & 0 & 1 \\ 1 & 0 & 0 & 0 & 0 & 0 \end{array}\right] \end{array}$$

As a check, we note that the arc set A lists eight arcs and that M contains eight 1s.

All the information about a digraph is contained in its adjacency matrix. Given the matrix, we can draw the digraph, as the following example demonstrates.

Example 1. Draw the digraph whose adjacency matrix is

$$M = \begin{array}{c} \\ v_1 \\ v_2 \\ v_3 \\ v_4 \\ v_5 \end{array} \begin{array}{c} \begin{array}{ccccc} v_1 & v_2 & v_3 & v_4 & v_5 \end{array} \\ \left[\begin{array}{ccccc} 0 & 1 & 0 & 1 & 0 \\ 0 & 0 & 0 & 1 & 0 \\ 1 & 1 & 0 & 0 & 1 \\ 0 & 0 & 0 & 0 & 1 \\ 1 & 0 & 0 & 0 & 0 \end{array}\right] \end{array}$$

Solution

Once we locate the five vertices of the digraph in our figure, we require arcs from v_1 to v_2 and from v_1 to v_4, to correspond to the 1s in the a_{12}- and a_{14}-locations; we require an arc from v_2 to v_4 because $a_{24} = 1$, and so on. The completed digraph is shown in Figure 7-40. Note that the eight 1s in the matrix are represented by eight arcs in this figure.

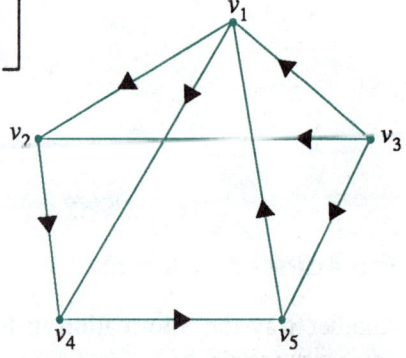

Figure 7-40

7.3 DIGRAPH MATRICES

In Section 7.2, we introduced the status of a vertex, which depends on the distances between vertices in a digraph. Recall that the distance from a vertex u to a vertex v in a digraph is the minimum of the lengths of all paths from u to v. Even for small digraphs, it is tedious to try to determine distance just by looking at the figure, as we did previously, since there can be several paths, of various lengths, between the same two vertices. That method becomes extremely undependable if the digraph represents an organization with many members. We will next show how we may use the adjacency matrix of a digraph more conveniently to compute the distances between vertices.

Matrices of Paths

We illustrate the method for computing distances using the digraph of Figure 7-41. The corresponding adjacency matrix, of course, identifies the five paths of length 1 in the digraph.

$$M = \begin{array}{c} \\ v_1 \\ v_2 \\ v_3 \\ v_4 \end{array} \begin{array}{c} \begin{array}{cccc} v_1 & v_2 & v_3 & v_4 \end{array} \\ \left[\begin{array}{cccc} 0 & 1 & 1 & 0 \\ 0 & 0 & 0 & 1 \\ 1 & 0 & 0 & 0 \\ 0 & 0 & 1 & 0 \end{array} \right] \end{array}$$

Both the digraph and its adjacency matrix M tell us that v_1 is distance one from both v_2 and v_3, v_2 is distance one from v_4, v_3 is distance one from v_1, and v_4 is distance one from v_3. Thus, the distance from v_1 to v_4, for instance, must be more than one; but we do not yet know what that distance is, or even whether v_4 is reachable from v_1 at all.

In Figure 7-41 we can see that there is a path of length two from v_1 to v_4 consisting of the arc from v_1 to v_2, followed by the arc from v_2 to v_4. The arc from v_1 to v_2 is represented by $a_{12} = 1$ in the adjacency matrix and the arc from v_2 to v_4 by $a_{24} = 1$. Arcs with initial v_1 are represented by 1s in the first row of M. A 1 in the fourth column of M represents an arc with terminal vertex v_4. The location of a 1 in the first row indicates the terminal vertex of an arc and the location of a 1 in the fourth column indicates the initial vertex of an arc. An arc with initial vertex v_1 and an arc with terminal vertex v_4 that have a vertex in common will occupy corresponding locations in their row and column. Therefore, if we multiply the first row of M times the fourth column of M (see Section 1.3), these arcs will contribute a term of the form $1 \cdot 1$ to the sum. If we multiply the first row of M and the fourth column of M, what happens is this:

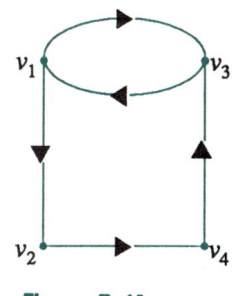

Figure 7-41

$$[0 \ 1 \ 1 \ 0] \begin{bmatrix} 0 \\ 1 \\ 0 \\ 0 \end{bmatrix} = 0 \cdot 0 + 1 \cdot 1 + 1 \cdot 0 + 0 \cdot 0 = 1$$

The 1 in the a_{14}-location of the matrix $M^2 = MM$ comes from multiplying a_{12} times a_{24}, so it represents the path of length two consisting of the arc from v_1 to v_2 followed by the arc from v_2 to v_4. When we complete the multiplication of M by itself:

$$M^2 = \begin{bmatrix} 0 & 1 & 1 & 0 \\ 0 & 0 & 0 & 1 \\ 1 & 0 & 0 & 0 \\ 0 & 0 & 1 & 0 \end{bmatrix} \begin{bmatrix} 0 & 1 & 1 & 0 \\ 0 & 0 & 0 & 1 \\ 1 & 0 & 0 & 0 \\ 0 & 0 & 1 & 0 \end{bmatrix} = \begin{bmatrix} 1 & 0 & 0 & 1 \\ 0 & 0 & 1 & 0 \\ 0 & 1 & 1 & 0 \\ 1 & 0 & 0 & 0 \end{bmatrix}$$

we obtain a matrix with six 1s. Looking back at Figure 7-41, we see that there is, for instance, a path from v_1 to itself obtained by combining the arc from v_1 to v_3 with the arc from v_3 back to v_1, and this appears as the 1 in the a_{11}-location of M^2. In this way, we can identify the six 1s in M^2 with the six paths of length two in Figure 7-3. Furthermore, we now know that the distance from v_1 to v_4 is two. Also, from the matrix M^2 we learn that the distance from v_2 to v_3 is two, as is the distance from v_3 to v_2 and the distance from v_4 to v_1.

An easy way to determine if there are any distances left to compute is to add the matrices M and M^2 since a 0 in the a_{ij}-location of the sum will indicate that there is no path of length zero or one between the vertices v_i and v_j. We find that

$$M + M^2 = \begin{bmatrix} 0 & 1 & 1 & 0 \\ 0 & 0 & 0 & 1 \\ 1 & 0 & 0 & 0 \\ 0 & 0 & 1 & 0 \end{bmatrix} + \begin{bmatrix} 1 & 0 & 0 & 1 \\ 0 & 0 & 1 & 0 \\ 0 & 1 & 1 & 0 \\ 1 & 0 & 0 & 0 \end{bmatrix} = \begin{bmatrix} 1 & 1 & 1 & 1 \\ 0 & 0 & 1 & 1 \\ 1 & 1 & 1 & 0 \\ 1 & 0 & 1 & 0 \end{bmatrix}$$

The 0s in the diagonal locations a_{22} and a_{44} do not signify anything of interest since the distance from a vertex to itself is zero. However, the 0s at the remaining locations, a_{21}, a_{34}, and a_{42}, tell us that we still must determine whether v_1 is reachable from v_2, v_4 is reachable from v_3, and v_2 is reachable from v_4 and, if so, what the distances are.

Since the nonzero entries in M^2 indicate paths of length two, it is not difficult to see that, to determine the paths of length three in the digraph, we should multiply M^2 by M to obtain the matrix M^3, which is given by

$$M^3 = M^2 M = \begin{bmatrix} 1 & 0 & 0 & 1 \\ 0 & 0 & 1 & 0 \\ 0 & 1 & 1 & 0 \\ 1 & 0 & 0 & 0 \end{bmatrix} \begin{bmatrix} 0 & 1 & 1 & 0 \\ 0 & 0 & 0 & 1 \\ 1 & 0 & 0 & 0 \\ 0 & 0 & 1 & 0 \end{bmatrix} = \begin{bmatrix} 0 & 1 & 2 & 0 \\ 1 & 0 & 0 & 0 \\ 1 & 0 & 0 & 1 \\ 0 & 1 & 1 & 0 \end{bmatrix}$$

The 1 in the a_{21}-location in the matrix M^3 tells us that there is a path of length three from v_2 to v_1. Thus, the distance from v_2 to v_1 is three, as is the distance from v_3 to v_4 and from v_4 to v_2. If we examine Figure 7-41, we find a path of length three from v_2 to v_1, namely, the arc from v_2 to v_4, then the arc from v_4 to v_3, and finally the arc from v_3 to v_1. However, even in such a simple digraph, it would not be easy to be sure we had calculated all the distances correctly, if we did not have the matrices M, M^2, and M^3 to guide us.

In the digraph of Figure 7-41, every vertex was reachable from all of the other vertices by paths of length at most 3. However, had we found that in any location a_{ij} there was a 0 in M^3 as well as in $M + M^2$, we would not have to calculate M^4. The reason is the following general fact about vertices in a digraph:

> If u and v are vertices of a digraph that has k vertices, then either the distance from u to v is less than or equal to $k-1$, or else v is not reachable from u.

Since the digraph in Figure 7-41 has $k = 4$ vertices, by calculating through $M^{k-1} = M^3$, we have identified all the paths that are necessary to calculate the distances between reachable vertices.

The preceding discussion applies to digraphs with any number of vertices and to paths of any length. If M is the adjacency matrix of a digraph with vertices $v_1, v_2, \ldots, v_k$ and if M^r is the rth power of M, then the entry in the a_{ij}-location of M^r is the number of paths of length r from v_i to v_j. Thus, the following is true:

> If there is a 0 in the a_{ij}-location of each of the matrices $M, M^2, \ldots, M^{r-1}$ or, equivalently, in the a_{ij}-location of the sum
> $$M + M^2 + \ldots + M^{r-1}$$
> and if the a_{ij}-location of M^r is not zero, then the distance from v_i to v_j is r. If there is a 0 in the a_{ij}-location of $M + M^2 + \ldots + M^{k-1}$, then v_j is not reachable from v_i.

Distance and Status

Recall from Section 7.2 that the status of the vertex u is the sum of the distances from u to all vertices reachable from u. Now that we know the distances between the vertices in Figure 7-41, we can use this information to calculate the status of any vertex. For example, to calculate the status of the vertex v_3 in the digraph of Figure 7-41, we note that the matrices told us that the distance from v_3 to v_1 is 1, the distance from v_3 to v_2 is 2, and the distance from v_3 to v_4 is 3. Of course the distance from v_3 to itself is 0, so we conclude that the status of v_3 is $1 + 2 + 3 = 6$.

538   GRAPHS AND NETWORKS

Example 2. A branch of a bank consists of two divisions: operations and credit. The operations division is run by an assistant manager (AMO) who gives orders to a head teller (HT), who, in turn, directs the work of two tellers (T1 and T2). The assistant manager of the credit division (AMC) supervises the work of two credit officers (CO1 and CO2). In addition, a secretary (S) in the credit division works for the assistant manager as well as for the two credit officers. Use adjacency matrices to calculate the status of the two assistant managers.

Solution

The lines of supervision in the two divisions of the bank branch are shown in Figure 7-42. The adjacency matrix of the operations division is

$$M = \begin{array}{c} \\ \text{AMO} \\ \text{HT} \\ \text{T1} \\ \text{T2} \end{array} \begin{array}{cccc} \text{AMO} & \text{HT} & \text{T1} & \text{T2} \\ \begin{bmatrix} 0 & 1 & 0 & 0 \\ 0 & 0 & 1 & 1 \\ 0 & 0 & 0 & 0 \\ 0 & 0 & 0 & 0 \end{bmatrix} \end{array}$$

Calculating M^2 is sufficient to determine all the distances from the vertex AMO.

$$M^2 = \begin{bmatrix} 0 & 1 & 0 & 0 \\ 0 & 0 & 1 & 1 \\ 0 & 0 & 0 & 0 \\ 0 & 0 & 0 & 0 \end{bmatrix} \begin{bmatrix} 0 & 1 & 0 & 0 \\ 0 & 0 & 1 & 1 \\ 0 & 0 & 0 & 0 \\ 0 & 0 & 0 & 0 \end{bmatrix} = \begin{bmatrix} 0 & 0 & 1 & 1 \\ 0 & 0 & 0 & 0 \\ 0 & 0 & 0 & 0 \\ 0 & 0 & 0 & 0 \end{bmatrix}$$

The distance from the assistant manager for operations to the head teller is one, and to each of the tellers is two, so the status of the assistant manager for operations is $1 + 2 + 2 = 5$. From the credit-division digraph we have the adjacency matrix

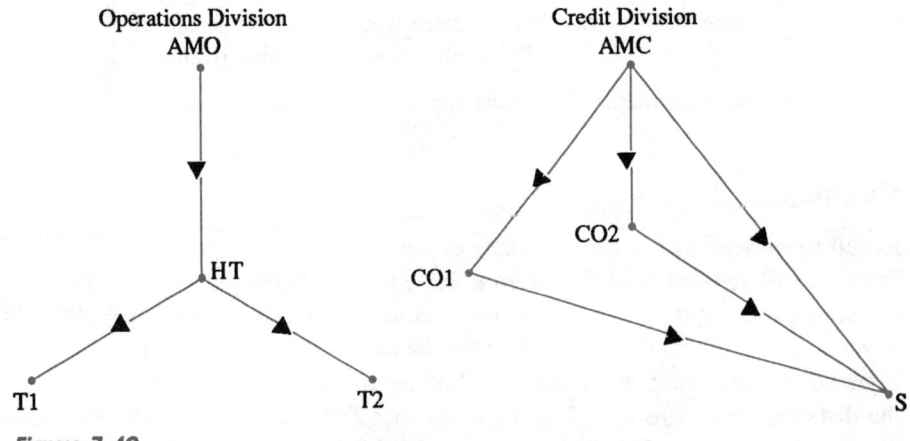

Figure 7-42

$$N = \begin{array}{c} \\ AMC \\ CO1 \\ CO2 \\ S \end{array} \begin{array}{c} \begin{array}{cccc} AMC & CO1 & CO2 & S \end{array} \\ \left[\begin{array}{cccc} 0 & 1 & 1 & 1 \\ 0 & 0 & 0 & 1 \\ 0 & 0 & 0 & 1 \\ 0 & 0 & 0 & 0 \end{array} \right] \end{array}$$

We see that the distance from the assistant manager for credit to each of the other employees of this division is one. Therefore, this assistant manager has less status than the other one, namely, $1 + 1 + 1 = 3$. ∎

Actually, it would have been less work to solve Example 2 using just the digraphs in Figure 7-42, without introducing the adjacency matrices. However, for large organizations, the matrix technique, implemented on a computer, is the only practical way to calculate status.

Reachability

Sometimes it is important to know whether the vertices of a graph are reachable from each other, but it is not necessary to know the distances between them. For example, if the vertices of a graph represent a computer-to-computer electronic-mail network, it does not matter whether A's message goes directly to B's computer or is rerouted through various stations of the network. The transmission is fast enough so that the length of the path makes little or no difference. As an example, take an electronic-mail network consisting of five universities, which we will label as A through E. Suppose that messages are routed by the following one-way links: from A to C and E; from B to A; from C to B, D, and E; from D to A, B, and E; from E to A, C, and D. The corresponding adjacency matrix is

$$M = \begin{array}{c} \\ A \\ B \\ C \\ D \\ E \end{array} \begin{array}{c} \begin{array}{ccccc} A & B & C & D & E \end{array} \\ \left[\begin{array}{ccccc} 0 & 0 & 1 & 0 & 1 \\ 1 & 0 & 0 & 0 & 0 \\ 0 & 1 & 0 & 1 & 1 \\ 1 & 1 & 0 & 0 & 1 \\ 1 & 0 & 1 & 1 & 0 \end{array} \right] \end{array}$$

We would like to verify that every university is reachable from every other one. We could do that by calculating the powers of M, that is, M^2, M^3, and M^4 and thus, in effect, list all possible paths. (We recall that since there are $k = 5$ vertices of the digraph, the distance from any vertex to any other is no more than $k - 1 = 4$.) Instead, we will use some information about matrices from Chapter 1 to reduce the effort. The amount of computation saved is rather modest in this small example, but if we wished to analyze a large communications network, the time saved in multiplying large matrices, even with a computer, can be considerable.

If we did calculate each of the powers through M^4, a simple way to verify that we can reach every university from every other one would be to find the sum of the matrices $M + M^2 + M^3 + M^4$, just as we did for the digraph of Figure 7-41. This will be the case provided that each a_{ij}-location (with i different from j) in the sum is nonzero.

If we use this idea with the matrices we calculated for the digraph of Figure 7-41, since there are just 4 vertices in that digraph we calculate

$$M + M^2 + M^3 = \begin{bmatrix} 0 & 1 & 1 & 0 \\ 0 & 0 & 0 & 1 \\ 1 & 0 & 0 & 0 \\ 0 & 0 & 1 & 0 \end{bmatrix} + \begin{bmatrix} 1 & 0 & 0 & 1 \\ 0 & 0 & 1 & 0 \\ 0 & 1 & 1 & 0 \\ 1 & 0 & 0 & 0 \end{bmatrix} + \begin{bmatrix} 0 & 1 & 2 & 0 \\ 1 & 0 & 0 & 0 \\ 1 & 0 & 0 & 1 \\ 0 & 1 & 1 & 0 \end{bmatrix}$$

$$= \begin{bmatrix} 1 & 2 & 3 & 1 \\ 1 & 0 & 1 & 1 \\ 2 & 1 & 1 & 1 \\ 1 & 1 & 2 & 0 \end{bmatrix}$$

and since the only 0s are on the main diagonal, we again see that every vertex is reachable from every other one in that digraph.

For a $k \times k$-adjacency matrix M, instead of the sum $M + M^2 + \ldots + M^{k-1}$, we can calculate the single matrix $(I + M)^{k-1}$, where I is the $k \times k$-identity matrix. To see why this calculation is equally effective in determining reachability, we use the distributive law for matrix multiplication and the property of the identity matrix given in Section 1.2:

$$(I + M)^2 = (I + M)(I + M) = I(I + M) + M(I + M) = I + M + M + M^2$$

$$(I + M)^3 = (I + M)(I + M)^2 = (I + M)(I + M + M + M^2)$$
$$= I + M + M + M^2 + M + M^2 + M^2 + M^3$$

In general, $(I + M)^{k-1}$ is a sum of the identity matrix and many copies of M, M^2, M^3, and so on, through M^{k-1}. Whenever there is a path from v_i to v_j of any length up to $k - 1$, it will contribute at least 1 to the a_{ij}-location of $(I + M)^{k-1}$. Thus, any 0s in $(I + M)^{k-1}$ would indicate that no such path exists and therefore, that v_j is not reachable from v_i. Since I contributes a 1 to every diagonal location, we have a simple condition for determining whether we can reach every vertex of the digraph from every other one: the matrix $(I + M)^{k-1}$ must not contain any 0s.

For the matrix of the five-university electronic-mail network, we calculate

$$I + M = \begin{bmatrix} 1 & 0 & 0 & 0 & 0 \\ 0 & 1 & 0 & 0 & 0 \\ 0 & 0 & 1 & 0 & 0 \\ 0 & 0 & 0 & 1 & 0 \\ 0 & 0 & 0 & 0 & 1 \end{bmatrix} + \begin{bmatrix} 0 & 0 & 1 & 0 & 1 \\ 1 & 0 & 0 & 0 & 0 \\ 0 & 1 & 0 & 1 & 1 \\ 1 & 1 & 0 & 0 & 1 \\ 1 & 0 & 1 & 1 & 0 \end{bmatrix} = \begin{bmatrix} 1 & 0 & 1 & 0 & 1 \\ 1 & 1 & 0 & 0 & 0 \\ 0 & 1 & 1 & 1 & 1 \\ 1 & 1 & 0 & 1 & 1 \\ 1 & 0 & 1 & 1 & 1 \end{bmatrix}$$

Then we find

$$(I+M)^2 = \begin{bmatrix} 1 & 0 & 1 & 0 & 1 \\ 1 & 1 & 0 & 0 & 0 \\ 0 & 1 & 1 & 1 & 1 \\ 1 & 1 & 0 & 1 & 1 \\ 1 & 0 & 1 & 1 & 1 \end{bmatrix} \begin{bmatrix} 1 & 0 & 1 & 0 & 1 \\ 1 & 1 & 0 & 0 & 0 \\ 0 & 1 & 1 & 1 & 1 \\ 1 & 1 & 0 & 1 & 1 \\ 1 & 0 & 1 & 1 & 1 \end{bmatrix} = \begin{bmatrix} 2 & 1 & 3 & 2 & 3 \\ 2 & 1 & 1 & 0 & 1 \\ 3 & 3 & 2 & 3 & 3 \\ 4 & 2 & 2 & 2 & 3 \\ 3 & 2 & 3 & 3 & 4 \end{bmatrix}$$

To complete the verification, we could calculate

$$(I+M)^4 = (I+M)^2(I+M)^2$$

However, that is quite a lot of work and is really unnecessary in this case. Recalling that $(I+M)^2 = I + M + M + M^2$, we can already see that every university in the network is reachable from all the others, except that, since there is a 0 in the a_{24}-location of $(I+M)^2$, we do not yet know whether the network will reach from B to D. Since we really only need to determine the entry in the a_{24}-location of $(I+M)^4$, we recall from Chapter 1 that we can just multiply the second row of $(I+M)^2$ by the fourth column of the same matrix:

$$\begin{bmatrix} 2 & 1 & 1 & 0 & 1 \end{bmatrix} \begin{bmatrix} 2 \\ 0 \\ 3 \\ 2 \\ 3 \end{bmatrix} = 4 + 0 + 3 + 0 + 3 = 10$$

This product is nonzero, so the network can reach every university from every other one. This conclusion is what we would expect; but the next example indicates how the same method can give us additional information about the network.

Example 3. If the computer at E fails and can no longer participate in the electronic mail network, is it still possible for the other four universities to communicate with each other?

Solution

Starting with the adjacency matrix of the original network,

$$\begin{array}{c} \\ A \\ B \\ C \\ D \\ E \end{array} \begin{array}{c} \begin{array}{ccccc} A & B & C & D & E \end{array} \\ \begin{bmatrix} 0 & 0 & 1 & 0 & 1 \\ 1 & 0 & 0 & 0 & 0 \\ 0 & 1 & 0 & 1 & 1 \\ 1 & 1 & 0 & 0 & 1 \\ 1 & 0 & 1 & 1 & 0 \end{bmatrix} \end{array}$$

removing university E from the network removes the fifth row, which represents

542   GRAPHS AND NETWORKS

transmissions from E, and the fifth column, which indicates messages sent to E. Therefore, the adjacency matrix is now

$$M = \begin{bmatrix} 0 & 0 & 1 & 0 \\ 1 & 0 & 0 & 0 \\ 0 & 1 & 0 & 1 \\ 1 & 1 & 0 & 0 \end{bmatrix}$$

and we compute

$$(I+M)^3 = \begin{bmatrix} 1 & 0 & 1 & 0 \\ 1 & 1 & 0 & 0 \\ 0 & 1 & 1 & 1 \\ 1 & 1 & 0 & 1 \end{bmatrix} \begin{bmatrix} 1 & 0 & 1 & 0 \\ 1 & 1 & 0 & 0 \\ 0 & 1 & 1 & 1 \\ 1 & 1 & 0 & 1 \end{bmatrix} \begin{bmatrix} 1 & 0 & 1 & 0 \\ 1 & 1 & 0 & 0 \\ 0 & 1 & 1 & 1 \\ 1 & 1 & 0 & 1 \end{bmatrix}$$

$$= (I+M)(I+M)^2 = \begin{bmatrix} 1 & 0 & 1 & 0 \\ 1 & 1 & 0 & 0 \\ 0 & 1 & 1 & 1 \\ 1 & 1 & 0 & 1 \end{bmatrix} \begin{bmatrix} 1 & 1 & 2 & 1 \\ 2 & 1 & 1 & 0 \\ 2 & 3 & 1 & 2 \\ 3 & 2 & 1 & 1 \end{bmatrix} = \begin{bmatrix} 3 & 4 & 3 & 3 \\ 3 & 2 & 3 & 1 \\ 7 & 6 & 3 & 3 \\ 6 & 4 & 4 & 2 \end{bmatrix}$$

The resulting matrix contains no 0s, so the network will function properly for the remaining four universities if university E's computer is down.   ∎

The Kula Ring

A similar technique can be used in analyzing the trade of the Kula Ring that we discussed in the introduction to this chapter. A digraph drawn from a simplified map of the trade route appears in Figure 7-43. The adjacency matrix for this digraph is

$$M = \begin{array}{c} \\ A \\ Tr \\ D \\ W \\ Tu \\ MB \end{array} \begin{array}{c} \begin{matrix} A & Tr & D & W & Tu & MB \end{matrix} \\ \begin{bmatrix} 0 & 1 & 0 & 1 & 0 & 1 \\ 1 & 0 & 1 & 0 & 0 & 0 \\ 1 & 1 & 0 & 0 & 0 & 0 \\ 0 & 0 & 0 & 0 & 1 & 0 \\ 0 & 0 & 0 & 0 & 0 & 1 \\ 0 & 0 & 0 & 0 & 1 & 0 \end{bmatrix} \end{array}$$

One of the questions that might be asked about this trade ring is: what happens if an island drops out? We now discuss one such case; other cases will be considered as Exercises 13 and 14.

Example 4. If the Marshall Bennett Islands (MB) dropped out of the ring, no longer contributing their traditional items to the trade, can the other islands of

7.3 DIGRAPH MATRICES   543

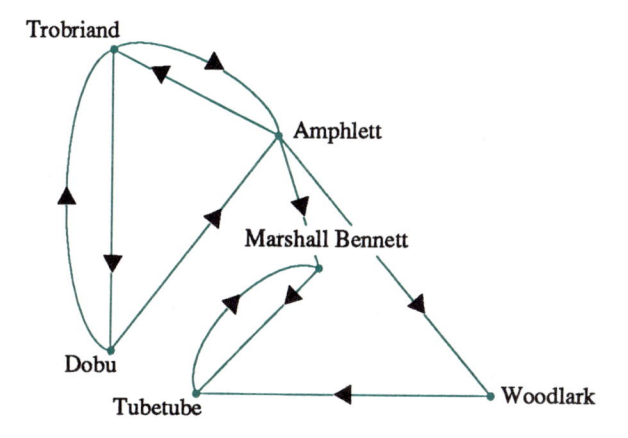

Figure 7-43

the Kula Ring all trade with one another, without the Marshall Bennetts as an intermediary?

Solution

We use the adjacency matrix obtained by deleting from the adjacency matrix of the Kula Ring the last row and last column, which concern trade from and to the Marshall Bennetts.

$$M = \begin{array}{c} \\ A \\ Tr \\ D \\ W \\ Tu \end{array} \begin{array}{c} A \quad Tr \quad D \quad W \quad Tu \\ \begin{bmatrix} 0 & 1 & 0 & 1 & 0 \\ 1 & 0 & 1 & 0 & 0 \\ 1 & 1 & 0 & 0 & 0 \\ 0 & 0 & 0 & 0 & 1 \\ 0 & 0 & 0 & 0 & 0 \end{bmatrix} \end{array}$$

First we find

$$(I+M)^2 = \begin{bmatrix} 1 & 1 & 0 & 1 & 0 \\ 1 & 1 & 1 & 0 & 0 \\ 1 & 1 & 1 & 0 & 0 \\ 0 & 0 & 0 & 1 & 1 \\ 0 & 0 & 0 & 0 & 1 \end{bmatrix} \begin{bmatrix} 1 & 1 & 0 & 1 & 0 \\ 1 & 1 & 1 & 0 & 0 \\ 1 & 1 & 1 & 0 & 0 \\ 0 & 0 & 0 & 1 & 1 \\ 0 & 0 & 0 & 0 & 1 \end{bmatrix} = \begin{bmatrix} 2 & 2 & 1 & 2 & 1 \\ 3 & 3 & 2 & 1 & 0 \\ 3 & 3 & 2 & 1 & 0 \\ 0 & 0 & 0 & 1 & 2 \\ 0 & 0 & 0 & 0 & 1 \end{bmatrix}$$

Then

$$(I+M)^4 = (I+M)^2(I+M)^2$$

$$= \begin{bmatrix} 2 & 2 & 1 & 2 & 1 \\ 3 & 3 & 2 & 1 & 0 \\ 3 & 3 & 2 & 1 & 0 \\ 0 & 0 & 0 & 1 & 2 \\ 0 & 0 & 0 & 0 & 1 \end{bmatrix} \begin{bmatrix} 2 & 2 & 1 & 2 & 1 \\ 3 & 3 & 2 & 1 & 0 \\ 3 & 3 & 2 & 1 & 0 \\ 0 & 0 & 0 & 1 & 2 \\ 0 & 0 & 0 & 0 & 1 \end{bmatrix} = \begin{array}{c} A \\ Tr \\ D \\ W \\ Tu \end{array} \begin{array}{c} A \quad Tr \quad D \quad W \quad Tu \\ \begin{bmatrix} 13 & 13 & 8 & 9 & 7 \\ 21 & 21 & 13 & 12 & 5 \\ 21 & 21 & 13 & 12 & 5 \\ 0 & 0 & 0 & 1 & 4 \\ 0 & 0 & 0 & 0 & 1 \end{bmatrix} \end{array}$$

The 0s of the matrix $(I + M)^4$ tell us that although Tubetube will continue to import goods from the other island groups, it will be unable to send anything out, unless it changes its trading patterns. We can also see that Woodlark would be severely effected if the Marshall Bennetts dropped out of the Kula Ring. ∎

Summary

Given a digraph with vertex set $V = \{v_1, v_2, \ldots, v_k\}$, the **adjacency matrix** of the digraph is the $k \times k$-matrix M with a 1 in the a_{ij}-location if the arc $a = (v_i, v_j)$ is in the digraph and with a 0 in the a_{ij}-location otherwise. For M^r, the rth power of M, the entry in the a_{ij}-location of M^r equals the number of paths of length r from v_i to v_j. If there is a 0 in the a_{ij}-location of all of the matrices $M, M^2, \ldots, M^{r-1}$ or, equivalently, in the a_{ij}-location of $M + M^2 + \ldots + M^{r-1}$, and if the entry in the a_{ij}-location of M^r is nonzero, then the distance from v_i to v_j is r. If the entry in the a_{ij}-location of $M + M^2 + \ldots + M^{k-1}$ is zero, then v_j is not reachable from v_i. To determine whether the vertex v_j is reachable from the vertex v_i without calculating the distance from v_i to v_j, compute $(I + M)^{k-1}$, where I is the $k \times k$-identity matrix. Then v_j is reachable from v_i if the entry in the a_{ij}-location of $(I + M)^{k-1}$ is nonzero and it is not reachable if that entry is zero.

Exercises

In Exercises 1 and 2, draw a pictorial representation of the digraph represented by the adjacency matrix.

1. $\begin{array}{c} \\ A \\ B \\ C \\ D \end{array} \begin{array}{c} A\ B\ C\ D \\ \begin{bmatrix} 0 & 1 & 1 & 1 \\ 0 & 0 & 0 & 1 \\ 1 & 1 & 0 & 1 \\ 1 & 1 & 1 & 0 \end{bmatrix} \end{array}$

2. $\begin{array}{c} \\ A \\ B \\ C \\ D \\ E \end{array} \begin{array}{c} A\ B\ C\ D\ E \\ \begin{bmatrix} 1 & 1 & 0 & 1 & 0 \\ 1 & 0 & 1 & 0 & 1 \\ 0 & 1 & 0 & 1 & 1 \\ 1 & 0 & 0 & 0 & 1 \\ 0 & 1 & 0 & 0 & 0 \end{bmatrix} \end{array}$

In Exercises 3 and 4, find the adjacency matrices of the digraphs in the figures.

3.

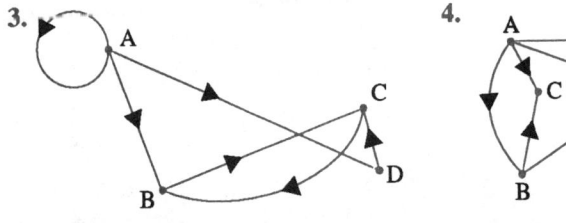

Figure 7-44

4.

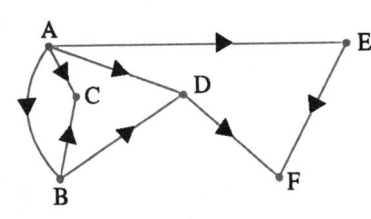

Figure 7-45

7.3 DIGRAPH MATRICES

For Exercises 5 through 8, use the adjacency matrix M of a digraph that represents daily flights of an airline between Armadale (A), Bostock (B), Cranford (C), and Dalton (D).

$$M = \begin{array}{c} \\ A \\ B \\ C \\ D \end{array} \begin{array}{c} \begin{array}{cccc} A & B & C & D \end{array} \\ \left[\begin{array}{cccc} 0 & 0 & 1 & 0 \\ 1 & 0 & 0 & 0 \\ 0 & 1 & 0 & 1 \\ 1 & 1 & 0 & 0 \end{array} \right] \end{array}$$

5. Compute M^2 and M^3.

6. Compute $M + M^2 + M^3$.

7. How many paths are there from Cranford to Armadale?

8. (a) What is the minimum number of flights required to travel from Dalton to Cranford?
   (b) List this sequence of flights.

In Exercises 9 through 11, we assume that because of anomalies in space, the four fantasy countries of Islandia, Gondor, Pern, and Xanth can only communicate with each other as follows:

    Islandia can send messages to Pern
    Pern can send messages to Gondor and Xanth
    Xanth can send messages to Islandia and Gondor

9. Find the adjacency matrix of the communication digraph.

10. Use adjacency matrices to show that Xanth cannot communicate directly with Pern, but can communicate through one intermediary.

11. Use adjacency matrices to show that there are two ways in which Islandia can communicate with Gondor: through one intermediary and through two intermediaries.

12. The administration of Maple College consists of the president, executive vice president, vice presidents for finance and for student services, and the provost. The president's decisions are transmitted to the executive vice president, who communicates them to the other vice presidents. The president also oversees the work of the provost. The board of governors of the college communicates its decisions to the president and to the provost. Use matrix methods to calculate the status of the President and of the Board of Governors of Maple College.

For Exercises 13 and 14 use the adjacency matrix corresponding to the digraph of Figure 7-43.

13. Which island groups would be able to send out their products if Tubetube dropped out of the Kula trade ring and the other five island groups remained in the ring?

14. If Amphlett dropped out of the Kula trade ring and the other five island groups remained, which two island groups could trade with each other but not with the rest of the ring?

7.4 Networks

In Section 7.1, we saw how graphs can be used to represent connections or relationships. In many applications it is necessary not only to join point *A* to point *B* but also to attach a number to the connection. For instance, it may be important to know not only that we can fly from Atlanta to Tallahassee, but also that it takes approximately 1 hour or that it costs $85.

> A graph, each of whose edges has a positive number associated with it, is called a **network**.

Routes on a map may be viewed as networks. Figure 7-46 shows a group of cities in northern Europe, the connections that can be made among them using the European equivalent of the U. S. interstate-highway system, and the corresponding distances in kilometers. Figure 7-46 is a network with seven vertices and nine edges.

Least-Cost Path

Suppose we want to find the shortest route between Utrecht and Brussels. Looking at the graph, we see that we can either go through Rotterdam and Antwerp or through Eindhoven and Antwerp. Adding the distances along the two paths, we find that if we go through Rotterdam the distance is 173 kilometers, while if we go through Eindhoven the distance is 179. Thus, the path Utrecht-Rotterdam-Antwerp-Brussels is the shortest. In general:

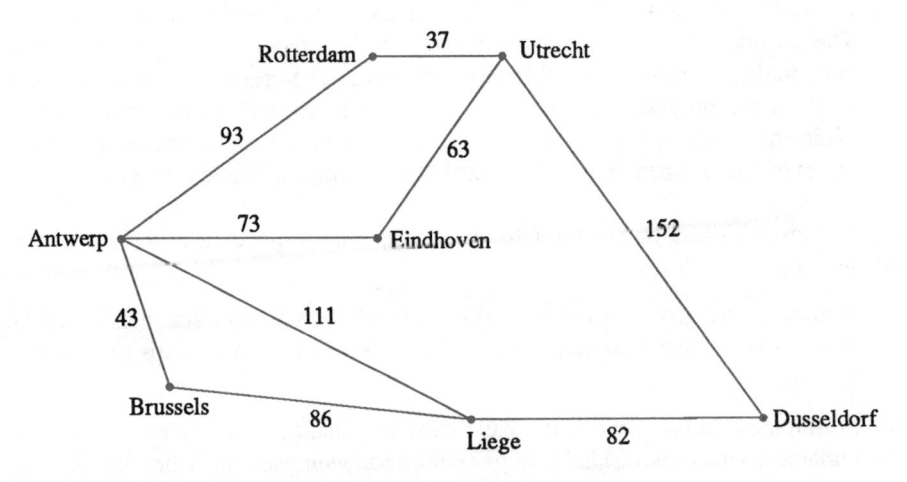

Figure 7-46

> The **length** of a path in a network is the sum of the numbers associated with the edges in the path. The shortest path between two vertices in a network is called the **least-cost path**.

We can easily determine the least-cost path in a small network because we can just find all possible paths, add the distances, and choose the shortest one. However, in a network with even a moderate number of vertices, the number of possible paths is very large. For instance, suppose a network with 15 vertices contains an edge between any two of them. Let us count the number of paths from a vertex A to another vertex B that pass through every vertex of the network. For the edge starting at A, we may go to any one of the remaining 13 vertices. There are 12 choices for the next edge, 11 choices for the edge after that, and so on, until we have a single choice for the last edge. According to the method for counting orderings in Section 3.3, there are 13! different paths of this type. Since 13! is more than 6 billion, there are that many different paths from A to B, not even counting those that pass through fewer vertices.

Finding the Least-Cost Path

There is a method that can be easily implemented on a computer to determine a least cost path through a network of any size[†]. To illustrate the method, we suppose that the network in Figure 7-47(a) represents the cost, in thousands of dollars, of shipping plastic valves from the factory (A), where they are manufactured, through three possible distribution centers (B, C, D) to their destination at a plumbing supply warehouse (E).

The method starts by labeling the vertex at A with a 0 since there is 0 cost to go from A to A; see Figure 7-47(b). Then we look at all the edges incident with A and observe that the edge from A to D has the least cost. We label D with that

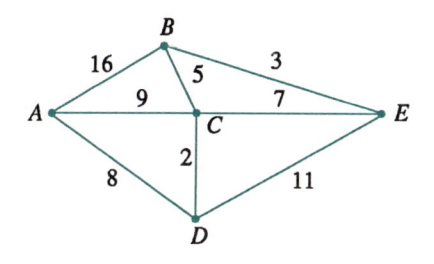

Figure 7-47(a)

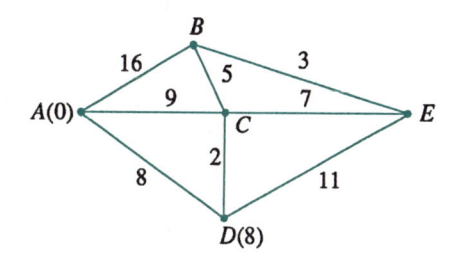

Figure 7-47(b)

[†]This Algorithm was constructed by E. Dijkstra, 1959. "A Note on Two Problems in Connection with Graphs," *Numerische Mathematik* 1:269–271.

548 GRAPHS AND NETWORKS

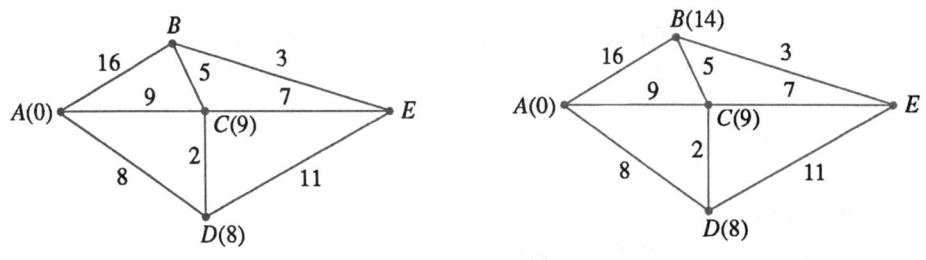

Figure 7-47(c)

Figure 7-47(d)

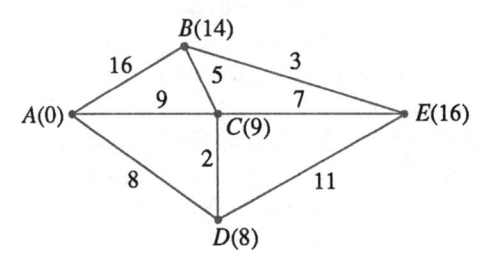

Figure 7-47(e)

cost, 8. Next, we look at all the edges linking a vertex that we have labeled, that is, A or D, with one of the remaining, unlabeled, vertices B, C, or E, and we compute the costs of the resulting paths starting from A, as shown in Table 7-48.

Labeled Vertices: A, D

Edge	Path from A	Cost
A–C	A–C	9*
A–B	A–B	16
D–C	A–D–C	8 + 2 = 10
D–E	A–D–E	8 + 11 = 19

Table 7-48

The least-cost path from A to an unlabeled vertex goes from A to C at a cost of 9, as we indicate by the asterisk. To indicate the least cost, we label C with a 9 as in Figure 7-47(c).

We again look at all edges linking a labeled vertex, now A, C, and D, to an unlabeled vertex, B or E, and compute the costs of the resulting paths starting at A in Table 7-49. To do this, we first look back at Table 7-48 and delete the edges A–C and D–C that end in C since that vertex is now labeled. Then we look at edges starting with the new vertex C (emphasized by boldface at the top of Table 7-49) and ending with unlabeled vertices. We determine the paths from A; because the least cost among these paths is 14 for A–C–B, we label B with 14 in Figure 7-47(d).

In forming Table 7-50, we delete from Table 7-49, the edges that end in the new labeled vertex B and add to the list the edge B–E that connects that vertex to

Labeled Vertices: A, C, D

Edge	Path from A	Cost
A–B	A–B	16
D–E	A–D–E	8 + 11 = 19
C–B	A–C–B	9 + 5 = 14*
C–E	A–C–E	9 + 7 = 16

Table 7-49

the remaining unlabeled vertex, E. The label 14 on vertex B in Figure 7-47(d) reminds us that the least-cost path from A to B has a cost of 14 thousand dollars. It is very important to notice that in Table 7-49, our choice of path from A to the labeled point B is the path of least cost previously established (in Table 7-48), that is, the path that passes through C. At each step, we go from A to a labeled vertex only by means of the least-cost path to that vertex, which must have been identified previously because the vertex was labeled.

Labeled Vertices: A, B, C, D

Edge	Path from A	Cost
C–E	A–C–E	9 + 7 = 16*
B–E	A–C–B–E	14 + 3 = 17
D–E	A–D–E	8 + 11 = 19

Table 7-50

From Table 7-50 we see that the least-cost path from A to E is A–C–E. Thus, the plastic valves should be shipped from factory A to the supply warehouse E through distribution center C, as shown in Figure 7-47(e), and the total cost will be 16 thousand dollars. We have found the least-cost path from A to E because at each step of the method we made the cost as low as possible.

This example was so small that we could have easily determined the least-cost path by just looking at the network. However, if we had a more complicated problem, such as trying to determine the route with the shortest driving time for sending a truck from Akron, Ohio to Albuquerque, New Mexico using the interstate-highway network, this method would be helpful because a computer could use it to determine the fastest route.

The next example illustrates the method for finding the least-cost path (here, in terms of time) in a somewhat more complicated network.

Example 1. As part of a study of information flow in a large organization, the network of Figure 7-51 represents the average number of days for a memo from one member of an organization to reach another member and to be acted upon. Find the fastest route for information about a new product to get from the general sales manager (A) to a salesperson (G).

550 GRAPHS AND NETWORKS

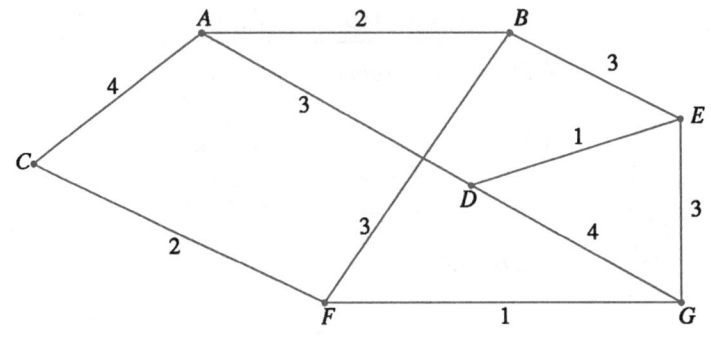

Figure 7-51

Solution

The quickest, that is least-cost, edge starting at A is A–B, so we begin the labeling with these two vertices, to construct Table 7-52 of edges between a labeled vertex, A or B, and an unlabeled vertex. We no longer display the cost computation.

Labeled Vertices: A, B

Edge	Path from A	Time
A–C	A–C	4
A–D	A–D	3*
B–E	A–B–E	5
B–F	A–B–F	5

Table 7-52

Thus, we label D, delete the edge ending in D from Table 7-52 and add the edges connecting D with unlabeled vertices to form Table 7-53.

Labeled Vertices: A, B, D

Edge	Path from A	Time
A–C	A–C	4*
B–E	A–B–E	5
B–F	A–B–F	5
D–E	A–D–E	4*
D–G	A–D–G	7

Table 7-53

Since this time there is a tie between two different vertices, C and E, for the shortest path from A, we label both vertices and treat both of them as in the previous tables.

Labeled Vertices: A, B, C, D, E

Edge	Path from A	Time
B–F	A–B–F	5*
D–G	A–D–G	7
C–F	A–C–F	6
E–G	A–D–E–G	7

Table 7-54

We emphasize that the indicated path from A to E in Table 7-54 was chosen because it was the least-cost path in the preceding table. Since we now label vertex F, the only unlabeled vertex is G and, taking care to use the least-cost paths from previous tables to determine the "path from A," we obtain Table 7-55.

Labeled Vertices: A, B, C, D, E, F

Edge	Path from A	Time
D–G	A–D–G	7
E–G	A–D–E–G	7
F–G	A–B–F–G	6*

Table 7-55

Thus, the quickest route for information from the sales manager A to salesperson G goes through B and F, at an average time of 6 days. ■

Communication Networks

Communication networks describe the paths among various locations that can communicate with each other. The paths might consist of closed circuit television, computer, or telephone connections; the locations could be universities, branch offices of a corporation or government offices. Often it is impractical to link every location directly to every other one. Just to link the central computer in each of 100 offices with every other one would require a total of

$$C_2^{100} = \frac{100 \cdot 99}{2 \cdot 1} = 4950$$

direct links, according to the combinations formula of Section 3.4. It is usually cheaper and more efficient to use a system based on paths from one office to another which go through intermediate offices. In finding these paths we not only want to enable every office to communicate with every other one, but to be able to do it as efficiently as possible, that is, with the least cost.

Minimal Spanning Trees

In order to discuss the least-cost communication problem, we introduce a few new terms. We assume that we are working with a connected graph, as in Section 7.1: a graph in which every two vertices can be joined by a path.

> A **cycle** in a graph is a path that starts and ends at the same vertex.

For instance, the path Liege–Antwerp–Brussels–Liege is a cycle in the road network of Figure 7-46.

> A **tree** is a connected graph that contains no cycles. For a graph G with vertex set V and edge set E, a **spanning tree** is a tree G' with the same vertex set V and with edge set E' a subset of E.

Using a spanning tree to determine which computers to connect gives an efficient way for the 100 offices to communicate. However, there may be many different spanning trees for the same graph. If the graph is a network in which each edge has a cost associated with it, then the **cost** of the spanning tree is the sum of the costs of its edges. A spanning tree of least cost is called a **minimal spanning tree**.

There is a labeling method for finding the minimal spanning tree of a communication network that is a variation of the method we have presented for finding the least-cost path in a network. However, a least-cost path may pass through only a few vertices of the network, whereas the tree must contain all the vertices. The idea of the method is to construct the minimal spanning tree to include, and thus label, one new vertex at each stage, at minimal cost, and to continue until all vertices of the network have been labeled. As in the preceding method, the edges of the tree are added only if they join a labeled vertex to one that is unlabeled. Since in order to close a cycle of labeled vertices we would have to include an edge between two labeled vertices, the method will produce a tree. We illustrate the detailed method by means of the network of Figure 7-56.

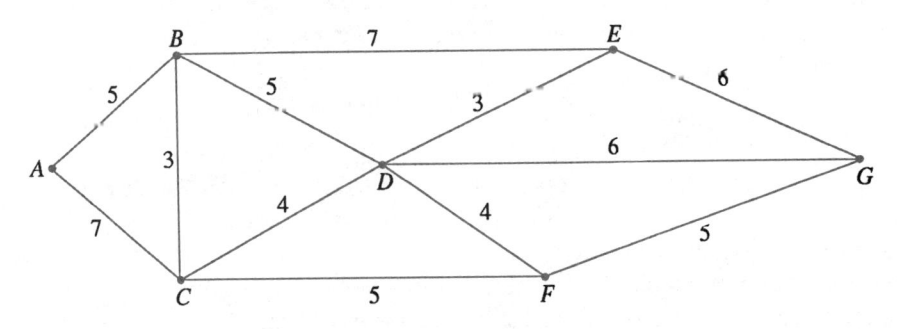

Figure 7-56

First we choose an edge that has the least cost. In Figure 7-56, one such edge, that costs 3 units, connects B to C. We label these two vertices and in Table 7-57 we list all the edges of the network connecting a labeled vertex (listed first) with an unlabeled vertex. The cost of the edge is listed next to it. We will again place an asterisk next to the lowest cost, and we also print the corresponding edge in boldface because we will want to pick it out easily when we construct the minimal spanning tree as the last step of the method. Since the construction will begin with the edge B–C, in order that we not lose sight of it, we write

$$B\text{–}C \quad 3*$$

Labeled Vertices: B, C

B–A 5	B–E 7	**C–D 4***	B–D 5	C–A 7	C–F 5

Table 7-57

Table 7-58 presents the remaining steps of the method. To begin, we look back to Table 7-57 and delete the edges B–D and C–D that end with the vertex D that we just labeled. We then add edges connecting D to unlabeled vertices. We repeat this operation, which is already familiar from the least-cost-path method, until all vertices are labeled in the final step of Table 7-58. In general, this form of the labeling method takes more steps than the form in which we originally met it, for least-cost paths. However, each individual step is simpler because we do not have to look back to the starting vertex to find the least-cost path.

Labeled Vertices: B, C, D

B–A 5	C–A 7	**D–E 3***	D–G 6	B–E 7	C–F 5	D–F 5

Labeled Vertices: B, C, D, E

B–A 5	C–F 5	D–G 6	C–A 7	**D–F 4***	E–G 6

Labeled Vertices: B, C, D, E, F

B–A 5*	D–G 6	F–G 5	C–A 7	E–G 6

Labeled Vertices: A, B, C, D, E, F

D–G 6	E–G 6	**F–G 5***

Table 7-58

The method concludes because all vertices are now labeled, with G added from the final step of Table 7-58. Note that in the next-to-last step there were two edges, B–A and F–G, with a minimal cost of 5, but we chose only one, B–A. We cannot speed up the process by labeling all vertices in a tie, as we did in the least-cost-path method, because if we did, we might form a cycle and therefore not construct a tree, as required. Using the boldface edges:

$$B\text{–}C \quad C\text{–}D \quad D\text{–}E \quad D\text{–}F \quad B\text{–}A \quad F\text{–}G$$

we have the tree pictured in Figure 7-59. From this figure we see that the total cost of the tree is $5 + 3 + 4 + 3 + 4 + 5 = 24$ (units).

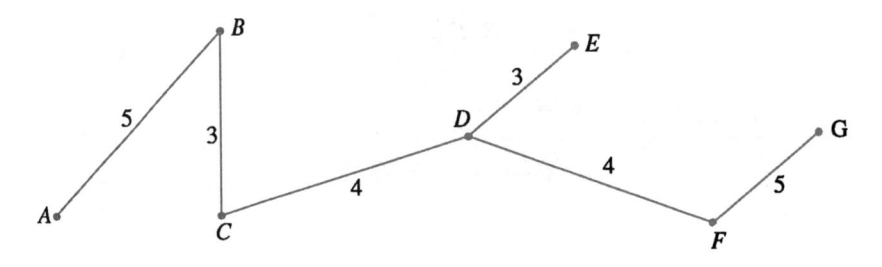

Figure 7-59

Example 2 demonstrates how a minimal spanning tree is used in a communications network.

Example 2. The diagram in Figure 7-60 shows the plans for upgrading a closed-circuit television network on a university campus. The cost of upgrading each link is shown as an integer on each edge. The costs are given in units of one thousand dollars; so the cost of upgrading the link between A and B would be $5000, whereas

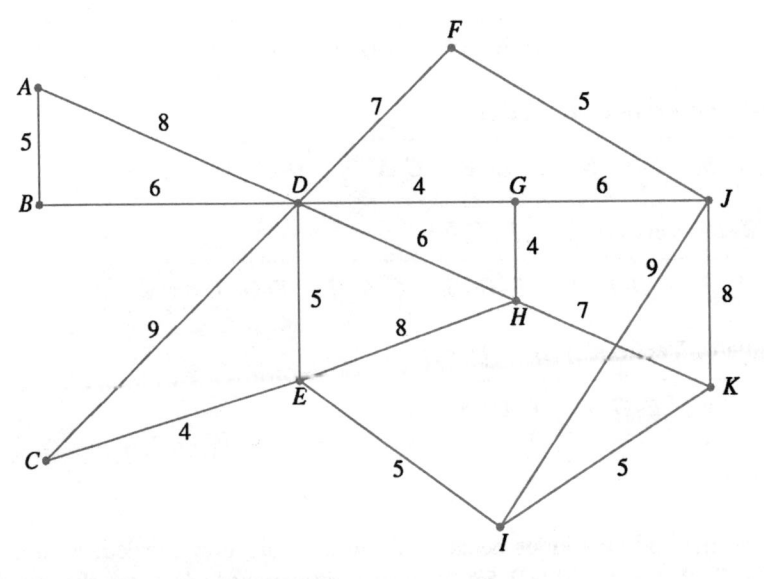

Figure 7-60

7.4 NETWORKS

the cost for the link between B and D would be $6000. The problem is to find a minimal spanning tree for the network, that is, to choose a set of links (edges) to be upgraded so that there is a path from any station (vertex) to any other station and so that the set of links chosen costs no more than any other such set of links.

Solution

There are two edges that have the least cost (4 units), namely C–E and D–G, and we choose

$$C\text{–}E \quad 4*$$

We proceed just as before, with the steps shown in Table 7-61.

Labeled Vertices: C, E

| C–D 9 | E–D 5* | E–H 8 | E–I 5 |

Labeled Vertices: C, D, E

| E–H 8 | D–A 8 | D–F 7 | D–H 6 | E–I 5 | D–B 6 | D–G 4* |

Labeled Vertices: C, D, E, G

| E–H 8 | D–A 8 | D–F 7 | G–H 4* | E–I 5 | D–B 6 | D–H 6 | G–J 6 |

Labeled Vertices: C, D, E, G, H

| E–I 5* | D–B 6 | G–J 6 | D–A 8 | D–F 7 | H–K 7 |

Labeled Vertices: C, D, E, G, H, I

| D–A 8 | D–F 7 | H–K 7 | I–K 5* | D–B 6 | G–J 6 | I–J 9 |

Labeled Vertices: C, D, E, G, H, I, K

| D–A 8 | D–F 7 | I–J 9 | D–B 6* | G–J 6 | K–J 8 |

Labeled Vertices: B, C, D, E, G, H, I, K

| D–A 8 | G–J 6 | K–J 8 | D–F 7 | I–J 9 | B–A 5* |

Labeled Vertices: $A, B, C, D, E, G, H, I, K$

| D–F 7 | G–J 6* | I–J 9 | K–J 8 |

Labeled Vertices: $A, B, C, D, E, G, H, I, J, K$

| D–F 7 | J–F 5* | I–J 9 | K–J 8 |

Table 7-61

556   GRAPHS AND NETWORKS

The minimal spanning tree is constructed by the boldface edges:

C–E, E–D, D–G, G–H, E–I, I–K, D–B, B–A, G–J, J–F

as shown in Figure 7-62. The figure presents the least expensive plan for upgrading the closed-circuit television network, which will cost $49,000. ∎

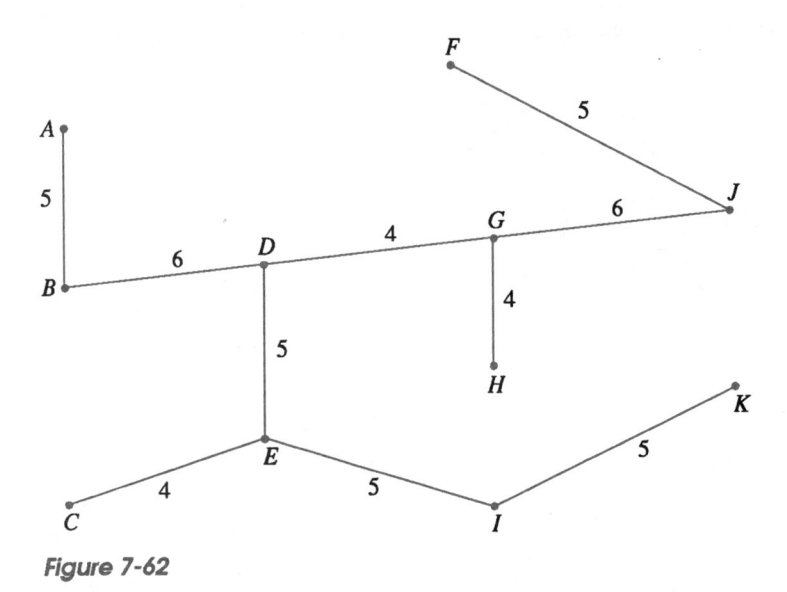

Figure 7-62

Summary

A graph, each of whose edges has a positive number associated with it, is called a **network**. The **length** of a path in a network is the sum of the numbers associated with the edges of the path. The shortest path between two vertices in a network is called the **least-cost path**. To find the least-cost path of a network from a vertex A to a vertex X, start by labeling A with a 0. Look at all the edges incident with A that are not loops, find the one with the smallest number associated with it, and label the vertex other than A with that number. At each succeeding stage of the method, list all edges connecting a labeled vertex with an unlabeled one and add the number associated with the edge to the label number. Choosing the edge or one of the edges for which the sum is smallest, label the unlabeled vertex of the edge with that sum. Continue until the destination vertex X is labelled. The label of X is the cost of the least-cost path. A **cycle** in a graph is a path that starts and ends at the same vertex. A **tree** is a connected graph that contains no cycles. For a graph G with vertex set V and edge set E, a **spanning tree** is a tree G' with the same vertex set V and with edge set E' a subset of E. If the graph is a network in which each edge has a cost associated with it, then the **cost** of the spanning tree is the sum of the costs of its edges. A spanning tree of least cost is called a **minimal spanning tree**. To construct a minimal spanning tree, first choose an edge or one of the edges that has the least cost. Label its two vertices to identify them (it is not necessary to use a number to label the vertex), and select this edge as a member of the minimal spanning tree. At each step of the method, list all edges connecting a vertex previously labeled with an unlabeled vertex and choose one edge with least cost for the minimal spanning tree. Continue until all vertices are labeled.

Exercises

In Exercises 1 through 6 use the algorithm to find a least-cost path between the specified vertices of the network in Figure 7-63.

1. A, E
2. A, I
3. F, H
4. F, E
5. C, F
6. C, I

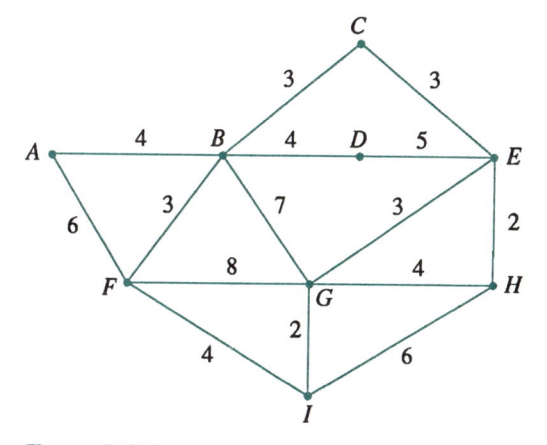

Figure 7-63

7. The We Move Anything Company has a contract to move some very heavy equipment from Martinsburg to Brownsville. Since the truck will move slowly, the company wishes to take the shortest possible route. From the graph in Figure 7-64, which shows the available routes with their distances, use the algorithm to choose a route for the company to use.

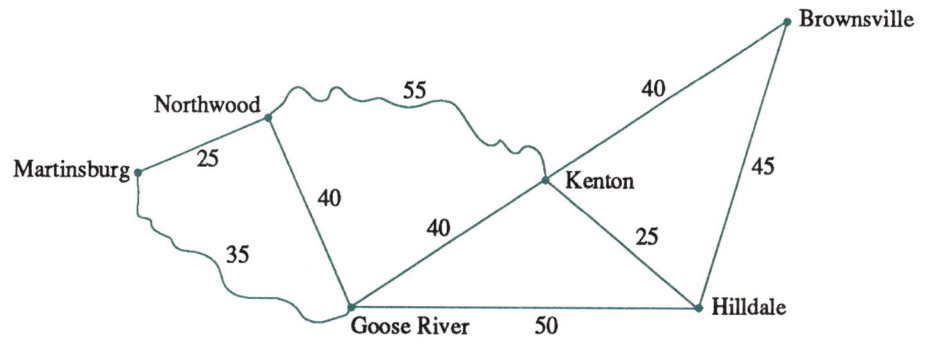

Figure 7-64

558 GRAPHS AND NETWORKS

In Exercises 8 through 11 state whether the graph shown is a tree and if it is not, indicate why not.

8.

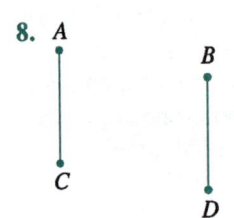

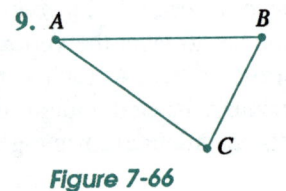

Figure 7-65

9.

Figure 7-66

10.

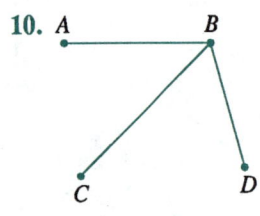

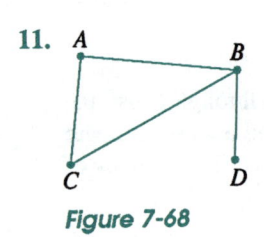

Figure 7-67

11.

Figure 7-68

In Exercises 12 and 13 use the algorithm in order to find a minimal spanning tree for the graph shown.

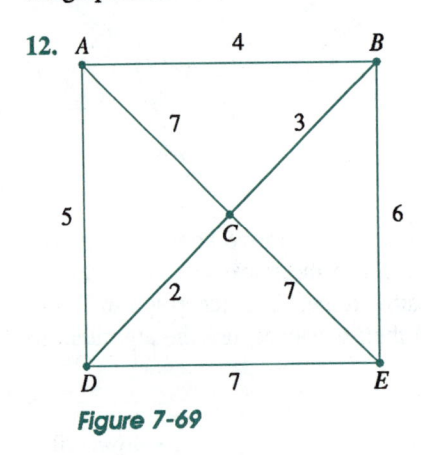

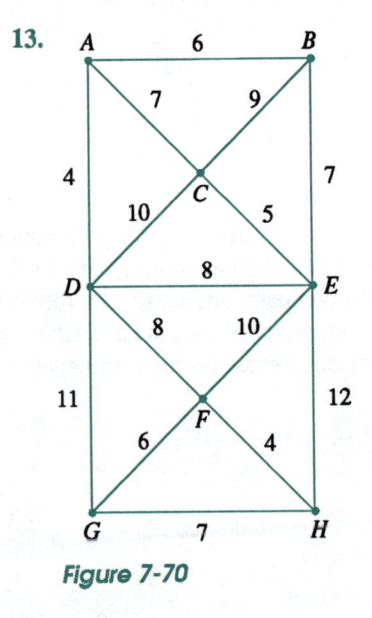

12.

Figure 7-69

13.

Figure 7-70

14. The Westside Investment Company has an office located in each of ten western cities: San Francisco (SF), Stockton (St), Los Angeles (LA), Fresno (Fr), Anaheim (An), San Diego (SD), Reno (Re), Las Vegas (LV), Phoenix (Ph), and Tucson (Tu). The company wants to establish a private communication system that will connect all its offices in the most economical way. Figure 7-71 shows in

schematic form the network of these offices. Find the minimal spanning tree that will connect these offices.

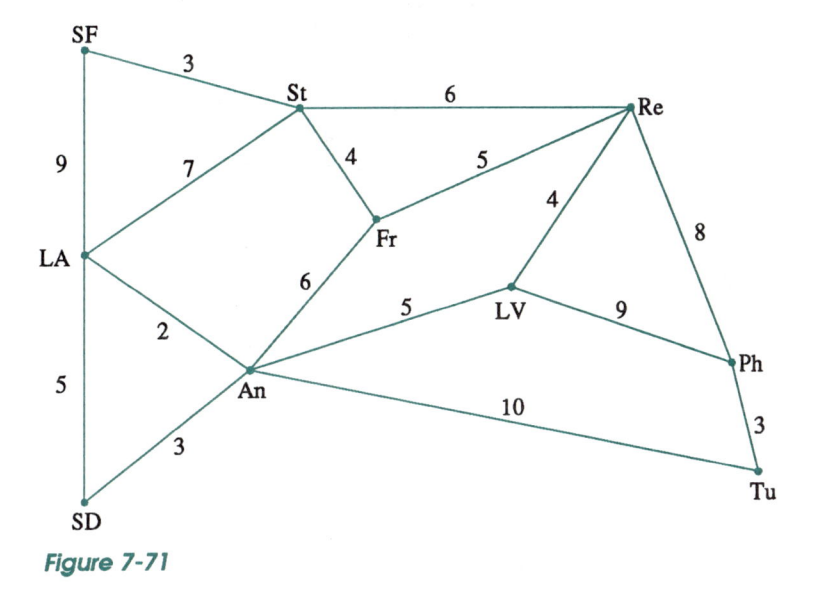

Figure 7-71

Review Exercises for Chapter 7

Exercises 1 through 3 concern the digraph in Figure 7-72 and can be answered by inspecting the figure.

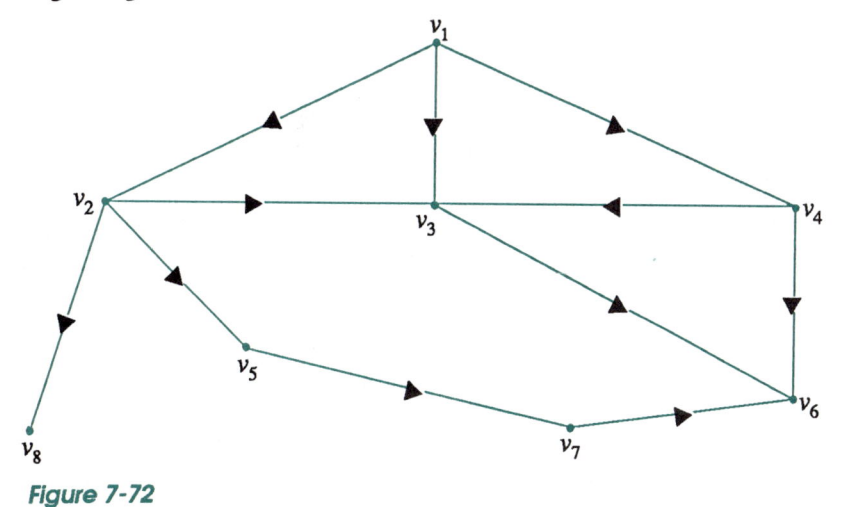

Figure 7-72

1. List the vertices reachable from v_2.
2. Calculate the distance from v_2 to v_6.
3. Calculate the status of v_5.

560   GRAPHS AND NETWORKS

4. A tasting panel tries seven types of tea, two at a time, and the majority preferences are:

   Lapsang over Matte, Gunpowder, Keemun, and Pekoe
   Matte over Oolong and Earl Grey
   Earl Grey over Lapsang and Pekoe
   Gunpowder over Earl Grey, Pekoe, and Keemun
   Keemun over Earl Grey, Matte, Pekoe, and Oolong
   Oolong over Lapsang, Earl Grey, and Gunpowder
   Pekoe over Matte and Oolong

   (a) Calculate the total number of triples in this tournament.
   (b) Calculate how many of those triples are cyclic.

In the graphs of Exercises 5 and 6, find an Euler circuit if one exists.

5.

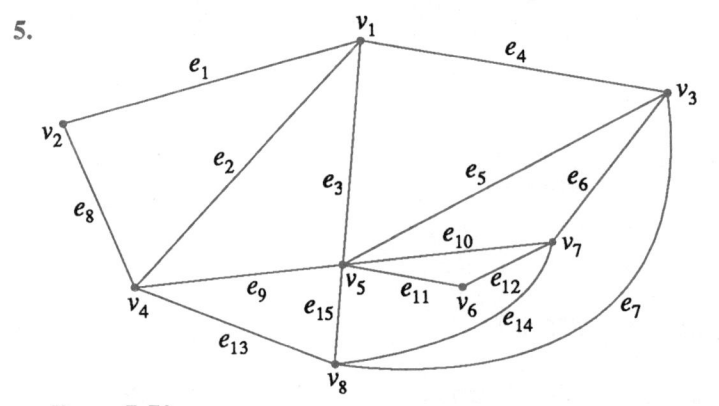

Figure 7-73

6.

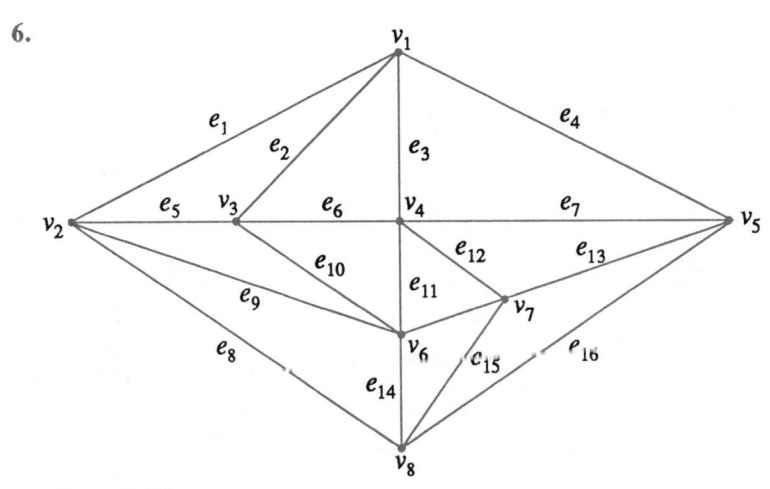

Figure 7-74

Exercises 7 through 9 concern the digraph in Figure 7-75.

7. Write the adjacency matrix of the digraph.

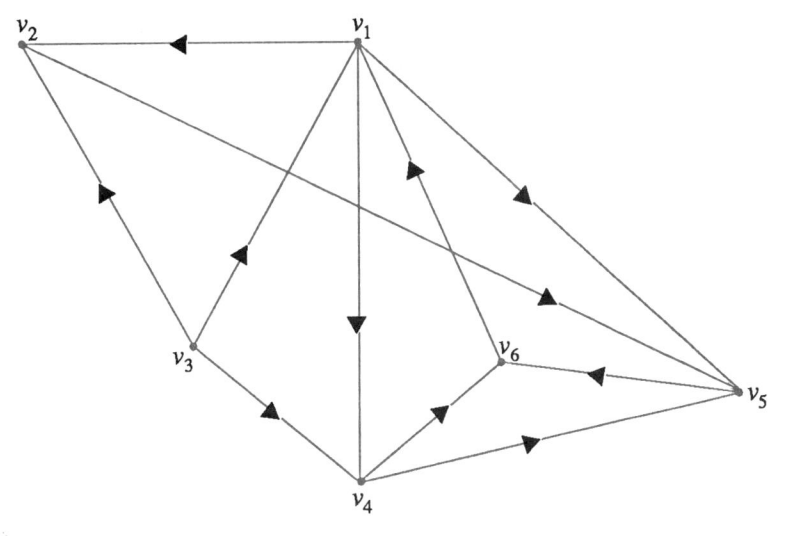

Figure 7-75

8. Use matrix methods to find all vertices whose distance from vertex v_1 is two.

9. Suppose vertex v_4 is removed from the digraph. Use matrix methods to determine which vertices are reachable from vertex v_1.

10. Draw a graph showing consanguinity (blood relationship) among the following nine people. (Constructing a family tree may be helpful.)

    Alice and Sally are sisters
    Alice marries John and they have two children, Joe and Kay
    Sally marries Fred and they have one child, Art
    Sally and Fred are divorced
    Fred marries Betty and they have one child, Tom

11. The White Rose Perfume Company wants to find the least expensive route for sending its flower essences from the flower fields (FF) to the perfume factory (PF). Figure 7-76 shows the possible routes and their associated costs. Use the appropriate algorithm to find the least-cost path.

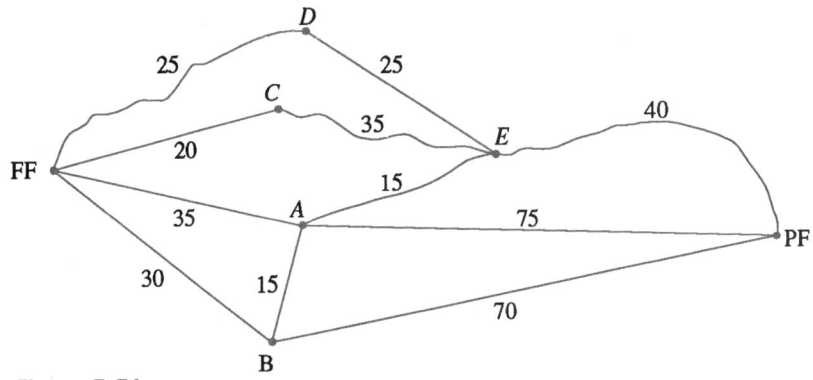

Figure 7-76

12. Figure 7-77 represents a communications network. Use the appropriate algorithm to construct a minimal spanning tree that connects the vertices of the network with each other at the least possible cost.

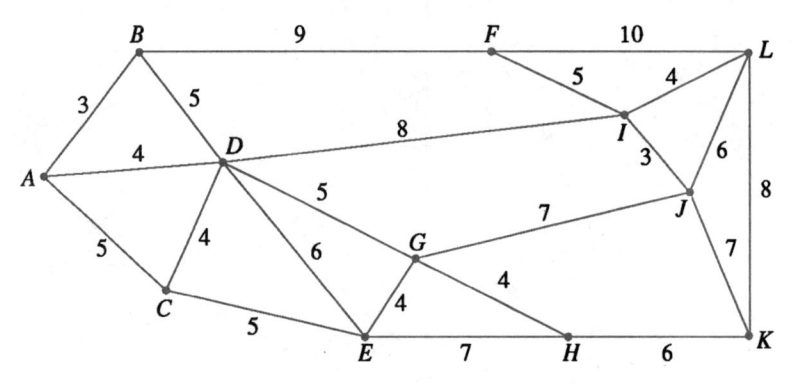

Figure 7-77

8

Mathematics of Finance

Home Mortgages

For most people, buying a house is the largest investment they will ever make. In comparison with prices in the mid-seventies, house prices have risen all over the country and in some areas this price rise has been spectacular. Because this has made it harder to sell houses, real estate professionals have come up with all sorts of "creative financing" to make it possible for more people to buy homes.

For many years, a mortgage usually meant a loan at a fixed rate of interest paid over a twenty- to thirty-year period. Now banks also offer adjustable-rate mortgages, which fluctuate with changes in national interest rates and graduated-rate mortgages, which usually start with a relatively low rate for the first year and then rise by agreed-upon amounts at regular intervals. Second mortgages, taken out to complete the financing or to make home improvements, are subject to an even greater variety of terms.

Whatever financing the buyers arrange, there are limitations on the amount of money that they can borrow. A typical rule for estimating whether a household can afford the house they want is the following: 28% of the household monthly gross income should be enough to pay the mortgage, real-estate taxes, and insurance. As an example, Jack and Judy Miller both work and have a combined gross income of $5000 a month, 28% of which is $1400. Suppose taxes and insurance for the sort of home they are considering will be about $250 a month. Then they can afford a mortgage with payments of $1150 a month. The size of the mortgage that the Millers can obtain for these payments will depend on the interest rate they are charged and other terms of the mortgage. We will compute mortgage amounts for the Millers in Section 8.5.

Of course, before the Millers can think of making monthly mortgage payments they must have saved money for a down payment on their home. For most mortgages, the cash down payment must be at least 10% of the price of the house, and often the bank will require 20% (or more) for a favorable interest rate. If that is the case, since few families can save the entire 20%, the Millers will probably have to obtain a second mortgage. In addition, the Millers will most likely be charged "points" (a loan-origination fee that is a percentage of the amount of the mortgage) and various other closing costs.

We can see that buying a house is not as simple as buying a loaf of bread or even buying a car and committing a part of your salary for the next 36 months to pay for it. Understanding more about the mathematics of finance helps in the intelligent management of money and particularly, in the important negotiations involved in buying a house.

8.1 Simple Interest

The mathematics of finance involves a considerable amount of arithmetic computation. We assume that you will use a calculator to do the necessary arithmetic. The simplest kind of calculator is all that you will need for Sections 8.1 and 8.2. From Section 8.3 on, it would be convenient to use a scientific calculator. However, we have included in the Appendix Table B which, together with a simple calculator, is all you will need for the rest of the chapter. The use of Table B will be explained in Section 8.3.

Computing Interest

The fee paid for a loan, that is, for the use of a sum of money over a period of time, is called the **interest** and the sum borrowed is called the **principal**. If the money is lent for a relatively short time (generally, a year or less), the fee is often in the form of *simple interest*. For a **simple interest** loan, the interest paid is a percentage of the principal. This percentage is called the **interest rate** of the loan. The interest rate is written either as a percent, such as 12% or 9.5%, or in the equivalent decimal form, .12 or .095, respectively.

> The relationship between the interest, I, the principal, P, and the interest rate, i, is expressed by the formula
>
> $$I = Pi$$
>
> where i is given in decimal form.

For example, suppose General Construction Corporation arranges to buy $600,000 worth of structural steel for an office building from United Steel Company. Under the terms of the contract, the full $600,000 amount will be due 30 days after delivery. Payment may be made 60 days after delivery provided General Construction pays United Steel a one-time penalty consisting of 1% simple interest on the $600,000. To find out how much the interest penalty would be, we convert the 1% interest rate to decimal form, $i = .01$, and find that the interest payment would be

$$I = Pi = 600,000(.01) = 6000$$

that is, $6000.

The General Construction treasurer has $600,000 available when the steel is delivered, so she could pay the bill at that time, or she could invest the money and pay the bill later. It is not to her advantage to pay the bill immediately because she can earn interest on this considerable sum for at least 30 days. Suppose she can lend the $600,000 to one borrower for 30 days, receiving simple interest at an annual rate of 7% or to another borrower for 60 days, receiving simple interest at an annual interest rate of 8.5%. If she chooses the 60-day loan, in addition to the $600,000 due she must pay $6000 interest to United Steel.

If the treasurer lends the money just for the initial 30 days, how much will her company earn from this investment? The interest rate for this loan was expressed in terms of an **annual interest rate** of 7%, or .07, which is the interest rate for an entire year. Expressing all interest rates over the same time period, one year, makes it easier to compare loan programs. We will use r to represent an *annual* interest rate.

The interest for the 30-day loan is still computed by $I = Pi$, where i means the interest rate for the duration of the loan, which is a fraction of a year. This interest rate is related to the annual rate by the formula $i = rt$, where r is the annual interest and t is the duration of the loan. To compute the interest the company will receive from this loan, we substitute rt for i and thus calculate

$$I = Prt$$

It is customary in the mathematics of finance to designate the number of days in the year by 360, rather than by the actual 365 (or 366 in a leap year). Similarly, a month is assumed to be 30 days. Thus, the fraction of a year for a 30-day loan is calculated to be

566   MATHEMATICS OF FINANCE

$$t = \frac{30}{360}$$

and the interest General Construction will receive from this loan comes to

$$I = Prt = 600{,}000(.07)\frac{30}{360} = 3500$$

Of course, the 60-day loan at an 8.5% annual interest rate produces more interest for General Construction:

$$I = Prt = 600{,}000(.085)\frac{60}{360} = 8500$$

However, recall that if General Construction does not pay the $600,000 within 30 days, it must pay $6000 interest. Therefore, General Construction will really earn only $8500 − $6000 = $2500 by investing the principal in this way. Thus, we conclude that the treasurer of General Construction should pay United Steel $600,000 on the 30th day. By lending out the money for 30 days, her company earns $3500. In contrast, if General Construction withholds payment to United Steel for 60 days, it can only earn a net of $2500.

Finding the Interest Rate

General Construction was able to lend money at a 7% or 8.5% annual rate, depending on the length of the loan. How much is United Steel charging General Construction for the use of the money? To make this comparison, we convert the 1% interest penalty to an annual interest rate in the following example.

Example 1. United Steel will charge General Construction at a simple-interest rate of 1% for delaying payment for 30 days. Calculate the annual rate for this loan.

Solution

We wish to determine the annual rate r when $i = .01$ and the time period is 30 days, so that $t = \frac{30}{360}$. Since $i = rt$, we have

$$.01 = r\left(\frac{30}{360}\right)$$

and solving the equation,

$$r = (.01)\left(\frac{360}{30}\right) = .12$$

We see that United Steel is charging interest at an annual rate of 12%.   ∎

When the interest payment is known, rather than the interest rate, the formula $I = Prt$ is useful in computing annual interest for purposes of comparison, as the next example demonstrates.

Example 2. The $350 premium on a homeowner's insurance policy can either be paid one year in advance or else in two installments: $175 at the beginning of the year and $190 (including a $15 service charge) after 6 months. Since the service charge is interest for keeping $175 for 6 months, what is the annual interest rate (to the nearest whole percent) for this loan.

Solution

The amount borrowed is $P = 175$, the interest charged is $I = 15$, and the time period is $t = \frac{1}{2}$ year. Substituting into the formula $I = Prt$, we obtain

$$15 = 175 \cdot r \cdot \frac{1}{2}$$

and thus the annual interest rate (rounded to two decimal places) is

$$r = \frac{30}{175} = .17$$

The homeowner will be paying interest at a 17% annual rate if he pays in two installments. ∎

Amount

The sum due at the end of the lending period is called the **amount** of the loan (also called the **future value**). The amount is represented by the symbol S; it consists of the principal P plus the interest I.

In symbols,

$$S = P + I = P + Prt$$

For instance, for the 30-day loan of $600,000 at an annual rate of 7% that the treasurer of General Construction was considering, the amount is

$$S = P + Prt = 600{,}000 + 600{,}000(.07)\frac{30}{360}$$
$$= 600{,}000 + 3500$$
$$= 603{,}500$$

Thus, the principal of $600,000 would be worth $603,500 after 30 days.

Bonds

Corporations and government bodies raise money by issuing bonds. In return for the payment of an initial sum, the **principal**, a **bond** obligates the issuer to pay a given amount, called the **face value** of the bond, at a **maturity date** several years in the future. Usually, the bondholder also receives periodic interest payments at a stated annual interest rate before the bond matures. The bondholder thus receives the face value of the bond plus a number of interest payments, each computed as simple interest.

Example 3. Pacific Gas and Electric Company has issued bonds that will mature in 14 years and that will yield 7.5 % annual interest, to be paid semiannually. Calculate the total amount paid on such a bond with a face value of $30,000 over the 14-year period.

Solution

Each semiannual payment is the interest on a principal of $30,000 for $\frac{1}{2}$ year at 7.5% annual interest rate, so it is calculated by the formula

$$Prt = 30,000(.075)\frac{1}{2} = 1125$$

In 14 years there will be 28 semiannual payments; thus, the total interest will be $I = 28 \cdot 1125 = 31,500$. We conclude that the total amount paid out on this bond is

$$S = P + I = 30,000 + 31,500 = 61,500$$ ∎

Present Value

Corporations can often make good use of cash that they are holding for a brief time by investing in short-term bank certificates of deposit. The certificates pay interest in the form of *compound interest*, which we will discuss in Section 8.3. However, compound interest can also be expressed in terms of an equivalent simple-interest rate and these investments treated as simple-interest loans.

The treasurer of General Construction must pay a payroll-tax bill of $75,000, due in 45 days. She wishes to invest exactly enough money in a 45-day bank certificate of deposit paying an 8% annual interest rate so that she has the $75,000 she needs to pay the tax bill when the loan matures. Thus, she wants to invest a sum P in the certificate so that the amount is precisely $S = \$75,000$.

Notice that this problem is the reverse of the ones we have been studying. Previously, we had a certain principal P to be borrowed and we were interested in determining S, the amount of the loan at maturity. Now we know that we want the amount to be $75,000 under the given time period and interest rate. When we think of the principal as the sum of money required to produce a given amount, it is called the **present value**.

The relationship between P (principal or present value, depending on how you think of it) and the amount S has already been described by an equation

8.1 SIMPLE INTEREST

$$S = P + Prt$$

Since we know S, the problem is to determine P, and it will be convenient to have a general formula for this purpose. Notice that we can factor P from the two terms on the right-hand side

$$S = P(1 + rt)$$

Dividing this equation by $1 + rt$ gives us the formula for the present value P

$$\boxed{P = \frac{S}{1 + rt}}$$

Applying this formula to the treasurer's problem of obtaining $S = 75{,}000$ after 45 days, that is $t = \frac{45}{360}$, then at an annual interest rate of $r = .08$, the present value is

$$P = \frac{S}{1+rt} = \frac{75{,}000}{1+(.08)\frac{45}{360}} = \frac{75{,}000}{1+.01} = 74{,}257.43$$

Thus, she should invest $74,257.43 in the certificate.

Notice that we rounded the answer to the nearest cent, which we will always do when the answer does not come out to a whole dollar.

Example 4. Bill owns a house that he estimates is worth $130,000 at present and he thinks the price will go up to $135,000 in 8 months. Bill can sell the house now and invest the money in a certificate of deposit that yields 7.5% simple interest or wait and sell the house in 8 months.

(a) Calculate the present value of $135,000 in 8 months at a 7.5% annual interest rate.

(b) Is it more profitable for Bill to sell now or should he wait 8 months?

Solution

(a) Since 8 months is $\frac{8}{12}$ of a year, it follows that $t = \frac{8}{12}$. The annual interest rate is $r = .075$. The present value of the $135,000 selling price Bill expects to get in 8 months is therefore calculated as

$$P = \frac{S}{1+rt} = \frac{135{,}000}{1+(.075)\frac{8}{12}} = 128{,}571.43$$

(b) By Part (a), in order to obtain $135,000 in 8 months, Bill would have to get $128,571.43 now for the house. Since the present selling price is $130,000, and $135,000 available 8 months in the future is worth considerably less now, it would be more profitable for Bill to sell the house immediately. ■

MATHEMATICS OF FINANCE

Summary

If a sum of money P, the **principal** is loaned at simple interest at a rate i given in decimal form, then the **interest** I charged for the loan is computed by $I = Pi$. If r is the **annual interest rate**, that is, the rate for one year, then the interest i for a time t expressed as a fraction of a year, is defined by $i = rt$ and therefore, the interest I charged for a loan with principal P at an annual interest rate r for time t is computed by $I = Prt$. If the duration of the loan is in days, then the time t is calculated by dividing the number of days by 360; if the duration is in months, then t is divided by 12. The **amount** S of the loan, the total to be paid back, is $S = P + I$. The sum of money that must be invested for a time t at an annual interest rate r in order to produce the amount S is called the **present value** of the investment, and it is computed by the formula

$$P = \frac{S}{1 + rt}$$

Exercises

1. Calculate the interest on a loan of $1200 for one month at an annual interest rate of 15%.

2. Calculate the semiannual interest payment for a $10,000 Exxon bond that pays interest at a 6.5% annual rate.

3. Calculate the amount of a loan of $125,000 at a 10% annual interest rate for 9 months.

4. A treasury bond with a face value of $50,000 matures in 3 years and pays interest semiannually at a 9% annual rate. Calculate the amount of the bond at maturity.

5. Calculate the present value of a loan that will produce $15,500 in 2 months if the annual interest rate is 7.5%.

6. Sam's Computer Shop signs a contract with a computer manufacturer to receive a shipment of a new model of its computer, with the full payment of $150,000 for the computers due in 120 days. Sam's can invest money at a 10% annual interest rate for 120 days. How much should it invest in order to have exactly $150,000 available to pay the bill when it is due?

7. Tickets for the antique-car show cost $20 at the door the day of the show, but can be purchased a month ahead of time at the group rate of $15 through a car club. A car-club member planning to go to the show buys her ticket at the door, figuring that she has the use of the $15 for a month, and the $5 difference in price represents the interest for the use of the $15. What is the annual interest rate for this loan?

8.1 SIMPLE INTEREST   571

8. While traveling, Ann obtains a cash advance on her credit card, which she pays back in full when her bill arrives 3 weeks later. At that time she must pay a service charge of 1% of the sum she borrowed, so that can be viewed as interest for 3 weeks (that is, 21 days). Calculate the corresponding annual interest rate for the cash advance.

9. Juan's Boutique was two months late in paying $800 it owed the state in sales taxes. The state sent Juan a bill for interest (including penalties) on the $800 at a 20% annual rate. What was the total on the bill?

10. Suppose that when the national debt was $1 trillion (= $1,000,000 million), the annual interest rate on the debt was 8%. Calculate the daily interest on the debt at that time, to the nearest $1 million

11. John's MasterCard monthly statement says he owes $1700 and that the annual interest rate on his card is 18%. If John sends MasterCard the minimum payment of $100 and does not use his card for the next month, how much will he owe on his next MasterCard statement?

12. Maria charges a tankful of gasoline for her truck on her credit card and pays the charge in full on her credit-card bill just in time to avoid a penalty, which turns out to be 1.5 months after she bought the gasoline. The service station charged 2% more for gasoline purchased by credit card than for cash, so that was, in effect, the interest she paid for borrowing the price of the gasoline for 1.5 months. Calculate the annual interest rate she paid by using the credit card to buy the gasoline.

13. The Major Investment Fund purchases a $100,000 noninterest-bearing bond (that is, a bond that gives no periodic interest payments) that will mature in 8 months and pay simple interest at a 9.5% annual interest rate. How much does the Fund pay for the bond?

14. The investment committee of Lake College buys 10,000 shares of American Express stock at a price of $27 a share plus a brokers fee of $6000. It sells all the stock one year later at a price of $33. In addition, it receives $2.90 a share in dividends during the year. The college pays an additional broker's fee of $15,000 when it sells the stock. If the committee views this investment in the stock as a loan of $270,000 to the stock market for one year, what annual interest rate did the college receive on its investment?

15. A farmer has a field of tomatoes that can be harvested in 90 days and sold for $70,000. An investor offers the farmer $65,000 immediately for the right to take possession of the tomatoes when they are harvested. Suppose the farmer can invest for 90 days at a 9% annual interest rate.
    (a) What is the present value of the $70,000?
    (b) Should the farmer accept the $65,000 offer?

16. Star Production Company needs to borrow $800,000 for 6 months to pay the production costs of making a television documentary. The EZ Finance Company will loan Star the money at an annual interest rate of 16%. The Last National Bank will

loan the money at an 11% annual rate, but it also charges a "loan initiation fee" of $20,000, which can be viewed as additional interest on the money. Which loan, from the finance company or from the bank, will cost Star less and how much will Star pay in interest for the less expensive loan?

17. Matthew can lease a stand near the beach for the summer, where he can rent out beach equipment, such as beach chairs and umbrellas. The lease costs $1000 and it must be paid at the beginning of the summer. Matthew spends $1200 to purchase the beach equipment. During the summer, he expects to make $3300 in rental charges and he can then sell the used equipment for $500 at the end of the 3-month summer season. He borrows the $2200 he needs for the lease and the equipment purchase at an annual interest rate of 14% and repays the loan 3 months later. Calculate Matthew's profit for the summer, that is, his income from rental charges and used-equipment sales, minus the amount of the loan.

18. The Rossi family will take a cruise this winter that costs $8000 payable upon departure. If they pay 60 days in advance, the price drops to $7800. They can invest money for 60 days at an annual interest rate of 8.5%.
    (a) What is the present value of $8000 for a 60-day loan at that interest rate?
    (b) Should they pay for the cruise 60 days in advance or pay at departure?

19. The Real Estate Partnership has the opportunity to buy a parcel of land near a new university for $1.6 million. It believes that it will be able to auction off the land one year later to a developer for a high-tech industrial park for $2.1 million. The partnership announces to its members that after paying the amount of the $1.6 million loan to purchase the land, they can expect a profit of $300,000. At what annual interest rate is the partnership assuming it will borrow the money?

20. The First National Bank loaned $200,000 to the Ajax Lumber Company for 90 days, charging 11.5% simple interest. After 30 days, First National sold the loan to Smith Investments for $202,500; so 60 days later Ajax paid the full amount of the loan to Smith rather than to First National.
    (a) Calculate the interest that Ajax will pay on the loan.
    (b) Calculate the annual interest rate that Smith Investments received by buying the Ajax loan.

8.2 Discount

When a retail merchant buys from a manufacturer, often the manufacturer will offer the retailer a reduction in the price of the manufactured goods in return for immediate payment in cash. Anticipating a busy Christmas sales season, in mid-September Supersound Stereo orders $70,0000 worth of stereophonic speakers from the manufacturer, who offers Supersound a 3% price reduction in return for an immediate cash payment. This means that Supersound can save

(.03)$70,000 = $2100 by paying cash immediately and that the bill will be $70,000 − $2100 = $67,900, instead of $70,000. Supersound does not have $67,900 available in cash in September, but is reasonably sure that pre-Christmas sales will make the money available by the end of November. It may therefore be a good idea for Supersound to borrow money now in order to pay the manufacturer in cash to take advantage of the price reduction, and then pay the loan back later from pre-Christmas receipts.

Notes and Discount

When the owner of Supersound Stereo goes to the a bank to borrow the money to pay the manufacturer for the speakers, rather than giving him a loan at simple interest, the bank may offer him a **noninterest-bearing note** for the required time at a specified **discount rate**. We will suppose that Atlantic Bank offers Supersound a 75-day noninterest-bearing note in the amount of $70,000 at a discount rate of 9%. This means that Atlantic will give Supersound somewhat less than $70,000 immediately and in 75 days Supersound must pay exactly $70,000 back to the bank. The sum of money the bank subtracts from $70,000 before giving the cash to Supersound, in effect the price Supersound pays for borrowing the money, is called the **discount** D and is calculated by the formula

$$D = Sdt$$

where S is the amount of the note, d is the discount rate, and the time t is computed as in Section 8.1. The sum of money that Supersound receives is called the **proceeds** P of the note. This is the amount minus the discount. Thus we write:

$$P = S - D = S - Sdt$$

For the note that the bank offers Supersound,

$$D = Sdt = 70,000(.09)\frac{75}{360} = 1312.50$$

$$P = S - D = 70,000 - 1312.50 = 68,687.50$$

Therefore, Atlantic Bank gives Supersound $68,687.50 immediately and Supersound must pay $70,000 to the bank 75 days later.

The bank note is a good idea for Supersound Stereo, assuming it correctly predicts pre-Christmas sales. It must pay $1312.50 (the discount) to Atlantic Bank, but Supersound will save $2100 through the price reduction for cash payment offered by the manufacturer.

Computing the Amount

Supersound Stereo took out a note for $70,000, for which the proceeds were $68,687.50, but it only needed $67,900 cash to pay the manufacturer of the stereo speakers. If Supersound wanted exactly $67,900 cash from the bank, what should the amount of a 75-day noninterest-bearing note with a 9% discount rate be? In this case, we know that we want the proceeds P to equal $67,900 and we have to compute S, the amount of the note. We solve the equation

$$P = S - Sdt$$

by factoring S

$$P = S(1 - dt)$$

and then by dividing by $1 - dt$ to get

$$S = \frac{P}{1 - dt}$$

Therefore, if Supersound wanted exactly $67,900 cash from the bank, the amount of a 75-day noninterest-bearing note with a 9% discount rate would be

$$S = \frac{67{,}900}{1 - (.09)\frac{75}{360}} = \frac{67{,}900}{.98125} = 69{,}197.45$$

It is unlikely that Atlantic Bank would issue a note in exactly that amount, but the owner of Supersound Stereo can see that if he borrows, for instance, $69,200 rather than the full $70,000, he will still have more than the $67,900 he needs to pay the manufacturer. If he does that, the cost of the note is a bit less since the discount is now:

$$D = Sdt = 69{,}200(.09)\frac{75}{360} = \$1297.50$$

Example 1. Worldwide Importing Company decides that the exchange rate between dollars and Japanese yen is presently very favorable, so it wishes to purchase $200,000 worth of yen that it can use for future purchases from Japan and thus protect itself from adverse currency fluctuations. It does not have $200,000 available at present, but expects to have considerably more than that amount available from receipts of past sales within 30 days. To obtain the $200,000 cash it needs immediately, it takes out a 30-day noninterest-bearing note from a bank, at a 10.5% discount rate. How much will Worldwide Importing have to pay the bank in 30 days?

Solution

The proceeds of this note are $P = \$200{,}000$, the discount rate is $r = .0105$, and the time $t = \frac{30}{360}$. The amount S of the note is given by

8.2 DISCOUNT   575

$$S = \frac{P}{1-dt} = \frac{200{,}000}{1-(.09)\frac{30}{360}} = \frac{200{,}000}{.9925} = 201{,}511.34$$

Thus, in 30 days, Worldwide must pay $201,511.34 to the bank.

Discounted Loans

When a lender sells an existing loan to an investor, the loan may be **discounted**, that is, offered to the investor at a discount rate d. This means that the amount of the loan is treated as that of a noninterest-bearing note and the investor pays the lender the proceeds of the note for the time t remaining on the loan at the discount rate d. The next example illustrates this type of loan resale transaction.

Example 2. At the time of sale, the Ready Real Estate Company gives the purchaser of a commercial property a simple-interest loan of $500,000 for one year at a 10% annual interest rate. Two months after the property is sold, Ready Real Estate discounts the loan to Super Savings and Loan at a 4% discount rate. How much does Super Savings pay to Ready Real Estate for the loan?

Solution

Super Savings will pay Ready Real Estate the proceeds of a noninterest-bearing note whose amount is equal to the amount of the simple-interest loan. The principal of that loan is $P = 500{,}000$ and the annual interest rate is $r = .10$. The simple-interest loan lasts a year, that is, $t = 1$; thus by the formula given in Section 8.1, the amount of this loan is

$$S = P + Prt = 500{,}000 + 500{,}000(.10)1 = 550{,}000$$

Two months after the initiation of the loan, the proceeds from a noninterest-bearing note for the remaining 10 months (so that $t = \frac{10}{12}$ for the note) in the amount of $550,000 at a 4% discount rate is

$$P = S - Sdt = 550{,}000 - 550{,}000(.04)\frac{10}{12}$$
$$= 550{,}000 - 18{,}333.33$$
$$= 531{,}666.67$$

Thus, Super Savings and Loan will then pay Ready Real Estate $531,666.67, and will receive $550,000 from the borrower ten months later.

Discount Rates and Annual Interest Rates

Suppose the owner of Supersound Stereo decides to shop around in an effort to borrow the $70,000 he needs for 75 days in order to pay cash to the speaker manufacturer and thus obtain the price reduction for an immediate cash payment.

Loyalty Savings offers him a 75-day simple-interest loan of the $70,000 principal at a 9% annual interest rate. We recall from the preceding section that this means that Loyalty will give Supersound Stereo $70,000 at once and 75 days later Supersound must pay Loyalty Savings back the $70,000 plus interest of

$$I = Prt = 70,000(.09)\frac{75}{360} = 1312.50$$

The total payment is therefore $71,312.50.

The interest figure of $1312.50 from Loyalty Savings will certainly sound familiar to the Supersound Stereo owner; this is precisely the discount he would pay Atlantic Bank for the $70,000 noninterest-bearing note. Should he conclude, therefore, that it doesn't matter whether he borrows the money from Loyalty or from Atlantic because the cost of the loan is the same $1312.50 in either case?

No, the loan from Loyalty Savings is a little more attractive to Supersound Stereo than the note from Atlantic Bank. The owner of Supersound Stereo observes that his company will receive $70,000 from Loyalty, rather than the discounted proceeds of $68,687.50, which amounts to $1312.50 less, from Atlantic Bank. If he borrowed only the principal of $68,687.50 from Loyalty Savings, as a 75-day simple-interest loan at a 9% annual interest rate, the amount of the loan would now be

$$S = P + Prt = 68,687.50 + 68,687.50(.09)\frac{75}{360}$$
$$= 68,687.50 + 1287.89$$
$$= 69,975.39$$

So, after 75 days he would pay back to Loyalty about $25 less than the $70,000 he would pay to Atlantic Bank. Thus, the Loyalty Savings loan is slightly more advantageous. Since the owner of Supersound Stereo often borrows money to cover short-term cash requirements, over many years even small differences in the cost of borrowing money can have a significant effect on the cost of running his business. Therefore, he would like a more systematic method of comparing the costs of various ways of borrowing money.

A good way to compare a noninterest-bearing note with a simple interest loan is to pretend that the proceeds of the note, which the borrower receives at the beginning of the loan, is the principal of a simple-interest loan, and the amount of the note that the borrower pays back at the end is the amount of a simple-interest loan. We then calculate the annual interest rate for such a loan. Thus, if we just think of the 75-day noninterest-bearing note from Atlantic Bank as a simple-interest loan with principal $P = 68,687.50$ and amount $S = 70,000$, then since $S = P + Prt$, we can write

$$70,000 = 68,687.50 + (68,687.50) \cdot r \cdot \frac{75}{360}$$

and solve for r:

$$(68,687.50) \cdot \frac{75}{360} r = 70,000 - 68,687.50 = 1312.50$$

$$r = \frac{1312.50}{68,687.50} \cdot \frac{360}{75} = .092 \text{ (approximately)}$$

Now the owner of Supersound Stereo has a more precise way of comparing the proposals of Atlantic Bank and Loyalty Savings: the noninterest-bearing note offered by Atlantic is equivalent to a simple interest loan at a 9.2% annual rate, compared to the 9% rate charged by Loyalty.

The next time Supersound Stereo has to borrow money, the owner would rather not have to repeat this analysis. He would prefer a general formula for expressing the discount rate d of a noninterest-bearing note as the annual interest rate of a simple-interest loan of the same duration. That interest rate is called the **effective interest rate** r of the noninterest-bearing note. We can obtain a formula for effective interest rates by following the steps we went through for Atlantic's note. Recall that the proceeds P of a noninterest-bearing note are calculated by

$$P = S - Sdt = S(1 - dt)$$

The borrower must pay the amount S at the end of the loan; so, if we think of S as the amount of a simple interest loan for the same duration t at an unknown annual interest rate r, then

$$S = P + I = P + Prt = P(1 + rt)$$

Since P represents the money the borrower receives at the start of the time period of the noninterest-bearing note, it may also be considered to be the principal of the corresponding simple-interest loan. Thus, the symbol P has the same meaning in both equations, and we can use the first equation to substitute $S(1 - dt)$ for P in the second one, obtaining

$$S = [S(1 - dt)](1 + rt)$$

Now we divide both sides of this equation by the (nonzero) amount S:

$$1 = (1 - dt)(1 + rt)$$

then multiply the terms on the right-hand side and simplify.

$$1 = 1 - dt + rt - rdt^2$$
$$dt = rt - rdt^2$$

We may divide all terms by the time t since the duration of the loan is not zero:

$$d = r - rdt = r(1 - dt)$$

Therefore, noninterest-bearing notes are related to simple-interest loans as follows:

> A noninterest-bearing note lasting time t at discount rate d costs the borrower the same as a simple-interest loan for the same duration with an annual interest rate of r, the effective interest rate of the note, where r is given by
>
> $$r = \frac{d}{1 - dt}$$

578 MATHEMATICS OF FINANCE

If the owner of Supersound Stereo knew this formula when he heard the terms of the simple-interest loan from Loyalty Savings, he could have easily converted the effective interest rate of the noninterest-bearing note from Atlantic Bank:

$$r = \frac{d}{1-dt} = \frac{.09}{1-(.09)\frac{75}{360}} = \frac{.09}{.098125} = .092 \text{ (approximately)}$$

He would have realized that the proposal from Loyalty, with its annual-interest rate of .09, was superior.

Notice that the relationship between the effective interest rate r and the discount rate d depends on t, the duration of the loan. For instance, for a 6-month noninterest-bearing note at a 9% discount rate, the effective interest rate is

$$r = \frac{.09}{1-(.09)\frac{1}{2}} = \frac{.09}{.955} = .094 \text{ (approximately)}$$

As this example suggests, for a given discount rate d, as the duration of the loan gets longer, the effective rate r becomes larger. For very short-term borrowing, a simple-interest loan and a noninterest-bearing note whose discount rate equals the annual interest rate of the loan cost about the same amount. But for longer-time periods, the noninterest-bearing note can be significantly more expensive.

Example 3. Peter needs to obtain a $2000 personal loan to pay some outstanding bills. His credit union will lend him the money at 13% simple interest. A bank offers him a noninterest-bearing note at a 12.5% discount rate.
(a) Which is the cheaper way for Peter to borrow the money if the loan is for 3 months?
(b) Which is the cheaper way for Peter to borrow the money if the loan is for 6 months?

Solution

(a) The effective interest rate corresponding to a 12.5% discount rate for 3 months is

$$r = \frac{.125}{1-(.125)\frac{1}{4}} = .129 \text{ (approximately)}$$

Thus, Peter should take the noninterest-bearing note offered by the bank because it is equivalent to a simple-interest loan at 12.9%, which is a little lower than the 13% charged by the credit union.

(b) The effective interest rate corresponding to a 12.5% discount rate for 6 months is

$$r = \frac{.125}{1-(.125)\frac{1}{2}} = .133 \text{ (approximately)}$$

which is higher than the 13% credit-union annual-interest rate. In this case, Peter would be better off with the credit-union loan. ∎

Summary

The **discount** D on an amount S at a **discount rate** d for a time t is calculated by $D = Std$. A **noninterest-bearing note** with discount rate d, time t, and amount S is a loan in which the borrower receives **proceeds** $P = S - D = S - Sdt$ and pays back S after time t has elapsed. If the proceeds, time, and discount rate of a noninterest-bearing note are known, the amount of the note is calculated by the formula

$$S = \frac{P}{1 - dt}$$

If a loan is **discounted** by the lender to an investor at a discount rate d, this means that the investor pays the lender the proceeds of a noninterest-bearing note with discount rate d for an amount equal to the amount of the loan and a time equal to the time remaining on the loan; the borrower pays the amount to the investor when the loan is due. A noninterest-bearing note with proceeds P for time t at a discount rate d will produce the same amount S as a simple-interest loan with the same principal P for the same time t with an annual interest rate r, given by

$$r = \frac{d}{1 - dt}$$

Here, r is called the **effective interest rate** of the noninterest-bearing note.

Exercises

1. Calculate the discount for a noninterest-bearing note in the amount of $3200 for 3 months at an 8% discount rate.

2. Calculate the proceeds of a noninterest-bearing note in the amount of $10,000 for 21 days at a 10.5% discount rate.

3. Calculate the amount of an 8-month noninterest-bearing note at a 9.5% discount rate if the proceeds of the note are $80,000.

4. Calculate the effective interest rate equivalent to a 10% discount rate on a noninterest-bearing note for 4 months.

580   MATHEMATICS OF FINANCE

5. The harvest is late this year, so the Northern Grape Juice Company has no grapes to press. It needs $40,000 immediately to meet its payroll. It receives this sum from a bank on a noninterest-bearing note for 2 months at a 10% discount rate. How much money will Northern Grape Juice have to pay back to the bank in 2 months?

6. To finance the initial costs of a new recycling facility, a bank grants the Eco-wise Corporation a noninterest-bearing note for 9 months in the amount of $450,000 at a 7.5% discount rate. Calculate how much the note will cost Eco-wise, that is, calculate the discount for this note.

7. Celebrity Publishing Company is sure that its latest book, the biography of a movie star, who has just appeared in a very successful movie, will be a best-seller, provided the book gets to the bookstores right away, while the movie is still playing in major cities. But the printer will only speed up the printing if Celebrity pays the full $300,000 cost of the print run immediately in cash. So, Celebrity obtains a 75-day noninterest-bearing note from a bank at a 13% discount rate with proceeds of $300,000 to pay for the printing. How much money will Celebrity Publishing have to pay the bank 75 days later?

8. Wong's Hardware buys $20,000 worth of power tools from a manufacturer who offers a 2% price reduction in return for an immediate cash payment. The store can borrow the money it needs to pay the manufacturer from a bank that offers a 1-month noninterest-bearing note at a 10.5% discount rate.
   (a) Calculate the amount of the note.
   (b) Will Wong's save money by borrowing from the bank to pay cash to the manufacturer?

9. In order to set up an office for the tax season, Alice's Tax Service borrows from a bank on a 90-day noninterest-bearing note at an 11.5% discount rate. If the amount of the note is $30,000, how much money does Alice's Tax Service receive as proceeds from the bank?

10. The Sturdy Storage Cabinets Company buys $200,000 worth of steel stock from a steel company and can get a 4% price reduction if it pays the steel company immediately. Sturdy can obtain the money it needs to pay the steel company through a noninterest-bearing note from a bank for 6 months at a 9.5% discount rate.
    (a) Calculate the amount of the note.
    (b) Will Sturdy save money by borrowing from the bank?

11. The Friendly Furniture Corporation sells $15,000 worth of beds and chairs to a motel owner. To finance this purchase, Friendly gives the owner an 8-month loan at a 10% annual simple-interest rate. After 5 months, Friendly discounts the loan to the National Finance Company at a 9% discount rate. How much does National Finance pay Friendly Furniture for the loan?

12. The Valley Bank will lend you money to finance your European vacation for 5 months at 12.5% simple annual interest. Alternately, you can go to the River Bank for a noninterest-bearing note for 5 months at an 11% discount rate.
    (a) Calculate the effective interest rate of the noninterest-bearing note.
    (b) Which bank will charge you more to borrow the money?

13. In March, the XYZ Holding Corporation gives its theme-park subsidiary an $8 million simple-interest loan for 90 days at a 12% annual interest rate to provide cash for the subsidiary's operations until attendance at the parks increases in the summer. However, 45 days later, XYZ finds it needs cash; so, it discounts the loan to a commercial bank at a 12.5% discount rate. How much does the bank pay XYZ for the loan?

14. Classic Investment Company decides to obtain cash from a loan it made to Steinberg Trucking Company 3 months ago. Steinberg's loan is a simple-interest loan for 8 months with a $55,000 principle and a 15% annual interest rate. Classic can discount Steinberg's loan to a bank at a 13.5% discount rate. Alternately, Classic could borrow money from another bank in the form of a simple-interest loan at 12.5%, where the amount of the new loan equals the amount of Steinberg's loan. Classic's loan is due when Steinberg's is, so Classic can use the money it gets from Steinberg to pay the bank.
    (a) Will Classic obtain more cash immediately by discounting Steinberg's loan or by taking out its own simple-interest loan?
    (b) How much cash will Classic obtain from the more advantageous choice?

8.3 Compound Interest

Tony's Restaurant has become so popular that Tony, the owner, often has to turn away customers. By the beginning of next year, some adjacent space will become available, and he will have the opportunity to expand his restaurant. However, it will cost him $1,000,000 for redecoration and new equipment. He is confident that if he can borrow that amount at the beginning of January, his restaurant's earnings will increase enough so that one year later he will be able to pay back the principal of the loan together with the interest the bank charges. Tony visits Valley Bank to arrange a one-year loan and is told that the annual interest rate for this sort of business loan is 11%.

Now let us look at Tony's loan from the point of view of the president of Valley Bank. She has $1,000,000 available to lend to Tony and considers him a good credit risk. If she offers him a simple-interest loan of $1,000,000 for one year at an 11% annual interest rate, we know from Section 8.1 that her bank will earn interest amounting to

$$I = Prt = (1,000,000)(.11)(1) = 110,000$$

But suppose that on January 1, instead of lending $1,000,000 to Tony for a year, Valley Bank lends the same amount to another business for just 3 months in a simple-interest loan at the same 11% annual interest rate. Since now $t = \frac{3}{12} = \frac{1}{4}$, at the end of March the bank would have an amount

$$S = P + Prt = P(1 + rt) = 1{,}000{,}000\left(1 + (.11)\frac{1}{4}\right)$$
$$= 1{,}000{,}000(1.0275)$$
$$= 1{,}027{,}500$$

that it could then use as the principal for another loan. Next, suppose that on April 1 the bank made a 3-month loan of $1,027,500 to some borrower at the same annual rate. Then at the end of June it would get back

$$1{,}027{,}500\left(1 + (.11)\frac{1}{4}\right) = \left[1{,}000{,}000\left(1 + (.11)\frac{1}{4}\right)\right]\left(1 + (.11)\frac{1}{4}\right)$$

or, writing the product in another way, and rounding to whole-dollar amounts,

$$1{,}000{,}000\left(1 + (.11)\frac{1}{4}\right)^2 = 1{,}055{,}756$$

We wrote out the computation this way to show a pattern: the original principal of $1,000,000 is multiplied by a power of

$$1 + rt = 1 + (.11)\frac{1}{4}$$

to calculate the last amount. Using the newly available principal to make another 3-month loan on July 1, the bank will then have at the end of September (since we multiply once more by $1 + rt$)

$$1{,}055{,}756\left(1 + (.11)\frac{1}{4}\right) = \left[1{,}000{,}000\left(1 + (.11)\frac{1}{4}\right)^2\right]\left(1 + (.11)\frac{1}{4}\right)$$
$$= 1{,}000{,}000\left(1 + (.11)\frac{1}{4}\right)^3$$
$$= 1{,}084{,}790$$

Repeating the process a final time with a loan on October 1, at the end of the year the bank would have

$$\left[1{,}000{,}000\left(1 + (.11)\frac{1}{4}\right)^3\right]\left(1 + (.11)\frac{1}{4}\right) = 1{,}000{,}000\left(1 + (.11)\frac{1}{4}\right)^4$$
$$= 1{,}114{,}621$$

Therefore, by making a series of loans for 3 months each and lending the amount of the previous loan each time, the bank would earn $114,621, rather than the $110,000 it would have received in interest from Tony for a simple interest loan for one year.

Of course, Valley Bank would not earn a clear additional profit of $4621 by making four 3-month simple-interest loans instead of lending the money to Tony for a year because it would incur administrative costs for arranging each loan. Furthermore, Valley Bank cannot be sure it would have borrowers seeking these exact amounts as they become available every 3 months. Nevertheless, the bank president's analysis of how her bank might lend out the money illustrates the reason why, if money were lent only at simple interest, it could be difficult for a borrower to obtain a long-term loan. It would be more profitable for the lender to make numerous short-term loans, lending out the amount received from each short-term loan as the principal on the next one. Therefore, interest on a long-term loan is generally charged as if the loan were a succession of short-term loans at simple interest. This type of interest is called **compound interest**. The home mortgages that we discussed at the beginning of this chapter illustrate the importance of long-term lending since they can last as long as 30 years. Consequently, mortgages are always compound-interest loans.

Computing the Amount

The president of Valley Bank offers Tony $1,000,000 for one year at an 11% annual interest rate with interest compounded quarterly, that is, every 3 months. That means the interest is calculated as we did it: at the end of each 3-month period the amount is used as the principal for another 3-month loan. So, at the end of the year Tony will owe the bank

$$1{,}000{,}000\left(1 + (.11)\frac{1}{4}\right)^4 = 1{,}114{,}621$$

as we have seen.

In general, to compute the **amount** S of a compound-interest loan, which, as in Section 8.1, means the sum of money, principal plus interest, that the borrower must pay back at the conclusion of the loan, we just follow the pattern of Tony's loan. The loan will be compounded m times a year ($m = 4$ for quarterly compounding), so the time of each short-term loan in $t = \frac{1}{m}$. The amount at the end of each interest period is calculated by multiplying the previous amount by

$$1 + rt = 1 + r \cdot \frac{1}{m} = 1 + \frac{r}{m}$$

Multiplying the principal P by $1 + \frac{r}{m}$ a total of n times:

The amount of the loan is

$$S = P\left(1 + \frac{r}{m}\right)^n$$

where r is the annual interest rate, m is the number of times each year that the interest is compounded, and n is the total number of times during the life of the loan that interest is compounded.

The number obtained by dividing the annual interest rate r by m, the number of times per year that interest is compounded is called the **interest rate per period**.

584 MATHEMATICS OF FINANCE

If this is denoted by i, then

$$i = \frac{r}{m}$$

In terms of interest rate per period, the amount of a compound interest loan is calculated by the simpler-looking formula

$$S = P(1 + i)^n$$

The duration of a loan is usually stated in years rather than in interest periods. If the loan is for y years and if there are m interest periods per year, then the total number of interest periods is $n = ym$.

Table B on page A15 gives the values of $(1 + i)^n = \left(1 + \frac{r}{m}\right)^n$ for various annual interest rates r and lengths of loans, in years, both for quarterly compounding ($m = 4$) and for monthly compounding ($m = 12$). In practice, compounding may take place at other time intervals such as daily, weekly, or semi-annually. However, varying the compounding periods does not affect the principles of the mathematics of finance so, for simplicity, we will discuss only monthly and quarterly compounding.

Example 1. Suppose the president of Valley Bank had offered Tony a compound-interest loan with principal $1,000,000 at an 11% annual interest rate, but with interest compounded monthly. Then how much would Tony owe the bank at the end of the year?

Solution

Monthly compounding means that $m = 12$, and we still have $r = .11$; thus, $i = \frac{.11}{12}$. (The unusual form of the fraction $\frac{.11}{12}$, that is, with a decimal number in the numerator, is customary in the mathematics of finance because it indicates the computation: divide .11 by 12.) There are now $n = 12$ interest periods. Using Table B or a calculator, we compute

$$S = P(1 + i)^n = 1{,}000{,}000\left(1 + \frac{.11}{12}\right)^{12}$$

$$= 1{,}000{,}000(1.115719)$$

$$= 1{,}115{,}719$$

Therefore, at the end of the year Tony must pay the bank $1,115,719. ∎

Example 2. Suppose Tony borrows $1,000,000 from Valley Bank, still at an 11% annual interest rate, but with interest compounded quarterly for a two-year period. How much would Tony owe the bank at the end of two years?

Solution

Since this is quarterly compounding, $m = 4$ and $i = \frac{.11}{4}$; but now $y = 2$, so that $n = 2 \cdot 4 = 8$. The amount is given by

$$S = P(1+i)^n = 1,000,000\left(1 + \frac{.11}{4}\right)^8$$
$$= 1,000,000(1.242381)$$
$$= 1,242,381 \qquad \blacksquare$$

The two preceding examples illustrate an important feature of compound interest. The one-year simple-interest loan would cost Tony $110,000 in interest, whereas quarterly compounding brought that up by more than $4600, to $114,621. Example 1 showed that monthly compounding would bring the cost up about $1000 more than quarterly compounding, to $115,719. A further increase to daily compounding would have relatively little additional effect. Using a calculator it is easy to show that with 360 interest periods in a year, the interest would be $116,259, which is a little more than $500 above the interest cost for the $1,000,000 loan with monthly compounding. However, the total length of the loan has a much greater effect on the amount of interest charged. Example 2 showed us that increasing the length of the loan with quarterly compounding from one year to two increased the interest Tony must pay to the bank from $114,621 to $242,381. Thus, doubling the length of the loan more than doubled the interest charge.

Present Value

In Section 8.1, we discussed the **present value** of a loan at a certain annual interest rate for a given time, that is, the sum of money that would have to be invested at that rate over this period of time in order to produce a given amount. This important concept is as appropriate for loans with compound interest as it is for simple-interest loans. In the compound-interest case, we know the amount S of the loan, the interest rate per period i (because we know the r, annual interest rate, and m, the number of times per year that interest is compounded), and we know n, the number of interest periods; but we do not know P, the principal. To find P, we solve the amount equation
$$S = P(1+i)^n$$
for the unknown P by dividing the equation by $(1+i)^n$.

$$\boxed{P = \frac{S}{(1+i)^n} = S(1+i)^{-n}}$$

Example 3. In 4 year's time, United Transport Company will need to replace a number of its vans, and it expects to require $150,000 for this purpose. It can buy a 4-year certificate of deposit from a bank in multiples of $1000 at an annual interest rate of 8.5% with interest compounded quarterly. Calculate how much money the company should invest in the certificate of deposit to have at least $150,000 available in 4 years.

Solution

The sum of money United Transport should invest equals the present value of the amount $S = \$150,000$ in 4 year's time with interest compounded quarterly ($m = 4$) at an annual rate of $r = 8.5\%$, so $i = \frac{.085}{4}$. During the $y = 4$ years duration of the loan, interest will be compounded $n = ym = 4 \cdot 4 = 16$ times. Therefore, the present value of $150,000 is

$$P = \frac{S}{(1+i)^n} = 150,000\left(1 + \frac{.085}{4}\right)^{-16}$$
$$= 150,000(.714310)$$
$$= 107,146.50$$

Since United Transport cannot purchase a certificate of deposit for exactly $107,146.50, but only in multiples of $1000, to be sure that it will have at least $150,000 after 4 years, it should obtain a certificate for $108,000. ∎

Effective Annual Rate

Recall that Valley Bank offered Tony a one-year loan with principal $P = \$1,000,000$ at an 11% annual interest rate, where interest is compounded quarterly. Suppose Tony finds that he can obtain a loan from River Savings and Loan at a 10.5% annual interest rate, compounded monthly. To determine whether the loan from River would be cheaper than that from Valley Bank, he could just calculate the amount of this loan and compare it to the amount $1,114,621 of the loan from Valley.

Alternatively, Tony could compare the terms, 11% compounded quarterly versus 10.5% compounded monthly, by means of a simple formula that converts them to the same type of loan. The procedure is the same one we used in Section 8.2 to compare noninterest-bearing notes. We know how much money the borrower receives at the start of the compound-interest loan, that is the principal P, and the amount S that must be paid back. We treat these as if they were for a simple-interest loan and calculate the corresponding annual interest rate, which is called the **effective annual rate** e—but this time for the compound-interest loan. We have seen that a loan with principal P at an annual interest rate r compounded m times a year, and thus, with periodic interest rate $i = \frac{r}{m}$, has an amount after one year equal to

$$S = P(1+i)^m$$

since interest will be compounded m times in the year. On the other hand, a simple-interest loan with principal P at an annual interest rate e that lasts one year has an amount given by the formula

$$S = P(1 + e \cdot 1) = P(1 + e)$$

Thus if we think of a one-year compound-interest loan as a one-year simple-interest loan with the same principal and amount, we may set the two expressions for S equal to each other:

$$P(1+i)^m = P(1+e)$$

Dividing the equation by the common nonzero factor P

$$(1+i)^m = 1 + e$$

we solve the equation to obtain the formula for the effective annual rate of the compound-interest loan, which is

$$\boxed{e = (1+i)^m - 1}$$

Thus, the effective annual interest rate of the loan from Valley Bank, in which $i = \frac{r}{m} = \frac{.11}{4}$, is

$$e = \left(1 + \frac{.11}{4}\right)^4 - 1 = 1.114621 - 1 = .115 \text{ (approximately)}$$

In other words, the compound-interest loan from Valley is equivalent to a simple-interest loan at an annual interest rate of about 11.5%. For River Savings and Loan, 10.5% compounded monthly, the interest rate per period is $i = \frac{.105}{12}$ and the effective annual rate is therefore

$$e = \left(1 + \frac{.105}{4}\right)^{12} - 1 = 1.110203 - 1 = .11 \text{ (approximately)}$$

This is significantly less expensive, so Tony should take this loan.

In Example 1, we calculated the amount of a loan with interest compounded monthly, and compared it with the amount for the same principal, time period, and annual interest rate, compounded quarterly. In Example 4, we carry out this comparison in another way.

Example 4. Suppose Valley Bank had proposed compounding the 11% annual interest rate on a monthly basis. What would be the effective annual rate of this loan?

Solution

Since we have 12 monthly payments in a year on a loan with an 11% annual interest rate, we calculate $i = \frac{.11}{12}$. We then have

$$e = \left(1 + \frac{.11}{12}\right)^{12} - 1 = 1.115719 - 1 = .116 \text{ (approximately)}$$

Therefore, this loan has an effective annual rate of approximately 11.6%, compared to the effective annual rate of about 11.5% with quarterly compounding that we observed previously. ∎

Effective Rates

The effective interest rate r of a noninterest-bearing note and the effective annual rate e of a compound-interest loan both depend on the same underlying concept:

treating another type of borrowing as if it were a simple-interest loan and computing the corresponding annual interest rate. The formulas

$$r = \frac{d}{1-dt} \quad \text{and} \quad e = (1+i)^m - 1$$

are, of course, different because different forms of borrowing are involved. But, in addition, the effective interest rate of a noninterest-bearing note depends on the duration t of the note, whereas the effective annual rate of a compound-interest loan is computed in terms of a loan duration of one year, which is why we prefer the terminology, effective *annual* rate. The reason for the difference in approach is that noninterest-bearing notes are generally for short-term borrowing, that is, for less than a year, whereas compound-interest loans are usually for a number of years. Whatever form of borrowing is involved, federal "truth in lending" laws require that the borrower be informed of the corresponding effective rate and be furnished with an estimate of the other costs connected with the loan.

Summary

If a loan with principal P pays compound interest at an annual interest rate r for y years, with interest compounded m times a year, then the **amount** S of that loan is

$$S = P(1+i)^n$$

where $i = \frac{r}{m}$ is the **interest rate per period** and $n = ym$ is the total number of interest periods. The **present value** P of a loan with amount S at interest per period i for n interest periods is

$$P = S(1+i)^{-n}$$

The **effective annual rate** e corresponding to an annual interest rate r compounded m times a year is the simple-interest rate that would produce the same amount from a given principal on a one-year loan. The effective annual rate is calculated by

$$e = (1+i)^m - 1$$

Exercises

1. Calculate the amount of a loan of $5000 for 3 years at 8% annual with interest compounded quarterly.

2. Calculate the present value of a loan for 5 years at 9.5% annual interest compounded monthly, if the amount of the loan is $40,000.

3. Calculate the effective annual rate corresponding to an annual interest rate of 14% compounded quarterly.

8.3 COMPOUND INTEREST   589

4. Susan's aunt offers to lend her $4000 to help pay for repairs on her home. They agree that after 2 years, Susan will repay her aunt the principal plus interest at a 10% annual rate, compounded quarterly. How much must Susan pay her aunt after 2 years?

5. The Toda family wins a large prize in a lottery. The family wishes to put part of the money in a bank certificate of deposit so that it will have at least $80,000 in 6 years to pay educational expenses. If the certificate pays 7.5% annual interest compounded monthly, to the nearest $100, how much of their winnings from the lottery should the Todas use for the certificate?

6. Ace Development Company estimates that it can sell a parcel of land that it owns either for $3 million now or for $3.5 million in three years.
   (a) If Ace Development can earn 8% annual interest compounded monthly by investing its money, what is the present value of $3.5 million in three years?
   (b) Should Ace Development sell the land now or wait for three years?

7. A graduate of Sunrise College promises to donate $500,000 in ten year's time. Instead, the Sunrise development office convinces him to donate $250,000 now so that the college can invest the $250,000 at an annual rate of 9.5% compounded quarterly. In ten years time, how much more money (that is, above $500,000) will Sunrise College have than if they had accepted his original offer?

8. General Research Corporation needs a relatively short-term loan of $2.3 million to cover unexpected developmental costs for a new computer product. A group of investors will lend General Research the money for one year at a 15% annual interest rate, compounded quarterly. Alternatively, General Research can obtain a two-year loan from a bank at an annual interest rate of 11.5% compounded monthly.
   (a) For which loan, from the investors or from the bank, will General Research pay more interest and how much more will that be?
   (b) For which loan, from the investors or from the bank, is the effective annual rate lower, and what is that rate?

9. A lottery awards a $1 million jackpot, payable in 20 annual payments of $50,000 each. It gives the winner the first payment of $50,000 and it sets up a fund investing the remaining $950,000 at a 7% annual rate, compounded quarterly. At the end of the year, how much money does the lottery have available in the fund to make the second payment to the winner?

10. Dependable Trucking Company plans to replace several of its older trucks in three years and it estimates that it will need $225,000 for that purpose. Dependable has cash on hand that it can invest in multiples of $1000 for three years at a 10.5% annual rate, compounded monthly. How much should Dependable invest now in order to have at least $225,000 available in three years to buy new trucks?

11. The Corner Toy Shop orders $4000 worth of computer games from a manufacturer who offers to reduce Corner's bill by 4% if it will pay the entire bill immediately in cash. Corner Toy will have enough money available to pay the manufacturer if it borrows $1000 for one year at 9% compounded quarterly.

Calculate the advantage to the Toy Shop of obtaining the loan by computing the cash reduction minus the cost in interest for the loan.

12. As part of the financing package to buy a home, the Garcia family borrows $30,000 from the seller of the house for three years at a 12.5% annual interest rate, compounded quarterly, with principal and interest due at the end of that period. How much will the Garcia family have to pay the seller when the loan becomes due?

13. Prestige Cars Inc. lends the purchaser of a restored classic Rolls Royce $75,000 at a 16% annual rate compounded monthly, with principal and interest due in three years. One year later, Prestige discounts the loan to Ace Finance Company at a 14% discount. That is, Ace pays Prestige the proceeds on a two-year non-interest-bearing note at a discount rate of 14% with the amount equal to the amount of the compound-interest loan to the purchaser, who then will pay that amount to Ace when the loan is due. How much does Ace pay Prestige for the loan?

14. When an early frost destroyed much of its crop this year, Orchard Fruit Company found it needed to borrow $500,000 for a year in order to stay in business until the next harvest. Orchard can obtain a one-year noninterest-bearing note at a 12% discount rate or a one-year loan at a 13.5% annual interest rate, compounded monthly.
    (a) Calculate the effective interest rate of the noninterest-bearing note.
    (b) Calculate the effective annual rate of the compound-interest loan.

8.4 Annuities

Banks sometimes promote regular savings by their customers through a special program, such as a Christmas club or a vacation club. In a vacation club, a person pays a fixed amount of money to the bank each month for a number of months and then, when it is time for vacation, the club member receives all the money, including interest, to pay for the vacation.

Examples of Annuities

A program in which payments of fixed sums of money are made at equally spaced intervals of time and on which compound interest is paid is called an **annuity**. Thus, a vacation club is an example of an annuity. Another example of an annuity is a retirement program. In this case, both the employer and the employee make regular payments into a fund that pays compound interest, so that when the employee retires, money is available for a retirement pension.

There are several types of annuities; but we will discuss just one very common type, called an "ordinary annuity." To illustrate how an ordinary annuity works,

8.4 ANNUITIES

we suppose Mary joins a bank's vacation club in August to save money to take a Mediterranean cruise in August of the following year. Throughout the life of the annuity, at the end of each month Mary will pay a fixed sum of money, called the **rent** of the annuity, to the bank. (The word "rent" is used in this way in the mathematics of finance; it is not the same usage as when rent is paid for an apartment.) The bank will pay compound interest on the money she deposits at an annual interest rate of 7.5%, compounded monthly. Mary decides on a monthly rent of $300, so she will put $300 into the vacation club at the end of each month from August of one year through July of the next year. Mary wishes to know how much money she will have available for the cruise.

If Mary just put $300 under a mattress at the end of each month from August through July of the next year, the 12 payments of $300 each would yield $3600 to pay for the cruise. In addition to greater safety, the advantage that the bank's vacation club has over the mattress lies in the interest that the bank will pay on Mary's money. For instance, Mary's first payment to the bank at the end of August is a loan to the bank of $300 for the 11 months, until the beginning of August of the next year. The bank pays interest at an annual rate of 7.5% compounded monthly, so the interest rate per period is

$$i = \frac{.075}{12} = .00625$$

As we discussed in the preceding section, since the initial payment will be compounded $n = 11$ times, the amount of the loan to the bank that consists of the first payment into the vacation club is

$$300\left(1 + \frac{.075}{12}\right)^{11} = 300(1.070939) = 321.28$$

If you are using Table B, you will find that there is no entry in the table of $\left(1 + \frac{r}{12}\right)^n$ for $n = 11$. However, you can use the properties of exponents to compute the value as follows.

$$\left(1 + \frac{.075}{12}\right)^{11} = \left(1 + \frac{.075}{12}\right)^{5}\left(1 + \frac{.075}{12}\right)^{6}$$
$$= (1.031643)(1.038091)$$
$$= 1.070939$$

The next payment into the vacation fund, at the end of September, stays in the bank for 10 months, so it is compounded 10 times. Thus, the amount of this loan is

$$300\left(1 + \frac{.075}{12}\right)^{10} = 300(1.064287) = 319.29$$

since

$$\left(1 + \frac{.075}{12}\right)^{10} = \left(1 + \frac{.075}{12}\right)^{5}\left(1 + \frac{.075}{12}\right)^{5} = 1.064287$$

Continuing in this manner, the payment at the end of June stays in the bank only one month, which makes the amount of that loan equal to

$$300\left(1 + \frac{.075}{12}\right)^1 = 300(1.00625) = 301.88$$

Finally, the payment at the end of July is given back immediately, so it returns just the $300. Of course, in real life the final payment is not made. However, it is included in the formal definition of the annuity.

The sum Mary will receive to go on her Mediterranean cruise is the total of the amounts of the 12 loans to the bank corresponding to the 12 payments. Table 8-1 supplies all the amounts, including the nine payments we skipped. The sum of the 12 amounts is $3726.36, so that is how much Mary will receive. Thus, by putting money in the vacation club, rather than under a mattress, Mary will have an additional $126.36 to spend for the cruise.

We will next see that we could have eliminated the tedious summing up of separate amounts in Table 8-1, by using a simple one-step formula.

Month of Payment	Amount
August	$321.28
September	319.29
October	317.30
November	315.33
December	313.37
January	311.43
February	309.49
March	307.57
April	305.66
May	303.76
June	301.88
July	300.00

Table 8-1

The Amount of an Ordinary Annuity

In an **ordinary annuity**, payments of fixed amounts R, the **rent** of the annuity, are made at regular intervals of time, with m payments over the course of a year. Interest is paid at an annual rate r, compounded m times a year; so interest rate per period is $i = \frac{r}{m}$. There are n interest periods during the life of the annuity and the rent R is paid at the end of each period. Since the first rent payment is not made until the end of the first time period, it collects interest only for the remaining $(n-1)$ interest periods. Therefore, by the discussion in Section 8.3, the first payment is a loan for which the amount is $R(1+i)^{n-1}$. The next payment has the same rent and interest per period, but it is in effect for one less period, so its amount is $R(1+i)^{n-2}$. Continuing in this way, we get to the next-to-last payment that is in effect for only 1 month, and in this case the amount returned is just $R(1+i)$. The final payment collects no interest, but the rent $R = R(1)$ is returned. The **amount** S of the annuity is the sum of the amounts of all the payments, that is,

$$S = R(1+i)^{n-1} + R(1+i)^{n-2} + \ldots + R(1+i) + R(1)$$

which we can simplify somewhat by factoring the common term R,

$$S = R[(1+i)^{n-1} + (1+i)^{n-2} + \ldots + (1+i) + 1]$$

The expression within the brackets is so important in the mathematics of finance that it is given its own symbol, $s_{\overline{n}|i}$, that is,

8.4 ANNUITIES

$$\boxed{s_{\overline{n}|i} = (1+i)^{n-1} + (1+i)^{n-2} + \ldots + (1+i) + 1}$$

The amount formula becomes

$$\boxed{S = R s_{\overline{n}|i}}$$

To calculate the amount S easily, we need a simple way to compute $s_{\overline{n}|i}$. To illustrate the computation of $s_{\overline{n}|i}$, we suppose that the General Manufacturing Company is in the habit of paying all its employees substantial New Year's bonuses. To help cover the cash needed for these bonuses, General Manufacturing sets up a one-year annuity with its bank with rent R of \$150,000 paid quarterly, for which the bank offers an 8% annual interest rate. Thus, there are 4 payments in the year and the interest rate per period is $i = \frac{.08}{4} = .02$. We would like to find a convenient way to calculate the amount of the annuity, that is,

$$S = (150{,}000)[(1+.02)^3 + (1+.02)^2 + (1+.02) + 1] = R s_{\overline{4}|.02}$$

We need a good computational formula for

$$s_{\overline{4}|.02} = (1.02)^3 + (1.02)^2 + (1.02) + 1$$

Notice what happens to $s_{\overline{4}|.02}$ when we multiply it by 1.02:

$$\begin{aligned} 1.02 s_{\overline{4}|.02} &= 1.02[(1.02)^3 + (1.02)^2 + (1.02) + 1] \\ &= (1.02)^4 + (1.02)^3 + (1.02)^2 + (1.02) \end{aligned}$$

We have an expression that contains most of the same terms as $s_{\overline{4}|.02}$ itself since the sum

$$(1.02)^3 + (1.02)^2 + (1.02)$$

appears in both $s_{\overline{4}|.02}$ and $(1.02)s_{\overline{4}|.02}$. Therefore, when we subtract $s_{\overline{4}|.02}$ from $(1.02)s_{\overline{4}|.02}$, most terms of the summands cancel each other and we are left with a very simple expression:

$$\begin{aligned} (1.02)s_{\overline{4}|.02} - s_{\overline{4}|.02} &= [(1.02)^4 + (1.02)^3 + (1.02)^2 + (1.02)] \\ &\quad - [(1.02)^3 + (1.02)^2 + (1.02) + 1] \\ &= (1.02)^4 - 1 \end{aligned}$$

Since

$$(1.02)s_{\overline{4}|.02} - s_{\overline{4}|.02} = (1.02)s_{\overline{4}|.02} - (1)s_{\overline{4}|.02} = (.02)s_{\overline{4}|.02}$$

we have shown that

$$(.02)s_{\overline{4}|.02} = (1.02)^4 - 1$$

594 MATHEMATICS OF FINANCE

and thus dividing the equation by .02 produces the simple computational formula we were seeking

$$s_{\overline{4}|.02} = \frac{(1.02)^4 - 1}{.02}$$

By using a scientific calculator or Table B, we calculate the amount of General Manufacturing's annuity to be

$$S = 150{,}000 s_{\overline{4}|.02} = 150{,}000 \frac{(1.02)^4 - 1}{.02}$$

$$= 150{,}000 \frac{\left(1 + \frac{.08}{4}\right)^4 - 1}{.02} = 150{,}000 \frac{(1.082432) - 1}{.02}$$

$$= 618{,}240$$

Thus, General Manufacturing will have $618,240 cash available from the annuity at the end of the year to pay the bonuses.

The General Formula for $s_{\overline{n}|i}$

Turning now to ordinary annuities in general, we will follow the model of General Manufacturing's annuity to find a simple way to calculate

$$s_{\overline{n}|i} = (1+i)^{n-1} + (1+i)^{n-2} + \ldots + (1+i) + 1$$

We multiply by $(1 + i)$ and see that

$$(1+i)s_{\overline{n}|i} = (1+i)[(1+i)^{n-1} + (1+i)^{n-2} + \ldots + (1+i) + 1]$$
$$= (1+i)^n + (1+i)^{n-1} + \ldots + (1+i)^2 + (1+i)$$

Therefore, when we cancel the common summands

$$(1+i)s_{\overline{n}|i} - s_{\overline{n}|i} = (1+i)^n + (1+i)^{n-1} + \ldots + (1+i)^2 + (1+i)$$
$$- [(1+i)^{n-1} + (1+i)^{n-2} + \ldots + (1+i) + 1]$$
$$= (1+i)^n - 1$$

Since

$$(1+i)s_{\overline{n}|i} - s_{\overline{n}|i} = (1+i)s_{\overline{n}|i} - (1)s_{\overline{n}|i} = i s_{\overline{n}|i}$$

we divide both sides of the equation

$$i s_{\overline{n}|i} = (1+i)^n - 1$$

by i to produce the general formula:

8.4 ANNUITIES 595

$$s_{\overline{n}|i} = \frac{(1+i)^n - 1}{i}$$

Recall Mary's vacation-club annuity, which had 12 payments of $300 each with interest compounded monthly at a 7.5% annual interest rate, so that $i = \frac{.075}{12} = .00625$. From Table B, we quickly compute the amount Mary will have for the cruise:

$$S = Rs_{\overline{n}|i} = 300 s_{\overline{12}|.00625} = 300 \frac{\left(1 + \frac{.075}{12}\right)^{12} - 1}{.00625}$$

$$= 300 \frac{(1.077633) - 1}{.00625}$$

$$= 3726.38$$

which is within a few cents of the answer we obtained before by summing the 12 amounts in Table 8-1. (Rounding off each month's payment to the nearest cent causes the slight difference in the total.) The reduction in effort by using the simple formula for $s_{\overline{n}|i}$ would be even more evident if we had to calculate the amount of an employee's retirement program that involved weekly payments over a span of 40 years.

Example 1. An employee earning $2800 a month is fired. The employee sues the company and 2 years later, the court orders the employee rehired and given a lump-sum payment consisting of all back pay plus interest at an annual rate of 6%, with the interest to be compounded monthly. How much is the lump sum payment to the employee?

Solution

In this case, $i = \frac{.06}{12} = .005$. The employee would have been paid a salary of $R = \$2800$ each month for 2 years, so $n = 12 \cdot 2 = 24$. Therefore,

$$S = Rs_{\overline{n}|i} = 2800 s_{\overline{24}|.005} = \frac{\left(1 + \frac{.06}{12}\right)^{24} - 1}{.005}$$

$$= 2800 \frac{1.127160 - 1}{.005}$$

$$= 71{,}209.60$$

The lump-sum payment to the fired employee is $71,209.60. ∎

Computing the Rent

When a much-loved professor died, a group of 34 of her former students decide to establish a perpetual scholarship fund in her memory. In order to have a significant amount of money to distribute each year from the income of the fund, the students will need a $300,000 endowment. They cannot afford that amount at present, but they decide to contribute every three months for 10 years to an annuity that will produce the $300,000 at the end of that time. A bank offers to set up the annuity with a 7% annual interest rate. If each former student contributes equally to the fund, what must his or her quarterly donation be?

In this example, we know the amount S of the annuity, $300,000, and we need to determine the quarterly rent R since each member of the group will be responsible for paying $\frac{R}{34}$ every 3 months. We may write the formula

$$S = Rs_{\overline{n}|i} = R\frac{(1+i)^n - 1}{i}$$

in the form

$$\frac{(1+i)^n - 1}{i} R = S$$

To solve for R, we multiply both sides of the equation by the reciprocal of the fraction, obtaining

$$\boxed{R = S\frac{i}{(1+i)^n - 1}}$$

For the scholarship fund, $i = \frac{.07}{4} = .0175$ and there are to be $n = 4 \cdot 10 = 40$ payments during the life of the annuity. Therefore, the total quarterly rent will be

$$R = 300{,}000 \frac{.0175}{\left(1 + \frac{.07}{4}\right)^{40} - 1}$$

$$= 300{,}000 \frac{.0175}{2.001597 - 1}$$

$$= 300{,}000(.017472097)$$

$$= 5241.63$$

and since

$$\frac{R}{34} = \frac{5241.63}{34} = 154.17$$

each student agrees to pay $154.17 every three months.

Sinking Funds

Example 3 of Section 8.3 concerned United Transport Company, which, in 4 year's time, will need $150,000 to purchase new vans. We calculated that if United Transport purchased a 4-year certificate of deposit from a bank with interest at an 8.5% annual rate, compounded quarterly, it would need a $108,000 certificate. United Transport may not wish to tie up over $100,000 throughout a four-year period in anticipation of the van purchase; yet it may still want to put money aside so that it will have $150,000 in four years. As an alternative to the certificate of deposit, United Transport might decide to set up a **sinking fund**, that is, an annuity to which it will make periodic payments in order to have the money available when it needs it. In the next example we calculate the rent on the sinking fund, that is, the sum United Transport would pay each quarter into the annuity.

Example 2. United Transport sets up a sinking fund that will pay $150,000 in four years. If the payments are made quarterly and if the fund pays at an 8.5% annual interest rate, calculate the rent for the sinking fund.

Solution

There are $n = 4 \cdot 4 = 16$ interest periods with interest per period $i = \frac{.085}{4} = .02125$. Using Table B, we see that the rent is

$$R = S \frac{i}{(1+i)^n - 1}$$

$$= 150{,}000 \frac{.02125}{\left(1 + \frac{.085}{4}\right)^{16} - 1}$$

$$= 150{,}000 \frac{.02125}{1.399952 - 1}$$

$$= 7969.71$$

United Transport will pay $7969.71 to the bank every 3 months to obtain $150,000 at the end of 4 years. ∎

Summary

In an **ordinary annuity**, payments of fixed sums R, the **rent** of the annuity, are made at regular intervals with m payments each year. Interest is paid at an annual rate r, compounded m times a year, so the interest rate per period is $i = \frac{r}{m}$. There are n interest periods during the life of the annuity and the rent R is paid at the end of each period. The **amount** S of the annuity, the value of the investment after the last payment, is calculated by

$$S = R s_{\overline{n}|i}$$

where

$$s_{\overline{n}|i} = \frac{(1+i)^n - 1}{i}$$

In order to save an amount S by means of an annuity of n payments at interest rate per period i, the rent must be

$$R = S \frac{i}{(1+i)^n - 1}$$

An annuity with rent calculated to produce a given amount at a given time in the future is known as a **sinking fund**.

Exercises

1. Calculate the amount of an annuity for which the rent is $50 a month for 5 years, and the interest is at a 9% annual rate, compounded monthly.

2. An annuity in the amount of $3000 will consist of payments every 3 months for 8 years and will pay interest at a 7% annual rate, compounded quarterly. What is the rent for this annuity?

3. In order to save enough money for a large-screen television set, the Walsh family starts a savings account with $100 and then puts $20 at the end of each month into the account, which pays interest at a 6% annual interest rate, compounded monthly. After 2 years, how much money does the Walsh family have in the account?

4. In the retirement program for employees of Metropolitan Hospital, the employee and the hospital each contribute to the program every month an amount equal to 5% of the employee's salary. The money earns interest at a 9.5% annual rate, compounded monthly. John is hired by the hospital at a salary of $3100 a month. How much money is in John's retirement account at the end of his first year of employment?

5. Mountain City must pay $500,000 interest on municipal bonds in 6 months. It will invest a sum of money each month at a 7.5% annual interest rate, compounded monthly, so that it will have exactly $500,000 when the payment becomes due. If Mountain City invests the same amount each month, to the nearest dollar, how much should it invest?

6. Star Storage Company estimates that an old warehouse will have to be replaced in 6 years at a cost of $6 million. It establishes a sinking fund, with payments every 3 months, on which it can receive interest at a 10% annual rate, compounded quarterly, in order to have the required sum at that time. What is the rent for the sinking fund?

7. A bank pays interest on savings accounts at a 5.5% annual rate, compounded quarterly. For 2 years, at the end of each 3-month period, Karen puts $275 in a

8.4 ANNUITIES 599

savings account and never withdraws any money. The bank also sells certificates of deposit that pay 6.5% annual interest, compounded quarterly. The money in the certificate cannot be withdrawn for 2 years. Lisa buys a $2000 certificate. At the end of the 2 years, which investor, Karen or Lisa, has more money, and how much more does she have?

8. An investor buys a bond with a $5000 face value that will mature in 9 years. The bonds pay interest quarterly at an 8.5% annual rate. Instead of spending the interest payments from the bonds, the investor sets up an annuity for 9 years with quarterly payments equal to the bond interest payments. The annuity receives interest at a 7% annual rate, compounded quarterly. When the bonds mature, how much money from the bonds and annuity combined does the investor have?

9. The president of AZ Industries Inc. is hired at the age of 58 and signs a deferred-compensation agreement that pays her a $2 million retirement bonus at the age of 65. AZ Industries establishes a sinking fund to cover the retirement bonus with the money invested at an annual interest rate of 9.5%, compounded monthly. How much must AZ Industries put into the sinking fund each month in order to have the $2 million in 7 years?

10. Every three months during his 10-year long playing career, a professional basketball player puts $10,000 into an investment program that pays interest at an 11.5% annual rate, compounded quarterly. When the player retires, how much money does he have in the investment program?

11. Aloft Airlines orders new planes. The order will be filled in 3 years and a payment of $150 million will be due at that time. Aloft Airlines can set up a sinking fund that will produce $150 million in three years in two different ways. First Commercial Bank offers Aloft an annual interest rate of 7% compounded monthly and National Bank pays 7.5% compounded quarterly. Aloft Airlines will choose the sinking fund that will require it to pay in less money each year. Determine which plan it will choose, First Commercial's or National's, and how much it will pay each year into the sinking fund it chooses.

12. A tennis player has $40,000 to invest as the result of winning a major tournament. She can put the money in a 10-year certificate of deposit paying interest at a 6.5% annual rate, compounded monthly. Instead, she could buy $40,000 worth of bonds paying interest at a 5% annual rate, with quarterly interest payments, and she can then put the interest payments in an annuity paying interest at a 7.5% annual rate, compounded quarterly. How much money will the tennis player have at the end of 10 years if
(a) she invests in the certificate?
(b) she buys the bonds and uses the interest payments for the annuity?

13. Quality Rug Emporium plans to expand its showroom in 5 years at an estimated cost of $600,000. Quality will set up a sinking fund with quarterly interest payments earning interest at an 11% annual rate, compounded quarterly. How much must Quality pay into the sinking fund each quarter?

600   MATHEMATICS OF FINANCE

14. In 1991, an employee earning $53,400 or more a year paid 7.65% of $53,400 to Social Security that year. If a 45-year old employee paid $\frac{1}{12}$ of that annual payment to Social Security each month and the government put the money into an annuity paying interest at a 7% annual rate, compounded monthly, and if the employee continued to pay just that same sum each month until retiring at the age of 70, what would be the amount of the annuity when the employee retired?

8.5 Amortization

Jack and Judy Miller want to buy a house and they need to know how much they can afford to pay. They have saved some money for a down payment and they will have to obtain a mortgage for the rest. In the introduction to this chapter, we estimated that they can afford mortgage payments of $1150 a month. How much money can they borrow from the bank?

Village Bank offers 30-year fixed-rate mortgages at a 10.5% annual interest rate. This means that the bank will give the Millers a sum of money A to apply to the cost of the house. Each month, the Millers will pay $1150 against the amount they owe Village Bank, which charges them interest at a 10.5% annual interest rate, compounded monthly, on the money still owed. The sum A is calculated so that after 30 years, that is, after $n = 30 \cdot 12 = 360$ payments, the Millers will have paid off the mortgage or, in the language of the mathematics of finance, they will have *amortized* their debt to the bank.

In general, the **amortization** of a debt means the payment of a fixed sum of money, the **rent** R, at periodic intervals until the money loaned together with the interest has been paid off. As in Section 8.4, we assume that interest is compounded when the payment is made.

The money A that Village Bank will lend to the Millers depends on the present value to the bank of each of those 360 payments of $1150 that the Millers will make in the future. We defined the present value of a future payment in Section 8.1 as the money that would have to be invested at the given interest rate to produce an amount equal to the payment at that particular time. Thus, the present value of the first payment the Millers make to Village Bank, one month after taking out the mortgage, is the present value of a simple-interest loan of $1150 borrowed for one month at a 10.5% annual interest rate. The present value of an amount S borrowed at simple interest for time t at an annual interest rate r is

$$\frac{S}{1 + rt}$$

For the Millers' first mortgage payment, the amount borrowed is the rent $R = 1150$. The length of time is $t = \frac{1}{12}$ of a year or, to put it another way, the Millers will be paying $m = 12$ times a year. As in Section 8.3, it will be convenient for reference

8.5 AMORTIZATION 601

to replace rt by $i = \frac{r}{m}$, the interest rate per period, which, for the mortgage is given by $i = \frac{.105}{12} = .00875$. The present value of the Millers' first mortgage payment is

$$\frac{R}{1+i} = \frac{1150}{1+.00875} = \frac{1150}{1.00875} = 1140.02$$

To determine the present value to Village Bank of the second mortgage payment, we can no longer use the simple-interest concept. The reason is that the bank compounds the interest every month, so the interest on the second $1150, which the Millers hold for two months, is charged as compound interest. The present value of a compound-interest loan with amount S at interest rate per period i compounded for n periods is

$$\frac{S}{(1+i)^n}$$

For the second payment, the amount is $R = 1150$ and $n = 2$. Therefore, by Table B, the present value of the Millers' second mortgage payment is

$$\frac{R}{(1+i)^2} = \frac{1150}{\left(1+\frac{.105}{12}\right)^2} = \frac{1150}{1.017577} = 1130.14$$

Section 8.3 tells us that each mortgage payment of rent R that the Millers will make has a different present value to Village Bank. Payment number k has present value

$$\frac{R}{(1+i)^k}$$

Thus, the last monthly payment, number 360, has present value of only

$$\frac{R}{(1+i)^{360}} = \frac{1150}{\left(1+\frac{.105}{12}\right)^{360}} = \frac{1150}{23.018509} = 49.96$$

That is, to obtain $1150 after 360 compounding periods at an interest rate per period of .00875, Village Bank need only invest $49.96.

The money A that Village Bank will be willing to lend the Millers in return for the 360 payments of $1150 each is the sum of the present values of all these payments, that is

$$A = \frac{1150}{(1+.00875)} + \frac{1150}{(1+.00875)^2} + \ldots + \frac{1150}{(1+.00875)^{360}}$$

$$= 1140.02 + 1130.14 + \ldots + 49.96$$

Of course, we do not want to have to calculate all 360 present values and then add them up, so we need a formula for computing A directly.

A Formula for Present Value

We have seen that if payments of rent R are made periodically to a lender for n periods at an interest rate per period of i, then the present value A to the lender is given by

$$A = \frac{R}{(1+i)} + \frac{R}{(1+i)^2} + \ldots + \frac{R}{(1+i)^n}$$

We factor the R that appears in every term on the right-hand side, obtaining

$$A = R\left[\frac{1}{(1+i)} + \frac{1}{(1+i)^2} + \ldots + \frac{1}{(1+i)^n}\right]$$

and we introduce another standard symbol from the mathematics of finance,

$$\boxed{a_{\overline{n}|i} = \frac{1}{(1+i)} + \frac{1}{(1+i)^2} + \ldots + \frac{1}{(1+i)^n}}$$

The definition of the present value of the mortgage becomes

$$\boxed{A = R a_{\overline{n}|i}}$$

We want a one-step computational formula for $a_{\overline{n}|i}$. Notice what happens when we multiply $a_{\overline{n}|i}$ by $(1+i)^n$ and then simplify:

$$(1+i)^n a_{\overline{n}|i} = (1+i)^n\left[\frac{1}{(1+i)} + \frac{1}{(1+i)^2} + \ldots + \frac{1}{(1+i)^n}\right]$$

$$= \frac{(1+i)^n}{(1+i)} + \frac{(1+i)^n}{(1+i)^2} + \ldots + \frac{(1+i)^n}{(1+i)^n}$$

$$= (1+i)^{n-1} + (1+i)^{n-2} + \ldots + 1$$

The last expression was represented by the symbol $s_{\overline{n}|i}$ in Section 8.4. Therefore, we see that

$$(1+i)^n a_{\overline{n}|i} = s_{\overline{n}|i}$$

Since we found a simple expression for $s_{\overline{n}|i}$, namely,

$$s_{\overline{n}|i} = \frac{(1+i)^n - 1}{i} = [(1+i)^n - 1]\left(\frac{1}{i}\right)$$

we have

$$(1+i)^n a_{\overline{n}|i} = [(1+i)^n - 1]\left(\frac{1}{i}\right)$$

8.5 AMORTIZATION

To solve for $a_{\overline{n}|i}$, we must divide both sides of the equation by $(1+i)^n$, which is the same as multiplying by $(1+i)^{-n}$.

$$a_{\overline{n}|i} = (1+i)^{-n}[(1+i)^n - 1]\left(\frac{1}{i}\right)$$

$$= [(1+i)^0 - (1+i)^{-n}]\left(\frac{1}{i}\right)$$

This is the formula that we were seeking:

$$a_{\overline{n}|i} = \frac{1-(1+i)^{-n}}{i}$$

The formula for $a_{\overline{n}|i}$ is customarily written in this form, using a term with a negative exponent. The calculation of $(1+i)^{-n}$, with $i = \frac{r}{n}$, is given in Table B. Now we will compute the size of the mortgage that the Millers can obtain. Recall that they will make $n = 360$ payments of $R = 1150$ and that the interest rate per period is $i = \frac{.105}{12} = .00875$. Then using Table B or a scientific calculator, we find that

$$A = Ra_{\overline{n}|i} = R\frac{1-(1+i)^{-n}}{i}$$

$$= 1150\frac{1-\left(1+\frac{.105}{12}\right)^{-360}}{.00875}$$

$$= 1150\left(\frac{1-.043443}{.00875}\right)$$

$$= 1150(109.3208)$$

$$= 125{,}718.92$$

Thus, the Millers' payments will amortize a mortgage of about $126,000. Village Bank will not finance more than 90% of the cost of the house, but if the Millers have at least $14,000 for the down payment, their income will permit them to buy a house that costs about $140,000.

Example 1. Jerry, who buys a used car from a dealer, pays $500 cash and $300 a month for three years. The dealer is charging Jerry interest at an annual rate of 16.5%. Calculate the cash value of Jerry's car when he buys it.

Solution

The $300-a-month payments that Jerry makes to the dealer will amortize a loan consisting of the cash value of the car less Jerry's $500 down payment. The interest rate per period on this loan is $i = \frac{.16}{12} = .013333$ and there will be $n = 3 \cdot 12 = 36$ payments.

Thus, the present value of the loan is

$$A = Ra_{\overline{36}|.01375} = 300\,\frac{1-\left(1+\frac{.16}{12}\right)^{-36}}{.013333}$$

$$= 300\,\frac{1-.620749}{.013333}$$

$$= 8533.36$$

Therefore, the total cash value of Jerry's car is $8533.56 + $500 = $9033.56 ∎

Computing the Rent

Unlike the Millers, the Singhs already have a house on which they are making mortgage payments. However, the house is not big enough for their growing family so, to pay for construction to expand their home, they will take out a second mortgage, a loan on their house in addition to the mortgage they obtained. Suppose they can borrow $25,000 on a 15-year second mortgage at a 12.5% annual interest rate. How much will the monthly payment on this second mortgage be?

The relationship between the amount borrowed for the second mortgage and the monthly rent is still given by

$$A = Ra_{\overline{n}|i}$$

where

$$a_{\overline{n}|i} = \frac{1-(1+i)^{-n}}{i}$$

This time we know the size of the debt, $A = 25{,}000$, and we wish to determine the rent. We also know that the interest rate per period is $i = \frac{.125}{12} = .010417$ and that there will be $n = 15 \cdot 12 = 180$ payments during the 15 year life of the mortgage. To find a formula for computing the rent on a mortgage, we solve the equation $A = Ra_{\overline{n}|i}$ for R by dividing by $a_{\overline{n}|i}$

$$R = \frac{A}{a_{\overline{n}|i}}$$

Since dividing by $a_{\overline{n}|i}$ is the same as multiplying by its reciprocal, that is, by

$$\frac{i}{1-(1+i)^{-n}}$$

we have

$$\boxed{R = A\,\frac{i}{1-(1+i)^{-n}}}$$

Thus, the monthly payments for the Singh's second mortgage will be

$$R = 25{,}000 \, \frac{.010417}{1 - \left(1 + \frac{.125}{12}\right)^{-180}}$$

$$= 25{,}000 \, \frac{.010417}{1 - .154849}$$

$$= 308.13$$

Example 2. Alice buys a boat that costs $8000. She makes a down payment of $1000 and pays off the rest in monthly payments over a 5-year period. If the installment loan for the boat is at a 14% annual interest rate, compounded monthly, what is the size of the monthly payments?

Solution

The loan is for $8000 − $1000 = $7000 and there will be $n = 5 \cdot 12 = 60$ monthly payments. The interest rate per period is $i = \frac{.14}{12} = .011667$, and therefore the monthly payment is

$$R = 7000 \, \frac{.011667}{1 - \left(1 + \frac{.14}{12}\right)^{-60}}$$

$$= 7000 \, \frac{.011667}{1 - .498601}$$

$$= 162.88 \qquad \blacksquare$$

Refinancing a Loan

Banks often offer borrowers a low "introductory" interest rate on a mortgage. This makes the mortgage more attractive because the monthly rent is relatively low during a time that the new home buyer has many additional expenses. Suppose that River Bank will lend the Millers $126,000 as a 30-year mortgage, just as the Village Bank will, but in place of the 10.5% annual rate that Village Bank charges, River Bank will give the Millers a 7% introductory rate for the first year, after which the rate will rise to 11%. The Millers need to know how much their mortgage payments will be during the first year and then how much they will be for the remaining 29 years.

During the first year, River Bank calculates the Miller's mortgage payments as though they had a 30-year mortgage for $126,000 at a 7% annual interest rate throughout the life of the mortgage, that is, with interest rate per period $i = \frac{.07}{12} = .005833$. Thus, each monthly payment during that year will be

$$R = 126{,}000 \, \frac{.005833}{1 - \left(1 + \frac{.07}{12}\right)^{-360}}$$

$$= 838.23$$

which is certainly considerably less than the $1150 a month the Millers would pay that first year to Village Bank.

However, at the end of the first year, the River Bank periodic interest rate goes up to $i = \frac{.11}{12} = .009167$ and there are 29 years, or $29 \cdot 12 = 348$ payments, still to go. But the Millers no longer owe the bank the full $126,000 since they paid part of it off during the first year. Under the River Bank mortgage plan, the Millers replace the original loan with what is really a new loan on the amount still owed, for 29 years at an 11% annual interest rate. When an existing loan that is not fully amortized is to be replaced by a new loan whose amount equals the amount still owed on the original loan, it is customary to say that the borrower will **refinance** the existing loan. To calculate the rent for the remaining 29 years of the mortgage, we need to know how much of the original $126,000 the Millers will still owe River Bank at the end of one year.

River Bank loaned $126,000 to the Millers at an interest rate per period of $i = \frac{.07}{12} = .005833$. If the Millers had paid nothing to the bank during the first year, according to Section 8.3 and using Appendix Table B, after the 12 compounding periods the amount of that loan, the principal plus the compounded interest, would have risen to

$$A(1+i)^n = 126{,}000\left(1 + \frac{.07}{12}\right)^{12} = 135{,}185.40$$

Of course the Millers did pay the bank $838.23 each month. If we view the Millers' 12 mortgage payments as payments into an annuity with the same interest rate per period, then the amount of this annuity, which River Bank should credit against their debt, is

$$S = R\frac{(1+i)^n - 1}{i} = 838.23 \frac{\left(1 + \frac{.07}{12}\right)^{12} - 1}{.005833} = 10{,}476.08$$

as shown in Section 8.4. Subtracting this sum from the amount of the loan, we find that after one year the Millers will still owe River Bank

$$\$135{,}185.40 - \$10{,}476.08 = \$124{,}709.32$$

Thus, their monthly mortgage payments for the final 29 years of the River Bank mortgage at the 11% annual interest rate will be

$$R = 124{,}709.32 \frac{.009167}{1 - \left(1 + \frac{.11}{12}\right)^{-348}}$$

$$= 124{,}709.32 \frac{.009167}{1 - .041775}$$

$$= 1193.05$$

If you are using Table B, you will not find an entry for $\left(1 + \frac{r}{12}\right)^{-n}$ with $n = 348$. You can again use the properties of exponents as follows.

$$\left(1 + \frac{.11}{12}\right)^{-348} = \left(1 + \frac{.11}{12}\right)^{-360}\left(1 + \frac{.11}{12}\right)^{12}$$

$$= (.037442)(1.115719)$$

$$= .041775$$

To summarize, if the Millers take the River Bank mortgage, they pay only $838.23 a month during the first year, but then the monthly mortgage payment rises to $1193.05.

You would probably expect the River Bank mortgage to be quite a bit more expensive than that of Village Bank over the 30-year period because River Bank charges a higher interest rate during most of that time, and you would be correct. To see how much more expensive, suppose the Millers borrow exactly $126,000 from Village Bank at a 10.5% annual interest rate. Then the monthly rent will be $1152.57, so that the total of the 360 payments will come to ($1152.57)360 = $414,925.20. Thus, to borrow $126,000 now, the Millers will have to pay Village Bank more than $400,000 during the next 30 years. But they would pay more to River Bank because 12 payments of $838.23 plus 348 payments of $1193.05 comes to $425,240.16.

If a loan is refinanced before it is fully amortized, it is necessary to calculate the amount of the debt remaining. We reason as we did in analyzing the River Bank loan. If periodic payments of size R are made on a debt of A dollars and if the interest rate per period is i, after k periods the amount of the original debt would have grown to $A(1 + i)^k$. On the other hand, the value of the k payments, viewed as an annuity with the same interest rate per period, would be $Rs_{\overline{k}|i}$. Therefore, the amount still to be paid is the difference:

$$\boxed{A(1 + i)^k - Rs_{\overline{k}|i} = A(1 + i)^k - R\frac{(1 + i)^k - 1}{i}}$$

Example 3. Jeff buys a condominium, taking out a 25-year mortgage of $50,000 at a 9.5% annual interest rate. After 5 years, Jeff decides that his income has grown sufficiently so that he can afford to pay off the mortgage more quickly. Therefore, he refinances the condominium with a 10-year mortgage at the same annual interest rate.

(a) How much does Jeff pay each month during the first 5 years?

(b) What is the size of Jeff's monthly mortgage payment after he refinances?

Solution

(a) The rent for a 25-year $50,000 mortgage at an interest rate per period of

608   MATHEMATICS OF FINANCE

$$i = \frac{.095}{12} = .007917$$

is given by

$$R = 50{,}000 \frac{.007917}{1 - \left(1 + \frac{.095}{12}\right)^{-300}} = 436.87$$

(b) Using the determination of the rent in Part (a), we calculate that the amount Jeff still owes the bank after 5 years, that is, after $k = 5 \cdot 12 = 60$ payments, is

$$A(1+i)^k - R\frac{(1+i)^k - 1}{i} = 50{,}000\left(1 + \frac{.095}{12}\right)^{60} - 436.87 \frac{\left(1 + \frac{.095}{12}\right)^{60} - 1}{.007917}$$

$$= 80{,}254.50 - 33{,}389.63$$

$$= 46{,}864.87$$

Since there are 120 payments in a 10-year loan, his new rent will be

$$R = 46{,}864.87 \frac{.007917}{1 - \left(1 + \frac{.095}{12}\right)^{-120}} = 606.45$$ ∎

Balloon Payments

It often happens that a debt is paid partly through amortization and the rest in a single cash payment, called a **balloon payment**. For example, a bank may be willing to finance only 80% of the cost of a house. If the purchaser has 10% of the price of the house in cash, the seller may agree to finance the other 10% for a few years, accepting relatively modest payments each month, but will then require that the remaining debt be paid off in a balloon payment. Our final example returns to the Miller's efforts to finance the home they want, to illustrate the balloon-payment concept.

Example 4. Jack and Judy Miller find the house they want to buy, but it will cost them $155,000. They have $15,500 cash for the required 10% down payment, but (as we have seen) the largest mortgage they can qualify for at a bank is only $126,000. In order to make the sale possible, the seller offers to lend them the remaining $13,500 at a 10% annual interest rate, compounded monthly. The Millers will pay the loan back by monthly payments of $200 for three years and then a balloon payment for the unpaid balance. Calculate the balloon payment the Millers must pay in three years.

Solution

The calculation of the balloon payment uses the same formula as that used to determine the amount still owed when a loan is refinanced. The amount of the debt is

$13,500 and the interest rate per period is $i = \frac{.10}{12} = .008333$, so after $k = 3 \cdot 12 = 36$ monthly payments of $200, the sum still owed for the balloon payment is

$$A(1+i)^k - R\frac{(1+i)^k - 1}{i} = 13{,}500\left(1 + \frac{.10}{12}\right)^{36} - 200\frac{\left(1 + \frac{.10}{12}\right)^{36} - 1}{.008333}$$

$$= 18{,}200.46 - 8{,}356.70$$

$$= 9{,}843.76$$

Thus, the Millers have agreed to a cash payment of $9,843.76 at the end of three years. ∎

If the Millers do not have $9,843.76 cash available to pay the balloon payment to the seller at the end of the two years, they will have to borrow the money, perhaps by taking a second mortgage on the house.

Summary

The **amortization** of a debt is the payment of a certain sum of money, the **rent** R, at fixed intervals until the debt, together with accumulated interest, has been paid off. The sum lent, A, is the present value of all payments. If the size of the debt is A, if the interest rate per period is i, and if the debt is paid off in n payments, each of size R, then

$$A = R a_{\overline{n}|i}$$

where

$$a_{\overline{n}|i} = \frac{1 - (1+i)^{-n}}{i}$$

If a debt with present value A is to be paid off in n payments with interest rate per payment i, then the rent can be calculated by

$$R = A \frac{i}{1 - (1+i)^{-n}}$$

If a debt of size A is amortized by periodic payments of rent R and if interest per period is at rate i, then after k payments the money still owed on the debt is

$$A(1+i)^k - R\frac{(1+i)^k - 1}{i}$$

A **balloon payment** is a cash payment on a loan equal to the amount that is not fully amortized.

Exercises

1. Compute the present value of a loan in which the borrower pays $10,000 every 3 months for 15 years, and the interest is at a 6.5% annual rate, compounded quarterly.

2. A debt of $100,000 at an annual interest rate of 8.5%, compounded quarterly, is to be paid off in 5 years, with a payment every 3 months. What will each payment be?

3. As part of the financing of a home, the seller lends the purchaser $35,000 at an annual interest rate of 13%, compounded monthly. The debt is to be paid off by payments of $500 a month for 4 years, to be followed by a balloon payment for the unpaid balance. Calculate the amount of the balloon payment.

4. Dan buys audio equipment for a $150 down payment together with monthly payments of $80 for 2 years. If the store is charging Dan interest at an 18% annual rate, compounded monthly, what is the present cash value of the stereo?

5. A house is purchased for $180,000 with a down payment of $40,000 and the rest to be paid monthly at a 10.5% annual interest rate, compounded monthly for 30 years. How large will each payment be?

6. Susan receives an inheritance in the form of a payment of $1000 every 3 months for 15 years. If she can invest the money received at an annual interest rate of 6% compounded quarterly, what is the present value of her inheritance?

7. The Perez family offer Mrs. Gray $250,000 for her home. They have $25,000 cash for a down payment and can obtain a $200,000 mortgage from a bank. In place of the remaining $25,000, Mrs. Gray will accept monthly payments of $400 for two years with interest on the unpaid balance, compounded monthly, at a 12.5% annual rate and the remainder of the debt payable as a balloon payment at the end of the 2 years. Calculate the balloon payment that Mrs. Gray will receive in 2 years.

8. The Sharks Swim Club builds a diving pool costing $80,000, to be financed by paying $30,000 in cash and by borrowing the rest at 11% annual interest, compounded quarterly, with payments every 3 months for 10 years. How large is each payment?

9. General Manufacturing Company borrows $7 million to pay for improvements to its manufacturing plant. The money will be paid back by payments every 3 months over 4 years at a 9.5% annual interest rate, compounded quarterly. At the end of 2 years, how much money does General Manufacturing Company still owe on this debt?

10. Larry buys a microwave oven costing $300 for a down payment of $50 and

monthly payments over 2 years at a 14.5% annual interest rate, compounded monthly. At the end of the first year, how much does Larry still owe on the oven?

11. The Green family buys a house with a 25-year mortgage for $200,000. Interest is charged at an annual rate of 8% for the first two years of the mortgage; then it rises to 10%.
    (a) What are the Greens' monthly payments during the first two years of the mortgage?
    (b) What are the Greens' monthly mortgage payments during the last 23 years?

12. The Chang family purchases a vacation cottage costing $60,000 for a $15,000 down payment and monthly payments over 15 years at an 11.5% annual interest rate, compounded monthly. What is the family's equity in the cottage ($60,000 minus the amount still owed) after 10 years?

13. Donna inherits $25,000 and decides to use it as a down payment on a condominium. Based on her salary, she can afford $700 a month for a mortgage. If she can obtain a 20-year mortgage at a 10% annual interest rate, compounded monthly, what is the most expensive condominium Donna can buy?

14. Linda buys a $12,000 automobile for a $1500 down payment and monthly payments to a finance company over 3 years at a 13.5% annual interest rate, compounded monthly. After 18 payments, Linda loses her job and cannot afford to make any more payments; so, the finance company takes possession of the automobile. How much did the automobile cost the finance company? That is, how much of Linda's debt was still unpaid?

15. A divorce settlement is to be paid in 40 quarterly payments of $5000 each. Calculate the present value of the settlement if money can be invested at a 9% annual interest rate, compounded quarterly.

16. In 1975, the Johnsons bought a house with a 25-year mortgage for $50,000 at an annual interest rate of 6%, compounded monthly. They sold the house in 1990 for $350,000. How much of the $350,000 did the Johnsons have left after they paid off the mortgage?

17. Jim can buy a new car for $26,000 by putting down $1750 and paying the rest monthly for 4 years at an annual interest rate of 13%, compounded monthly. Instead, Jim can keep his old car and deposit the down payment in an account that pays a 5.5% annual interest rate, compounded monthly. Furthermore, each month he will deposit an amount equal to the monthly installment on the new car. How much money would Jim have in the bank at the end of 3 years if he keeps his old car?

18. Mike bought a house with an $80,000 mortgage for 30 years at a 13.5% annual interest rate, compounded monthly. After 10 years, he refinanced the house,

taking a mortgage (on the amount still owed) for the last 20 years at a 10% annual rate.
(a) What was Mike's monthly mortgage payment during the first 10 years?
(b) What was the size of the mortgage payment after he refinanced?

19. To finance a new shopping mall, Northern Development Corporation obtains a loan for $3 million at a 9.5% annual interest rate, compounded monthly. The payments will be $30,000 a month for 5 years, with the unpaid balance due as a balloon payment at the end of that time. Two years after the purchase of the mall, Northern Development sets up a sinking fund at a bank to pay the balloon payment when it becomes due. The bank pays interest at an 8% annual rate, compounded quarterly, and the payments into the sinking fund will be made quarterly. How much must be paid into the sinking fund every 3 months?

Review Exercises for Chapter 8

1. Ajax Management Company manages an office building that presently earns a profit of $6000 a month. However, the building is in need of substantial repairs, so Ajax wants to use the $6000 monthly profits to pay a second mortgage in order to finance the repairs. If Ajax can get a 4-year second mortgage at an 11% annual interest rate, compounded monthly, how much money can it raise in this way?

2. White's Department Store announces that store credit-card holders need not pay anything on their Christmas purchases until April. However, in the meantime, interest accrues at a 20% annual rate, compounded monthly. Sam charges $800 worth of Christmas presents at White's. When he starts to pay for them 4 months later, what is the total that Sam owes the department store?

3. Jose borrows $2000 from Neighborhood Finance Corporation with an 8-month noninterest-bearing note at an 18% annual discount rate.
   (a) How much money does he receive from Neighborhood Finance?
   (b) Calculate the effective interest rate of the note.

4. The winner of a "$1 million" lottery payoff is paid $12,500 every 3 months for 20 years. If the winner could invest money at an annual interest rate of 8% compounded quarterly, what is the present value of the "$1 million" lottery payoff?

5. Central Investments Inc. is considering the purchase of a $3 million issue of local government bonds that will pay $45,000 interest every 3 months until the bonds mature and the principal is repaid at the end of 5 years. Alternatively, Central can obtain a $3 million certificate of deposit at a 7% annual interest rate, compounded monthly.
   (a) If Central puts the $45,000 quarterly payments into a 5-year annuity paying interest at a 6.5% annual rate, compounded quarterly, how much money will Central have at the end of the 5 years?
   (b) If Central chooses the certificate instead, how much money will it have at the end of 5 years?

6. Banks offering certificates of deposit publish both the annual interest rate and the "yield," which is the effective annual rate. If a bank offers certificates at a 7.5% annual interest rate, compounded quarterly, what does it advertise as the yield?

7. The registration fee for the annual meeting of the American Mathematical Society is $135. However, if a member preregisters and sends in the money 3 months in advance, the fee is only $105. Suppose a member of the Society can invest money at simple interest at a 9% annual rate.
   (a) Calculate the present value of $135 in 3 months.
   (b) How much does a Society member save over the present value by paying the $105 fee now?

8. In 1975, Judy obtained a 30-year mortgage for $70,000 at an 11.5% annual interest rate, compounded monthly. In 1980, she added a 15-year second mortgage of $10,000 at a 14% annual interest rate, compounded monthly. Then in 1990, when interest rates were lower, Judy decided to consolidate the loans into a single mortgage.
   (a) Calculate Judy's monthly rent between 1975 and 1980.
   (b) Calculate Judy's total monthly rent for the two mortgages from 1980 to 1990.
   (c) When Judy took out the new mortgage in 1990, what was the total she owed on the two existing mortgages?

9. The management committee of the Seaside Condominium Complex has just paid $60,000 to carpet the public areas of the complex, and it estimates that it will have to do it again in 10 years. The committee sets up a sinking fund for 10 years at an 8% annual interest rate, compounded monthly, to cover the cost of recarpeting, with rent to be paid by the condominium owners each month. The committee estimates that the cost of recarpeting will rise 6% a year, compounded quarterly, so the cost of recarpeting is the same as the amount of a $60,000 loan for 10 years at a 6% annual interest rate, compounded quarterly.
   (a) How much will it cost to recarpet in 10 years?
   (b) How much will the condo's monthly rent be for the sinking fund?

10. In December, World Travel Services can charter a cruise ship for the summer at a very favorable price of $200,000. World can either borrow the $200,000 for 6 months through a noninterest-bearing note at a 10% discount rate or take out a loan for $200,000 at an 11% annual interest rate, compounded monthly, and amortize the loan over 6 months.
    (a) How much would World Travel have to pay the holder of the noninterest-bearing note at the end of the 6 months?
    (b) How much would World pay for the loan, that is, what is 6 times the monthly rent?

11. When fire destroys part of the Ace Manufacturing Company's plant, the insurance company immediately pays Ace $3 million for repairs. It will take time to repair the damage and there are no bills to pay at once, so the Ace treasurer invests the $3 million in a 4-month simple-interest loan at an annual interest rate of 11.5%. How much money does Ace have at the end of 4 months?

12. Mrs. Patel takes out a 12-year second mortgage on her home for $20,000 from First Local Bank at a 12.5% annual interest rate, compounded monthly. After 5 years, First Local sells Mrs. Patel's loan to an investor at par, that is, for the sum still owed on the loan. How much does the investor pay First Local for Mrs. Patel's loan?

13. While Brad has a part in a Broadway play, he puts $500 a month into a bank account paying interest at a 7% annual interest rate, compounded monthly. He does this every month of the 2-year run of the play. How much money does Brad have in the account when the play closes?

14. Sue wins $2,000,000 in the lottery, payable in 80 quarterly installments of $25,000 each. She is offered $1,100,000 cash for the rights to these payments. If Sue can invest money at a 7.5% annual rate, compounded quarterly, would she have more money 20 years later if she accepts the cash offer and invests the $1,100,000 now, or if she invests the lottery payments as she receives them? Explain.

Appendices

Appendix 1: Set Theory and Logic

Set Theory

In probability, graph theory, and other branches of mathematics the objects studied are collectively known as **sets**. For example, a probability problem may be concerned with the set of all 5-card hands that can be dealt in a poker game. The set of vertices of a graph may represent a set of American cities, and the set of edges of the graph, intercity highways. The set of all objects considered in a particular context is called the **universal set,** and is often denoted by $\mathcal{U}$. In probability studies, $\mathcal{U}$ represents the **sample space**, introduced in Section 3.6.

Venn Diagram

It is convenient to represent sets by means of a **Venn diagram**, in which the universal set is shown as a large rectangle within which other sets are depicted as circles or ovals. The diagram for a single set A is illustrated in Figure A-1. Here, the set of objects that are in A is represented by the circle; the set A' of objects of $\mathcal{U}$ that are *not* in A, known as the **complement** of A, consists of the points inside the rectangle, but outside of the circle.

It is sometimes necessary to refer to the set with no elements, called the **empty set.** For example, if the universal set consists of the cities of the United

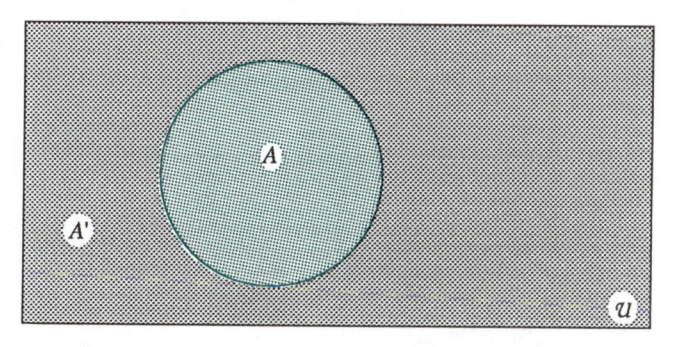

Figure A-1

States, one description of the empty set would be the set of cities with a population of over one billion people.

Unions and Intersections

Venn diagrams representing two sets, A and B, illustrate the relationship between them.

> If every element of A is an element of B, then A is said to be a **subset** of B.

This relationship is shown in Figure A-2, in which the circle for the subset A lies within that for B.

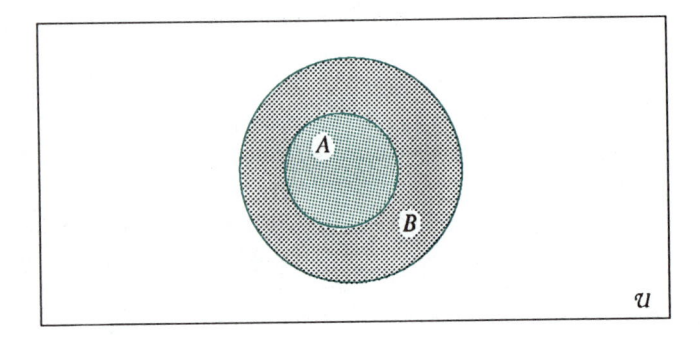

Figure A-2

In the general case represented in Figure A-3, neither A nor B is necessarily a subset of the other.

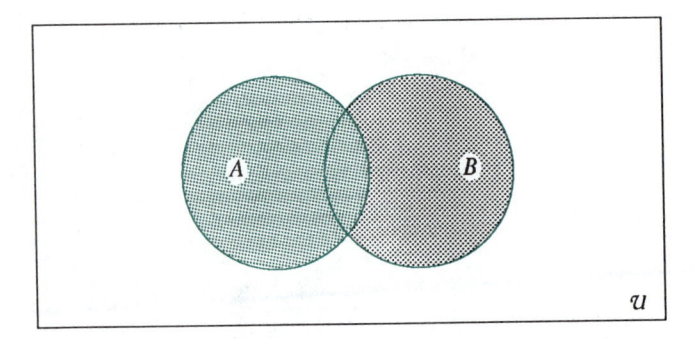

Figure A-3

> The set of objects that are in at least one of the sets A *or* B is called the **union** of A and B, written $A \cup B$, and is represented by the colored region of Figure A-4.

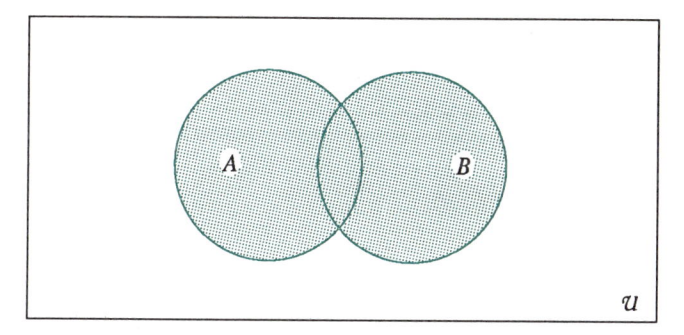

Figure A-4

> The set of objects that are in both sets *A and B*, called the **intersection** of A and B, written $A \cap B$, is shown by the colored region of Figure A-5. If A and B do not have objects in common that is, if $A \cap B$ is the empty set, A and B are said to be **disjoint**.

A Venn diagram showing disjoint sets is shown in Figure A-6.

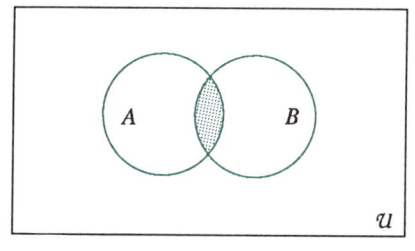

Figure A-5

Figure A-6

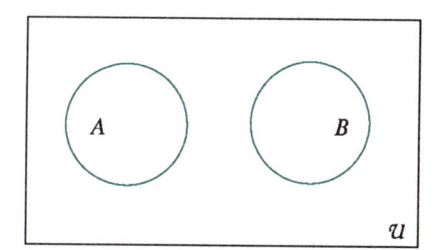

> Sets A and B are **equal** if they contain precisely the same elements; in this case, we write $A = B$.

Venn diagrams can be useful in establishing the equality of sets, as Example 1 illustrates.

Example 1. Show that for any sets **A** and **B**, the sets $A' \cup B$ and $(A \cap B')'$ are equal, that is, $A' \cup B = (A \cap B')'$.

Solution

In Figure A-7, we represent the sets A and B in a Venn diagram and number the non-overlapping regions of the diagram. The set A consists of region 1, the subset of elements of A that are also in B, and region 2, the subset of elements of A that are not in B. The subset of B consisting of elements not in A is numbered 3. The subset of $\mathcal{U}$ of elements not in $A \cup B$ is numbered 4.

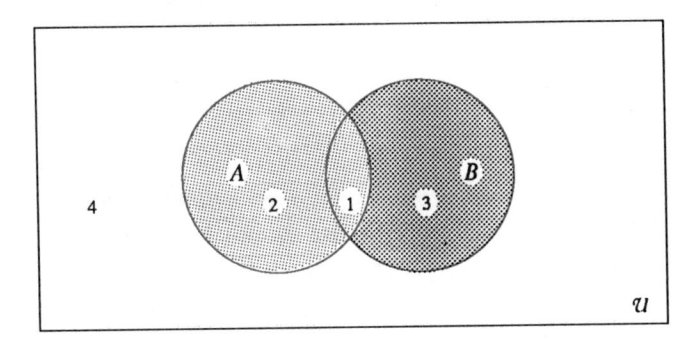

Figure A-7

In Figure A-8, which shows $A' \cup B$, we see that A' consists of regions 3 and 4. The union with B adds only region 1 since region 3 is already included in A'. On the other hand, in Figure A-9, we observe that A consists of regions 1 and 2, whereas B' is made up of regions 2 and 4. Therefore, $A \cap B'$ is the region they have in common, namely, 2. The complement of region 2, that is, $(A \cap B')'$, consists of the other regions, that is 1, 3, and 4. Thus, in both cases, the set in question is represented by regions 1, 3, and 4. Therefore, the sets are equal, that is,

$$A' \cup B = (A \cap B')'$$

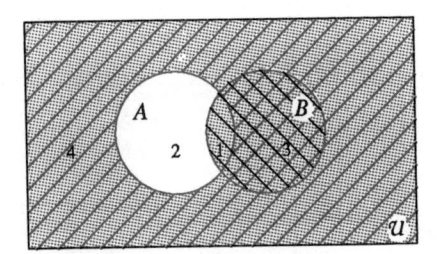

Figure A-8

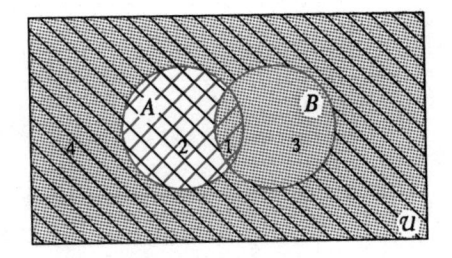

Figure A-9

Logic

Logic is the science that investigates the principles of reasoning. Every statement analyzed in logic is either strictly defined or is accepted without definition. We begin by considering the following sentences.

1. All cats are black.
2. Stop running.
3. Where are the keys?
4. Luciano Pavarotti is a famous tenor.
5. J.R.R.Tolkien was a professor at Oxford University and he wrote popular fantasy.

Not all these sentences are "statements" that can be analyzed by the methods of logic. A **statement** is defined as a meaningful declarative sentence that is regarded as being either true or false, but not both true and false. Looking at the five given sentences, we see that Sentence 1 is a statement that is false. Sentence 2 is not a statement because it is a command. Sentence 3 is not a statement because it is a question. Sentences 4 and 5 are both true statements. Sentence 4 is an example of a **simple statement**; it cannot be broken up into anything "simpler." Sentence 5 is called a **compound statement** because it is made up of two statements, which we label p and q:

p: J.R.R. Tolkien was a professor at Oxford University
q: he wrote popular fantasy

In logic, statements are represented by the letters p, q, r, s, etc. A compound statement always has at least one **connective**, which is a combining or modifying expression, such as *and, or, not, if ... then, neither ... nor,* and so on. We will restrict our discussion to three fundamental connectives: **conjunction** (*p and q*) represented by the symbol $p \wedge q$, **disjunction** (*p or q*) symbolized by $p \vee q$, and **negation** (*not p*) symbolized by $\sim p$. Thus, Statement 5 is of the form $p \wedge q$.

Truth Tables

Every statement has a **truth value**, either true (T) or false (F). The truth value of a compound statement depends on the truth values of its parts. The truth values of the fundamental connectives are defined by means of the following **truth table** (Table A-10).

(1)	(2)	(3)	(4)	(5)
p	q	$p \wedge q$	$p \vee q$	$\sim p$
T	T	T	T	F
T	F	F	T	F
F	T	F	T	T
F	F	F	F	T

Table A-10

From column (3) of the truth table we see that $p \wedge q$ (*p and q*) is true when both statements p, q are true and is false in the other three cases. The statement $p \vee q$ (*p or q*) is considered to be true except in the case in which both p and q are false (column (4)). Finally, $\sim p$ (*not p*) and p have opposite values; when one is true the other is false (column (5)).

Truth Tables and Venn Diagrams

There is a correspondence between the truth table for statements p and q (Table A-10) and the Venn diagram for sets A and B (Figure A-7). By examining the (horizontal) rows of the truth table with the value T and the regions of the Venn diagram we see that

> $p \wedge q$ corresponds to $A \cap B$
> $p \vee q$ corresponds to $A \cup B$
> $\sim p$ corresponds to A'

For example, $A \cup B$ corresponds to regions 1, 2, and 3 of the Venn diagram; $p \vee q$ is true in rows 1, 2, and 3 of the truth table.

Truth Tables for Compound Statements

To determine whether a more complicated compound statement is true or false, we use a truth table and the truth values of the fundamental connectives given in Table A-10.

Example 2. Construct a truth table for the compound statement:

Tracy will be elected class president or Jennifer will not be elected treasurer.

Solution

We have two simple statements

p: Tracy will be elected class president
q: Jennifer will be elected treasurer

The negation of q is then

$\sim q$: Jennifer will not be elected treasurer

Statement p is connected with $\sim q$ by "or" ($\vee$). Thus, the given statement can be represented symbolically as $p \vee (\sim q)$. We construct our truth table (A-11) from Table A-10.

(1) p	(2) q	(3) $\sim q$	(4) $p \vee (\sim q)$
T	T	F	T
T	F	T	T
F	T	F	F
F	F	T	T

Table A-11

SET THEORY AND LOGIC A7

The final column, (4), is obtained by comparing columns (1) and (3). Since our statement is of the form

$$p \vee (\sim q)$$

column (4) contains a T in any row in which either column (1) or column (3) has a T. Thus, column (4) contains T's in rows 1, 2, and 4. The given statement is regarded as true except in the case in which Tracy will not be elected president, and yet Jennifer will be elected treasurer. ∎

Example 3. Construct a truth table for the compound statement

Susan did not come yesterday and she did not collect her library books.

Solution

We define statements p and q by

p: Susan came yesterday
q: she did collect her library books

We see that the first part of the statement, that is,

Susan did not come yesterday

should be designated as $\sim p$ and the second part,

she did not collect her library books

as $\sim q$. The connective *and* in the given statement is represented by $\wedge$, so that symbolically, we write the complete statement as $(\sim p) \wedge (\sim q)$. To construct the required truth table, we first list all the possible combinations of truth values for p and q. Then we put into the table the truth values for $\sim p$ and $\sim q$, using Table A-10; we obtain Table A-12.

Finally, we insert the truth values for $(\sim p) \wedge (\sim q)$ in Table A-13. We can find these values by comparing the entries for $\sim p$ and $\sim q$ with the definition of conjunction in Table A-10.

(1)	(2)	(3)	(4)
p	q	$\sim p$	$\sim q$
T	T	F	F
T	F	F	T
F	T	T	F
F	F	T	T

Table A-12

(1)	(2)	(3)	(4)	(5)
p	q	$\sim p$	$\sim q$	$(\sim p) \wedge (\sim q)$
T	T	F	F	F
T	F	F	T	F
F	T	T	F	F
F	F	T	T	T

Table A-13

Thus, the compound statement

Susan did not come yesterday and she did not collect her library books

A8 APPENDICES

is true only when both

p: Susan came yesterday

and

q: Susan did collect her library books

are false. ∎

Equivalent Statements

If we are only considering one statement p, our truth table need only have two rows. Thus, if we use Table A-10 to construct a truth table for the statement $\sim(\sim p)$, we obtain Table A-14.

We can see that $\sim(\sim p)$ and p have the same truth values, that is $\sim(\sim p)$ is true when p is true, and false when p is false. Two statements are logically **equivalent** if they have the same truth tables. Thus, p and $\sim(\sim p)$ are equivalent statements. A statement may be replaced by an equivalent statement in any logical expression.

Equivalent statements can be represented by an equation using the symbol $\equiv$. For instance, we have just shown that

(1)	(2)	(3)
p	$\sim p$	$\sim(\sim p)$
T	F	T
F	T	F

Table A-14

$$\boxed{\sim(\sim p) \equiv p}$$

Example 4. In Parts (a) and (b), determine whether the following statements are equivalent.

(a) $\sim(p \vee (\sim q))$ and $(\sim p) \wedge q$
(b) $p \wedge (\sim q)$ and $(\sim p) \vee (\sim q)$

Solution

Using Table A-10 we form truth tables A-15 for Part (a) and A-16 for Part (b).

(1)	(2)	(3)	(4)	(5)	(6)	(7)
p	q	$\sim p$	$\sim q$	$p \vee (\sim q)$	$\sim(p \vee (\sim q))$	$(\sim p) \wedge q$
T	T	F	F	T	F	F
T	F	F	T	T	F	F
F	T	T	F	F	T	T
F	F	T	T	T	F	F

Table A-15

SET THEORY AND LOGIC A9

(1)	(2)	(3)	(4)	(5)	(6)
p	q	$\sim p$	$\sim q$	$p \wedge (\sim q)$	$(\sim p) \vee (\sim q)$
T	T	F	F	F	F
T	F	F	T	T	T
F	T	T	F	F	T
F	F	T	T	F	T

Table A-16

From Table A-15 we can see that the expressions in Part (a) are equivalent because the truth values in columns (6) and (7) are identical. From Table A-16 we see that the expressions in Part (b) are not equivalent because the truth values in columns (5) and (6) are different, that is, they differ in at least one row. ■

De Morgan's Laws

De Morgan's Laws (for statements) tell us how to negate conjunctions and disjunctions.

1. $\sim(p \wedge q) \equiv (\sim p) \vee (\sim q)$
2. $\sim(p \vee q) \equiv (\sim p) \wedge (\sim q)$

In words, Law (1) states that

"not (p and q)" is equivalent to "not p" or "not q"

Law (2) states that

"not (p or q)" is equivalent to "not p" and "not q"

To verify the first of De Morgan's Laws, concerning the equivalence of the statements $\sim(p \wedge q)$ and $(\sim p) \vee (\sim q)$, we construct Table A-17.

(1)	(2)	(3)	(4)	(5)	(6)	(7)
p	q	$\sim p$	$\sim q$	$p \wedge q$	$\sim(p \wedge q)$	$(\sim p) \vee (\sim q)$
T	T	F	F	T	F	F
T	F	F	T	F	T	T
F	T	T	F	F	T	T
F	F	T	T	F	T	T

Table A-17

Noting that columns (6) and (7) are identical, we have demonstrated that

$$\sim(p \wedge q) \equiv (\sim p) \vee q$$

We will leave the proof of Law (2) as Exercise 21.

De Morgan's Laws can also be stated for sets. Thus

1. $(A \cap B)' = A' \cup B'$
2. $(A \cup B)' = A' \cap B'$

To visualize the second law for sets, see Figures A-18 and A-19 and note that the colored regions representing $(A \cup B)'$ and $A' \cap B'$ are the same.

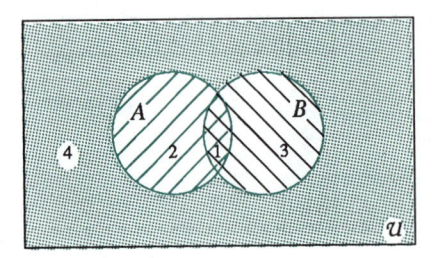

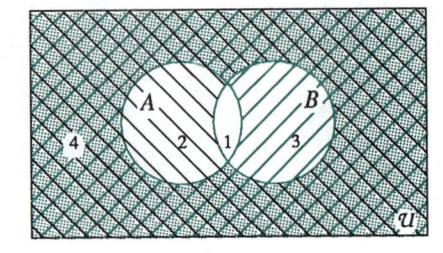

Figure A-18

Figure A-19

Summary

Collections of objects are known as **sets**. The set of all objects considered for some purpose is called the **universal set,** and is often denoted by $\mathcal{U}$. A **Venn diagram** is a diagram in which the universal set is represented by a large rectangle within which other sets are depicted as circles or ovals. The set A' of objects of $\mathcal{U}$ that are *not* in A is called the **complement** of A. The set containing no element is called the **empty set**. The set $A \cup B$, of objects that are in at least one of the sets A or B is called the **union** of A and B. The set $A \cap B$, of objects that are in both sets A and B is called the **intersection** of A and B. When A and B do not have objects in common, they are said to be **disjoint**.

A **statement** is defined as a meaningful declarative sentence that is either true or false, but not both true and false. A **compound statement** is made up of two or more **simple statements**, combined or modified by means of **connectives**. The connective **conjunction** (p and q) is represented by $p \wedge q$, **disjunction** (p or q) by $p \vee q$, and **negation** (not p) by $\sim p$. Every statement has a **truth value**—either true (T) or false (F). The truth value of a compound statement depends on the truth values of its parts; to determine this truth value we construct a **truth table**. Two statements are logically **equivalent** if they have the same truth tables. **De Morgan's Laws** for statements are

1. $\sim(p \wedge q) \equiv (\sim p) \vee (\sim q)$
2. $\sim(p \vee q) \equiv (\sim p) \wedge (\sim q)$

and for sets are

1. $(A \cap B)' = A' \cup B'$
2. $(A \cup B)' = A' \cap B'$

Exercises

Exercises 1–5 concern the Hi-Tech Computer consultants, who are listed by country of birth as follows:

Name	Born	Name	Born
Mr. White	USA	Ms. Taylor	Australia
Ms. Wong	USA	Mr. Campos	Mexico
Ms. Goldberg	USA	Mr. Nguyen	Vietnam
Ms. Kjos	USA	Mr. Sanchez	USA
Ms. Singh	India		

The universal set $\mathcal{U}$ of the Venn diagram, Figure A-20, represents the set of Hi-Tech Computer consultants.

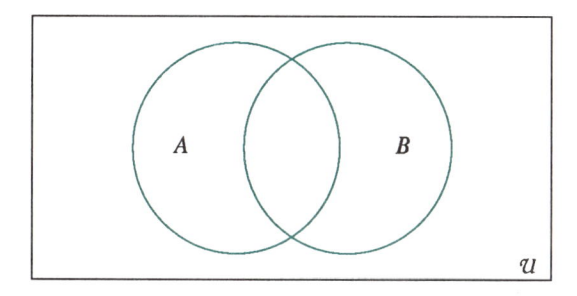

Figure A-20

The subset A represents the foreign-born employees and B represents the female employees. List the Hi-Tech consultants in the following subsets of $\mathcal{U}$.

1. B
2. A
3. $A \cup B$
4. $A \cap B$
5. A'

6. Show that for any sets A and B, $A \cup B' = (A' \cap B)'$.

7. Which of the following are statements?
   (a) Jack and Jill went up the hill.
   (b) When are you graduating?
   (c) 7 + 8 is 17.
   (d) Take your seats!
   (e) John Adams was the second President of the United States.
   (f) Did she sell sea shells?
   (g) Mary was famous for being contrary.

Construct truth tables for the statements given in Exercises 8 through 10.

8. $(\sim p) \vee q$
9. $p \wedge (\sim q)$
10. $\sim(p \wedge q) \vee (\sim q)$

A12   APPENDICES

Translate the statements in Exercises 11 through 14 into symbols and construct the corresponding truth tables.

11. Maria will arrive in July and will see Matthew.

12. It is raining today and it will not snow tomorrow.

13. Jose will come or Susan will call him.

14. Tom will bring the book or Harry will not go to school.

In Exercises 15 through 20, determine by using truth tables whether or not the following expressions are equivalent.

15. $p \vee (p \wedge q)$ and p

16. $p \wedge (p \vee q)$ and p

17. $\sim(p \vee q)$ and $(\sim p) \wedge (\sim q)$

18. $\sim((\sim p) \wedge q)$ and $p \vee q$

19. $\sim((\sim p) \vee (\sim q))$ and $p \wedge q$

20. $\sim(p \vee q)$ and $(\sim p) \wedge q$

21. Use a truth table to show that the second De Morgan Law (for statements) is true. That is, show that
$$\sim(p \vee q) \equiv (\sim p) \wedge (\sim q)$$

22. Using Venn diagrams show that the first De Morgan Law is true for sets. That is, show that
$$(A \cap B)' = A' \cup B'$$

Appendix 2: Tables

Table A

z	.00	.01	.02	.03	.04	.05	.06	.07	.08	.09
0.0	.0000	.0040	.0080	.0120	.0160	.0199	.0239	.0279	.0319	.0359
0.1	.0398	.0438	.0478	.0517	.0557	.0596	.0636	.0675	.0714	.0753
0.2	.0793	.0832	.0871	.0910	.0948	.0987	.1026	.1064	.1103	.1141
0.3	.1179	.1217	.1255	.1293	.1331	.1368	.1406	.1443	.1480	.1517
0.4	.1554	.1591	.1628	.1664	.1700	.1736	.1772	.1808	.1844	.1879
0.5	.1915	.1950	.1985	.2019	.2054	.2088	.2123	.2157	.2190	.2224
0.6	.2257	.2291	.2324	.2357	.2389	.2422	.2454	.2486	.2517	.2549
0.7	.2580	.2611	.2642	.2673	.2704	.2734	.2764	.2794	.2823	.2852
0.8	.2881	.2910	.2939	.2967	.2995	.3023	.3051	.3078	.3106	.3133
0.9	.3159	.3186	.3212	.3238	.3264	.3289	.3315	.3340	.3365	.3389
1.0	.3413	.3438	.3461	.3485	.3508	.3531	.3554	.3577	.3599	.3621
1.1	.3643	.3665	.3686	.3708	.3729	.3749	.3770	.3790	.3810	.3830
1.2	.3849	.3869	.3888	.3907	.3925	.3944	.3962	.3980	.3997	.4015
1.3	.4032	.4049	.4066	.4082	.4099	.4115	.4131	.4147	.4162	.4177
1.4	.4192	.4207	.4222	.4236	.4251	.4265	.4279	.4292	.4306	.4319
1.5	.4332	.4345	.4357	.4370	.4382	.4394	.4406	.4418	.4429	.4441
1.6	.4452	.4463	.4474	.4484	.4495	.4505	.4515	.4525	.4535	.4545
1.7	.4554	.4564	.4573	.4582	.4591	.4599	.4608	.4616	.4625	.4633
1.8	.4641	.4649	.4656	.4664	.4671	.4678	.4686	.4693	.4699	.4706
1.9	.4713	.4719	.4726	.4732	.4738	.4744	.4750	.4756	.4761	.4767
2.0	.4772	.4778	.4783	.4788	.4793	.4798	.4803	.4808	.4812	.4817
2.1	.4821	.4826	.4830	.4834	.4838	.4842	.4846	.4850	.4854	.4857
2.2	.4861	.4864	.4868	.4871	.4875	.4878	.4881	.4884	.4887	.4890
2.3	.4893	.4896	.4898	.4901	.4904	.4906	.4909	.4911	.4913	.4916
2.4	.4918	.4920	.4922	.4925	.4927	.4929	.4931	.4932	.4934	.4936
2.5	.4938	.4940	.4941	.4943	.4945	.4946	.4948	.4949	.4951	.4952
2.6	.4953	.4955	.4956	.4957	.4959	.4960	.4961	.4962	.4963	.4964
2.7	.4965	.4966	.4967	.4968	.4969	.4970	.4971	.4972	.4973	.4974
2.8	.4974	.4975	.4976	.4977	.4977	.4978	.4979	.4979	.4980	.4981
2.9	.4981	.4982	.4982	.4983	.4984	.4984	.4985	.4985	.4986	.4986
3.0	.4987	.4987	.4987	.4988	.4988	.4989	.4989	.4989	.4990	.4990

Table B

Part 1. Table of $\left(1 + \dfrac{r}{4}\right)^n$

n/r	0.050	0.055	0.060	0.065	0.070	0.075	0.080
2	1.025156	1.027689	1.030225	1.032764	1.035306	1.037852	1.040400
3	1.037971	1.041820	1.045678	1.049546	1.053424	1.057311	1.061208
4	1.050945	1.056145	1.061364	1.066602	1.071859	1.077136	1.082432
8	1.104486	1.115442	1.126493	1.137639	1.148882	1.160222	1.171659
12	1.160755	1.178068	1.195618	1.213408	1.231439	1.249716	1.268242
16	1.219890	1.244211	1.268986	1.294222	1.319929	1.346114	1.372786
20	1.282037	1.314067	1.346855	1.380420	1.414778	1.449948	1.485947
24	1.347351	1.387845	1.429503	1.472358	1.516443	1.561791	1.608437
40	1.643619	1.726771	1.814018	1.905559	2.001597	2.102349	2.208040
60	2.107181	2.269092	2.443220	2.630471	2.831816	3.048297	3.281031
80	2.701485	2.981737	3.290663	3.631154	4.006392	4.419872	4.875439

n/r	0.085	0.090	0.095	0.100	0.105	0.110	0.115
2	1.042952	1.045506	1.048064	1.050625	1.053189	1.055756	1.058327
3	1.065114	1.069030	1.072956	1.076891	1.080835	1.084790	1.088753
4	1.087748	1.093083	1.098438	1.103813	1.109207	1.114621	1.120055
8	1.183196	1.194831	1.206567	1.218403	1.230341	1.242381	1.254523
12	1.287019	1.306050	1.325339	1.344889	1.364703	1.384784	1.405135
16	1.399952	1.427621	1.455803	1.484506	1.513738	1.543509	1.573829
20	1.522795	1.560509	1.599110	1.638616	1.679049	1.720428	1.762775
24	1.656417	1.705767	1.756523	1.808726	1.862413	1.917626	1.974406
40	2.318904	2.435189	2.557152	2.685064	2.819206	2.959874	3.107377
60	3.531215	3.800135	4.089167	4.399790	4.733585	5.092251	5.477607
80	5.377316	5.930145	6.539028	7.209568	7.947922	8.760854	9.655791

n/r	0.120	0.125	0.130	0.140	0.150
2	1.060900	1.063477	1.066056	1.071225	1.076406
3	1.092727	1.096710	1.100703	1.108718	1.116771
4	1.125509	1.130982	1.136476	1.147523	1.158650
8	1.266770	1.279121	1.291578	1.316809	1.342471
12	1.425761	1.446664	1.467847	1.511069	1.555454
16	1.604706	1.636151	1.668173	1.733986	1.802228
20	1.806111	1.850458	1.895838	1.989789	2.088152
24	2.032794	2.092835	2.154574	2.283328	2.419438
40	3.262038	3.424195	3.594201	3.959260	4.360379
60	5.891603	6.336329	6.814023	7.878091	9.105134
80	10.640891	11.725110	12.918284	15.675738	19.012903

Part 2. Table of $\left(1 + \dfrac{r}{4}\right)^{-n}$

n/r	0.050	0.055	0.060	0.065	0.070	0.075	0.080
2	0.975461	0.973057	0.970662	0.968275	0.965898	0.963529	0.961169
3	0.963418	0.959859	0.956317	0.952792	0.949285	0.945795	0.942322
4	0.951524	0.946840	0.942184	0.937557	0.932959	0.928388	0.923845
8	0.905398	0.896506	0.887711	0.879013	0.870412	0.861904	0.853490
12	0.861509	0.848847	0.836387	0.824125	0.812058	0.800182	0.788493
16	0.819746	0.803722	0.788031	0.772665	0.757616	0.742879	0.728446
20	0.780009	0.760996	0.742470	0.724417	0.706825	0.689680	0.672971
24	0.742197	0.720542	0.699544	0.679183	0.659438	0.640291	0.621721
40	0.608413	0.579116	0.551262	0.524780	0.499601	0.475658	0.452890
60	0.474568	0.440705	0.409296	0.380160	0.353130	0.328052	0.304782
80	0.370167	0.335375	0.303890	0.275395	0.249601	0.226251	0.205110

n/r	0.085	0.090	0.095	0.100	0.105	0.110	0.115
2	0.958817	0.956474	0.954140	0.951814	0.949497	0.947188	0.944888
3	0.938866	0.935427	0.932005	0.928599	0.925210	0.921838	0.918482
4	0.919331	0.914843	0.910383	0.905951	0.901545	0.897166	0.892813
8	0.845169	0.836938	0.828798	0.820747	0.812783	0.804906	0.797115
12	0.776990	0.765667	0.754524	0.743556	0.732760	0.722134	0.711675
16	0.714310	0.700466	0.686906	0.673625	0.660616	0.647874	0.635393
20	0.656687	0.640816	0.625348	0.610271	0.595575	0.581251	0.567287
24	0.603713	0.586247	0.569306	0.552875	0.536938	0.521478	0.506482
40	0.431238	0.410646	0.391060	0.372431	0.354710	0.337852	0.321815
60	0.283189	0.263149	0.244549	0.227284	0.211256	0.196377	0.182561
80	0.185966	0.168630	0.152928	0.138705	0.125819	0.114144	0.103565

n/r	0.120	0.125	0.130	0.140	0.150
2	0.942596	0.940312	0.938037	0.933511	0.929017
3	0.915142	0.911818	0.908510	0.901943	0.895438
4	0.888487	0.884187	0.879913	0.871442	0.863073
8	0.789409	0.781787	0.774247	0.759412	0.744895
12	0.701380	0.691246	0.681270	0.661783	0.642899
16	0.623167	0.611191	0.599458	0.576706	0.554869
20	0.553676	0.540407	0.527471	0.502566	0.478892
24	0.491934	0.477821	0.464129	0.437957	0.413319
40	0.306557	0.292039	0.278226	0.252572	0.229338
60	0.169733	0.157820	0.146756	0.126934	0.109828
80	0.093977	0.085287	0.077410	0.063793	0.052596

Part 3. Table of $\left(1 + \dfrac{r}{12}\right)^n$

n/r	0.050	0.055	0.060	0.065	0.070	0.075	0.080
2	1.008351	1.009188	1.010025	1.010863	1.011701	1.012539	1.013378
3	1.012552	1.013813	1.015075	1.016338	1.017602	1.018867	1.020134
4	1.016771	1.018460	1.020151	1.021843	1.023538	1.025235	1.026935
5	1.021008	1.023128	1.025251	1.027378	1.029509	1.031643	1.033781
6	1.025262	1.027817	1.030378	1.032943	1.035514	1.038091	1.040673
12	1.051162	1.056408	1.061678	1.066972	1.072290	1.077633	1.083000
24	1.104941	1.115998	1.127160	1.138429	1.149806	1.161292	1.172888
36	1.161472	1.178949	1.196681	1.214672	1.232926	1.251446	1.270237
48	1.220895	1.245451	1.270489	1.296020	1.322054	1.348599	1.375666
60	1.283359	1.315704	1.348850	1.382817	1.417625	1.453294	1.489846
120	1.647009	1.731076	1.819397	1.912184	2.009661	2.112065	2.219640
180	2.113704	2.277584	2.454094	2.644201	2.848947	3.069452	3.306921
240	2.712640	2.996626	3.310204	3.656447	4.038739	4.460817	4.926803
300	3.481290	3.942672	4.464970	5.056198	5.725418	6.482880	7.340176
360	4.467744	5.187388	6.022575	6.991798	8.116497	9.421534	10.935730

n/r	0.085	0.090	0.095	0.100	0.105	0.110	0.115
2	1.014217	1.015056	1.015896	1.016736	1.017577	1.018417	1.019259
3	1.021401	1.022669	1.023939	1.025209	1.026480	1.027753	1.029026
4	1.028636	1.030339	1.032045	1.033752	1.035462	1.037174	1.038888
5	1.035922	1.038067	1.040215	1.042367	1.044522	1.046681	1.048844
6	1.043260	1.045852	1.048450	1.051053	1.053662	1.056276	1.058895
12	1.088391	1.093807	1.099248	1.104713	1.110203	1.115719	1.121259
24	1.184595	1.196414	1.208345	1.220391	1.232552	1.244829	1.257222
36	1.289302	1.308645	1.328271	1.348182	1.368383	1.388879	1.409672
48	1.403265	1.431405	1.460098	1.489354	1.519184	1.549598	1.580608
60	1.527301	1.565681	1.605009	1.645309	1.686603	1.728916	1.772272
120	2.332647	2.451357	2.576055	2.707041	2.844630	2.989150	3.140948
180	3.562653	3.838043	4.134593	4.453920	4.797761	5.167988	5.566613
240	5.441243	6.009152	6.636061	7.328074	8.091918	8.935015	9.865552
300	8.310413	9.408415	10.650941	12.056945	13.647852	15.447889	17.484440
360	12.692499	14.730576	17.094862	19.837399	23.018509	26.708098	30.987181

n/r	0.120	0.125	0.130	0.135	0.140
2	1.020100	1.020942	1.021784	1.022627	1.023469
3	1.030301	1.031577	1.032853	1.034131	1.035410
4	1.040604	1.042322	1.044043	1.045765	1.047490
5	1.051010	1.053180	1.055353	1.057530	1.059710
6	1.061520	1.064150	1.066786	1.069427	1.072074
12	1.126825	1.132416	1.138032	1.143674	1.149342
24	1.269735	1.282366	1.295118	1.307991	1.320987
36	1.430769	1.452172	1.473886	1.495916	1.518266
48	1.612226	1.644463	1.677330	1.710841	1.745007
60	1.816697	1.862216	1.908857	1.956645	2.005610
120	3.300387	3.467849	3.643733	3.828460	4.022471
180	5.995802	6.457884	6.955364	7.490939	8.067507
240	10.892554	12.025975	13.276792	14.657109	16.180270
300	19.788466	22.394964	25.343491	28.678761	32.451308
360	35.949641	41.704262	48.377089	56.114160	65.084661

n/r	0.145	0.150	0.155	0.160	0.165
2	1.024313	1.025156	1.026000	1.026844	1.027689
3	1.036690	1.037971	1.039253	1.040536	1.041820
4	1.049216	1.050945	1.052676	1.054410	1.056145
5	1.061894	1.064082	1.066273	1.068468	1.070667
6	1.074726	1.077383	1.080046	1.082715	1.085388
12	1.155035	1.160755	1.166500	1.172271	1.178068
24	1.334107	1.347351	1.360721	1.374219	1.387845
36	1.540940	1.563944	1.587281	1.610957	1.634975
48	1.779841	1.815355	1.851563	1.888477	1.926112
60	2.055779	2.107181	2.159847	2.213807	2.269092
120	4.226227	4.440213	4.664940	4.900941	5.148777
180	8.688187	9.356334	10.075557	10.849737	11.683046
240	17.860991	19.715494	21.761665	24.019222	26.509903
300	36.718246	41.544120	47.001870	53.173919	60.153398
360	75.484592	87.540995	101.516858	117.716787	136.493572

n/r	0.170	0.180	0.190	0.200	0.210
2	1.028534	1.030225	1.031917	1.033611	1.035306
3	1.043105	1.045678	1.048256	1.050838	1.053424
4	1.057882	1.061364	1.064853	1.068352	1.071859
5	1.072869	1.077284	1.081714	1.086158	1.090617
6	1.088068	1.093443	1.098841	1.104260	1.109702
12	1.183892	1.195618	1.207451	1.219391	1.231439
24	1.401600	1.429503	1.457938	1.486915	1.516443
36	1.659342	1.709140	1.760389	1.813130	1.867407
48	1.964482	2.043478	2.125583	2.210915	2.299599
60	2.325733	2.443220	2.566537	2.695970	2.831816
120	5.409036	5.969323	6.587114	7.268255	8.019183
180	12.579975	14.584368	16.906072	19.594998	22.708854
240	29.257669	35.632816	43.390065	52.827531	64.307303
300	68.045538	87.058800	111.362218	142.421445	182.106467
360	158.255782	212.703781	285.815282	383.963963	515.692058

Part 4. Table of $\left(1 + \dfrac{r}{12}\right)^{-n}$

n/r	0.050	0.055	0.060	0.065	0.070	0.075	0.080
2	0.991718	0.990896	0.990075	0.989254	0.988435	0.987616	0.986799
3	0.987603	0.986375	0.985149	0.983924	0.982702	0.981482	0.980264
4	0.983506	0.981875	0.980248	0.978624	0.977003	0.975386	0.973772
5	0.979425	0.977395	0.975371	0.973351	0.971337	0.969327	0.967323
6	0.975361	0.972936	0.970518	0.968107	0.965704	0.963307	0.960917
12	0.951328	0.946604	0.941905	0.937232	0.932583	0.927960	0.923361
24	0.905025	0.896059	0.887186	0.878404	0.869712	0.861110	0.852596
36	0.860976	0.848213	0.835645	0.823268	0.811079	0.799076	0.787255
48	0.819071	0.802922	0.787098	0.771593	0.756399	0.741510	0.726921
60	0.779205	0.760050	0.741372	0.723161	0.705405	0.688092	0.671210
120	0.607161	0.577675	0.549633	0.522962	0.497596	0.473470	0.450523
180	0.473103	0.439062	0.407482	0.378186	0.351007	0.325791	0.302396
240	0.368645	0.333709	0.302096	0.273490	0.247602	0.224174	0.202971
300	0.287250	0.253635	0.223966	0.197777	0.174660	0.154252	0.136237
360	0.223827	0.192775	0.166042	0.143025	0.123206	0.106140	0.091443

n/r	0.085	0.090	0.095	0.100	0.105	0.110	0.115
2	0.985982	0.985167	0.984353	0.983539	0.982727	0.981916	0.981105
3	0.979048	0.977833	0.976621	0.975411	0.974203	0.972997	0.971792
4	0.972161	0.970554	0.968950	0.967350	0.965752	0.964158	0.962568
5	0.965324	0.963329	0.961340	0.959355	0.957375	0.955401	0.953431
6	0.958534	0.956158	0.953789	0.951427	0.949071	0.946722	0.944380
12	0.918788	0.914238	0.909713	0.905212	0.900736	0.896283	0.891854
24	0.844171	0.835831	0.827578	0.819410	0.811325	0.803323	0.795404
36	0.775613	0.764149	0.752859	0.741740	0.730789	0.720005	0.709385
48	0.712624	0.698614	0.684885	0.671432	0.658248	0.645329	0.632668
60	0.654750	0.638700	0.623049	0.607789	0.592908	0.578397	0.564248
120	0.428698	0.407937	0.388190	0.369407	0.351540	0.334543	0.318375
180	0.280690	0.260549	0.241862	0.224521	0.208431	0.193499	0.179642
240	0.183782	0.166413	0.150692	0.136462	0.123580	0.111919	0.101363
300	0.120331	0.106288	0.093888	0.082940	0.073272	0.064734	0.057194
360	0.078787	0.067886	0.058497	0.050410	0.043443	0.037442	0.032271

n/r	0.120	0.125	0.130	0.135	0.140	0.145	0.150
2	0.980296	0.979488	0.978680	0.977874	0.977069	0.976264	0.975461
3	0.970590	0.969390	0.968192	0.966995	0.965801	0.964609	0.963418
4	0.960980	0.959396	0.957815	0.956238	0.954663	0.953092	0.951524
5	0.951466	0.949506	0.947550	0.945600	0.943654	0.941713	0.939777
6	0.942045	0.939717	0.937395	0.935080	0.932772	0.930470	0.928175
12	0.887449	0.883068	0.878710	0.874375	0.870063	0.865774	0.861509
24	0.787566	0.779809	0.772130	0.764531	0.757010	0.749565	0.742197
36	0.698925	0.688624	0.678478	0.668487	0.658646	0.648954	0.639409
48	0.620260	0.608101	0.596185	0.584508	0.573064	0.561848	0.550856
60	0.550450	0.536995	0.523874	0.511079	0.498601	0.486434	0.474568
120	0.302995	0.288363	0.274444	0.261202	0.248603	0.236618	0.225214
180	0.166783	0.154849	0.143774	0.133495	0.123954	0.115099	0.106879
240	0.091806	0.083153	0.075319	0.068226	0.061804	0.055988	0.050722
300	0.050534	0.044653	0.039458	0.034869	0.030815	0.027234	0.024071
360	0.027817	0.023978	0.020671	0.017821	0.015365	0.013248	0.011423

n/r	0.160	0.165	0.170	0.180	0.190	0.200	0.210
2	0.973857	0.973057	0.972258	0.970662	0.969070	0.967482	0.965898
3	0.961043	0.959859	0.958676	0.956317	0.953965	0.951622	0.949285
4	0.948398	0.946840	0.945285	0.942184	0.939096	0.936021	0.932959
5	0.935919	0.933997	0.932080	0.928260	0.924459	0.920677	0.916913
6	0.923604	0.921329	0.919060	0.914542	0.910050	0.905583	0.901143
12	0.853045	0.848847	0.844672	0.836387	0.828191	0.820081	0.812058
24	0.727686	0.720542	0.713471	0.699544	0.685900	0.672534	0.659438
36	0.620749	0.611630	0.602648	0.585090	0.568056	0.551532	0.535502
48	0.529527	0.519181	0.509040	0.489362	0.470459	0.452301	0.434858
60	0.451711	0.440705	0.429972	0.409296	0.389630	0.370924	0.353130
120	0.204042	0.194221	0.184876	0.167523	0.151812	0.137585	0.124701
180	0.092168	0.085594	0.079491	0.068567	0.059150	0.051033	0.044036
240	0.041633	0.037722	0.034179	0.028064	0.023047	0.018930	0.015550
300	0.018806	0.016624	0.014696	0.011486	0.008980	0.007021	0.005491
360	0.008495	0.007326	0.006319	0.004701	0.003499	0.002604	0.001939

Answers to Odd-Numbered Exercises

Chapter 1

Section 1.1, page 13

1. (a) $\begin{bmatrix} -1 \\ 1 \end{bmatrix}$
   (b) 2×4
   (c) 2nd row, 4th column

3. (a) 0
   (b) 5×1
   (c) 3rd row, 1st column

5. Many possible answers

7. Equal

9. Equal

11. $x = 4, y = -1$

13. $x = -1, y = 1$

15. $x = -1, y = 5$

17. $\begin{bmatrix} 1 & -\frac{1}{2} & -2 \end{bmatrix}$

19. Cannot be added; different sizes

21. $\begin{bmatrix} -3 \\ -2 \end{bmatrix}$

23. $\begin{bmatrix} -3 & -1 \\ 0 & 1 \\ 1 & 1 \end{bmatrix}$

25. $\begin{bmatrix} -4 \\ -3 \\ -4 \end{bmatrix}$

27. (a) $\begin{bmatrix} 0 & 8 & -8 \\ -4 & 8 & 4 \\ 4 & 0 & 0 \end{bmatrix}$
    (b) $\begin{bmatrix} 2 & 4 & -4 \\ 3 & 2 & 1 \\ -1 & -4 & -6 \end{bmatrix}$
    (c) $\begin{bmatrix} 3 & -7 & 7 \\ 11 & -10 & -5 \\ -8 & -6 & -9 \end{bmatrix}$

29. $x = -6, y = -2$

31. (a) $\begin{bmatrix} 82 & 36 \\ 108 & 56 \\ 162 & 95 \\ 158 & 65 \end{bmatrix}$
    (b) $\begin{bmatrix} 2 & 6 \\ -2 & -4 \\ -2 & 5 \\ -2 & 1 \end{bmatrix}$
    (c) $\begin{bmatrix} 41 & 18 \\ 54 & 28 \\ 81 & 47.5 \\ 79 & 32.5 \end{bmatrix}$

    (d) $\begin{bmatrix} 46.2 & 23.1 \\ 58.3 & 28.6 \\ 88.0 & 55.0 \\ 85.8 & 36.3 \end{bmatrix}$

Section 1.2, page 30

1. AB is 1×4; BA is not defined.
3. Both are 2×2.
5. AB is not defined; BA is 1×4.
7. Neither is defined.
9. AB is 4×3; BA is not defined.
11. $\begin{bmatrix} 1 \end{bmatrix}$
13. $\begin{bmatrix} 0 & -1 & 2 & -2 \end{bmatrix}$
15. $\begin{bmatrix} 2a+b-c & a+c & -a \end{bmatrix}$
17. $\begin{bmatrix} 1 & 0 & 2 \\ 2 & 0 & 4 \\ -1 & 0 & -2 \\ -2 & 0 & -4 \end{bmatrix}$
19. $\begin{bmatrix} -x+2y \\ 3x \\ -x-z \end{bmatrix}$
21. $\begin{bmatrix} 2 \\ 4 \\ -9 \end{bmatrix}$
23. $\begin{bmatrix} -w + \frac{1}{2}x + y \\ \frac{1}{3}w + 2x - z \\ -\frac{2}{3}x + y + -\frac{2}{3}z \end{bmatrix}$
25. $AB = \begin{bmatrix} 0 & -2 \\ 0 & 0 \end{bmatrix}$, $BA = \begin{bmatrix} 0 & 2 \\ 0 & 0 \end{bmatrix}$
27. $AB = \begin{bmatrix} -12 & -6 \\ 6 & 0 \end{bmatrix}$, $BA = \begin{bmatrix} -12 & -6 \\ 6 & 0 \end{bmatrix}$
29. $\begin{bmatrix} 1 & 2 \\ -1 & 0 \\ 2 & 3 \end{bmatrix}$
31. $\begin{bmatrix} 2 & -2 & 0 \\ 0 & 2 & 2 \end{bmatrix}$
33. $\begin{bmatrix} 0 & 0 & 0 \\ 0 & 0 & 0 \end{bmatrix}$
35. (a) $\begin{bmatrix} 100 & 200 & 400 & 100 \end{bmatrix}$
    (b) $\begin{bmatrix} 46 \\ 34 \\ 15 \\ 10 \end{bmatrix}$
    (c) $18,400
37. (a) $\begin{bmatrix} 160{,}000 & 90{,}000 & 80{,}000 \\ 50 & 20 & 10 \end{bmatrix}$
    (b) $\begin{bmatrix} 1 \\ 3 \\ 1 \end{bmatrix}$
    (c) $510,000; 120 shares
39. (a) $\begin{bmatrix} \frac{1}{5} & \frac{2}{5} & \frac{1}{5} \\ 0 & 1 & \frac{1}{2} \\ \frac{1}{10} & 0 & 0 \\ \frac{1}{8} & \frac{1}{8} & \frac{1}{16} \end{bmatrix}$
    (b) $\begin{bmatrix} 2 \\ 1 \\ 3 \end{bmatrix}$
    (c) $\frac{7}{5}$ loaves of bread; $\frac{5}{2}$ quarts of milk; $\frac{1}{5}$ pound of coffee; $\frac{9}{16}$ pound of cheese

Section 1.3, page 46

1. $\begin{bmatrix} 2 & 2 \\ 5 & 2 \end{bmatrix}$

3. $\begin{bmatrix} 2 & -3 \\ -2 & 4 \end{bmatrix}$

5. $\begin{bmatrix} -1 & 3 \\ 1 & -2 \end{bmatrix}$

7. $\begin{bmatrix} 4 & -1 \\ -3 & 1 \end{bmatrix}$

9. $\begin{bmatrix} \frac{1}{2} & -\frac{1}{2} \\ -\frac{1}{2} & \frac{3}{2} \end{bmatrix}$

11. $\begin{bmatrix} -\frac{2}{5} & \frac{1}{5} \\ \frac{1}{5} & \frac{2}{5} \end{bmatrix}$

13. $\begin{bmatrix} 2 & -1 \\ 3 & -2 \end{bmatrix}$

15. $\begin{bmatrix} \frac{2}{3} \\ \frac{2}{3} \end{bmatrix}$

17. $x = 3, y = -4$

19. $x = 1, y = 2$

21. $x = -\frac{5}{4}, y = -\frac{7}{4}$

23. HURRY

Section 1.4, page 62

1. $\begin{bmatrix} 3 & 2 & 1 & | & 1 \\ 1 & 1 & -3 & | & -2 \\ 2 & -2 & 1 & | & 1 \end{bmatrix}$

3. $\begin{bmatrix} 1 & 1 & 1 & 1 & | & 1 \\ -1 & 0 & 1 & -1 & | & 0 \\ 0 & 2 & 3 & 1 & | & -1 \\ 1 & 1 & 0 & -1 & | & 1 \end{bmatrix}$

5. $\begin{bmatrix} 1 & -2 & 0 & | & 2 \\ \frac{1}{3} & 1 & 2 & | & 0 \\ -1 & 1 & 2 & | & -1 \end{bmatrix}$

7. $\begin{bmatrix} 1 & -2 & 1 & | & 1 \\ 0 & -2 & 3 & | & 5 \\ 1 & -1 & 1 & | & -1 \end{bmatrix}$

9. $\begin{bmatrix} 1 & 0 & 0 & | & \frac{1}{3} \\ 0 & 1 & 0 & | & -\frac{8}{3} \\ 0 & 0 & 1 & | & 2 \end{bmatrix}$

11. Multiply the third row by $\frac{1}{3}$.

13. Interchange the second and third rows.
15. Interchange the first and third rows.
17. Interchange the second and third rows.

19. $x = -3, y = 6, z = -4$

21. $x = \frac{5}{2}, y = -\frac{1}{2}, z = 1$

23. $x = -\frac{3}{2}, y = \frac{5}{2}, z = -\frac{3}{2}$

25. $x = -\frac{1}{3}, y = \frac{2}{3}, z = 0$

27. $w = 3, x = -6, y = -9, z = 5$

A26   ANSWERS TO ODD-NUMBERED EXERCISES

Section 1.5, page 77

1. $\begin{bmatrix} 1 & 2 & | & 1 & 0 \\ 0 & -3 & | & -2 & 1 \end{bmatrix}$

3. $\begin{bmatrix} 1 & 0 & 0 & 1 & | & \frac{1}{4} & -\frac{1}{2} & 2 & 0 \\ 0 & 1 & 0 & -3 & | & \frac{1}{8} & \frac{7}{2} & -3 & 0 \\ 0 & 0 & 1 & 1 & | & \frac{1}{8} & -\frac{1}{2} & 1 & 0 \\ 0 & 0 & 1 & 2 & | & 1 & -1 & 0 & 1 \end{bmatrix}$

5. $\begin{bmatrix} 2 & 3 \\ -1 & -2 \end{bmatrix}$

7. $\begin{bmatrix} -1 & 1 & 2 \\ -3 & 2 & 4 \\ 1 & -1 & -1 \end{bmatrix}$

9. $\begin{bmatrix} \frac{5}{8} & \frac{1}{4} & -\frac{1}{8} \\ \frac{1}{4} & \frac{1}{2} & -\frac{1}{4} \\ -\frac{3}{4} & -\frac{1}{2} & \frac{3}{4} \end{bmatrix}$

11. $\begin{bmatrix} 0 & -\frac{2}{3} & \frac{1}{2} & -\frac{2}{3} \\ 0 & \frac{1}{3} & 0 & -\frac{1}{3} \\ -\frac{1}{2} & 0 & 0 & 0 \\ -\frac{3}{2} & -\frac{1}{3} & -\frac{1}{2} & \frac{2}{3} \end{bmatrix}$

13. $\begin{bmatrix} 1 \\ \frac{1}{2} \\ -\frac{1}{2} \end{bmatrix}$

15. $\begin{bmatrix} -1 \\ 1 \\ -1 \end{bmatrix}$

17. (a) $x = -\frac{5}{4}, y = \frac{3}{4}, z = -\frac{3}{2}$
    (b) $x = -\frac{1}{4}, y = \frac{3}{4}, z = \frac{1}{2}$

19. (a) $w = 3, x = -3, y = 2, z = -1$
    (b) $w = 1, x = -2, y = 1, z = 1$

21. (a) $w = -1, x = -3, y = -5, z = 6$
    (b) $w = \frac{1}{2}, x = 1, y = 1, z = -\frac{1}{2}$

Section 1.6, page 94

1. $x = -\frac{2}{3}, y = -\frac{1}{3}$

3. No solution

5. No solution

7. No solution

9. $x = \frac{3}{4}, y = -\frac{1}{2}, z = \frac{3}{2}$

11. $x = \frac{3}{2}, y = \frac{3}{2}, z = -1$

13. No solution

15. Infinitely many solutions; $w = -\frac{2}{3}; x = 0, y = 1, z = \frac{4}{3}$;
    $w = -\frac{5}{3}, x = 1, y = 1, z = \frac{4}{3}$; and so on

17. Infinitely many solutions; $w = 2, x = 2, y = 0, z = -1$; $w = 0, x = 4, y = -2, z = -1$; and so on
19. No solution

Section 1.7, page 104

1. Inconsistent
3. Inconsistent
5. Eight red cards, five black cards
7. Last year: 5 vice presidents, 10 division managers, 10 assistant managers
   This year: 4 vice presidents, 6 division managers, 8 assistant managers
9. Thursday: 3 attendants, 4 mechanics; Friday: 5 attendants, 3 mechanics; Saturday: 4 attendants, 1 mechanic
11. SEND MONEY

Section 1.8, page 112

1. (a) 0.293
   (b) 58,600
   (c) 0
   (d) 2200
   (e) $\begin{bmatrix} 0.707 & 0 & 0 \\ -0.014 & 0.793 & -0.017 \\ -0.044 & -0.010 & 0.784 \end{bmatrix}$
   (f) Ag = 45.248, Mfg = 18.788, Energy = 15.544

3. (a) $\begin{bmatrix} 0.702 & -0.002 & 0 & 0 \\ -0.088 & 0.788 & 0 & -0.002 \\ -0.010 & 0 & 0.950 & -0.006 \\ -0.029 & -0.003 & -0.004 & 0.970 \end{bmatrix}$
   (b) vehicles = 7770.804, steel = 5025.669, glass = 213.568, rubber = 1219.905

Review Exercises for Chapter 1, page 115

1. $\begin{bmatrix} 3 & 2 & -1 \\ 1 & 4 & 2 \\ 1 & 1 & -1 \end{bmatrix}$
3. [0]
5. $\begin{bmatrix} 2 & 0 & 2 \\ -2 & 0 & -2 \\ 2 & 0 & 2 \end{bmatrix}$
7. $\begin{bmatrix} -4 & -11 & -2 \\ -3 & -2 & -1 \\ 2 & -3 & 3 \end{bmatrix}$
9. $\begin{bmatrix} 0 & \frac{1}{2} \\ 1 & \frac{1}{2} \end{bmatrix}$
11. $\begin{bmatrix} 3 \\ -2 \end{bmatrix}$
13. $\begin{bmatrix} -2 \\ 2 \end{bmatrix}$
15. $X = \begin{bmatrix} 2 \\ -5 \end{bmatrix}$
17. No solution
19. Infinitely many solutions; $x = -\frac{1}{3}, y = \frac{2}{3}, z = 0$; $x = 0, y = 0, z = 1$; and so on
21. $18,040
23. 10 inexpensive models, 5 standard models, 0 deluxe models

Chapter 2

Section 2.1, page 127

1. (a) Yes (b) Yes (c) No
5. (a) No (b) Yes (c) No (d) Yes
9. (a) 210 (b) 3600 (c) 285
11. Minimize: Sugar = $z = 1.25x + y$
    Subject to: $x + y \geq 5$
    $x - y \geq 1$
    $x, y \geq 0$

3. (a) Yes (b) Yes (c) No (d) No
7. (a) 8 (b) 4 (c) 4

13. Maximize: Profit = $z = 200x + 30y$
    Subject to: $x \leq 100$
    $15x + 6y \leq 2000$
    $x + y \leq 310$
    $x, y \geq 0$

Section 2.2, page 143

1. (a) Yes (b) Yes (c) No (d) No (e) Yes (f) Yes

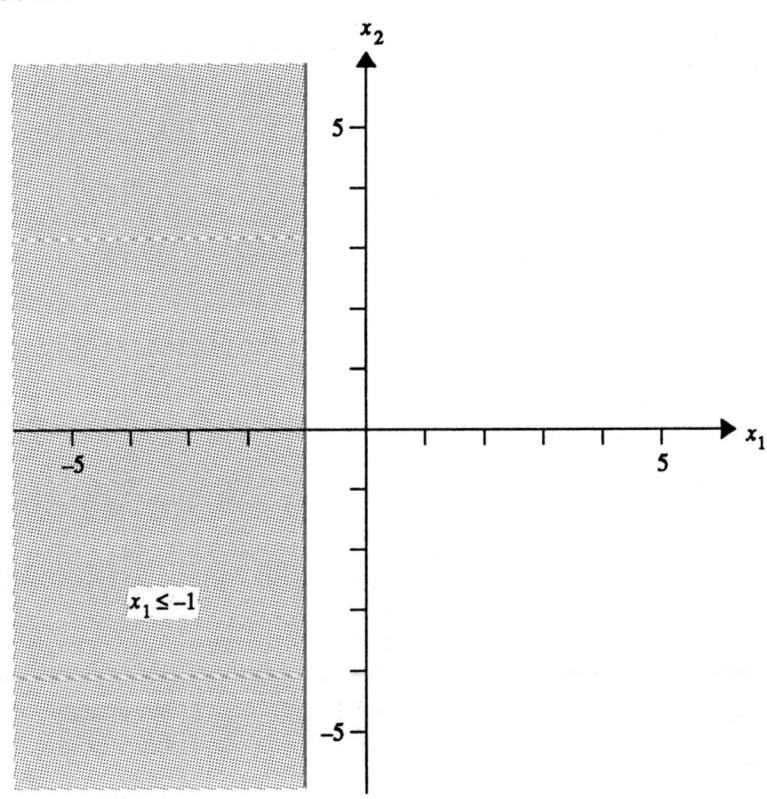

3. Figure A-1

5. Figure A-2

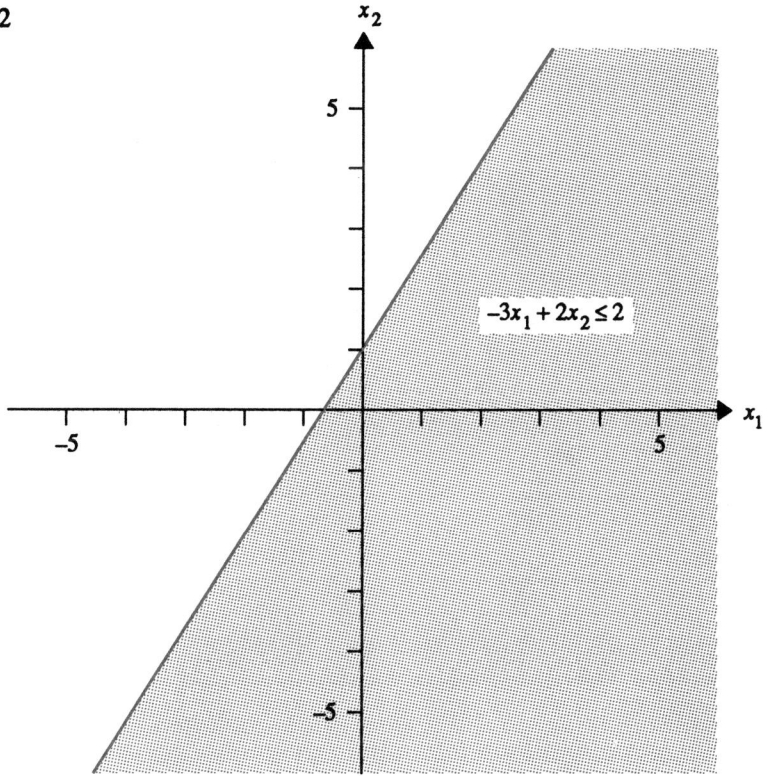

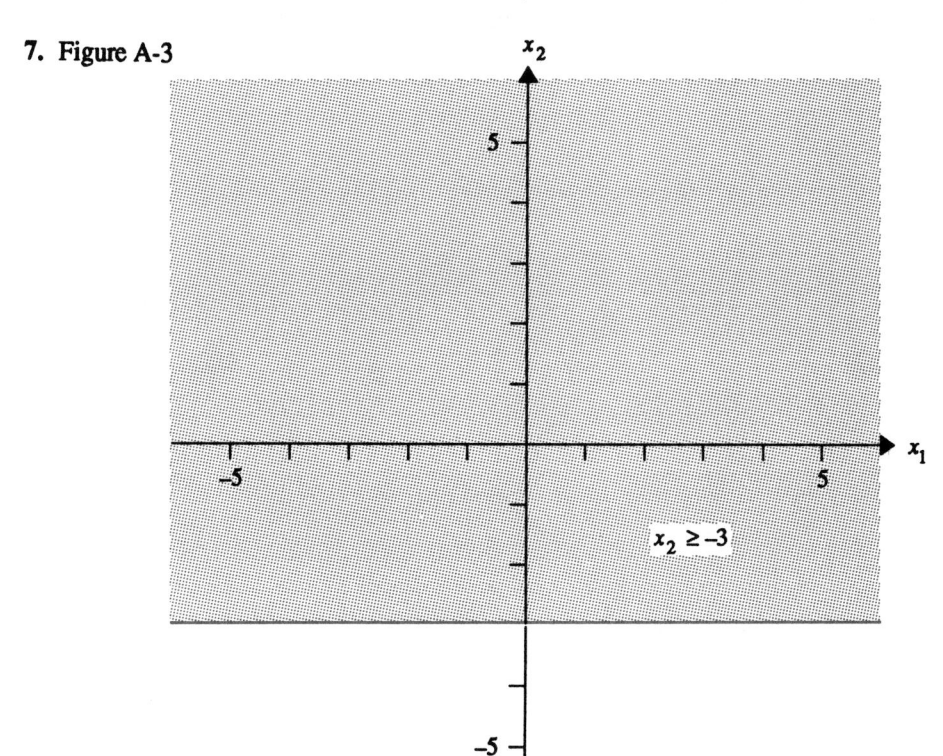

7. Figure A-3

9. Figure A-4

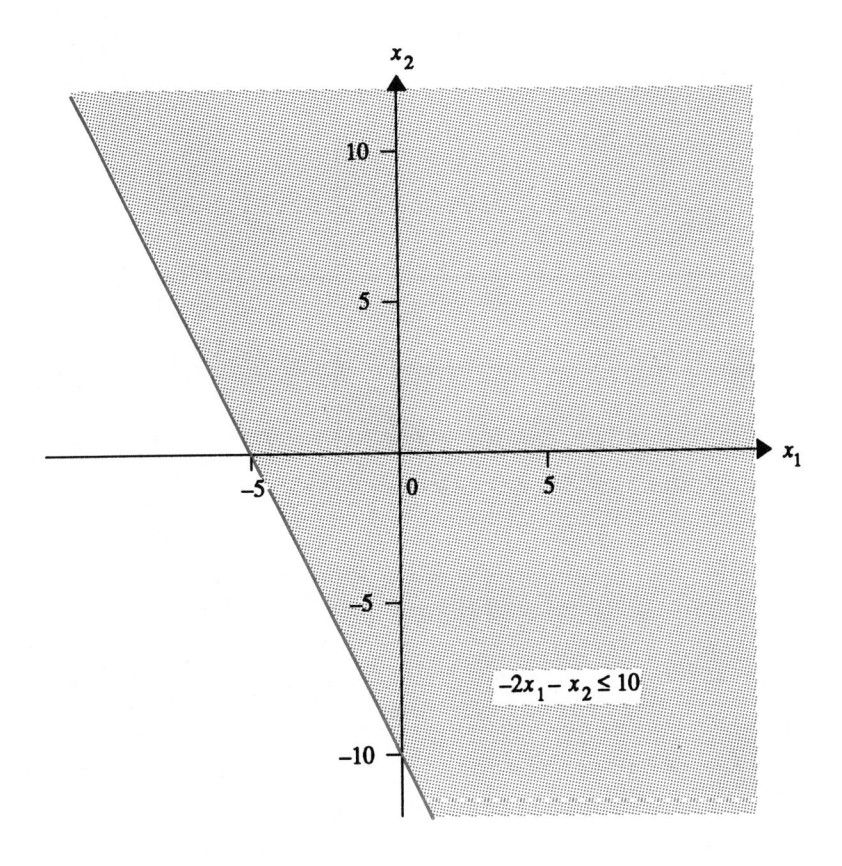

11. Figure A-5

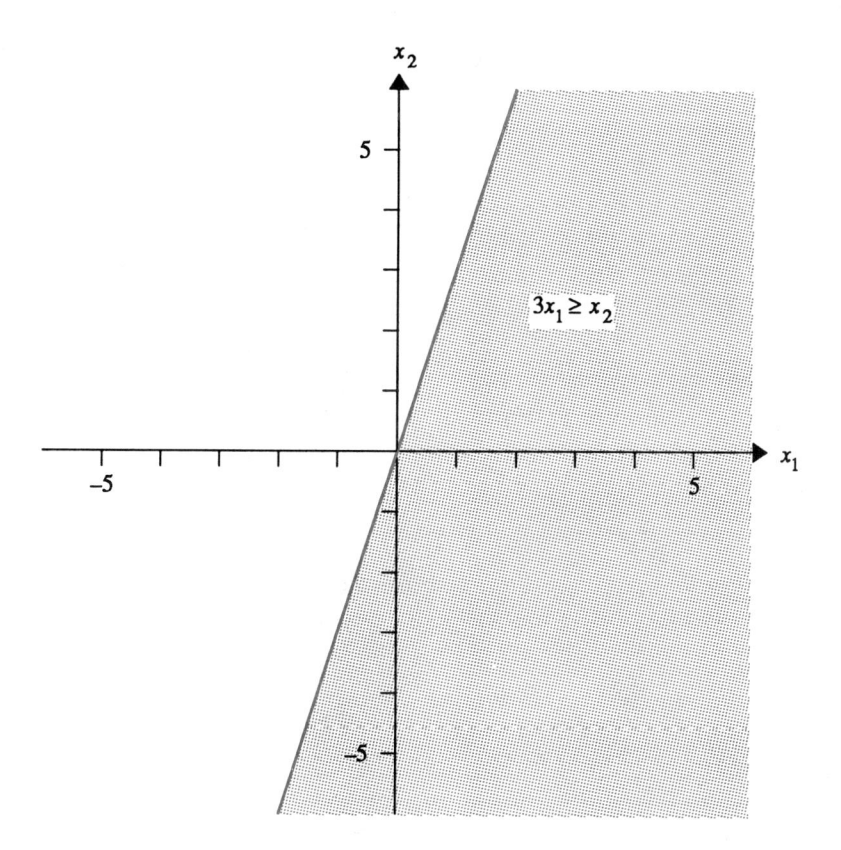

A32   ANSWERS TO ODD-NUMBERED EXERCISES

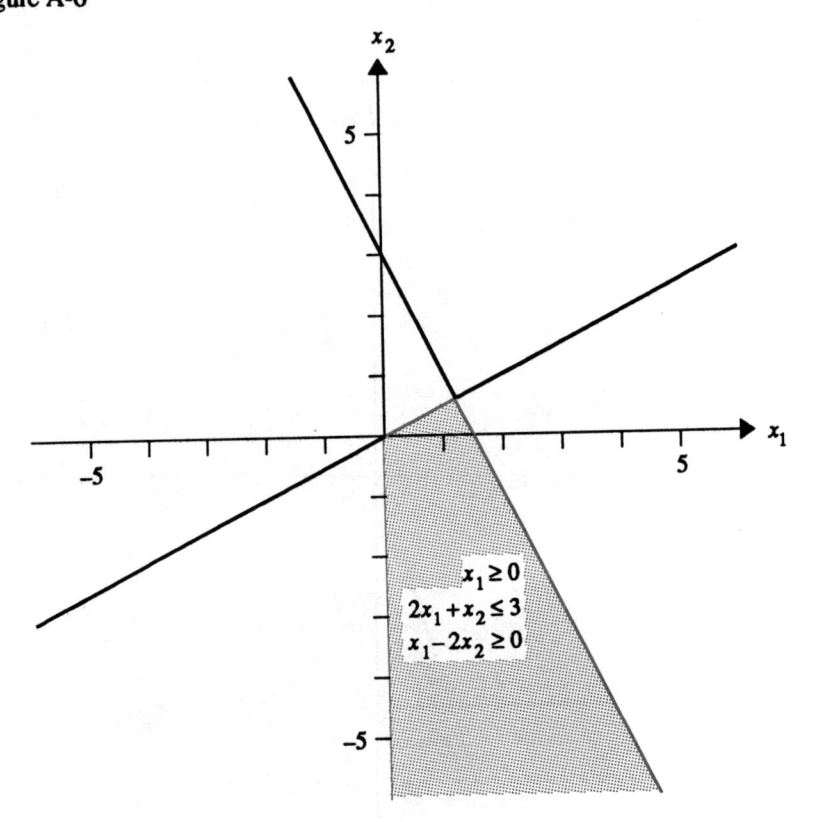

13. Figure A-6

$x_1 \geq 0$
$2x_1 + x_2 \leq 3$
$x_1 - 2x_2 \geq 0$

15. Figure A-7

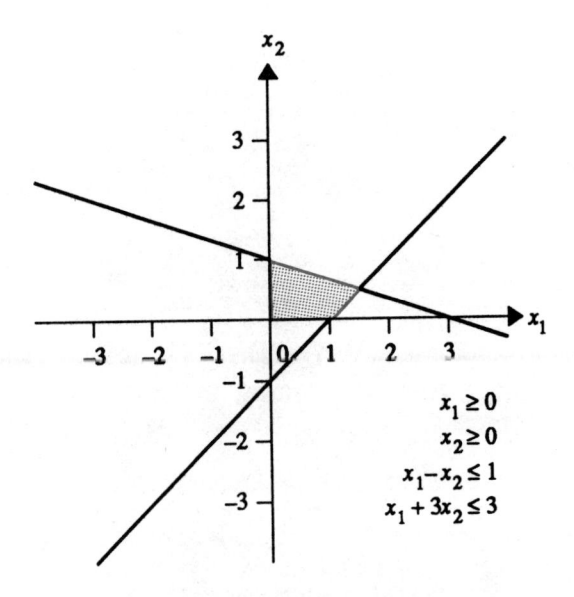

$x_1 \geq 0$
$x_2 \geq 0$
$x_1 - x_2 \leq 1$
$x_1 + 3x_2 \leq 3$

17. Figure A-8

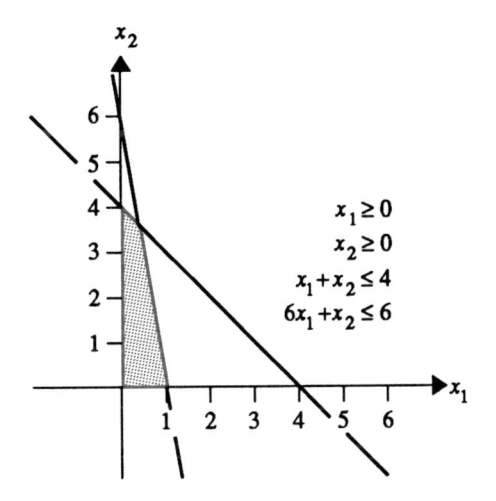

$x_1 \geq 0$
$x_2 \geq 0$
$x_1 + x_2 \leq 4$
$6x_1 + x_2 \leq 6$

19. Figure A-9

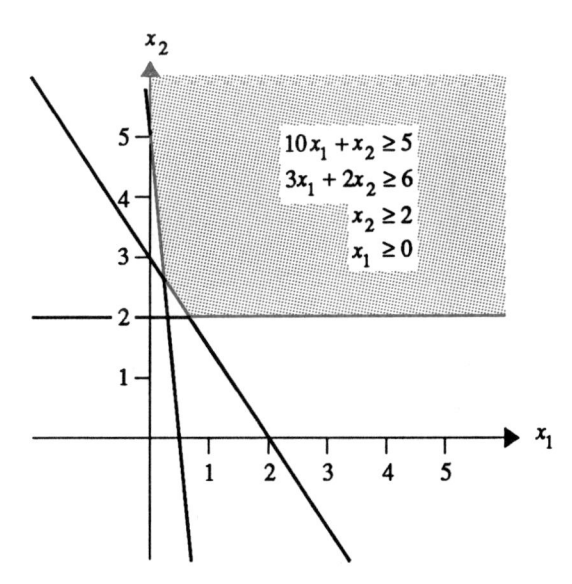

$10x_1 + x_2 \geq 5$
$3x_1 + 2x_2 \geq 6$
$x_2 \geq 2$
$x_1 \geq 0$

21. Figure A-10

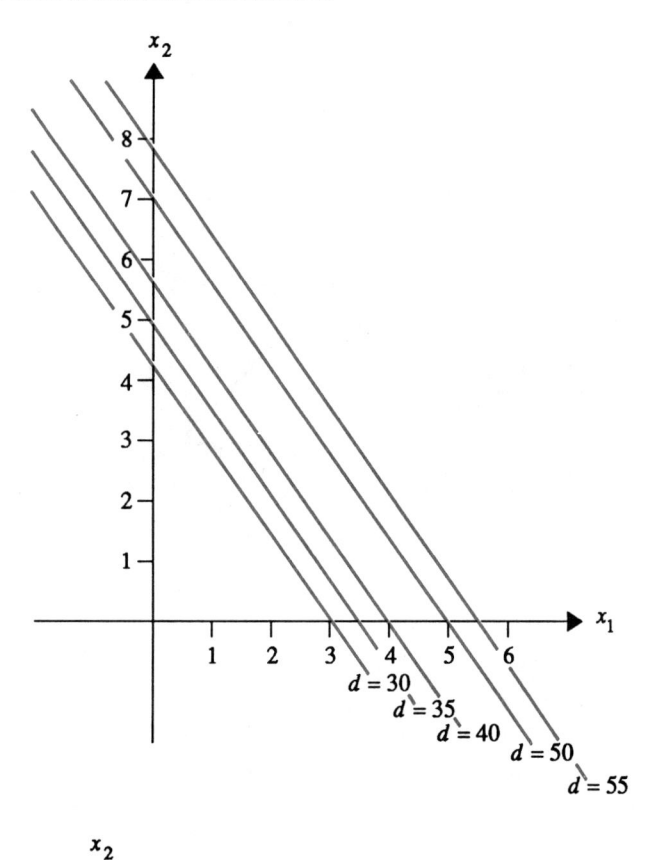

23. Figure A-11

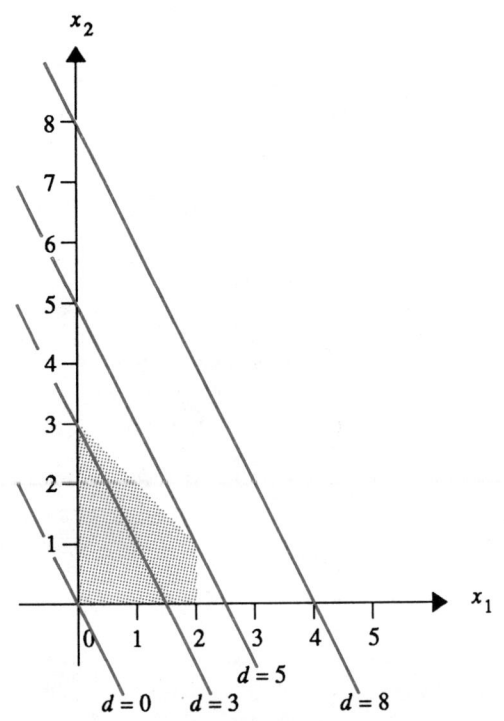

25. $x_1 = 0, x_2 = 2$ **27.** $x_1 = \frac{8}{3}, x_2 = \frac{1}{3}$

29. Smaller smelter: 8 hours, larger smelter: 11 hours

31. None under diesel fuel, 10 hours under natural gas

33. $375 million in mortgages, $125 million in business loans

Section 2.3, page 160

1. Neither **3.** Neither **5.** $A \geq B$

7. $\begin{bmatrix} -1 \\ -2 \\ 0 \\ \frac{1}{2} \end{bmatrix}$ **9.** $\begin{bmatrix} 1 & -1 & 1 & 2 \\ 2 & 0 & 2 & -1 \end{bmatrix}$ **11.** $\begin{bmatrix} 1 & 2 & -2 \end{bmatrix}$

13. $X = \begin{bmatrix} x_1 \\ x_2 \end{bmatrix}$, $A = \begin{bmatrix} 1 & 1 \\ 6 & 1 \end{bmatrix}$, $B = \begin{bmatrix} 4 \\ 6 \end{bmatrix}$, $C = \begin{bmatrix} 3 & 4 \end{bmatrix}$

15. $X = \begin{bmatrix} x_1 \\ x_2 \end{bmatrix}$, $A = \begin{bmatrix} 2 & 3 \\ 1 & 2 \\ 3 & 1 \end{bmatrix}$, $B = \begin{bmatrix} 10 \\ 7 \\ 12 \end{bmatrix}$, $C = \begin{bmatrix} 5 & 9 \end{bmatrix}$

17. $X = \begin{bmatrix} x_1 \\ x_2 \\ x_3 \end{bmatrix}$, $A = \begin{bmatrix} 1 & 1 & 1 \\ 2 & 0 & 1 \\ 0 & 2 & 1 \end{bmatrix}$, $B = \begin{bmatrix} 100 \\ 40 \\ 60 \end{bmatrix}$, $C = \begin{bmatrix} 10 & 12 & 5 \end{bmatrix}$

19. $X = \begin{bmatrix} x_1 \\ x_2 \\ x_3 \\ x_4 \end{bmatrix}$, $A = \begin{bmatrix} 1 & 0 & 0 & 0 \\ 0 & 1 & -1 & 0 \\ 0 & 1 & 1 & 1 \end{bmatrix}$, $B = \begin{bmatrix} 12 \\ 0 \\ 15 \end{bmatrix}$, $C = \begin{bmatrix} 2 & 3 & 1 & 1 \end{bmatrix}$

21. $X = \begin{bmatrix} x_1 \\ x_2 \\ x_3 \end{bmatrix}$, $A = \begin{bmatrix} 1 & 1 & 1 \\ 0 & 0 & -1 \\ 1 & -0.5 & 0 \\ 1 & 0 & -1 \end{bmatrix}$, $B = \begin{bmatrix} 100 \\ -15 \\ 0 \\ 0 \end{bmatrix}$, $C = \begin{bmatrix} 0.09 & 0.075 & 0.05 \end{bmatrix}$

23. $X = \begin{bmatrix} x_1 \\ x_2 \\ x_3 \\ x_4 \end{bmatrix}$, $A = \begin{bmatrix} 50 & 80 & 30 & 200 \\ 1 & 0 & 0 & 0 \\ 0 & 1 & 0 & 0 \\ 0 & 0 & 1 & 0 \\ 0 & 0 & 0 & 1 \end{bmatrix}$, $B = \begin{bmatrix} 1500 \\ 5 \\ 5 \\ 5 \\ 5 \end{bmatrix}$, $C = \begin{bmatrix} 1 & \frac{3}{2} & \frac{1}{2} & 2 \end{bmatrix}$

A36 ANSWERS TO ODD-NUMBERED EXERCISES

25. Maximize: $u = 5y_1 + 4y_2 + 2y_3 + y_4$
    Subject to: $y_1 + 3y_2 + 3y_3 + y_4 \leq 2$
    $4y_1 + y_2 + y_4 \leq 5$
    $y_1, y_2, y_3, y_4 \geq 0$

27. Minimize: $u = 9y_1 + 5y_2$
    Subject to: $2y_1 + y_2 \geq 0$
    $y_1 + 4y_2 \geq 3$
    $3y_1 \geq 4$
    $y_1, y_2 \geq 0$

29. Minimize: $u = 50y_1 + 100y_2$
    Subject to: $y_1 + 2y_2 \geq 1$
    $y_1 \geq 2$
    $y_1 + 3y_2 \geq 0$
    $y_1 - 2y_2 \geq 0$
    $y_1 + 5y_2 \geq 3$
    $y_1, y_2 \geq 0$

31. Maximize: $u = 15y_1 + 12y_2 + 9y_4$
    Subject to: $-y_1 + 3y_2 + y_3 + 3y_4 \leq 1$
    $2y_1 + y_2 - 4y_3 + 2y_4 \leq 0$
    $3y_1 + y_2 - 2y_3 + y_4 \leq 0$
    $y_1, y_2, y_3, y_4 \geq 0$

Section 2.4, page 176

1. $\begin{bmatrix} 2 & 1 & -2 & | & -1 \\ 0 & -1 & 0 & | & 5 \\ 0 & 5 & 2 & | & -1 \end{bmatrix}$

3. $\begin{bmatrix} 3 & 0 & 1 & | & 0 \\ 0 & 1 & 1 & | & 1 \\ 0 & 0 & -1 & | & 0 \end{bmatrix}$

5. $\begin{bmatrix} 2 & 1 & 0 & 1 & 1 & | & 2 \\ -7 & 0 & 2 & -3 & -2 & | & -5 \end{bmatrix}$

7. $\begin{bmatrix} -4 & -4 & -2 & -2 & 0 & | & -2 \\ 1 & 2 & 2 & 0 & -2 & | & 1 \end{bmatrix}$

9. $\begin{bmatrix} 0 & 2 & 2 & 0 & | & 1 \\ 2 & 0 & 2 & 0 & | & 1 \\ 0 & 0 & -2 & -4 & | & 5 \end{bmatrix}$

11. $x = -\frac{1}{17}, y = -\frac{12}{17}$

13. No solutions

15. Infinitely many solutions; $x = -\frac{2}{3}, y = -\frac{5}{3}, z = 0$; $x = -3, y = -2, z = 1$; and so on

17. Infinitely many solutions; $x_1 = -\frac{1}{2}, x_2 = -\frac{1}{2}, x_3 = \frac{1}{2}, x_4 = 0$; $x_1 = -1, x_2 = 1, x_3 = 2, x_4 = 1$; and so on

19. Infinitely many solutions; $x_1 = -1, x_2 = 2, x_3 = 0, x_4 = 0, x_5 = 2$; $x_1 = -2, x_2 = 0, x_3 = 1, x_4 = 0, x_5 = 2$; and so on

21. $x_1 = 0, x_2 = -\frac{1}{6}, x_3 = -\frac{4}{3}, x_4 = -\frac{2}{3}$

23. $x_1 = -\frac{5}{3}, x_2 = \frac{7}{6}, x_3 = 0, x_4 = \frac{1}{3}$

25. $x_1 = 0, x_2 = \frac{3}{2}, x_3 = \frac{15}{8}, x_4 = 0, x_5 = -\frac{1}{6}$

Section 2.5, page 196

1. $\begin{bmatrix} \frac{3}{5} \\ \frac{8}{5} \end{bmatrix}$

3. $\begin{bmatrix} 2 \\ 1 \end{bmatrix}$

5. $x_1 = \frac{5}{3}, x_2 = 1$

7. $x_1 = 2, x_2 = 1$

9. $\frac{1}{2}$ hour fast walking; $\frac{1}{4}$ hour strolling; $\frac{1}{4}$ hour talking

Section 2.6, page 210

1. $\begin{bmatrix} 0 \\ 2 \end{bmatrix}$

3. $\begin{bmatrix} 1 \\ \frac{1}{2} \\ 1 \end{bmatrix}$

5. $\begin{bmatrix} 2 \\ 8 \\ 0 \end{bmatrix}$

7. $x_1 = \frac{1}{2}, x_2 = 1$

9. $x_1 = 0, x_2 = 0, x_3 = \frac{3}{2}, x_4 = 1$

11. 500 square feet of vegetables, 500 square feet of flowers
13. 2 hours jogging, 4 hours bicycling, 2 hours swimming
15. (a) 36 chairs, 32 tables;
    (b) Machine work-hour = $10, assembly work-hour = $17.50, chair limit = 0, table limit = 0

Section 2.7, page 216

1. Not enough water in periods 2 and 3
3. 60 acres of cotton, 90 acres of potatoes, profit = $30,420

Review Exercises for Chapter 2, page 216

1. Maximize: $u = 5y_1$
   Subject to: $y_1 + y_2 \leq 4$
   $-y_1 + y_2 \leq 1$
   $y_1 - y_2 \leq 1$
   $y_1 \leq 2$
   $y_1, y_2 \geq 0$

3. Extreme points: $(0, 2), (2, 0), \left(\frac{16}{9}, \frac{10}{9}\right), (0, 0)$; Solution: $x_1 = \frac{16}{9}, x_2 = \frac{10}{9}$

   See Figure A-12 on page 444.

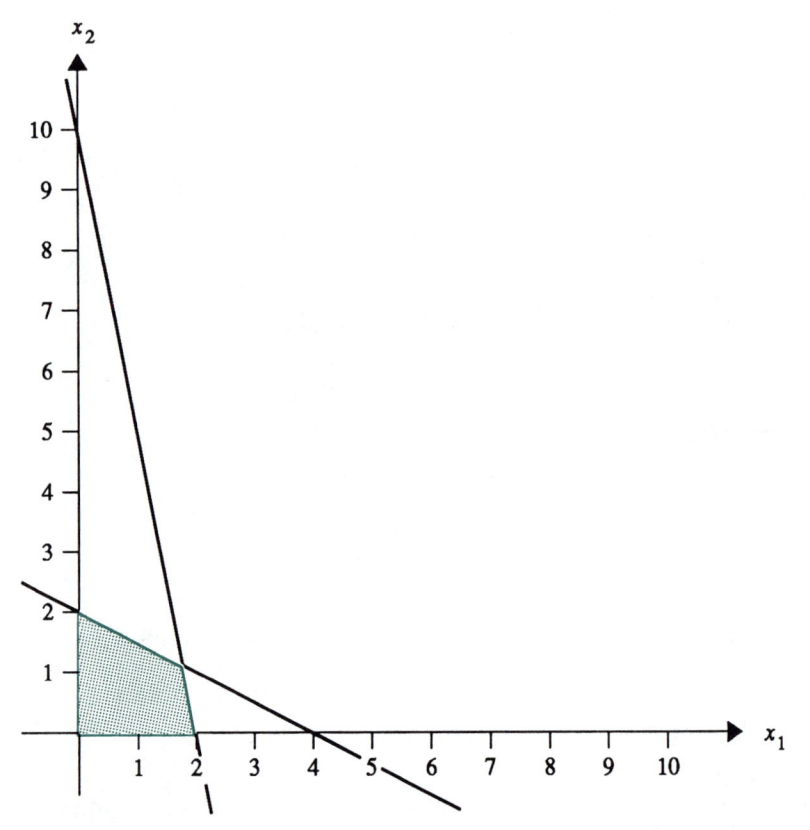

Figure A-12

5. $\begin{bmatrix} 3 \\ 2 \end{bmatrix}$

7. $\begin{bmatrix} 0 \\ 1 \\ \frac{1}{2} \\ 2 \end{bmatrix}$

9. $x_1 = \frac{2}{3}$, $x_2 = \frac{2}{3}$, $x_3 = \frac{1}{3}$

11. $\frac{3}{8}$ pound of rice, $\frac{5}{8}$ pound of vermicelli

Chapter 3

Section 3.1, page 227

1. .25   **3.** .6667   **5.** .2   **7.** .1538
9. .2933   **11.** .4884   **13.** .0263   **15.** .3158
17. .2075   **19.** .5888   **21.** .0476   **23.** 25
25. Thundercloud

CHAPTER 3

Section 3.2, page 235

1. 105
3. 312
5. 32
7. 20
9. 36
11. 30,000
13. 720
15. 216
17. 120
19. 15,600

Section 3.3, page 243

1. (a) 120 (b) 5040
3. (a) 720 (b) 12 (c) 10 (d) 715
5. 120
7. 6,652,800
9. 6720
11. 479,001,600
13. 210
15. 3024
17. 2160

Section 3.4, page 251

1. (a) 21 (b) 84
3. 5040
5. 6
7. 10
9. 30,240
11. 23,100
13. 90
15. 31,500
17. $C_7^{52} C_7^{45}$

Section 3.5, page 256

1. 5005
3. .40
5. 15
7. .0833
9. .676
11. .000000153
13. 120
15. 22,100
17. 1144
19. 52
21. .0024
23. 672

Section 3.6, page 265

1. .7105
3. .4231
5. .66
7. .164
9. .62
11. .1585
13. .15
15. .55

Section 3.7, page 272

1. .65
3. .1111
5. .20
7. .15
9. .38
11. .35
13. .35
15. .60
17. 120
19. .1667

Section 3.8, page 283

1. .40
3. .75
5. .3333
7. .66
9. .0909
11. .4286
13. .29
15. .60
17. Yes
19. .25
21. .15
23. Independent
25. Not independent
27. .8333

Section 3.9, page 291

1. .348
3. .32
5. .0375
7. .0278
9. .0833
11. .2778
13. .755
15. .05
17. .000000595
19. .005
21. .1875
23. .16

Review Exercises for Chapter 3, page 293

1. 27,216　　3. 55,272　　5. .36　　7. (a) Yes　(b) No
9. 990　　11. .55　　13. .41　　15. 96
17. 35　　19. .0192　　21. .4808　　23. .6875
25. .6122　　27. .7143　　29. .38

Chapter 4

Section 4.1, page 305

1. .795　　3. .1825　　5. .0078　　7. .0367　　9. .365

Section 4.2, page 312

1. .50　　3. .8333　　5. .1509　　7. .5833　　9. .9259　　11. .5063

Section 4.3, page 320

1. (a) X = Number of aces
   (b) 2

3. (a) X = Number of legal packages
   (b) 5

5. (a) X = Number of units that fail inspection
   (b) 2

7. $p(0) = .30, p(1) = .50, p(20 = .15, p(3) = .05$; See Figure A-13.

9. $p(2) = .03, p(3) = .06, p(4) = .08, p(5) = .11, p(6) = .14, p(7) = .17, p(8) = .14, p(9) = .11, p(10) = .08, p(11) = .06, p(12) = .03$; See Figure A-14

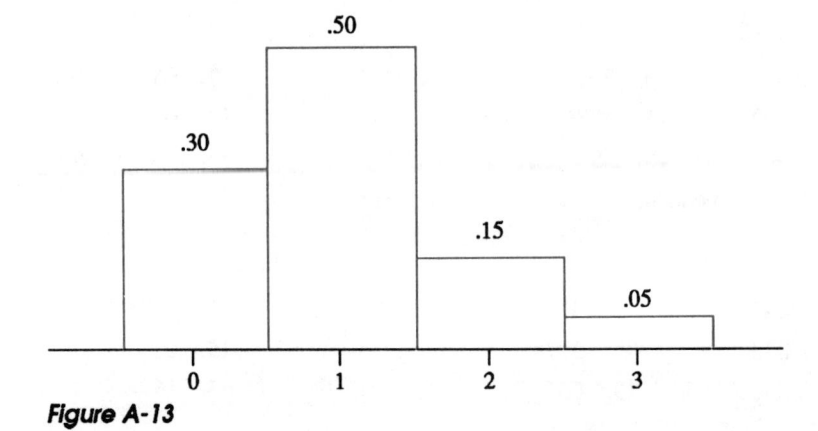

Figure A-13

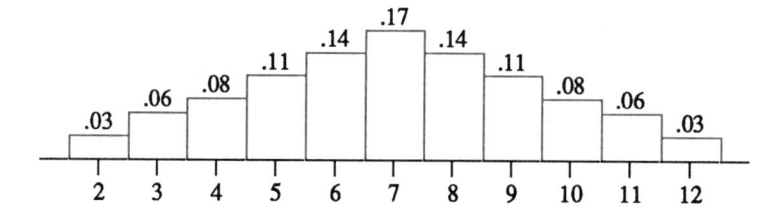

Figure A-14

Section 4.4, page 329

1. $\mu = 2.6, \sigma^2 = 2.14$
3. $\mu = .49, \sigma^2 = .4499$
5. 3.975 points
7. $\mu = \$240, \sigma^2 = 70{,}400$
9. −$0.33
11. −$0.53

Section 4.5, page 338

1. .2188
3. (a) $p(0) = .729, p(1) = .243, p(2) = .027, p(3) = .000001$
   (b) See Figure A-15.
   (c) .3
   (d) .27

5. .1382
7. (a) .655 (b) 4.8
9. .9782
11. .8906
13. .05
15. .83
17. (a) $p(0) = (1-p)^3, p(1) = 3p(1-p)^2, p(2) = 3p^2(1-p), p(3) = p^3$

Section 4.6, page 350

1. .4834
3. .6808
5. .0019
7. .7823
9. 25%
11. 24%
13. 98%
15. 80%

Section 4.7, page 359

1. .0967
3. .2578
5. .1056
7. .0013
9. .2033
11. .0336
13. .1190
15. .1251
17. .8554
19. .9920

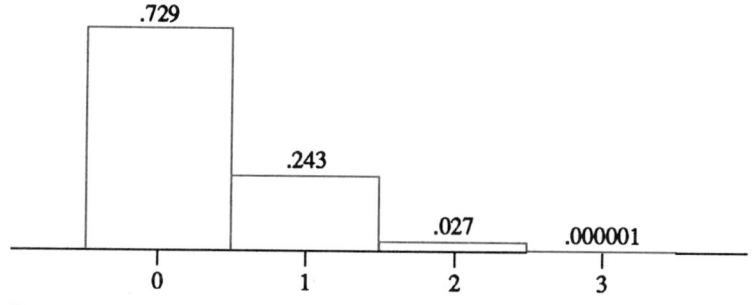

Figure A-15

Review Exercises for Chapter 4, page 361

1. .1814   3. 1   5. .40   7. .2075   9. .68
11. $p(0) = .0370, p(1) = .2222, p(2) = .4444, p(3) = .2963$

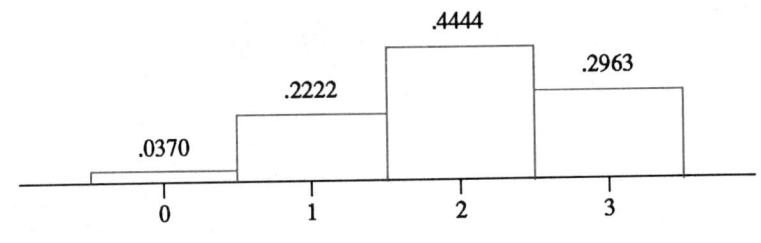

Figure A-16

13. .75   15. .624
17. $p(0) = .03, p(1) = .22, p(2) = .47, p(3) = .28$

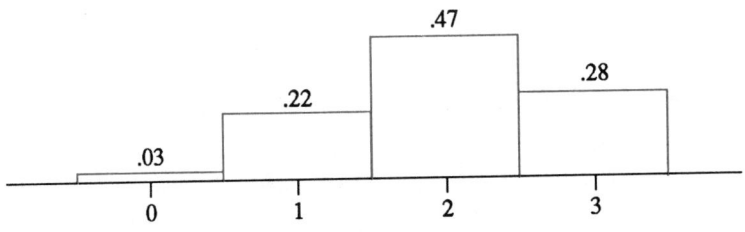

Figure A-17

19. .62   21. .096

Chapter 5

Section 5.1, page 374

1. Republican 58%   3. .47   5. .2375
7. 27.5% large, 55% medium, 17.5% small

Section 5.2, page 386

1. $S^{(1)} = \begin{bmatrix} .11 \\ .89 \end{bmatrix}$, $S^{(2)} = \begin{bmatrix} .072 \\ .928 \end{bmatrix}$, $S^{(3)} = \begin{bmatrix} .0644 \\ .9356 \end{bmatrix}$

3. $S^{(1)} = \begin{bmatrix} 0 \\ 0.4 \\ 0.6 \end{bmatrix}$, $S^{(2)} = \begin{bmatrix} .24 \\ .28 \\ .48 \end{bmatrix}$, $S^{(3)} = \begin{bmatrix} .24 \\ .256 \\ .504 \end{bmatrix}$

5. .275, .2925, .3048

7. $\begin{bmatrix} .33424 \\ .66576 \end{bmatrix}$

9. 39.25%

17. P_1 and P_3 are regular; P_2 is not.

Section 5.3, page 398

1. $\begin{bmatrix} .5 \\ .5 \end{bmatrix}$

3. $\begin{bmatrix} .59375 \\ .40624 \end{bmatrix}$

5. .8824

7. .625

9. .3194 win, .4444 draw, .2361 lose

Section 5.4, page 411

1. (a) $\begin{bmatrix} 1 & 0 & .2 \\ 0 & 1 & .6 \\ 1 & 0 & .2 \end{bmatrix}$ (b) $R = \begin{bmatrix} .2 \\ .6 \end{bmatrix}$, $Q = \begin{bmatrix} .2 \end{bmatrix}$ (c) $m = 1$

3. (a) $\begin{bmatrix} 1 & .2 & .3 \\ 0 & .3 & .6 \\ 0 & .5 & .1 \end{bmatrix}$ (b) $R = \begin{bmatrix} .2 & .3 \end{bmatrix}$, $Q = \begin{bmatrix} .3 & .6 \\ .5 & .1 \end{bmatrix}$

   (c) $m = 1$

5. (a) $\begin{bmatrix} 1 & 0 & .7 \\ 0 & 1 & .1 \\ 0 & 0 & .2 \end{bmatrix}$ (b) $R = \begin{bmatrix} .7 \\ .1 \end{bmatrix}$, $Q = \begin{bmatrix} .2 \end{bmatrix}$ (c) $m = 1$

7. (a) $\begin{bmatrix} 1 & .2 & 0 \\ 0 & .5 & .4 \\ 0 & .3 & .6 \end{bmatrix}$ (b) $R = \begin{bmatrix} .2 & 0 \end{bmatrix}$, $Q = \begin{bmatrix} .5 & .4 \\ .3 & .6 \end{bmatrix}$

   (c) $m = 2$

9. (a) $\begin{bmatrix} 1 & 0.9 & 0 & 0 \\ 0 & 0.1 & 0 & 0.3 \\ 0 & 0 & 0.6 & 0 \\ 0 & 0 & 0.4 & 0.7 \end{bmatrix}$ (b) $R = \begin{bmatrix} .9 & 0 & 0 \end{bmatrix}$, $Q = \begin{bmatrix} 0.1 & 0 & 0.3 \\ 0 & 0.6 & 0 \\ 0 & 0.4 & 0.7 \end{bmatrix}$

   (c) $m = 3$

Section 5.5, page 422

1. (a) 1.67 (b) .167   3. .64   5. (a) 4 (b) 3

A44   ANSWERS TO ODD-NUMBERED EXERCISES

7. .8667   **9.** (a) Compute P^2 (b) 22.5

Review Exercises for Chapter 5, page 424

1. (a) $\begin{bmatrix} .15 & .60 \\ .85 & .40 \end{bmatrix}$ (b) .41

3. 14% excellent, 26% good, 38% fair, 22% poor

5. 50%   **7.** .72

Chapter 6

Section 6.1, page 435

1.

		Customs	
		Mountains	River
Smugglers	Mountains	−2	1
	Rivers	1	−2

3.

		Market		
		Decline	Steady	Increase
Investor	Conservative	8,000	10,500	11,500
	Speculative	3,000	9,500	16,000

5.

		Highway	
		Far	Near
Developer	Housing	40	10
	Shopping	60	−40
	Offices	−30	75
	Industrial	−50	100

7.

		You		
		Nickel	Dime	Quarter
Me	Nickel	−5	5	20
	Dime	5	10	15
	Quarter	20	15	−25

9.

		Antibiotic			
		I	II	III	IV
Flu	A	0.80	0.30	0	0
	B	0	0.90	0	0.50
	C	0.60	0.20	0.90	0.20

Section 6.2, page 443

1. Row 3, column 1; value = 1
3. Row 3, column 1 or 3; value = –1
5. Row 2, column 2; value = 2
7. Row 1 or 3, column 3 or 4; value = –2
9. R chooses –1 or 1, C chooses 1; value = 0
11. Mexican border
13. Home
15. Guilty

Section 6.3, page 460

1. 0
3. .5
5. $\frac{21}{100}$
7. (a) $-\frac{2}{3}$
   (b) –2
   (c) $\frac{2}{3}$
   (d) 0
9. (a) –3
   (b) $-\frac{1}{2}$
   (c) $-\frac{5}{3}$

Section 6.4, page 471

1. $P^* = \begin{bmatrix} \frac{1}{2} & \frac{1}{2} \end{bmatrix}$, $Q^* = \begin{bmatrix} \frac{1}{4} \\ \frac{3}{4} \end{bmatrix}$; $v = \frac{1}{2}$

3. Row 2, column 1; $v = 1$
5. Mrs. Adams: $\frac{4}{7}$ on *Sun*, $\frac{3}{7}$ on *Review*; Miss Butler: $\frac{5}{7}$ on *Sun*, $\frac{2}{7}$ on *Review*. Mrs. Adams will gain $\frac{1}{7}$ of a percent.
7. Kicker: right $\frac{4}{7}$, left $\frac{3}{7}$; Goalkeeper: right $\frac{1}{2}$, left $\frac{1}{2}$; probability .5
9. Spicy Chicken should locate downtown and Happy Hen in the suburbs.
11. NewsTV: $\frac{1}{7}$ general news, $\frac{6}{7}$ sports news; AllNews: $\frac{9}{14}$ general news, $\frac{5}{14}$ sports news; $\frac{5}{7}$ point gain for AllNews

Section 6.5, page 487

1. $P^* = \begin{bmatrix} \frac{1}{3} & \frac{2}{3} \end{bmatrix}$, $Q^* = \begin{bmatrix} \frac{1}{3} \\ \frac{2}{3} \end{bmatrix}$; $v = \frac{5}{3}$

3. $P^* = \begin{bmatrix} \frac{1}{2} & \frac{1}{2} \end{bmatrix}$, $Q^* = \begin{bmatrix} \frac{1}{2} \\ \frac{1}{2} \\ 0 \end{bmatrix}$; $v = \frac{3}{2}$

5. $P^* = \begin{bmatrix} \frac{1}{5} & \frac{4}{5} \end{bmatrix}$, $Q^* = \begin{bmatrix} \frac{3}{5} \\ \frac{2}{5} \end{bmatrix}$; $v = -\frac{3}{5}$

7. Any row, column 3; $v = 1$ 9. Row 2, column 3; $v = -1$

11. $P^* = \begin{bmatrix} \frac{2}{3} & \frac{1}{3} & 0 \end{bmatrix}$, $Q^* = \begin{bmatrix} \frac{1}{3} \\ \frac{2}{3} \end{bmatrix}$; $v = \frac{1}{6}$

13. Amy and Bill up $\frac{1}{2}$, down $\frac{1}{2}$; value = 0

15. Jeff penny, Matt dime; value is one cent to Matt.

17. Offense: run $\frac{11}{13}$, pass $\frac{2}{13}$; defense: run $\frac{7}{13}$, pass $\frac{6}{13}$; value = $\frac{38}{13}$

19. Both players: paper $\frac{1}{3}$, rock $\frac{1}{3}$, scissors $\frac{1}{3}$

Section 6.6, page 500

1. If $t \leq -1$, then row 1, column 1, $v = -1$; if $-1 \leq t \leq 3$, then row 2, column 1, $v = t$; if $t \geq 3$, then row 2, column 2, $v = 3$

3. If $t \leq 1$, then row 1, column 2, $v = 1$

5. If $t \leq 3$, then

$$P^* = \begin{bmatrix} \frac{3-t}{6-t} & \frac{3}{6-t} \end{bmatrix}, \quad Q^* = \begin{bmatrix} \frac{4}{6-t} \\ \frac{2-t}{6-t} \end{bmatrix}; \quad v = \frac{6+t}{6-t}$$

if $t \geq 3$, then row 2, column 2, $v = 3$

7. (a) $P^* = \begin{bmatrix} \frac{3}{5} & \frac{2}{5} \end{bmatrix}$, $Q^* = \begin{bmatrix} 0 \\ \frac{3}{5} \\ \frac{2}{5} \end{bmatrix}$; $v = \frac{11}{5}$ (b) $1 \leq 2, 1 \leq 3, t \leq 1$

(c) $t \leq 4$

9. (a)

$$P^* = \begin{bmatrix} \frac{3}{7} & \frac{4}{7} \end{bmatrix}, \quad Q^* = \begin{bmatrix} 0 \\ 0 \\ \frac{4}{7} \\ \frac{3}{7} \end{bmatrix}; \quad v = -\frac{9}{7}$$

(b) $0 \le 1, -1 \le -1, -1 \le 0, t \le -3$ \qquad (c) $t \le -\frac{5}{3}$

Review Exercises for Chapter 6, page 503

1.
$$P^* = \begin{bmatrix} \frac{1}{2} & \frac{1}{2} \end{bmatrix}, \quad Q^* = \begin{bmatrix} \frac{1}{6} \\ \frac{5}{6} \end{bmatrix}; \quad v = -\frac{3}{2}$$

3.
$$P^* = \begin{bmatrix} \frac{1}{2} & \frac{1}{2} \end{bmatrix}, \quad Q^* = \begin{bmatrix} 0 \\ \frac{1}{2} \\ \frac{1}{2} \\ 0 \end{bmatrix}; \quad v = -1$$

5. Row 2, column 2; $v = -1$ **7.** $\frac{19}{8}$ **9.** Deluxe $\frac{4}{7}$, regular 0, economy $\frac{3}{7}$

11. Allen supports, Baker opposes, Allen loses 2000 votes to Baker.

13. $\frac{1}{3}$ type I, $\frac{2}{3}$ type II

15. If $t \le 0$, then row 2, column 1, $v = 0$; if $0 \le t \le 1$, then row 3, column 1, $v = t$; if $t \ge 1$, then row 3, column 2, $v = 1$

Chapter 7

Note: If the answer to an exercise is a figure, there are many correct ways to draw it. Therefore, the figures given as answers in this chapter are intended only as illustrations.

Section 7.1, page 519

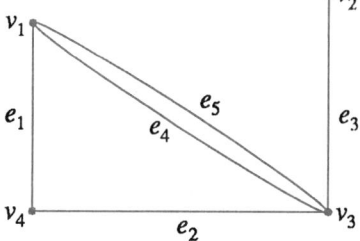

1. Figure A-18

A48 ANSWERS TO ODD-NUMBERED EXERCISES

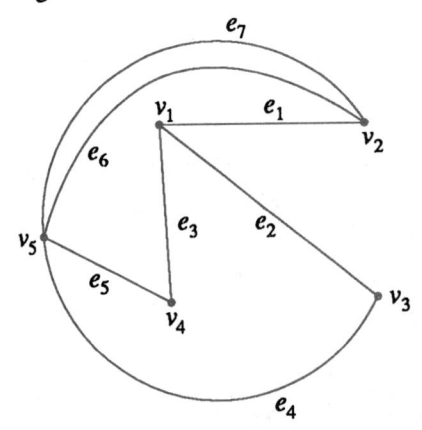

3. Figure A-19

5. (a) $V = \{v_1, v_2, v_3, v_4, v_5, v_6, v_7, v_8\}$
(b) $E = \{e_1, e_2, e_3, e_4, e_5, e_6, e_7, e_8, e_9\}$, where $e_1 = \{v_1, v_2\}$, $e_2 = \{v_2, v_4\}$, $e_3 = \{v_1, v_3\}$, $e_4 = \{v_3, v_4\}$, $e_5 = \{v_5, v_6\}$, $e_6 = \{v_5, v_7\}$, $e_7 = \{v_6, v_8\}$
$e_8 = \{v_7, v_8\}$, $e_9 = \{v_7, v_8\}$
(c) Parallel edges: e_8, e_9 (d) Not connected
(e) Vertex: v_1 v_2 v_3 v_4 v_5 v_6 v_7 v_8
  Degree: 2 2 2 2 2 2 3 3

7. (a) $V = \{v_1, v_2, v_3, v_4, v_5, v_6\}$
(b) $E = \{e_1, e_2, e_3, e_4, e_5, e_6, e_7, e_8, e_9, e_{10}, e_{11}\}$, where $e_1 = \{v_1, v_2\}$, $e_2 = \{v_1, v_3\}$, $e_3 = \{v_2, v_3\}$, $e_4 = \{v_2, v_5\}$, $e_5 = \{v_3, v_6\}$, $e_6 = \{v_3, v_4\}$, $e_7 = \{v_2, v_4\}$,
$e_8 = \{v_4, v_5\}$, $e_9 = \{v_4, v_6\}$, $e_{10} = \{v_5, v_6\}$, $e_{11} = \{v_5, v_5\}$
(c) Loop e_{11} (d) Connected
(e) Vertex: v_1 v_2 v_3 v_4 v_5 v_6
  Degree: 2 4 4 4 5 3

9. Figure A-20

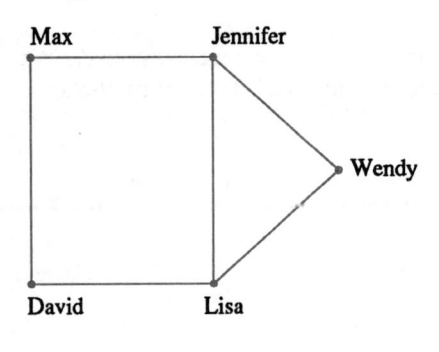

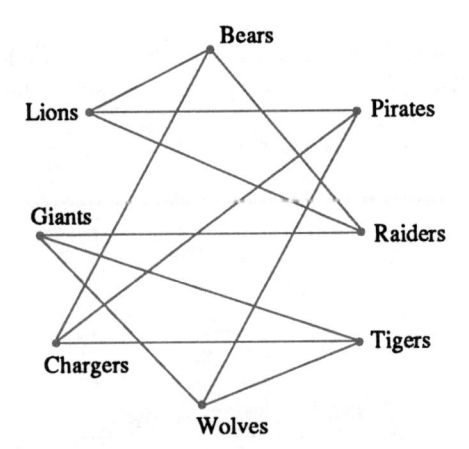

11. Figure A-21

13. No   **15.** Yes; for instance: $e_2e_4e_7e_9e_8e_3e_1e_6e_{10}e_5$

17. (a) Figure A-22

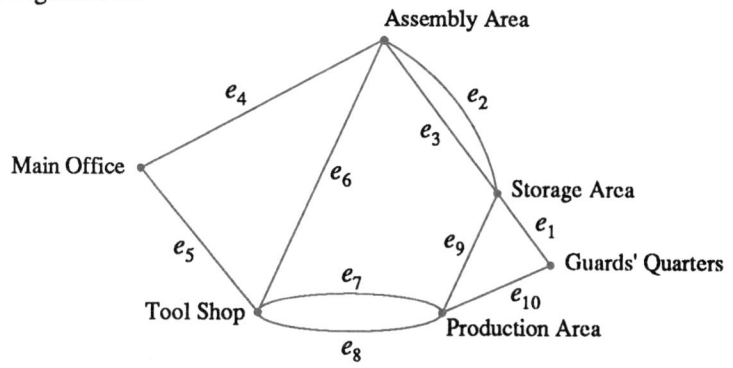

Assembly Area, e_4, e_2, e_3, Main Office, e_6, Storage Area, e_1, e_5, e_9, Guards' Quarters, e_7, e_{10}, Tool Shop, e_8, Production Area

(b) For instance: $e_{10}e_9e_3e_4e_5e_8e_7e_6e_2e_1$

Section 7.2, page 531

1. Vertex: A B C D E
Status: 6 1 3 0 1

3. Vertex: A B C D E F
       4 3 3 1 2 0

5. (a) (1, 1, 1)
(b) No transitive triples
(c) 1 cyclic triple

7. Figure A-23

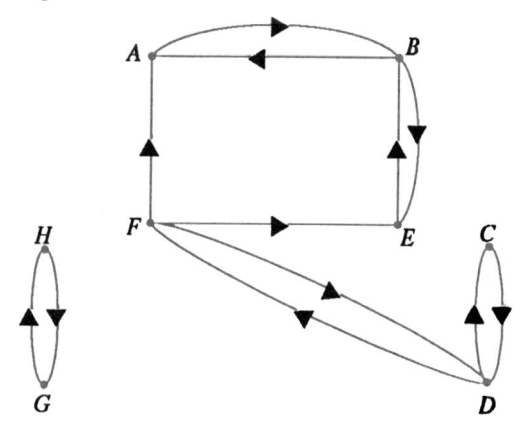

9. Figure A-24

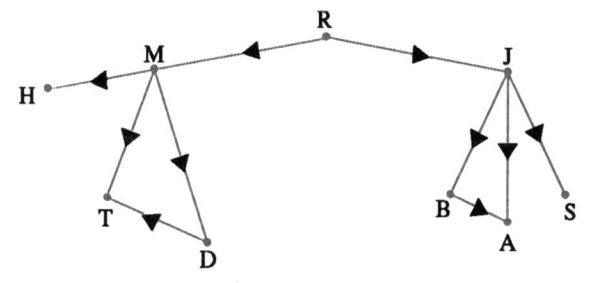

Vertex: R M J D B H T A S
Status: 14 3 3 1 1 0 0 0 0

11. (a) $(3, 2, 2, 2, 1)$
(b) 4 cyclic triples

Section 7.3, Page 544

1. Figure A-25

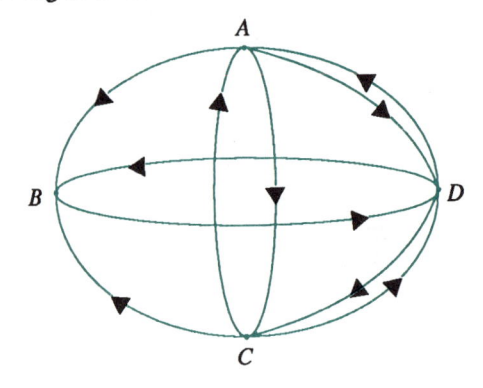

3.
$$\begin{array}{c c} & \begin{array}{c c c c} A & B & C & D \end{array} \\ \begin{array}{c} A \\ B \\ C \\ D \end{array} & \left[\begin{array}{c c c c} 1 & 1 & 0 & 1 \\ 0 & 0 & 1 & 0 \\ 0 & 1 & 0 & 0 \\ 0 & 0 & 1 & 0 \end{array} \right] \end{array}$$

5. $M^2 = \begin{bmatrix} 0 & 1 & 0 & 1 \\ 0 & 0 & 1 & 0 \\ 2 & 1 & 0 & 0 \\ 1 & 0 & 1 & 0 \end{bmatrix}$
$M^3 = \begin{bmatrix} 2 & 1 & 0 & 0 \\ 0 & 1 & 0 & 1 \\ 1 & 0 & 2 & 0 \\ 0 & 1 & 1 & 1 \end{bmatrix}$

7. 3

9.
$$\begin{array}{c c} & \begin{array}{c c c c} I & G & P & X \end{array} \\ \begin{array}{c} I \\ G \\ P \\ X \end{array} & \left[\begin{array}{c c c c} 0 & 0 & 1 & 0 \\ 0 & 0 & 0 & 0 \\ 0 & 1 & 0 & 1 \\ 1 & 1 & 0 & 0 \end{array} \right] \end{array}$$

11. The number in the a_{12}-location of M^2 is 1. The number in the a_{12}-location of M^3 is 1.

13. Woodlark and Marshall Bennett

Section 7.4, page 557

1. A–B–C–E; cost = 10

3. F–I–H or F–I–G–H; cost = 10

5. C–B–F; cost = 6

7. Martinsburg–Goose River–Kenton–Brownsville; distance = 115

9. No; contains a cycle. **11.** No; contains a cycle.

13. A–B, A–D, C–E, D–F, F–G, F–H, and either A–C or B–E; cost = 40

Review Exercises for Chapter 7, page 559

1. $v_2, v_3, v_5, v_6, v_7, v_8$

3. 3

5. For instance: $e_8 e_1 e_3 e_9 e_2 e_4 e_6 e_{12} e_{11} e_{15} e_7 e_5 e_{10} e_{14} e_{13}$

7.

$$\begin{array}{c c} & \begin{array}{c c c c c c} v_1 & v_2 & v_3 & v_4 & v_5 & v_6 \end{array} \\ \begin{array}{c} v_1 \\ v_2 \\ v_3 \\ v_4 \\ v_5 \\ v_6 \end{array} & \left[\begin{array}{c c c c c c} 0 & 1 & 0 & 1 & 1 & 0 \\ 0 & 0 & 0 & 0 & 1 & 0 \\ 1 & 1 & 0 & 1 & 0 & 0 \\ 0 & 0 & 0 & 0 & 1 & 1 \\ 0 & 0 & 0 & 0 & 0 & 1 \\ 1 & 0 & 0 & 0 & 0 & 0 \end{array} \right] \end{array}$$

9. All except v_3

11. FF–D–E–PF or FF–A–E–PF; cost = 90

Chapter 8

Note: Answers were calculated using an 8-place calculator and Table B, where appropriate.

Section 8.1, page 570

1. $15
3. $134,375
5. $15,308.64
7. .04
9. $826.67
11. $1624
13. $94,043.89
15. (a) $68,459.66 (b) No
17. $1523
19. .125

Section 8.2, page 579

1. $64
3. $85,409.25
5. $40,677.97
7. $308,351.18
9. $29,137.50
11. $15,640
13. $8,111,250

Section 8.3, page 588

1. $6341.21
3. .148
5. $51,100
7. $139,288
9. $1,018,266
11. $66.92
13. $86,991.84

Section 8.4, page 598

1. $3771.21
3. $621.36
5. $82,040
7. Karen; $33.56 more
9. $16,854.80
11. National; $45,051,120
13. $22,903.02

Section 8.5, page 610

1. $381,440
3. $27,445.03
5. $1280.62
7. $21,216.23
9. $3,827,644.40
11. (a) $1543.60 (b) $1801.92
13. $97,540
15. $130,967.55
17. $27,463.25
19. $188,064.63

Review Exercises for Chapter 8, page 612

1. $232,148.28
3. (a) $1760 (b) .205
5. (a) $4,053,470.70 (b) $4,252,875
7. (a) $132.03 (b) $27.03
9. (a) $108,841.08 (b) $594.93
11. $3,115,000
13. $12,840.59

Appendix 1

Exercises, page A11

1. Ms. Wong, Ms. Goldberg, Ms. Kjos, Ms. Singh, Ms. Taylor

3. Ms. Wong, Ms. Goldberg, Ms. Kjos, Ms. Singh, Ms. Taylor, Mr. Campos, Mr. Nguyen

5. Mr. White, Ms. Wong, Ms. Goldberg, Ms. Kjos, Mr. Sanchez

7. (a), (c), (e), (g)

9.

p	q	$(\sim q)$	$p \wedge (\sim q)$
T	T	F	F
T	F	T	T
F	T	F	F
F	F	T	F

11. p: Maria will arrive in July.
    q: Maria will see Matthew.

p	q	$p \wedge q$
T	T	T
T	F	F
F	T	F
F	F	F

13. p: Jose will come.
    q: Susan will call him.

p	q	$p \vee q$
T	T	T
T	F	T
F	T	T
F	F	F

15. Equivalent
17. Equivalent
19. Equivalent

Index

Absorbing Markov chain processes, 401–402, 411, 421
Absorbing state, 401–402, 406–407, 409, 411, 413, 416, 421
Acceptance number, 351, 358
Acceptance sampling, 351, 355–356, 358
Acre-inches, 118n, 123, 205–207, 213, 216
Addition of matrices, 7, 13
Adjacency matrix, 534, 544
Amortization, 600, 609
Amount, 567, 570
    with compound interest, 583, 585, 588
Annual interest rate, 565, 570
Annuities
    amounts of, 592–593, 597–598
    defined, 590
    ordinary, 592, 597
    rents of, 592, 596–597
Arcs, 524, 530
Associative property, 38n
Augmented matrices, 50, 61, 67–68, 77, 85, 94, 164–165
Average, 12

Balloon payment, 608, 610
Basic feasible solution, 181–182, 184, 188, 195
Basic solution, 173, 176, 195
Basic variables, 173, 175–176, 184
Basis, 173–174, 176, 188, 195, 198
Bayes' Theorem, 309, 312
Bill of demands, 2, 107
Bimatrix, 434
Bimatrix games, 433
Binomial experiments, 332, 338, 352, 359
Binomial probability distributions, 336, 338, 351–353, 359
Bonds, 568
Brand share, 364–365, 376–377, 379–380

Canonical form, 204

Cards, 296
Central Limit Theorem, 341–342
Chance moves, 434
Circuit, 514–515, 519
Climatological Data, 220, 264–265, 279, 281–283, 291, 293–294, 296
Climatological record, 220
Coalitions, 434
Column matrices, 5, 13, 152, 450
Columns, 4, 13
Combinations, 245–250, 253–254
Commutative property, 8–9, 13, 20–21
Complement, A1, A10
Compound interest, 583
Conditional probability, 276–278, 283, 307, 312, 366, 371, 413, 415, 421
Conjunction, A5, A9, A10
Connected graphs, 510, 519, 552, 556
Connectives, A5
Coordinates, 129, 140, 143
Corners, 138
Counting rules
    for ordered k-tuples, 234–235, 238, 255–256
    for ordered pairs, 232–233, 247, 249
    for union, 261–262
Credible interval, 221–222
Cycles, 552, 556
Cyclic triples, 527, 531
    counting, 528–531
Cycling, 189

Degree, 514, 519
Demand matrices, 4, 64, 109
De Morgan's Laws, A9-A10
Determinants, 41–42, 46, 76–77, 418
Diagonal entries, 77
Diagonal location, 26, 53–54, 61
Digraphs, 524, 530, 534
Directed paths, 525, 530
Disconnected graphs, 510, 519
Discounted loans, 575, 579
Discount rate, 573, 577, 579

A53

Discounts, 573, 579
Disjoint events, 267–270, 272, 290, 300, 413
Disjoint sets, A3
Disjunction, A5, A9
Distance (directed), 525, 530, 537, 544
Distributive laws, 27–29, 65, 390, 416, 481, 540
Domination, 494, 499, 500
Drug testing, 297–298, 311–312
Dual
    of maximization problem, 153–154, 156–157, 160, 206
    of minimization problem, 155, 157, 160, 197–198, 209
Duality Theorem, 156–157, 159–160, 198–199, 477, 480
Dual simplex algorithm, 497, 499

Economic planning, 1–3, 106
Edges, 508, 518
Effective annual rate, 586–588
Effective interest rate, 577, 579
Empty set, A1, A3, A10
Encoding matrices, 43, 103
Entries, 5, 13
Equal sets, A3
Equilibrium, 379–382, 385–386, 389, 391–392, 394, 397
Equivalent statements, A8–A10
Estimate of probability, 222, 227
Euler circuits, 512, 514, 519
    construction of, 515–516
Events, 261–263, 265
Expected value, 322–324, 326, 328, 416, 421, 448
    of binomial distribution, 336–337, 353
    in histogram, 324–325
Experiments, 253, 331
Extreme points, 138–140, 143, 184, 195

Face value, 568
Factorial, 238, 241–242
Failure, 253, 268–269, 292, 331, 338
False negatives, 298, 311–312
False positives, 298, 311–312
Farm management, 118–119, 212–216
Favorable outcomes, 223–224, 227
Feasible set, 120, 126, 132, 138–139, 148, 184, 189, 195
Feasible solutions, 120, 126

Functions
    linear, 122, 126, 134–135
    objective, 123, 127, 139–140, 148, 182, 198, 206, 209, 212
Fundamental matrices, 416, 421
Future value, 567

Gauss-Jordan method, 48, 53–61, 80, 93–94, 98, 164–165, 169, 176
    for finding equilibrium, 392, 394, 398
    for inverses, 77, 419
    rules for, 53–54, 61–62
Graphs, 508, 518
    of linear equations, 80–84
    of linear inequalities, 129–131, 136, 143

Half-planes, 131, 143
Histograms, 317–318, 320, 334–335, 341, 349, 353–354, 356

Identity matrices, 26–27, 29, 64, 390, 392, 398, 404, 415, 540
Incident, 508, 518
Independent events, 280, 283, 288–290, 331, 338, 448
Inequalities
    linear, 120, 126, 129–132, 148, 204, 209
    of matrices, 146–147, 149, 159
Initial vertex, 524, 530
Input-output equations, 64–65, 108–110
Input-output matrices, 3, 5, 64, 109, 112
Intensity matrices, 4, 64, 109
Interest, 564–565, 570
Interest rate, 564–565, 570
    finding the, 566–567
    per period, 583–584, 588, 592, 601–602
Intersection
    of events, 280, 287, 290, 300, 306–307, 413
    of lines, 81–82
    of sets, A3, A10
Interval forecast, 221
Inverses
    of matrices, 37, 46, 64, 77, 112, 416, 419
    of two-by-two matrices, 38–42, 46, 417

Keno, 322

Least-cost path, 547–549, 556
Length of a path, 544, 547, 556
Leontief model, 1–2, 64, 106–112
   bill of demands for, 2, 107
   input-output equation of, 64–65, 108–110
   and sectors, 2, 106, 112
Level curves, 135–138, 143
Linear equations, 34, 96, 119–120, 126, 131, 200, 204
Linear functions, 122, 126, 134–135
Linear inequalities, 120, 126, 129–132, 148, 204, 209
   solutions of, 120, 126, 132
   systems of, 146
Linear polynomials, 119, 126
Linear programming problems (LP problems), 123, 126–127, 137, 140, 143, 204, 499
   and feasible set, 120, 126, 132, 138–139, 184, 189, 195
   and geometry, 129–142
   of maximization, 149, 159, 179, 197–198, 203, 206, 209, 214
   of minimization, 151, 160, 197–199, 206, 209
   and objective functions, 123, 127, 139–140, 148, 182, 198, 206, 209, 212
   and variables, 122–123, 126
Linear systems, 32, 47, 80, 90, 163, 169
   and inverses of matrices, 66–67
   matrix form of, 45, 61, 176, 391
   matrix methods for, 80–94
   nonsingular n-by-n, 47–61
   operations on, 48–50
Lines, 80–84
Location, 5, 13
Loops, 509, 518

Main diagonal, 26, 29
Marginal value, 205–209, 214–216
Markov chain processes, 378, 382, 386, 397
   absorbing, 401–402, 411, 421
   regular, 384, 386, 409
Matrices
   addition of, 7–8, 13
   applications of, 96–103
   augmented, 50, 61, 67–68, 77, 85, 94, 164–165
   canonical form of, 204
   and codes, 43, 102–103
   coefficient, 35, 45, 48, 61, 148
   columns of, 4, 13
   definition of, 3, 13
   demand, 4, 64, 109
   determinants of, 41–42, 46, 76–77, 417
   diagonal entries of, 77
   diagonal locations in, 26, 53–54, 61
   distributive laws for, 27–29, 65, 390
   encoding, 43, 103
   entries of, 5, 13
   equality of, 6, 13
   fundamental, 416, 421
   identity, 26–27, 29, 64, 390, 392, 398, 404, 415
   inequalities, 146–147, 149, 159
   input-output, 3, 5, 64, 109, 112
   intensity, 4, 64, 109
   inverses of, 37, 46, 64, 77, 112, 416, 419
   inverses of two-by-two, 38–42, 46, 66, 417
   locations in, 5, 13
   and main diagonal, 26, 29
   multiplication of, 17–18, 20–21, 26, 29, 187, 368, 381
   nonsingular, 37, 46, 48, 61, 64, 77, 85, 416
   positive, 383–384, 386, 407, 409, 411, 474, 481–482, 486
   and probability, 365–374
   regular transition, 384, 386, 392
   row operations on, 51–54, 61–63, 94
   row-reduced form of, 93–94
   rows of, 4, 13
   scalar multiplication for, 10, 13
   singular, 37, 46, 64, 76–77, 83, 85
   sizes of, 5, 13
   square, 5, 13
   subtraction of, 7, 13
   transition, 367, 370, 374, 378, 382, 386, 394
   transposes of, 152–155, 160, 198–199
   of unknowns, 36
   zero, 6, 8, 13, 390, 404
Matrix codes, 43–44, 102–103
Matrix equations, 9, 71, 370, 414–416
   for finding equilibrium, 390–392, 398
   Gauss-Jordan method for, 80, 94
   input-output, 64
   for linear systems, 36–38, 45, 48, 50, 61
   and solutions using inverses, 48, 77, 85

INDEX

Matrix form of system, 35
Matrix games, 430, 434, 458–459
Matrix multiplication, 21, 29, 187, 368, 535
    applications of, 24–25
    associative property of, 38
    and distributive laws, 27–29, 65, 390, 416
    by matrix of unknowns, 24
    noncommutativity of, 20–21, 26
    row-column, 17–18, 29
    size restrictions for, 20, 29
Matrix of unknowns, 36
Maturity date, 568
Maximization problems, 149, 159, 179, 197–198, 203, 206, 209, 214, 475, 478
Maximum, 122–123, 126
Mean, 343, 349, 352, 359
Minimal spanning tree, 552–554, 557
Minimization problems, 151, 160, 197–199, 206, 209, 476
Minimum, 122–123, 126
Mixed strategies, 446, 449, 450, 459
Mortgages, 563–564

Negation, A5, A9–A10
Networks, 546, 556
Nonabsorbing states, 402, 407, 409, 411, 416, 421
Nonbasic variables, 173, 176, 184
Noninterest-bearing notes, 573, 574, 577, 579
Nonsingular matrices, 37, 46, 48, 61, 64, 77, 85, 416
Nonstrictly determined games, 443
Normal curves, 342
Normal distributions, 341, 348, 351–352, 358
n-person games, 434

Objective functions, 123, 127, 139–140, 148, 182, 198, 206, 209, 212
Odds, 225–226
Optimum pure strategy, 438, 439, 443
Optimum strategy, 457, 459, 478, 487
Ordered k-tuples, 233, 235, 247
Ordered pairs, 230
Ordering, 237–238

Parallel edges, 509, 518
Parameters, 491, 500
Parametrized matrix games, 491, 493, 500
Partitions, 303, 305, 308–309, 312, 365–366, 369, 371

Paths, 509, 519
Payoff, 429, 434, 447, 450, 459
Payoff matrix, 429, 434
Permutations, 241–242, 246–247, 256
Pivot(ing), 165-167, 174-176
    column, 165, 167, 176, 184, 188
    element, 165, 167, 176, 184, 188, 195
    without fractions, 166–167
    row, 165, 167, 176, 184, 188
    rules for, 167, 176
Present value, 568–569, 570
    with compound interest, 585, 588
    formula for, 602, 609
Principal, 564–565, 570
Probability
    computation of, 223–224, 227, 253, 256, 260
    conditional, 276–278, 283, 307, 312, 366, 371, 413, 415, 421
    and counting, 223, 229
    of disjoint events, 267–270, 272, 290, 300, 413
    distributions, 317–320, 323, 328
    estimate, 222, 227
    and experience, 222
    of intersection, 280, 287, 290, 300, 306–307, 413
    and matrices, 365–373
    and odds, 224–226
    subjective, 225
    trees, 301–302, 304, 310–311
    types of, 226, 263–264
    of union, 262–265, 267
Proceeds, 573, 579
Pure strategies, 429, 454, 458

Random variables, 315, 320, 328, 342
Reachability, 525, 530
    using adjacency matrices, 537, 539, 544
Refinancing loans, 607
Rent, 604, 609
Roulette, 258–259
Row-column multiplication, 17–18, 29
Row matrices, 5, 13, 152, 450
Row operations, 51–54, 61–62, 94
Row-reduced form, 93–94
Rows, 4, 13
R-Q form, 405–407, 409, 411, 414–415, 421

Saddle points, 439, 441, 443, 458–459, 467, 492
Sample space, 261–263, 265, 328, A1

Scalar multiplication, 10, 13
Scalar product, 10
Score, 526, 530
Score sequence, 526, 530
Sector, 2, 106, 112
Security level, 438, 443, 452, 459
Sensitivity analysis, 495, 499, 500
Sets, A1, A6, A10
Side payments, 434
Simple interest, 564, 570, 577
Simplex algorithm, 179, 182, 188–189, 195, 198, 203, 209, 215, 479, 486, 500
Simplex tableau, 182, 188, 195, 203, 215, 479, 486, 493, 495
Singular matrices, 37, 46, 64, 76–77, 83, 85
Sinking funds, 597, 598
Slack variables, 180, 186, 188, 195, 198–199, 204, 206, 209, 214
Solutions
    basic, 173, 176, 195
    basic feasible, 181–182, 184, 188, 195
    of equations, 9–10
    of linear equations, 34
    of linear inequalities, 120, 126, 132
    of systems of linear equations, 34–35, 163–165, 169–172
Spanning tree, 552, 556
Square matrices, 5, 13
Standard deviation, 345, 349, 352, 359
State, 371, 374, 378, 382, 385, 389, 391, 399
    absorbing, 401–402, 406–407, 409–411, 413, 416, 421
    nonabsorbing, 402, 407, 409, 411, 416, 421
Statements
    compound, A5, A6, A10
    simple, A5, A6, A10
Status, 525, 530, 537
Strictly determined game, 439, 443, 458–459
    solution of, 440, 443
Subjective probability, 225
Subsets, A2
Subtraction of matrices, 7, 13
Success, 253, 268–269, 292, 331, 338
Systems of linear equations, 34–36

Target, 165, 167, 176
Terminal vertex, 524, 530

Tournaments, 526, 530
Transition matrices, 367, 370, 374, 378, 382, 386, 394
Transitive triples, 527, 530
Transpose, 152–155, 160, 198–199
Tree diagrams, 230, 255
Trees, 552, 556
Triples, 233
Truth tables, A5, A8, A10
    for compound statements, A6
    and Venn diagrams, A6
Truth values, A5, A10
Two-by-two games, 461, 464–465, 471, 493
    equalizing strategies for, 463, 465–466, 471
    optimum solutions of, 467, 471
    payoff matrices for, 464–465
    security levels for, 462
    strictly determined, 470
Two-person games, 429, 434

Union
    of events, 261–265, 267
    of sets, A2, A10
Universal set, A1, A10

Value (of a game), 439, 443, 457, 459, 480, 482, 487
Variables
    basic, 173, 175–176, 184
    dependent, 122, 126, 186, 188
    independent, 122, 126, 148, 155, 195, 198
    and LP-problems, 123
    nonbasic, 173, 176, 184
    random, 315, 320, 328, 342
    slack, 180, 186, 188, 195, 198–199, 204, 206, 209, 214
Variance, 326, 328–329
    of binomial distribution, 337, 353
    computation of, 327–328
Venn diagrams, 261, A1, A10
Vertices, 508, 518

Weather, 219–222

Zero matrices, 6, 8, 13, 391, 404
Zero-sum games, 429, 435